T0203854

HANDBOOK
OF
MATHEMATICS
AND
STATISTICS
FOR THE
ENVIRONMENT

HANDBOOK

OF

MATHEMATICS

AND

STATISTICS

FOR THE

ENVIRONMENT

Frank R. Spellman

Nancy E. Whiting

CRC Press
Taylor & Francis Group
Boca Raton London New York

CRC Press is an imprint of the
Taylor & Francis Group, an **informa** business

CRC Press
Taylor & Francis Group
6000 Broken Sound Parkway NW, Suite 300
Boca Raton, FL 33487-2742

First issued in paperback 2019

© 2014 by Taylor & Francis Group, LLC
CRC Press is an imprint of Taylor & Francis Group, an Informa business

No claim to original U.S. Government works

ISBN-13: 978-1-4665-8637-6 (hbk)
ISBN-13: 978-0-367-86813-0 (pbk)

Library of Congress Cataloging-in-Publication Data

Spellman, Frank R.
  Handbook of mathematics and statistics for the environment / authors Frank R. Spellman and Nancy E. Whiting.
    pages cm
  Includes bibliographical references and index.
    ISBN 978-1-4665-8637-6 (alk. paper)
    1. Environmental engineering--Mathematics--Handbooks, manuals, etc. 2. Environmental engineering--Statistical methods--Handbooks, manuals, etc. I. Whiting, Nancy E. II. Title.

TD145.S6765 2013
333.701′51--dc23
                                                                            2013018633

Visit the Taylor & Francis Web site at
http://www.taylorandfrancis.com

and the CRC Press Web site at
http://www.crcpress.com

Contents

Preface...xix
Authors..xxi

SECTION I *Fundamental Conversions, Computations, Modeling, and Algorithms*

Chapter 1   Introduction 3

    1.1   Setting the Stage 3
    1.2   SI Units 3
    1.3   Conversion Factors 4
    1.4   Conversion Factors: Practical Examples 15
           1.4.1   Weight, Concentration, and Flow....................................16
           1.4.2   Water/Wastewater Conversion Examples.........................18
           1.4.3   Temperature Conversions 24
    1.5   Conversion Factors: Air Pollution Measurements 25
           1.5.1   Conversion from ppm to $\mu g/m^3$ 26
           1.5.2   Conversion Tables for Common Air Pollution Measurements...........27
    1.6   Soil Test Results Conversion Factors...29
    1.7   Greenhouse Gas Emission Numbers to Equivalent Units 31
           1.7.1   Electricity Reductions (Kilowatt-Hours)..........................31
           1.7.2   Passenger Vehicles per Year...31
           1.7.3   Gallons of Gasoline Consumed...32
           1.7.4   Therms of Natural Gas..32
           1.7.5   Barrels of Oil Consumed...32
           1.7.6   Tanker Trucks Filled with Gasoline..................................33
           1.7.7   Home Electricity Use 33
           1.7.8   Home Energy Use 34
           1.7.9   Number of Tree Seedlings Grown for 10 Years................34
           1.7.10 Acres of U.S. Forests Storing Carbon for 1 Year 35
           1.7.11 Acres of U.S. Forest Preserved from Conversion to Cropland..........36
           1.7.12 Propane Cylinders Used for Home Barbecues...................38
           1.7.13 Railcars of Coal Burned..38
           1.7.14 Tons of Waste Recycled Instead of Landfilled.................39
           1.7.15 Coal-Fired Power Plant Emissions for 1 Year..................39
    1.8   Units of Derivatives and Integrals...39
    References and Recommended Reading 39

Chapter 2   Basic Math Operations 41

    2.1   Introduction 41
    2.2   Basic Math Terminology and Definitions 41
           2.2.1   Keywords..42
    2.3   Sequence of Operations..42
           2.3.1   Rules for Sequence of Operations 43
           2.3.2   Examples of Sequence of Operations................................43

2.4   Percent 44
2.5   Significant Digits 47
2.6   Powers and Exponents 49
2.7   Averages (Arithmetic Mean) 50
2.8   Ratio 52
2.9   Dimensional Analysis 55
        2.9.1   Basic Operation: Division of Units 55
        2.9.2   Basic Operation: Divide by a Fraction 55
        2.9.3   Basic Operation: Cancel or Divide Numerators and Denominators 56
2.10   Threshold Odor Number (TON) 59
2.11   Geometrical Measurements 60
        2.11.1   Definitions 60
        2.11.2   Relevant Geometric Equations 61
        2.11.3   Geometrical Calculations 61
2.12   Force, Pressure, and Head Calculations 69
        2.12.1   Force and Pressure 69
        2.12.2   Head 70
2.13   Review of Advanced Algebra Key Terms/Concepts 75
References and Recommended Reading 77

Chapter 3   Environmental Modeling 79
3.1   Introduction 79
3.2   Basic Steps for Developing an Effective Model 79
3.3   What Are Models Used For? 80
3.4   Media Material Content 81
        3.4.1   Material Content: Liquid Phases 82
3.5   Phase Equilibrium and Steady State 85
3.6   Math Operations and Laws of Equilibrium 86
        3.6.1   Solving Equilibrium Problems 86
        3.6.2   Laws of Equilibrium 87
3.7   Chemical Transport Systems 89
3.8   Final Word on Environmental Modeling 91
References and Recommended Reading 91

Chapter 4   Algorithms and Environmental Engineering 93
4.1   Introduction 93
4.2   Algorithms: What Are They? 93
4.3   Expressing Algorithms 94
4.4   General Algorithm Applications 94
4.5   Environmental Practice Algorithm Applications 96
4.6   Dispersion Models 96
4.7   Screening Tools 97
References and Recommended Reading 98

Chapter 5   Quadratic Equations 99
5.1   Quadratic Equations and Environmental Practice 99
        5.1.1   Key Terms 99
5.2   Quadratic Equations: Theory and Application 100

5.3   Derivation of the Quadratic Equation Formula.. 101
5.4   Using the Quadratic Equation 102
References and Recommended Reading.. 102

Chapter 6   Trigonometric Ratios.. 103

6.1   Trigonometric Functions and the Environmental Practitioner..................... 103
6.2   Trigonometric Ratios or Functions... 103
References and Recommended Reading.. 105

SECTION II   Statistics, Risk Measurement, and Boolean Algebra

Chapter 7   Statistics Review 109

7.1   Statistical Concepts 109
    7.1.1   Probability and Statistics... 110
7.2   Measure of Central Tendency... 111
7.3   Symbols, Subscripts, Basic Statistical Terms, and Calculations.................. 111
    7.3.1   Mean... 112
    7.3.2   Median.. 113
    7.3.3   Mode... 113
    7.3.4   Range.. 113
7.4   Distribution.. 115
7.5   Standard Deviation 117
7.6   Coefficient of Variation 118
7.7   Standard Error of the Mean.. 118
7.8   Covariance... 119
7.9   Simple Correlation Coefficient 120
7.10   Variance of a Linear Function.. 121
7.11   Sampling Measurement Variables... 123
    7.11.1   Simple Random Sampling.. 123
    7.11.2   Stratified Random Sampling 126
7.12   Sampling—Discrete Variables 129
    7.12.1   Random Sampling... 129
    7.12.2   Sample Size 130
7.13   Cluster Sampling for Attributes.. 130
    7.13.1   Transformations... 131
7.14   Chi-Square Tests.. 131
    7.14.1   Test of Independence... 131
    7.14.2   Test of a Hypothesized Count 132
    7.14.3   Bartlett's Test of Homogeneity of Variance 133
7.15   Comparing Two Groups by the t Test... 135
    7.15.1 t Test for Unpaired Plots.. 135
    7.15.2 t Test for Paired Plots 137
7.16   Comparison of Two or More Groups by Analysis of Variance..................... 139
    7.16.1   Complete Randomization... 139
    7.16.2   Multiple Comparisons.. 141
    7.16.3   Randomized Block Design... 144
    7.16.4   Latin Square Design.. 147

7.16.5 Factorial Experiments 149
7.16.6 Split-Plot Design.. 154
7.16.7 Missing Plots 159
7.17 Regression... 161
7.17.1 Simple Linear Regression 161
7.17.2 Multiple Regression.. 166
7.17.3 Curvilinear Regressions and Interactions 171
7.17.4 Group Regressions 173
7.17.5 Analysis of Covariance in a Randomized Block Design 176
References and Recommended Reading 179

Chapter 8 Computation and Measurement of Risk.. 181
8.1 Introduction 181
8.2 Frequency Measures.. 182
8.2.1 Ratio 182
8.2.2 Proportion.. 184
8.2.3 Rate.. 187
8.3 Morbidity Frequency Measures... 188
8.3.1 Incidence Proportion or Risk 188
8.3.2 Incidence Rate or Person-Time Rate 190
8.3.3 Prevalence 193
8.4 Mortality Frequency Measures 196
8.4.1 Mortality Rate 196
8.4.2 Death-to-Case Ratio.. 201
8.4.3 Case-Fatality Rate... 202
8.4.4 Proportionate Mortality.. 203
8.4.5 Years of Potential Life Lost..................................... 204
8.5 Natality (Birth) Measures.. 205
8.6 Measures of Association.. 206
8.6.1 Risk Ratio 206
8.6.2 Rate Ratio 208
8.6.3 Odds Ratio.. 209
8.7 Measures of Public Health Impact 210
8.7.1 Attributable Proportion 210
8.7.2 Vaccine Efficacy or Vaccine Effectiveness 211
References and Recommended Reading 211

Chapter 9 Boolean Algebra.. 213
9.1 Why Boolean Algebra?.. 213
9.1.1 Fault-Tree Analysis.. 214
9.1.2 Key Terms 216
9.2 Technical Overview... 216
9.2.1 Commutative Law 216
9.2.2 Distributive Law... 217
9.2.3 Identity and Inverse Variables.................................. 217
9.3 Boolean Synthesis.. 217
References and Recommended Reading 219

SECTION III   Economics

Chapter 10 Environmental Economics 223

　　10.1   Environmental Practice and Economics 223
　　　　　10.1.1 Key Terms 224
　　10.2   Capital-Recovery Factor (Equal-Payment Series)....................224
　　10.3   Uniform Series Present Worth (Value) Factor...........................225
　　10.4   Future Value 226
　　10.5   Annual Payment (Uniform Series Sinking Fund) 226
　　10.6   Present Value of Future Dollar Amount.....................................227
　　10.7   Future Value of a Present Amount 228
　　References and Recommended Reading 228

SECTION IV   Engineering Fundamentals

Chapter 11 Fundamental Engineering Concepts 231

　　11.1   Introduction 231
　　11.2   Resolution of Forces 232
　　11.3   Slings 235
　　　　　11.3.1 Rated Sling Loads 238
　　11.4   Inclined Plane 238
　　11.5   Properties of Materials 241
　　　　　11.5.1 Friction 245
　　　　　11.5.2 Specific Gravity...246
　　　　　11.5.3 Force, Mass, and Acceleration 246
　　　　　11.5.4 Centrifugal and Centripetal Forces.................................246
　　　　　11.5.5 Stress and Strain..247
　　11.6   Materials and Principles of Mechanics 247
　　　　　11.6.1 Statics 247
　　　　　11.6.2 Dynamics 248
　　　　　11.6.3 Hydraulics and Pneumatics—Fluid Mechanics.............248
　　　　　11.6.4 Welds..248
　　　　　11.6.5 Moment 248
　　　　　11.6.6 Beams and Columns..249
　　　　　11.6.7 Bending Moment..250
　　　　　11.6.8 Radius of Curvature (Beam) 252
　　　　　11.6.9 Column Stress 252
　　　　　11.6.10 Beam Flexure 252
　　　　　11.6.11 Reaction Force (Beams) 253
　　　　　11.6.12 Buckling Stress (Wood Columns)...................................254
　　　　　11.6.13 Floors...254
　　11.7   Principles of Electricity 255
　　　　　11.7.1 Nature of Electricity 256
　　　　　11.7.2 Simple Electrical Circuit 257
　　　　　11.7.3 Ohm's Law 259

11.7.4 Electrical Power 261
11.7.5 Electrical Energy (Kilowatt-Hours) 263
11.7.6 Series DC Circuit Characteristics 264
11.7.7 Parallel DC Circuits 274
11.7.8 Series–Parallel Circuits 284
11.7.9 Conductors 285
11.7.10 Magnetic Units 288
11.7.11 AC Theory 289
11.7.12 Inductance 298
References and Recommended Reading 303

SECTION V   Soil Mechanics

Chapter 12 Fundamental Soil Mechanics 307

12.1   Introduction 307
12.2   Soil: What Is It? 309
12.3   Soil Basics 310
       12.3.1 Soil for Construction 310
12.4   Soil Characteristics 311
       12.4.1 Soil Weight–Volume and Space–Volume Relationships 311
       12.4.2 Specific Gravity 313
12.5   Soil Particle Characteristics 314
12.6   Soil Stress and Strain 315
12.7   Soil Compressibility 316
12.8   Soil Compaction 316
12.9   Soil Failure 316
12.10 Soil Physics 319
12.11 Structural Failure 319
References and Recommended Reading 321

SECTION VI   Biomass Basic Computations

Chapter 13 Forest-Based Biomass Basic Computations 325

13.1   Introduction 325
       13.1.1 Forestry Service and Mathematics: The Interface 326
13.2   Forest Biomass Computations and Statistics 328
13.3   Sample Forest-Based Biomass Computations 328
       13.3.1 Lower and Higher Heating Values 328
       13.3.2 Effect of Fuel Moisture on Wood Heat Content 329
       13.3.3 Forestry Volume Unit to Biomass Weight Considerations 330
       13.3.4 Stand-Level Biomass Estimation 332
       13.3.5 Biomass Equations 333
13.4   Timber Scaling and Log Rules 335
       13.4.1 Theory of Scaling 335
       13.4.2 Authorized Log Rules 336
       13.4.3 Wood Density and Weight Ratios 339
References and Recommended Reading 339

SECTION VII   Fundamental Science Computations

Chapter 14 Fundamental Chemistry and Hydraulics..343

    14.1   Fundamental Chemistry 343
        14.1.1 Density and Specific Gravity 344
        14.1.2 Water Chemistry Fundamentals.......................................346
    14.2   Fundamental Hydraulics 369
        14.2.1 Principles of Water Hydraulics..369
        14.2.2 Basic Pumping Calculations..374
        14.2.3 Calculating Head Loss 375
        14.2.4 Calculating Head..375
        14.2.5 Calculating Horsepower and Efficiency..........................376
        14.2.6 Pump Efficiency and Brake Horsepower (BHP)...............376
    References and Recommended Reading 378

SECTION VIII   Environmental Health and Safety Computations

Chapter 15 Basic Calculations for Occupational Safety and Environmental Health
Professionals..381

    15.1   Industrial Hygiene: What Is It? 381
    15.2   Workplace Stressors 383
        15.2.1 Chemical Stressors...383
        15.2.2 Physical Stressors 384
        15.2.3 Biological Stressors 384
        15.2.4 Ergonomic Stressors...385
    15.3   Environmental Health and Safety Terminology.............................385
    15.4   Indoor Air Quality..395
        15.4.1 Sources of Indoor Air Pollutants.....................................396
    15.5   Air Pollution Fundamentals 396
        15.5.1 Six Common Air Pollutants 397
        15.5.2 Gases 399
        15.5.3 Particulate Matter...404
        15.5.4 Pollution Emission Measurement Parameters...................405
        15.5.5 Standard Corrections..405
        15.5.6 Atmospheric Standards 406
        15.5.7 Air Monitoring and Sampling...407
    15.6   Ventilation 409
        15.6.1 Definitions 410
        15.6.2 Concept of Ventilation...412
        15.6.3 Common Ventilation Measurements................................416
        15.6.4 Good Practices 417
        15.6.5 Facts, Concepts, and Example Math Problems 418
    15.7   Noise...422
        15.7.1 Definitions 422
        15.7.2 Occupational Noise Exposure 426
        15.7.3 Determining Workplace Noise Levels 427
        15.7.4 Engineering Control for Industrial Noise.........................429
        15.7.5 Noise Units, Relationships, and Equations.......................429

15.8   Thermal Stress..433
        15.8.1   Causal Factors 433
        15.8.2   Thermal Comfort 433
        15.8.3   Body's Response to Heat..434
        15.8.4   Work-Load Assessment..435
        15.8.5   Sampling Methods 435
        15.8.6   Heat Stress Sample Measurement and Calculation.......................437
        15.8.7   Cold Hazards 437
15.9   Radiation...438
        15.9.1   Definitions 438
        15.9.2   Ionizing Radiation..440
        15.9.3   Effective Half-Life 440
        15.9.4   Alpha Radiation..441
        15.9.5   Beta Radiation..441
        15.9.6   Gamma Radiation and X-Rays..442
        15.9.7   Radioactive Decay Equations..443
        15.9.8   Radiation Dose 444
References and Recommended Reading 444

SECTION IX   Math Concepts: Air Pollution Control

Chapter 16 Gas Emission Control...449

16.1   Introduction 449
16.2   Definitions 449
16.3   Absorption 452
        16.3.1   Solubility 453
        16.3.2   Equilibrium Solubility and Henry's Law 454
        16.3.3   Material (Mass) Balance 456
        16.3.4   Packed Towers and Plate Towers..459
16.4   Adsorption 469
        16.4.1   Adsorption Steps 470
        16.4.2   Adsorption Forces—Physical and Chemical 470
        16.4.3   Adsorption Equilibrium Relationships..471
        16.4.4   Factors Affecting Adsorption..474
16.5   Incineration...478
        16.5.1   Factors Affecting Incineration for Emission Control....................478
        16.5.2   Incineration Example Calculations 481
16.6   Condensation 483
        16.6.1   Contact Condenser Calculations 484
        16.6.2   Surface Condenser Calculations 485
References and Recommended Reading 489

Chapter 17 Particulate Emission Control...491

17.1   Particulate Emission Control Basics 491
        17.1.1   Interaction of Particles with Gas..491
        17.1.2   Particulate Collection..492
17.2   Particulate Size Characteristics and General Characteristics 493

17.2.1 Aerodynamic Diameter 493
17.2.2 Equivalent Diameter... 493
17.2.3 Sedimentation Diameter.. 493
17.2.4 Cut Diameter 493
17.2.5 Dynamic Shape Factor 494
17.3 Flow Regime of Particle Motion 494
17.4 Particulate Emission Control Equipment Calculations 499
17.4.1 Gravity Settlers.. 499
17.4.2 Cyclones 504
17.4.3 Electrostatic Precipitators 510
17.4.4 Baghouses (Fabric Filters)... 516
References and Recommended Reading 525

Chapter 18 Wet Scrubbers for Emission Control... 527

18.1 Introduction 527
18.1.1 Wet Scrubbers 527
18.2 Wet Scrubber Collection Mechanisms and Efficiency (Particulates)............ 529
18.2.1 Collection Efficiency 529
18.2.2 Impaction... 529
18.2.3 Interception.. 530
18.2.4 Diffusion.. 530
18.2.5 Calculation of Venturi Scrubber Efficiency 531
18.3 Wet Scrubber Collection Mechanisms and Efficiency (Gaseous Emissions) ...541
18.4 Assorted Venturi Scrubber Example Calculations..................................... 542
18.4.1 Calculations for Scrubber Design of a Venturi Scrubber................. 542
18.4.2 Spray Tower Calculations... 548
18.4.3 Packed Tower Calculations 549
18.4.4 Packed Column Height and Diameter Calculations......................... 551
References and Recommended Reading 555

SECTION X Math Concepts: Water Quality

Chapter 19 Running Waters... 559

19.1 Balancing the Aquarium 559
19.1.1 Sources of Stream Pollution... 560
19.2 Is Dilution the Solution? 561
19.2.1 Dilution Capacity of Running Waters 562
19.3 Discharge Measurement 562
19.4 Time of Travel 563
19.5 Dissolved Oxygen.. 563
19.5.1 Dissolved Oxygen Correction Factor 565
19.6 Biochemical Oxygen Demand... 565
19.6.1 BOD Test Procedure.. 565
19.6.2 Practical BOD Calculation Procedure 566
19.7 Oxygen Sag (Deoxygenation).. 568
19.8 Stream Purification: A Quantitative Analysis 569
References and Recommended Reading 572

Chapter 20 Still Waters 573

    20.1   Still Water Systems 574
    20.2   Still Water System Calculations 575
          20.2.1 Still Water Body Morphometry Calculations 575
    20.3   Still Water Surface Evaporation 578
          20.3.1 Water Budget Model 578
          20.3.2 Energy Budget Model 579
          20.3.3 Priestly–Taylor Equation 579
          20.3.4 Penman Equation 580
          20.3.5 DeBruin–Keijman Equation 580
          20.3.6 Papadakis Equation 580
    References and Recommended Reading 580

Chapter 21 Groundwater 581

    21.1   Groundwater and Aquifers 581
          21.1.1 Groundwater Quality 583
          21.1.2 Groundwater under the Direct Influence of Surface Water 583
    21.2   Aquifer Parameters 583
          21.2.1 Aquifer Porosity 583
          21.2.2 Specific Yield (Storage Coefficient) 584
          21.2.3 Permeability 584
          21.2.4 Transmissivity 584
          21.2.5 Hydraulic Gradient and Head 584
          21.2.6 Flow Lines and Flow Nets 585
    21.3   Groundwater Flow 585
    21.4   General Equations of Groundwater Flow 586
          21.4.1 Steady Flow in a Confined Aquifer 587
          21.4.2 Steady Flow in an Unconfined Aquifer 587
    References and Recommended Reading 587

Chapter 22 Water Hydraulics 589

    22.1   Introduction 589
    22.2   Basic Concepts 589
          22.2.1 Stevin's Law 591
          22.2.2 Density and Specific Gravity 591
          22.2.3 Force and Pressure 593
          22.2.4 Hydrostatic Pressure 594
          22.2.5 Head 594
    22.3   Flow and Discharge Rate: Water in Motion 596
          22.3.1 Area and Velocity 597
          22.3.2 Pressure and Velocity 598
    22.4   Bernoulli's Theorem 598
          22.4.1 Bernoulli's Equation 599
    22.5   Calculating Major Head Loss 600
          22.5.1 *C* Factor 601
    22.6   Characteristics of Open-Channel Flow 601
          22.6.1 Laminar and Turbulent Flow 601
          22.6.2 Uniform and Varied Flow 602

22.6.3 Critical Flow...602
22.6.4 Parameters Used in Open Channel Flow 602
22.7 Open-Channel Flow Calculations 603
References and Recommended Reading 604

Chapter 23 Water Treatment Processes 605

23.1 Introduction 605
23.2 Water Source and Storage Calculations 605
    23.2.1 Water Source Calculations 606
    23.2.2 Vertical Turbine Pump Calculations 609
23.3 Water Storage 612
    23.3.1 Water Storage Volume Calculations..613
    23.3.2 Copper Sulfate Dosing 613
23.4 Coagulation, Mixing, and Flocculation...615
    23.4.1 Coagulation 615
    23.4.2 Mixing 615
    23.4.3 Flocculation 616
    23.4.4 Coagulation and Flocculation General Calculations......................616
    23.4.5 Determining Percent Strength of Solutions 619
    23.4.6 Dry Chemical Feeder Calibration 621
    23.4.7 Determining Chemical Usage 623
23.5 Sedimentation Calculations...625
    23.5.1 Tank Volume Calculations 625
    23.5.2 Detention Time...626
    23.5.3 Surface Overflow Rate 627
    23.5.4 Mean Flow Velocity 627
    23.5.5 Weir Overflow Rate..628
    23.5.6 Percent Settled Biosolids 629
    23.5.7 Determining Lime Dosage (mg/L)..630
    23.5.8 Determining Lime Dosage (lb/day)...632
    23.5.9 Determining Lime Dosage (Grams per Minute).............................633
    23.5.10 Particle Settling (Sedimentation) 633
    23.5.11 Overflow Rate (Sedimentation)...636
23.6 Water Filtration Calculations...638
    23.6.1 Flow Rate through a Filter (gpm)...638
    23.6.2 Filtration Rate...640
    23.6.3 Unit Filter Run Volume (UFRV)...641
    23.6.4 Backwash Rate 642
    23.6.5 Backwash Rise Rate 643
    23.6.6 Volume of Backwash Water Required (gal).....................................644
    23.6.7 Required Depth of Backwash Water Tank (ft) 644
    23.6.8 Backwash Pumping Rate (gpm)...645
    23.6.9 Percent Product Water Used for Backwashing...............................645
    23.6.10 Percent Mudball Volume 646
    23.6.11 Filter Bed Expansion 647
    23.6.12 Filter Loading Rate 648
    23.6.13 Filter Medium Size 648
    23.6.14 Mixed Media 649
    23.6.15 Head Loss for Fixed Bed Flow...649

        23.6.16 Head Loss through a Fluidized Bed..651
        23.6.17 Horizontal Washwater Troughs..652
        23.6.18 Filter Efficiency..653
   23.7   Water Chlorination Calculations...653
        23.7.1 Chlorine Disinfection..654
        23.7.2 Determining Chlorine Feed Rate...654
        23.7.3 Calculating Chlorine Dose, Demand, and Residual 655
        23.7.4 Breakpoint Chlorination Calculations...656
        23.7.5 Calculating Dry Hypochlorite Feed Rate.......................................658
        23.7.6 Calculating Hypochlorite Solution Feed Rate................................659
        23.7.7 Percent Strength of Solutions..660
        23.7.8 Calculating Percent Strength Using Dry Hypochlorite...................660
        23.7.9 Calculating Percent Strength Using Liquid Hypochlorite..............661
   23.8   Chemical Use Calculations...662
   23.9   Chlorination Chemistry...662
   References and Recommended Reading..664

SECTION XI   Math Concepts: Wastewater

Chapter 24 Wastewater Calculations 667
   24.1   Introduction 667
   24.2   Preliminary Treatment 667
        24.2.1 Screening Removal Calculations 667
        24.2.2 Screening Pit Capacity Calculations 669
        24.2.3 Head Loss through Bar Screen Calculations 670
        24.2.4 Grit Removal Calculations 670
        24.2.5 Grit Channel Velocity Calculations 672
        24.2.6 Required Settling Time Calculations 672
        24.2.7 Required Grit Channel Length Calculations.................................673
        24.2.8 Velocity of Scour Calculations..673
   24.3   Primary Treatment 674
        24.3.1 Process Control Calculations 674
        24.3.2 Surface Loading Rate (Surface Settling Rate/
                  Surface Overflow Rate) Calculations 675
        24.3.3 Weir Overflow Rate (Weir Loading Rate) Calculations 675
        24.3.4 Primary Sedimentation Basin Calculations 676
   24.4   Biosolids Pumping Calculations...677
        24.4.1 Percent Total Solids Calculations...677
        24.4.2 BOD and SS Removal Calculations 677
   24.5   Trickling Filter...678
        24.5.1 Trickling Filter Process Calculations...679
   24.6   Rotating Biological Contactors...682
        24.6.1 RBC Process Control Calculations...682
   24.7   Activated Biosolids...687
        24.7.1 Activated Biosolids Process Control Calculations.........................687
   24.8   Oxidation Ditch Detention Time 701
   24.9   Treatment Ponds 702
        24.9.1 Treatment Pond Parameters 702
        24.9.2 Treatment Pond Process Control Calculations..............................703

24.9.3 Aerated Ponds..704
24.10 Chemical Dosage..705
    24.10.1 Chemical Feed Rate...706
    24.10.2 Chlorine Dose, Demand, and Residual...........................707
    24.10.3 Hypochlorite Dosage 708
    24.10.4 Chemical Solutions..709
    24.10.5 Mixing Solutions of Different Strength 710
    24.10.6 Solution Mixtures and Target Percent Strength............711
    24.10.7 Chemical Solution Feeder Setting (gpd)........................712
    24.10.8 Chemical Feed Pump: Percent Stroke Setting 713
    24.10.9 Chemical Solution Feeder Setting (mL/min).................713
    24.10.10 Chemical Feed Rate Calibration....................................713
    24.10.11 Average Use Calculations 716
24.11 Biosolids Production and Pumping 717
    24.11.1 Process Residuals 717
    24.11.2 Primary and Secondary Solids Production Calculations...............717
    24.11.3 Primary Clarifier Solids Production Calculations 717
    24.11.4 Secondary Clarifier Solids Production Calculation 718
    24.11.5 Percent Solids Calculations..718
    24.11.6 Biosolids Pumping Calculations 719
24.12 Biosolids Thickening..722
    24.12.1 Gravity/Dissolved Air Flotation Thickener Calculations 722
    24.12.2 Concentration Factor Calculations 724
    24.12.3 Air-to-Solids Ratio Calculations 724
    24.12.4 Recycle Flow in Percent Calculations 724
    24.12.5 Centrifuge Thickening Calculations 725
24.13 Stabilization...725
    24.13.1 Biosolids Digestion...725
    24.13.2 Aerobic Digestion Process Control Calculations 726
    24.13.3 Aerobic Tank Volume Calculations................................727
    24.13.4 Anaerobic Digestion Process Control Calculations 727
24.14 Biosolids Dewatering and Disposal..731
    24.14.1 Biosolids Dewatering 731
    24.14.2 Pressure Filtration Calculations 732
    24.14.3 Rotary Vacuum Filter Dewatering Calculations 736
    24.14.4 Sand Drying Bed Calculations.......................................738
    24.14.5 Biosolids Disposal Calculations 740
24.15 Wastewater Laboratory Calculations..744
    24.15.1 Wastewater Lab 744
    24.15.2 Composite Sampling Procedures and Calculations 745
    24.15.3 Biochemical Oxygen Demand Calculations...................746
    24.15.4 Mole and Molarity Calculations....................................748
    24.15.5 Settleability (Activated Biosolids Solids) Calculations....................750
    24.15.6 Settleable Solids Calculations 751
    24.15.7 Biosolids Total Solids, Fixed Solids and
         Volatile Solids Calculations 752
    24.15.8 Wastewater Suspended Solids and
         Volatile Suspended Solids Calculations 753
    24.15.9 Biosolids Volume Index and Biosolids
         Density Index Calculations 754
References and Recommended Reading...755

SECTION XII Math Concepts: Stormwater Engineering

Chapter 25 Stormwater Calculations 759

25.1 Introduction 759
25.2 Stormwater Terms and Acronyms 759
25.3 Hydrologic Methods 765
    25.3.1 Precipitation 767
25.4 Runoff Hydrographs 772
25.5 Runoff and Peak Discharge 772
25.6 Calculation Methods 773
    25.6.1 Rational Method 773
    25.6.2 Modified Rational Method 777
    25.6.3 *TR-55* Estimating Runoff Method 779
    25.6.4 *TR-55* Graphical Peak Discharge Method 787
    25.6.5 *TR-55* Tabular Hydrograph Method 788
25.7 General Stormwater Engineering Calculations 790
    25.7.1 Detention, Extended Detention, and
            Retention Basin Design Calculations 790
    25.7.2 Allowable Release Rates 791
    25.7.3 Storage Volume Requirement Estimates 791
    25.7.4 Graphical Hydrograph Analysis—SCS Methods 792
    25.7.5 *TR-55* Storage Volume for Detention Basins (Shortcut Method) 794
    25.7.6 Graphical Hydrograph Analysis, Modified
            Rational Method, Critical Storm Duration 796
    25.7.7 Modified Rational Method, Critical
            Storm Duration—Direct Solution 799
    25.7.8 Stage–Storage Curve 806
    25.7.9 Water Quality and Channel Erosion
            Control Volume Calculations 808
  References and Recommended Reading 814

Index 817

Preface

For many years, first as students and later as instructors, we observed undergraduate and graduate students in several environmental disciplines who had been required as undergraduates to take basic college-level math, calculus, and statistics courses. Those courses have mostly emphasized how to prove theorems but have neglected real-world applications. Most of the time, the students have come out of such courses with little or no appreciation of how to apply mathematical operations in their own work. Based on these observations, we developed our *Environmental Engineer's Mathematics Handbook* in 2005, and it was well received by users. However, as with anything composed of the written or digital word, there are critics and room for improvement. This is the case with the original handbook. Even though it was an industry-wide best seller hailed on its first publication as a masterly account written in an engaging, highly readable, and useable style, the original handbook had room for improvement. Specifically, we found that the original handbook was too narrow in scope. Environmental practice is a broad area; thus, the guiding principle behind development of the original handbook's replacement edition, *Handbook of Mathematics and Statistics for the Environment*, was to broaden the scope of presentation.

This replacement book assembles, explains, and integrates in a single text the fundamental math operations performed by environmental practitioners for air; water; wastewater; solid and hazardous wastes; biosolids; environmental health, safety, and welfare; environmental science; environmental economics; stormwater operations; environmental laboratory operations; auditing; risk management; environmental monitoring; nuclear medicine; and environmental planning and managing. *Handbook of Mathematics and Statistics for the Environment* offers the reader an unusual approach to presenting environmental math concepts, one that emphasizes the relationship between the principles in natural processes and those employed in environmental processes.

This text covers in detail the environmental principles, practices, and math operations involved in the design and operation of conventional environmental works (i.e., water and wastewater treatment works) and presents environmental modeling tools and environmental algorithm examples. The arrangement of the material lends itself to several different specific environmental specialties and several different forma course formats.

Major subjects covered in this book include

- Math concepts review
- Modeling
- Algorithms
- Air pollution control calculations
- Water assessment and control calculations

New material in this edition includes

- Quadratic equations, trigonometric ratios, statistics review, Boolean algebra
- Environmental economics
- Fundamental engineering and environmental concepts
- Environmental health computations and solutions
- Basic electricity for environmental practitioners
- Lab computations
- Greenhouse gas emission computation procedures
- Fisheries pond and tank calculations

- Calculations for carbon sequestration by trees
- Forest biomass calculations
- U.S. Department of Agriculture calculations

We emphasize concepts, definitions, descriptions, and derivations and provide a touch of the natural world blended with common sense in our approach. This book is intended to be a combination textbook and reference tool for those practitioners involved in the protection of the four environmental media: air, weather, land resources, and biota.

Authors

Frank R. Spellman, PhD, is a retired assistant professor of environmental health at Old Dominion University, Norfolk, Virginia, and the author of more than 83 books covering topics ranging from concentrated animal feeding operations (CAFOs) to all areas of environmental science and occupational health. Many of his texts are readily available online, and several have been adopted for classroom use at major universities throughout the United States, Canada, Europe, and Russia; two have been translated into Spanish for South American markets. Dr. Spellman has been cited in more than 850 publications. He serves as a professional expert witness for three law groups and as an incident/accident investigator for the U.S. Department of Justice and a northern Virginia law firm. In addition, he consults on homeland security vulnerability assessments for critical infrastructures including water/wastewater facilities nationwide and conducts pre-Occupational Safety and Health Administration (OSHA)/Environmental Protection Agency audits throughout the country. Dr. Spellman receives frequent requests to co-author with well-recognized experts in several scientific fields; for example, he was a contributing author to the prestigious text *The Engineering Handbook*, 2nd ed. (CRC Press). Dr. Spellman lectures on wastewater treatment, water treatment, homeland security, and safety topics and teaches water/wastewater operator short courses at Virginia Tech (Blacksburg, Virginia). Recently, he traced and documented the ancient water distribution system at Machu Pichu, Peru, and surveyed several drinking water resources in Amazonia-Coco, Ecuador. Dr. Spellman also studied and surveyed two separate potable water supplies in the Galapagos Islands and while there researched Darwin's finches. He holds a BA in public administration, a BS in business management, and an MBA, an MS, and a PhD in environmental engineering.

Nancy E. Whiting is a freelance technical writer in water/wastewater, environmental technology, quality management systems, security, occupational safety and health, and international traffic in arms compliance.

Section I

Fundamental Conversions, Computations, Modeling, and Algorithms

It is the mark of an instructed mind to rest satisfied with the degree of precision which the nature of the subject permits and not to see an exactness where only an approximation of the truth is possible.

—**Aristotle** (*Nicomachean Ethics*)

1 Introduction

I thought I was mathematically dysfunctional. But I am not unique—69 out of every 9 people are also dysfunctional.

—Frank R. Spellman (2005)

1.1 SETTING THE STAGE

It is general knowledge that mathematics is the study of numbers and counting and measuring, but its associated collaterals are less recognized. Simply, mathematics is more than numbers; it also involves the study of number patterns and relationships. It is also a means of communicating ideas. Perhaps, however, mathematics, more than anything, is a way of reasoning that is unique to human beings. No matter how we describe or define mathematics, one thing is certain—without an understanding of mathematical units and conversion factors, one might as well delve into the mysteries of deciphering hieroglyphics while blindfolded and lacking the sense of touch and reason.

1.2 SI UNITS

The units most commonly used by environmental engineering professionals are based on the complicated English system of weights and measures; however, bench work is usually based on the metric system, or International System of Units (SI), due to the convenient relationship between milliliters (mL), cubic centimeters (cm^3), and grams (g). The SI is a modernized version of the metric system established by international agreement. The metric system of measurement was developed during the French Revolution and was first promoted in the United States in 1866. In 1902, proposed congressional legislation requiring the U.S. government to use the metric system exclusively was defeated by a single vote. Although we use both systems in this text, the SI provides a logical and interconnected framework for all measurements in engineering, science, industry, and commerce. The metric system is much simpler to use than the existing English system, because all of its units of measurement are divisible by 10.

Before listing the various conversion factors commonly used in environmental engineering it is important to describe the prefixes commonly used in the SI system. These prefixes are based on the power of 10. For example, a "kilo" means 1000 grams, and a "centimeter" means 1/100 of 1 meter. The 20 SI prefixes used to form decimal multiples and submultiples of SI units are given in Table 1.1. Multiple prefixes may not be used. The prefix *names* shown in Table 1.1 are combined with the unit name "gram," and the prefix *symbols* are used with the unit symbol "g." With this exception, any SI prefix may be used with any SI unit, including the degree Celsius and its symbol °C.

■ EXAMPLE 1.1

10^{-6} kg = 1 mg (one milligram) is acceptable, but not 10^{-6} kg = 1 μkg (one microkilogram).

TABLE 1.1

SI Prefixes

Factor	Name	Symbol	Factor	Name	Symbol
10^{24}	Yotta	Y	10^{-1}	Deci	d
10^{21}	Zetta	Z	10^{-2}	Centi	c
10^{18}	Exa	E	10^{-3}	Milli	m
10^{15}	Peta	P	10^{-6}	Micro	μ
10^{12}	Tera	T	10^{-9}	Nano	n
10^{9}	Giga	G	10^{-12}	Pico	p
10^{6}	Mega	M	10^{-15}	Femto	f
10^{3}	Kilo	k	10^{-18}	Atto	a
10^{2}	Hecto	h	10^{-21}	Zepto	z
10^{1}	Deka	da	10^{-24}	Yocto	y

■ **EXAMPLE 1.2**

Consider the height of the Washington Monument. We may write it as 169,000 mm, 16,900 cm, 169 m, or 0.169 km, using the units of millimeter (prefix "milli," symbol "m"), centimeter (prefix "centi," symbol "c"), or kilometer (prefix "kilo," symbol "k").

1.3 CONVERSION FACTORS

Conversion factors are given alphabetically in Table 1.2 and are listed by unit category in Table 1.3.

■ **EXAMPLE 1.3**

Problem: Find degrees in Celsius of water at 72°F.

Solution:

$$°C = (F - 32) \times 5/9 = (72 - 32) \times 5/9 = 22.2$$

DID YOU KNOW?

The Fibonacci sequence is the following sequence of numbers:

1, 1, 2, 3, 5, 8, 13, 21, 34, 55, 89, 144, …

Or, alternatively,

0.1, 1, 2, 3, 5, 8, 13, 21, 34, 55, 89, 144, …

Note that each term from the third one onward is *the sum of the previous two*. Another point to notice is that, if you divide each number in the sequence by the next number, beginning with the first, an interesting thing appears to be happening:

1/1 = 1, 1/2 = 0.5, 2/3 = 0.66666, 3/5 = 0.6,
5/8 = 0.625, 8/13 = 0.61538, 13/21 = 0.61904, …

Note that the first of these ratios appear to be converging to a number just a bit larger than 0.6.

TABLE 1.2
Alphabetical List of Conversion Factors

Factor	Metric (SI) or English Conversions
°C	$(5/9)[(°F) - 32°]$
°F	$(9/5)[(°C) + 32°]$
1°C (expressed as an interval)	$33.8°F = (9/5)(°F)$
	1.8°R (degrees Rankine)
	1.0 K (degrees Kelvin)
1°F (expressed as an interval)	$0.556°C = (5/9)°C$
	1.0°R (degrees Rankine)
	0.556 K (degrees Kelvin)
1 atm (atmosphere)	1.013 bars
	10.133 N/cm^2 (newtons/square centimeter)
	33.90 ft of H$_2$O (feet of water)
	101.325 kPa (kilopascals)
	1,013.25 mbar (millibars)
	psia (pounds/square inch absolute)
	760 torr
	760 mmHg (millimeters of mercury)
1 bar	0.987 atm (atmospheres)
	1×10^6 dynes/cm^2 (dynes/square centimeter)
	33.45 ft of H$_2$O (feet of water)
	1×10^5 Pa (pascals)
	750.06 torr
	750.06 mmHg (millimeters of mercury)
1 Bq (becquerel)	1 radioactive disintegration/second
	2.7×10^{-11} Ci (curie)
	2.7×10^{-8} mCi (millicurie)
1 BTU (British Thermal Unit)	252 cal (calories)
	1055.06 j (joules)
	10.41 L atm (liter-atmosphere)
	0.293 Wh (watt-hours)
1 cal (calorie)	3.97×10^{-3} BTUs (British Thermal Units)
	4.18 j (joules)
	0.0413 L atm (liter-atmospheres)
	1.163×10^{-3} Wh (watt-hours)
1 Ci (curie)	3.7×10^{10} radioactive disintegrations/second
	3.7×10^{10} Bq (becquerel)
	1000 mCi (millicurie)
1 cm (centimeter)	0.0328 ft (feet)
	0.394 in. (inches)
	10,000 μm (microns, micrometers)
	100,000,000 Å = 10^8 Å (Ångstroms)
1 cm^2 (square centimeter)	1.076×10^{-3} ft^2 (square feet)
	0.155 in.^{2} (square inches)
	1×10^{-4} m^2 (square meters)
1 cm^3 (cc, cubic centimeter)	3.53×10^{-5} ft^3 (cubic feet)
	0.061 in.^{3} (cubic inches)
	2.64×10^{-4} gal (gallons)
	52.18 L (liters)
	52.18 mL (milliliters)

(continues)

TABLE 1.2 (continued)
Alphabetical List of Conversion Factors

Factor	Metric (SI) or English Conversions
1 day	24 hr (hours)
	1440 min (minutes)
	86,400 sec (seconds)
	0.143 wk (weeks)
	2.738×10^{-3} yr (years)
1 dyne	1×10^{-5} N (newton)
1 erg	1 dyn•cm (dyne-centimeter)
	1×10^{-7} j (joules)
	2.78×10^{-11} Wh (watt-hours)
1 eV (electron volt)	1.602×10^{-12} ergs
	1.602×10^{-19} j (joules)
1 fps (feet per second)	1.097 kmph (kilometers/hour)
	0.305 mps (meters/second)
	0.01136 mph (miles/hour)
1 ft (foot)	30.48 cm (centimeters)
	12 in. (inches)
	0.3048 m (meters)
	1.65×10^{-4} NM (nautical miles)
	1.89×10^{-4} mi (statute miles)
1 ft^2 (square foot)	2.296×10^{-5} acres
	9.296 cm^2 (square centimeters)
	144 in.^{2} (square inches)
	0.0929 m^2 (square meters)
1 ft^3 (cubic foot)	28.317 cm^3 (cc, cubic centimeters)
	1,728 in.^{3} (cubic inches)
	0.0283 m^3 (cubic meters)
	7.48 gal (gallons)
	28.32 L (liters)
	29.92 qt (quarts)
1 g (gram)	0.001 kg (kilogram)
	1000 mg (milligrams)
	$1,000,000$ ng = 10^6 ng (nanograms)
	2.205×10^{-3} lb (pounds)
1 g/cm^3 (grams per cubic centimeter)	62.43 lb/ft^3 (pounds/cubic foot)
	0.0361 lb/in.^{3} (pounds/cubic inch)
	8.345 lb/gal (pounds/gallon)
1 gal (gallon)	3785 cm^3 (cc, cubic centimeters)
	0.134 ft^3 (cubic feet)
	231 in.^{3} (cubic inches)
	3.785 L (liters)
1 Gy (gray)	1 j/kg (joules/kilogram)
	100 rad
	1 Sv (sievert), unless modified through division by an appropriate factor, such as Q or N
1 hp (horsepower)	745.7 j/sec (joules/sec)

TABLE 1.2 (continued)
Alphabetical List of Conversion Factors

Factor	Metric (SI) or English Conversions
1 hr (hour)	0.0417 days
	60 min (minutes)
	3600 sec (seconds)
	5.95×10^{-3} wk (weeks)
	1.14×10^{-4} yr (years)
1 in. (inch)	2.54 cm (centimeters)
	1000 mil
1 in.^{3} (cubic inch)	16.39 cm^3 (cc, cubic centimeters)
	16.39 mL (milliliters)
	5.79×10^{-4} ft^3 (cubic feet)
	1.64×10^{-5} m^3 (cubic meters)
	4.33×10^{-3} gal (gallons)
	0.0164 L (liters)
	0.55 fl oz. (fluid ounces)
1 inch of water	1.86 mmHg (millimeters of mercury)
	249.09 Pa (pascals)
	0.0361 psi (lb/in.^{2})
1 j (joule)	9.48×10^{-4} BTUs (British Thermal Units)
	0.239 cal (calories)
	10,000,000 ergs = 1×10^7 ergs
	9.87×10^{-3} L atm (liter-atmospheres)
	1.0 N-m (newton-meters)
1 kcal (kilocalories)	3.97 BTUs (British Thermal Units)
	1000 cal (calories)
	4186.8 j (joules)
1 kg (kilogram)	1000 g (grams)
	2205 lb (pounds)
1 km (kilometer)	3280 ft (feet)
	0.54 NM (nautical miles)
	0.6214 mi (statute miles)
1kW (kilowatt)	56.87 BTU/min (British Thermal Units per minute)
	1.341 hp (horsepower)
	1000 j/sec (joules per second)
1 kWh (kilowatt-hour)	3412.14 BTU (British Thermal Units)
	3.6×10^6 j (joules)
	859.8 kcal (kilocalories)
1 L (liter)	1000 cm^3 (cc, cubic centimeters)
	1 dm^3 (cubic decimeters)
	0.0353 ft^3 (cubic feet)
	61.02 in.^{3} (cubic inches)
	0.264 gal (gallons)
	1000 mL (milliliters)
	1.057 qt (quarts)
1 lb (pound)	453.59 g (grams)
	16 oz. (ounces)

(continues)

TABLE 1.2 (continued)
Alphabetical List of Conversion Factors

Factor	Metric (SI) or English Conversions
1 lb/ft^3 (pounds per cubic foot)	16.02 g/L (grams/liter)
1 lb/in.^{3} (pounds per cubic inch)	27.68 g/cm^3 (grams/cubic centimeter)
	1728 lb/ft^3 (pounds/cubic feet)
1 m (meter)	1×10^{10} Å (Ångstroms)
	100 cm (centimeters)
	3.28 ft (feet)
	39.37 in. (inches)
	1×10^{-3} km (kilometers)
	1000 mm (millimeters)
	1,000,000 μm = 1×10^6 μm (micrometers)
	1×10^9 nm (nanometers)
1 m^2 (square meter)	10.76 ft^2 (square feet)
	1550 in.^{2} (square inches)
1 m^3 (cubic meter)	1,000,000 cm^3 = 10^6 cm^3 (cc, cubic centimeters)
	33.32 ft^3 (cubic feet)
	61,023 in.^{3} (cubic inches)
	264.17 gal (gallons)
	1000 L (liters)
1 mCi (millicurie)	0.001 Ci (curie)
	3.7×10^{10} radioactive disintegrations/second
	3.7×10^{10} Bq (becquerel)
1 mi (statute mile)	5280 ft (feet)
	1.609 km (kilometers)
	1609.3 m (meters)
	0.869 NM (nautical miles)
	1760 yd (yards)
1 mi^2 (square mile)	640 acres
	2.79×10^7 ft^2 (square feet)
	2.59×10^6 m^2 (square meters)
1 min (minute)	6.94×10^{-4} days
	0.0167 hr (hours)
	60 sec (seconds)
	9.92×10^{-5} wk (weeks)
	1.90×10^{-6} yr (years)
1 mmHg (mm of mercury)	1.316×10^{-3} atm (atmosphere)
	0.535 in H$_2$O (inches of water)
	1.33 mb (millibars)
	133.32 Pa (pascals)
	1 torr
	0.0193 psia (pounds/square inch absolute)
1 mph (miles per hour)	88 fpm (feet/minute)
	1.61 kmph (kilometers/hour)
	0.447 mps (meters/second)
1 mps (meters per second)	196.9 fpm (feet/minute)
	3.6 kmph (kilometers/hour)
	2.237 mph (miles/hour)
1 N (newton)	1×10^5 dynes

TABLE 1.2 (continued)
Alphabetical List of Conversion Factors

Factor	Metric (SI) or English Conversions
1 N-m (newton-meter)	1.00 j (joules)
1 NM (nautical mile)	6076.1 ft (feet)
	1.852 km (kilometers)
	1.15 mi (statute miles)
	2025.4 yd (yards)
1 Pa (pascal)	9.87×10^{-6} atm (atmospheres)
	4.015×10^{-3} in. H_2O (inches of water)
	0.01 mb (millibars)
	7.5×10^{-3} mmHg (milliliters of mercury)
1 ppm (parts per million)	1.00 mL/m³ (milliliters/cubic meter)
	1.00 mg/kg (milligrams/kilogram)
1 psi (pounds/square inch)	0.068 atm (atmospheres)
	27.67 in H_2O (inches of water)
	68.85 mb (millibars)
	51.71 mmHg (millimeters of mercury)
	6894.76 Pa (pascals)
1 qt (quart)	946.4 cm³ (cc, cubic centimeters)
	57.75 in.³ (cubic inches)
	0.946 L (liters)
1 rad	100 ergs/g (ergs/gram)
	0.01 Gy (gray)
	1 rem, unless modified through division by an appropriate factor, such as Q or N
1 rem	1 rad, unless modified through division by an appropriate factor, such as Q or N
1 Sv (sievert)	1 Gy, unless modified through division by an appropriate factor, such as Q or N
1 torr	1.33 mb (millibars)
1 W (watt)	3.41 BTU/hr (British Thermal Units/hour)
	1.341×10^{-3} hp (horsepower)
	52.18 j/sec (joules/second)
1 week	7 days
	168 hr (hours)
	10,080 min (minutes)
	6.048×10^5 sec (seconds)
	0.0192 yr (years)
1 Wh (watt-hour)	3.412 BTUs (British Thermal Units)
	859.8 cal (calories)
	3600 j (joules)
	35.53 L atm (liter-atmosphere)
1 yd³ (cubic yard)	201.97 gal (gallons)
	764.55 L (liters)
1 yr (year)	365.25 days
	8766 hr (hours)
	5.26×10^5 min (minutes)
	3.16×10^7 sec (seconds)
	52.18 weeks

TABLE 1.3

Conversion Factors by Unit Category

Factor	Metric (SI) or English Conversions
	Units of Length
1 cm (centimeter)	0.0328 ft (feet)
	0.394 in. (inches)
	10,000 µm (microns, micrometers)
	100,000,000 Å = 10^8 Å (Ångstroms)
1 ft (foot)	30.48 cm (centimeters)
	12 in. (inches)
	0.3048 m (meters)
	1.65×10^{-4} NM (nautical miles)
	1.89×10^{-4} mi (statute miles)
1 in. (inch)	2.54 cm (centimeters)
	1000 mils
1 km (kilometer)	3280.8 ft (feet)
	0.54 NM (nautical miles)
	0.6214 mi (statute miles)
1 m (meter)	1×10^{10} Å (Ångstroms)
	100 cm (centimeters)
	3.28 ft (feet)
	39.37 in. (inches)
	1×10^{-3} km (kilometers)
	1000 mm (millimeters)
	1,000,000 µm = 1×10^6 µm (microns, micrometers)
	1×10^9 nm (nanometers)
1 NM (nautical mile)	6076.1 ft (feet)
	1.852 km (kilometers)
	1.15 mi (statute miles)
	2025.4 yd (yards)
1 mi (statute mile)	5280 ft (feet)
	1.609 km (kilometers)
	1690.3 m (meters)
	0.869 NM (nautical miles)
	1760 yd (yards)
	Units of Area
1 cm² (square centimeter)	1.076×10^{-3} ft² (square feet)
	0.155 in.² (square inches)
	1×10^{-4} m² (square meters)
1 ft² (square foot)	2.296×10^{-5} acres
	929.03 cm² (square centimeters)
	144 in.² (square inches)
	0.0929 m² (square meters)
1 m² (square meter)	10.76 ft² (square feet)
	1550 in.² (square inches)
1 mi² (square mile)	640 acres
	2.79×10^7 ft² (square feet)
	2.59×10^6 m² (square meters)

TABLE 1.3 (continued)
Conversion Factors by Unit Category

Factor	Metric (SI) or English Conversions
	Units of Volume
1 cm³ (cubic centimeter)	3.53×10^{-5} ft³ (cubic feet)
	0.061 in.³ (cubic inches)
	2.64×10^{-4} gal (gallons)
	0.001 L (liters)
	1.00 mL (milliliters)
1 ft³ (cubic foot)	28,317 cm³ (cc, cubic centimeters)
	1728 in³ (cubic inches)
	0.0283 m³ (cubic meters)
	7.48 gal (gallons)
	28.32 L (liters)
	29.92 qt (quarts)
1 in.³ (cubic inch)	16.39 cm³ (cc, cubic centimeters)
	16.39 mL (milliliters)
	5.79×10^{-4} ft³ (cubic feet)
	1.64×10^{-5} m³ (cubic meters)
	4.33×10^{-3} gal (gallons)
	0.0164 L (liters)
	0.55 fl oz. (fluid ounces)
1 m³ (cubic meter)	1,000,000 cm³ = 10^6 cm³ (cc, cubic centimeters)
	35.31 ft³ (cubic feet)
	61,023 in³ (cubic inches)
	264.17 gal (gallons)
	1000 L (liters)
1 yd³ (cubic yards)	201.97 gal (gallons)
	764.55 L (liters)
1 gal (gallon)	3785 cm³ (cc, cubic centimeters)
	0.134 ft³ (cubic feet)
	231 in³ (cubic inches)
	3.785 L (liters)
1 L (liter)	1000 cm³ (cc, cubic centimeters)
	1 dm³ (cubic decimeters)
	0.0353 ft³ (cubic feet)
	61.02 in.³ (cubic inches)
	0.264 gal (gallons)
	1000 mL (milliliters)
	1.057 qt (quarts)
1 qt (quart)	946.4 cm³ (cc, cubic centimeters)
	57.75 in.³ (cubic inches)
	0.946 L (liters)
	Units of Mass
1 g (grams)	0.001 kg (kilograms)
	1000 mg (milligrams)
	1,000,000 mg = 10^6 ng (nanograms)
	2.205×10^{-3} lb (pounds)

(continues)

TABLE 1.3 (continued)

Conversion Factors by Unit Category

Factor	Metric (SI) or English Conversions
1 kg (kilogram)	1000 g (grams)
	2.205 lb (pounds)
1 lbs (pound)	453.59 g (grams)
	16 oz. (ounces)

Units of Time

1 day	24 hr (hours)
	1440 min (minutes)
	86,400 sec (seconds)
	0.143 weeks
	2.738×10^{-3} yr (years)
1 hr (hours)	0.0417 days
	60 min (minutes)
	3600 sec (seconds)
	5.95×10^{-3} yr (years)
1 hr (hour)	0.0417 days
	60 min (minutes)
	3600 sec (seconds)
	5.95×10^{-3} weeks
	1.14×10^{-4} yrs (years)
1 min (minutes)	6.94×10^{-4} days
	0.0167 hr (hours)
	60 sec (seconds)
	9.92×10^{-5} weeks
	1.90×10^{-6} yr (years)
1 week	7 days
	168 hr (hours)
	10,080 min (minutes)
	6.048×10^{5} sec (seconds)
	0.0192 yr (years)
1 yr (year)	365.25 days
	8,766 hr (hours)
	5.26×10^{5} min (minutes)
	3.16×10^{7} sec (seconds)
	52.18 weeks

Units of the Measure of Temperature

°C	$(5/9)[(°F) - 32°]$
1°C (expressed as an interval)	$33.8°F = (9/5)(°F)$
	1.8°R (degrees Rankine)
	1.0 K (degrees Kelvin)
°F Fahrenheit)	$(9/5)[(°C) + 32°]$
1°F (expressed as an interval)	$0.556°C = (5/9)°C$
	1.0°R (degrees Rankine)
	0.556 K (degrees Kelvin)

Units of Force

1 dyne	1×10^{-5} N (newtons)
1 nt (newton)	1×10^{5} dynes

TABLE 1.3 (continued)
Conversion Factors by Unit Category

Factor	Metric (SI) or English Conversions
Units of Work or Energy	
1 BTU (British Thermal Unit)	252 cal (calories)
	1055.06 j (joules)
	10.41 L atm (liter-atmospheres)
	0.293 Wh (watt-hours)
1 cal (calories)	3.97×10^{-3} BTUs (British Thermal Units)
	4.18 j (joules)
	0.0413 L atm (liter-atmospheres)
	1.163×10^{-3} Wh (watt-hours)
1 eV (electron volt)	1.602×10^{-12} ergs
	1.602×10^{-19} j (joules)
1 erg	1 dyne-centimeter
	1×10^{-7} j (joules)
	2.78×10^{-11} Wh (watt-hours)
1 j (joule)	9.48×10^{-4} BTUs (British Thermal Units)
	0.239 cal (calories)
	$10,000,000$ ergs $= 1 \times 10^{7}$ ergs
	9.87×10^{-3} L atm (liter-atmospheres)
	1.00 N-m (newton-meters)
1 kcal (kilocalorie)	3.97 BTUs (British Thermal Units)
	1000 cal (calories)
	4,186.8 j (joules)
1 kWh (kilowatt-hour)	3412.14 BTU (British Thermal Units)
	3.6×10^{6} j (joules)
	859.8 kcal (kilocalories)
1 N-m (newton-meter)	1.00 j (joules)
	2.78×10^{-4} Wh (watt-hours)
1 Wh (watt-hour)	3.412 BTUs (British Thermal Units)
	859.8 cal (calories)
	3,600 j (joules)
	35.53 L atm (liter-atmospheres)
Units of Power	
1 hp (horsepower)	745.7 j/sec (joules/sec)
1 kW (kilowatt)	56.87 BTU/min (British Thermal Units/minute)
	1.341 hp (horsepower)
	1000 j/sec (joules/sec)
1 W (watt)	3.41 BTU/hr (British Thermal Units/hour)
	1.341×10^{-3} hp (horsepower)
	1.00 j/sec (joules/second)
Units of Pressure	
1 atm (atmosphere)	1.013 bars
	10.133 N/cm^2 (newtons/square centimeters)
	33.90 ft of H_2O (feet of water)
	101.325 kPa (kilopascals)
	14.70 psia (pounds per square inch absolute)
	760 torr
	760 mmHg (millimeters of mercury)

(continues)

TABLE 1.3 (continued)
Conversion Factors by Unit Category

Factor	Metric (SI) or English Conversions
1 bar	0.987 atm (atmospheres)
	1×10^6 dynes/cm^2 (dynes/square centimeter)
	33.45 ft of H$_2$O (feet of water)
	1×10^5 Pa (pascals)
	750.06 torr
	750.06 mmHg (millimeters of mercury)
1 inch of water	1.86 mmHg (millimeters of mercury)
	249.09 Pa (pascals)
	0.0361 psi (lb/in.^{2})
1 mmHg (millimeter of	1.316×10^{-3} atm (atmospheres)
mercury)	0.535 in H$_2$O (inches of water)
	1.33 mb (millibars)
	133.32 Pa (pascals)
	1 torr
	0.0193 psia (pounds per square inch absolute)
1 pascal	9.87×10^{-6} atm (atmospheres)
	4.015×10^{-3} in H$_2$O (inches of water)
	0.01 mb (millibars)
	7.5×10^{-3} mmHg (millimeters of mercury)
1 psi (pounds per square inch)	0.068 atm (atmospheres)
	27.67 in H$_2$O (inches of water)
	68.85 mb (millibars)
	51.71 mmHg (millimeters of mercury)
	6,894.76 Pa (pascals)
1 torr	1.33 mb (millibars)

Units of Velocity or Speed

1 fps (feet per second)	1.097 kmph (kilometers/hour)
	0.305 mps (meters/second)
	0.01136 mph (miles/hours)
1 mps (meters per second)	196.9 fpm (feet/minute)
	3.6 kmph (kilometers/hour)
	2.237 mph (miles/hour)
1 mph (miles per hour)	88 fpm (feet/minute)
	1.61 kmph (kilometers/hour)
	0.447 mps (meters/second)

Units of Density

1 g/cm^3 (grams per cubic	62.43 lb/ft^3 (pounds/cubic foot)
centimeter)	0.0361 lb/in.^{3} (pounds/cubic inch)
	8.345 lb/gal (pounds/gallon)
1 lb/ft^3 (pounds/cubic foot)	16.02 g/L (grams/liter)
1 lb/in^2 (pounds/cubic inch)	27.68 g/cm^3 (grams/cubic centimeter)
	1.728 lb/ft^3 (pounds/cubic foot)

Units of Concentration

1 ppm (parts/million-volume)	1.00 mL/m^3 (milliliters/cubic meter)
1 ppm (wt)	1.00 mg/kg (milligrams/kilograms)

TABLE 1.3 (continued)
Conversion Factors by Unit Category

Factor	Metric (SI) or English Conversions
	Radiation and Dose Related Units
1 Bq (becquerel)	1 radioactive disintegration/second
	2.7×10^{-11} Ci (curie)
	2.7×10^{-8} (millicurie)
1 Ci (curie)	3.7×10^{10} radioactive disintegration/second
	3.7×10^{10} Bq (becquerel)
	1000 mCi (millicurie)
1 Gy (gray)	1 j/kg (joule/kilogram)
	100 rad
	1 Sv, unless modified through division by an appropriate factor, such as Q or N
1 mCi (millicurie)	0.001 Ci (curie)
	3.7×10^{10} radioactive disintegrations/second
	3.7×10^{10} Bq (becquerel)
1 rad	100 ergs/g (ergs/gram)
	0.01 Gy (gray)
	1 rem, unless modified through division by an appropriate factor, such as Q or N
1 rem	1 rad, unless modified through division by an appropriate factor, such as Q or N
1 Sv (sievert)	1 Gy, unless modified through division by an appropriate factor, such as Q or N

1.4 CONVERSION FACTORS: PRACTICAL EXAMPLES

Sometimes we have to convert between different units. Suppose that a 60-inch piece of pipe is attached to an existing 6-foot piece of pipe. Joined together, how long are they? Obviously, we cannot find the answer to this question by adding 60 to 6, because the two lengths are given in different units. Before we can add the two lengths, we must convert one of them to the units of the other. Then, when we have two lengths in the same units, we can add them.

To perform this conversion, we need a *conversion factor*. In this case, we have to know how many inches make up a foot: 12 inches. Knowing this, we can perform the calculation in two steps:

1. 60 in. is really 60/12 = 5 ft
2. 5 ft + 6 ft = 11 ft

DID YOU KNOW?

Units and dimensions are not the same concepts. *Dimensions* are concepts such as time, mass, length, or weight. *Units* are specific cases of dimensions, such as hour, gram, meter, or pound. You can *multiply* and *divide* quantities with different units: 4 ft × 8 lb = 32 ft-lb, but you can *add* and *subtract* terms only if they have the same units. So, 5 lb + 8 kg = **no way!**

From the example above, it can be seen that a conversion factor changes known quantities in one unit of measure to an equivalent quantity in another unit of measure. In making the conversion from one unit to another, we must know two things:

1. The exact number that relates the two units
2. Whether to multiply or divide by that number

When making conversions, confusion over whether to multiply or divide is common; on the other hand, the number that relates the two units is usually known and thus is not a problem. Understanding the proper methodology—the "mechanics"—to use for various operations requires practice and common sense.

Along with using the proper mechanics (and practice and common sense) in making conversions, probably the easiest and fastest method of converting units is to use a conversion table. The simplest conversion requires that the measurement be multiplied or divided by a constant value. For instance, if the depth of wet cement in a form is 0.85 ft, multiplying by 12 in. per foot converts the measured depth to inches (10.2 in.). Likewise, if the depth of the cement in the form is measured as 16 in., dividing by 12 in. per ft converts the depth measurement to feet (1.33 ft).

1.4.1 Weight, Concentration, and Flow

Using Table 1.4 to convert from one unit expression to another and *vice versa* is good practice; however, in making conversions to solve process computations in water treatment operations, for example, we must be familiar with conversion calculations based upon a relationship between weight, flow or volume, and concentration. The basic relationship is

$$\text{Weight} = \text{Concentration} \times (\text{Flow or Volume}) \times \text{Factor} \qquad (1.1)$$

DID YOU KNOW?

Many environmental health professionals choose to work for the Food Safety and Inspection Service (FSIS) at the federal level or for other food inspection services at the state and local government levels. Their objective, of course, is to ensure that meat, meat food, poultry, and poultry food products distributed in interstate commerce or locally are wholesome, not adulterated, and properly marked, labeled, and packaged. One obstacle that inspection personnel encounter in ascertaining restricted ingredient compliance with the regulations is that calculations for allowable ingoing amounts could be based on one of five different weights. These different weights vary according to the type of ingredient, type of product, and reason for using the ingredient in the product. The five weights (or bases for restricted ingredient calculations) are

- *Green weight*—Weight of the meat and/or poultry byproduct (meat block) component at formulation
- *Formulated weight*—Total weight after meat products are added to sausage products at the time of formulation, excluding the water and ice
- *Finished weight*—Total weight of the entire meat or poultry product, including breading and butter
- *Projected finished weight*—Total weight of meat or poultry, including nonfat dry milk, soy flours, and cereals added
- *Weight of the fat content*—Weight of the fat content of a fresh meat or poultry product

TABLE 1.4
Conversion Table

To Convert	Multiply by	To Get
Feet	12	Inches
Yards	3	Feet
Yards	36	Inches
Inches	2.54	Centimeters
Meters	3.3	Feet
Meters	100	Centimeters
Meters	1000	Millimeters
Square yards	9	Square feet
Square feet	144	Square inches
Acres	43,560	Square feet
Cubic yards	27	Cubic feet
Cubic feet	1728	Cubic inches
Cubic feet (water)	7.48	Gallons
Cubic feet (water)	62.4	Pounds
Acre-feet	43,560	Cubic feet
Gallons (water)	8.34	Pounds
Gallons (water)	3.785	Liters
Gallons (water)	3785	Milliliters
Gallons (water)	3785	Cubic centimeters
Gallons (water)	3785	Grams
Liters	1000	Milliliters
Days	24	Hours
Days	1440	Minutes
Days	86,400	Seconds
Million gallons/day	1,000,000	Gallons/day
Million gallons/day	1.55	Cubic feet/second
Million gallons/day	3.069	Acre-feet/day
Million gallons/day	36.8	Acre-inches/day
Million gallons/day	3785	Cubic meters/day
Gallons/minute	1440	Gallons/day
Gallons/minute	63.08	Liters/minute
Pounds	454	Grams
Grams	1000	Milligrams
Pressure (psi)	2.31	Head (feet of water)
Horsepower	33,000	Foot-pounds/minute
Horsepower	0.746	Kilowatts

Table 1.5 summarizes weight, volume, and concentration calculations. With practice, many of these calculations become second nature to users.

The following conversion factors are used extensively in environmental engineering (e.g., water and wastewater operations):

- 7.48 gallons = 1 cubic foot (ft^3)
- 3.785 liters = 1 gallon
- 454 grams = 1 pound
- 1000 mL = 1 liter
- 1000 mg = 1 gram
- 1 ft^3/sec (cfs) = 0.6465 MGD

TABLE 1.5
Weight, Volume, and Concentration Calculations

To Calculate	Formula
Pounds	Concentration (mg/L) × Tank Volume (MG) × 8.34 lb/mg/L/MG
Pounds/day	Concentration (mg/L) × Flow (MGD) × 8.34 lb/mg/L/MG
Million gallons/day	$\dfrac{\text{Quantity (lb/day)}}{\text{Concentration (mg/L)} \times 8.34 \text{ lb/mg/L/MG}}$
Milligrams/liter	$\dfrac{\text{Quantity (lb)}}{\text{Tank volume (MG)} \times 8.34 \text{ lb/mg/L/MG}}$
Kilograms/liter	Concentraiton (mg/L) × Volume, (MG) × 3.785 lb/mg/L/MG
Kilograms/day	Concentration (mg/L) × Flow (MGD) × 3.785 lb/mg/L/MG
Pounds/dry ton	Concentration (mg/kg) × 0.002 lb/dry ton/mg/kg

Note: *Density* (also called *specific weight*) is mass per unit volume and may be written as lb/ft^3, lb/gal, g/mL, or g/m^3. If we take a fixed-volume container, fill it with a fluid, and weigh it, we can determine the density of the fluid (after subtracting the weight of the container).

- 1 gallon of water weighs 8.34 pounds; the density is 8.34 lb/gal
- 1 milliliter of water weighs 1 gram; the density is 1 g/mL
- 1 cubic foot of water weighs 62.4 pounds; the density is 62.4 lb/gal
- 8.34 lb/gal = milligrams per liter, which is used to convert dosage in mg/L into lb/day/MGD (e.g., 1 mg/L × 10 MGD × 8.34 lb/gal = 83.4 lb/day)
- 1 psi = 2.31 feet of water (head)
- 1 foot head = 0.433 psi
- °F = 9/5(°C + 32)
- °C = 5/9(°F − 32)
- Average water usage, 100 gallons/capita/day (gpcd)
- Persons per single family residence, 3.7

1.4.2 WATER/WASTEWATER CONVERSION EXAMPLES

Use Tables 1.4 and 1.5 to make the conversions indicated in the following example problems. Other conversions are presented in appropriate sections of the text.

■ EXAMPLE 1.4

Convert cubic feet to gallons.

$$\text{Gallons} = \text{Cubic feet (ft}^3) \times \text{gal/ft}^3$$

Problem: How many gallons of biosolids can be pumped to a digester that has 3600 ft^3 of volume available?

Solution:

$$\text{Gallons} = 3600 \text{ ft}^3 \times 7.48 \text{ gal/ft}^3 = 26{,}928 \text{ gal}$$

■ **EXAMPLE 1.5**

Convert gallons to cubic feet.

$$\text{Cubic feet (ft}^3) = \frac{\text{Gallons}}{7.48 \text{ gal/ft}^3}$$

Problem: How many cubic feet of biosolids are removed when 18,200 gal are withdrawn?

Solution:

$$\text{Cubic feet} = \frac{18{,}200 \text{ gal}}{7.48 \text{ gal/ft}^3} = 2433 \text{ ft}^3$$

■ **EXAMPLE 1.6**

Convert gallons to pounds.

$$\text{Pounds (lb)} = \text{Gallons} \times 8.34 \text{ lb/gal}$$

Problem: If 1650 gal of solids are removed from the primary settling tank, how many pounds of solids are removed?

Solution:

$$\text{Pounds} = 1650 \text{ gal} \times 8.34 \text{ lb/gal} = 13{,}761 \text{ lb}$$

■ **EXAMPLE 1.7**

Convert pounds to gallons.

$$\text{Gallons (gal)} = \frac{\text{Pounds (lb)}}{8.34 \text{ lb/gal}}$$

Problem: How many gallons of water are required to fill a tank that holds 7540 lb of water?

Solution:

$$\text{Gallons} = \frac{7540 \text{ lb}}{8.34 \text{ lb/gal}} = 904 \text{ gal}$$

■ **EXAMPLE 1.8**

Convert milligrams/liter to pounds.

Key Point: For plant operations, concentrations in milligrams per liter or parts per million determined by laboratory testing must be converted to quantities of pounds, kilograms, pound per day, or kilograms per day.

$$\text{Pounds (lb)} = \text{Concentration (mg/L)} \times \text{Volume (MG)} \times 8.34 \text{ lb/mg/L/MG}$$

Problem: The solids concentration in the aeration tank is 2580 mg/L. The aeration tank volume is 0.95 MG. How many pounds of solids are in the tank?

Solution:

$$\text{Pounds} = 2580 \text{ mg/L} \times 0.95 \text{ MG} \times 8.34 \text{ lb/mg/L/MG} = 20{,}441.3 \text{ lb}$$

■ **EXAMPLE 1.9**

Convert milligrams per liter to pounds per day.

$$\text{Pounds/day} = \text{Concentration (mg/L)} \times \text{Flow (MGD)} \times 8.34 \text{ lb/mg/L/MG}$$

Problem: How many pounds of solids are discharged per day when the plant effluent flow rate is 4.75 MGD and the effluent solids concentration is 26 mg/L?

Solution:

$$\text{Pounds/day} = 26 \text{ mg/L} \times 4.75 \text{ MGD} \times 8.34 \text{ lb/mg/L/MG} = 1030 \text{ lb/day}$$

■ **EXAMPLE 1.10**

Convert milligrams per liter to kilograms per day.

$$\text{Kilograms/day} = \text{Concentration (mg/L)} \times \text{Volume (MG)} \times 3.785 \text{ kg/mg/L/MG}$$

Problem: The effluent contains 26 mg/L of BOD_5. How many kilograms per day of BOD_5 are discharged when the effluent flow rate is 9.5 MGD?

Solution:

$$\text{Kilograms/day} = 26 \text{ mg/L} \times 9.5 \text{ MG} \times 3.785 \text{ kg/mg/L/MG} = 934 \text{ kg/day}$$

■ **EXAMPLE 1.11**

Convert pounds to milligrams per liter.

$$\text{Concentration (mg/L)} = \frac{\text{Quantity (lb)}}{\text{Volume (MG)} \times 8.34 \text{ lb/mg/L/MG}}$$

Problem: The aeration tank contains 89,990 pounds of solids. The volume of the aeration tank is 4.45 MG. What is the concentration of solids in the aeration tank in milligrams per liter?

Solution:

$$\text{Concentration} = \frac{89,990 \text{ lb}}{4.45 \text{ MG} \times 8.34 \text{ lb/mg/L/MG}} = 2425 \text{ mg/L}$$

■ **EXAMPLE 1.12**

Convert pounds per day to milligrams per liter.

$$\text{Concentration (mg/L)} = \frac{\text{Quantity (lb/day)}}{\text{Volume (MGD)} \times 8.34 \text{ lb/mg/L/MG}}$$

Problem: The disinfection process uses 4820 pounds per day of chlorine to disinfect a flow of 25.2 MGD. What is the concentration of chlorine applied to the effluent?

Solution:

$$\text{Concentration} = \frac{4820 \text{ lb/day}}{25.2 \text{ MGD} \times 8.34 \text{ lb/mg/L/MG}} = 22.9 \text{ mg/L}$$

■ EXAMPLE 1.13

Convert pounds to flow in million gallons per day.

$$\text{Flow (MGD)} = \frac{\text{Quantity (lb/day)}}{\text{Concentration (mg/L)} \times 8.34 \text{ lb/mg/L/MG}}$$

Problem: 9640 pounds of solids must be removed from the activated biosolids process per day. The waste activated biosolids concentration is 7699 mg/L. How many million gallons per day of waste activated biosolids must be removed?

Solution:

$$\text{Flow} = \frac{9640 \text{ lb/day}}{7699 \text{ mg/L} \times 8.34 \text{ lb/mg/L/MG}} = 0.15 \text{ MGD}$$

■ EXAMPLE 1.14

Convert million gallons per day (MGD) to gallons per minute (gpm).

$$\text{Flow (gpm)} = \frac{\text{Flow (MGD)} \times 1{,}000{,}000 \text{ gal/MG}}{1440 \text{ min/day}}$$

Problem: The current flow rate is 5.55 MGD. What is the flow rate in gallons per minute?

Solution:

$$\text{Flow} = \frac{5.55 \text{ MGD} \times 1{,}000{,}000 \text{ gal/MG}}{1440 \text{ min/day}} = 3854 \text{ gpm}$$

■ EXAMPLE 1.15

Convert million gallons per day (MGD) to gallons per day (gpd).

$$\text{Flow (gpd)} = \text{Flow (MGD)} \times 1{,}000{,}000 \text{ gal/MG}$$

Problem: The influent meter reads 28.8 MGD. What is the current flow rate in gallons per day?

Solution:

$$\text{Flow} = 28.8 \text{ MGD} \times 1{,}000{,}000 \text{ gal/MG} = 28{,}800{,}000 \text{ gpd}$$

■ EXAMPLE 1.16

Convert million gallons per day (MGD) to cubic feet per second (cfs).

$$\text{Flow (cfs)} = \text{Flow (MGD)} \times 1.55 \text{ cfs/MGD}$$

Problem: The flow rate entering the grit channel is 2.89 MGD. What is the flow rate in cubic feet per second?

Solution:

$$\text{Flow} = 2.89 \text{ MGD} \times 1.55 \text{ cfs/MGD} = 4.48 \text{ cfs}$$

■ **EXAMPLE 1.17**

Convert gallons per minute (gpm) to million gallons per day (MGD).

$$\text{Flow (MGD)} = \frac{\text{Flow (gpm)} \times 1440 \text{ min/day}}{1{,}000{,}000 \text{ gal/MG}}$$

Problem: The flow meter indicates that the current flow rate is 1469 gpm. What is the flow rate in million gallons per day?

Solution:

$$\text{Flow} = \frac{1469 \text{ gpm} \times 1440 \text{ min/day}}{1{,}000{,}000 \text{ gal/MG}} = 2.12 \text{ MGD (rounded)}$$

■ **EXAMPLE 1.18**

Convert gallons per day (gpd) to million gallons per day (MGD).

$$\text{Flow (MGD)} = \frac{\text{Flow (gal/day)}}{1{,}000{,}000 \text{ gal/MG}}$$

Problem: The totalizing flow meter indicates that 33,444,950 gal of wastewater have entered the plant in the past 24 hr. What is the flow rate in million gallons per day?

Solution:

$$\text{Flow} = \frac{33{,}444{,}950 \text{ gal/day}}{1{,}000{,}000 \text{ gal/MG}} = 33.44 \text{ MGD}$$

■ **EXAMPLE 1.19**

Convert flow in cubic feet per second (cfs) to million gallons per day (MGD).

$$\text{Flow (MGD)} = \frac{\text{Flow (cfs)}}{1.55 \text{ cfs/MG}}$$

Problem: The flow in a channel is determined to be 3.89 cubic feet per second (cfs). What is the flow rate in million gallons per day (MGD)?

Solution:

$$\text{Flow} = \frac{3.89 \text{ cfs}}{1.55 \text{ cfs/MG}} = 2.5 \text{ MGD}$$

■ **EXAMPLE 1.20**

Problem: The water in a tank weighs 675 lb. How many gallons does it hold?

Solution: Water weighs 8.34 lb/gal; therefore,

$$\frac{675 \text{ lb}}{8.34 \text{ lb/gal}} = 80.9 \text{ gal}$$

■ **EXAMPLE 1.21**

Problem: A liquid chemical weighs 62 lb/ft³. How much does a 5-gal can of it weigh?

Solution: Solve for specific gravity, determine lb/gal, and multiply by 5:

$$\text{Specific gravity} = \frac{\text{Weight of chemical (lb/ft}^3)}{\text{Weight of water (lb/ft}^3)} = \frac{62 \text{ lb/ft}^3}{62.4 \text{ lb/ft}^3} = 0.99$$

$$0.99 = \frac{\text{Weight of chemical (lb/gal)}}{8.34 \text{ lb/gal}}$$

$$\text{Weight of chemical} = 8.26 \text{ lb/gal}$$

$$8.26 \text{ lb/gal} \times 5 \text{ gal} = 41.3 \text{ lb}$$

■ **EXAMPLE 1.22**

Problem: A wooden piling with a diameter of 16 in. and a length of 16 ft weighs 50 lb/ft³. If it is inserted vertically into a body of water, what vertical force is required to hold it below the water surface?

Solution: If this piling had the same weight as water, it would rest just barely submerged. Find the difference between its weight and that of the same volume of water—that is the weight needed to keep it down:

$$\begin{array}{r} 62.4 \text{ lb/ft}^3 \text{ (water)} \\ - 50.0 \text{ lb/ft}^3 \text{ (piling)} \\ \hline 12.4 \text{ lb/ft}^3 \text{ difference} \end{array}$$

$$\text{Volume of piling} = 0.785 \times (1.33)^2 \times 16 \text{ ft} = 22.21 \text{ ft}^3$$

$$12.4 \text{ lb/ft}^3 \times 22.21 \text{ ft}^3 = 275.4 \text{ lb}$$

■ **EXAMPLE 1.23**

Problem: A liquid chemical with a specific gravity (SG) of 1.22 is pumped at a rate of 40 gpm. How many pounds per day are being delivered by the pump?

Solution: Solve for pounds pumped per minute, then change to pounds/day.

$$8.34 \text{ lb/gal water} \times 1.22 \text{ SG liquid chemical} = 10.2 \text{ lb/gal liquid}$$

$$40 \text{ gal/min} \times 10.2 \text{ lb/gal} = 408 \text{ lb/min}$$

$$408 \text{ lb/min} \times 1440 \text{ min/day} = 587{,}520 \text{ lb/day}$$

■ **EXAMPLE 1.24**

Problem: A cinder block weighs 70 lb in air. When immersed in water, it weighs 40 lb. What are the volume and specific gravity of the cinder block?

Solution: The cinder block displaces 30 lb of water; solve for cubic feet of water displaced (equivalent to volume of cinder block).

$$\frac{30 \text{ lb water displaced}}{62.4 \text{ lb/ft}^3} = 0.48 \text{ ft}^3 \text{ water displaced}$$

Cinder block volume is 0.48 ft³, which weighs 70 lb; thus,

$$\frac{70 \text{ lb}}{0.48 \text{ ft}^3} = 145.8 \text{ lb/ft}^3 \text{ density of cinder block}$$

$$\text{Specific gravity} = \frac{\text{Density of cinder block}}{\text{Density of water}} = \frac{145.8 \text{ lb/ft}^3}{62.4 \text{ lb/ft}^3} 2.34$$

1.4.3 TEMPERATURE CONVERSIONS

Two commonly used methods used to make temperature conversions. We have already demonstrated the following methods:

- $°C = 5/9(°F - 32)$
- $°F = 9/5(°C) + 32$

■ EXAMPLE 1.25

Problem: At a temperature of 4°C, water is at its greatest density. What is that temperature in degrees Fahrenheit?

Solution:

$$9/5(°C) + 32 = 9/5(4) + 32 = 7.2 + 32 = 39.2°F$$

The difficulty arises when one tries to recall these formulas from memory. Probably the easiest way to recall these important formulas is to remember these basic steps for both Fahrenheit and Celsius conversions:

1. Add 40°.
2. Multiply by the appropriate fraction (5/9 or 9/5).
3. Subtract 40°.

Obviously, the only variable in this method is the choice of 5/9 or 9/5 in the multiplication step. To make the proper choice, you must be familiar with the two scales. The freezing point of water is 32° on the Fahrenheit scale and 0° on the Celsius scale. The boiling point of water is 212° on the Fahrenheit scale and 100° on the Celsius scale.

Note: At the same temperature, higher numbers are associated with the Fahrenheit scale and lower numbers with the Celsius scale. This important relationship helps you decide whether to multiply by 5/9 or 9/5.

Now look at a few conversion problems to see how the three-step process works.

■ EXAMPLE 1.26

Problem: Suppose that we wish to convert 240°F to Celsius.

Solution: Using the three-step process, we proceed as follows:

1. Add 40°

$$240° + 40° = 280°$$

2. $280°$ must be multiplied by either 5/9 or 9/5. Because the conversion is to the Celsius scale, we will be moving to a number *smaller* than 280. Through reason and observation, obviously, if 280 were multiplied by 9/5, the result would be almost the same as multiplying by 2, which would double 280 rather than make it smaller. If we multiply by 5/9, the result will be about the same as multiplying by 1/2, which would cut 280 in half. Because in this problem we wish to move to a smaller number, we should multiply by 5/9:

$$(5/9)(280°) = 156.0°C$$

3. Now subtract $40°$.

$$156.0°C - 40.0°C = 116.0°C$$

Therefore, $240°F = 116.0°C$.

■ EXAMPLE 1.27

Problem: Convert $22°C$ to Fahrenheit.

Solution:

1. Add $40°$:

$$22° + 40° = 62°$$

2. Because we are converting from Celsius to Fahrenheit, we are moving from a smaller to a larger number, and 9/5 should be used in the multiplications:

$$(9/5)(62°) = 112°$$

3. Subtract 40:

$$112° - 40° = 72°$$

Thus, $22°C = 72°F$.

Obviously, knowing how to make these temperature conversion calculations is useful, but it is generally more practical to use a temperature conversion table.

1.5 CONVERSION FACTORS: AIR POLLUTION MEASUREMENTS

The recommended units for reporting air pollutant emissions are commonly stated in metric system whole numbers. If possible, the reported units should be the same as those that are actually being measured. For example, weight should be recorded in grams, and volume of air should be recorded in cubic meters. When the analytical system is calibrated in one unit, the emissions should be reported in the same units of the calibration standard. For example, if a gas chromatograph is calibrated with a 1-ppm standard of toluene in air, then the emissions monitored by the system should also be reported in ppm. Finally, if the emission standard is defined in a specific unit, the monitoring system should be selected to monitor in that unit. Tables 1.6 and 1.7 illustrate the conversion for various volumes to attain 1 part per million (ppm) and also to illustrate conversion for parts per million in proportion and percent.

The preferred reporting units for the following types of emissions should be

- Nonmethane organic and volatile organic compound emissions   ppm, ppb
- Semi-volatile organic compound emissions $\mu g/m^3$, mg/m^3
- Particulate matter (TSP/PM10) emissions $\mu g/m^3$
- Metal compound emissions ng/m^3

TABLE 1.6

Conversion for Various Volumes to Attain One Part Per Million

Amount of Active Ingredient	Unit of Volume	Parts per Million
2.71 pounds	Acre-foot	1 ppm
1.235 grams	Acre-foot	1 ppm
1.24 kilograms	Acre-foot	1 ppm
0.0283 grams	Cubic foot	1 ppm
1 milligram	Liter	1 ppm
8.34 pounds	Million gallons	1 ppm
1 gram	Cubic meter	1 ppm
0.0038 grams	Gallon	1 ppm
3.8 grams	Thousand gallons	1 ppm

TABLE 1.7

Conversion for Parts per Million in Proportion and Percent

Parts per Million	Proportion	Percent	Parts per Million	Proportion	Percent
0.1	1:10,000,000	0.00001	25.0	1:40,000	0.0025
0.5	1:2,000,000	0.00005	50.0	1:20,000	0.005
1.0	1:1,000,000	0.0001	100.0	1:10,000	0.01
2.0	1:500,000	0.0002	200.0	1:5,000	0.02
3.0	1:333,333	0.0003	250.0	1:4,000	0.025
5.0	1:200,000	0.0005	500.0	1:2,000	0.05
7.0	1:142,857	0.0007	1550.0	1:645	0.155
10.0	1:100,000	0.001	5000.0	1:200	0.5
15.0	1:66,667	0.0015	10,000.0	1:100	1.0

1.5.1 Conversion from ppm to µg/m³

Often, the environmental practitioner must be able to convert from ppm to µg/m³. Following is an example of how one would perform that conversion using sulfur dioxide (SO_2) as the monitored constituent.

■ EXAMPLE 1.28

The expression "parts per million" is without dimensions; that is, no units of weight or volume are specifically designated. Using the format of other units, the expression may be written:

$$\frac{\text{Parts}}{\text{Million parts}}$$

"Parts" are not defined. If cubic centimeters replace parts, we obtain:

$$\frac{\text{Cubic centimeters}}{\text{Million cubic centimeters}}$$

Similarly, we might write pounds per million pounds, tons per million tons, or liters per million liters. In each expression, identical units of weight or volume appear in both the numerator and denominator and may be canceled out, leaving a dimensionless term. An analog of parts per million is the more familiar term "percent." Percent can be written as

$$\frac{\text{Parts}}{\text{Hundred parts}}$$

To convert from parts per million by volume (μL/L) to μg/m^3 at standard temperature (25°C) and standard pressure (760 mmHg), known as STP, it is necessary to know the molar volume at the given temperature and pressure and the molecular weight of the pollutant. At 25°C and 760 mmHg, 1 mole of any gas occupies 24.46 L.

Problem: 2.5 ppm by volume of sulfur dioxide (SO$_2$) was reported as the atmospheric concentration. What is this concentration in micrograms (μg) per cubic meter (m^3) at 25°C and 760 mmHg? What is the concentration in μg/m^3 at 37°C and 752 mmHg?

Note: This example problem points out the need for reporting temperature and pressure when the results are present on a weight to volume basis.

Solution: Let parts per million equal μL/L, then 2.5 ppm = 2.5 μL/L. The molar volume at 25°C and 760 mmHg is 24.46 L, and the molecular weight of SO$_2$ is 64.1 g/mole.

At 25°C and 760 mmHg (STP):

$$\frac{2.5\ \mu L}{L} \times \frac{1\ \mu mole}{24.46\ \mu L} \times \frac{64.1\ \mu g}{\mu mole} \times \frac{1000\ L}{m^3} = \frac{6.66 \times 10^3\ \mu g}{m^3}$$

At 37°C and 752 mmHg:

$$24.46\ \mu L \left(\frac{310°K}{298°K} \times \frac{760\ mmHg}{752\ mmHg} \right) = 25.72\ \mu L$$

$$\frac{2.5\ \mu L}{L} \times \frac{1\ \mu mole}{25.72\ \mu L} \times \frac{64.1\ \mu g}{\mu mole} \times \frac{1000\ L}{m^3} = \frac{6.2 \times 10^3\ \mu g}{m^3}$$

1.5.2 Conversion Tables for Common Air Pollution Measurements

To assist the environmental engineer in converting from one set of units to another, the following conversion factors for common air pollution measurements and other useful information are provided in Tables 1.8 through 1.12. These conversion tables provide factors for

- Atmospheric gases
- Atmospheric pressure
- Velocity
- Atmospheric particulate matter
- Concentration

TABLE 1.8

Atmospheric Gases

To Convert from	to	Multiply by
Milligram per cubic meter (mg/m³)	Micrograms per cubic meter (μm/m³)	1000.0
	Micrograms per liter (μg/L)	1.0
	ppm by volume (20°C)	24.04/molecular weight of gas
	ppm by weight	0.8347
	Pounds per cubic foot (lb/ft³)	62.43×10^{-9}
Micrograms per cubic foot (μm/ft³)	Milligrams per cubic foot (mg/ft³)	0.001
	Micrograms per liter (μg/L)	0.001
	ppm by volume (20° C)	0.02404/molecular weight of gas
	ppm by weight	834.7×10^{-6}
	Pounds per cubic foot (lb/ft³)	62.43×10^{-12}
Micrograms/liter (μm/L)	Milligrams per cubic meter (mg/m³)	1.0
	Micrograms per cubic meter (μg/m³)	1000.0
	ppm by volume (20°C)	24.04/molecular weight of gas
	ppm by weight	0.8347
	Pounds per cubic ft (lb/ft³)	62.43×10^{-9}
ppm by volume (20°C)	Milligrams per cubic meter (mg/m³)	Molecular weight of gas/24.04
	Micrograms per cubic meter (μg/m³)	Molecular weight of gas/0.02404
	Micrograms per liter (μg/L)	Molecular weight of gas/24.04
	ppm by weight	Molecular weight of gas/28.8
	Pounds per cubic ft (lb/ft³)	Molecular weight of gas/385.1 $\times 10^6$
ppm by weight	Milligrams per cubic meter (mg/m³)	1.198
	Micrograms per cubic meter (μg/m³)	1.198×10^3
	Micrograms per liter (μg/L)	1.198
	ppm by volume (20°C)	28.8/molecular weight of gas
	Pounds per cubic ft (lb/ft³)	7.48×10^{-6}
Pounds per cubic foot (lb/ft³)	Milligrams per cubic meter (mg/m³)	16.018×10^6
	Micrograms per cubic meter (μg/m³)	16.018×10^9
	Micrograms per liter (μg/L)	16.018×10^6
	ppm by volume (20°)	385.1×10^6/molecular weight of gas
	ppm by weight	133.7×10^3

Following is a list of conversions from ppm to $\mu g/m^3$ (at 25°C and 760 mmHg) for several common air pollutants:

- ppm $SO_2 \times 2620 = \mu g/m^3$ SO_2 (sulfur dioxide)
- ppm $CO \times 1150 = \mu g/m^3$ CO (carbon monoxide)
- ppm $CO_x \times 1.15 = mg/m^3$ CO (carbon dioxide)
- ppm $CO_2 \times 1.8 = mg/m^3$ CO_2 (carbon dioxide)
- ppm $NO \times 1230 = \mu g/m^3$ NO (nitrogen oxide)
- ppm $NO_2 \times 1880 = \mu g/m^3$ NO_2 (nitrogen dioxide)
- ppm $O_2 \times 1960 = \mu g/m^3$ O_3 (ozone)
- ppm $CH_4 \times 655 = \mu g/m^3$ CH_4 (methane)
- ppm $CH_4 \times 655 = mg/m^3$ CH_4 (methane)
- ppm $CH_3SH \times 2000 = \mu g/m^3$ CH_3SH (methyl mercaptan)
- ppm $C_3H_8 \times 1800 = \mu g/m^3$ C_3H_8 (propane)
- ppm $C_3H_8 \times 1.8 = mg/m^3$ C_3H_8 (propane)

TABLE 1.9
Atmospheric Pressure

To Convert from	to	Multiply by
Atmospheres	Millimeters of mercury	760.0
	Inches of mercury	29.92
	Millibars	1013.2
Millimeters of mercury	Atmospheres	1.316×10^{-3}
	Inches of mercury	39.37×10^{-3}
	Millibars	1.333
Inches of mercury	Atmospheres	0.03333
	Millimeters of mercury	25.4005
	Millibars	33.35
Millibars	Atmospheres	0.000987
	Millimeters of mercury	0.75
	Inches of mercury	0.30
Sampling Pressures		
Millimeters of mercury	Inches of water (60°C)	0.5358
Inches of mercury	Inches of water (60°C)	13.609
Inches of water	Millimeters of mercury (0°C)	1.8663
	Inches of mercury (0°C)	73.48×10^{-2}

TABLE 1.10
Velocity

To Convert from	to	Multiply by
Meters/second (m/sec)	Kilometers/hour (km/hr)	3.6
	Feet/second (fps)	3.281
	Miles/hour (mph)	2.237
Kilometers/hour (km/hr)	Meters/second (m/sec)	0.2778
	Feet/second (fps)	0.9113
	Miles/hour (mph)	0.6241
Feet/hour (ft/hr)	Meters/second (m/sec)	0.3048
	Kilometers/hour (km/hr)	1.0973
	Miles/hour (mph)	0.6818
Miles/hour (mph)	Meters/second (m/sec)	0.4470
	Kilometers/hour (km/hr)	1.6093
	Feet/second (fps)	1.4667

- ppm F^- × 790 = $\mu g/m^3$ F^- (fluoride)
- ppm H_2S × 1400 = $\mu g/m^3$ H_2S (hydrogen sulfide)
- ppm NH_3 × 696 = $\mu g/m^3$ NH_3 (ammonia)
- ppm HCHO × 1230 = $\mu g/m^3$ HCHO (formaldehyde)

1.6   SOIL TEST RESULTS CONVERSION FACTORS

Soil test results can be converted from parts per million (ppm) to pounds per acre by multiplying ppm by a conversion factor based on the depth to which the soil was sampled. Because a slice of soil 1 acre in area and 3 inches deep weighs approximately 1 million pounds, the conversion factors given in Table 1.13 can be used.

TABLE 1.11

Atmospheric Particulate Matter

To Convert from	to	Multiply by
Milligrams/cubic meter (mg/m³)	Grams/cubic foot (g/ft³)	283.2×10^{-6}
	Grams/cubic meter (g/m³)	0.001
	Micrograms/cubic meter (μg/m³)	1000.0
	Micrograms/cubic foot (μg/ft³)	28.32
	Pounds/1000 cubic feet (lb/1000 ft³)	62.43×10^{-6}
Grams/cubic foot (g/ft³)	Milligrams/cubic meter (mg/m³)	35.3145×10^{3}
	Grams/cubic meter (g/m³)	35.314
	Micrograms/cubic meter (μg/m³)	35.314×10^{3}
	Micrograms/cubic foot (μg/ft³)	1.0×10^{6}
	Pounds/1000 cubic feet (lb/1000 ft³)	2.2046

TABLE 1.12

Concentration

To Convert from	to	Multiply by
Grams/cubic meter (g/m³)	Milligrams/cubic meter (mg/m³)	1000.0
	Grams/cubic foot (g/ft³)	0.02832
	Micrograms/cubic foot (μg/ft³)	1.0×10^{6}
	Pounds/1000 cubic feet (lb/1000 ft³)	0.06243
Micrograms/cubic meter (μg/m³)	Milligrams/cubic meter (mg/m³)	0.001
	Grams/cubic foot (g/ft³)	28.43×10^{-9}
	Grams/cubic meter (g/m³)	1.0×10^{-6}
	Micrograms/cubic foot (μg/ft³)	0.02832
	Pounds/1000 cubic feet (lb/1000 ft³)	62.43×10^{-9}
Micrograms/cubic foot (μg/ft³)	Milligrams/cubic meter (mg/m³)	35.314×10^{-3}
	Grams/cubic foot (g/ft³)	1.0×10^{-6}
	Grams/cubic meter (g/m³)	35.314×10^{-6}
	Micrograms/cubic foot (μg/ft³)	35.314
	Pounds/1000 cubic feet (lb/1000 ft³)	2.2046×10^{-6}
Pounds/1000 cubic feet (lb/1000 ft³)	Milligrams/cubic meter (mg/m³)	16.018×10^{3}
	Grams/cubic foot (g/ft³)	0.35314
	Micrograms/cubic meter (μg/m³)	16.018×10^{6}
	Grams/cubic meter (g/m³)	16.018
	Micrograms/cubic foot (μg/ft³)	353.14×10^{2}

TABLE 1.13

Soil Test Conversion Factors

Soil Sample Depth (inches)	Multiply ppm by
3	1
6	2
7	2.33
8	2.66
9	3
10	3.33
12	4

STANDARD CONVERSIONS FOR MANUAL CALCULATIONS

Volume	Weight	Length
1 gal = 3.78 L	1 lb = 453 g or 0.453 kg	1 in. = 2.54 cm
1 L = 0.26 gal	1 kg = 2.2 lb	1 cm = 0.39 in.
1 tsp = 5 mL		3.28 ft = 1 m

1.7 GREENHOUSE GAS EMISSION NUMBERS TO EQUIVALENT UNITS

This section describes the calculations used to convert greenhouse gas emission numbers—an area of increasing concern for environmental practitioners—into different types of equivalent units.

1.7.1 ELECTRICITY REDUCTIONS (KILOWATT-HOURS)

The U.S. Environmental Protection Agency's Greenhouse Gas Equivalencies Calculator uses the Emissions & Generation Resource Integrated Database (eGRID) of U.S. annual non-baseload CO_2 output emission rates to convert reductions of kilowatt-hours into avoided units of carbon dioxide emissions. Most users of the Equivalencies Calculator who seek equivalencies for electricity-related emissions want to know equivalencies for emissions reductions due to energy efficiency or renewable entry programs. These programs are not generally assumed to affect base-load emissions (the emissions from power plants that run all the time), but rather non-baseload generation (power plants that are brought online as necessary to meet demand). For that reason, the Equivalencies Calculator uses a non-baseload emissions rate (USEPA 2012a).

1.7.1.1 Emission Factor

$$7.0555 \times 10^{-4} \text{ metric tons } CO_2/\text{kWh}$$

Note: This calculation does not include any greenhouse gases other than CO_2, and it does not include line losses.

1.7.2 PASSENGER VEHICLES PER YEAR

Passenger vehicles are defined as two-axle, four-tire vehicles, including passenger cars, vans, pickup trucks, and sport/utility vehicles. In 2010, the weighted average combined fuel economy of cars and light trucks was 21.6 miles per gallon (FHWA, 2012). The average vehicle miles traveled in 2010 was 11,489 miles per year. In 2010, the ratio of carbon dioxide emissions to total greenhouse gas emissions (including carbon dioxide, methane, and nitrous oxide, all expressed as carbon dioxide equivalents) for passenger vehicles was 0.985 (USEPA, 2013a). The amount of carbon dioxide emitted per gallon of motor gasoline burned was 8.92×10^{-3} metric tons, as calculated in the Section 1.7.3.

To determine the annual greenhouse gas emissions per passenger vehicle, the following methodology was used: Vehicle miles traveled (VMT) was divided by average gas mileage to determine gallons of gasoline consumed per vehicle per year. Gallons of gasoline consumed was multiplied by carbon dioxide per gallon of gasoline to determine carbon dioxide emitted per vehicle per year. Carbon dioxide emissions were than divided by the ratio of carbon dioxide emissions to total vehicle greenhouse gas emissions to account for vehicle methane and nitrous oxide emissions.

1.7.2.1 Calculation

Due to rounding, performing the calculations given in the equations below may not return the exact results shown.

$$(8.92 \times 10^{-3} \text{ metric tons } CO_2 \text{ per gal gasoline}) \times (11{,}489 \text{ VMT car/truck average})$$

$$\times (1/21.6 \text{ miles per gal car/truck average}) \times [(1 \; CO_2, \; CH_4, \text{ and } N_2O)/0.985 \; CO_2]$$

$$= 4.8 \text{ metric tons } CO_2 \text{ emissions per vehicle per year.}$$

1.7.3 GALLONS OF GASOLINE CONSUMED

To obtain the number of grams of CO_2 emitted per gallon of gasoline combusted, the heat content of the fuel per gallon is multiplied by the kg CO_2 per heat content of the fuel. The average heat content per gallon of gasoline is 0.125 mmbtu/gallon and the average emissions per heat content of gasoline is 71.35 kg CO_2/mmbtu (USEPA 2012b). The fraction oxidized to CO_2 is 100% (IPCC, 2006).

1.7.3.1 Calculation

Due to rounding, performing the calculations given in the equations below may not return the exact results shown.

$$(0.125 \text{ mmbtu/gal}) \times (71.35 \text{ kg } CO_2 \text{ per mmbtu}) \times (1 \text{ metric ton}/1000 \text{ kg})$$

$$= 8.92 \times 10^{-3} \text{ metric tons } CO_2 \text{ per gal of gasoline.}$$

1.7.4 THERMS OF NATURAL GAS

Carbon dioxide emissions per therm are determined by multiplying heat content times the carbon coefficient times the fraction oxidized times the ratio of the molecular weight of carbon dioxide to that of carbon (44/12). The average heat content of natural gas is 0.1 mmbtu per therm, and the average carbon coefficient of natural gas is 14.47 kg carbon per mmbtu (USEPA, 2013a). The fraction oxidized to CO_2 is 100% (IPCC, 2006).

Note: When using this equivalency, please keep in mind that it represents the CO_2 equivalency for natural gas burned as a fuel, not natural gas released to the atmosphere. Direct methane emissions released to the atmosphere (without burning) are about 21 times more powerful than CO_2 in terms of their warming effect on the atmosphere.

1.7.4.1 Calculation

Due to rounding, performing the calculations given in the equations below may not return the exact results shown.

$$(0.1 \text{ mmbtu}/1 \text{ therm}) \times (14.47 \text{ kg C per mmbtu}) \times (44 \text{ g } CO_2 \text{ per } 12 \text{ g C})$$

$$\times (1 \text{ metric ton}/1000 \text{ kg})$$

$$= 0.005 \text{ metric tons } CO_2 \text{ per therm.}$$

1.7.5 BARRELS OF OIL CONSUMED

Carbon dioxide emissions per barrel of crude oil are determined by multiplying heat content times the carbon coefficient times the fraction oxidized times the ratio of the molecular weight of carbon dioxide to that of carbon (44/12). The average heat content of crude oil is 5.80 mmbtu per barrel, and the average carbon coefficient of crude oil is 20.31 kg carbon per mmbtu (USEPA, 2013a). The fraction oxidized to CO_2 is 100% (IPCC, 2006).

1.7.5.1   Calculation

Due to rounding, performing the calculations given in the equations below may not return the exact results shown.

$$(5.80 \text{ mmbtu/barrel}) \times (20.31 \text{ kg C per mmbtu}) \times (44 \text{ g } CO_2 \text{ per } 12 \text{ g C}) \times (1 \text{ metric ton/ } 1000 \text{ kg})$$
$$= 0.43 \text{ metric tons } CO_2 \text{ per barrel.}$$

1.7.6   Tanker Trucks Filled with Gasoline

Carbon dioxide emissions per barrel of gasoline are determined by multiplying the heat content times the carbon dioxide coefficient times the fraction oxidized times the ratio of the molecular weight of carbon dioxide to that of carbon (44/12). A barrel equals 42 gallons. A typical gasoline tanker truck contains 8500 gallons. The average heat content of conventional motor gasoline is 0.125 mmbtu/gal, and the average carbon coefficient of motor gasoline is 71.35 kg CO_2 (USEPA, 2012). The fraction oxidized to CO_2 is 100% (IPCC, 2006).

1.7.6.1   Calculation

Due to rounding, performing the calculations given in the equations below may not return the exact results show.

$$(0.125 \text{ mmbtu/gal}) \times (71.35 \text{ kg } CO_2 \text{ per mmbtu}) \times (1 \text{ metric ton/1000 kg})$$
$$= 8.92 \times 10^{-3} \text{ metric tons } CO_2 \text{ per gallon.}$$
$$(8.92 \times 10^{-3} \text{ metric tons } CO_2 \text{ per gallon}) \times (8500 \text{ gal per tanker truck})$$
$$= 75.82 \text{ metric tons } CO_2 \text{ per tanker truck.}$$

1.7.7   Home Electricity Use

The U.S. Department of Energy's Residential Energy Consumption Surveys defines a single-family home as follows: A housing unit, detached or attached, that provides living space for one home or family. Attached houses are considered single-family houses as long as they are not divided into more than one housing unit and they have an independent outside entrance. A single-family house is contained within walls extending from the basement (or the ground floor, if there is no basement) to the roof. A mobile home with one or more rooms added is classified as a single-family home. Townhouses, row-houses, and duplexes are considered single-family attached housing units, as long as there is no home living above another one within the walls extending from the basement to the roof to separate the units. In 2009, there were 113.6 million homes in the United States; of those, 71.8 million were single-family detached homes and 6.7 million were single-family attached homes for a total of 78.9 million single-family homes nationally (USEIA, 2009). On average, each single-family home consumed 11,319 kWh of delivered electricity. The national average carbon dioxide output rate for electricity generated in 2009 was 1216 lb CO_2 per megawatt-hour (USEPA, 2012), which translates to about 1301 lb CO_2 per megawatt-hour for delivered electricity (assuming 7% in transmission and distribution losses). Annual single-family home electricity consumption is multiplied by the carbon dioxide emission rate (per unit of electricity delivered) to determine annual carbon dioxide emissions per home.

1.7.7.1   Calculation

Due to rounding, performing the calculations give in the equations below may not return the exact results shown.

$$(11,319 \text{ kWh per home} \times 1301.31 \text{ lb } CO_2 \text{ per megawatt-hour delivered})$$
$$\times (1 \text{ mWh/1000 kWh}) \times (1 \text{ metric ton/2204.6 lb})$$
$$= 6.68 \text{ metric tons } CO_2 \text{ per home.}$$

1.7.8   Home Energy Use

The average carbon dioxide coefficient of natural gas is 0.0544 kg CO_2 per cubic foot (USEPA 2013a), and the fraction oxidized to CO_2 is 100% (IPCC, 2006). The average carbon dioxide coefficient of distillate fuel oil is 429.61 kg CO_2 per 42-gallon barrel (USEPA, 2013a), and the fraction oxidized to CO_2 is 100% (IPCC, 2006). The average carbon dioxide coefficient of liquefied petroleum gases is 219.3 kg CO_2 per 42-gallon barrel (USEPA, 2013a), and the fraction oxidized is 100% (IPCC, 2006). The average carbon dioxide coefficient of kerosene is 426.31 kg CO_2 per 42-gallon barrel (USEPA, 2013a), and the fraction oxidized to CO_2 is 100% (IPCC, 2006). Total single-family home electricity, natural gas, distillate fuel oil, and liquefied petroleum gas consumption figures were converted from their various units to metric tons of CO_2 and added together to obtain total CO_2 emissions per home.

1.7.8.1   Calculation

Due to rounding, performing the calculations given in the equations below may not return the exact results shown.

1. *Delivered electricity:* (11,319 kWh per home) × (1301.31 lb CO_2 per megawatt-hour delivered) × (1 mWh/1000 kWh) × (1 metric ton/2204.6 lb) = 6.8 metric tons CO_2 per home.
2. *Natural gas:* (66,000 ft³ per home) × (00544 kg CO_2 per cubic foot) × (1/1,000 kg/metric ton) = 3.59 metric tons CO_2 per home.
3. *Liquid petroleum gas:* (464 gal per home) × (1/42 barrels/gal) × (219.3 kg CO_2 per barrel) × (1/1000 kg/metric ton) = 2.42 metric tons CO_2 per home.
4. *Fuel oil:* (551 gal per home) × (1/42 barrels/gallon) × (429.61 kg CO_2 per barrel) × (1/1000 kg/metric ton) = 5.64 metric tons CO_2 per home.
5. *Kerosene:* (108 gal per home) × (1/42 barrels/gallon) × (426.32 kg CO_2 per barrel) × (1/1000 kg/metric ton) = 1.10 tons CO_2 per home.

Total CO_2 emissions for energy use per single-family home, then, is equal to 6.68 metric tons CO_2 for electricity + 3.59 metric tons CO_2 for natural gas + 2.42 metric tons CO_2 for liquid petroleum gas + 5.64 metric tons CO_2 for fuel oil + 1.10 metric tons CO_2 for kerosene = 19.43 metric tons CO_2 per home per year.

1.7.9   Number of Tree Seedlings Grown for 10 Years

A medium-growth coniferous tree, planted in an urban setting and allowed to grow for 10 years sequesters 23.2 lb of carbon. This estimate is based on the following assumptions:

- Medium-growth coniferous trees are raised in a nursery for one year until they become 1 inch in diameter at 4.5 feet above the ground (the size of tree purchased in a 15-gallon container).
- The nursery-grown trees are then planted in a suburban/urban setting; the trees are not densely planted.
- The calculation takes into account "survival factors" developed by the U.S. Department of Energy. For example, after 5 years (1 year in the nursery and 4 in the urban setting), the probability of survival is 68%; after 10 years, the probability declines to 59%. For each year, the sequestration rate (in pounds per tree) is multiplied by the survival factor to yield a probability-weighted sequestration rate. These values are summed over the 10-year period, beginning from the time of planting, to derive the estimate of 23.2 lb of carbon per tree.

DID YOU KNOW?

Forest land in the United States includes land that is at least 10% stocked with trees of any size, or, in the case of stands dominated by certain western woodland species for which stocking parameters are not available, at least 5% crown cover by trees of any size. Timberland is defined as unreserved productive forest land producing or capable of producing crops of industrial wood. Productivity is at a minimum rate of 20 ft^3 of industrial wood per acre per year. The remaining portion of forest land is classified as "reserved forest land," which is forest withdrawn from timber use by statute or regulation, or "other forest land," which includes forests on which timber is growing at a rate less than 20 ft^3 per acre per year (Smith et. al., 2010).

Please note the following caveats to these assumptions:

- Although most trees take 1 year in a nursery to reach the seedling stage, trees grown under different conditions and trees of certain species may take longer—up to 6 years.
- Average survival rates in urban areas are based on broad assumptions, and the rates will vary significantly depending upon site conditions.
- Carbon sequestration depends on growth rate, which varies by location and other conditions.
- This method estimates only direct sequestration of carbon and does not include the energy savings that result from buildings being shaded by urban tree cover.

To convert to units of metric tons CO_2 per tree, multiply by the ratio of the molecular weight of carbon dioxide to that of carbon (44/12) and the ratio of metric tons per pound (1/2204.6).

1.7.9.1 Calculation

Due to rounding, performing the calculations given in the equations below may not return the exact results shown.

(23.2 lb C per tree) × (44 units CO_2 ÷ 12 units C) × (1 metric ton ÷ 2204.6 lb) = 0.039 metric ton CO_2 per urban tree planted.

1.7.10 ACRES OF U.S. FORESTS STORING CARBON FOR 1 YEAR

Growing forests accumulate and store carbon. Through the process of photosynthesis, trees remove CO_2 from the atmosphere and store it as cellulose, lignin, and other compounds. The rate of accumulation is equal to growth minus removals (i.e., harvest for the production of paper and wood) minus decomposition. In most U.S. forests, growth exceeds removals and decomposition, so that amount of carbon stored nationally is increasing overall.

1.7.10.1 Calculation for U.S. Forests

The *Inventory of U.S. Greenhouse Gas Emissions and Sinks* (USEPA, 2013a) provides data on the net change in forest carbon stocks and forest area. Net changes in carbon attributed to harvested wood products are not included in the calculation.

Annual net change in carbon stocks per area in year n = (Carbon stocks$_{(t+1)}$ − Carbon stocks$_t$) ÷ (Area of land remaining in the same land-use category)

1. Determine the carbon stock change between years by subtracting carbon stocks in year t from carbon stocks in year $(t + 1)$. (This includes carbon stocks in the above-ground biomass, below-ground biomass, dead wood, litter, and soil organic carbon pools.)
2. Determine the annual net change in carbon stocks (i.e., sequestration) per area by dividing the carbon stock change in U.S. forests from step 1 by the total area of U.S. forests remaining in forests in year $(n + 1)$ (i.e., the area of land that did not change land-use categories between the time periods).

Applying these calculations to data developed by the USDA Forest Service for the *Inventory of U.S. Greenhouse Gas Emissions and Sinks* yields a result of 150 metric tons of carbon per hectare (or 61 metric tons of carbon per acre)[*] for the carbon stock density of U.S forests in 2010, with an annual net change in carbon stock per area in 2010 of 0.82 metric tons of carbon sequestered per hectare per year (or 0.33 metric tons of carbon sequestered per acre per year). These values include carbon in the five forest pools of above-ground biomass, below-ground biomass, deadwood, litter, and soil organic carbon, and they are based on state-level Forest Inventory and Analysis (FIA) data. Forest carbon stocks and carbon stock change are based on the stock difference methodology and algorithms described by Smith et al. (2010).

1.7.10.2 Conversion Factors for Carbon Sequestered Annually by 1 Acre of Average U.S. Forest

Due to rounding, performing the calculations given in the equations below may not return the exact results shown. In the following calculation, negative values indicate carbon sequestration.

$(-0.33$ metric ton C per acre/year$) \times (44$ units $CO_2 \div 12$ units C$) = -1.22$ metric ton CO_2 sequestered annually by one acre of average U.S. forest.

Note that this is an estimate for "average" U.S. forests in 2010 (i.e., for U.S. forests as a whole in 2010). Significant geographical variations underlie the national estimates, and the values calculated here might not be representative of individual regions of states. To estimate carbon sequestered for additional acres in one year, simply multiply the number of acres by 1.22 metric tons CO_2 per acre/year. From 2000 to 2010, the average annual sequestration per area was 0.73 metric tons C per hectare/year (or 0.30 metric tons C per acre/year) in the United States, with a minimum value of 0.36 metric tons C per hectare/year (or 0.15 metric tons C per acre/year) in 2000, and a maximum value of 0.83 metric tons C per hectare/year (or 0.34 metric tons C per acre/year) in 2006.

1.7.11 ACRES OF U.S. FOREST PRESERVED FROM CONVERSION TO CROPLAND

The carbon stock density of U.S. forests in 2010 was 150 metric tons of carbon per hectare (or 61 metric tons of carbon per acre) (USEPA, 2013a). This estimate is composed of the five carbon pools of above-ground biomass (52 metric tons C per hectare), below-ground biomass (10 metric tons C per hectare), dead wood (9 metric tons C per hectare), litter (17 metric tons per C hectare), and soil organic carbons (62 metric tons C per hectare).

The *Inventory of U.S. Greenhouse Gas Emissions and Sinks* estimates soil carbon stock changes using U.S.-specific equations and data from the USDA Natural Resource Inventory and the CENTURY biogeochemical model (USEPA, 2013a). When calculating carbon stock changes in biomass due to conversion from forestland to cropland, the IPCC guidelines indicate that the average carbon stock change is equal to the carbon stock change due to removal of biomass from the outgoing land use (i.e., forestland) plus the carbon stocks from one year of growth in the incoming land use (i.e., cropland), or the carbon in biomass immediately after the conversion minus the carbon in biomass prior to the conversion plus the carbon stocks from one year of growth in the incoming

---

[*] 1 hectare = 10,000 m²; 100 m by 100 m; 2.47 acres.

land use (i.e., cropland) (IPCC, 2006). The carbon stock in annual cropland biomass after 1 year is 5 metric tons carbon per hectare, and the carbon content of dry above-ground biomass is 45% (IPCC, 2006). Therefore, the carbon stock in cropland after 1 year of growth is estimated to be 2.25 metric tons carbon per hectare (or 0.91 metric tons carbon per acre).

The averaged reference soil carbon stock (for high-activity clay, low-activity clay, and sandy soils for all climate regions in the United States) is 40.83 metric tons carbon per hectare (USEPA 2013a). Carbon stock change in soils is time dependent, with a default time period for transition between equilibrium soil organ carbon values of 20 years for mineral soils in cropland systems (IPCC, 2006). Consequently, it is assumed that the change in equilibrium mineral soil organic carbon will be annualized over 20 years to represent the annual flux. The IPCC (2006) guidelines indicate that there are insufficient data to provide a default approach or parameters to estimate carbon stocks in perennial cropland.

1.7.11.1 Calculations for Converting U.S. Forests to U.S. Cropland

Annual change in biomass carbon stocks on land converted to other land-use category:

$$\Delta C_B = \Delta C_G + C_{Conversion} - \Delta C_L$$

where

ΔC_B = Annual change in carbon stocks in biomass due to growth on land converted to another land-use category (i.e., 2.25 metric tons C per hectare).

ΔC_G = Annual increase in carbon stocks in biomass due to growth on land converted to another land-use category (i.e., 2.25 metric tons C per hectare).

$C_{Conversion}$ = Initial change in carbon stocks in biomass on land converted to another land-use category; the sum of the carbon stocks in above-ground, below-ground, deadwood, and litter biomass (–88.47 metric tons C per hectare). Immediately after conversion from forestland to cropland, biomass is assumed to be zero, as the land is cleared of all vegetation before planting crops.

ΔC_L = Annual decrease in biomass stocks due to losses from harvesting, fuel wood gathering, and disturbances on land converted to other land-use category (assumed to be zero).

Therefore, $\Delta C_B = \Delta C_G + C_{Conversion} - \Delta C_L$ = –86.22 metric tons carbon per hectare per year of biomass carbon stocks are lost when forestland is converted to cropland.

Annual change in organic carbon stocks in mineral soils

$$\Delta C_{Mineral} = (SOC_O - SOC_{(O-T)}) \div D$$

where

$\Delta C_{Mineral}$ = Annual change in carbon stocks in mineral soils.

SOC_O = Soil organic carbon stock in last year of inventory time period (i.e., 40.83 mt C per hectare).

$SOC_{(O-T)}$ = Solid organic carbon stock at beginning of inventory time period (i.e., 62 mt C per hectare).

D = Time dependence of stock change factors which is the default time period for transition between equilibrium SOC values (i.e., 20 years for cropland systems).

Therefore, $\Delta C_{Mineral}$ $(SPC_O - SOC_{(O-T)}) \div D = (40.83 - 62) \div 20 = -1.06$ metric tons C per hectare per year of soil organic C are lost. Consequently, the change in carbon density from converting forestland to cropland would be –86.22 metric tons of C per hectare per year of biomass plus –1.06 metric tons C per hectare per year of soil organic C, equaling a total loss of 87.28 metric tons C per hectare per year (or –35.32 metric tons C per acre per year). To convert to carbon dioxide, multiply by the ratio of the molecular weight of carbon dioxide to that of carbon (44/12), to yield a value of –320.01 metric tons CO_2 per hectare per year (or –129.51 metric tons CO_2 per acre per year).

1.7.11.2 Conversion Factor for Carbon Sequestered Annually by 1 Acre of Forest Preserved from Conversion to Cropland

Due to rounding, performing the calculations given in the equations below may not return the exact results shown. Negative values indicate CO_2 that is *not* emitted.

(–35.32 metric tons C per acre per year) × (44 units CO_2 ÷ 12 units C) = –129.51 metric tons CO_2 per acre per year.

To estimate CO_2 not emitted when an acre of forest is preserved from conversion to cropland, simply multiply the number of acres of forest not converted by –129.51 metric tons CO_2 per acre per year. Note that this calculation method assumes that all of the forest biomass is oxidized during clearing (i.e., one of the burned biomass remains as charcoal or ash). Also note that this estimate only includes mineral soil carbon stocks, as most forests in the contiguous United States are growing on mineral soils. In the case of mineral soil forests, soil carbon stocks could be replenished or even increased, depending on the starting stocks, how the agricultural lands are managed, and the time frame over which lands are managed.

1.7.12 PROPANE CYLINDERS USED FOR HOME BARBECUES

Propane is 81.7% carbon. The fraction oxidized is 100% (IPCC, 2006; USEPA, 2013a). Carbon dioxide emissions per pound of propane were determined by multiplying the weight of propane in a cylinder times the carbon content percentage times the fraction oxidized times the ratio of the molecular weight of carbon dioxide to that of carbon (44/12). Propane cylinders vary with respect to size; for the purpose of this equivalency calculation, a typical cylinder for home use was assumed to contain 18 pounds of propane.

1.7.12.1 Calculation

Due to rounding, performing the calculations given in the equations below may not return the exact results shown.

(18 lb propane/1 cylinder) × (0.817 lb C per lb propane) × (0.4536 kg/lb) × (44 kg CO_2 per 12 kg C) × (1 metric ton/1000 kg) = 0.024 metric tons CO_2 per cylinder.

1.7.13 RAILCARS OF COAL BURNED

The average heat content of coal in 2009 was 27.56 mmbtu per metric ton. The average carbon coefficient of coal in 2009 was 25.34 kg carbon per mmbtu (USEPA, 2011). The fraction oxidized to CO_2 is 100% (IPCC, 2006). Carbon dioxide emissions per ton of coal were determined by multiplying heat content times the carbon coefficient times the fraction oxidized times the ratio of the molecular weight of carbon dioxide to that of carbon (44/12). The amount of coal in an average railcar was assumed to be 100.19 short tons, or 90.89 metric tons (Hancock and Sreekanth, 2001).

1.7.13.1 Calculation

Due to rounding, performing the calculations given in the equations below may not return the exact results shown.

(27.56 mmbtu/metric ton coal) × (25.34 kg C per mmbtu) × (44g CO_2 per 12g C) × (90.89 metric tons coal per railcar) × (1 metric ton/1000 kg) = 232.74 metric tons CO_2 per railcar.

1.7.14   TONS OF WASTE RECYCLED INSTEAD OF LANDFILLED

To develop the conversion factor for recycling rather than landfilling waste, emission factors from the USEPA's Waste Reduction Model (WARM) were used (USEPA, 2013a). These emission factors were developed following a life-cycle assessment methodology using estimation techniques developed for national inventories of greenhouse gas emissions. According to WARM, the net emission reduction from recycling mixed recyclables (e.g., paper, metals, plastics), compared with a baseline in which the materials are landfilled, is 0.73 metric tons of carbon equivalent per short ton. This factor was then converted to metric tons of carbon dioxide equivalent by multiplying by 44/12, the molecular weight ratio of carbon dioxide to carbon.

1.7.14.1   Calculation

Due to rounding, performing the calculation given in the equation below may not return the exact results show.

(0.73 metric tons of carbon equivalent per ton) × (44 g CO_2 per 12 g C) = 2.67 metric tons CO_2 equivalent per ton of waste recycled instead of landfilled.

1.7.15   COAL-FIRED POWER PLANT EMISSIONS FOR 1 YEAR

In 2009, a total of 457 power plants used coal to generate at least 95% of their electricity (USEPA, 2012a). These plants emitted 1,614,625,638.1 metric tons of CO_2 in 2009. Carbon dioxide emissions per power plant were calculated by dividing the total emissions from power plants whose primary source of fuel was coal by the number of power plants.

1.7.15.1   Calculation

Due to rounding, performing the calculations given in the equations below may not return the exact results shown.

(1,614,625,638.1 metric tons of CO_2) × (1/457 power plants) = 3,533,098 metric tons CO_2/power plant.

1.8   UNITS OF DERIVATIVES AND INTEGRALS

If x = meters and t = seconds, the units of these quantities.

$$\frac{dx}{dt}; \quad \frac{d^2x}{dt^2}; \quad \left(\frac{dx}{dt}\right)^2; \quad y = \int_{t_1}^{t} x^2 dt$$

are m s^{-1}, m s^{-2}, m^{2}s^{-2}, and m^{2}s, respectively.

REFERENCES AND RECOMMENDED READING

FHWA (2010). *Highway Statistics 2010*. Office of Highway Policy Information, Federal Highway Administration, U.S. Department of Transportation, Washington, DC (http://www.fhwa.dot.gov/policy-information/statistics/2010/index.cfm).

Hancock, K. and Sreekanth, A. (2001). Conversion of weight of freight to number of railcars. *Transportation Research Record*, 1768, 1–10.

IPCC. (2006). *2006 IPCC Guidelines for National Green House Gas Inventories*. Intergovernmental Panel on Climate Change, Geneva, Switzerland.

Smith, J.L., Heath, L., and Nichols, M. (2010). *U.S. Forest Carbon Calculation Tool User's Guide: Forestland Carbon Stocks and Net Annual Stock Change*, General Technical Report NRS-13 revised. U.S. Department of Agriculture Forest Service, Northern Research Station, St. Paul, MN.

USEIA. (1998). *Method for Calculating Carbon Sequestration by Trees in Urban and Suburban Settings*. U.S. Energy Information Administration, Washington, DC.

USEIA. (2009). *2009 Residential Energy Consumption Survey*, Table CE2.6, Fuel Expenditures Totals and Averages, U.S. Homes. U.S. Energy Information Administration, Washington, DC.

USEPA. (2011). *Inventory of U.S. Greenhouse Gas Emissions and Sinks (MMT CO_2 Equivalents): Fast Facts 1990–2009*. U.S. Environmental Protection Agency, Washington, DC (http://www.epa.gov/climatechange/emissions/usinventoryreport.html).

USEPA. (2012a). *eGrid2012 Version 1.0 Year 2009 Summary Tables*. U.S. Environmental Protection Agency, Washington, DC (http://www.epa.gov/cleanenergy/documents/egridzips/eGRID2012V1_0_year09_SummaryTables.pdf).

USEPA. (2013a). *Inventory of U.S. Greenhouse Gas Emissions and Sinks: 1990–2011*. U.S. Environmental Protection Agency, Washington, DC (http://www.epa.gov/climatechange/ghgemissions/usinventoryreport.html).

USEPA. (2013b). *Waste Reduction Model (WARM)*. U.S. Environmental Protection Agency, Washington, DC (http://epa.gov/epawaste/conserve/tools/warm/index.html).

2 Basic Math Operations

Note everything that counts can be counted, and not everything that can be counted counts.

—Albert Einstein

2.1 INTRODUCTION

Most calculations required by wastewater operators and engineers (as with many others) start with the basics, such as addition, subtraction, multiplication, division, and sequence of operations. Although many of the operations are fundamental tools within each operator's toolbox, it is important to reuse these tools on a consistent basis to remain sharp in their use. Wastewater operators should master basic math definitions and the formation of problems; daily operations require calculation of percentage, average, simple ratio, geometric dimensions, threshold odor number, force, pressure, and head, and, at the higher levels of licensure, the use of dimensional analysis and advanced math operations.

2.2 BASIC MATH TERMINOLOGY AND DEFINITIONS

The following basic definitions will aid in understanding the material that follows.

- An *integer*, or an *integral number*, is a whole number; thus, 1, 2, 3, 4, 5, 6, 7, 8, 9, 10, 11, and 12 are the first 12 positive integers.
- A *factor*, or *divisor*, of a whole number is any other whole number that exactly divides it; thus, 2 and 5 are factors of 10.
- A *prime number* in math is a number that has no factors except itself and 1; examples of prime numbers are 1, 3, 5, 7, and 11.
- A *composite number* is a number that has factors other than itself and 1. Examples of composite numbers are 4, 6, 8, 9, and 12.
- A *common factor*, or *common divisor*, of two or more numbers is a factor that will exactly divide each of them. If this factor is the largest factor possible, it is called the *greatest common divisor*. Thus, 3 is a common divisor of 9 and 27, but 9 is the greatest common divisor of 9 and 27.
- A *multiple* of a given number is a number that is exactly divisible by the given number. If a number is exactly divisible by two or more other numbers, it is a common multiple of them. The least (smallest) such number is called the *lowest common multiple*. Thus, 36 and 72 are common multiples of 12, 9, and 4; however, 36 is the lowest common multiple.
- An *even number* is a number exactly divisible by 2; thus, 2, 4, 6, 8, 10, and 12 are even integers.
- An *odd number* is an integer that is not exactly divisible by 2; thus, 1,3, 5, 7, 9, and 11 are odd integers.

- A *product* is the result of multiplying two or more numbers together; thus, 25 is the product of 5 × 5. Also, 4 and 5 are factors of 20.
- A *quotient* is the result of dividing one number by another; for example, 5 is the quotient of 20 ÷ 4.
- A *dividend* is a number to be divided, and a *divisor* is a number that divides; for example, in 100 ÷ 20 = 5, 100 is the dividend, 20 is the divisor, and 5 is the quotient.
- *Area* is the area of an object, measured in square units.
- *Base* is a term used to identify the bottom leg of a triangle, measured in linear units.
- *Circumference* is the distance around an object, measured in linear units. When determined for other than circles, it may be called the *perimeter* of the figure, object, or landscape.
- *Cubic units* are measurements used to express volume, cubic feet, cubic meters, etc.
- *Depth* is the vertical distance from the bottom of the tank to the top. This is normally measured in terms of liquid depth and given in terms of *sidewall depth* (SWD), measured in linear units.
- *Diameter* is the distance from one edge of a circle to the opposite edge passing through the center, measured in linear units.
- *Height* is the vertical distance from the base or bottom of a unit to the top or surface.
- *Linear units* are measurements used to express distances: feet, inches, meters, yards, etc.
- *Pi* (π) is a number in calculations involving circles, spheres, or cones ($\pi = 3.14$).
- *Radius* is the distance from the center of a circle to the edge, measured in linear units.
- *Sphere* is a container shaped like a ball.
- *Square units* are measurements used to express area, square feet, square meters, acres, etc.
- *Volume* is the capacity of the unit (how much it will hold), measured in cubic units (cubic feet, cubic meters) or in liquid volume units (gallons, liters, million gallons).
- *Width* is the distance from one side of the tank to the other, measured in linear units.

2.2.1 KEYWORDS

- *Of* means to multiply.
- *And* means to add.
- *Per* means to divide.
- *Less than* means to subtract.

2.3 SEQUENCE OF OPERATIONS

Mathematical operations such as addition, subtraction, multiplication, and division are usually performed in a certain order or sequence. Typically, multiplication and division operations are done prior to addition and subtraction operations. In addition, mathematical operations are also generally performed from left to right using this hierarchy. The use of parentheses is also common to set apart operations that should be performed in a particular sequence.

Consider the expression 2 + 3 × 4. You might answer 20 or, if you know the rules, you would provide the correct answer of 14. The preceding expression may be written as 2 + (3 × 4), but the brackets are unnecessary if you know the rules, as multiplication has precedence even without the parentheses.

Note: It is assumed that the reader has a fundamental knowledge of basic arithmetic and math operations. Thus, the purpose of the following section is to provide a brief review of the mathematical concepts and applications frequently employed by environmental practitioners.

2.3.1   RULES FOR SEQUENCE OF OPERATIONS

Rule 1

In a series of additions, the terms may be placed in any order and grouped in any way; thus, $4 + 3 = 7$ and $3 + 4 = 7$; $(4 + 3) + (6 + 4) = 17$, $(6 + 3) + (4 + 4) = 17$, and $[6 + (3 + 4)] + 4 = 17$.

Rule 2

In a series of subtractions, changing the order or the grouping of the terms may change the result; thus, $100 - 30 = 70$, but $30 - 100 = -70$, and $(100 - 30) - 10 = 60$, but $100 - (30 - 10) = 80$.

Rule 3

When no grouping is given, the subtractions are performed in the order written, from left to right; thus, $100 - 30 - 15 - 4 = 51$ (by steps, $100 - 30 = 70$, $70 - 15 = 55$, $55 - 4 = 51$).

Rule 4

In a series of multiplications, the factors may be placed in any order and in any grouping; thus, $[(2 \times 3) \times 5] \times 6 = 180$ and $5 \times [2 \times (6 \times 3)] = 180$.

Rule 5

In a series of divisions, changing the order or the grouping may change the result; thus, $100 \div 10 = 10$ but $10 \div 100 = 0.1$, and $(100 \div 10) \div 2 = 5$ but $100 \div (10 \div 2) = 20$. Again, if no grouping is indicated, the divisions are performed in the order written, from left to right; thus, $100 \div 10 \div 2$ is understood to mean $(100 \div 10) \div 2$.

Rule 6

In a series of mixed mathematical operations, the convention is as follows: Whenever no grouping is given, multiplications and divisions are to be performed in the order written, then additions and subtractions in the order written.

2.3.2   EXAMPLES OF SEQUENCE OF OPERATIONS

In a series of additions, the terms may be placed in any order and grouped in any way:

$$4 + 6 = 10 \text{ and } 6 + 4 = 10$$

$$(4 + 5) + (3 + 7) = 19, (3 + 5) + (4 + 7) = 19, \text{ and } [7 + (5 + 4)] + 3 = 19$$

In a series of subtractions, changing the order or the grouping of the terms may change the result:

$$100 - 20 = 80, \text{ but } 20 - 100 = -80$$

$$(100 - 30) - 20 = 50, \text{ but } 100 - (30 - 20) = 90$$

When no grouping is given, the subtractions are performed in the order written—from left to right:

$$100 - 30 - 20 - 3 = 47$$

or by steps:

$$100 - 30 = 70, 70 - 20 = 50, 50 - 3 = 47$$

In a series of multiplications, the factors may be placed in any order and in any grouping:

$$[(3 \times 3) \times 5] \times 6 = 270 \text{ and } 5 \times [3 \times (6 \times 3)] = 270$$

In a series of divisions, changing the order or the grouping may change the result:

$$100 \div 10 = 10, \text{ but } 10 \div 100 = 0.1$$

$$(100 \div 10) \div 2 = 5, \text{ but } 100 \div (10 \div 2) = 20$$

If no grouping is indicated, the divisions are performed in the order written—from left to right:

$$100 \div 5 \div 2 \text{ is understood to mean } (100 \div 5) \div 2$$

In a series of mixed mathematical operations, the rule of thumb is that, whenever no grouping is given, multiplications and divisions are to be performed in the order written, then additions and subtractions in the order written.

Consider the following classic example of sequence of operations (Stapel, 2012):

Problem: Simplify $4 - 3[4 - 2(6 - 3)] \div 2$.

Solution:

$$4 - 3\,[4 - 2(6 - 3)] \div 2$$
$$4 - 3[4 - 2(3)] \div 2$$
$$4 - 3[4 - 6] \div 2$$
$$4 - 3[-2] \div 2$$
$$4 + 6 \div 2$$
$$4 + 3 = 7$$

2.4 PERCENT

The word "percent" means "by the hundred." Percentage is usually designated by the symbol %; thus, 15% means 15 percent or 15/100 or 0.15. These equivalents may be written in the reverse order: 0.15 = 15/100 = 15%. In wastewater treatment, percent is frequently used to express plant performance and for control of biosolids treatment processes. When working with percent, the following key points are important:

- *Percents* are another way of expressing a part of a whole.
- Percent means "by the hundred," so a percentage is the number out of 100. To determine percent, divide the quantity we wish to express as a percent by the total quantity, then multiply by 100:

$$\text{Percent } (\%) = \frac{\text{Part}}{\text{Whole}} \tag{2.1}$$

For example, 22 percent (or 22%) means 22 out of 100, or 22/100. Dividing 22 by 100 results in the decimal 0.22:

$$22\% = \frac{22}{100} = 0.22$$

- When using percentage in calculations (such as when used to calculate hypochlorite dosages and when the percent available chlorine must be considered), the percentage must be converted to an equivalent decimal number; this is accomplished by dividing the percentage by 100. For example, calcium hypochlorite (HTH) contains 65% available chlorine. What is the decimal equivalent of 65%? Because 65% means 65 per hundred, divide 65 by 100: 65/100, which is 0.65.
- Decimals and fractions can be converted to percentages. The fraction is first converted to a decimal, then the decimal is multiplied by 100 to get the percentage. For example, if a 50-foot-high water tank has 26 feet of water in it, how full is the tank in terms of the percentage of its capacity?

$$\frac{26 \text{ ft}}{50 \text{ ft}} = 0.52 \text{ (decimal equivalent)}$$

$$0.52 \times 100 = 52$$

Thus, the tank is 52% full.

■ EXAMPLE 2.1

Problem: The plant operator removes 6500 gal of biosolids from the settling tank. The biosolids contain 325 gal of solids. What is the percent solids in the biosolids?

Solution:

$$\text{Percent} = \frac{325 \text{ gal}}{6500 \text{ gal}} \times 100 = 5\%$$

■ EXAMPLE 2.2

Problem: Convert 65% to decimal percent.

Solution:

$$\text{Decimal Percent} = \frac{\text{Percent}}{100} = \frac{65}{100} = 0.65$$

■ EXAMPLE 2.3

Problem: Biosolids contains 5.8% solids. What is the concentration of solids in decimal percent?

Solution:

$$\text{Decimal percent} = \frac{5.8\%}{100} = 0.058$$

Note: Unless otherwise noted, all calculations in the text using percent values require the percent to be converted to a decimal before use.

Key Point: To determine what quantity a percent equals, first convert the percent to a decimal then multiply by the total quantity:

$$\text{Quantity} = \text{Total} \times \text{Decimal percent} \qquad (2.2)$$

■ **EXAMPLE 2.4**

Problem: Biosolids drawn from the settling tank are 5% solids. If 2800 gal of biosolids are withdrawn, how many gallons of solids are removed?

Solution:

$$\text{Gallons} = \frac{5\%}{100} \times 2800 \text{ gal} = 140 \text{ gal}$$

■ **EXAMPLE 2.5**

Problem: Convert 0.55 to percent.

Solution:

$$0.55 = \frac{55}{100} = 55\%$$

To convert 0.55 to 55%, we simply move the decimal point two places to the right.

■ **EXAMPLE 2.6**

Problem: Convert 7/22 to a decimal percent to a percent.

Solution:

$$\frac{7}{22} = 0.318 = 0.318 \times 100 = 31.8\%$$

■ **EXAMPLE 2.7**

Problem: What is the percentage of 3 ppm?

Note: Because 1 liter of water weighs 1 kg (1000 g = 1,000,000 mg), milligrams per liter is parts per million (ppm)

Solution: Because 3 parts per million (ppm) = 3 mg/L:

$$3 \text{ mg/L} = \frac{3 \text{ mg}}{1 \text{ L} \times 1,000,000 \text{ mg/L}} \times 100\%$$

$$= \frac{3}{10,000}\% = 0.0003\%$$

■ **EXAMPLE 2.8**

Problem: How many mg/L is a 1.4% solution?

Solution:

$$1.4\% = \frac{1.4}{100} \times 1,000,000 \text{ mg/L (the weight of 1 L water to } 10^6) = 14,000 \text{ mg/L}$$

■ **EXAMPLE 2.9**

Problem: Calculate pounds per million gallons for 1 ppm (1 mg/L) of water.

Solution: Because 1 gal of water = 8.34 lb,

$$1\,ppm = \frac{1\,gal}{10^6\,gal} = \frac{1\,gal \times 8.34\,lb/gal}{1,000,000\,gal} = 8.34\,lb/1,000,000\,gal$$

■ **EXAMPLE 2.10**

Problem: How many pounds of activated carbon (AC) are needed with 42 lb of sand to make the mixture 26% AC?

Solution: Let x be the weight of AC; thus,

$$\frac{x}{42+x} = 0.26$$

$$x = 0.26(42+x) = 10.92 + 0.26x$$

$$x = \frac{10.92}{0.74} = 14.76\,lb$$

■ **EXAMPLE 2.11**

Problem: A pipe is laid at a rise of 140 mm in 22 m. What is the grade?

Solution:

$$Grade = \frac{140\,mm}{22\,m} \times 100\% = \frac{140\,mm}{22\,m \times 1000\,mm} \times 100\% = 0.64\%$$

■ **EXAMPLE 2.12**

Problem: A motor is rated as 40 horsepower (hp). However, the output horsepower of the motor is only 26.5 hp. What is the efficiency of the motor?

Solution:

$$Efficiency = \frac{hp\,output}{hp\,input} \times 100\% = \frac{26.5\,hp}{40\,hp} \times 100\% = 66\%$$

2.5 SIGNIFICANT DIGITS

When rounding numbers, the following key points are important:

- Numbers are rounded to reduce the number of digits to the right of the decimal point. This is done for convenience, not for accuracy.
- A number is rounded off by dropping one or more numbers from the right and adding zeroes if necessary to place the decimal point. If the last figure dropped is 5 or more, increase the last retained figure by 1. If the last digit dropped is less than 5, do not increase the last retained figure. If the digit 5 is dropped, round off preceding digit to the nearest *even* number.

RULE

Significant figures are those numbers that are known to be reliable. The position of the decimal point does not determine the number of significant figures.

■ **EXAMPLE 2.13**

Problem: Round off the following numbers to one decimal.

Solution:
   34.73 = 34.7
   34.77 = 34.8
   34.75 = 34.8
   34.45 = 34.4
   34.35 = 34.4

■ **EXAMPLE 2.14**

Problem: Round off 10,546 to 4, 3, 2, and 1 significant figures.

Solution:
   10,546 = 10,550 to 4 significant figures
   10,546 = 10,500 to 3 significant figures
   10,546 = 11,000 to 2 significant figures
   10,547 = 10,000 to 1 significant figure

When determining significant figures, the following key points are important:

1. The concept of significant figures is related to rounding.
2. It can be used to determine where to round off.

Key Point: No answer can be more accurate than the least accurate piece of data used to calculate the answer.

■ **EXAMPLE 2.15**

Problem: How many significant figures are in a measurement of 1.35 in.?

Solution: Three significant figures: 1, 3, and 5.

■ **EXAMPLE 2.16**

Problem: How many significant figures are in a measurement of 0.000135?

Solution: Again, three significant figures: 1, 3, and 5. The three zeros are used only to place the decimal point.

■ **EXAMPLE 2.17**

Problem: How many significant figures are in a measurement of 103,500?

Solution: Four significant figures: 1, 0, 3, and 5. The remaining two zeros are used to place the decimal point.

■ EXAMPLE 2.18

Problem: How many significant figures are in 27,000.0?

Solution: There are six significant figures: 2, 7, 0, 0, 0, 0. In this case, the .0 in 27,000.0 means that the measurement is precise to 1/10 unit. The zeros indicate measured values and are not used solely to place the decimal point.

2.6   POWERS AND EXPONENTS

In working with powers and exponents, the following key points are important:

- *Powers* are used to identify *area*, as in square feet, and *volume*, as in cubic feet.
- Powers can also be used to indicate that a number should be squared, cubed, etc. This later designation is the number of times a number must be multiplied times itself.
- If all of the factors are alike, as $4 \times 4 \times 4 \times 4 = 256$, the product is called a *power*. Thus, 256 is a power of 4, and 4 is the *base* of the power. A power is a *product* obtained by using a base a certain number of times as a factor.
- Instead of writing $4 \times 4 \times 4 \times 4$, it is more convenient to use an *exponent* to indicate that the factor 4 is used as a factor four times. This exponent, a small number placed above and to the right of the base number, indicates how many times the base is to be used as a factor. Using this system of notation, the multiplication $4 \times 4 \times 4 \times 4$ is written as 4^4. The ^{4} is the exponent, showing that 4 is to be used as a factor 4 times.
- These same consideration apply to letters (a, b, x, y, etc.) as well; for example:

$$z^2 = z \times z$$

$$z^4 = z \times z \times z \times z$$

Note: When a number or letter does not have an exponent, it is considered to have an exponent of one.

■ EXAMPLE 2.19

Problem: How is the term 2^3 written in expanded form?

Solution: The power (exponent) of 3 means that the base number (2) is multiplied by itself three times:

$$2^3 = 2 \times 2 \times 2$$

POWERS OF 1	POWERS OF 10
$1^0 = 1$	$10^0 = 1$
$1^1 = 1$	$10^1 = 10$
$1^2 = 1$	$10^2 = 100$
$1^3 = 1$	$10^3 = 1000$
$1^4 = 1$	$10^4 = 10,000$

■ **EXAMPLE 2.20**

Problem: How is the term $(3/8)^2$ written in expanded form?

Note: When parentheses are used, the exponent refers to the entire term within the parentheses.

Solution: In this example, $(3/8)^2$ means:

$$(3/8)^2 = (3/8 \times 3/8)$$

Key Point: When a negative exponent is used with a number or term, a number can be rewritten using a positive exponent:

$$6^{-3} = 1/6^3$$

Another example is

$$11^{-5} = 1/11^5$$

■ **EXAMPLE 2.21**

Problem: How is the term 8^{-3} written in expanded form?

Solution:

$$8^{-3} = \frac{1}{8^3} = \frac{1}{8 \times 8 \times 8}$$

Key Point: A number or letter written as, for example, 3^0 or X^0 does not equal 3×1 or $X \times 1$, but simply 1.

2.7 AVERAGES (ARITHMETIC MEAN)

Whether we speak of harmonic mean, geometric mean, or arithmetic mean, each represents the "center," or "middle," of a set of numbers. They capture the intuitive notion of a "central tendency" that may be present in the data. In statistical analysis, an "average of data" is a number that indicates the middle of the distribution of data values.

An *average* is a way of representing several different measurements as a single number. Although averages can be useful in that they tell us "about" how much or how many, they can also be misleading, as we demonstrate below. You will find two kinds of averages in environmental engineering calculations: the *arithmetic mean* (or simply *mean*) and the *median*.

■ **EXAMPLE 2.22**

Problem: The operator of a waterworks or wastewater treatment plant takes a chlorine residual measurement every day; part of the operator's log is shown below. Find the mean.

> **DEFINITION**
>
> The *mean* (what we usually refer to as an *average*) is the total of values of a set of observations divided by the number of observations. We simply add up all of the individual measurements and divide by the total number of measurements we took.

Monday	0.9 mg/L
Tuesday	1.0 mg/L
Wednesday	0.9 mg/L
Thursday	1.3 mg/L
Friday	1.1 mg/L
Saturday	1.4 mg/L
Sunday	1.2 mg/L

Solution: Add up the seven chlorine residual readings: 0.9 + 1.0 + 0.9 + 1.3 + 1.1 + 1.4 + 1.2. = 7.8. Next, divide by the number of measurements—in this case, 7:

$$7.8 \div 7 = 1.11$$

The mean chlorine residual for the week was 1.11 mg/L.

■ EXAMPLE 2.23

Problem: A water system has four wells with the following capacities: 115 gpm (gallons per minute), 100 gpm, 125 gpm, and 90 gpm. What is the mean?

Solution:

$$115 \text{ gpm} + 100 \text{ gpm} + 125 \text{ gpm} + 90 \text{ gpm} = 430$$

$$430 \div 4 = 107.5 \text{ gpm}$$

■ EXAMPLE 2.24

Problem: A water system has four storage tanks. Three of them have a capacity of 100,000 gal each, while the fourth has a capacity of 1 million gal. What is the mean capacity of the storage tanks?

Solution: The mean capacity of the storage tanks is

$$100,000 + 100,000 + 100,000 + 1,000,000 = 1,300,000$$

$$1,300,000 \div 4 = 325,000 \text{ gal}$$

Notice that no tank in Example 2.24 has a capacity anywhere close to the mean.

■ EXAMPLE 2.25

Problem: Effluent biochemical oxygen demand (BOD) test results for the treatment plant during the month of August are shown below:

Test 1	22 mg/L
Test 2	33 mg/L
Test 3	21 mg/L
Test 4	13 mg/L

What is the average effluent BOD for the month of August?

Solution:

$$22 + 33 + 21 + 13 = 89$$

$$89 \div 4 = 22.3 \text{ mg/L}$$

■ **EXAMPLE 2.26**

Problem: For the primary influent flow, the following composite-sampled solids concentrations were recorded for the week:

Monday	310 mg/L SS
Tuesday	322 mg/L SS
Wednesday	305 mg/L SS
Thursday	326 mg/L SS
Friday	313 mg/L SS
Saturday	310 mg/L SS
Sunday	320 mg/L SS
Total	2206 mg/L SS

What is the average SS?

Solution:

$$\text{Average SS} = \frac{\text{Sum of all measurements}}{\text{Number of measurements used}}$$

$$= \frac{2206 \text{ mg/L SS}}{7} = 315.1 \text{ mg/L SS}$$

2.8 RATIO

A *ratio* is the established relationship between two numbers; it is simply one number divided by another number. For example, if someone says, "I'll give you four to one the Redskins over the Cowboys in the Super Bowl," what does that person mean? Four to one, or 4:1, is a ratio. If someone gives you four to one, it's his or her $4 to your $1. As another more pertinent example, if an average of 3 cubic feet (ft³) of screenings are removed from each million gallons (MG) of wastewater treated, the ratio of screenings removed to treated wastewater is 3:1. Ratios are normally written using a colon (such as 2:1) or as a fraction (such as 2/1). When working with ratios, the following key points are important to remember.

- One place where fractions are used in calculations is when ratios are used, such as calculating solutions.
- A ratio is usually stated in the form A is to B as C is to D, which can be written as two fractions that are equal to each other:

$$\frac{A}{B} = \frac{C}{D}$$

- Cross-multiplying solves ratio problems; that is, we multiply the left numerator (A) by the right denominator (D) and say that the product is equal to the left denominator (B) times the right numerator (C):

$$A \times D = B \times C \text{ (or, } AD = BC)$$

- If one of the four items is unknown, dividing the two known items that are multiplied together by the known item that is multiplied by the unknown solves the ratio. For example, if 2 lb of alum are needed to treat 500 gal of water, how many pounds of alum will we need to treat 10,000 gal? We can state this as a ratio: "2 lb of alum is to 500 gal of water as x lb of alum is to 10,000 gal of water." This is set up in this manner:

$$\frac{1 \text{ lb alum}}{500 \text{ gal water}} = \frac{x \text{ lb alum}}{10,000 \text{ gal water}}$$

Cross-multiplying,

$$500 \times x = 1 \times 10,000$$

Transposing,

$$\frac{1 \times 10,000}{500} = 20 \text{ lb alum}$$

To calculate proportion, suppose, for example, that 5 gal of fuel costs \$5.40. What will 15 gal cost?

$$\frac{5 \text{ gal}}{\$5.40} = \frac{15 \text{ gal}}{\$y}$$

$$5 \text{ gal} \times y = 15 \text{ gal} \times \$5.40 = 81$$

$$y = \frac{81}{5} = \$16.20$$

■ **EXAMPLE 2.27**

Problem: If a pump will fill a tank in 20 hr at 4 gpm, how long will it take a 10-gpm pump to fill the same tank?

Solution: First, analyze the problem. Here, the unknown is some number of hours. But, should the answer be larger or smaller than 20 hr? If a 4-gpm pump can fill the tank in 20 hr, a larger (10-gpm) pump should be able to complete the filling in less than 20 hr. Therefore, the answer should be less than 20 hours. Now set up the proportion:

$$\frac{x \text{ hr}}{20 \text{ hr}} = \frac{4 \text{ gpm}}{10 \text{ gpm}}$$

$$x = \frac{(4 \times 20)}{10} = 8 \text{ hr}$$

■ **EXAMPLE 2.28**

Problem: Solve for the unknown value x in the problem given below.

Solution:

$$\frac{36}{180} = \frac{x}{4450}$$

$$\frac{4450 \times 36}{180} = x = 890$$

■ **EXAMPLE 2.29**

Problem: Solve for the unknown value x in the problem given below.

$$\frac{3.4}{2} = \frac{6}{x}$$

Solution:

$$3.4 \times x = 2 \times 6$$

$$x = \frac{2 \times 6}{3.4} = 3.53$$

■ **EXAMPLE 2.30**

Problem: 1 lb of chlorine is dissolved in 65 gal of water. To maintain the same concentration, how many pounds of chlorine would have to be dissolved in 150 gal of water?

Solution:

$$\frac{1 \text{ lb}}{65 \text{ gal}} = \frac{x \text{ lb}}{150 \text{ gal}}$$

$$65 \times x = 1 \times 150$$

$$x = \frac{1 \times 150}{65} = 2.3 \text{ lb}$$

■ **EXAMPLE 2.31**

Problem: It takes 5 workers 50 hr to complete a job. At the same rate, how many hours would it take 8 workers to complete the job?

Solution:

$$\frac{5 \text{ workers}}{8 \text{ workers}} = \frac{x \text{ hr}}{50 \text{ hr}}$$

$$x = \frac{5 \times 50}{8} = 31.3 \text{ hr}$$

■ **EXAMPLE 2.32**

Problem: If 1.6 L of activated sludge (biosolids) with volatile suspended solids (VSS) of 1900 mg/L are mixed with 7.2 L of raw domestic wastewater with BOD of 250 g/L, what is the food/microorganisms (F/M) ratio?

Solution:

$$\frac{F}{M} = \frac{\text{Amount of BOD}}{\text{Amount of VSS}} = \frac{250 \text{ mg/L} \times 7.2 \text{ L}}{1900 \text{ mg/L} \times 1.6 \text{ L}} = \frac{0.59}{1} = 0.59$$

2.9   DIMENSIONAL ANALYSIS

Dimensional analysis is a problem-solving method that uses the fact that any number or expression can be multiplied by 1 without changing its value. It is a useful technique used to check if a problem is set up correctly. In using dimensional analysis to check a math setup, we work with the dimensions (units of measure) only—not with numbers.

An example of dimensional analysis that is common to everyday life is the unit pricing found in many hardware stores. A shopper can purchase a 1-lb box of nails for 98¢ at a local hardware store, but a nearby warehouse store sells a 5-lb bag of the same nails for $3.50. The shopper will analyze this problem almost without thinking about it. The solution calls for reducing the problem to the price per pound. The pound is selected without much thought because it is the unit common to both stores. The shopper will pay 70¢ a pound for the nails at the warehouse store but 98¢ at the local hardware store. Implicit in the solution to this problem is knowing the unit price, which is expressed in dollars per pound ($/lb).

Note: Unit factors may be made from any two terms that describe the same or equivalent amounts of what we are interested in; for example, we know that 1 inch = 2.54 centimeters.

In order to use the dimensional analysis method, we must know how to perform three basic operations.

2.9.1   Basic Operation: Division of Units

To complete a division of units, always ensure that all units are written in the same format; it is best to express a horizontal fraction (such as gal/ft²) as a vertical fraction.

Horizontal to vertical

$$\text{gal/ft}^3 \text{ to } \frac{\text{gal}}{\text{ft}^3}$$

$$\text{psi to } \frac{\text{lb}}{\text{in.}^2}$$

The same procedures are applied in the following examples.

$$\text{ft}^3/\text{min becomes } \frac{\text{ft}^3}{\text{min}}$$

$$\text{s/min becomes } \frac{\text{s}}{\text{min}}$$

2.9.2   Basic Operation: Divide by a Fraction

We must know how to divide by a fraction. For example,

$$\frac{\left(\dfrac{\text{lb}}{\text{day}}\right)}{\left(\dfrac{\text{min}}{\text{day}}\right)} \text{ becomes } \frac{\text{lb}}{\text{day}} \times \frac{\text{day}}{\text{min}}$$

In the above, notice that the terms in the denominator were inverted before the fractions were multiplied. This is a standard rule that must be followed when dividing fractions.

Another example is

$$\frac{mm^2}{\left(\dfrac{mm^2}{m^2}\right)} \text{ becomes } mm^2 \times \frac{m^2}{mm^2}$$

2.9.3 BASIC OPERATION: CANCEL OR DIVIDE NUMERATORS AND DENOMINATORS

We must know how to cancel or divide terms in the numerator and denominator of a fraction. After fractions have been rewritten in the vertical form and division by the fraction has been re-expressed as multiplication, as shown above, then the terms can be canceled (or divided) out.

Key Point: For every term that is canceled in the numerator of a fraction, a similar term must be canceled in the denominator and *vice versa*, as shown below:

$$\frac{kg}{\cancel{d}} \times \frac{\cancel{d}}{min} = \frac{kg}{min}$$

$$\cancel{mm^2} \times \frac{m^2}{\cancel{mm^2}} = m^2$$

$$\frac{gal}{min} \times \frac{ft^3}{gal} = \frac{ft^3}{min}$$

Question: How do we calculate units that include exponents?

Answer: When written with exponents, such as ft^3, a unit can be left as is or put in expanded form, (ft)(ft)(ft), depending on other units in the calculation. The point is that it is important to ensure that square and cubic terms are expressed uniformly (e.g., sq ft, ft^2 cu ft, ft^3). For dimensional analysis, the latter system is preferred.

For example, to convert a volume of 1400 ft^3 to gallons, we will use 7.48 gal/ft^3 in the conversions. The question becomes do we multiply or divide by 7.48? In this instance, it is possible to use dimensional analysis to answer this question of whether we multiply or divide by 7.48.

To determine if the math setup is correct, only the dimensions are used. First, try dividing the dimensions:

$$\frac{ft^3}{gal/ft^3} = \frac{ft^3}{\left(\dfrac{gal}{ft^3}\right)}$$

Multiply the numerator and denominator to get

$$\frac{ft^6}{gal}$$

So, by dimensional analysis, we have determined that if we divide the two dimensions (ft^3 and gal/ft^3) then the units of the answer are ft^6/gal, not gal. It is clear that division is not the right approach to making this conversion.

What would have happened if we had multiplied the dimensions instead of dividing?

$$\text{ft}^3 \times (\text{gal/ft}^3) = \text{ft}^3 \times \left(\frac{\text{gal}}{\text{ft}^3} \right)$$

Multiply the numerator and denominator to obtain

$$\frac{\text{ft}^3 \times \text{gal}}{\text{ft}^3}$$

and cancel common terms to obtain

$$\frac{\cancel{\text{ft}^3} \times \text{gal}}{\cancel{\text{ft}^3}}$$

Obviously, by multiplying the two dimensions (ft³ and gal/ft³), the answer will be in gallons, which is what we want. Thus, because the math setup is correct, we would then multiply the numbers to obtain the number of gallons:

$$(1400 \text{ ft}^3) \times (7.48 \text{ gal/ft}^3) = 10,472 \text{ gal}$$

Now, let's try another problem with exponents. We wish to obtain an answer in square feet. If we are given the two terms—70 ft³/s and 4.5 ft/s—is the following math setup correct?

$$(70 \text{ ft}^3/\text{s}) \times (4.5 \text{ ft/s})$$

First, only the dimensions are used to determine if the math setup is correct. By multiplying the two dimensions, we get

$$(\text{ft}^2/\text{s}) \times (\text{ft/s}) = \frac{\text{ft}^3}{\text{s}} \times \frac{\text{ft}}{\text{s}}$$

Multiply the terms in the numerators and denominators of the fraction:

$$\frac{\text{ft}^3 \times \text{ft}}{\text{s} \times \text{s}} = \frac{\text{ft}^4}{\text{s}^2}$$

Obviously, the math setup is incorrect because the dimensions of the answer are not square feet; therefore, if we multiply the numbers as shown above, the answer will be wrong.

Let's try division of the two dimensions instead:

$$(\text{ft}^3/\text{s}) = \frac{\left(\dfrac{\text{ft}^3}{\text{s}} \right)}{\left(\dfrac{\text{ft}}{\text{s}} \right)}$$

Invert the denominator and multiply to get

$$= \frac{\text{ft}^3}{\text{s}} \times \frac{\text{s}}{\text{ft}} = \frac{(\text{ft} \times \text{ft} \times \text{ft}) \times \text{s}}{\text{s} \times \text{ft}} = \frac{(\text{ft} \times \text{ft} \times \cancel{\text{ft}}) \times \cancel{\text{s}}}{\cancel{\text{s}} \times \cancel{\text{ft}}} = \text{ft}^2$$

Because the dimensions of the answer are square feet, this math setup is correct; therefore, by dividing the numbers as was done with units, the answer will also be correct.

$$\frac{70 \text{ ft}^3/\text{s}}{4.5 \text{ ft/s}} = 15.56 \text{ ft}^2$$

■ EXAMPLE 2.33

Problem: We are given two terms, 5 m/s and 7 m², and the answer to be obtained should be in cubic meters per second (m³/s). Is multiplying the two terms the correct math setup?

Solution:

$$(\text{m/s}) \times (\text{m}^2) = \frac{\text{m}}{\text{s}} \times \text{m}^2$$

Multiply the numerators and denominator of the fraction:

$$= \frac{\text{m} \times \text{m}^2}{\text{s}} = \frac{\text{m}^3}{\text{s}}$$

Because the dimensions of the answer are cubic meters per second (m³/s), the math setup is correct; therefore, multiply the numbers to get the correct answer:

$$5 \text{ m/s} \times 7 \text{ m}^2 = 35 \text{ m}^3/\text{s}$$

■ EXAMPLE 2.34

Problem: The flow rate in a water line is 2.3 ft³/s. What is the flow rate expressed as gallons per minute?

Solution: Set up the math problem and then use dimensional analysis to check the math setup:

$$(2.3 \text{ ft}^3/\text{s}) \times (7.48 \text{ gal/ft}^3) \times (60 \text{ s/min})$$

Dimensional analysis can be used to check the math setup:

$$(\text{ft}^3/\text{s}) \times (\text{gal/ft}^3) \times (\text{s/min}) = \frac{\text{ft}^3}{\text{s}} \times \frac{\text{gal}}{\text{ft}^3} \times \frac{\text{s}}{\text{min}} = \frac{\text{ft}^3}{\text{s}} \times \frac{\text{gal}}{\text{ft}^3} \times \frac{\text{s}}{\text{min}} = \frac{\text{gal}}{\text{min}}$$

The math setup is correct as shown above; therefore, this problem can be multiplied out to get the answer in correct units:

$$(2.3 \text{ ft}^3/\text{s}) \times (7.48 \text{ gal/ft}^3) \times (60 \text{ s/min}) = 1032.24 \text{ gal/min}$$

■ EXAMPLE 2.35

Problem: During an 8-hr period, a water treatment plant treated 3.2 million gallons of water. What is the plant total volume treated per day, assuming the same treatment rate?

Solution:

$$\frac{3.2 \text{ million gal}}{8 \text{ hr}} \times \frac{24 \text{ hr}}{\text{day}} = \frac{3.2 \times 24}{8} \text{ MGD} = 9.6 \text{ MGD}$$

■ **EXAMPLE 2.36**

Problem: One million gallons per day equals how many cubic feet per second (cfs)?

Solution:

$$1 \text{ MGD} = \frac{10^6}{1 \text{ day}} = \frac{10^6 \text{ gal} \times 0.1337 \text{ ft}^3/\text{gal}}{1 \text{ day} \times 86,400 \text{ s/day}} = \frac{133,700}{86,400} = 1.547 \text{ cfs}$$

■ **EXAMPLE 2.37**

Problem: A 10-gal empty tank weighs 4.6 lb. What is the total weight of the tank filled with 6 gal of water?

Solution:

$$\text{Weight of water} = 6 \text{ gal} \times 8.34 \text{ lb/gal} = 50.04 \text{ lb}$$

$$\text{Total weight} = 50.04 + 4.6 \text{ lb} = 54.6 \text{ lb}$$

■ **EXAMPLE 2.38**

Problem: The depth of biosolids applied to the biosolids drying bed is 10 in. What is the depth in centimeters (2.54 cm = 1 in.)?

Solution:

$$10 \text{ in.} = 10 \times 2.54 \text{ cm} = 25.4 \text{ cm}$$

2.10   THRESHOLD ODOR NUMBER (TON)

The environmental practitioner responsible for water supplies soon discovers that taste and odor are the most common customer complaints. Odor is typically measured and expressed in terms of a *threshold odor number* (TON), the ratio by which the sample has to be diluted with odor-free water for the odor to become virtually unnoticeable. In 1989, the USEPA issued a Secondary Maximum Contaminant Level (SMCL) of 3 TON for odor.

Note: Secondary Maximum Contaminant Levels are parameters not related to health.

When a dilution is used, a number can be devised in clarifying odor.

$$\text{TON (threshold odor number)} = \frac{V_T + V_P}{V_T} \tag{2.3}$$

where
  V_T = Volume tested
  V_P = Volume of dilution with odor-free distilled water

  For $V_P = 0$, TON = 1 (lowest value possible)
  For $V_P = V_T$, TON = 2
  For $V_P = 2V_T$, TON = 3
  $\vdots$

■ **EXAMPLE 2.39**

Problem: The first detectable odor is observed when a 50-mL sample is diluted to 200 mL with odor-free water. What is the TON of the water sample?

Solution:

$$\text{TON} = \frac{200}{V_T} = \frac{200 \text{ mL}}{50 \text{ mL}} = 4$$

2.11 GEOMETRICAL MEASUREMENTS

Wastewater treatment plants consist of a series of tanks and channels. Proper design and operational control requires the engineer and operator to perform several process control calculations. Many of these calculations include parameters such as the circumference or perimeter, area, or volume of the tank or channel as part of the information necessary to determine the result. Many process calculations require computation of surface areas. To aid in performing these calculations, the following definitions and relevant equations used to calculate areas and volumes for several geometric shapes are provided.

Environmental practitioners involved in fisheries, water/wastewater treatment plants, and other operations dealing with tanks, basins, and ponds operations must know the area and volume of all tanks, basins, and ponds they deal with. For example, in water and wastewater treatment plant operations, the plant configuration usually consists of a series of tanks and channels. Proper design and operational control require the environmental practitioner and plant operator to perform several process control calculations. Many of these calculations require parameters such as circumference or perimeter, surface area, or volume of a tank or channel. Moreover, in fisheries operations, exact measurements of area and volume are essential to calculate stocking rates and chemical applications. Stocking fish in a pond of uncertain area can result in poor production, disease, and possibly death. Chemical treatments can be ineffective if the volume or area is underestimated and can be potentially lethal if they are overestimated (Masser and Jensen, 1991). To aid in performing these calculations, the following definitions and relevant equations used to calculate areas and volumes for several geometric shapes are provided.

2.11.1 DEFINITIONS

Area—The area of an object, measured in square units.

Base—The term used to identity the bottom leg of a triangle, measured in linear units.

Circumference—The distance around an object, measured in linear units. When determined for other than circles, it may be called the *perimeter* of the figure, object, or landscape.

Cubic units—Measurements used to express volume, cubic feet, cubic meters, etc.

Depth—The vertical distance from the bottom the tank to the top. It is normally measured in terms of liquid depth and given in terms of *sidewall depth* (SWD), measured in linear units.

Diameter—The distance, measured in linear units, from one edge of a circle to the opposite edge passing through the center.

Height—The vertical distance, measured in linear units, from one end of an object to the other.

Length—The distance, measured in linear units, from one end of an object to the other.

Linear units—Measurements used to express distance (e.g., feet, inches, meters, yards).

Pi (π)—A number in the calculations involving circles, spheres, or cones (π = 3.14).

Radius—The distance, measured in linear units, from the center of a circle to the edge.

Sphere—A container shaped like a ball.

Square units—Measurements used to express area (e.g., square feet, square meters, acres).

Volume—The capacity of a unit (how much it will hold), measured in cubic units (e.g., cubic feet, cubic meters) or in liquid volume units (e.g., gallons, liters, million gallons).

Width—The distance from one side of the tank to the other, measured in linear units.

2.11.2 RELEVANT GEOMETRIC EQUATIONS

Circumference C of a circle:	$C = \pi d = 2\pi r$
Perimeter P of a square with side a:	$P = 4a$
Perimeter P of a rectangle with sides a and b:	$P = 2a + 2b$
Perimeter P of a triangle with sides a, b, and c:	$P = a + b + c$
Area A of a circle with radius r ($d = 2r$):	$A = \pi d^2/4 = \pi r^2$
Area A of duct in square feet when d is in inches:	$A = 0.005454d^2$
Area A of a triangle with base b and height h:	$A = 0.5bh$
Area A of a square with sides a:	$A = a^2$
Area A of a rectangle with sides a and b:	$A = ab$
Area A of an ellipse with major axis a and minor axis b:	$A = \pi ab$
Area A of a trapezoid with parallel sides a and b and height h:	$A = 0.5(a + b)h$
Area A of a duct in square feet when d is in inches:	$A = \pi d^2/576 = 0.005454d^2$
Volume V of a sphere with a radius r ($d = 2r$):	$V = 1.33\pi r^3 = 0.1667\pi d^3$
Volume V of a cube with sides a:	$V = a^3$
Volume V of a rectangular solid (sides a and b and height c):	$V = abc$
Volume V of a cylinder with a radius r and height H:	$V = \pi r^2 h = \pi d^2 h/4$
Volume V of a pyramid:	$V = 0.33$

2.11.3 GEOMETRICAL CALCULATIONS

2.11.3.1 Perimeter and Circumference

On occasion, it may be necessary to determine the distance around grounds or landscapes. To measure the distance around property, buildings, and basin-like structures, it is necessary to determine either perimeter or circumference. The *perimeter* is the distance around an object; a border or outer boundary. *Circumference* is the distance around a circle or circular object, such as a clarifier. Distance is a linear measurement that defines the distance (or length) along a line. Standard units of measurement such as inches, feet, yards, and miles and metric units such as centimeters, meters, and kilometers are used.

The perimeter (P) of a rectangle (a four-sided figure with four right angles) is obtained by adding the lengths (L_i) of the four sides (see Figure 2.1):

$$\text{Perimeter} = L_1 + L_2 + L_3 + L_4 \tag{2.4}$$

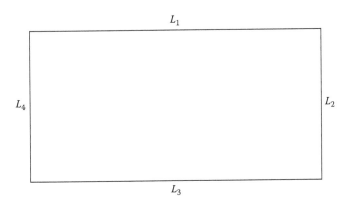

FIGURE 2.1 Perimeter.

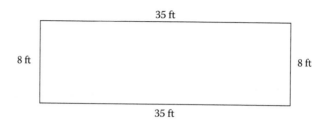

FIGURE 2.2   Perimeter of a rectangle for Example 2.40.

■ **EXAMPLE 2.40**

Problem: Find the perimeter of the rectangle shown in Figure 2.2.

Solution:

$$P = 35 \text{ ft} + 8 \text{ ft} + 35 \text{ ft} + 8 \text{ ft} = 86 \text{ ft}$$

■ **EXAMPLE 2.41**

Problem: What is the perimeter of a rectangular field if its length is 100 ft and its width is 50 ft?

Solution:

$$P = (2 \times \text{length}) + (2 \times \text{width}) = (2 \times 100 \text{ ft}) + (2 \times 50 \text{ ft}) = 200 \text{ ft} + 100 \text{ ft} = 300 \text{ ft}$$

■ **EXAMPLE 2.42**

Problem: What is the perimeter of a square with 8-in. sides?

Solution:

$$P = (2 \times \text{length}) + (2 \times \text{width})$$

$$= (2 \times 8 \text{ in.}) + (2 \times 8 \text{ in.}) = 16 \text{ in.} + 16 \text{ in.} = 32 \text{ in.}$$

The circumference is the length of the outer border of a circle. The circumference is found by multiplying pi (π) times the *diameter* (*D*) (a straight line passing through the center of a circle, or the distance across the circle; see Figure 2.3):

$$C = \pi \times D \qquad\qquad\qquad (2.5)$$

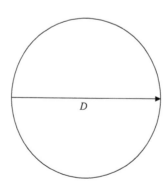

FIGURE 2.3   Diameter of circle.

where
    C = Circumference.
    π = pi = 3.1416.
    D = Diameter.

Use this calculation if, for example, the circumference of a circular tank must be determined.

■ EXAMPLE 2.43

Problem: Find the circumference of a circle that has a diameter of 25 feet (π = 3.14)

Solution:

$$C = \pi \times 25 \text{ ft}$$

$$C = 3.14 \times 25 \text{ ft} = 78.5 \text{ ft}$$

■ EXAMPLE 2.44

Problem: A circular chemical holding tank has a diameter of 18 m. What is the circumference of this tank?

Solution:

$$C = \pi \times 18 \text{ m}$$

$$C = 3.14 \times 18 \text{ m} = 56.52 \text{ m}$$

■ EXAMPLE 2.45

Problem: An influent pipe inlet opening has a diameter of 6 ft. What is the circumference of the inlet opening in inches?

Solution:

$$C = \pi \times 6 \text{ ft}$$

$$C = 3.14 \times 6 \text{ ft} = 18.84 \text{ ft}$$

2.11.3.2 Area

For area measurements in water/wastewater operations, three basic shapes are particularly important—namely, circles, rectangles, and triangles. Area is the amount of surface an object contains or the amount of material it takes to cover the surface. The area on top of a chemical tank is called the *surface area*. The area of the end of a ventilation duct is called the *cross-sectional area* (the area at right angles to the length of ducting). Area is usually expressed in square units, such as square inches (in.^{2}) or square feet (ft^2). Land may also be expressed in terms of square miles (sections) or acres (43,560 ft^2) or in the metric system as hectares. In fisheries operations, pond stocking rates, limiting rates, and other important management decisions are based on surface area (Masser and Jensen, 1991).

If contractor's measurements or country field offices of the U.S Department of Agricultural Soil Conservation Service do not have records on basin, lake, or pond measurements, then surveying basins, tanks, lagoons, and ponds using a transit is the most accurate way to determine area. Less accurate but acceptable methods of measuring basin or pond area are chaining and pacing. Inaccuracies in these methods come from mismeasurements and measurement over uneven or sloping terrain. Measurements made on flat or level areas are the most accurate.

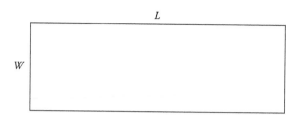

FIGURE 2.4 Rectangle.

Chaining uses a surveyor's chain or tape of known length. Stakes are placed at each end of the tape. The stakes are used to set or locate the starting point for each progressive measurement and to maintain an exact count of the number of times the tape was moved. Sight down the stakes to keep the measurement in a straight line. The number of times the tape is moved multiplied by the length of the tape equals the total distance.

Pacing uses the average distance of a person's pace or stride. To determine your pace length, measure a 100-foot distance and pace it, counting the number of strides. Pace in a comfortable and natural manner. Repeat the procedure several times and get an average distance for your stride. It is good practice to always pace a distance more than once and average the number of paces (Masser and Jennings, 1991). The formula for calculating distances from pacing is

$$\text{Distance (ft)} = \text{Total number of paces} \times \text{Length of average pace}$$

A *rectangle* is a two-dimensional box. The area of a rectangle is found by multiplying the length (*L*) times width (*W*) (see Figure 2.4).

$$\text{Area} = L \times W \tag{2.6}$$

■ EXAMPLE 2.46

Problem: Find the area of the rectangle shown in Figure 2.5.

Solution:

$$\text{Area} = L \times W = 14 \text{ ft} \times 6 \text{ ft} = 84 \text{ ft}^2$$

To find the area of a circle, we need to introduce a new term, the *radius*, which is represented by *r*. The circle shown in Figure 2.6 has a radius of 6 ft. The radius is any straight line that radiates from the center of the circle to some point on the circumference. By definition, all radii (plural of radius) of the same circle are equal. The surface area of a circle is determined by multiplying π times the radius squared:

$$A = \pi \times r^2 \tag{2.7}$$

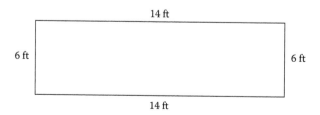

FIGURE 2.5 Area of a rectangle for Example 2.46.

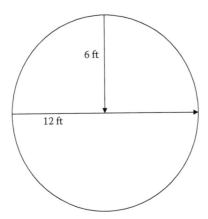

FIGURE 2.6   Area of a circle for Example 2.47.

where
   A = Area.
   π = pi = 3.14.
   r = Radius of circle = one half of the diameter.

■ EXAMPLE 2.47

Problem: What is the area of the circle shown in Figure 2.6?

Solution:

$$\text{Area of circle} = \pi \times r^2 = \pi \times 6^2 = 3.14 \times 36 = 113 \text{ ft}^2$$

If we are assigned to paint a water storage tank, we must know the surface area of the walls of the tank to determine how much paint is required. In this case, we need to know the area of a circular or cylindrical tank. To determine the surface area of the tank, we need to visualize the cylindrical walls as a rectangle wrapped around a circular base. The area of a rectangle is found by multiplying the length by the width; in the case of a cylinder, the width of the rectangle is the height of the wall, and the length of the rectangle is the distance around the circle (circumference).

Thus, the area (A) of the side wall of a circular tank is found by multiplying the circumference of the base ($C = \pi \times D$) times the height of the wall (H):

$$A = \pi \times D \times H \tag{2.8}$$

$$A = \pi \times 20 \text{ ft} \times 25 \text{ ft} = 3.14 \times 20 \text{ ft} \times 25 \text{ ft} = 1570 \text{ ft}^2$$

To determine the amount of paint needed, remember to add the surface area of the top of the tank, which is 314 ft². Thus, the amount of paint needed must cover 1570 ft² + 314 ft² = 1884 ft². If the tank floor should be painted, add another 314 ft².

Many ponds are watershed ponds that have been built by damming valleys. These ponds are irregular in shape. If no good records exist on the pond, then a reasonable estimate can be made by chaining or pacing off the pond margins and using the following procedures to calculate area:

1. Draw the general shape of the pond on graph paper.
2. Draw a rectangle over the pond shape that would approximate the area of the pond if some water was eliminated and placed onto an equal amount of land. This will give you a rectangle on which to base the calculation of area (see Figure 2.7).

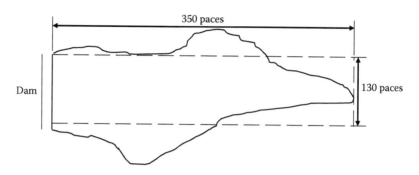

FIGURE 2.7   Irregularly shaped pond.

3. Mark the corners of the rectangle (from the drawing) on the ground around the pond and chain or pace its length and width. For example, a length of 375 paces and a width of 130 paces and a pace length of 2.68 (for example) would be equal to 1005 ft (375 paces × 2.68 ft/pace) by 348.4 ft.
4. Multiply the length times width to get the approximate pond area. For example, 1005 ft × 348.4 ft = 350,142 ft² or 8.04 acres (350,142 ÷ 43,500).

2.11.3.3   Volume

Volume is the amount of space occupied by or contained in an object (see Figure 2.8). It is expressed in cubic units, such as cubic inches (in.³), cubic feet (ft³), or acre-feet (1 acre-foot = 43,560 ft³). The volume (V) of a rectangular object is obtained by multiplying the length times the width times the depth or height:

$$V = L \times W \times H \qquad (2.9)$$

where
    V = Volume.
    L = Length.
    W = Width.
    H (or D) = Height (or depth).

■ **EXAMPLE 2.48**

Problem: A unit rectangular process basin has a length of 15 ft, width of 7 ft, and depth of 9 ft. What is the volume of the basin?

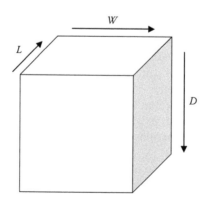

FIGURE 2.8   Volume.

TABLE 2.1

Volume Formulas

Sphere volume	=	$(\pi/6) \times (\text{diameter})^3$
Cone volume	=	$1/3 \times (\text{volume of a cylinder})$
Rectangular tank volume	=	$(\text{Area of rectangle}) \times (D \text{ or } H)$
	=	$(L \times W) \times (D \text{ or } H)$
Cylinder volume	=	$(\text{Area of cylinder}) \times (D \text{ or } H)$
	=	$\pi r^2 \times (D \text{ or } H)$

Solution:

$$V = L \times W \times D = 15 \text{ ft} \times 7 \text{ ft} \times 9 \text{ ft} = 945 \text{ ft}^3$$

For wastewater operators, representative surface areas are most often rectangles, triangles, circles, or a combination of these. Practical volume formulas used in water/wastewater calculations are given in Table 2.1.

When determining the volume of round pipe and round surface areas, the following examples are helpful.

■ EXAMPLE 2.49

Problem: Find the volume of a 3-in. round pipe that is 300 ft long.

Solution:

1. Change the diameter (D) of the duct from inches to feet by dividing by 12:

$$D = 3 \div 12 = 0.25 \text{ ft}$$

2. Find the radius (r) by dividing the diameter by 2:

$$r = 0.25 \text{ ft} \div 2 = 0.125$$

3. Find the volume (V):

$$V = L \times \pi \times r^2$$

$$V = 300 \text{ ft} \times 3.14 \times 0.0156 = 14.72 \text{ ft}^2$$

■ EXAMPLE 2.50

Problem: Find the volume of a smokestack that is 24 in. in diameter (entire length) and 96 in. tall.

Solution: First find the radius of the stack. The radius is one half the diameter, so 24 in. ÷ 2 = 12 in. Now find the volume:

$$V = H \times \pi \times r^2$$

$$V = 96 \text{ in.} \times \pi \times (12 \text{ in.})^2$$

$$V = 96 \text{ in.} \times \pi \times (144 \text{ in.}^2) = 43,407 \text{ ft}^3$$

To determine the volume of a cone and sphere, we use the following equations and examples.

Volume of cone

$$\text{Volume of cone} = \frac{\pi}{12} \times \text{Diameter} \times \text{Diameter} \times \text{Height} \qquad (2.10)$$

Note that

$$\frac{\pi}{12} = \frac{3.14}{12} = 0.262$$

Key Point: The diameter used in the formula is the diameter of the base of the cone.

■ EXAMPLE 2.51

Problem: The bottom section of a circular settling tank has the shape of a cone. How many cubic feet of water are contained in this section of the tank if the tank has a diameter of 120 ft and the cone portion of the unit has a depth of 6 ft?

Solution:

$$\text{Volume (ft}^3) = 0.262 \times 120 \text{ ft} \times 120 \text{ ft} \times 6 \text{ ft} = 22,637 \text{ ft}^3$$

Volume of sphere

$$\text{Volume of sphere} = \frac{\pi}{6} \times \text{Diameter} \times \text{Diameter} \times \text{Diameter} \qquad (2.11)$$

Note that

$$\frac{\pi}{6} = \frac{3.14}{6} = 0.524$$

■ EXAMPLE 2.52

Problem: What is the volume (ft³) of a spherical gas storage container with a diameter of 60 ft?

Solution:

$$\text{Volume (ft}^3) = 0.524 \times 60 \text{ ft} \times 60 \text{ ft} \times 60 \text{ ft} = 113,184 \text{ ft}^3$$

Circular process and various water and chemical storage tanks are commonly found in water/wastewater treatment. A circular tank consists of a circular floor surface with a cylinder rising above it (see Figure 2.9). The volume of a circular tank is calculated by multiplying the surface area times the height of the tank walls.

■ EXAMPLE 2.53

Problem: If a tank is 20 feet in diameter and 25 feet deep, how many gallons of water will it hold?

Hint: In this type of problem, calculate the surface area first, multiply by the height, and then convert to gallons.

Solution:

$$r = D \div 2 = 20 \text{ ft} \div 2 = 10 \text{ ft}$$

$$A = \pi \times r^2 = \pi \times 10 \text{ ft} \times 10 \text{ ft} = 314 \text{ ft}^2$$

$$V = A \times H = 314 \text{ ft}^2 \times 25 \text{ ft} = 7850 \text{ ft}^3 \times 7.48 \text{ gal/ft}^3 = 58,718 \text{ gal}$$

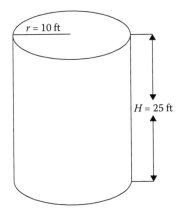

FIGURE 2.9   Circular or cylindrical water tank.

2.12   FORCE, PRESSURE, AND HEAD CALCULATIONS

Before we review calculations involving force, pressure, and head, we must first define these terms:

- *Force*—The push exerted by water on any confined surface. Force can be expressed in pounds, tons, grams, or kilograms.
- *Pressure*—The force per unit area. The most common way of expressing pressure is in pounds per square inch (psi).
- *Head*—The vertical distance or height of water above a reference point. Head is usually expressed in feet. In the case of water, head and pressure are related.

2.12.1   FORCE AND PRESSURE

Figure 2.10 helps to illustrate these terms. A cubical container measuring 1 foot on each side can hold 1 cubic foot of water. A basic fact of science states that 1 cubic foot of water weights 62.4 lb and contains 7.48 gal. The force acting on the bottom of the container is 62.4 lb/ft². The area of the bottom in square inches is

$$1 \text{ ft}^2 = 12 \text{ in.} \times 12 \text{ in.} = 144 \text{ in.}^2$$

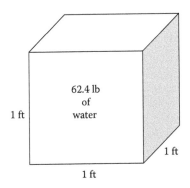

FIGURE 2.10   One cubic foot of water weighs 62.4 lb.

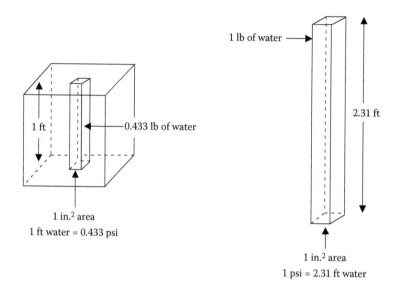

FIGURE 2.11   Relationship between pressure and head.

Therefore, the pressure in pounds per square inch (psi) is

$$\frac{62.4 \text{ lb/ft}^2}{1 \text{ ft}^2} = \frac{62.4 \text{ lb/ft}^2}{144 \text{ in.}^2/\text{ft}^2} = 0.433 \text{ lb/in.}^2 \text{ (psi)}$$

If we use the bottom of the container as our reference point, the head would be 1 foot. From this, we can see that 1 foot of head is equal to 0.433 psi—an important parameter to remember. Figure 2.11 illustrates some other important relationships between pressure and head.

Note: Force acts in a particular direction. Water in a tank exerts force down on the bottom and out of the sides. Pressure, however, acts in all directions. A marble at a water depth of 1 foot would have 0.433 psi of pressure acting inward on all sides.

Using the preceding information, we can develop Equations 2.12 and 2.13 for calculating pressure and head:

$$\text{Pressure (psi)} = 0.433 \times \text{Head (ft)} \tag{2.12}$$

$$\text{Head (ft)} = 2.31 \times \text{Pressure (psi)} \tag{2.13}$$

2.12.2   HEAD

Head is the vertical distance the water must be lifted from the supply tank or unit process to the discharge. The total head includes the vertical distance the liquid must be lifted (*static head*), the loss to friction (*friction head*), and the energy required to maintain the desired velocity (*velocity head*):

$$\text{Total head} = \text{Static head} + \text{Friction head} + \text{Velocity head} \tag{2.14}$$

2.12.2.1   Static Head

Static head is the actual vertical distance the liquid must be lifted.

$$\text{Static head} = \text{Discharge elevation} - \text{Supply elevation} \qquad (2.15)$$

■ EXAMPLE 2.54

Problem: The supply tank is located at elevation 108 ft. The discharge point is at elevation 205 ft. What is the static head in feet?

Solution:

$$\text{Static head (ft)} = 205 \text{ ft} - 108 \text{ ft} = 97 \text{ ft}$$

2.12.2.2   Friction Head

Friction head is the equivalent distance of the energy that must be supplied to overcome friction. Engineering references include tables showing the equivalent vertical distance for various sizes and types of pipes, fittings, and valves. The total friction head is the sum of the equivalent vertical distances for each component:

$$\text{Friction head (ft)} = \text{Energy losses due to friction} \qquad (2.16)$$

2.12.2.3   Velocity Head

Velocity head is the equivalent distance of the energy consumed in achieving and maintaining the desired velocity in the system:

$$\text{Velocity head (ft)} = \text{Energy losses to maintain velocity} \qquad (2.17)$$

2.12.2.4   Total Dynamic Head (Total System Head)

$$\text{Total head} = \text{Static head} + \text{Friction head} + \text{Velocity head} \qquad (2.18)$$

2.12.2.5   Pressure and Head

The pressure exerted by water/wastewater is directly proportional to its depth or head in the pipe, tank, or channel. If the pressure is known, the equivalent head can be calculated:

$$\text{Head (ft)} = \text{Pressure (psi)} \times 2.31 \text{ ft/psi} \qquad (2.19)$$

■ EXAMPLE 2.55

Problem: The pressure gauge on the discharge line from the influent pump reads 75.3 psi. What is the equivalent head in feet?

Solution:

$$\text{Head (ft)} = 75.3 \times 2.31 \text{ ft/psi} = 173.9 \text{ ft}$$

2.12.2.6 Head and Pressure

If the head is known, the equivalent pressure can be calculated by

$$\text{Pressure (psi)} = \frac{\text{Head (ft)}}{2.31 \text{ ft/psi}} \tag{2.20}$$

■ **EXAMPLE 2.56**

Problem: A tank is 15 ft deep. What is the pressure in psi at the bottom of the tank when it is filled with wastewater?

Solution:

$$\text{Pressure (psi)} = \frac{15 \text{ ft}}{2.31 \text{ ft/psi}} = 6.49 \text{ psi}$$

Before we look at a few example problems dealing with force, pressure, and head, it is important to review the key points related to force, pressure, and head:

1. By definition, water weighs 62.4 lb/ft^3.
2. The surface of any one side of a 1-ft^3 cube contains 144 in.^{2} (12 in. × 12 in. = 144 in.^{2}); therefore, the cube contains 144 columns of water that are 1 ft tall and 1 inch square.
3. The weight of each of these pieces can be determined by dividing the weight of the water in the cube by the number of square inches:

$$\text{Weight} = \frac{62.4 \text{ lb}}{144 \text{ in.}^2} = 0.433 \text{ lb/in.}^2 \text{ or } 0.433 \text{ psi}$$

4. Because this is the weight of one column of water 1 ft tall, the true expression would be 0.433 pounds per square inch per foot of head, or 0.433 psi/ft.

Note: 1 foot of head = 0.433 psi.

In addition to remembering the important parameter that 1 ft of head = 0.433 psi, it is important to understand the relationship between pressure and feet of head—in other words, how many feet of head 1 psi represents. This is determined by dividing 1 ft by 0.433 psi:

$$\text{Feet of head} = \frac{1 \text{ ft}}{0.433 \text{ psi}} = 2.31 \text{ ft/psi}$$

If a pressure gauge reads 12 psi, the height of the water necessary to represent this pressure would be 12 psi × 2.31 ft/psi = 27.7 feet.

Key Point: Both of the above conversions are commonly used in water/wastewater treatment calculations; however, the most accurate conversion is 1 ft = 0.433 psi. This is the conversion we use throughout this text.

■ **EXAMPLE 2.57**

Problem: Convert 40 psi to feet head.

Solution:

$$\frac{40 \text{ psi}}{1} \times \frac{\text{ft}}{0.433 \text{ psi}} = 92.4 \text{ ft}$$

■ **EXAMPLE 2.58**

Problem: Convert 40 ft to psi.

Solution:

$$40\frac{\text{ft}}{1} \times \frac{0.433 \text{ psi}}{1 \text{ ft}} = 17.32 \text{ psi}$$

As the above examples demonstrate, when attempting to convert psi to feet we *divide* by 0.433, and when attempting to convert feet to psi we *multiply* by 0.433. The above process can be most helpful in clearing up confusion about whether to multiply or divide; however, there is another approach that may be easier for many operators to use. Notice that the relationship between psi and feet is almost 2 to 1. It takes slightly more than 2 feet to make 1 psi; therefore, in a problem where the data are provided in pressure and the result should be in feet, the answer will be at least twice as large as the starting number. For instance, if the pressure were 25 psi, we intuitively know that the head is over 50 feet, so we must divide by 0.433 to obtain the correct answer.

■ **EXAMPLE 2.59**

Problem: Convert a pressure of 45 psi to feet of head.

Solution:

$$45\frac{\text{psi}}{1} \times \frac{1 \text{ ft}}{0.433 \text{ psi}} = 104 \text{ ft}$$

■ **EXAMPLE 2.60**

Problem: Convert 15 psi to feet.

Solution:

$$15\frac{\text{psi}}{1} \times \frac{1 \text{ ft}}{0.433 \text{ psi}} = 34.6 \text{ ft}$$

■ **EXAMPLE 2.61**

Problem: Between the top of a reservoir and the watering point, the elevation is 125 feet. What will the static pressure be at the watering point?

Solution:

$$125\frac{\text{psi}}{1} \times \frac{1 \text{ ft}}{0.433 \text{ psi}} = 288.7 \text{ ft}$$

■ **EXAMPLE 2.62**

Problem: Find the pressure (psi) in a tank 12 ft deep at a point 5 ft below the water surface.

Solution:

$$\text{Pressure (psi)} = 0.433 \times 5 \text{ ft} = 2.17 \text{ psi}$$

■ **EXAMPLE 2.63**

Problem: A pressure gauge at the bottom of a tank reads 12.2 psi. How deep is the water in the tank?

Solution:

$$\text{Head (ft)} = 2.31 \times 12.2 \text{ psi} = 28.2 \text{ ft}$$

■ **EXAMPLE 2.64**

Problem: What is the pressure (static pressure) 4 miles beneath the ocean surface?

Solution: Change miles to feet, then to psi:

$$5280 \text{ ft/mile} \times 4 = 21{,}120 \text{ ft}$$

$$\frac{21{,}120 \text{ ft}}{2.31 \text{ ft/psi}} = 9143 \text{ psi}$$

■ **EXAMPLE 2.65**

Problem: A 150-ft-diameter cylindrical tank contains 2.0 MG water. What is the water depth? At what pressure would a gauge at the bottom read in psi?

Solution:

1. Change MG to cubic feet:

$$\frac{2{,}000{,}000 \text{ gal}}{7.48} = 267{,}380 \text{ ft}^3$$

2. Using volume, solve for depth:

$$\text{Volume} = 0.785 \times D^2 \times \text{depth}$$
$$267{,}380 \text{ ft}^3 = 0.785 \times (150)^2 \times \text{depth}$$
$$\text{Depth} = 15.1 \text{ ft}$$

■ **EXAMPLE 2.66**

Problem: The pressure in a pipe is 70 psi. What is the pressure in feet of water? What is the pressure in psf?

Solution:

1. Convert pressure to feet of water:

$$70 \text{ psi} \times 2.31 \text{ ft/psi} = 161.7 \text{ ft of water}$$

2. Convert psi to psf:

$$70 \text{ psi} \times 144 \text{ in.}^2/\text{ft}^2 = 10{,}080 \text{ psf}$$

■ EXAMPLE 2.67

Problem: The pressure in a pipeline is 6476 psf. What is the head on the pipe?

Solution:

$$\text{Head on pipe} = \text{Feet of pressure}$$
$$\text{Pressure} = \text{Weight} \times \text{height}$$
$$6476 \text{ psf} = 62.4 \text{ lb/ft}^3 \times \text{height}$$
$$\text{Height} = 104 \text{ ft}$$

2.13 REVIEW OF ADVANCED ALGEBRA KEY TERMS/CONCEPTS

Advanced algebraic operations (linear, linear differential, and ordinary differential equations) have in recent years become an essential part of the mathematical background required by environmental engineers, among others. It is not the intent here to provide complete coverage of the topics (environmental practitioners are normally well grounded in these critical foundational areas), but it is important to review the key terms and relevant concepts. Key definitions include the following:

Algebraic multiplicity of an eigenvalue—The algebraic multiplicity of eigenvalue c of matrix A is the number of times the factor $(t - c)$ occurs in the characteristic polynomial of A.

Basis for a subspace—A basis for subspace W is a set of vectors $\{\mathbf{v}_1, \ldots, \mathbf{v}_k\}$ in W such that
1. $\{\mathbf{v}_1, \ldots, \mathbf{v}_k\}$ is linearly independent, and
2. $\{\mathbf{v}_1, \ldots, \mathbf{v}_k\}$ spans W.

Characteristic polynomial of a matrix—The characteristic polynomial of $n \times n$ matrix A is the polynomial in t given by the formula $\det(A - tI)$.

Column space of a matrix—The subspace spanned by the columns of the matrix considered as a set of vectors (also see row space).

Consistent linear system—A system of linear equations is consistent if it has at least one solution.

Defective matrix—Matrix A is defective if A has an eigenvalue whose geometric multiplicity is less than its algebraic multiplicity.

Diagonalizable matrix—A matrix is diagonalizable if it is similar to a diagonal matrix.

Dimension of a subspace—The dimension of subspace W is the number of vectors in any basis of W. (If W is the subspace $\{\mathbf{0}\}$, then we say that its dimension is 0.)

Echelon form of a matrix—A matrix is in row echelon form if
1. All rows that consist entirely of zeros are grouped together at the bottom of the matrix, and
2. The first (counting left to right) nonzero entry in each nonzero row appears in a column to the right of the first nonzero entry in the preceding row (if there is a preceding row).

Eigenspace of a matrix—The eigenspace associated with the eigenvalue c of matrix A is the null space of $A - cI$.

Eigenvalue of a matrix—An eigenvalue of matrix A is scalar c such that $A\mathbf{x} = c\mathbf{x}$ holds for some nonzero vector $\mathbf{x}$.

Eigenvector of a matrix—An eigenvector of square matrix A is a nonzero vector $\mathbf{x}$ such that $A\mathbf{x} = c\mathbf{x}$ holds for some scalar c.

Elementary matrix—A matrix that is obtained by performing an elementary row operation on an identity matrix.

Equivalent linear systems—Two systems of linear equations in *n* unknowns are equivalent if they have the same set of solutions.

Geometric multiplicity of an eigenvalue—The geometric multiplicity of eigenvalue *c* of matrix *A* is the dimension of the eigenspace of *c*.

Homogeneous linear system—A system of linear equations $A\mathbf{x} = \mathbf{b}$ is homogeneous if $\mathbf{b} = 0$.

Inconsistent linear system—A system of linear equations is inconsistent if it has no solutions.

Inverse of a matrix—Matrix *B* is an inverse for matrix *A* if $AB = BA = I$.

Invertible matrix—A matrix is invertible if it has no inverse.

Least-squares solution of a linear system—A least-squares solution to a system of linear equations $A\mathbf{x} = \mathbf{b}$ is a vector $\mathbf{x}$ that minimizes the length of the vector $A\mathbf{x} - \mathbf{b}$.

Linear combination of vectors—Vector $\mathbf{v}$ is a linear combination of the vectors $\mathbf{v}_1, \ldots, \mathbf{v}_k$ if there exist scalars $a_1, \ldots, a_k$ such that $\mathbf{v} = a_1\mathbf{v}_1 + \ldots + a_k\mathbf{v}_k$.

Linear dependence relation for a set of vectors—A linear dependence relation for the set of vectors $\{\mathbf{v}_1, \ldots, \mathbf{v}_k\}$ is an equation of the form $a_1\mathbf{v}_1 + \ldots + a_k\mathbf{v}_k = 0$, where the scalars $a_1, \ldots, a_k$ are zero.

Linearly dependent set of vectors—The set of vectors $\{\mathbf{v}_1, \ldots, \mathbf{v}_k\}$ is linearly dependent if the equation $a_1\mathbf{v}1 + \ldots + a_k\mathbf{v}_k = 0$ has a solution where not all the scalars $a_1, \ldots, a_k$ are zero (i.e., if $\{\mathbf{v}_1, \ldots, \mathbf{v}_k\}$ satisfies a linear dependence relation).

Linearly independent set of vectors—The set of vectors $\{\mathbf{v}_1, \ldots, \mathbf{v}_k\}$ is linearly independent if the only solution to the equation $a_1\mathbf{v}_1 + \ldots + a_k\mathbf{v}_k = 0$ is the solution where all the scalars $a_1, \ldots, a_k$ are zero (i.e., if $\{\mathbf{v}_1, \ldots, \mathbf{v}_k\}$ does not satisfy any linear dependence relation).

Linear transformation—A linear transformation from *V* to *W* is a function *T* from *V* to *W* such that
1. $T(\mathbf{u} + \mathbf{v}) = T(\mathbf{u}) + T(\mathbf{v})$ for all vectors $\mathbf{u}$ and $\mathbf{v}$ in *V*.
2. $T(a\mathbf{v}) = aT(\mathbf{v})$ for all vectors $\mathbf{v}$ in *V* and all scalars *a*.

Nonsingular matrix—Square matrix *A* is nonsingular if the only solution to the equation $A\mathbf{x} = 0$ is $\mathbf{x} = 0$.

Null space of a matrix—The null space of *m*×*n* matrix *A* is the set of all vectors $\mathbf{x}$ in R^n such that $A\mathbf{x} = 0$.

Null space of a linear transformation—The null space of linear transformation *T* is the set of vectors $\mathbf{v}$ in its domain such that $T(\mathbf{v}) = 0$.

Nullity of a linear transformation—The nullity of linear transformation *T* is the dimension of its null space.

Nullity of a matrix—The dimension of its null space.

Orthogonal complement of a subspace—The orthogonal complement of subspace *S* of R^n is the set of all vectors $\mathbf{v}$ in R^n such that $\mathbf{v}$ is orthogonal to every vector in *S*.

Orthogonal set of vectors—A set of vectors in R^n is orthogonal if the dot product of any two of them is 0.

Orthogonal matrix—Matrix *A* is orthogonal if *A* is invertible and its inverse equals its transpose; that is, $A^{-1} = A^T$.

Orthogonal linear transformation—Linear transformation *T* from *V* to *W* is orthogonal if $T(\mathbf{v})$ has the same length as $\mathbf{v}$ for all vectors $\mathbf{v}$ in *V*.

Orthonormal set of vectors—A set of vectors in R^n is orthonormal if it is an orthogonal set and each vector has length 1.

Range of a linear transformation—The range of linear transformation *T* is the set of all vectors $T(\mathbf{v})$, where $\mathbf{v}$ is any vector in its domain.

Rank of a matrix—The rank of matrix *A* is the number of nonzero rows in the reduced row echelon form of *A*; that is, the dimension of the row space of *A*.

Rank of a linear transformation—The rank of a linear transformation (and hence of any matrix regarded as a linear transformation) is the dimension of its range. Note that a theorem tells us that the two definitions of rank of a matrix are equivalent.

Reduced row echelon form of a matrix—A matrix is in reduced row echelon form if
1. The matrix is in row echelon form.
2. The first nonzero entry in each nonzero row is the number 1.
3. The first nonzero entry in each nonzero row is the only nonzero entry in its column.

Row equivalent matrices—Two matrices are row equivalent if one can be obtained from the other by a sequence of elementary row operations.

Row operations—The elementary row operations performed on a matrix are
1. Interchange two rows.
2. Multiply a row by a nonzero scalar.
3. Add a constant multiple of one row to another.

Row space of a matrix—The subspace spanned by the rows of the matrix considered as a set of vectors.

Similar matrices—Matrices A and B are similar if there is a square invertible matrix S such that $S^{-1}AS = B$.

Singular matrix—Square matrix A is singular if the equation $A\mathbf{x} = 0$ has a nonzero solution for $\mathbf{x}$.

Span of a set of vectors—The span of the set of vectors $\{\mathbf{v}_1, \ldots, \mathbf{v}_k\}$ is the subspace V consisting of all linear combinations of $\mathbf{v}_1, \ldots, \mathbf{v}_k$. One also says that the subspace V is spanned by the set of vectors $\{\mathbf{v}_1, \ldots, \mathbf{v}_k\}$ and that this set of vectors spans V.

Subspace—A subset W of R^n is a subspace of R^n if
1. The zero vector is in W.
2. $\mathbf{x} + \mathbf{y}$ is in W whenever $\mathbf{x}$ and $\mathbf{y}$ are in W.
3. $a\mathbf{x}$ is in W whenever $\mathbf{x}$ is in W and a is any scalar.

Symmetric matrix—Matrix A is symmetric if it equals its transpose; that is, $A = A^T$.

REFERENCES AND RECOMMENDED READING

Masser, M.P. and Jensen, J.W. (1991). *Calculating Area and Volume of Ponds and Tanks*, USDA Grant 89-38500-4516. Southern Regional Aquaculture Center, Washington, DC.

Stapel, E. (2013). *The Order of Operations: More Examples*. Purplemath, http://www.purplemath.com/modules/orderops2.htm.

3 Environmental Modeling

Who has measured the waters in the hollow of the hand, or with the breadth of his hand marked off the heavens? Who has held the dust of the earth in a basket, or weighed the mountains on the scales and the hills in a balance?

—Isaiah 40:12

3.1 INTRODUCTION

There is a growing interest in the field of environmental monitoring and quantitative assessment of environmental problems. For some years now, the results of environmental models and assessment analyses have been influencing environmental regulation and policies. These results are widely cited by politicians in forecasting consequences of such greenhouse gas emissions as carbon dioxide (CO_2) and in advocating dramatic reductions of energy consumption at local, national, and international levels. For this reason, and because environmental modeling is often based on extreme conceptual and numerical intricacy and uncertain validity, environmental modeling has become one of the most controversial topics of applied mathematics.

Having said this, environmental modeling continues to be widely used in environmental practice, with its growth limited only by the imagination of the modelers. Environmental problem-solving techniques incorporating the use of modeling are widely used in watershed management, surface water monitoring, flood hazard mapping, climate modeling, and groundwater modeling, among others. It is important to keep in mind, however, that modelers often provide models for product developers who use the results to describe what their products are based on and why.

This chapter does not provide a complete treatment of environmental modeling. For the reader who desires such a treatment, we highly recommend Nirmalakhandan (2002) and NIST (2012). Much of the work presented in this chapter is modeled after these works. Here, we present an overview of quantitative operations implicit to environmental modeling processes.

3.2 BASIC STEPS FOR DEVELOPING AN EFFECTIVE MODEL

The basic steps used for model building are the same across all modeling methods. The details vary somewhat from method to method, but an understanding of the common steps, combined with the typical underlying assumptions needed for the analysis, provides a framework in which the results from almost any method can be interpreted and understood. The basic steps of the model-building process are

1. Model selection
2. Model fitting
3. Model validation

These three basic steps are used iteratively until an appropriate model for the data has been developed. In the model selection step, plots of the data, process knowledge, and assumptions about the process are used to determine the form of the model to be fit to the data. Then, using the selected model and possibly information about the data, an appropriate model-fitting method is used to estimate the unknown parameters in the model. When the parameter estimates have been made, the model is then carefully assessed to see if the underlying assumptions of the analysis appear plausible. If the assumptions seem valid, the model can be used to answer the scientific or engineering questions that prompted the modeling effort. If the model validation identifies problems with the current model, however, then the modeling process is repeated using information from the model validation step to select or fit an improved model.

The three basic steps of process modeling described in the paragraph above assume that the data have already been collected and that the same dataset can be used to fit all of the candidate models. Although this is often the case in mode-building situations, one variation on the basic model-building sequence comes up when additional data are needed to fit a newly hypothesized model based on a model fit to the initial data. In this case, two additional steps, experimental design and data collection, can be added to the basic sequence between model selection and mode-fitting.

3.3   WHAT ARE MODELS USED FOR?

Models are used for four main purposes:

1. Estimation
2. Prediction
3. Calibration
4. Optimization

A brief explanation of the different uses of models is provided below (NIST, 2012):

- *Estimation*—The goal of estimation is to determine the value of the regression function (i.e., the average value of the response variable) for a particular combination of the values of the predictor variables. Regression function values can be estimated for any combination of predictor variable values, including values for which no data have been measured or observed. Function values estimated for points within the observed space of predictor variable values are sometimes called *interpolations*. Estimation of regression function values for points outside the observed space of predictor variable values, called *extrapolations*, are sometimes necessary but require caution.
- *Prediction*—The goal of prediction is to determine either
  1. The value of a new observation of the response variable, or
  2. The values of a specified proportion of all future observations of the response variable for a particular combination of the values of the predictor variables. Predictions can be made for any combination of predictor variable values, including values for which no data have been measured or observed. As in the case of estimation, predictions made outside the observed space of predictor variable values are sometimes necessary but require caution.
- *Calibration*—The goal of calibration is to quantitatively relate measurements made using one measurement system to those of another measurement system. This is done so that measurements can be compared in common units or to tie results from a relative measurement method to absolute units.
- *Optimization*—Optimization is performed to determine the values of process inputs that should be used to obtain the desired process output. Typical optimization goals might be to maximize the yield of a process, to minimize the processing time required to fabricate a product, or to hit a target product specification with minimum variation in order to maintain specified tolerances.

3.4 MEDIA MATERIAL CONTENT

Media material content is a measure of the material contained in a bulk medium, quantified by the ratio of the amount of material present to the amount of the medium. The terms *mass*, *moles*, or *volume* can be used to quantify the amounts. Thus, the ratio can be expressed in several forms such as mass or moles of material per volume of medium, resulting in mass or molar concentration; moles of material per mole of medium, resulting in mole fraction; and volume of material per volume of medium, resulting in volume fraction.

When dealing with mixtures of materials and media, the use of different forms of measures in the ratio to quantify material content may become confusing. With regard to mixtures, the ratio can be expressed in concentration units. The *concentration* of a chemical (liquid, gaseous, or solid) substance expresses the amount of substance present in a mixture. There are many different ways to express concentration.

Chemists use the term *solute* to describe the substance of interest and the term *solvent* to describe the material in which the solute is dissolved. For example, in a can of soft drink (a solution of sugar in carbonated water), there are approximately 12 tablespoons of sugar (the solute) dissolved in the carbonated water (the solvent). In general, the component that is present in the greatest amount is the solvent.

Some of the more common concentration units are

1. *Mass per unit volume.* Some concentrations are expressed in milligrams per milliliter (mg/mL) or milligrams per cubic centimeter (mg/cm^3). Note that 1 mL = 1 cm^3 and that a cubic centimeter is sometimes denoted as a "cc." Mass per unit volume is handy when discussing how soluble a material is in water or a particular solvent—for example, "the solubility of substance x is 4 grams per liter."

2. *Percent by mass.* Also called weight percent or percent by weight, this is simply the mass of the solute divided by the total mass of the solution and multiplied by 100%:

$$\text{Percent by mass} = \frac{\text{Mass of component}}{\text{Mass of solution}} = 100\% \qquad (3.1)$$

The mass of the solution is equal to the mass of the solute plus the mass of the solvent. For example, a solution consisting of 30 g of sodium chloride and 70 g of water would be 30% sodium chloride by mass: [(30 g NaCl)/(30 g NaCl + 70 g water)] × 100% = 30%. To avoid confusion as to whether a solution is percent by weight or percent by volume, "w/w" (for weight to weight) is often added after the concentration—for example, "10% potassium iodide solution in water (w/w)."

3. *Percent by volume.* Also called volume percent or volume/volume percent, this is typically only used for mixtures of liquids. Percent by volume is simply the volume of the solute divided by the sum of the volumes of the other components multiplied by 100%. If we mix 30 mL of ethanol and 70 mL of water, the percent ethanol by volume will be 30%, but the total volume of the solution will *not* be 100 mL (although it will be close), because ethanol and water molecules interact differently with each other than they do with themselves. To avoid confusion as to whether we have a percent by weight or percent by volume solution, we can label this mixture as "30% ethanol in water (v/v)," where v/v stands for "volume to volume."

4. *Molarity.* Molarity is the number of moles of solute dissolved in 1 liter of solution. For example, a quantity of 90 g of glucose (molar mass = 180 g/mol) is equal to (90 g)/(180 g/mol) = 0.50 moles of glucose. If we place this glucose in a flask and add water until the total volume is 1 liter, we would have a 0.5 molar solution. Molarity is usually denoted with

M (e.g., a 0.50-M solution). Recognize that molarity is moles of solute per liter of solution, not per liter of solvent. Also recognize that molarity changes slightly with temperature because the volume of a solution changes with temperature.

5. *Molality.* Molality is used for calculations of colligative properties; it is the number of moles of solute dissolved in 1 kilogram of solvent. Notice the two key differences between molarity and molality. Molality uses mass rather than volume and uses solvent instead of solution:

$$\text{Molality} = \frac{\text{Moles of solute}}{\text{Kilograms of solution}} \tag{3.2}$$

Unlike molarity, molality is independent of temperature because mass does not change with temperature. If we were to place 90 g of glucose (0.50 mol) in a flask and then add 1 kg of water we would have a 0.50-molal solution. Molality is usually denoted with a small m (e.g., a 0.50-m solution).

6. *Parts per million.* Parts per million (ppm) works like percent by mass but is more convenient when there is only a small amount of solute present. Parts per million is defined as the mass of the component in solution divided by the total mass of the solution multiplied by 10^6 (one million):

$$\text{Parts per million} = \frac{\text{Mass of component}}{\text{Mass of solution}} \times 1{,}000{,}000 \tag{3.3}$$

A solution with a concentration of 1 ppm has 1 gram of substance for every million grams of solution. Because the density of water is 1 g/mL and we are adding such a tiny amount of solute, the density of a solution at such a low concentration is approximately 1 g/mL. Therefore, in general, 1 ppm implies 1 mg of solute per liter of solution. Finally, recognize that 1% = 10,000 ppm. Therefore, something that has a concentration of 300 ppm could also be said to have a concentration of (300 ppm)/(10,000 ppm/percent) = 0.03% percent by mass.

7. *Parts per billion.* Parts per billion (ppb) works like above, but we multiply by 1 billion (10^9) (be aware that the word "billion" has different meanings in different countries). A solution with 1 ppb of solute has 1 μg (10^{-6}) of material per liter.

8. *Parts per trillion.* Parts per trillion (ppt) works like parts per million and parts per billion except that we multiply by 1 trillion (10^{12}). There are few, if any, solutes that are harmful at concentrations as low as 1 ppt.

The following examples can help in formalizing these different forms. In these examples, subscripts for components are $i = 1, 2, 3, \ldots, n$, and subscripts for phases are g = gas, a = air, l = liquid, w = water, and s = solids and soil.

3.4.1 Material Content: Liquid Phases

Mass concentration, molar concentration, or mole fraction can be used to quantify material content in liquid phases:

$$\text{Mass concentration of component } i \text{ in water} = p_{i,w} = \frac{\text{Mass of material } i}{\text{Volume of water}} \tag{3.4}$$

$$\text{Molar concentration of component } i \text{ in water} = C_{i,w} = \frac{\text{Moles of material } i}{\text{Volume of water}} \tag{3.5}$$

Because moles of material = mass/molecular weight (MW), mass concentrations ($p_{i,w}$) are related by the following:

$$C_{i,w} = \frac{p_{i,w}}{MW_i} \qquad (3.6)$$

For molarity M, $[X]$ is the molar concentration of X.

The mole fraction (X) of a single chemical in water can be expressed as follows:

$$\text{Mole fraction } X = \frac{\text{Moles of component/chemical}}{\text{Total moles of solution (moles of chemical + moles of water)}} \qquad (3.7)$$

For dilute solutions, the moles of chemical in the denominator of the above equation can be ignored in comparison to the moles of water (n_w) and can be approximated by

$$X = \frac{\text{Moles of chemical}}{\text{Moles of water}} \qquad (3.8)$$

If X is less than 0.02, an aqueous solution can be considered dilute. On a mass basis, similar expressions can be formulated to yield mass fractions. Mass fractions can also be expressed as a percentage or as other ratios such as parts per million (ppm) or parts per billion (ppb).

The mole fraction of a component in a solution is simply the number of moles of that component divided by the total moles of all of the components. We use the mole fraction because the sum of the individual fractions should equal 1. This constraint can reduce the number of variables when modeling mixtures of chemicals. Mole fractions are strictly additive. The sum of the mole fractions of all components is equal to 1. Mole fraction X_i of component i in an n-component mixture is defined as follows:

$$X_i = \frac{\text{Moles of } i}{\left(\sum_{1}^{n} n_i\right) + n_w} \qquad (3.9)$$

$$\text{The sum of all mole fractions} = \left(\sum_{1}^{n} x_w\right) = 1 \qquad (3.10)$$

For dilute solutions of multiple chemicals (as in the case of single-chemical systems), mole fraction X_i of component i in an n-component mixture can be approximated by the following:

$$X = \frac{\text{Moles of } i}{n_w} \qquad (3.11)$$

Note that the preceding ratio is known as an *intensive property* because it is independent of the system and the mass of the sample. An intensive property is any property that can exist at a point in space. Temperature, pressure, and density are good examples. On the other hand, an *extensive property* is any property that depends on the size (or extent) of the system under consideration. Volume is an example. If we double the length of all edges of a solid cube, the volume increases by a factor of eight. Mass is another. The same cube will undergo an eightfold increase mass when the length of the edges is doubled.

Note: The material content in solid and gas phases is different from those in liquid phases. For example, the material content in solid phases is often quantified by a ratio of masses and is expressed as ppm or ppb. The material content in gas phases is often quantified by a ratio of moles or volumes and is expressed as ppm or ppb. It is preferable to report gas-phase concentrations at standard temperature and pressure (STP; 0°C and 769 mmHg or 273 K and 1 atm).

■ **EXAMPLE 3.1**

Problem: A certain chemical has a molecular weight of 80. Derive the conversion factors to quantify the following:

1. 1 ppm (volume/volume) of the chemical in air in molar and mass concentration form.
2. 1 ppm (mass ratio) of the chemical in water in mass and molar concentration form.
3. 1 ppm (mass ratio) of the chemical in soil in mass ratio form.

Solution:

1. *Gas phase*—The volume ratio of 1 ppm can be converted to the mole or mass concentration form using the assumption of ideal gas, with a molar volume of 22.4 L/g mol at STP conditions (273 K and 1.0 atm.).

$$1 \text{ ppm}_v = \frac{1 \text{ m}^3 \text{ chemical}}{1,000,000 \text{ m}^3 \text{ of air}}$$

$$1 \text{ ppm}_v \equiv \frac{1 \text{ m}^3 \text{ chemical}}{1,000,000 \text{ m}^3 \text{ of air}} \left(\frac{\text{mol}}{22.4 \text{ L}} \right) \left(\frac{1000 \text{ L}}{\text{m}^3} \right) \equiv 4.46 \times 10^{-5} \text{ mol/m}^3$$

$$\equiv 4.46 \times 10^{-5} \text{ mol/m}^3 \left(\frac{80 \text{ g}}{\text{gmol}} \right) \equiv 0.0035 \text{ g/m}^3 \equiv 3.5 \text{ mg/m}^3 \equiv 3.5 \text{ μg/L}$$

   The general relationship is 1 ppm = (MW/22.4) mg/m³.

2. *Water phase*—The mass ratio of 1 ppm can be converted to mole or mass concentration form using the density of water, which is 1 g/cm³ at 4°C and 1 atm:

$$1 \text{ ppm} = \frac{1 \text{ g chemical}}{1,000,000 \text{ g of water}}$$

$$1 \text{ ppm} \equiv \frac{1 \text{ g chemical}}{1,000,000 \text{ g of water}} \left(1 \text{ g/cm}^3 \right) \left(1,000,000 \text{ cm}^3/\text{m}^3 \right) \equiv 1 \text{ g/m}^3 \equiv 1 \text{ mg/L}$$

$$\equiv 1 \text{ g/m}^3 \left(\frac{\text{mol}}{80 \text{ g}} \right) \equiv 0.0125 \text{ mol/m}^3$$

3. *Soil phase*—The conversion is direct:

$$1 \text{ ppm} = \frac{1 \text{ g chemical}}{1,000,000 \text{ g of soil}}$$

$$1 \text{ ppm} = \frac{1 \text{ g chemical}}{1,000,000 \text{ g of soil}} \left(\frac{1000 \text{ g}}{\text{kg}} \right) \left(\frac{1000 \text{ mg}}{\text{g}} \right) = 1 \text{ mg/kg}$$

■ EXAMPLE 3.2

Problem: Analysis of a water sample from a pond gave the following results: volume of sample = 2 L, concentration of suspended solids in the sample = 15 mg/L, concentration of dissolved chemical = 0.01 mol/L, and concentration of the chemical adsorbed onto the suspended solids = 400 μg/g solids. If the molecular weight of the chemical is 125, determine the total mass of the chemical in the sample.

Solution:

Dissolved concentration = Molar concentration × MW:

$$0.001 \text{ mol/L} \times 125 \text{ g/mol} = 0.125 \text{ g/L}$$

Dissolved mass in sample = Dissolved concentration × Volume:

$$(125 \text{ g/L}) \times (2 \text{ L}) = 0.25 \text{ g}$$

Mass of solids in sample = Concentration of solids × Volume:

$$(25 \text{ mg/L} \times (2 \text{ L}) = 50 \text{ mg} = 0.05 \text{ g}$$

Adsorbed mass in sample = Adsorbed concentration × Mass of solids:

$$(400 \text{ μg/g}) \times (0.05 \text{ g}) \times \left(\frac{1 \text{ g}}{10^6 \text{ μg}} \right) = 0.00020 \text{ g}$$

Thus, the total mass of chemical in the sample = 0.25 g + 0.00020 g = 0.25020 g.

3.5 PHASE EQUILIBRIUM AND STEADY STATE

The concept of phase equilibrium (balance of forces) is an important one in environmental modeling. In the case of mechanical equilibrium, consider the following example. A cup sitting on a table top remains at rest because the downward force exerted by the Earth's gravity action on the cup's mass (this is what is meant by the "weight" of the cup) is exactly balanced by the repulsive force between atoms that prevents two objects from simultaneously occupying the same space, acting in this case between the table surface and the cup. If you pick up the cup and raise it above the tabletop, the additional upward force exerted by your arm destroys the state of equilibrium as the cup moves upward. If one wishes to hold the cup at rest above the table, it is necessary to adjust the upward force to exactly balance the weight of the cup, thus restoring equilibrium.

For more pertinent examples (chemical equilibrium, for example) consider the following. Chemical equilibrium is a dynamic system in which chemical changes are taking place in such a way that there is no overall change in the composition of the system. In addition to partial ionization, equilibrium situations include simple reactions such as when the air in contact with a liquid is saturated with the liquid's vapor, meaning that the rate of evaporation is equal to the rate of condensation. When a solution is saturated with a solute, this means that the rate of dissolving is just equal to the rate of precipitation from solution. In each of these cases, both processes continue. The equality of rate creates the illusion of static conditions. The point is that no reaction actually goes to completion.

Equilibrium is best described by the principle of Le Chatelier, which sums up the effects of changes in any of the factors influencing the position of equilibrium. It states that a system in equilibrium, when subjected to a stress resulting from a change in temperature, pressure, or concentration and causing the equilibrium to be upset, will adjust its position of equilibrium to relieve the stress and reestablish equilibrium.

What is the difference between steady state and equilibrium? Steady state implies no changes with passage of time. Likewise, equilibrium can also imply no change of state with passage of time. In many situations, this is the case—the system is not only at steady state but also at equilibrium. However, this is not always the case. In some cases, where the flow rates are steady but the phase contents, for example, are not being maintained at the equilibrium values, the system is at steady state but not at equilibrium.

3.6 MATH OPERATIONS AND LAWS OF EQUILIBRIUM

Earlier we observed that no chemical reaction goes to completion. There are qualitative consequences of this insight that go beyond the purpose of this text, but in this text we are interested in the basic quantitative aspects of equilibria. The chemist usually starts with the chemistry of the reaction and fully utilizes chemical intuition before resorting to mathematical techniques. That is, science should always precede mathematics in the study of physical phenomena. Note, however, that most chemical problems do not require exact, closed-form solutions, and the direct application of mathematics to a problem can lead to an impasse. Several basic math operations and fundamental laws from physical chemistry and thermodynamics serve as the tools, blueprints, and foundational structures of mathematical models. They can be used and applied to environmental systems under certain conditions to solve a variety of problems. Many laws serve as important links between the state of a system, its chemical properties, and its behavior. As such, some of the basic math operations used to solve basic equilibrium problems and laws essential for modeling the fate and transport of chemicals in natural and engineered environmental systems are reviewed in the following sections.

3.6.1 SOLVING EQUILIBRIUM PROBLEMS

In the following math operations, we provide examples of the various forms of combustion of hydrogen to yield water to demonstrate the solution of equilibrium problems. Let's first consider the reaction at 1000.0 K where all constituents are in the gas phase and the equilibrium constant is 1.15×10^{10} atm^{-1}. This reaction is represented by the following equation and equilibrium constant expression (Equation 3.12).

$$2H_2 \text{ (g)} + O_2 \text{ (g)} = 2H_2O \text{ (g)}$$

$$K = [H_2O]^2/[H_2]^2[O_2] \qquad (3.12)$$

where concentrations are given as partial pressures in atm. Observe that K is very large; consequently, the concentration of water is large and/or the concentration of at least one of the reactants is very small.

■ EXAMPLE 3.3

Problem: Consider a system at 1000.0 K in which 4.00 atm of oxygen is mixed with 0.500 atm of hydrogen and no water is initially present. Note that oxygen is in excess and hydrogen is the limiting reagent. Because the equilibrium constant is very large, virtually all of the hydrogen is converted to water, yielding $[H_2O] = 0.500$ atm and $[O_2] = 4.000 - 0.5(0.500) = 3.750$ atm. The final concentration of hydrogen, a small number, is an unknown, the only unknown.

Solution: Using the equilibrium constant expression, we obtain

$$1.15 \times 10^{10} = (0.500)^2/[H_2]^2(3.750)$$

from which we determine that $[H_2] = 2.41 \times 10^{-6}$ atm. Because this is a small number, our initial approximation is satisfactory.

■ EXAMPLE 3.4

Problem: Again, consider a system at 1000.0 K, where 0.250 atm of oxygen is mixed with 0.500 atm of hydrogen and 2000 atm of water.

Solution: Again, the equilibrium constant is very large and the concentration of least reactants must be reduced to a very small value.

$$[H_2O] = 2.000 + 0.500 = 2.500 \text{ atm}$$

In this case, oxygen and hydrogen are present in a 1:2 ratio, the same ratio given by the stoichiometric coefficients. Neither reactant is in excess, and the equilibrium concentrations of both will be very small values. We have two unknowns but they are related by stoichiometry. Because neither product is in excess and one molecule of oxygen is consumed for two of hydrogen, the ratio $[H2]/[O2] = 2/1$ is preserved during the entire reaction and $[H_2] = 2[O_2]$:

$$1.15 \times 10^{10} = 2.500^2/(2[O_2])^2[O_2]$$

$$[O2] = 5.14 \times 10^{-4} \text{ atm and } [H2] = 2[O2] = 1.03 \times 10^{-3} \text{ atm}$$

3.6.2 LAWS OF EQUILIBRIUM

Some of the laws essential for modeling the fate and transport of chemicals in natural and engineered environmental system include the following:

- Ideal gas law
- Dalton's law
- Raoult's law
- Henry's law

3.6.2.1 Ideal Gas Law

An ideal gas is defined as one in which all collisions between atoms or molecules are perfectly elastic and in which there are no intermolecular attractive forces. One can visualize it as collections of perfectly hard spheres that collide but otherwise do not interact with each other. In such a gas, all the internal energy is in the form of kinetic energy and any change in the internal energy is accompanied by a change in temperature. An ideal gas can be characterized by three state variables: absolute pressure (P), volume (V), and absolute temperature (T). The relationship between them may be deduced from kinetic theory and is called the *ideal gas law*:

$$P \times V = n \times R \times T = N \times k \times T \tag{3.13}$$

where
  P = Absolute pressure.
  V = Volume.
  n = Number of moles.
  R = Universal gas constant = 8.3145 J/mol·K or 0.821 L·atm/mol·K.
  T = Temperature.
  N = Number of molecules.
  k = Boltzmann constant = 1.38066×10^{-23} J/K = R/N_A, where N_A is Avogadro's number (6.0221 $\times 10^{23}$).

Note: At standard temperature and pressure (STP), the volume of 1 mol of ideal gas is 22.4 L, a volume called the *molar volume of a gas*.

■ **EXAMPLE 3.5**

Problem: Calculate the volume of 0.333 mol of gas at 300 K under a pressure of 0.950 atm.

Solution:

$$V = \frac{n \times R \times T}{P} = \frac{0.333 \text{ mol} \times 0.0821 \text{ L} \cdot \text{atm/mol} \cdot \text{K} \times 300 \text{ K}}{0.959 \text{ atm}} = 8.63 \text{ L}$$

Most gases in environmental systems can be assumed to obey this law. The ideal gas law can be viewed as arising from the kinetic pressure of gas molecules colliding with the walls of a container in accordance with Newton's laws, but there is also a statistical element in the determination of the average kinetic energy of those molecules. The temperature is taken to be proportional to this average kinetic energy; this invokes the idea of kinetic temperature.

3.6.2.2 Dalton's Law

Dalton's law states that the pressure of a mixture of gases is equal to the sum of the pressures of all of the constituent gases alone. Mathematically, this can be represented as

$$P_{Total} = P_1 + P_2 + \ldots + P_n \tag{3.14}$$

where
P_{Total} = Total pressure.
$P_1, \ldots$ = Partial pressure.

and

$$\text{Partial } P = \frac{n_j \times R \times T}{V} \tag{3.15}$$

where n_j is the number of moles of component j in the mixture.

Note: Although Dalton's law explains that the total pressure is equal to the sum of all of the pressures of the parts, this is only absolutely true for ideal gases, but the error is small for real gases.

■ **EXAMPLE 3.6**

Problem: The atmospheric pressure in a lab is 102.4 kPa. The temperature of a water sample is 25°C at a pressure of 23.76 torr. If we use a 250-mL beaker to collect hydrogen from the water sample, what are the pressure of the hydrogen and the moles of hydrogen using the ideal gas law?

Solution:
1. Make the following conversions—A torr is 1 mm of mercury at standard temperature. In kilopascals, that would be 3.17 (1 mmHg = 7.5 kPa). Convert 250 mL to 0.250 L and 25°C to 298 K.
2. Use Dalton's law to find the hydrogen pressure:

$$P_{Total} = P_{Water} + P_{Hydrogen}$$

$$102.4 \text{ kPa} = 3.17 \text{ kPa} + P_{Hydrogen}$$

$$P_{Hydrogen} = 99.23 \text{ kPa}$$

3. Recall that the ideal gas law is

$$P \times V = n \times R \times T$$

where P is pressure, V is volume, n is the number of moles, R is the ideal gas constant (8.31 L·kPa/mol·K or 0.821 L·atm/mol·K), and T is temperature. Therefore,

$$99.2 \text{ kPa} \times 0.250 \text{ L} = n \times 8.31 \text{ L·kPa/mol·K} \times 298 \text{ K}$$

Rearranged:

$$n = 99.2 \text{ kPa} \times 0.250 \text{ L}/8.31 \text{ L·kPa/mol·K}/298 \text{ K}$$

$$n = 0.0100 \text{ mol or } 1.00 \times 10^{-2} \text{ mol hydrogen}$$

3.6.2.3 Raoult's Law

Raoult's law states that the vapor pressure of mixed liquids is dependent on the vapor pressures of the individual liquids and the molar fraction of each present. Accordingly, for concentrated solutions where the components do not interact, the resulting vapor pressure (P) of component a in equilibrium with other solutions can be expressed as

$$P = x_a \times P_a \tag{3.16}$$

where
P = Resulting vapor pressure.
x_a = Mole fraction of component a in solution.
P_a = Vapor pressure of pure a at the same temperature and pressure as the solution.

3.6.2.4 Henry's Law

Henry's law states that the mass of a gas that dissolves in a definite volume of liquid is directly proportional to the pressure of the gas, provided the gas does not react with the solvent. A formula for Henry's law is

$$P = H \times x \tag{3.17}$$

P is the partial pressure of a gas above the solution, H is Henry's constant, and x is the solubility of a gas in the solution phase.

Henry's law constant (H) is a partition coefficient usually defined as the ratio of the concentration of a chemical in air to its concentration in water at equilibrium. Henry's law constants generally increase with increased temperature, primarily due to the significant temperature dependency of chemical vapor pressures. Solubility is much less affected by the changes in temperature that are normally found in the environment (Hemond and Fechner-Levy, 2000). H can be expressed either in a dimensionless form or with units. Table 3.1 lists the Henry's law constants for some common environmental chemicals.

3.7 CHEMICAL TRANSPORT SYSTEMS

In environmental modeling, environmental practitioners have a fundamental understanding of the phenomena involved with the transport of certain chemicals through the various components of the environment. The primary transport mechanism at the microscopic level is molecular *diffusion* driven by concentration gradients; whereas, mixing and bulk movement of the medium are

TABLE 3.1
Henry's Law Constants (H)

Chemical	Henry's Law Constant	
	(atm × m³/mol)	(dimensionless)
Aroclor 1254	2.7×10^{-3}	1.2×10^{-1}
Aroclor 1260	7.1×10^{-3}	3.0×10^{-1}
Atrazine	3×10^{-9}	1×10^{-7}
Benzene	5.5×10^{-3}	2.4×10^{-1}
Benz[a]anthracene	5.75×10^{-6}	2.4×10^{-4}
Carbon tetrachloride	2.3×10^{-2}	9.7×10^{-1}
Chlorobenzene	3.7×10^{-3}	1.65×10^{-1}
Chloroform	4.8×10^{-3}	2.0×10^{-1}
Cyclohexane	0.18	7.3
1,1-Dichloroethane	6×10^{-3}	2.4×10^{-1}
1,2-Dichloroethane	10^{-3}	4.1×10^{-2}
cis-1,2-Dichloroethene	3.4×10^{-3}	0.25
trans-1,2-Dichlorethene	6.7×10^{-3}	0.23
Ethane	4.9×10^{-1}	20
Ethanol	6.3×10^{-6}	—
Ethylbenzene	8.7×10^{-3}	3.7×10^{-1}
Lindane	4.8×10^{-7}	2.2×10^{-5}
Methane	0.66	27
Methylene chloride	3×10^{-3}	1.3×10^{-1}
n-Octane	2.95	121
Pentachlorophenol	3.4×10^{-6}	1.5×10^{-4}
n-Pentane	1.23	50.3
Perchloroethane	8.3×10^{-3}	3.4×10^{-1}
Phenanthrene	3.5×10^{-5}	1.5×10^{-3}
Toluene	6.6×10^{-3}	2.8×10^{-1}
1,1,1-Trichloroethane (TCA)	1.8×10^{-2}	7.7×10^{-1}
Trichloroethene (TCE)	1×10^{-2}	4.2×10^{-1}
o-Xylene	5.1×10^{-3}	2.2×10^{-1}
Vinyl chloride	2.4	99

Source: Adapted from Lyman, W.J. et al., *Handbook of Chemical Property Estimation Methods*, American Chemical Society, Washington, DC, 1990.

the primary transport mechanisms at the macroscopic level. Transport by molecular diffusion and mixing is referred to as *dispersive transport*; transport by bulk movement of the medium is referred to as *advective transport*. Advective and dispersive transports are fluid-element driven. Advection, for example, is the movement of dissolved solute with flowing groundwater. The amount of contaminant being transported is a function of its concentration in the groundwater and the quantity of groundwater flowing, and advection will transport contaminants at different rates in each stratum. Diffusive transport, on the other hand, is the process by which a contaminant in water will move from an area of greater concentration toward an area where it is less concentrated. Diffusion will occur as long as a concentration gradient exists, even if the fluid is not moving, and as a result a contaminant may spread away from the place where it is introduced into a porous medium.

3.8   FINAL WORD ON ENVIRONMENTAL MODELING

In this chapter, we have provided a basic survey of some of the basic math and science involved in environmental modeling. In today's computer age, environmental engineers have the advantage of choosing from a wide variety of mathematical models available. These models enable environmental engineers and students with minimal computer programming skills to develop computer-based mathematical models for natural and engineered environmental systems. Commercially available syntax-free authoring software can be adapted to create customized, high-level models of environmental phenomena in groundwater, air, soil, aquatic, and atmospheric systems. We highly recommend that aspiring environmental engineering students take full advantage of college-level computer modeling courses. Without such a background, the modern environmental engineer's technical toolbox is missing a vital tool.

REFERENCES AND RECOMMENDED READING

Harter, H.L. (1983). Least squares, in Kotz, S. and Johnson, N.L., Eds., *Encyclopedia of Statistical Sciences*. John Wiley & Sons, New York.

Hemond, F.H. and Fechner-Levy, E.J. (2000). *Chemical Fate and Transport in the Environment*, 2nd ed. Academic Press, San Diego.

Lyman, W.J., Reehl, W.R., and Rosenblatt, D.H. (1990). *Handbook of Chemical Property Estimation Methods*. American Chemical Society, Washington, DC.

Nirmalakhandan, N. (2002). *Modeling Tools for Environmental Engineers and Scientists*. CRC Press, Boca Raton, FL.

NIST. (2012). *Engineering Statistics*. Technology Administration, U.S. Commerce Department, Washington, DC.

4 Algorithms and Environmental Engineering

Algorithm—The word comes from the Persian author Abu Ja'far Mohammed ibn Musa al-Khomwarizmi, who wrote a book with arithmetic rules dating from about 825 AD.

4.1 INTRODUCTION

In Chapter 3, we observed that environmental modeling has become an important tool within the environmental engineer's well-equipped toolbox. Continuing with that analogy, we can say that if a skilled handyperson's toolbox usually contains a socket and ratchet set and several different sized wrench attachments, then the well-equipped environmental practitioner's toolbox includes a number of environmental models (sockets and ratchets) with a varying set of algorithms (socket wrench attachments). Although a complete treatment or discussion of algorithms is beyond the scope of this book, we do explain what algorithms are and provide examples of their applications in cyberspace. For those interested in a more complete discussion of algorithms, many excellent texts on the general topic are available. We list several of these resources in the References and Recommended Reading section at the end of the chapter.

4.2 ALGORITHMS: WHAT ARE THEY?

An *algorithm* is a specific mathematical calculation procedure, a computable set of steps to achieve a desired result. More specifically, "an algorithm is any well-defined computational procedure that takes some value, or set of values, as input and produces some value, or set of values, as output" (Cormen et al., 2002). In other words, an algorithm is a recipe for an automated solution to a problem. A computer model may contain several algorithms. The word "algorithm" is derived from the name of a ninth-century Persian mathematician: al-Khomwarizmi.

Algorithms should not be confused with computations. Whereas an algorithm is a systematic method for solving problems, and computer science is the study of algorithms (although the algorithm was developed and used long before any device resembling a modern computer was available), the act of executing an algorithm—that is, manipulating data in a systematic manner—is called *computation*. For example, the following algorithm for finding the greatest common divisor of two given whole numbers (attributed to Euclid *ca.* 300 BC, thus known for millennia) may be stated as follows:

- Set a and b to the values A and B, respectively.
- Repeat the following sequence of operations until b has value 0:
  1. Let r take the value of a mod b.
  2. Let a take the value of b.
  3. Let b take the value of r.
- The greatest common divisor of A and B is the final value of a.

Note: The operation a mod b gives the remainder obtained upon dividing a by b.

Here, the problem—finding the greatest common divisor of two numbers—is specified by stating what is to be computed; the problem statement itself does not require that any particular algorithm be used to compute the value. Such method-independent specifications can be used to define the meaning of algorithms: the meaning of an algorithm is the value that it computes.

Several methods can be used to compute the required value; Euclid's method is just one. The chosen method assumes a set of standard operations (such as basic operations on the whole number and a means to repeat an operation) and combines these operations to form an operation that computes the required value. Also, it is not at all obvious to the vast majority of people that the proposed algorithm does actually compute the required value. That is one reason why a study of algorithms is important—to develop methods that can be used to establish what a proposed algorithm achieves.

4.3 EXPRESSING ALGORITHMS

Although an in-depth discussion of algorithms is beyond the scope of this text, the analysis of algorithms often requires us to draw upon a body of mathematical operations. Some of these operations are as simple as high-school algebra, but others may be less familiar to the average environmental engineer. Along with learning how to manipulate asymptotic notations and solving recurrences, several other concepts and methods must be learned to analyze algorithms.

Methods for evaluating bounding summations, for example, occur frequently in the analysis of algorithms and are used when an algorithm contains an iterative control construct such as a *while* or *for* loop. In this case, the running time can be expressed as the sum of the times spent on each execution of the body of the loop. Many of the formulas commonly used in analyzing algorithms can be found in any calculus text. In addition, in order to analyze many algorithms, we must be familiar with the basic definitions and notations for sets, relations, functions, graphs, and trees. A basic understanding of elementary principles of counting (permutations, combinations, and the like) is important as well. Most algorithms used in environmental engineering require no probability for their analysis; however, a familiarity with these operations can be useful.

Because mathematical and scientific analyses (and many environmental engineering functions) are so heavily based on numbers, computation has tended to be associated with numbers; however, this need not be the case. Algorithms can be expressed using any formal manipulation system—that is, any system that defines a set of entities and a set of unambiguous rules for manipulating those entities. For example, *SKI calculus* consists of three combinators (entities) called, coincidentally, S, K, and I. The computation rules for the calculus are

1. $Sfgx \rightarrow fx(gx)$
2. $Kxy \rightarrow x$
3. $Ix \rightarrow x$

where f, g, x, and y are strings of the three entities. SKI calculus is computationally complete; that is, any computation that can be performed using any formal system can be performed using SKI calculus. (Equivalently, all algorithms can be expressed using SKI calculus.) Not all systems of computation are equally as powerful, though; some problems that can be solved using one system cannot be solved using another. Further, it is known that problems exist that cannot be solved using any formal computation system.

4.4 GENERAL ALGORITHM APPLICATIONS

Practical applications of algorithms are ubiquitous. All computer programs are expressions of algorithms, where the instructions are expressed in computer language being used to develop the program. Computer programs are described as expressions of algorithms, as an algorithm is a general technique for achieving some purpose and can be expressed in a number of different ways.

Algorithms exist for many purposes and are expressed in many different ways. Examples of algorithms include recipes in cook books, servicing instructions in the manual of a computer, knitting patterns, digital instructions for a welding robot indicating where each weld should be made, or cyber-speak for any system used in cyberspace.

Algorithms can be used in sorting operations—for example, to reorder a list into some defined sequence. It is possible to express an algorithm as instructions given to a human who has a similar requirement to reorder some list—for example, to sort a list of tax records into a sequence determined by the date of birth on the record. These instructions could employ the *insertion sort algorithm*, the *bubble sort algorithm*, or one of many other available algorithms. Thus, an algorithm, as a general technique for expressing the process of completing a defined task, is independent of the precise manner in which it is expressed.

Sorting is by no means the only application for which algorithms have been developed. Practical applications of algorithms include the following examples:

- Internet routing (e.g., single-source shortest paths)
- Search engine (e.g., string matching)
- Public-key cryptography and digital signatures (e.g., number-theoretic algorithms)
- Allocating scarce resources in the most beneficial way (e.g., linear programming)

Algorithms are at the core of most technologies used in contemporary computers:

- Hardware design uses algorithms.
- The design of any GUI relies on algorithms.
- Routing in networks relies heavily on algorithms.
- Compilers, interpreters, and assemblers make extensive use of algorithms.

A few classic algorithms are commonly used to illustrate the function, purpose, and applicability of algorithms. One of these classics is known as the *Byzantine Generals* (Black, 2012). Briefly, this algorithm is about the problem of reaching a consensus among distributed units if some of them give misleading answers. The problem is couched in terms of generals deciding on a common plan of attack. Some traitorous generals may lie about whether they will support a particular plan and what other generals told them. What decision-making algorithm should the generals use to reach a consensus through only an exchange of messages? What percentage of liars can the algorithm tolerate and still correctly determine a consensus?

Another classic algorithm that is used to illustrate how an algorithm can be applied to real-world situations (because of its general usefulness and because it is easy to explain to just about anyone) is the *Traveling Salesman* problem. The Traveling Salesman problem is the most notorious NP-complete problem; that is, no polynomial-time algorithm has yet been discovered for an NP-complete problem, nor has anyone yet been able to prove that no polynomial-time algorithm can exist for any one of them. For the Traveling Salesman problem, imagine that a traveling salesman has to visit each of a given set of cities by car, but he can only stop in each city one time. In Figure 4.1A, find the shortest possible route that visits each city once and returns to the origin city (Figure 4.1B).

We have pointed out some of the functions that algorithms can perform, but the question arises: "Can every problem be solved algorithmically?" The simple and complex answer is *no*. For example, for some problems no generalized algorithmic solution can possibly exist (they are unsolvable). Also, some problems—*NP-complete problems*—have no known efficient solutions; that is, it is unknown if efficient algorithms exist for these problems. If an efficient algorithm exists for any one of them, then efficient algorithms exist for all of them (e.g., Traveling Salesman problem). Finally, problems exist that we simply do not know how to solve algorithmically. From this discussion, it should be apparent that computer science is not simply about word processing and spreadsheets.

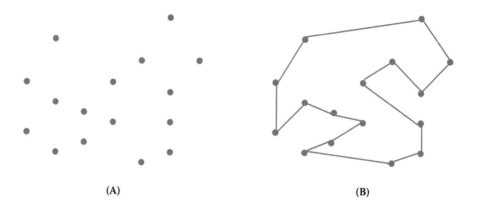

(A)   (B)

FIGURE 4.1   Traveling Salesman problem.

4.5   ENVIRONMENTAL PRACTICE ALGORITHM APPLICATIONS

Although algorithms can be used in transportation applications (e.g., Traveling Salesman problem), many of their most important applications are in environmental engineering functions. For example, consider a robot arm assigned to weld all the metal parts on an automobile in an assembly line. The shortest path that visits each weld point exactly once would be the most efficient one for the robot. A similar application arises in minimizing the amount of time taken by a design engineer or draftsperson to draw a given structure. Algorithms have found widespread application in all branches of environmental practice. In environmental engineering, for example, the U.S. Environmental Protection Agency (USEPA) uses computer models relying upon various algorithms to monitor chemical spill and ultimate fate data. In the following, we provide selected model summary descriptions of applications used in dispersion modeling. Specifically, we discuss how the USEPA (and others) employ preferred or recommended models (i.e., refined models that are recommended for a specific type of regulatory application) in monitoring air quality (i.e., ambient pollutant concentrations and their temporal and spatial distribution). Further information on this important topic can be found at USEPA (2003).

4.6   DISPERSION MODELS*

- *BLP* (buoyant line and point source model) is a Gaussian plume dispersion model designed to handle unique modeling problems associated with aluminum reduction plants and other industrial sources where plume rise and downwash effects from stationary line sources are important.
- *CALINE3* is a steady-state Gaussian dispersion model designed to determine air pollution concentrations at receptor locations downwind of "at-grade," fill," "bridge," and "cut section" highways located in relatively uncomplicated terrain.
- *CALPUFF* is a multi-layer, multi-species, non-steady-state puff dispersion model that simulates the effects of time- and space-varying meteorological conditions on pollution transport, transformation, and removal. CALPUFF can be applied on scales of tens to hundreds of kilometers. It includes algorithms for subgrid scale effects (such as terrain impingement), as well as longer range effects (such as pollutant removal due to wet scavanging and dry deposition, chemical transformation, and visibility effects of particulate matter concentrations).

---

* The algorithm-based models described in this section are currently listed in Appendix A of the *Guidelines on Air Quality Models* (Appendix W of 40 CFR Part 51).

- *CTDMPLUS* (Complex Terrain Dispersion Model Plus Algorithms for Unstable Situations) is a refined point-source Gaussian air quality model for use in all stability conditions for complex terrain (i.e., terrain exceeding the height of the stack being modeled as contrasted with simple terrain, which is defined as an area where terrain features are all lower in elevation than the top of the stack of the source). The model contains, in its entirety, the technology of CTDM for stable and neutral conditions.
- *ISC3* (Industrial Source Complex Model) is a steady-state Gaussian plume model that can be used to assess pollutant concentrations from a wide variety of sources associated with an industrial complex. This model can account for the following: settling and dry deposition of particles; downwash; point, area, line, and volume sources; plume rise as a function of downwind distance; separation of point sources; and limited terrain adjustment. ISC3 operates in both long-term and short-term modes.
- *OCD* (Offshore and Coastal Dispersion Model) is a straight-line Gaussian model developed to determine the impact of offshore emissions from point, area, or line sources on the air quality of coastal regions. OCD incorporates overwater plume transport and dispersion as well as changes that occur as the plume crosses the shoreline. Hourly meteorological data are needed from both offshore and onshore locations.

4.7 SCREENING TOOLS

Screening tools are relatively simple analysis techniques to determine if a given source is likely to pose a threat to air quality. Concentration estimates from screening techniques precede a refined modeling analysis and are conservative. Following are some of the screening tools available:

- *CAL3QHC/CAL3QHCR* (CALINE3 with queuing and hot-spot calculations) is a CALINE3-based CO model with a traffic model to calculate delays and queues that occur at signalized intersections; CAL3QHCR requires local meteorological data.
- *COMPLEX 1* is a multiple point-source screening technique with terrain adjustment that incorporates the plume impaction algorithm of the VALLEY model.
- *CTSCREEN* (Complex Terrain Screening model) is a Gaussian plume dispersion model designed as a screening technique for regulatory application to plume impaction assessments in complex terrain. CTSCREEN is a screening version of the CTDMPLUS model.
- *LONGZ* is a steady-state Gaussian plume formulation for both urban and rural areas in flat or complex terrain to calculate long-term (seasonal and/or annual) ground-level ambient air concentrations attributable to emissions from up to 14,000 arbitrarily placed sources (stack, buildings, and area sources).
- *SCREEN3* is a single-source Gaussian plume model that provides maximum ground-level concentrations for point, area, flare, and volume sources, as well as concentrations in the cavity zone and concentrations due to inversion break-up and shoreline fumigation. SCREEN3 is a screening version of the ISC3 model.
- *SHORTZ* is a steady-state bivariate Gaussian plume formulation model for both urban and rural areas in flat or complex terrain to calculate ground-level ambient air concentrations. It can calculate 1-hr, 2-hr, 3-hr, etc., average concentrations due to emissions from stacks, buildings, and area sources for up to 300 arbitrarily placed sources.
- *VALLEY* is a steady-state, complex-terrain, univariate Gaussian plume dispersion algorithm designed for estimating either 24-hour or annual concentrations resulting from emissions from up to 50 (total) point and area sources.
- *VISCREEN* calculates the potential impact of a plume of specified emissions for specific transport and dispersion conditions.

REFERENCES AND RECOMMENDED READING

Black, P.E. (2009). Byzantine generals, in *Dictionary of Algorithms and Data Structures*. U.S. National Institute of Standards and Technology, Washington, DC, http://xlinux.nist.gov/dads/HTML/byzantine.html.

Cormen, T.H., Leiserson, C.E., Rivest, R.L., and Stein, C. (2002). *Introduction to Algorithms*, 2nd ed. Prentice-Hall, New Delhi.

Gusfield, D. (1997). *Algorithms on Strings, Trees, and Sequences: Computer Science and Computational Biology*. Cambridge University Press, Cambridge, U.K.

Lafore, R. (2002). *Data Structures and Algorithms in Java*, 2nd ed. Sams Publishing, Indianapolis, IN.

Mitchell, T.M., (1997). *Machine Learning*. McGraw-Hill, New York.

Poynton, C. (2003). *Digital Video and HDTV Algorithms and Interfaces*. Morgan Kaufmann, Burlington, MA.

USEPA. (2003). *Technology Transfer Network Support Center for Regulatory Air Models*. U.S. Environmental Protection Agency, Washington, DC, http://www.epa.gov/scram001/.

5 Quadratic Equations

$$ax^2 + bx + c = 0$$

$$x = \frac{-b \pm \sqrt{b^2 - 4ac}}{2a}$$

When studying a discipline that does not include mathematics, one thing is certain: The discipline under study has little or nothing to do with environmental practice.

5.1 QUADRATIC EQUATIONS AND ENVIRONMENTAL PRACTICE

A logical question at this point might be why is the quadratic equation important in environmental practice? The logical answer is that the quadratic equation is used in environmental practice to find solutions to problems primarily dealing with length and time determinations. Stated differently: The quadratic equation is a tool, an important tool that belongs in every environmental practitioner's toolbox.

To the student of mathematics, this explanation might seem somewhat strange. Math students know, for example, that there will be two solutions to a quadratic equation. In environmental disciplines such as environmental engineering, many times only one solution is meaningful. For example, if we are dealing with a length, a negative solution to the equation may be mathematically possible but is not the solution we would use. Negative time, obviously, would also pose the same problem.

So what is the point? The point is that we often need to find a solution to certain mathematical problems. In environmental problems involving the determination of length and time using quadratic equations, we will end up with two answers. In some instances, a positive answer and a negative answer may result. One of these answers is usable; thus, we would use it. Real engineering is about modeling situations that occur naturally and using the model to understand what is happening or maybe to predict what will happen in future. The quadratic equation is often used in modeling because it is a beautifully simple curve (Bourne, 2013).

5.1.1 KEY TERMS

- a is the coefficient of x^2.
- b is the coefficient of x.
- c is a number in the quadratic equation (not a coefficient of any x term).
- Simple equations are equations in which the unknown appears only in the first degree.
- Pure quadratic equations are equations in which the unknown appears only in the second degree.
- Affected quadratic equations are equations containing the first and second degree of an unknown.

5.2 QUADRATIC EQUATIONS: THEORY AND APPLICATION

The equation $6x = 12$ is a form of equation familiar to most of us. In this equation the unknown appears only in the first degree, so it is a simple equation or linear equation. Those experienced in mathematics know that not all equations reduce to this form. For instance, when an equation has been reduced, the result may be an equation in which the square of the unknown equals some number, as in $x^2 = 5$. In this equation, the unknown appears only in the second degree, so it is a pure quadratic equation. In some cases, when an equation is simplified and reduced, the resulting equation contains the square and first power of the unknown, which equal some number, such as $x^2 - 5x = 24$. An equation containing the first and second degree of an unknown is an affected quadratic equation.

Quadratic equations, and certain other forms, can be solved with the aid of factoring. The procedure for solving a quadratic equation by factoring is as follows:

1. Collect all terms on the left and simplify to the form $ax^2 + bx + c = 0$.
2. Factor the quadratic expression.
3. Set each factor equal to zero.
4. Solve the resulting linear equations.
5. Check the solution in the original equation.

■ **EXAMPLE 5.1**

Problem: Solve $x^2 - x - 12 = 0$.

Solution:

1. Factor the quadratic expression.

$$(x - 4)(x + 3) = 0$$

2. Set each factor equal to zero.

$$x - 4 = 0 \quad x + 3 = 0$$

3. Solve the resulting linear equations.

$$x = 4 \quad x = -3$$

Thus, the roots are $x = 4$ and $x = -3$.

4. Check the solution in the original equation.

$$(4)^2 - 4 - 12 = 0 \quad (-3)^2 - (-3) - 12 = 0$$
$$0 = 0 \qquad\qquad\qquad 0 = 0$$

Many times factoring is either too time consuming or not possible. The formula shown below is called the quadratic formula. It expresses the quadratic equation in terms of its coefficients. The quadratic formula allows us to quickly solve for x with no factoring.

$$x = \frac{-b \pm \sqrt{b^2 - 4ac}}{2a} \tag{5.1}$$

To use the quadratic equation, just substitute the appropriate coefficients into the equation and solve.

5.3   DERIVATION OF THE QUADRATIC EQUATION FORMULA

The equation $ax^2 + bx + c = 0$, where a, b, and c are any numbers, positive or negative, represents any quadratic equation with one unknown. When this general equation is solved, the solution can be used to determine the unknown value in any quadratic equation. The solution follows.

■ **EXAMPLE 5.2**

Problem: Solve $ax^2 + bx + c = 0$ for x.

Solution:

1. Subtract c from both members:

$$ax^2 + bx = -c$$

2. Divide both members by a:

$$x^2 + \frac{b}{a}(x) = -\frac{c}{a}$$

3. Add $(b/2a)^2$ to both sides:

$$x^2 + \left(\frac{b}{a}\right)(x) + \left(\frac{b}{2a}\right)^2 = -\frac{c}{a} + \left(\frac{b}{2a}\right)^2$$

4. Complete the square:

$$\left(x + \frac{b}{2a}\right)^2 = -\frac{c}{a} + \left(\frac{b}{2a}\right)^2$$

5. Take the square root of both members:

$$x + \frac{b}{2a} = \pm\sqrt{-\frac{c}{a} + \left(\frac{b}{2a}\right)^2}$$

6. Subtract $b/2a$ from both members:

$$x = -\frac{b}{2a} \pm \sqrt{-\frac{c}{a} + \left(\frac{b}{2a}\right)^2}$$

Thus, the following quadratic formula is obtained:

$$x = \frac{-b \pm \sqrt{b^2 - 4ac}}{2a}$$

5.4   USING THE QUADRATIC EQUATION

■ EXAMPLE 5.3

Problem: After conducting a study and deriving an equation representing time, we arrive at the following equation:

$$x^2 = 5x + 6 = 0$$

Solution: All like terms have been combined and the equation is set to equal zero. Use the quadratic formula to solve the problem:

$$x = \frac{-b \pm \sqrt{b^2 - 4ac}}{2a}$$

From our equation, $a = 1$ (the coefficient of x^2), $b = -5$ (the coefficient of x), and $c = 6$ (the constant or third term). Substituting these coefficients in the quadratic formula:

$$x = \frac{-(-5) \pm \sqrt{(-5)^2 - 4(1)(6)}}{2(1)}$$

$$x = \frac{5 \pm \sqrt{25 - 4}}{2}$$

$$x = \frac{5 \pm 1}{2}$$

$$x = 3, 2$$

Note: The roots may not always be rational (integers), but the procedure is the same.

REFERENCES AND RECOMMENDED READING

Bourne, M. (2013). *Quadratic Equations.* Interactive Mathematics, http://www.intmath.com/quadratic-equations/quadratic-equations-intro.php.

6 Trigonometric Ratios

$$\sin A = a/c \quad \cos A = b/c \quad \tan A = a/b$$

We owe a lot to the Indians, who taught us how to count, without which no worthwhile scientific discovery could have been made.

—Albert Einstein

Trigonometry is the branch of mathematics that is used to compute unknown angles and sides of triangles. The word *trigonometry* is derived from the Greek words for triangle and measurement. Trigonometry is based on the principles of geometry. Many problems require the use of geometry and trigonometry.

— Smith and Peterson (2007)

6.1 TRIGONOMETRIC FUNCTIONS AND THE ENVIRONMENTAL PRACTITIONER

Typically, environmental practitioners are called upon to make calculations involving the use of various trigonometric functions. Consider slings, for example; they are commonly used with cranes, derricks, and hoists to lift a load and move it to the desired location. For the environmental professional responsible for safety and health, knowledge of the properties and limitations of the sling, the type and condition of material being lifted, the weight and shape of the object being lifted, the angle of the sling to the load being lifted, and the environment in which the lift is to be made are all important considerations to be evaluated before the safe transfer of material can take place. Later, we put many of the following principles to work in determining sling load and working load on a ramp (inclined plane)—that is, to solve force-type problems. For now, we discuss the basic trigonometric functions used to make such calculations.

6.2 TRIGONOMETRIC RATIOS OR FUNCTIONS

In trigonometry, all computations are based on certain ratios (i.e., trigonometric functions). The trigonometric ratios or functions are sine, cosine, tangent, cotangent, secant, and cosecant. It is important to understand the definition of the ratios given in Table 6.1 and defined in terms of the lines shown in Figure 6.1.

Note: In a right triangle, the side opposite the right angle is the longest side. This side is called the *hypotenuse*. The other two sides are the *legs*.

TABLE 6.1

Definition of Trigonometric Ratios

Sine of angle A	$\dfrac{\text{Measure of leg opposite angle } A}{\text{Measure of hypotenuse}}$	$\sin A = a/c$
Cosine of angle A	$\dfrac{\text{Measure of leg adjacent to angle } A}{\text{Measure of hypotenuse}}$	$\cos A = b/c$
Tangent of angle A	$\dfrac{\text{Measure of leg opposite angle } A}{\text{Measure of leg adjacent to angle } A}$	$\tan A = a/b$

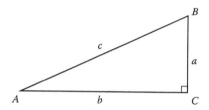

FIGURE 6.1   Right triangle.

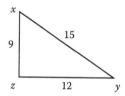

FIGURE 6.2   Illustration for Examples 6.1 and 6.2.

■ **EXAMPLE 6.1**

Problem: Find the sine, cosine, and tangent of angle Y in Figure 6.2.

Solution:

$$\sin Y = \frac{\text{Opposite leg}}{\text{Hypotenuse}} = \frac{9}{15} = 0.60$$

$$\cos Y = \frac{\text{Adjacent leg}}{\text{Hypotenuse}} = \frac{12}{15} = 0.80$$

$$\tan Y = \frac{\text{Opposite leg}}{\text{Adjacent leg}} = \frac{9}{12} = 0.75$$

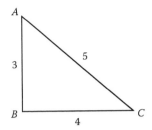

FIGURE 6.3   Illustration for Example 6.3.

■ **EXAMPLE 6.2**

Problem: Using Figure 6.2, find the measure of angle x to the nearest degree.

Solution:

$$\sin x = \frac{\text{Opposite leg}}{\text{Hypotenuse}} = \frac{12}{15} = 0.8$$

Use a scientific calculator to find the angle measure with a sine of 0.8.

Enter: 0.8 [2nd] or [INV]
Result: 53.13010235

So, the measure of angle $x = 53°$.

■ **EXAMPLE 6.3**

Problem: For the triangle shown in Figure 6.3, find sin C, cos C, and tan C.

Solution:

$$\sin C = 2/5; \cos C = 4/5; \tan C = 3/4$$

REFERENCES AND RECOMMENDED READING

McKeague, M. and Charles, P. (1998). *Algebra with Trigonometry for College Students*. Saunders College Publishing, Philadelphia, PA.
Smith, R.D. and Peterson, J.C. (2009). *Mathematics for Machine Technology*, 6th ed. Delmar, Clifton Park, NY.

Section II

Statistics, Risk Measurement, and Boolean Algebra

No aphorism is more frequently repeated in connection with field trials, than that we must ask Nature few questions, or, ideally, one question, at a time. The writer is convinced that this view is wholly mistaken.

—**Ronald Fisher (1926)**

7 Statistics Review

There are three kinds of lies: lies, damned lies, and statistics.

—**Benjamin Disraeli**

To the uninitiated it may often appear that the statistician's primary function is to prevent or at least impede the progress of research. And even those who suspect that statistical methods may be more boon than bane are at times frustrated in their efforts to make use of the statistician's wares.

—**Frank Freese (1967)**

7.1 STATISTICAL CONCEPTS*

Despite the protestation of Disraeli and the wisdom of Freese, environmental practice includes the study of and use of statistical analysis of the results. The principal concept of statistics is that of variation. Variation is often found when conducting typical environmental health functions requiring the use of *biostatistics*, where a wide range of statistics are applied to an even wider range of topics in biology, such as toxicological or biological sampling protocols for air contamination, and other environmental functions applied to agriculture, forestry, fisheries, and other specialized areas. This chapter provides environmental practitioners with a survey of the basic statistical and data analysis techniques that can be used to address many of the problems that they will encounter on a daily basis. It covers the data analysis process, from research design to data collection, analysis, reaching conclusions, and, most importantly, the presentation of findings.

Finally, it is important to point out that statistics can be used to justify the implementation of a program, identify areas that need to be addressed, or evaluate the impact that various environmental health and safety programs might have on losses and accidents. A set of occupational health and safety data (or other data) is only useful if it is analyzed properly. Better decisions can be made when the nature of the data is properly characterized. For example, the importance of using statistical data when selling an environmental health and safety plan or some other type of environmental operation and trying to win over those who control the purse strings cannot be overemphasized.

With regard to Freese's opening statement, much of the difficulty is due to not understanding the basic objectives of statistical methods. We can boil these objectives down to two:

1. Estimation of population parameters (values that characterize a particular population)
2. Testing hypotheses about these parameters

A common example of the first is estimation of the coefficients a and b in the linear relationship $Y = a + bX$. To accomplish this objective one must first define the population involved and specify the parameters to be estimated. This is primarily the research worker's job. The statistician helps devise efficient methods of collecting the data and calculating the desired estimates.

---

* Much of the information in this chapter is modeled after Freese, F., *Elementary Statistical Methods for Foresters*, Handbook 317, U.S. Department of Agriculture, Washington, DC, 1967.

Unless the entire population is examined, an estimate of a parameter is likely to differ to some degree from the population value. The unique contribution of statistics to research is that it provides ways of evaluating how far off the estimate may be. This is ordinarily done by computing confidence limits, which have a known probability of including the true value of the parameter. For example, the mean diameter of the trees in a pine plantation may be estimated from a sample as being 9.2 inches, with 95% confidence limits of 8.8 and 9.6 inches. These limits (if properly obtained) tell us that, unless a 1-in-20 chance has occurred in sampling, the true mean diameter is somewhere between 8.8 and 9.6 inches.

The second basic objective in statistics is to test some hypothesis about the population parameters. A common example is a test of the hypothesis that the regression coefficient in the linear model

$$Y = a + bX$$

has some specified value (say zero). Another example is a test of the hypothesis that the difference between the means of two populations is zero.

Again, it is the research worker who should formulate meaningful hypotheses to be tested, not the statistician. This task can be tricky. The beginner would do well to work with the statistician to be sure that the hypothesis is put in a form that can be tested. Once the hypothesis is set, it is up to the statistician to work out ways of testing it and to devise efficient procedures for obtaining the data (Freese, 1969).

7.1.1 PROBABILITY AND STATISTICS

Those who work with probabilities are commonly thought to have an advantage when it comes to knowing, for example, the likelihood of tossing coins heads up six times in a row, the chances of a crapshooter making several consecutive winning throws ("passes"), and other such useful bits of information. It is fairly well known that statisticians work with probabilities; thus, they are often associated with having the upper hand, so to speak, on predicting outcomes in games of chance. However, statisticians also know that this assumed edge they have in games of chance is often dependent on other factors.

The fundamental role of probability in statistical activities is often not appreciated. In putting confidence limits on an estimated parameter, the part played by probability is fairly obvious. Less apparent to the neophyte is the operation of probability in the testing of hypotheses. Some of them say with derision, "You can prove anything with statistics" (remember what Disraeli said about statistics). Anyway, the truth is, you can *prove* nothing; you can at most compute the probability of something happening and let the researcher draw his own conclusions.

Let's return to our game of chance to illustrate this point. In the game of craps, the probability of a shooter winning (making a pass) is approximately 0.493—assuming, of course, a perfectly balanced set of dice and a honest shooter. Suppose now that you run up against a shooter who picks up the dice and immediately makes seven passes in a row! It can be shown that if the probability of making a single pass is really 0.493, then the probability of seven or more consecutive passes is about 0.007 (or 1 in 141). This is where the job of statistics ends; you can draw your own conclusions about the shooter. If you conclude that the shooter is pulling a fast one, then in statistical terms you are rejecting the hypothesis that the probability of the shooter making a single pass is 0.493.

In practice, most statistical tests are of this nature. A hypothesis is formulated and an experiment is conducted or a sample is selected to test it. The next step is to compute the probability of the experimental or sample results occurring by chance if the hypothesis is true. If this probability is less than some preselected value (perhaps 0.05 or 0.01), then the hypothesis is rejected. Note that nothing has been proved—we haven't even proved that the hypothesis is false. We merely inferred this because of the low probability associated with the experiment or sample results.

Our inferences may be incorrect if we are given inaccurate probabilities. Obviously, reliable computation of these probabilities requires knowledge of how the variable we are dealing with is distributed (that is, what the probability is of the chance occurrence of different values of the variable). Accordingly, if we know that the number of beetles caught in light traps follows what is called the *Poisson distribution* we can compute the probability of catching X or more beetles. But, if we assume that this variable follows the Poisson distribution when it actually follows the negative binomial distribution, then our computed probabilities may be in error.

Even with reliable probabilities, statistical tests can lead to the wrong conclusions. We will sometimes reject a hypothesis that is true. If we always test at the 0.05 level, we will make this mistake on the average of 1 time in 20. We accept this degree of risk when we select the 0.05 level of testing. If we are willing to take a bigger risk, we can test at the 0.10 or the 0.25 level. If we are not willing to take this much risk, we can test at the 0.01 or 0.001 level.

Researchers can make more than one kind of error. In addition to rejecting a hypothesis that is true (a Type I error), one can make the mistake of not rejecting a hypothesis that is false (a Type II error). In crapshooting, it is a mistake to accuse an honest shooter of cheating (Type I error—rejecting a true hypothesis), but it is also a mistake to trust a dishonest shooter (Type II error—failure to reject a false hypothesis).

The difficulty is that for a given set of data, reducing the risk of one kind of error increases the risk of the other kind. If we set 15 straight passes as the critical limit for a crapshooter, then we greatly reduce the risk of making a false accusation (probability about 0.00025). But in doing so we have dangerously increased the probability of making a Type II error—failure to detect a phony. A critical step in designing experiments is the attainment of an acceptable level of probability of each type of error. This is usually accomplished by specifying the level of testing (i.e., probability of an error of the first kind) and then making the experiment large enough to attain an acceptable level of probability for errors of the second kind.

It is beyond the scope of this book to go into basic probability computations, distribution theory, or the calculation of Type II errors, but anyone who uses statistical methods should be fully aware that he or she is dealing primarily with probabilities (not necessarily lies or damnable lies) and not with immutable absolutes. Remember, 1-in-20 chances do actually occur—about one time out of twenty.

7.2   MEASURE OF CENTRAL TENDENCY

When we talk about statistics, it is usually because we are estimating something with incomplete knowledge. Maybe we can only afford to test 1% of the items we are interested in and we want to say something about the properties of the entire lot, or perhaps we must destroy the sample to test it. In that case, 100% sampling is not feasible if someone is supposed to get the items back after we are done with them. The questions we are usually trying to answer are "What is the central tendency of the item of interest?" and "How much dispersion about this central tendency can we expect?" Simply, the average or averages that can be compared are measures of central tendency or central location of the data.

7.3   SYMBOLS, SUBSCRIPTS, BASIC STATISTICAL TERMS, AND CALCULATIONS

In statistics, *symbols* such as X, Y, and Z are used to represent different sets of data. Hence, if we have data for five companies, we might let

X = company income
Y = company materials expenditures
Z = company savings

Subscripts are used to represent individual observations within these sets of data. Thus, X_i represents the income of the *i*th company, where *i* takes on the values 1, 2, 3, 4, and 5. Using this notation, X_1, X_2, X_3, X_4, and X_5 stand for the incomes of the first company, the second company, and so on. The data are arranged in some order, such as by size of income, the order in which the data were gathered, or any other way suitable to the purposes or convenience of the investigator.

The subscript *i* is a variable used to index the individual data observations; therefore, X_i, Y_i, and Z_i represent the income, materials expenditures, and savings of the *i*th company. For example, X_2 represents the income of the second company, Y_2 the materials expenditures of the second company, and Z_5 the savings of the fifth company.

Suppose that we have data for two different samples: the net worths of 100 companies and the test scores of 30 students. To refer to individual observations in these samples, we can let X_i denote the net worth of the *i*th company, where *i* assumes values from 1 to 100 (as indicated by the notation *i* = 1, 2, 3, ..., 100.) We can also let Y_j denote the test score of the *j*th student, where *j* = 1, 2, 3, ..., 20. The different subscript letters make it clear that different sample are involved. Letters such as *X, Y,* and *Z* generally represent the different variables or types of measurements involved, whereas subscripts such as *i, j, k,* and *l* designate individual observations (Hamburg, 1987).

Next, we turn our attention to the method of expressing summations of sets of data. Suppose we want to add a set of four observations, denoted X_1, X_2, X_3, and X_4. A convenient way of designating this addition is

$$\sum_{i=1}^{4} X_i = X_1 + X_2 + X_3 + X_4$$

where the symbol Σ (Greek capital "sigma") means the "sum of." Thus, the following

$$\sum_{i=1}^{4} X_i$$

is read "the sum of the *X* values going from 1 to 4." For example, if $X_1 = 5$, $X_2 = 1$, $X_3 = 8$, and $X_4 = 6$, then

$$\sum_{i=1}^{4} X_i = 5 + 1 + 8 + 6 = 20$$

In general, if there are *n* observations, we write

$$\sum_{i=1}^{n} X_i = X_1 + X_2 + X_3 + \ldots + X_n$$

Basic statistical terms include mean or average, median, mode, and range. The following is an explanation of each of these terms.

7.3.1 MEAN

Mean is one of the most familiar and commonly estimated population parameters. It is the total of the values of a set of observations divided by the number of observations. Given a random sample, the population mean is estimated by

$$\bar{X} = \frac{\sum_{i=1}^{n} X_i}{n}$$

where X_i is the observed value of the ith unit in the sample, n is the number of units in the sample, and

$$\sum_{i=1}^{n} X_1$$

means to sum up all n of the X values in the sample.

If there are N units in the population, the total of the X values over all units in the population would be estimated by

$$\hat{T} = N\bar{X}$$

The circumflex (^) over the T is frequently used to indicate an estimated value as opposed to the true but unknown population value. It should be noted that this estimate of the mean is used for a simple random sample. It may not be appropriate if the units included in the sample are not selected entirely at random.

7.3.2 MEDIAN

The median is the value of the central item when the data are arrayed in size.

7.3.3 MODE

The mode is the observation that occurs with the greatest frequency and thus is the most "fashionable" value.

7.3.4 RANGE

The range is the difference between the values of the highest and lowest terms.

■ EXAMPLE 7.1

Problem: Given the following laboratory results for the measurement of dissolved oxygen (DO) in water, find the mean, mode, median, and range.

6.5 mg/L, 6.4 mg/L, 7.0 mg/L, 6.9 mg/L, 7.0 mg/L

Solution: To find the mean:

$$\bar{X} = \frac{\sum_{i=1}^{n} X_i}{n} = \frac{6.5 \text{ mg/L} + 6.4 \text{ mg/L} + 7.0 \text{ mg/L} + 6.0 \text{ mg/L} + 7.0 \text{ mg/L}}{5} = 6.58 \text{ mg/L}$$

The mode is 7.0 mg/L, the number that appears most often.

Now arrange the measurements in order:

6.4 mg/L, 6.5 mg/L, 6.9 mg/L, 7.0 mg/L, 7.0 mg/L.

The median is 6.9 mg/L, the central value, and the range is 0.6 mg/L (7.0 mg/L – 6.4 mg/L).

The importance of using statistically valid sampling methods cannot be overemphasized. Several different methodologies are available. A careful review of these methods (with an emphasis on designing appropriate sampling procedures) should be made before computing analytic results. Using appropriate sampling procedures along with careful sampling techniques will provide basic data that are accurate. The need for statistics in environmental practice is driven by the discipline itself. Environmental studies often deal with entities that are variable. If there were no variation in collected data, then there would be no need for statistical methods.

Over a given time interval there will always be some variation in sampling analyses. Usually, the average and the range yield the most useful information. For example, in evaluating the indoor air quality (IAQ) in a factory, a monthly summary of air-flow measurements, operational data, and laboratory tests for the factory would be used. Another example is when a work center or organization evaluates its monthly on-the-job reports of accidents and illnesses, where a monthly summary of reported injuries, lost-time incidents, and work-caused illnesses would be used.

In the preceding section, we used the term *sample* and the scenario sampling to illustrate the use and definition of mean, mode, median, and range. Though these terms are part of the common terminology used in statistics, the term *sample* in statistics has its own unique meaning. There is a difference between the term *sample* and the term *population*. In statistics, we most often obtain data from a sample and use the results from the sample to describe an entire population. The population of a sample signifies that one has measured a characteristic for everyone or everything that belongs to a particular group. For example, if one wishes to measure that characteristic of the population defined as environmental professionals, one would have to obtain a measure of that characteristic for every environmental professional possible. Measuring a population is difficult, if not impossible.

We use the term *subject* or *case* to refer to a member of a population or sample. There are statistical methods for determining how many cases must be selected in order to have a credible study. *Data*, another important term, are the measurements taken for the purposes of statistical analysis. Data can be classified as either *qualitative* or *quantitative*. Qualitative data deal with characteristics of the individual or subject (e.g., gender of a person or the color of a car), whereas quantitative data describe a characteristic in terms of a number (e.g., the age of a horse or the number of lost-time injuries an organization had over the previous year). Along with common terminology, the field of statistics also generally uses some common symbols. Statistical notation uses Greek letters and algebraic symbols to convey meaning about the procedures that one should follow to complete a particular study or test. Greek letters are used as statistical notation for a population, while English letters are used for statistical notation for a sample. Table 7.1 summarizes some of the more common statistical symbols, terms and procedures used in statistical operations.

TABLE 7.1

Commonly Used Statistical Symbols and Procedures

Term or Procedure	Population Symbol	Sample Notation
Mean	$\bar{\mu}$	$\bar{x}$
Standard deviation	σ	s
Variance	σ^2	s^2
Number of cases	N	n
Raw umber or value	X	x
Correlation coefficient	R	r

Procedure	Symbol
Sum of	Σ
Absolute value of x	$|x|$
Factorial of n	$n!$

7.4 DISTRIBUTION

An environmental professional conducting a research study collects data, and a group of raw data is obtained, but to make sense out of the data they must be organized into a meaningful format. The formatting begins by putting the data into some logical order, then grouping the data. Before the data can be compared to other data it must be organized. Organized data are referred to *distributions*. When confronted with masses of ungrouped data (listings of individual values), it is difficult to generalize about the information the masses contain. However, if a frequency distribution of the figures is formed, then many features become readily discernible. A frequency distribution records the number of cases that fall into each class of the data.

■ EXAMPLE 7.2

Problem: An environmental health and safety professional gathered data on the medical costs of 24 on-the-job injury claims for a given year. The raw data collected are shown in the table below:

$60	$1500	$85	$120
$110	$150	$110	$340
$2000	$3000	$550	$560
$4500	$85	$2300	$200
$120	$880	$1200	$150
$650	$220	$150	$4600

Solution: To develop a frequency distribution, the investigator took the values of the claims and placed them in order. Then the investigator counted the frequency of occurrences for each value as shown in Table 7.2. In order to develop a frequency distribution, groupings were formed using the values in Table 7.2, ensuring that each group had an equal range. The safety engineer grouped the data into ranges of 1000. The lowest range and highest range were determined by the data. Because it was decided to group by thousands, values fell in the ranges of $0 to $4999, and the distribution ended with this. The frequency distribution for the data appears in Table 7.3.

7.4.1 NORMAL DISTRIBUTION

When large amounts of data are collected on certain characteristics, the data and subsequent frequency can follow a distribution that is bell shaped in nature—the *normal distribution*. Normal distributions are a very important class of statistical distribution. As stated, all normal distributions are symmetric and have bell-shaped curves with a single peak (see Figure 7.1).

TABLE 7.2
Value and Frequency of Claims

Value	Frequency	Value	Frequency
$60	1	$650	1
$85	2	$880	1
$110	2	$1200	1
$120	2	$1500	1
$150	3	$2000	1
$200	1	$2300	1
$220	1	$3000	1
$340	1	$4500	1
$550	1	$4600	1
$560	1	Total	24

TABLE 7.3
Frequency Distribution

Range	Frequency
$0–$999	17
$1000–1999	2
$2000–2999	2
$3000–3999	1
$4000–4999	2
Total	24

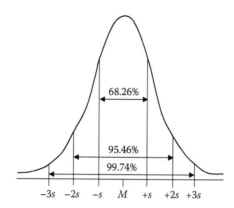

FIGURE 7.1 Normal distribution curve showing the frequency of a measurement.

To speak specifically of any normal distribution, two quantities have to be specified: the mean μ (pronounced "mu"), where the peak of the density occurs, and the standard deviation σ (sigma). Different values of μ and σ yield different normal density curves and hence different normal distributions. Although there are many normal curves, they all share an important property that allows us to treat them in a uniform fashion. All normal density curves satisfy the following property, which is often referred to as the *empirical rule*:

- 68% of the observations fall within 1 standard deviation of the mean; that is, between μ − σ and μ + σ.
- 95% of the observations fall within 2 standard deviations of the mean; that is, between μ − 2σ and μ + 2σ.
- 98% of the observations fall within 3 standard deviations of the mean; that is, between μ − 3σ and μ + 3σ.

Thus, for a normal distribution, almost all values lie within 3 standard deviations of the mean (see Figure 7.1). It is important to stress that the rule applies to all normal distributions. Also remember that it applies *only* to normal distributions.

Note: Before applying the empirical rule it is a good idea to identify the data being described and the value of the mean and standard deviation. A sketch of a graph summarizing the information provided by the empirical rule should also be made.

■ EXAMPLE 7.3

Problem: The scores for all high school seniors taking the math section of the Scholastic Aptitude Test (SAT) in a particular year had a mean of 490 (μ = 490) and a standard deviation of 100 (σ = 100). The distribution of SAT scores is bell shaped.

1. What percentage of seniors scored between 390 and 590 on this SAT test?
2. One student scored 795 on this test. How did this student do compared to the rest of the scores?
3. A rather exclusive university admits only students who received among the highest 16% of the scores on this test. What score would a student need on this test to be qualified for admittance to this university?

The data being described are the math SAT scores for all seniors taking the test in one year. A bell-shaped curve summarizing the percentages given by the empirical rule is shown in Figure 7.2.

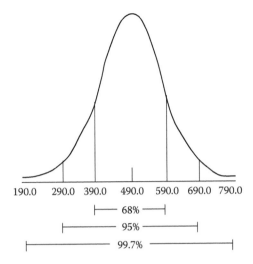

190.0   290.0   390.0   490.0   590.0 690.0 790.0

├─── 68% ───┤

├─────── 95% ───────┤

├────────── 99.7% ──────────┤

FIGURE 7.2   Sample Scholastic Aptitude Test (SAT) math percentages given by the empirical rule.

Solution:

1. From Figure 7.2, about 68% of seniors scored between 390 and 590 on this SAT test.
2. Because about 99.7% of the scores are between 190 and 790, a score of 795 is excellent. This is one of the highest scores on this test.
3. Because about 68% of the scores are between 390 and 590, this leaves 32% of the scores outside the interval. Because a bell-shaped curve is symmetric, one-half of these scores, or 16%, are on each end of the distribution.

7.5   STANDARD DEVIATION

The standard deviation, s or σ (sigma), is often used as an indicator of precision. The standard deviation is a measure of the variation (spread) in the set of observations; that is, it gives us some idea as to whether most of the individuals in a population are close to the mean or spread out. In order to a gain better understanding of the benefits derived from using statistical methods in safety engineering, it is appropriate to consider some of the basic theory of statistics. In any set of data, the true value (mean) will lie in the middle of all of the measurements taken. This is true, providing the sample size is large and only random error is present in the analysis. In addition, the measurements will show a normal distribution as shown in Figure 7.1. In Figure 7.1, 68.26% of the results fall between $M + s$ and $M - s$, 95.46% of the results lie between $M + 2s$ and $M - 2s$, and 99.74% of the results lie between $M + 3s$ and $M - 3s$. Therefore, if they are precise, then 68.26% of all the measurements should fall between the true value estimated by the mean, plus the standard deviation and the true value minus the standard deviation. The following equation is used to calculate the sample standard deviation:

$$s = \sqrt{\frac{\sum (X - \bar{X})^2}{n - 1}}$$

where
  s = Standard deviation.
  Σ = Means to sum the values from X to X_n.
  X = Measurements from X to X_n.
  $\bar{X}$ = Mean.
  n = Number of samples.

TABLE 7.4

Calculations for Example 7.4

X	$X - \bar{X}$	$(X - \bar{X})^2$
9.5	−0.5	0.25
10.5	0.5	0.25
10.1	0.1	0.01
9.9	−0.1	0.01
10.6	0.6	0.36
9.5	−0.5	0.25
11.5	1.5	2.25
9.5	−0.5	0.25
10.0	0	0
9.4	−0.6	0.36
		3.99

■ EXAMPLE 7.4

Problem: Calculate the standard deviation (σ) of the following dissolved oxygen values:

$$9.5, 10.5, 10.1, 9.9, 10.6, 9.5, 11.5, 9.5, 10.0, 9.4$$

$$\bar{X} = 10.0$$

Solution: See Table 7.4.

$$\sigma = \sqrt{\frac{\sum (X - \bar{X})^2}{n-1}} = \sqrt{\frac{3.99}{10-1}} = 0.67$$

7.6 COEFFICIENT OF VARIATION

In nature, populations with large means often show more variation than populations with small means. The coefficient of variation (C) facilitates comparison of variability in different sized means. It is the ratio of the standard deviation to the mean. A standard deviation of 2 for a mean coefficient of variation would be 0.20 or 20% in each case. If we have a standard deviation of 1.414 and a mean of 9.0, the coefficient of variation would be estimated by

$$C = \frac{s}{\bar{X}} = \frac{1.414}{9.0} = 0.157, \text{ or } 15.7\%$$

7.7 STANDARD ERROR OF THE MEAN

There is usually variation among the individual units of a population. Again, the standard deviation is a measure of this variation. Because the individual units vary, variation may also exist among the means (or any other estimates) computed from samples of these units. Take, for example, a population with a true mean of 10. If we were to select four units at random, they might have a sample mean of 8. Another sample of four units from the same population might have a mean of 11, another 10.5, and so forth. Clearly it would be desirable to know the variation likely to be encountered among the means of samples from this population. A measure of the variation among sample means is the standard error of the mean. It can be thought of as a standard deviation among sample means;

it is a measure of the variation among sample means, just as the standard deviation is a measure of the variation among individuals. The standard error of the mean may be used to compute confidence limits for a population mean.

The computation of the standard error of the mean (often symbolized by $s_{\bar{x}}$) depends on the manner in which the sample was selected. For simple random sampling without replacement (i.e., a given unit cannot appear in the sample more than once) from a population having a total of N units the formula for the estimated standard error of the mean is

$$s_{\bar{x}} = \sqrt{\frac{s^2}{n}\left(1 - \frac{n}{N}\right)}$$

In a forestry example, if we had $n = 10$ and found that $s = 1.414$ and $s^2 = 2$ in the population that contains 1000 trees, then the estimated mean diameter ($\bar{X} = 9.0$) would have a standard error of

$$s_{\bar{x}} = \sqrt{\frac{2}{10}\left(1 - \frac{10}{1000}\right)} = \sqrt{0.198} = 0.445$$

Note: The term $(1 - n/N)$ is called the *finite population correction*, or fpc. The fpc is used when the sampling fraction (the number of elements or respondents sampled relative to the population) becomes large. The fpc is used in the calculation of the standard error of the estimate. If the value of the fpc is close to 1, it will have little impact and can be safely ignored.

7.8   COVARIANCE

Very often, each unit of a population will have more than a single characteristic. In forestry practice, for example, trees may be characterized by their height, diameter, and form class (amount of taper). The covariance is a measure of the association between the magnitudes of two characteristics. If there is little or no association, the covariance will be close to zero. If the large values of one characteristic tend to be associated with the small values of another characteristic, the covariance will be negative. If the large values of one characteristic tend to be associated with the large values of another characteristic, the covariance will be positive. The population covariance of X and Y is often symbolized by σ_{xy}; the sample estimate by s_{xy}.

Let's return to a forestry practice example. Suppose that the diameter (inches) and age (years) have been obtained for a number of randomly selected trees. If we symbolize diameter by Y and age by X, then the sample covariance of diameter and age is given by

$$s_{xy} = \frac{\sum XY - \frac{\left(\sum X\right)\left(\sum Y\right)}{N}}{n - 1}$$

This is equivalent to the formula

$$s_{xy} = \frac{\sum (X - \bar{X})(Y - \bar{Y})}{n - 1}$$

If $n = 12$ and the Y and X values were as follows:

$$Y = 4 + 9 + 7 + 7 + 5 + 10 + 9 + 6 + 8 + 6 + 4 + 11 = 86$$

$$X = 20 + 40 + 30 + 45 + 25 + 45 + 30 + 40 + 20 + 35 + 25 + 40 = 395$$

DID YOU KNOW?

The computed value of a statistic such as the correlation efficient depends on which particular units were selected for the sample. Such estimates will vary from sample to sample. More important, they will usually vary from the population value which we try to estimate.

Then

$$s_{xy} = \frac{(4)(20)+(9)(40)+...+(11)(40)-\dfrac{(86)(395)}{12}}{12-1} = \frac{2960-2830.83}{11} = 11.74$$

The positive covariance is consistent with the well-known and economically unfortunate fact that larger diameters tend to be associated with older ages.

7.9 SIMPLE CORRELATION COEFFICIENT

The magnitude of the covariance, like that of the standard deviation, is often related to the size of the variables themselves. Units with large X and Y values tend to have larger covariances than units with small X and Y values. Also, the magnitude of the covariance depends on the scale of measurement; in the previous example, if the diameter had been expressed in millimeters instead of inches, the covariance would have been 298.196 instead of 11.74. The simple correlation coefficient, a measure of the degree of linear association between two variables, is free of the effects of scale of measurement. It can vary from between −1 and +1. A correlation of 0 indicates that there is no linear association (although there may be a very strong nonlinear association). A correlation of +1 or −1 would suggest a perfect linear association. As for the covariance, a positive correlation implies that the large values of X are associated with the large values of Y. If the large values of X are associated with the small values of Y, then the correlation is negative.

The population correlation coefficient is commonly symbolized by ρ (rho) and the sample-based estimate r. The population correlation coefficient is defined to be

$$\rho = \frac{\text{Covariance of } X \text{ and } Y}{\sqrt{(\text{Variance of } X)(\text{Variance of } Y)}}$$

For a simple random sample, the sample correlation coefficient is computed as follows:

$$r = \frac{s_{xy}}{s_x \cdot s_y} = \frac{\sum xy}{\sqrt{\left(\sum x^2\right)\left(\sum y^2\right)}}$$

where
s_{xy} = Sample covariance of X and Y.
s_x = Sample standard deviation of X.
s_y = Sample standard deviation of Y.
$\sum xy$ = Corrected sum of XY products:

$$\sum XY - \frac{\left(\sum X\right)\left(\sum Y\right)}{n}$$

$\sum x^2 =$ Corrected sum of squares for X:

$$\sum X^2 - \frac{\left(\sum X\right)^2}{n}$$

$\sum y^2 =$ Corrected sum of squares for Y:

$$\sum Y^2 - \frac{\left(\sum Y\right)^2}{n}$$

For the values used to illustrate the covariance we have:

$$\sum xy = (4)(20) + (9)(40) + \ldots + (11)(40) - \frac{(86)(395)}{12} = 129.1667$$

$$\sum y^2 = 4^2 + 9^2 + \ldots + 11^2 - \frac{(86)^2}{12} = 57.6667$$

$$\sum x^2 = 20^2 + 40^2 + \ldots + 40^2 - \frac{(395)^2}{12} = 922.9167$$

So:

$$r = \frac{129.1667}{\sqrt{(57.6667)(922.9167)}} = \frac{129.1667}{230.6980} = 0.56$$

7.10   VARIANCE OF A LINEAR FUNCTION

Routinely we combine variables or population estimates in a linear function. For example, if the mean timber volume per acre has been estimated as $\overline{X}$, then the total volume on M acres with be $M\overline{X}$; the estimate of total volume is a linear function of the estimated mean volume. If the estimate of cubic volume per acre in sawtimber is $\overline{X}_1$ and that of pulpwood above the sawtimber top is $\overline{X}_2$, then the estimate of total cubic foot volume per acre is $\overline{X}_1 + \overline{X}_2$. If on a given tract the mean volume per half-acre is $\overline{X}_1$ for spruce and the mean volume per quarter-acre is $\overline{X}_2$ for yellow birch, then the estimated total volume per acre of spruce and birch would be $2\overline{X} + 4\overline{X}_2$. In general terms, a linear function of three variables (say X_1, X_2, and X_3) can be written as

$$L = a_1 X_1 + a_2 X_2 + a_3 X_3$$

where a_1, a_2, and a_3 are constants.

If the variances are s_1^2, s_2^2, and s_3^2 (for X_1, X_2 and X_3, respectively) and the covariances are $s_{1,2}$, $s_{1,3}$, and $s_{2,3}$, then the variance of L is given by

$$s_L^2 = a_1^2 s_1^2 + a_2^2 s_2^2 + a_3^2 s_3^2 + 2\left(a_1 a_2 s_{1,2} + a_1 a_3 s_{1,3} + a_2 a_3 s_{2,3}\right)$$

The standard deviation (or standard error) of L is simply the square root of this. The extension of the rule to cover any number of variables should be fairly obvious.

■ **EXAMPLE 7.5**

Problem: The sample mean volume per forested acre for a 10,000-acre tract is $X = 5680$ board feet with a standard error of $s_{\bar{x}} = 632$ (so $s_{\bar{x}}^2 = 399{,}424$). The estimated total volume is

$$L = 10{,}000(\bar{X}) = 56{,}800{,}000 \text{ board feet}$$

The variance of this estimate would be

$$s_L^2 = (10{,}000)^2\left(s_{\bar{x}}^2\right) = 39{,}942{,}400{,}000{,}000$$

Since the standard error of an estimate is the square root of its variance, the standard error of the estimated total is

$$s_L = \sqrt{s_L^2} = 6{,}320{,}000$$

■ **EXAMPLE 7.6**

Problem: In 1995, a random sample of 40 1/4-acre circular plots was used to estimate the cubic foot volume of a stand of pine. Plot centers were monumented for possible relocation at a later time. The mean volume per plot was $\bar{X}_1 = 225$ ft^3. The plot variance was $s_{x1}^2 = 8281$ so that the variance of the mean was $s_{\bar{x}1}^2 = 8281/40 = 207.025$. In 2000, a second inventory was made using the same plot centers. This time, however, the circular plots were only 1/10 acre. The mean volume per plot was $\bar{X}_2 = 122$ ft^3. The plot variance was $s_{x2}^2 = 6084$, so the variance of the mean was $s_{\bar{x}2}^2 = 152.100$. The covariance of initial and final plot volumes was $s_{x1,2} = 4259$, making the covariance of the means $s_{\bar{x}1,\bar{x}2} = 4259/40 = 106.475$.

Solution: The net periodic growth per acre would be estimated as

$$G = 10\bar{X}_2 - 4\bar{X}_1 = 10(122) - 4(225) = 320 \text{ ft}^3/\text{acre}$$

By the rule for linear functions the variance of G would be

$$s_G^2 = (10)^2 s_{\bar{x}2}^2 + (-4)^2 s_{\bar{x}1}^2 + 2(10)(-4)s_{\bar{x}1,\bar{x}2}$$

$$= 100(152.100) + 16(207.025) - 80(106.475)$$

$$= 10{,}004.4$$

In this example there was a statistical relationship between the 2000 and 1995 means because the same plot locations were used in both samples. The covariance of the means ($s_{\bar{x}1,\bar{x}2}$) is a measure of this relationship. If the 2000 plots had been located at random rather than at the 1995 locations, the two means would have been considered statistically independent and their covariance would have been set at zero. In this case the equation for the variance of the net periodic growth per acre (G) would reduce to

$$s_G^2 = (10)^2 s_{\bar{x}2}^2 + (-4)^2 s_{\bar{x}1}^2 = 100(152.100) + 16(207.025) = 18{,}522.40$$

7.11 SAMPLING MEASUREMENT VARIABLES

7.11.1 SIMPLE RANDOM SAMPLING

Most environmental practitioners are familiar with *simple random sampling*. As in any sampling system, the aim is to estimate some characteristic of a population without measuring all of the population units. In a simple random sample of size n, the units are selected so that every possible combination of n units has an equal chance of being selected. If sampling is without replacement, then at any stage of the sampling each unused unit should have an equal chance of being selected.

7.11.1.1 Sample Estimates of the Population Mean and Total

From a population of $N = 100$ units, $n = 20$ units were selected at random and measured. Sampling was without replacement—once a unit had been included in the sample it could not be selected again. The unit values were

$$10 \quad 9 \quad 10 \quad 9 \quad 11 \quad 16 \quad 11 \quad 7 \quad 12 \quad 12 \quad 11 \quad 3 \quad 5 \quad 11 \quad 14 \quad 8 \quad 13 \quad 12 \quad 20 \quad 10$$

Sum of all 20 random units = 214

From this sample we estimate the population mean as

$$\bar{X} = \frac{\sum X}{n} = \frac{214}{20} = 10.7$$

A population of $N = 100$ units having a mean of 10.7 would then have an estimated total of

$$\hat{T} = N\bar{X} = 100(10.7) = 1070$$

7.11.1.2 Standard Errors

The first step in calculating a standard error is to obtain an estimate of the population variance (σ^2) or standard deviation (σ). As noted in a previous section, the standard deviation for a simple random sample (like our example here) is estimated by

$$s = \sqrt{\frac{\sum X^2 - \frac{\left(\sum X\right)^2}{n}}{n-1}} = \sqrt{\frac{10^2 + 16^2 + \ldots + 10^2 - \frac{214^2}{20}}{19}} = \sqrt{13.4842} = 3.672$$

For sampling without replacement, the standard error of the mean is

$$s_{\bar{x}} = \sqrt{\frac{s^2}{n}\left(1 - \frac{n}{N}\right)} = \sqrt{\frac{13.4842}{20}\left(1 - \frac{20}{100}\right)} = \sqrt{0.539368} = 0.734$$

From the formula for the variance of a linear function we can find the variance of the estimated total:

$$s_{\hat{T}}^2 = N^2 s_{\bar{x}}^2$$

The standard error of the estimated total is the square root of this, or

$$s_{\hat{T}} = N s_{\bar{x}} = 100(0.734) = 73.4$$

7.11.1.3 Confidence Limits

Sample estimates are subject to variation. How much they vary depends primarily on the inherent variability of the population (Var^2) and on the size of the sample (n) and of the population (N). The statistical way of indicating the reliability of an estimate is to establish confidence limits. For estimates made from normally distributed populations, the confidence limits are given by

$$\text{Estimate} \pm t \text{ (standard error)}$$

For setting confidence limits on the mean and total we already have everything we need except for the value of t, and that can be obtained from a table of the t distribution.

In the previous example, the sample of $n = 20$ units had a mean of $\overline{X} = 10.7$ and a standard error of $s_{\overline{x}} = 0.734$. For 95% confidence limits on the mean we would use a t value (from a t table) of 0.05 and (also from a t table) 19 degrees of freedom. As $t_{0.05} = 2.093$, the confidence limits are given by

$$\overline{X} \pm (t)(s_{\overline{x}}) = 10.7 \pm (2.093)(0.734) = 9.16 \text{ to } 12.24$$

This says that, unless a 1-in-20 chance has occurred in sampling, the population mean is somewhere between 9.16 and 12.24. It does not say where the mean of future samples from this population might fall, nor does it say where the mean may be if mistakes have been made in the measurements.

For 99% confidence limits, we find $t_{0.01} = 2.861$ (with 19 degrees of freedom), so the limits are

$$10.7 \pm (2.861)(0.734) = 8.6 \text{ to } 12.8$$

These limits are wider, but they are more likely to include the true population mean. For the population total the confidence limits are

95% limits = $1070 \pm (2.093)(73.4) = 916$ to 1224
99% limits = $1070 \pm (2.861)(73.4) = 860$ to 1280

For large samples ($n > 60$), the 95% limits are closely approximated by

$$\text{Estimate} \pm 2 \text{ (standard error)}$$

and the 99% limits by

$$\text{Estimate} \pm 2.6 \text{ (standard error)}$$

7.11.1.4 Sample Size

Samples cost money. So do errors. The aim in planning a survey should be to take enough observations to obtain the desired precision—no more, no less. The number of observations needed in a simple random sample will depend on the precision desired and the inherent variability of the population being sampled. Because sampling precision is often expressed in terms of the confidence interval on the mean, it is not unreasonable in planning a survey to say that in the computed confidence interval

$$\overline{X} \pm t s_{\overline{x}}$$

we would like to have the $ts_{\overline{x}}$ equal to or less than some specified value E, unless a 1-in-20 (or 1-in-100) chance has occurred in the sample. That is, we want

$$ts_{\overline{x}} = E$$

or, because

$$s_{\bar{x}} = \frac{s}{\sqrt{n}}$$

we want

$$t\left(\frac{s}{\sqrt{n}}\right) = E$$

Solving this for n gives the desired sample size:

$$n = \frac{t^2 s^2}{E^2}$$

To apply this equation we need to have an estimate (s^2) of the population variance and a value for Student's t at the appropriate level of probability. The variance estimate can be a real problem. One solution is to make the sample survey in two stages. In the first state, n_1 random observations are made and from these an estimate of the variance (s^2) is computed. This value is then plugged into the sample size equation:

$$n = \frac{t^2 s^2}{E^2}$$

where t has $n_1 - 1$ degrees of freedom and is selected from the appropriate table. The computed value of n is the total size of sample needed. As we have already observed n_1 units, this means that we will have to observed $(n - n_1)$ additional units.

   If pre-sampling as described above is not feasible then it will be necessary to make a guess at the variance. Assuming our knowledge of the population is such that the guessed variance (s^2) can be considered fairly reliable, then the size of sample (n) needed to estimate the mean to within $\pm E$ units is approximately

$$n = \frac{4s^2}{E^2}$$

for 95% confidence and

$$n = \frac{20s^2}{3E^2}$$

for 99% confidence.

Less reliable variance estimates could be doubled (as a safety factor) before applying these equations. In many cases, the variance estimate may be so poor as to make the sample size computation just so much statistical window dressing.

When sampling is without replacement (as it is in most forest sampling situations) the sample size estimates given above apply to populations with an extremely large number (N) of units so that the sampling fraction (n/N) is very small. If the sampling fraction is not small (say $n/N = 0.05$), then the sample size estimates should be adjusted. This adjusted value of n is

$$n_a = \frac{n}{1 + \dfrac{n}{N}}$$

Suppose that we plan to use quarter-acre plots in a survey and estimate the variance among plot volumes to be $s^2 = 160{,}000$. If the error limit is $E = 5000$ feet per acre, we must convert the variance to an acre basis or the error to a quarter-acre basis. To convert a quarter-acre volume to an acre basis we multiply by 4, and to convert a quarter-acre variance to an acre variance we multiply by 16. Thus, the variance would be 2,560,000 and the sample-size formula would be

$$n = \frac{t^2(2{,}560{,}000)}{(500)^2} = t^2(10.24)$$

Alternatively, we can leave the variance alone and convert the error statement from an acre to a quarter-acre basis ($E = 125$). Then, the sample-size formula is

$$n = \frac{t^2(160{,}000)}{(125)^2} = t^2(10.24), \text{ as before}$$

7.11.2 STRATIFIED RANDOM SAMPLING

In *stratified random sampling*, a population is divided into subpopulations (strata) of known size, and a simple random sample of at least two units is selected in each subpopulation. This approach has several advantages. For one thing, if there is more variation between subpopulations than within them, the estimate of the population mean will be more precise than that given by a simple random sample of the same size. Also, it may be desirable to have separate estimates for each subpopulation (e.g., in timber types or administrative subunits). In addition, it may be administratively more efficient to sample by subpopulations.

■ EXAMPLE 7.7

Problem: A 500-acre forested area was divided into three strata on the basis of timber type. A simple random sample of 0.2-acre plots was taken in each stratum, and the means, variances, and standard errors were computed by the formulas for a simple random sample. These results, along with the size (N_h) of each stratum (expressed in number of 0.1-acre plots), are shown in Table 7.5.

Solution: The squared standard error of the mean for stratum h is computed by the formula given for the simple random sample

$$s_{\bar{x}_h}^2 = \frac{s_h^2}{n_h}\left(1 - \frac{n_h}{N_h}\right)$$

TABLE 7.5
Data for Example 7.7

Type	Stratum Number (h)	Stratum Size (N_h)	Sample Size (n_h)	Stratum Mean (X_h)	Within-Stratum Variance (s_h^2)	Squared Standard Error of the Mean ($s_{\bar{x}_h}^2$)
Pine	1	1350	30	251	10,860	353.96
Upland hardwoods	2	700	15	164	9680	631.50
Bottom-land hardwoods	3	450	10	110	3020	265.29
Sum		2500				

Thus, for stratum 1 (pine type),

$$s_{\bar{x}}^2 = \frac{10,860}{30}\left(1 - \frac{30}{1350}\right) = 353.96$$

Where the sampling fraction (n_h/N_h) is small, the fpc can be omitted.
With these data, the population mean is estimated by

$$\bar{X}_{st} = \sum \frac{N_h \bar{X}_h}{N}$$

where $N = \Sigma N_h$.
For this example we have

$$\bar{X}_{st} = \frac{N_1 \bar{X}_1 + N_2 \bar{X}_2 + N_3 \bar{X}_4}{N} = \frac{(1350)(251) + (700)(164) + (450)(110)}{2500} = 201.26$$

The formula for the standard error of the stratified mean is cumbersome but not complicated:

$$s_{\bar{x}_{st}} = \sqrt{\frac{1}{N^2}\left[\sum N_h^2 s_{\bar{x}_h}^2\right]} = \sqrt{\frac{(1350)^2(353.96) + (700)^2(631.50) + (450)^2(295.29)}{(2500)^2}} = 12.74$$

If the sample size is fairly large, the confidence limits on the mean are given by

$$95\% \text{ confidence limits} = X_{st} \pm 2s_{\bar{x}_{st}}$$
$$99\% \text{ confidence limits} = X_{st} \pm 2.6s_{\bar{x}_{st}}$$

There is no simple way of compiling the confidence limits for small samples.

7.11.2.1 Sample Allocation

If a sample of n units is taken, how many units should be selected in each stratum? Among several possibilities, the most common procedure is to allocate the sample in proportion to the size of the stratum; in a stratum having 2/5 of the units of the population, we would take 2/5 of the samples. In the population discussed in the previous example, the proportional allocation of the 55 sample units was as follows:

Stratum	Relative Size (N_h/N)	Sample Allocation
1	0.54	29.7 or 30
2	0.28	15.4 or 15
3	0.18	9.9 or 10
Sums	1.00	55

Some other possibilities are equal allocation, allocation proportional to estimated value, and optimum allocation. In optimum allocation, an attempt is made to get the smallest standard error possible for a sample of n units. This is done by sampling more heavily in the state having a larger variation. The equation for optimum allocation is

$$n_h = \left(\frac{N_h s_h}{\sum N_h s_h} \right) n$$

Optimum allocation obviously requires estimates of the within-stratum variances—information that may be difficult to obtain. A refinement of optimum allocation is to take sampling cost differences into account and allocate the sample so as to get the most information per dollar. If the cost per sampling unit in stratum h is c_h, the equation is

$$n_h = \left[\frac{\dfrac{N_h s_h}{\sqrt{c_h}}}{\sum \left(\dfrac{N_h s_h}{\sqrt{c_h}} \right)} \right] n$$

7.11.2.2 Sample Size

To estimate the size of sample to take for a specified error at a given level of confidence, it is first necessary to decide on the method of allocation. Ordinarily, proportional allocation is the simplest and perhaps the best choice. With proportional allocation, the size of sample needed to be within $\pm E$ units of the true value at the 0.05 probability level can be approximated by

$$n = \frac{N \left(\sum N_h s_h^2 \right)}{\dfrac{N^2 E^2}{4} + \sum N_h s_h^2}$$

For the 0.01 probability level, use 6.76 in place of 4.

■ EXAMPLE 7.8

Problem: Assume that prior to sampling a 500-acre forest we decided to estimate the mean volume per acre to within ± 100 cubic feet per acre unless a 1-in-20 chance occurs in sampling. As we plan to sample with 0.2-acre plots, the error specification should be put on a 0.2-acre basis. Therefore, $E = 20$. From previous sampling, the stratum variances for 0.2-acre volumes are estimated to be

$$s_1^2 = 8000 \quad s_2^2 = 10,000 \quad s_3^2 = 5000$$

Therefore,

$$n = \frac{2500 \left[(1350)(8000) + (700)(10,000) + (450)(5000) \right]}{\dfrac{(2500)^2 (20)^2}{4} + \left[(1350)(8000) + (700)(10,000) + (450)(5000) \right]} = 77.7, \text{ or } 78$$

The 78 sample units would now be allocated to the strata by the formula

$$n_h = \left(\frac{N_h}{N}\right)n$$

giving $n_1 = 42$, $n_2 = 22$, and $n_3 = 14$.

7.12 SAMPLING—DISCRETE VARIABLES

7.12.1 RANDOM SAMPLING

The sampling methods discussed in the previous sections apply to data that are on a continuous or nearly continuous scale of measurement. These methods may not be applicable if each unit observed is classified as alive or dead, germinated or not germinated, infected or not infected. Data of this type may follow what is known as the binomial distribution. They require slightly different statistical techniques.

As an illustration, suppose that a sample of 1000 seeds was selected at random and tested for germination. If 480 of the seeds germinated, the estimated viability for the lot would be

$$\bar{p} = \frac{480}{1000} = 0.48, \text{ or } 48\%$$

For large samples (say, $n > 250$) with proportions greater than 0.20 but less than 0.80, approximate confidence limits can be obtained by first computing the standard error of $\bar{p}$ by the equation

$$s_{\bar{p}} = \sqrt{\frac{\bar{p}(1-\bar{p})}{(n-1)}\left(1-\frac{n}{N}\right)}$$

Then, the 95% confidence limits are given by

$$\bar{p} \pm \left[2(s_{\bar{p}}) + \frac{1}{2n}\right]$$

Applying this to the above example (and ignoring the fpc) we get

$$s_{\bar{p}} = \sqrt{\frac{(0.48)(0.52)}{999}} = 0.0158$$

And the 95% confidence interval is given by the following:

$$48 \pm \left[2(0.0158) + \frac{1}{2(1000)}\right] = 0.448 \text{ to } 0.512$$

The 99% confidence limits are approximated by

$$\bar{p} \pm \left(2.6s_{\bar{p}} + \frac{1}{2n}\right)$$

7.12.2 SAMPLE SIZE

An appropriate table can be used to estimate the number of units that would have to be observed in a simple random sample in order to estimate a population proportion with some specified precision. Suppose, for example, that we wanted to estimate the germination percent for a population to within ±10% (or 0.10) at the 95% confidence level. The first step is to guess about what the proportion of seed germination will be. If a good guess is not possible, then the safest course is to guess $\bar{p} = 0.59$, as this will give the maximum sample size.

Next, pick any of the sample sizes given in the appropriate table (e.g., 10, 15, 20, 30, 50, 100, 250, and 1000) and look at the confidence interval for the specified value of $\bar{p}$. Inspection of these limits will tell whether or not the precision will be met with a sample of this size or if a larger or smaller sample would be more appropriate.

Thus, if we guess $\bar{p} = 0.2$, then in a sample of $n = 50$ we would expect to observe $(0.2)(50) = 10$, and the table says that the 95% confidence limits on $\bar{p}$ would be 0.10 and 0.34. Since the upper limit is not within 0.10 of $\bar{p}$, a larger sample would be needed. For a sample of $n = 100$ the limits are 0.13 to 0.29. Because both of these values are within 0.10 of $\bar{p}$, a sample of 100 would be adequate.

If the table indicates the need for a sample of over 250, the size can be approximated by

$$n \approx \frac{4(\bar{p})(1-\bar{p})}{E^2}, \text{ for 95\% confidence}$$

or,

$$n \approx \frac{20(\bar{p})(1-\bar{p})}{3E^2}, \text{ for 99\% confidence}$$

where E is the precision with which $\bar{p}$ is to be estimated (expressed in same for as $\bar{p}$, either percent or decimal).

7.13   CLUSTER SAMPLING FOR ATTRIBUTES

Simple random sampling of discrete variables is often difficult or impractical. In estimating tree plantation survival, for example, we could select individual trees at random and examine them, but it wouldn't make much sense to walk down a row of planted trees in order to observe a single member of that row. It would usually be more reasonable to select rows at random and observe all of the trees in the selected row.

Seed viability is often estimated by randomly selecting several lots of 100 or 200 seeds each and recording for each lot the percentage of the seeds that germinate. These are examples of *cluster sampling*; the unit of observation is the cluster rather than the individual tree or single seed. The value attached to the unit is the proportion having a certain characteristic rather than the simple fact of having or not having that characteristic. If the clusters are large enough (say, over 100 individuals per cluster) and nearly equal in size, the statistical methods that have been described for measurement variables can often be applied. Thus, suppose that the germination percent of a seedlot is estimated by selecting $n = 10$ sets of 200 seed each and observing the germination percent for each set:

Set	1	2	3	4	5	6	7	8	9	10	Sum
Germination percent (p)	78.5	82.0	86.0	80.5	74.5	78.0	79.0	81.0	80.5	83.5	803.5

then the mean germination percent is estimated by

$$\bar{p} = \frac{\sum p}{n} = \frac{803.5}{10} = 80.35\%$$

The standard deviation of p is

$$s_p = \sqrt{\frac{\sum p^2 - \frac{\left(\sum p\right)^2}{n}}{(n-1)}} = \sqrt{\frac{(78.5)^2 + \ldots (83.5)^2 - \frac{(803.5)^2}{10}}{9}} = \sqrt{10.002778} = 3.163$$

And the standard error for $\bar{p}$ (ignoring the fpc) is

$$s_{\bar{p}} = \sqrt{\frac{s_p^2}{n}\left(1 - \frac{n}{N}\right)} = \sqrt{\frac{10.002778}{10}} = 1.000$$

Note that n and N in these equations refer to the number of clusters, not to the number of individuals. The 95% confidence interval, computed by the procedure for continuous variables is

$$p \pm (t_{0.05})(s_{\bar{p}}), \text{ where } t \text{ has } (n-1) = 9 \text{ degrees of freedom}$$

$$80.35 \pm (2.262)(1.000) = 78.1 \text{ to } 82.6$$

7.13.1 TRANSFORMATIONS

The above method of computing confidence limits assumes that the individual percentages follow something close to a normal distribution with homogeneous variance (i.e., same variance regardless of the size of the percent). If the clusters are small (say, less than 100 individuals per cluster) or some of the percentages are greater than 80 or less than 20, the assumptions may not be valid and the computed confidence limits will be unreliable. In such cases, it may be desirable to compute the transformation

$$y = \arcsin \sqrt{percent}$$

and to analyze the transformed variable.

7.14 CHI-SQUARE TESTS

7.14.1 TEST OF INDEPENDENCE

Individuals are often classified according to two (or more) distinct systems. A tree can be classified as to species and at the same time according to whether it is infected or not infected with some disease. A milacre plot can be classified as to whether or not it is stocked with adequate reproduction and whether it is shaded or not shaded. Given such a cross-classification, it may be desirable to know whether the classification of an individual according to one system is independent of its classification by the other system. In the species-infection classification, for example, independence of species and infection would be interpreted to mean that there is no difference in infection rate among species (i.e., infection rate does not depend on species).

The hypothesis that two or more systems of classification are independent can be tested by chi-square. The procedure can be illustrated by a test of three termite repellents. A batch of 1500 wooden stakes was divided at random into three groups of 500 each, and each group received a different termite-repellent treatment. The treated stakes were driven into the ground, with the treatment at any particular stake location being selected at random. Two years later the stakes were examined for termites. The number of stakes in each classification is shown in the following 2×3 (two rows by three columns) contingency table:

	Group I	Group II	Group III	Subtotals
Attacked by termites	193	148	210	551
Not attacked	307	352	390	949
Subtotals	500	500	500	1500

If the data in the table can be symbolized as shown below:

	Group I	Group II	Group III	Subtotals
Attacked by termites	a_1	a_2	a_3	A
Not attacked	b_1	b_2	b_3	B
Subtotals	T_1	T_2	T_3	G

then the test of independence is made by computing

$$\chi^2 = \frac{1}{(A)(B)} \sum_{i=1}^{3} \left(\frac{(a_i B - b_i A)^2}{T_i} \right)$$

$$= \frac{1}{(551)(949)} \left[\frac{((193)(949) - (307)(551))^2}{500} + \ldots + \frac{((210)(949) - (290)(551))^2}{500} \right]$$

$$= 17.66$$

The result is compared to the appropriate tabular accumulative distribution of chi-square values of χ^2 with $(c-1)$ degrees of freedom, where c is the number of columns in the table of data. If the computed value exceeds the tabular value given in the 0.05 column, then the difference among treatments is said to be significant at the 0.05 level (i.e., we reject the hypothesis that attack classification is independent of termite-repellent treatment).

For illustrative purposes, in this example, we say that the computed value of 17.66 (2 degrees of freedom) exceeds the tabular value in the 0.01 column, so the difference in rate of attack among treatments is said to be significant at the 1% level. Examination of the data suggests that this is primarily due to the lower rate of attack on the Group II stakes.

7.14.2 TEST OF A HYPOTHESIZED COUNT

A geneticist hypothesized that, if a certain cross were made, the progeny would be of four types, in the following proportions:

$$A = 0.48, \quad B = 0.32, \quad C = 0.12, \quad D = 0.08$$

The actual segregation of 1225 progeny is shown below, along with the numbers expected according to the hypothesis:

Type	A	B	C	D	Total
Number (X_i)	542	401	164	118	1225
Expected (m_i)	588	392	147	98	1225

As the observed counts differ from those expected, we might wonder if the hypothesis is false. Or, can departures as large as this occur strictly by chance?

The chi-square test is

$$\chi^2 = \sum_{i=1}^{k} \left(\frac{(X_i - m_i)^2}{m_i} \right), \text{ with } (k-1) \text{ degrees of freedom}$$

where

k = Number of groups recognized.
X_i = Observed count for the ith.
M_i = Count expected in the ith group if the hypothesis is true.

For the above data,

$$\chi^2_{3dt} = \frac{(542-588)^2}{588} + \frac{(401-392)^2}{392} + \frac{(164-147)^2}{147} + \frac{(118-98)^2}{98} = 9.85$$

This value exceeds the tabular χ^2 with 3 degrees of freedom at the 0.05 level (i.e., it is greater than 7.81). Hence, the hypothesis would be rejected (if the geneticist believed in testing at the 0.05 level).

7.14.3 BARTLETT'S TEST OF HOMOGENEITY OF VARIANCE

Many of the statistical methods described later are valid only if the variance is homogeneous (i.e., variance within each of the populations is equal). The t test of the following section assumes that the variance is the same for each group, and so does the analysis of variance. The fitting of an unweighted regression as described in the last section also assumes that the dependent variable has the same degree of variability (variance) for all levels of the independent variables.

Bartlett's test offers a means of evaluating this assumption. Suppose that we have taken random samples in each of four groups and obtained variances (s^2) of 84.2, 63.8, 88.6, and 72.1 based on samples of 9, 21, 5, and 11 units, respectively. We would like to know if these variances could have come from populations all having the same variance. The quantities needed for Bartlett's test are tabulated here:

Group	Variance (s^2)	($n-1$)	Corrected Sum of Squares (SS)	$1/(n-1)$	log s^2	($n-1$)(log s^2)
1	84.2	8	673.6	0.125	1.92531	15.40248
2	63.8	20	1276.0	0.050	1.80482	36.09640
3	88.6	5	443.0	0.200	1.94743	9.73715
4	72.1	10	721.0	0.100	1.85794	18.57940
Sums	—	43	3113.6	0.475	—	79.81543

where the number of groups (k) = 4, and the corrected sum of squares (SS) is

$$\left(\sum X^2 - \frac{\left(\sum X\right)^2}{n} \right) = (n-1)s^2$$

From this we compute the pooled within-group variance:

$$\bar{s}^2 = \frac{\sum SS_i}{\sum (n_i - 1)} = \frac{3113.6}{43} = 72.4093$$

and

$$\log \bar{s}^2 = 1.85979$$

Then the test for homogeneity is

$$\chi^2_{(k-1)df} = (2.3026)\left[\left(\log \bar{s}^2\right)\left(\sum (n_i - 1)\right) - \sum (n_i - 1)\left(\log s_i^2\right)\right]$$

$$\chi^2_{3df} = (2.3026)\left[(1.85979)(43) - 79.81543\right]$$

$$= 0.358$$

This value of χ^2 is now compared with the value of χ^2 in an accumulative distribution of chi-square for the desired probability level. A value greater than that given in the table would lead us to reject the homogeneity assumption.

Note: The original form of this equation used natural logarithms in place of the common logarithms shown here. The natural log of any number is approximately 2.3026 times its common log—hence, the constant of 2.3026 in the equation. In computations, common logarithms are usually more convenient than natural logarithms.

The χ^2 value given by the above equation is biased upward. If χ^2 is nonsignificant, the bias is not important. However, if the computed χ^2 is just a little above the threshold value for significance, a correction for bias should be applied. The correction (C) is

$$C = \frac{3(k-1) + \left[\sum\left(\dfrac{1}{n_i - 1}\right) - \dfrac{1}{\sum (n_i - 1)}\right]}{3(k-1)}$$

The corrected value of χ^2 is then

$$\frac{3(4-1) + \left(0.475 - \dfrac{1}{43}\right)}{3(4-1)} = 1.0502$$

DID YOU KNOW?

According to Ernst Mayr (1970, 2002), *races* are distinct, generally divergent populations within the same species with relatively small morphological and genetic differences. The populations can be described as ecological races if they arise from adaptations to different local habitats or geographic races when they are geographically isolated. If sufficiently different, two or more races can be identified as subspecies, which is an official biological taxonomy unit subordinate to species. If not, they are denoted as races, which means that a formal rank should not be given to the group or taxonomists are unsure whether or not a formal rank should be given. Again, according to Mayr, "a subspecies is a geographical race that is sufficiently different taxonomically to be worthy of a separate name" (pp. 89–94).

7.15 COMPARING TWO GROUPS BY THE *t* TEST

7.15.1 *T* Test for Unpaired Plots

An individual unit in a population may be characterized in a number of different ways. A single tree, for example, can be described as alive or dead, hardwood or softwood, infected or not infected, and so forth. When dealing with observations of this type, we usually want to estimate the proportion of a population having a certain attribute. Or, if there are two or more different groups, we will often be interested in testing whether or not the groups differ in the proportions of individuals having the specified attribute. Some methods of handling these problems have been discussed in previous sections.

Alternatively, we might describe a tree by a measurement of some characteristic such as its diameter, height, or cubic volume. For this measurement type of observation we may wish to estimate the mean for a group as discussed in the section on sampling for measurement variables. If there are two or more groups we will frequently want to test whether or not the group means are different. Often the groups will represent types of treatment that we wish to compare. Under certain conditions, the *t* or *F* tests may be used for this purpose.

Both of these tests have a wide variety of applications. For the present we will confine our attention to tests of the hypothesis that there is no difference between treatment (or group) means. The computational routine depends on how the observations have been selected or arranged. The first illustration of a *t* test of the hypothesis that there is no difference between the means of two treatments assumes that the treatments have been assigned to the experimental units completely at random. Except for the fact that there are usually (but not necessarily) an equal number of units or "plots" for each treatment, there is no restriction on the random assignment of treatments.

In this example the "treatments" were two races of white pine which were to be compared on the basis of their volume production over a specified period of time. Twenty-two square, 1-acre plots were staked out for the study; 11 of these were selected entirely at random and planted with seedlings of race A. The remaining 11 were planted with seedlings of race B. After the prescribed time period the pulpwood volume (in cords—a stack of wood 4 ft wide by 4 ft high by 8 ft in length) was determined for each plot:

Race A			Race B		
11	5	9	9	6	9
8	10	11	9	13	8
10	8	11	6	5	6
8	8		10	7	
	Sum = 99			Sum = 88	
	Average = 9.0			Average = 8.0	

To test the hypothesis that there is no difference between the race means (sometimes referred to as a null hypothesis—general or default position) we compute

$$t = \frac{\bar{X}_A - \bar{X}_B}{\sqrt{\dfrac{s^2(n_A + n_B)}{(n_A)(n_B)}}}$$

where

$\bar{X}_A$ and $\bar{X}_B$ = Arithmetic means for groups A and B.

n_A and n_B = Number of observations in groups A and B (n_A and n_B do not have to be the same).

s^2 = Pooled within-group variance (calculation shown below).

To compute the pooled within-group variance, we first get the corrected sum of squares (*SS*) within each group:

$$SS_A = \sum X_A^2 - \frac{\left(\sum X_A\right)^2}{n_A} = 11^2 + 8^2 + \ldots + 11^2 - \frac{(99)^2}{11} = 34$$

$$SS_B = \sum X_B^2 - \frac{\left(\sum X_B\right)^2}{n_B} = 9^2 + 9^2 + \ldots + 6^2 - \frac{(88)^2}{11} = 54$$

Then the pooled variance is

$$s^2 = \frac{SS_A + SS_B}{(n_A - 1) + (n_B - 1)} = \frac{88}{20} = 4.4$$

Hence,

$$t = \frac{9.0 - 8.0}{\sqrt{4.4\left(\frac{11 + 11}{(11)(11)}\right)}} = \frac{1.0}{\sqrt{0.800000}} = 1.118$$

This value of *t* has $(n_A - 1) + (n_B - 1)$ degrees of freedom. If it exceeds the tabular value (from a distribution of *t* table) at a specified probability level, we would reject the hypothesis. The difference between the two means would be considered significant (larger than would be expected by chance if there is actually no difference). In this case, tabular *t* with 20 degrees of freedom at the 0.05 level is 2.086. Since our sample value is less than this, the difference is not significant at the 0.05 levels.

One of the unfortunate aspects of the *t* test and other statistical methods is that almost any kind of numbers can be plugged into the equations. But, if the numbers and methods of obtaining them do not meet certain requirements, then the result may be a fancy statistical facade with nothing behind it. In a handbook of this scope it is not possible to make the reader aware of all of the niceties of statistical usage, but a few words of warning are certainly appropriate.

A fundamental requirement in the use of most statistical methods is that the experimental material be a random sample of the population to which the conclusions are to be applied. In the *t* test of white pine races, the plots should be a sample of the sites on which the pine are to be grown, and the planted seedlings should be a random sample representing the particular race. A test conducted in one corner of an experimental forest may yield conclusions that are valid only for that particular area or sites that are about the same. Similarly, if the seedlings of a particle race are the progeny of a small number of parents, their performance may be representative of those parents only, rather than of the race.

In addition to assuming that the observations for a given race are a valid sample of the population of possible observations, the *t* test described above assumes that the population of such observations follows the normal distribution. With only a few observations, it is usually impossible to determine whether or not this assumption has been met. Special studies can be made to check on the distribution, but often the question is left to the judgment and knowledge of the research worker.

Finally, the *t* test of unpaired plots assumes that each group (or treatment) has the same population variance. Since it is possible to compute a sample variance for each group, this assumption can be checked with Bartlett's test for homogeneity of variance. Most statistical textbooks present variations of the *t* test that may be used if the group variances are unequal.

7.15.1.1   Sample Size

If there is a real difference of D feet between the two races of white pine, how many replicates (plots) would be needed to show that it is significant? To answer this, we first assume that the number of replicates will be the same for each group $(n_A = n_B = n)$. The equation for t can then be written as

$$t = \frac{D}{\sqrt{\frac{2s^2}{n}}} \quad \text{or} \quad n = \frac{2t^2s^2}{D^2}$$

Next we need an estimate of the within-group variance, s^2. As usual, this must be determined from previous experiments, or by special study of the populations.

■ EXAMPLE 7.9

Problem: Suppose that we plan to test at the 0.05 level and wish to detect a true difference of $D = 1$ cord if it exists. From previous tests, we estimate $s^2 = 5.0$. Thus, we have

$$n = \frac{2t^2s^2}{D^2} = 2t^2\left(\frac{5.0}{1.0}\right)$$

Here we hit a snag. In order to estimate n we need a value for t, but the value of t depends on the number of degrees of freedom, which depends on n. The situation calls for an iterative solution, which is a mathematical procedure that generates a sequence of improving approximate solutions for a class of problems—in other words, a fancy name for trial and error. We start with a guessed value for n, say $n_0 = 20$. As t has $(n_A - 1) + (n_B - 1) = 2(n - 1)$ degrees of freedom, we'll use $t = 2.025$ (which is equal to $t_{0.05}$ with 38 degrees of freedom) and compute

$$n_1 = 2(2.025)^2\left(\frac{5.0}{1.0}\right) = 41$$

The proper value of n will be somewhere between n_0 and n_1—much closer to n_1 than to n_0. We can now make a second guess at n and repeat the process. If we try $n_2 = 38$, t will have $2(n - 1) = 74$ degrees of freedom and $t_{0.05} = 1.992$. Thus,

$$n_3 = 2(1.992)^2\left(\frac{5.0}{1.0}\right) = 39.7$$

Thus, n appears to be over 39 and we will use $n = 40$ plots for each group, or a total of 80 plots.

7.15.2 t Test for Paired Plots

A second test was made of the two races of white pine. It also had 11 replicates of each race, but instead of the two races being assigned completely at random over the 22 plots, the plots were grouped into 11 pairs and a different race was randomly assigned to each member of a pair. The cordwood volumes at the end of the growth period were as follows:

						Plot Pair							
	1	**2**	**3**	**4**	**5**	**6**	**7**	**8**	**9**	**10**	**11**	**Sum**	**Mean**
Race A	12	8	8	11	10	9	11	11	13	10	7	110	10.0
Race B	10	7	8	9	11	6	10	11	10	8	9	99	9.0
$d_i = A_i - B_i$	2	1	0	2	−1	3	1	0	3	2	−2	11	1.0

As before, we wish to test the hypothesis that there is no real difference between the race means. The value of t when the plots have been paired is

$$t = \frac{\bar{X}_A - \bar{X}_B}{\sqrt{\frac{s_d^2}{n}}} = \frac{\bar{d}}{\sqrt{s_d^2}}, \text{ with } (n-1) \text{ degrees of freedom}$$

where

n = Number of pairs of plots.
s_d^2 = Variance of the individual differences between A and B.

$$s_d^2 = \frac{\sum d_i^2 - \frac{\left(\sum d_i\right)^2}{n}}{(n-1)} = \frac{2^2 + 1^2 + \ldots + (-2)^2 - \frac{11^2}{11}}{10} = 2.6$$

So, in this example we find

$$t_{10df} = \frac{10.0 - 9.0}{\sqrt{2.6/11}} = 2.057$$

When this value of 2.057 is compared to the tabular value of t in a distribution of t table ($t_{0.05}$ with 10 degrees of freedom = 2.228), we find that the difference is not significant at the 0.05 level. That is, a sample means difference of 1 cord or more could have occurred by chance more than one time in twenty even if there is no real difference between the race means. Usually such an outcome is not regarded as sufficiently strong evidence to reject the hypothesis.

"The method of paired observations is a useful technique. Compared with the standard two-sample t test, in addition to the advantage that we do not have to assume that the two samples are independent, we also need not assume that the variances of the two samples are equal" (Hamburg, 1987, p. 304). Moreover, the paired test will be more sensitive (capable of detecting smaller real differences) than the unpaired test whenever the experimental units (plots, in this case) can be grouped into pairs such that the variation between pairs is appreciably larger than the variation within pairs. The basis for paring plots may be geographic proximity or similarity in any other characteristic that is expected to affect the performance of the plot. In animal-husbandry studies, litter mates are often paired, and where patches of human skin are the plots, the left and right arms may constitute the pair. If the experimental units are very homogeneous, then there may be no advantage in pairing.

7.15.2.1 Number of Replicates

The number (n) of plot pairs needed to detect a true mean difference of size D is

$$n = \frac{t^2 s_d^2}{D^2}$$

7.16 COMPARISON OF TWO OR MORE GROUPS BY ANALYSIS OF VARIANCE

7.16.1 Complete Randomization

A planter wanted to compare the effects of five site-preparation treatments on the early height growth of planted pine seedlings. He laid out 25 plots and applied each treatment to 5 randomly selected plots. The plots were then hand planted and at the end of 5 years the height for all pines was measured and an average height computed for each plot. The plot averages (in feet) were as follows:

	Treatments					
	A	B	C	D	E	Total
	15	16	13	11	14	
	14	14	12	13	12	
	12	13	11	10	12	
	13	15	12	12	10	
	13	14	10	11	11	
Sums	67	72	58	57	59	313
Treatment means	13.4	14.4	11.6	11.4	11.8	12.52

Looking at the data we see that there are differences among the treatment means: A and B have higher averages than C, D, and E. Soils and planting stock are seldom completely uniform, however, so we would expect some differences even if every plot had been given exactly the same site-preparation treatment. The question is, can differences as large as this occur strictly by chance if there is actually no difference among treatments? If we decide that the observed differences are larger than might be expected to occur strictly by chance, then the inference is that the treatment means are not equal. Statistically speaking, we reject the hypothesis of no difference among treatment means.

Problems like this are neatly handled by an analysis of variance. To make this analysis, we need to fill in a table like the following:

Source of Variation	Degrees of Freedom	Sums of Squares	Mean Squares
Treatments	4		
Error	20		
Total	24		

7.16.1.1 Source of Variation

There are a number of reasons why the height growth of these 25 plots might vary, but only one can be definitely identified and evaluated—that attributable to treatments. The unidentified variation is assumed to represent the variation inherent in the experimental material and is labeled error. Thus, total variation is being divided into two parts: one part attributable to treatments, and the other unidentified and called error.

7.16.1.2 Degrees of Freedom

Degrees of freedom are difficult to explain in non-statistical language. In the simpler analyses of variance, however, they are not difficult to determine. For the total, the degrees of freedom are one less than the number of observations; there are 25 plots, so the total has 24 degrees of freedom. For the sources, other than error, the degrees of freedom are one less than the number of classes or groups recognized in the source. Thus, in the source labeled "Treatments," there are five groups (five treatments), so there will be four degrees of freedom for treatments. The remaining degrees of freedom (24 – 4 = 20) are associated with the error term.

7.16.1.3 Sums of Squares

There is a sum of squares associated with every source of variation. These *SS* are easily calculated as follows. First we need what is known as a "correction term," or *CT*. This is simply

$$CT = \frac{\left(\sum_{}^{n} X\right)^2}{n} = \frac{313^2}{25} = 3918.76, \text{ where } \sum^{n} \text{ is the sum of } n \text{ items}$$

Then the total sum of squares is

$$\text{Total } \underset{24df}{SS} = \sum^{n} X^2 - CT = \left(15^2 + 14^2 + \ldots + 11^2\right) - CT = 64.24$$

The sum of squares attributable to treatments is

$$\text{Treatment } \underset{4df}{SS} = \frac{\sum^{n} (\text{treatment totals})^2}{\text{No. of plots per treatment}} - CT$$

$$= \frac{67^2 + 72^2 + \ldots + 59^2}{5} - CT = \frac{19,767}{5} - CT = 34.64$$

Note that in both *SS* calculations the number of items squared and added was one more than the number of degrees of freedom associated with the sum of squares. The number of degrees of freedom just below the *SS* and the numbers of items to be squared and added over the *n* value provided a partial check as to whether the proper totals are being used in the calculation—the degrees of freedom must be one less than the number of items.

Note also that the divisor in the treatment *SS* calculation is equal to the number of individual items that go to make up each of the totals being squared in the numerator. This was also true in the calculation of total *SS*, but there the divisor was 1 and did not have to be shown. Note further that the divisor times the number over the summation sign ($5 \times 5 = 25$ for treatments) must always be equal to the total number of observations in the test—another check.

The sum of squares for error is obtained by subtracting the treatment *SS* from total *SS*. A good habit to get into when obtaining sums of squares by subtraction is to perform the same subtraction using degrees of freedom. In the more complex designs, doing this provides a partial check on whether the right items are being used.

7.16.1.4 Mean Squares

The mean squares are now calculated by dividing the sums of squares by the associated degrees of freedom. It is not necessary to calculate the mean squares for the total. The items that have been calculated are entered directly into the analysis table, which at the present stage would look like this:

Source of Variation	Degrees of Freedom	Sums of Squares	Mean Squares
Treatments	4	34.64	8.66
Error	20	29.60	1.48
Total	24	64.25	

An *F* test of treatments (used to reject the null hypothesis) is now made by dividing the mean square for treatments by the mean square for error. In this case,

$$F = \frac{8.66}{1.48} = 5.851$$

Fortunately, critical values of the F ratio have been tabulated for frequently used significance levels analogous to the χ^2 distribution. Thus, the result, 5.851, can be compared to the appropriate value of F in the table. The tabular F for significance at the 0.05 level is 2.87 and that for the 0.01 level is 4.43. As the calculated value of F exceeds 4.43, we conclude that the difference in height growth between treatments is significant at the 0.01 level. (More precisely, we reject the hypothesis that there is no difference in mean height growth between the treatments.) If F had been smaller than 4.43 but larger than 2.87, we would have said that the difference is significant at the 0.05 level. If F had been less than 2.87, we would have said that the difference between treatments is not significant at the 0.05 level. Researchers should select their own levels of significance (preferably in advance of the study), keeping in mind that significance at the α (alpha) level, for example, means this: If there is actually no difference among treatments, then the probability of getting chance differences as large as those observed is α or less.

7.16.1.5   *t* Test vs. the Analysis of Variance

If only two treatments are compared, the analysis of variance of a completely randomized design and the t test of unpaired plots lead to the same conclusion. The choice of test is strictly one of personal preference, as may be verified by applying the analysis of variance to the data used to illustrate the t test of unpaired plots. The resulting F value will be equal to the square of the value of t that was obtained (i.e., $F = t^2$). Like the t test, the F test is valid only if the variable observed is normally distributed and if all groups have the same variance.

7.16.2   MULTIPLE COMPARISONS

In the example illustrating the completely randomized design, the difference among treatments was found to be significant at the 0.01 probability level. This is interesting as far as it goes, but usually we will want to take a closer look at the data, making comparisons among various combinations of the treatments. Suppose, for example, that A and B involve some mechanical form of site preparation while C, D, and E are chemical treatments. We might want to test whether the average of A and B together differ from the combined average of C, D, and E. Or, we might wish to test whether A and B differ significantly from each other. When the number of replications (n) is the same for all treatments, such comparisons are fairly easy to define and test.

The question of whether the average of treatments A and B differs significantly from the average of treatments C, D, and E is equivalent to testing whether the linear contrast

$$\bar{Q} = \left(3\bar{A} + 3\bar{B}\right) - \left(2\bar{C} + 2\bar{D} + 2\bar{E}\right)$$

differs significantly from zero ($\bar{A}$ = the mean for treatment A, etc.). Note that the coefficients of this contrast sum to zero ($3 + 3 - 2 - 2 - 2 = 0$) and are selected so as to put the two means in the first group on an equal basis with the three means in the second group.

7.16.2.1   *F* Test with Single Degree of Freedom

A comparison specified in advance of the study (on logical grounds and before examination of the data) can be tested by an F test with single degree of freedom. For the linear contrast

$$\hat{Q} = a_1\bar{X}_1 + a_2\bar{X}_2 + a_3\bar{X}_3 + \ldots$$

among means based on the same number (n) of observations, the sum of squares has one degree of freedom and is computed as

$$SS_{1df} = \frac{n\hat{Q}^2}{\sum a_i^2}$$

This sum of squares divided by the mean square for error provides an F test of the comparison. Thus, in testing A and B vs. C, D, and E we have

$$\hat{Q} = 3(13.4) + 3(14.4) - 2(11.6) - 2(11.4) - 2(11.8) = 13.8$$

and

$$SS_{1df} = \frac{5(13.8)^2}{3^3 + 3^3 + (-2)^2 + (-2)^2 + (-2)^2} = \frac{952.20}{30} = 31.74$$

Then, dividing by the error mean square gives the F value for testing the contrast:

$$F = \frac{31.74}{1.48} = 21.446 \text{ with 1 and 20 degrees of freedom}$$

This exceeds the tabular value of F (4.35) at the 0.05 probability level. If this is the level at which we decided to test, we would reject the hypothesis that the mean of treatments A and B does not differ from the mean of treatments C, D, and E.

If $\hat{Q}$ is expressed in terms of the treatment totals rather than their means so that

$$\hat{Q}_T = a_1\left(\sum X_1\right) + a_2\left(\sum X_2\right) + \ldots$$

then the equation for the single degree of freedom sum of squares is

$$SS_{1df} = \frac{\hat{Q}_T^2}{n\left(\sum a_i^2\right)}$$

The results will be the same as those obtained with the means. For the test of A and B vs. C, D, and E,

$$\hat{Q}_T = 3(67) + 3(72) - 2(58) - 2(57) - 2(59) = 69$$

And,

$$SS_{1df} = \frac{69^2}{5\left[3^2 + 3^2 + (-2)^2 + (-2)^2 + (-2)^2\right]} = \frac{4761}{150} = 31.74, \text{ as before}$$

Working with the totals saves the labor of computing means and avoids possible rounding errors.

7.16.2.2 Scheffe's Test

Quite often we will want to test comparisons that were not anticipated before the data were collected. If the test of treatments was significant, such unplanned comparisons can be tested by the method of Scheffe, or Scheffe's test. Named after the American statistician Henry Scheffe, the Scheffe test adjusts significant levels in a linear regression analysis to account for multiple comparisons. It is particularly useful in analysis of variance and in constructing simultaneous bands for regressions involving basic functions. When there are n replications of each treatment, k degrees of freedom for treatment, and v degrees of freedom for error, any linear contrast among the treatment means

$$\hat{Q} = a_1\bar{X}_1 + a_2\bar{X}_2 + \dots$$

is tested by computing

$$F = \frac{n\hat{Q}^2}{k\left(\sum a_i^2\right)(\text{Error mean square})}$$

This value is then compared to the tabular value of F with k and v degrees of freedom. For example, to test treatment B against the means of treatments C and E we would have

$$\hat{Q} = \left[2\bar{B} - (\bar{C} + \bar{E})\right] = \left[2(14.4) - 11.6 - 11.8\right] = 5.4$$

And,

$$F = \frac{5(5.4)^2}{(4)\left[2^2 + (-1)^2 + (-1)^2\right](1.48)} = 4.105, \text{ with 4 and 20 degrees of freedom}$$

This figure is larger than the tabular value of F (2.87), so in testing at the 0.05 level we would reject the hypothesis that the mean for treatment B did not differ from the combined average of treatments C and E.

For a contrast (Q_T) expressed in terms of treatment totals, the equation for F becomes

$$F = \frac{\hat{Q}_T^2}{nk\left(\sum a_i^2\right)(\text{Error mean square})}$$

7.16.2.3 Unequal Replications

If the number of replications is not the same for all treatments, then for the linear contrast

$$\hat{Q} = a_1\bar{X}_1 + a_2\bar{X}_2 + \dots$$

The sum of squares in the single degree of freedom F test is given by

$$SS_{1df} = \frac{\hat{Q}_2}{\left(\dfrac{a_1^2}{n_1} + \dfrac{a_2^2}{n_2} + \dots\right)}$$

where n_i is the number of replications on which $\bar{X}_i$ is based.

With unequal replication, the F value in Scheffe's test is computed by the equation

$$F = \frac{\hat{Q}^2}{k\left(\dfrac{a_1^2}{n_1} + \dfrac{a_2^2}{n_2} + \ldots\right)\text{(Error mean square)}}$$

Selecting the coefficients (a_i) for such contrasts can be tricky. When testing the hypothesis that there is no difference between the means of two groups of treatments, the positive coefficients are usually

$$\text{Positive } a_i = \frac{n_i}{p}$$

where p is the total number of plots in the group of treatments with positive coefficients. The negative coefficients are

$$\text{Negative } a_j = \frac{n_j}{m}$$

where m is the total number of plots in the group of treatments with negative coefficients.

To illustrate, if we wish to compare the mean of treatments A, B, and C with the mean of treatments D and E and there are two plots of treatment A, three of B, five of C, three of D, and two of E, then $p = 2 + 3 + 5 = 10$, $m = 3 + 2 = 5$, and the contrast would be

$$\hat{Q} = \left(\frac{2}{10}\bar{A} + \frac{3}{10}\bar{B} + \frac{5}{10}\bar{C}\right) - \left(\frac{3}{5}\bar{D} + \frac{2}{5}\bar{E}\right)$$

7.16.3　RANDOMIZED BLOCK DESIGN

There are two basic types of the two-factor analysis of variance: *completely randomized design* (discussed in the previous section) and *randomized block design*. In the completely randomized design, the error mean square is a measure of the variation among plots treated alike. It is in fact an average of the within-treatment variances, as may easily be verified by computation. If there is considerable variation among plots treated alike, the error mean square will be large and the F test for a given set of treatments is less likely to be significant. Only large differences among treatments will be detected as real and the experiment is said to be insensitive.

Often the error can be reduced (thus giving a more sensitive test) by use of a randomized block design in place of complete randomization. In this design, similar plots or plots that are close together are grouped into blocks. Usually the number of plots in each block is the same as the number of treatments to be compared, though there are variations having two or more plots per treatment in each block. The blocks are recognized as a source of variation that is isolated in the analysis. A general rule in randomized block design is to "block what you can, randomize what you can't." In other words, blocking is used to remove the effects of nuisance variables or factors. Nuisance factors are those that may affect the measured result but are not of primary interest. For example, in applying a treatment, nuisance factors might be the time of day the experiment was run, the room temperature, or the specific operator who prepared the treatment (Addelman, 1969, 1970).

As an example, a randomized block design with five blocks was used to test the height growth of cottonwood cuttings from four selected parent trees. The field layout looked like this:

D	B		B	C		A	D		B	A		C	D
C	A		A	D		B	C		C	D		A	B

I　　　　　　　　II　　　　　　　　III　　　　　　　　IV　　　　　　　　V

Each plot consisted of a planting of 100 cuttings of the clone assigned to that plot. When the trees were 5 years old the heights of all survivors were measured and an average computed for each plot. The plot averages (in feet) by clones and blocks are summarized below:

Block	Clone				Block Totals
	A	B	C	D	
I	18	14	12	16	60
II	15	15	16	13	59
III	16	15	8	15	54
IV	14	12	10	12	48
V	12	14	9	14	49
Clone totals	75	70	55	70	270
Clone means	15	14	11	14	—

The hypothesis to be tested is that clones do not differ in mean height. In this design there are two identifiable sources of variation—that attributable to clones and that associated with blocks. The remaining portion of the total variation is used as a measure of experimental error. The outline of the analysis is therefore as follows:

Source of Variation	Degrees of Freedom	Sums of Squares	Mean Squares
Blocks	4		
Clones	3		
Error	12		
Total	19		

The breakdown in degrees of freedom and computation of the various sums of squares follow the same pattern as in the completely randomized design. Total degrees of freedom (19) are one less than the total number of plots. Degrees of freedom for clones (three) are one less than the number of clones. With five blocks, there will be four degrees of freedom for blocks. The remaining 12 degrees of freedom are associated with the error term.

Sums-of-squares calculations proceed as follows:

1. $CT = \dfrac{\left(\sum\limits^{20} X\right)^2}{n} = \dfrac{270^2}{20} = 3645$

2. Total $\underset{19df}{SS} = \sum\limits^{20} X^2 - CT = \left(18^2 + 15^2 + \ldots + 14^2\right) - CT = 3766 - 3645 = 121$

3. Clone $\underset{3df}{SS} = \dfrac{\sum\limits^{4}(\text{Clone totals}^2)}{\text{No. of plots per clone}} - CT = \dfrac{75^2 + 70^2 + 55^2 + 70^2}{5} - CT = 3690 - 3645 = 45$

4. Block $SS_{4df} = \dfrac{\sum\limits^{5}(\text{Block totals}^2)}{\text{No. of plots per block}} - CT = \dfrac{60^2 + 59^2 + \ldots + 49^2}{4} - CT = 3675.5 - 3645 = 30.5$

5. Error $SS_{12df} = \text{Total } SS_{19df} - \text{Clone } SS_{3df} - \text{Block } SS_{4df} = 45.5$

Note that in obtaining the error SS by subtraction, we get a partial check on ourselves by subtracting clone and block df's from the total df to see if we come out with the correct number of error df. If these don't check, we have probably used the wrong sums of squares in the subtraction.

Mean squares are again calculated by dividing the sums of squares by the associated number of degrees of freedom. Tabulating the results of these computations

Source of Variation	Degrees of Freedom	Sums of Squares	Mean Squares
Blocks	4	30.5	7.625
Clones	3	45.0	15.000
Error	12	45.5	3.792
Total	19	121.0	

F for clones is obtained by dividing the clone means square by the error mean square. In this case $F = 15.000/3792 = 3.956$. As this is larger than the tabular F of 3.49 (obtained from a distribution of F table) ($F_{0.05}$ with 3 and 12 degrees of freedom) we conclude that the difference between clones is significant at the 0.05 level. The significance appears to be due largely to the low value of C as compared to A, B, and D.

Comparisons among clone means can be made by the methods previously described. For example, to test the prespecified (i.e., before examining the data) hypothesis that there is no difference between the mean of clone C and the combined average A, B, and D we would have:

$$SS_{1df} \text{ for } (\text{A} + \text{B} + \text{D vs. C}) = \frac{5(3\bar{C} - \bar{A} - \bar{B} - \bar{D})^2}{3^2 + (-1)^2 + (-1)^2 + (-1)^2} = \frac{5(-10)^2}{12} = 41.667$$

Then,

$$F = \frac{41.667}{3.792} = 10.988$$

Tabular F at the 0.01 level with 1 and 12 degrees of freedom is 9.33. As calculated F is greater than this, we conclude that the difference between C and the average of A, B, and D is significant at the 0.01 level.

The sum of squares for this single-degree-of-freedom comparison (41.667) is almost as large as that for clones (45.0) with three degrees of freedom. This result suggests that most of the clonal variation is attributable to the low value of C, and that comparisons between the other three means are not likely to be significant.

DID YOU KNOW?

With only two treatments, the analysis of variance of a randomized block design is equivalent to the t test of paired replicates. The value of F will be equal to the value of t^2 and the inferences derived from the tests will be the same. The choice of tests is a matter of personal preference.

There is usually no reason for testing blocks, but the size of the block mean square relative to the mean square for error does give an indication of how much precision was gained by blocking. If the block mean square is large (at least two or three times as large as the error mean square) the test is more sensitive than it would have been with complete randomization. If the block mean square is about equal to or only slightly larger than the error mean square, the use of blocks has not improved the precision of the test. The block mean square should not be appreciably smaller than the error mean square. If it is, the method of conducting the study and the computations should be re-examined.

In addition to the assumption of homogeneous variance and normality, the randomized block design assumes that there is no interaction between treatments and blocks; that is, that differences among treatments are about the same in all blocks. Because of this assumption, it is not advisable to have blocks that differ greatly, as they may cause an interaction with treatments.

7.16.4   LATIN SQUARE DESIGN

In the randomized block design the purpose of blocking is to isolate a recognizable extraneous source of variation. If successful, blocking reduces the error mean square and gives a more sensitive test than could be obtained by complete randomization. In some situations, however, we have a two-way source of variation that cannot be isolated by blocks alone. In an agricultural field, for example, fertility gradients may exist both parallel to and at right angles to plowed rows. Simple blocking isolates only one of these sources of variation, leaving the other to swell the error term and reduce the sensitivity of the test.

When such a two-way source of extraneous variation is recognized or suspected, the Latin square design may be helpful. In this design, the total number of plots or experimental units is made equal to the square of the number of treatments. In forestry and agricultural experiments, the plots are often (but not always) arranged in rows and columns, with each row and column having a number of poles equal to the number of treatments being tested. The rows represent different levels of one source of extraneous variation while the columns represent different levels of the other source of extraneous variation. Thus, before the assignment of treatments, the field layout of a Latin square for testing five treatments might look like this:

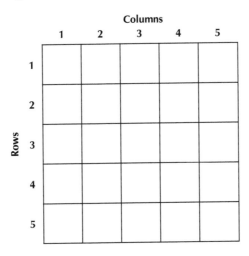

Treatments are assigned to plots at random, but with the very important restriction that a given treatment cannot appear more than once in any row or any column. An example of a field layout of a Latin square for testing five treatments is given below. The letters represent the assignment of five treatments (which here are five species of hardwoods). The numbers show the average 5-year height growth by plots. The tabulation below shows the totals for rows, columns, and treatments:

Rows	Columns 1	2	3	4	5
1	C 13	A 21	B 16	E 16	D 14
2	A 18	B 15	D 17	C 17	E 15
3	D 17	C 15	E 15	A 15	B 18
4	E 18	D 18	C 16	B 14	A 16
5	B 17	E 16	A 25	D 19	C 14

Row	Σ	Column	Σ	Treatment	Σ	$\overline{X}$
1	80	1	83	A	95	19
2	82	2	85	B	80	16
3	80	3	89	C	75	15
4	82	4	81	D	85	16
Total	415		415		415	16.6

The partitioning of degrees of freedom, the calculation of sums of squares, and the subsequent analysis follow much the same pattern illustrated previously for randomized blocks.

$$CT = \frac{\left(\sum^{25} X\right)^2}{n} = \frac{415^2}{25} = \frac{172{,}225}{25} = 6889.0$$

$$\text{Total } \underset{24df}{SS} = \sum^{25} X^2 - CT = 7041 - CT = 152.0$$

$$\text{Row } \underset{4df}{SS} = \frac{\sum^{5}(\text{Row totals}^2)}{\text{No. of plots per row}} - CT = \frac{34{,}529}{5} - CT = 16.8$$

$$\text{Column } \underset{4df}{SS} = \frac{\sum^{5}(\text{Column totals}^2)}{\text{No. of plots per column}} - CT = \frac{34{,}525}{5} - CT = 16.0$$

$$\text{Species } \underset{4df}{SS} = \frac{\sum^{5}(\text{Species totals}^2)}{\text{No. of plots per species}} - CT = \frac{34{,}675}{5} - CT = 46.0$$

$$\text{Error } \underset{12df}{SS} = \text{Total } \underset{24df}{SS} - \text{Row } \underset{4df}{SS} - \text{Column } \underset{4df}{SS} - \text{Species } \underset{4df}{SS} = 73.2$$

The analysis of variance is as follows:

Source of Variation	Degrees of Freedom	Sums of Squares	Mean Squares
Rows	4	16.8	4.2
Columns	4	16.0	4.0
Species	4	46.0	11.5
Error	12	73.2	6.1
Total	24	152.0	

$$F \text{ (for species)} = \frac{11.5}{6.1} = 1.885$$

As the computed value of F is less than the tabular value of F at the 0.05 level (with 4/12 degrees of freedom), the differences among species are considered nonsignificant.

The Latin square design can be used whenever there is a two-way heterogeneity that cannot be controlled simply by blocking. In greenhouse studies, distance from a window could be treated as a row effect while distance from the blower or heater might be regarded as a column effect. Though the plots are often physically arranged in rows or columns, this is not required. When testing the use of materials in a manufacturing process where different machines and machine operators will be involved, the variation between machines could be treated as a row effect and the variation due to operators could be treated as a column effect.

The Latin square should not be used if an interaction between rows and treatments or columns and treatments is suspected.

7.16.5 FACTORIAL EXPERIMENTS

In environmental practice, knowledge that interactions between elements of the environment occur and an understanding of what their influence or impact on the environment is, or can be, is important. Consider a comparison of corn yields following three rates or levels of nitrogen fertilization indicating that the yields depended on how much phosphorus was used along with the nitrogen. The differences in yield were smaller when no phosphorus was used than when the nitrogen applications were accompanied by 100 lb/acre of phosphorus. In statistics, this situation is referred to as an interaction between nitrogen and phosphorus. Another example is when leaf litter was removed from the forest floor, the catch of pine seedlings was much greater than when the litter was not removed, but for red oak the reverse was true—the seedling catch was lower where litter was removed. Thus, species and litter treatment were interacting.

Interactions are important in the interpretation of study results. In the presence of an interaction between species and litter treatment it obviously makes no sense to talk about the effects of litter removal without specifying the species. The nitrogen–phosphorus interaction means that it may be misleading to recommend a level of nitrogen without mentioning the associated level of phosphorus.

Factorial experiments are aimed at evaluating known or suspected interactions. In these experiments, each factor to be studied is tested at several levels and each level of a factor is tested at all possible combinations of the levels of the other factors. In a planting test involving three species of trees and four methods of preplanting site preparation, each method will be applied to each species, and the total number of treatment combinations will be 12. In a factorial test of the effects of two nursery treatments on the survival of four species of pine planted by three different methods, there would be 24 ($2 \times 4 \times 3 = 24$) treatment combinations.

The method of analysis can be illustrated by a factorial test of the effects of three levels of nitrogen fertilization (0, 100, and 200 lb/acre) on the growth of three species (A, B, and C) of planted pine. The nine possible treatment combinations were assigned at random to nine plots in each of three blocks. Treatments were evaluated on the basis of average annual height growth in inches per year over a 3-year period. Field layout and plot data were as follows (with subscripts denoting nitrogen levels: $0 = 0$, $1 = 100$, $2 = 200$):

C_2 17	B_2 18	C_0 37
A_0 45	B_0 24	C_1 20
B_1 21	A_2 24	A_1 17

I

B_2 23	C_2 20	A_1 18
B_1 18	A_2 14	C_0 43
A_0 40	C_1 25	B_0 35

II

C_1 19	C_2 21	A_0 37
B_1 19	A_2 17	A_1 28
B_0 29	C_0 39	B_2 15

III

Preliminary analysis of the nine combinations (temporarily ignoring their factorial nature) is made just as though this were a straight randomized block design (which is exactly what it is). (See table below.)

Summary of Plot Data

Species	Nitrogen Level	Blocks I	Blocks II	Blocks III	Nitrogen Subtotals	Species Totals
A	0	45	40	37	122	
	1	17	18	28	63	
	2	24	14	17	55	
	Block subtotals	86	72	82		240
B	0	24	35	29	88	
	1	21	18	19	58	
	2	18	23	15	56	
	Block subtotals	63	76	63		202
C	0	37	43	39	119	
	1	20	25	19	64	
	2	17	20	21	58	
	Block subtotals	74	88	79		241
All species	0	106	118	105	329	
	1	58	61	66	185	
	2	59	57	53	169	
	Totals	223	236	224	**683**	

$$CT = \frac{\left(\sum_{}^{27} X\right)^2}{27} = \frac{683^2}{27} = 17,277.3704$$

$$\text{Total } \underset{26df}{SS} = \sum^{26} X^2 - CT = (45^2 + 17^2 + \ldots + 21^2) - CT = 2275.63$$

$$\text{Block } \underset{2df}{SS} = \frac{\sum^{3}(\text{Block totals}^2)}{\text{No. of plots per block}} - CT = \frac{(223^2 + 236^2 + 224^2)}{9} - CT = 11.6296$$

$$\text{Treatment } \underset{8df}{SS} = \frac{\overset{9}{\sum}(\text{Treatment totals}^2)}{\text{No. of plots per treatment}} - CT = \frac{\left(122^2 + 63^2 + 58^2\right)}{3} - CT = 1970.2963$$

$$\text{Error } \underset{16df}{SS} = \text{Total } \underset{26df}{SS} - \text{Treatment } \underset{8df}{SS} - \text{Block } \underset{2df}{SS} = 293.7037$$

Tabulating these in the usual form:

Source of Variation	Degrees of Freedom	Sums of Squares	Mean Squares
Blocks	2	11.6296	5.8148
Treatments	8	1970.2963	246.2870
Error	16	293.7037	18.3565
Total	26	2275.6296	

$$F_{8/16df} = \frac{246.2870}{18.3565} = 13.417, \text{ significant at } 0.01$$

The next step is to analyze the components of the treatment variability. How do the species compare? What is the effect of fertilization? And does fertilization affect all species the same way (i.e., is there a species-nitrogen interaction)? To answer these questions we have to partition the degrees of freedom and sums of squares associated with treatments. This is easily done by summarizing the data for the nine combinations in a two-way table:

		Nitrogen Levels		
Species	0	1	2	Totals
A	122	63	55	240
B	88	58	56	202
C	119	64	58	241
Totals	329	185	169	683

The nine individual values will be recognized as those that entered into the calculation of the treatment SS. Keeping in mind that each entry in the body of the table is the sum of three plot values and that the species and nitrogen totals are each the sum of nine plots, the sums of squares for species, nitrogen, and the species–nitrogen interaction can be computed as follows:

$$\text{Treatment } \underset{8df}{SS} = 1970.2963, \text{ as previously calculated}$$

$$\text{Species } \underset{2df}{SS} = \frac{\overset{3}{\sum}(\text{Species totals}^2)}{\text{No. of plots per species}} - CT = \frac{\left(240^2 + 202^2 + 241^2\right)}{9} - CT$$

$$= \frac{156,485}{9} - CT = 109.8518$$

$$\text{Nitrogen } \underset{2df}{SS} = \frac{\overset{3}{\sum}(\text{Nitrogen totals}^2)}{\text{No. of plots per level of nitrogen}} - CT$$

$$= \frac{\left(329^2 + 185^2 + 169^2\right)}{9} - CT = \frac{171{,}027}{9} - CT = 1725.6296$$

$$\text{Species–nitrogen interation } \underset{4df}{SS} = \text{Treatment } \underset{8df}{SS} - \text{Species } \underset{2df}{SS} - \text{Nitrogen } \underset{2df}{SS} = 134.8149$$

The analysis now becomes:

Source of Variation	Degrees of Freedom	Sums of Squares	Mean Squares	F
Blocks	2	11.6296	5.8148	—
Treatments	8	1970.2963	246.2870	13.417[b]
Species	2[a]	109.8518[a]	54.9259	2.992[c]
Nitrogen	2[a]	1725.6296[a]	862.8148	47.003[b]
Species-nitrogen	4[a]	134.8149[a]	33.7037	1.836[c]
Error	16	293.7037	18.3565	—
Total	26	2275.6296	—	—

[a] These figures are a partitioning of the degrees of freedom and sum of squares for treatments and are therefore not included in the total at the bottom of the table.
[b] Significant at the 0.01 level.
[c] Not significant.

The degrees of freedom for simple interactions can be obtained in two ways. The first way is by subtracting the degrees of freedom associated with the component factors (in this case, two for species and two for nitrogen levels) from the degrees of freedom associated with all possible treatment combinations (eight in this case). The second way is to calculate the interaction degrees of freedom as the product of the component factor degrees of freedom (in this case, $2 \times 2 = 4$). Do it both ways as a check. The F values for species, nitrogen, and the species–nitrogen interaction are calculated by dividing their mean squares by the mean square for error. The analysis indicates a significant difference among levels of nitrogen, but no difference between species and no species–nitrogen interaction.

As before, a prespecified comparison among treatment means can be tested by breaking out the sum of squares associated with that comparison. To illustrate the computations, we will test nitrogen vs. no nitrogen and also 100 lb vs. 200 lb of nitrogen.

$$\text{Nitrogen vs. no nitrogen } \underset{1df}{SS} = \frac{9\left[2\left(\dfrac{329}{9}\right) - 1\left(\dfrac{185}{9}\right) - 1\left(\dfrac{169}{9}\right)\right]^2}{2^2 + 1^2 + 1^2}$$

$$= \frac{[2(329) - 185 - 169]^2}{9(6)} = 1711.4074$$

In the numerator, the mean for the zero level of nitrogen is multiplied by 2 to give it equal weight with the mean of levels 1 and 2 with which it is compared. The 9 is the number of plots on which each mean is based. The $(2^2 + 1^2 + 1^2)$ in the denominator is the sum of squares of the coefficients used in the numerator.

$$100 \text{ lb vs. } 200 \text{ lb } \underset{1df}{SS} = \frac{9\left[1\left(\dfrac{189}{9}\right) - 1\left(\dfrac{169}{9}\right)\right]^2}{1^2 + 1^2}$$

$$= \frac{[185 - 169]^2}{9(2)} = 14.2222$$

Note that these two sums of squares (1711.4075 and 14.2222), each with one degree of freedom, add up to the sum of squares for nitrogen (1,725.6296) with two degrees of freedom. This additive characteristic holds true only if the individual degree of freedom comparisons selected are orthogonal (i.e., independent). When the number of observations is the same for all of the treatments, then the orthogonality of any two comparisons can be checked in the following manner.

First, tabulate the coefficients and check to see that for each comparison the coefficients sum to zero:

	Nitrogen Level		
Comparison	1	2	Sum
$2N_0$ vs. $N_1 + N_2$	2	–	–	0
N_1 vs. N_2	0	+	–	0
Product of coefficients	0	–	–	0

Then, for two comparisons to be orthogonal the sum of the products of corresponding coefficients must be zero. Any sum of squares can be partitioned in a similar manner, with the number of possible orthogonal individual degree of freedom comparisons being equal to the total number of degrees of freedom with which the sum of squares is associated.

The sum of squares for species can also be partitioned into two orthogonal single-degree-of-freedom comparisons. If the comparisons were specified before the data were examined, we might make single degree of freedom tests of the difference between B and the average of A and C and also of the difference between A and C. The method is the same as that illustrated in the comparison of nitrogen treatments. The calculations are as follows:

$$2B \text{ vs. } (A+C) \underset{1df}{SS} = \frac{9\left[1\left(\dfrac{240}{9}\right) + 1\left(\dfrac{241}{9}\right) - 2\left(\dfrac{202}{9}\right)\right]^2}{1^2 + 1^2 + 2^2}$$

$$= \frac{[240 + 241 - 2(202)]^2}{9(6)} = 109.7963$$

$$A \text{ vs. } C \underset{1df}{SS} = \frac{9\left[1\left(\dfrac{241}{9}\right) - 1\left(\dfrac{240}{9}\right)\right]^2}{1^2 + 1^2} = \frac{[241 - 240]^2}{9(2)} = 0.0555$$

These comparisons are orthogonal, so the sums of squares each with one degree of freedom add up to the species SS with two degrees of freedom.

Note that in computing the sums of squares for the single-degree-of-freedom comparisons, the equations have been restated in terms of treatment totals rather than means. This often simplifies the computations and reduces the errors due to rounding.

With the partitioning the analysis has become:

Source of Variation	Degrees of Freedom	Sum of Squares	Mean Square	F
Blocks	2	11.6296	5.8148	—
Species	2	109.8518	54.9259	2.992[a]
2B vs. (A + C)	1	109.7963	109.7963	5.981[b]
A vs. C	1	0.0555	0.0555	—
Nitrogen	2	1725.6296	862.8148	47.003[c]
$2N_0$ vs. $(N_1 + N_2)$	1	1711.4074	1711.4074	93.232[c]
N_1 vs. N_2	1	14.2222	14.2222	—
Species × nitrogen interaction	4	134.8149	33.7037	1.836[a]
Error	16	293.7037	18.3565	—
Total	26	2275.6296		—

[a] Not significant.
[b] Significant at the 0.05 level.
[c] Significant at the 0.01 level.

We conclude that species B is poorer than A or C and that there is no difference in growth between A and C. We also conclude that nitrogen adversely affected growth and that 100 lb was about as bad as 200 lb. The nitrogen effect was about the same for all species (i.e., no interaction).

It is worth repeating that the comparisons to be made in an analysis should, whenever possible, be planned and specified prior to an examination of the data. A good procedure is to outline the analysis, putting in all the times that are to appear in the first two columns (source of variation and degrees of freedom) of the table.

The factorial experiment, it will be noted, is not an experimental design. It is, instead, a way of selecting treatments; given two or more factors each at two or more levels, the treatments are all possible combinations of the levels of each factor. If we have three factors with the first at four levels, the second at two levels, and the third at three levels, we will have $4 \times 2 \times 3 = 24$ factorial combinations or treatments. Factorial experiments may be conducted in any of the standard designs. The randomized block and split plot design are the most common for factorial experiments in forest research.

7.16.6 Split-Plot Design

When two or more types of treatment are applied in factorial combinations, it may be that one type can be applied on relatively small plots while the other type is best applied to larger plots. Rather than make all plots of the size needed for the second type, a split-plot design can be employed. In this design, the major (large-plot) treatments are applied to a number of plots with replication accomplished through any of the common designs (such as complete randomization, randomized blocks, Latin square). Each major plot is then split into a number of subplots, equal to the number of minor (small-plot) treatments. Minor treatments are assigned at random to subplots within each major plot.

As an example, a test was to be made of direct seeding of loblolly pine at six different dates on burned and unburned seedbeds. To get typical burn effects, major plots 6 acres in size were selected. There were to be four replications of major treatments in randomized blocks. Each major plot was divided into six 1-acre subplots for seeding at six dates. The field layout was somewhat as follows (blocks denoted by Roman numerals, burning treatment by capital letters, day of seeding by small letters):

One pound of seed was sowed on each 1-acre subplot. Seedling counts were made at the end of the first growing season. Results were as follows:

Date	I A	I BA	II BA		III BA		IV BA		Date Subtotals B		Date Totals
a	900	880	810	1100	760	960	1040	1040	3510	3980	7490
b	880	1050	1170	1240	1060	1110	910	1120	4020	4520	8540
c	1530	1140	1160	1270	1390	1320	1540	1080	5620	4810	10430
d	1970	1360	1890	1510	1820	1490	2140	1270	7820	5630	13450
e	1960	1270	1670	1380	1310	1500	1480	1450	6420	5600	12020
f	830	150	420	380	570	420	760	270	2580	1220	3800
Major plot totals	8070	5850	7120	6880	6910	6800	7870	6230	29,970	25,760	—
Block totals	13,920		14,000		13,710		14,100		—		55,730

The correction term and total sum of squares are calculated using the 48 subplot values:

$$CT = \frac{(\text{Grand total of all subplots})^2}{\text{Total number of subplots}} = \frac{55,730^2}{48} = 64,704,852$$

$$\text{Total } \underset{47df}{SS} = \sum^{48}(\text{Subplot values})^2 - CT = \left(900^2 + 880^2 + \ldots + 270^2\right) - CT = 9,339,648$$

Before partitioning the total sum of squares into its components, it may be instructive to ignore subplots for the moment, and examine the major plot phase of the study. The major phase can be viewed as a straight randomized block design with two burning treatments in each of four blocks. The analysis would be as follows:

Source of Variation	Degrees of Freedom
Blocks	3
Burning	1
Error (major plots)	3
Major plots	7

Now, looking at the subplots, we can think of the major plots as blocks. From this standpoint, we would have a randomized block design with six dates of treatment in each of eight blocks (major plots) for which the analysis is as follows:

Source of Variation	Degrees of Freedom
Major plots	7
Dates	5
Remainder	35
Subplots (= Total)	47

In this analysis, the remainder is made up of two components. One of these is the burning–date interaction, with five degrees of freedom. The rest, with 30 degrees of freedom, is called the *subplot error*. Thus, the complete breakdown of the split-plot design is as follows:

Source of Variation	Degrees of Freedom
Blocks	3
Burning	1
Major plot error	3
Total major plot	7
Date	5
Total date	5
Burning X date	5
Subplot error	30
Total remainder	35
Total	47

The various sums of squares are obtained in an analogous manner. We first compute the following:

$$\text{Major plot } \underset{7df}{SS} = \frac{\sum\limits^{8}(\text{Major plot totals}^2)}{\text{No. of subplots per major plot}} - CT$$

$$= \frac{\left(8070^2 + \ldots + 6230^2\right)}{6} - CT = 647,498$$

$$\text{Block } \underset{3df}{SS} = \frac{\sum\limits^{4}(\text{Block totals}^2)}{\text{No. of subplots per block}} - CT$$

$$= \frac{\left(13,920^2 + \ldots + 14,100^2\right)}{12} - CT = 6856$$

$$\text{Burning } \underset{1df}{SS} = \frac{\sum\limits^{2}(\text{Burning treatment totals}^2)}{\text{No. of subplots per burning treatment}} - CT$$

$$= \frac{\left(29,970^2 + 25,760^2\right)}{24} - CT = 369,252$$

$$\text{Major plot error } \underset{3df}{SS} = \text{Major plot } \underset{7df}{SS} - \text{Block } \underset{3df}{SS} - \text{Burning } \underset{1df}{SS} = 271,390$$

$$\text{Subplot } \underset{40df}{SS} = \text{Total } \underset{47df}{SS} - \text{Major plot } \underset{7df}{SS} = 8{,}692{,}150$$

$$\text{Date } \underset{5df}{SS} = \frac{\sum\limits^{6}(\text{Date totals}^2)}{\text{No. of subplots per date}} - CT = \frac{\left(7490^2 + \ldots + 3800^2\right)}{8} - CT = 7{,}500{,}086$$

To get the sum of squares for the interaction between date and burning we resort to a factorial experiment device—the two-way table of the treatment combination totals:

Burning	Date						Burning Subtotals
	a	b	c	d	e	f	
A	3510	4020	5620	7820	6420	2580	29,970
B	3980	4520	4810	5630	5600	1220	25,760
Date subtotals	7490	8540	10,430	13,450	12,020	3800	55,730

$$\text{Date–burning subclass } \underset{11df}{SS} = \frac{\sum\limits^{12}(\text{Date–burning combination totals}^2)}{\text{No. of subplots per date–burning combination}} - CT$$

$$= \frac{\left(3510^2 + \ldots + 1220^2\right)}{4} - CT = 8{,}555{,}723$$

$$\text{Date–burning interaction } \underset{5df}{SS} = \text{Date–burning subclass } \underset{11df}{SS} - \text{Date } \underset{5df}{SS} - \text{Burning } \underset{1df}{SS} = 686{,}385$$

$$\text{Subplot error } \underset{30df}{SS} = \text{Subplot } \underset{40df}{SS} - \text{Date } \underset{5df}{SS} - \text{Date–burning interaction } \underset{5df}{SS} = 505{,}679$$

Thus the completed analysis table is

Source of Variation	Degrees of Freedom	Sums of Squares	Mean Squares
Blocks	3	6856	—
Burning	1	369,252	369,252
Major plot error	3	271,390	90,463
Date	5	7,500,086	1,500,017
Date–burning interaction	5	686,385	137,277
Subplot error	30	505,679	16,856
Total	47	9,339,648	

The F test for burning is

$$F_{1/3df} = \frac{\text{Burning mean square}}{\text{Major plot error mean square}} = \frac{369{,}252}{90{,}463} = 4.082 \text{ (not significant at 0.05 level)}$$

The F test for dates is

$$F_{5/30df} = \frac{\text{Date mean square}}{\text{Subplot error mean square}} = \frac{1,500,017}{16,856} = 88.99 \text{ (significant at 0.01 level)}$$

And for the date–burning interaction,

$$F_{5/30df} = \frac{\text{Date–burning interaction mean square}}{\text{Subplot error mean square}} = \frac{137,277}{16,856} = 8.14 \text{ (significant at 0.01 level)}$$

Note that the major plot error is used to test the sources above the dashed line in the table, while the subplot error is used for the sources below the line. Because the subplot error is a measure of random variation *within* major plots it will usually be smaller than the major plot error, which is a measure of the random variation between major plots. In addition to being smaller, the subplot error will generally have more degrees of freedom than the major plot error, and for these reasons the sources below the dashed line will usually be tested with greater sensitivity than the sources above the line. This fact is important; in planning a split-plot experiment the designer should try to get the items of greatest interest below the line rather than above. Rarely will the major plot error be appreciably smaller than the subplot error. If it is, the conduct of the study and the computations should be carefully examined. If desired, the subplots can also be split for a third level of treatment, producing a split-split-plot design. The calculations follow the same general pattern but are more involved. A split-split-plot design has three separate error terms. For comparisons among major or subplot treatments, F tests with a single degree of freedom may be made in the usual manner. Comparisons among major plot treatments should be tested against the major plot error mean square, while subplot treatment comparisons are tested against the subplot error. In addition, it is sometimes desirable to compare the means of two treatment combinations. This can get tricky, for the variation among such means may contain more than one source of error. A few of the more common cases are discussed below.

In general, the t test for comparing two equally replicated treatment means is

$$t = \frac{\text{Mean difference}}{\text{Standard error of the mean difference}} = \frac{\bar{D}}{s_{\bar{D}}}$$

1. For the difference between two major treatment means:

$$s_{\bar{D}} = \sqrt{\frac{2(\text{Major plot error mean square})}{(m)(R)}}$$

where R is the number of replications of major treatments, and m is the number of subplots per major plot; t has degrees of freedom equal to the degree of freedom for the major plot error.

2. For the difference between two minor treatment means:

$$s_{\bar{D}} = \sqrt{\frac{2(\text{Subplot error mean square})}{(R)(M)}}$$

where M is the number of major plot treatments; t has degrees of freedom equal to the degree of freedom for the subplot error.

3. For the difference between two minor treatments within a single major treatment:

$$s_{\bar{D}} = \sqrt{\frac{2(\text{Subplot error mean square}}{R}}$$

where the degrees of freedom for t is equal to the degrees of freedom for the subplot error.

4. For the difference between the means of two major treatments at a single level of a minor treatment, or between the means of two major treatments at different levels of a minor treatment:

$$s_{\bar{D}} = \sqrt{2\left[\frac{(m-1)(\text{Subplot error mean square}) + (\text{Major plot error mean square}}{(m)(R)}\right]}$$

In this case, t will not follow the t distribution. A close approximation to the value of t required for significance at the a level is given by

$$t = \frac{(m-1)(\text{Subplot error mean square})t_m + (\text{Major plot error mean square})t_M}{(m-1)(\text{Subplot error mean square}) + (\text{Major plot error mean square})}$$

where
t_m = Tabular value of t at the α level for degrees of freedom equal to the degrees of freedom for the subplot error.
t_M = Tabular value of t at the α level for degrees of freedom equal to the degrees of freedom for the major plot error.

Other symbols are as previously defined.

7.16.7 Missing Plots

A mathematician who had developed a complex electronic computer program for analyzing a wide variety of experimental designs was asked how he handled missing plots. His disdainful reply was, "We tell our research workers not to have missing plots." This is good advice. But it is sometimes hard to follow, and particularly so in forest, environmental, and ecological research, where close control over experimental material is difficult and studies may run for several years. The likelihood of plots being lost during the course of a study should be considered when selecting an experimental design. Lost plots are least troublesome in the simple designs. For this reason, complete randomization and randomized blocks may be preferable to the more intricate designs when missing data can be expected.

In the complete randomization design, loss of one or more plots causes no computational difficulties. The analysis is made as though the missing plots never existed. Of course, a degree of freedom will be lost from the total and error terms for each missing plot and the sensitivity of the test will be reduced. If missing plots are likely, the number of replications should be increased accordingly. In the randomized block design, completion of the analysis will usually require an estimate of the values for the missing plots. A single missing value can be estimated by

$$X = \frac{bB + tT - G}{(b-1)(t-1)}$$

where

b = Number of blocks.
B = Total of all other units in the block with a missing plot.
t = Number of treatments.
T = Total of all other units that received the same treatment as the missing plot.
G = Total of all observed units.

If more than one plot is missing, the customary procedure is to insert guessed values for all but one of the missing units, which is then estimated by the above formula. This estimate is used in obtaining an estimated value for one of the guessed plots, and so on through each missing unit. Then the process is repeated, with the first estimates replacing the guessed values. The cycle should be repeated until the new approximations differ little from the previous estimates.

The estimated values are now applied in the usual analysis-of-variance calculations. For each missing unit one degree of freedom is deducted from the total and from the error term.

A similar procedure is used with the Latin square design, but the formula for a missing plot is

$$X = \frac{r(R + C + T) - 2G}{(r-1)(r-2)}$$

where

r = Number of rows.
R = Total of all observed units in the row with the missing plot.
C = Total of all observed units in the column with the missing plot.
T = Total of all observed units in the missing plot treatment.
G = Grand total of all observed units.

With the split-plot design, missing plots can cause trouble. A single missing subplot value can be estimated by the following equation:

$$X = \frac{rP + m\left(T_{ij}\right) - \left(T_i\right)}{(r-1)(m-1)}$$

where

r = Number of replications of major plot treatments.
P = Total of all observed subplots in the major plot having a missing subplot.
m = Number of subplot treatments.
T_{ij} = Total of all subplots having the same treatment combination as the missing unit.
T_i = Total of all subplots having the same major plot treatment as the missing unit.

For more than one missing subplot the iterative process described for randomized blocks must be used. In the analysis, one degree of freedom will be deducted from the total and subplot error terms for each missing subplot.

When data for missing plots are estimated, the treatment mean square for all designs is biased upwards. If the proportion of missing plots is small, the bias can usually be ignored. Where the proportion is large, adjustments can be made as described in the standard references on experimental designs.

7.17 REGRESSION

7.17.1 SIMPLE LINEAR REGRESSION

An environmental researcher had an idea that she could tell how well a loblolly pine was growing from the volume of the crown. Very simple: big crown—good growth, small crown—poor growth. But she couldn't say how big and how good, or how small and how poor. What she needed was regression analysis, which would allow her to express a relationship between tree growth and crown volume in an equation. Given a certain crown volume, she could use the equation to predict what the tree growth was. To gather data, she ran parallel survey lines across a large tract that was representative of the area in which she was interested. The lines were 5 chains apart. At each 2-chain mark along the lines, she measured the nearest loblolly pine of at least 5.6 inches diameter at breast height (d.b.h.; 4.5 ft above the forest floor on the uphill side of the tree) for crown volume and basal area growth over the past 10 years.

A portion of the data is printed below to illustrate the methods of calculation. Crown volume in hundreds of cubic feet is labeled X and basal area growth in square feet is labeled Y. Now, what can we tell the environmental researcher about the relationship?

X Crown Volume	Y Growth	X Crown Volume	Y Growth	X Crown Volume	Y Growth
22	0.36	53	0.47	51	0.41
6	0.09	70	0.55	75	0.66
93	0.67	5	0.07	6	0.18
62	0.44	90	0.69	20	0.21
84	0.72	46	0.42	36	0.29
14	0.24	36	0.39	50	0.56
52	0.33	14	0.09	9	0.13
69	0.61	60	0.54	2	0.10
104	0.66	103	0.74	21	0.18
100	0.80	43	0.64	17	0.17
41	0.47	22	0.50	87	0.63
85	0.60	75	0.39	97	0.66
90	0.51	29	0.30	33	0.18
27	0.14	76	0.61	20	0.06
18	0.32	20	0.29	96	0.58
48	0.21	29	0.38	61	0.42
37	0.54	30	0.53		
67	0.70	59	0.58		
56	0.67	70	0.62		
31	0.42	81	0.66		
17	0.39	93	0.69		
7	0.25	99	0.71		
2	0.06	14	0.14		
Totals				3,050	26.62
Means ($n = 62$)				49.1935	0.42935

Often, the first step is to plot the field data on coordinate paper (see Figure 7.3). This is done to provide some visual evidence of whether the two variables are related. If there is a simple relationship, the plotted points will tend to form a pattern (a straight line or curve). If the relationship is very strong, the pattern will generally be distinct. If the relationship is weak, the points will be more spread out and the pattern less definite. If the points appear to fall pretty much at random, there may be no simple relationship or one that is so very poor as to make it a waste of time to fit any regression.

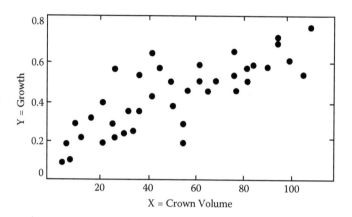

FIGURE 7.3   Plotting of growth (F) over crown volume (X).

The type of pattern (straight line, parabolic curve, exponential curve, etc.) will influence the regression model to be fitted. In this particular case, we will assume a simple straight-line relationship. After selecting the model to be fitted, the next step will be to calculate the corrected sums of squares and products. In the following equations, capital letters indicate uncorrected values of the variables; lowercase letters will be used for the corrected values.

The corrected sum of squares for Y is

$$\sum y^2 = \sum_n Y^2 - \frac{\left(\sum_n Y\right)^2}{n} = \left(0.36^2 + 0.09^2 + \ldots + 0.42^2\right) - \frac{26.62^2}{62} = 2.7826$$

The corrected sum of squares for X is

$$\sum x^2 = \sum_n X^2 - \frac{\left(\sum_n X\right)^2}{n} = \left(22^2 + 6^2 + \ldots + 61^2\right) - \frac{3050^2}{62} = 59,397.6775$$

The corrected sum of products is

$$\sum xy = \sum_n (XY) - \frac{\left(\sum_n X\right)\left(\sum_n Y\right)}{n} = \left[(22)(0.36) + (6)(0.09) + \ldots + (61)(0.42)\right] - \frac{(3050)(26.62)}{62}$$
$$= 354.1477$$

The general form of equation for a straight line is $Y = a + bX$. In this equation, a and b are constants or regression coefficients that must be estimated. According to the principle of least squares, the best estimates of these coefficients are

$$b = \frac{\sum xy}{\sum x^2} = \frac{354.1477}{59,397.6775} = 0.005962$$

$$a = \bar{Y} - b\bar{X} = 0.42935 - (0.005962)(49.1935) = 0.13606$$

Substituting these estimates in the general equation gives

$$\hat{Y} = 0.13606 + 0.005962X$$

where $\hat{Y}$ is used to indicate that we are dealing with an estimated value of Y.

With this equation we can estimate the basal area growth for the past 10 years ($\hat{Y}$) from the measurements of the crown volume X. Because Y is estimated from a known value of X, it is called the dependent variable and X is the independent variable. In plotting on graph paper, the values of Y are usually (purely by convention) plotted along the vertical axis (ordinate) and the values of x along the horizontal axis (abscissa).

7.17.1.1 How Well Does the Regression Line Fit the Data?

A regression line can be thought of as a moving average. It gives an average value of Y associated with a particular value of X. Of course, some values of Y will be above the regression line (moving average) and some below, just as some values of Y are above or below the general average of Y. The corrected sum of squares for Y (i.e., Σy^2) estimates the amount of variation of individual values of Y about the mean value of Y. A regression equation is a statement that part of the observed variation in Y (estimated by Σy^2) is associated with the relationship of Y to X. The amount of variation in Y that is associated with the regression on X is the reduction or regression sum of squares:

$$\text{Reduction } SS = \frac{\left(\sum xy\right)^2}{\sum x^2} = \frac{(354.1477)^2}{59,397.6775} = 2.1115$$

As noted above, the total variation in Y is estimated by $\Sigma y^2 = 2.7826$ (as previously calculated). The part of the total variation in Y that is not associated with the regression is the residual sum of squares:

$$\text{Residual } SS = \sum y^2 - \text{Reduction } SS = 2.7826 - 2.1115 = 0.6711$$

In analysis of variance we used the unexplained variation as a standard for testing the amount of variation attributable to treatments. We can do the same in regression. What's more, the familiar F test will serve.

Source of Variation	Degrees of Freedom[a]	Sums of Squares	Mean Squares[b]
Due to regression $\left[= \dfrac{(\Sigma xy)^2}{\Sigma x^2} \right]$	1	2.1115	2.1115
Residual (unexplained)	60	0.6711	0.01118
Total ($= \Sigma y^2$)	61	2.7826	

[a] As there are 62 values of Y, the total sum of squares has 61 degrees of freedom. The regression of Y on X has one degree of freedom. The residual degrees of freedom are obtained by subtraction.
[b] Mean square is, as always, equal to sum of squares/degrees of freedom.

The regression is tested by

$$F = \frac{\text{Regression mean square}}{\text{Residual mean square}} = \frac{2.1115}{0.01118} = 188.86$$

Because the calculated F is much greater than tabular $F_{0.01}$ with 1/60 degree of freedom, the regression is deemed significant at the 0.01 level.

Before we fitted a regression line to the data, Y had a certain amount of variation about its mean ($\overline{Y}$). Fitting the regression was, in effect, an attempt to explain part of this variation by the linear association of Y with X. But even after the line had been fitted, some variation was unexplained—that of Y about the regression line. When we tested the regression line above, we merely showed that the part of the variation in Y that is explained by the fitted line is significantly greater than the part that the line left unexplained. The test did not show that the line we fitted gives the best possible description of the data (a curved line might be even better), nor does it mean that we have found the true mathematical relationship between the two variables. There is a dangerous tendency to ascribe more meaning to a fitted regression than is warranted.

It might be noted that the residual sum of squares is equal to the sum of the squared deviations of the observed values of Y from the regression line. That is,

$$\text{Residual } SS = \sum\left(Y - \hat{Y}\right)^2 = \sum(Y - a - bX)^2$$

The principle of least squares says that the best estimates of the regression coefficients (a and b) are those that make this sum of squares a minimum.

7.17.1.2 Coefficient of Determination

The coefficient of determination, denoted R^2, is used in the context of statistical models whose main purpose is the prediction of future outcomes on the basis of other related information. Stated differently, the coefficient of determination is a ratio that measures how well a regression fits the sample data:

$$\text{Coefficient of determination} = \frac{\text{Reduction } SS}{\text{Total } SS} = \frac{2.1115}{2.7826} = 0.758823$$

When someone says, "76% of variation in Y was associated with X," she means that the coefficient of determination was 0.76. Note that R^2 is most often seen as a number between 0 and 1.0, used to describe how well a regression lien fit a set of data. An R^2 near 1.0 indicates that a regression line fits the data well, while an R^2 closer to 0 indicates that a regression line does not fit the data very well.

The coefficient of determination is equal to the square of the correlation coefficient:

$$\frac{\text{Reduction } SS}{\text{Total } SS} = \frac{\left(\sum xy\right)^2 / \sum x^2}{\sum y^2} = \frac{\left(\sum xy\right)^2}{\left(\sum x^2\right)\left(\sum y^2\right)} = r^2$$

In fact, most present-day users of regression refer to R^2 values rather than to coefficients of determination.

7.17.1.3 Confidence Intervals

Because it is based on sample data, a regression equation is subject to sample variation. Confidence limits (i.e., a pair of numbers used to estimate a characteristic of a population) on the regression line can be obtained by specifying several values over the range of X and computed by

$$\hat{Y} \pm t \sqrt{(\text{Residual mean square})\left(\frac{1}{n} + \frac{\left(X_0 - \overline{X}\right)^2}{\sum x^2}\right)}$$

where X_0 = a selected value of X, and degrees of freedom for t equal the degrees of freedom for residue mean square. In the example we had

$$\hat{Y} = 0.13606 + 0.005962X$$

Residual mean square = 0.01118 with 60 degrees of freedom

$$n = 62$$

$$\bar{X} = 49.1935$$

$$\sum x^2 = 59,397.6775$$

So, if we pick $X_0 = 28$ we have $\hat{Y} = 0.303$, and 95% confidence limits

$$\hat{Y} \pm t \sqrt{(\text{Residual mean square}) \left(1 + \frac{1}{n} + \frac{(X_0 - \bar{X})^2}{\sum x^2} \right)}$$

For other values of X_0 we would get:

		95% Limits	
X_0	$\hat{Y}$	Lower	Upper
8	0.184	0.139	0.229
49.1935	0.429	0.402	0.456
70	0.553	0.521	0.585
90	0.673	0.629	0.717

Note that these are confidence limits for the regression of Y on X (see Figure 7.4). They indicate the limits within which the true mean of Y for a given X will lie unless a 1-in-20 chance has occurred. The limits do not apply to a single predicted value of Y. The limits within which a single Y might lie are given by

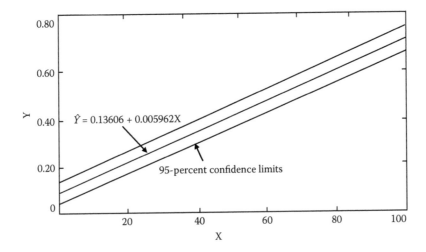

FIGURE 7.4   Confidence limits for the regression of Y on X.

$$\hat{Y} \pm t \sqrt{(\text{Residual mean square}) \left(1 + \frac{1}{n} + \frac{\left(X_0 - \bar{X}\right)^2}{\sum x^2}\right)}$$

In addition to assuming that the relationship of Y to X is linear, the above method of fitting assumes that the variance of Y about the regression line is the same at all levels of X (the assumption of homogeneous variance or homoscedasticity—that is, the property of having equal variances). The fitting does not assume nor does it require that the variation of Y about the regression line follow the normal distribution. However, the F test does assume normality, and so does the use of t for the computation of confidence limits.

There is also an assumption of independence of the errors (departures from regression) of the sample observations. The validity of this assumption is best ensured by selecting the sample units at random. The requirements of independence may not be met if successive observations are made on a single unit or if the units are observed in clusters. For example, a series of observations of tree diameter made by means of a growth band would probably lack independence.

Selecting the sample units so as to get a particular distribution of the X values does not violate any of the regression assumptions, provided the Y values are a random sample of all Y values associated with the selected values of X. Spreading the sample over a wide range of X values will usually increase the precision with which the regression coefficients are estimated. This device must be used with caution, however, for if the Y values are not random then the regression coefficients and mean square residual may be improperly estimated.

7.17.2 MULTIPLE REGRESSION

It frequently happens that a variable (Y) in which we are interested is related to more than one independent variable. If this relationship can be estimated, it may enable us to make more precise predictions of the dependent variable than would be possible by a simple linear regression. This brings us up against multiple regression, which describes the changes in a dependent variable associated with changes in one or more independent variables; it is a little more work but no more complicated than a simple linear regression.

The calculation methods can be illustrated with the following set of hypothetical data from an environmental study relating the growth of even-aged loblolly–shortleaf pine stands to the total basal area (X_1), the percentage of the basal area in loblolly pine (X_2), and loblolly pine site index (X_3).

Y	X_1	X_2	X_3
65	41	79	75
78	90	48	83
85	53	67	74
50	42	52	61
55	57	52	59
59	32	82	73
82	71	80	72
66	60	65	66
113	98	96	99
86	80	81	90
104	101	78	86
92	100	59	88
96	84	84	93
65	72	48	70
81	55	93	85

77	77	68	71
83	98	51	84
97	95	82	81
90	90	70	78
87	93	61	89
74	45	96	81
70	50	80	77
75	60	76	70
75	68	74	76
93	75	96	85
76	82	58	80
71	72	58	68
61	46	69	65
Sums	2206	1987	2003	2179
Means ($n = 28$)	78.7857	70.9643	71.5387	77.8214

With these data we would like to fit an equation of the form

$$Y = a + b_1X_1 + b_2X_2 + b_3X_3$$

According to the principle of least squares, the best estimates of the X coefficients can be obtained by solving the set of least squares normal equations.

b_1 equation: $\left(\sum x_1^2\right)b_1 + \left(\sum x_1x_2\right)b_2 + \left(\sum x_1x_3\right)b_3 = \sum x_1y$

b_2 equation: $\left(\sum x_1x_2\right)b_1 + \left(\sum x_2^2\right)b_2 + \left(\sum x_2x_3\right)b_3 = \sum x_2y$

b_3 equation: $\left(\sum x_1x_3\right)b_1 + \left(\sum x_2x_3\right)b_2 + \left(\sum x_3^2\right)b_3 = \sum x_3y$

where

$$\sum x_ix_j = \sum X_iY_j - \frac{\left(\sum X_i\right)\left(\sum X_j\right)}{n}$$

Having solved for the X coefficients (b_1, b_2, and b_3), we obtain the constant term by solving

$$a = \bar{Y} - b_1\bar{X}_1 + b_2\bar{X}_2 - b_3\bar{X}_3$$

Derivation of the least squares normal equations requires a knowledge of differential calculus. However, for the general linear mode with a constant term

$$Y = a + b_1X_1 + b_2X_2 + \ldots b_kX_k$$

the normal equations can be written quite mechanically once their pattern has been recognized. Every term in the first row contains an x_1, every term in the second row an x_2, and so forth down to the kth row, every term of which will have an x_k. Similarly, every term in the first column has an x_1 and a b_1, every term in the second column has an x_2 and a b_2, and so on through the kth column, every term of which has an x_k and a b_k. On the right side of the equations, each term has a y times the x that is appropriate for a particular row. So, for the general linear model given above, the normal equations are

b_1 equation: $\left(\sum x_1^2\right)b_1 + \left(\sum x_1 x_2\right)b_2 + \left(\sum x_1 x_3\right)b_3 + \ldots + \left(\sum x_1 x_k\right)b_k = \sum x_1 y$

b_2 equation: $\left(\sum x_1 x_2\right)b_1 + \left(\sum x_2^2\right)b_2 + \left(\sum x_2 x_3\right)b_3 + \ldots + \left(\sum x_2 x_k\right)b_k = \sum x_2 y$

b_3 equation: $\left(\sum x_1 x_3\right)b_1 + \left(\sum x_2 x_3\right)b_2 + \left(\sum x_3^2\right)b_3 + \ldots + \left(\sum x_3 x_k\right)b_k = \sum x_3 y$

$\vdots$

b_k equation: $\left(\sum x_1 x_k\right)b_1 + \left(\sum x_2 x_k\right)b_2 + \left(\sum x_3 x_k\right)b_3 + \ldots + \left(\sum x_k^2\right)b_k = \sum x_k y$

Given the X coefficients, the constant term can be computed as

$$a = \bar{Y} - b_1 \bar{X}_1 - b_2 \bar{X}_2 - \ldots - b_k \bar{X}_k$$

Note that the normal equations for the general linear model include the solution for the simple linear regression

$$\left(\sum x_1^2\right)b_1 = \sum x_1 y$$

Hence,

$$b_1 = \left(\sum x_1 y\right)\Big/\sum x_1^2$$

In fact, all of this section on multiple regression can be applied to the simple linear regression as a special case.

The corrected sums of squares and products are computed in the familiar manner:

$$\sum y^2 = \sum Y^2 - \frac{\left(\sum Y\right)^2}{n} = \left(65^2 + \ldots + 61^2\right) - \frac{(2206)^2}{28} = 5974.7143$$

$$\sum x_1^2 = \sum X_1^2 - \frac{\left(\sum X_1\right)^2}{n} = \left(41^2 + \ldots + 46^2\right) - \frac{(1987)^2}{28} = 11{,}436.9643$$

$$\sum x_1 y = \sum X_1 Y - \frac{\left(\sum X_1\right)\left(\sum Y\right)}{n} = (41)(65) + \ldots + (46)(61) - \frac{(1987)(2206)}{28} = 6428.7858$$

Similarly,

$$\sum x_1 x_2 = -1171.4642$$

$$\sum x_1 x_3 = 3458.8215$$

$$\sum x_2^2 = 5998.9643$$

$$\sum x_2 x_3 = 1789.6786$$

$$\sum x_2 y = 2632.2143$$

$$\sum x_3^2 = 2606.1072$$

$$\sum x_3 y = 3327.9286$$

Putting these values in the normal equations gives:

$$11,436.9643b_1 - 1171.4642b_2 + 3458.8215b_3 = 6428.7858$$
$$-1171.4642b_1 + 5998.9643b_2 + 1789.6786b_3 = 2632.2143$$
$$3458.8215b_1 - 1789.6786b_2 + 2606.1072b_3 = 3327.9286$$

These equations can be solved by any of the standard procedures for simultaneous equations. One approach (applied to the above equations) is as follows:

1. Divide through each equation by the numerical coefficient of b_1.

$$b_1 - 0.102,427,897b_2 + 0.302,424,788b_3 = 0.562,105,960$$
$$b_1 - 5.120,911,334b_2 - 1.527,727,949b_3 = -2.246,943,867$$
$$b_1 + 0.517,424,389b_2 + 0.753,466,809b_3 = 0.962,156,792$$

2. Subtract the second equation from the first and the third from the first so as to leave two equations in b_2 and b_3.

$$5.018,483,437b_2 + 1.830,152,737b_3 = 2.809,049,827$$
$$-0.619,852,286b_2 - 0.451,042,021b_3 = -0.400,050,832$$

3. Divide through each equation by the numerical coefficient of b_2.

$$b^2 + 0.364,682,430b_3 = 0.559,740,779$$
$$b^2 + 0.727,660,494b_2 = 0.645,397,042$$

4. Subtract the second of these equations from the first, leaving one equation in b_3.

$$-0.362,978,064b_3 = -0.085,656,263$$

5. Solve for b_3.

$$b_3 = \frac{-0.085,656,263}{-0.326,978,064} = 0.235,981,927$$

6. Substitute this value of b_3 in one of the equations (the first one, for example) of step 3 and solve for b_2.

$$b_2 + (0.364,682,43)(0.381,927) = 0.59,740,779$$
$$b_2 = 0.473,682,316$$

7. Substitute the solutions for b_2 and b_3 in one of the equations (the first one, for example) of step 1, and solve for b_1.

$$b_1 - (0.102,427,897)(0.473,682,316) + (0.302,424,788)(0.235,981,927) = 0.562,105,960$$
$$b_1 = 0.539,257,459$$

8. As a check, add up the original normal equations and substitute the solutions for b_1, b_2, and b_3.

$$13,724.3216b_1 + 6,617.1787b_2 + 7,854.6073b_3 = 12,388.9287$$

$$12,388.92869 \approx 12,388.9287$$

Given the values of b_1, b_2, and b_3 we can now compute

$$a = \bar{Y} - b_1\bar{X}_1 - b_2\bar{X}_2 - b_3\bar{X}_3 = -11.7320$$

Thus, after rounding of the coefficients, the regression equation is

$$\hat{Y} = -11.732 + 0.539X_1 + 0.474X_2 + 0.236X_3$$

It should be noted that in solving the normal equations more digits have been carried than would be justified by the rules for number of significant digits. Unless this is done, the rounding errors may make it difficult to check the computations.

7.17.2.1 Tests of Significance

Tests of significance refer to the methods of inference used to support or reject claims based on sample data. To test the significance of the fitted regression, the outline for the analysis of variance is

Source of Variation	Degrees of Freedom
Reduction due to regression on X_1, X_2, and X_3	3
Residuals	24
Total	27

The degrees of freedom for the total are equal to the number of observations minus 1. The total sum of squares is

$$\text{Total } SS = \sum y^2 = 5974.7143$$

The degrees of freedom for the reduction are equal to the number of independent variables fitted, in this case 3. The reduction sum of squares for any least squares regression is

$$\text{Reduction } SS = \sum (\text{Estimated coefficients}) \, (\text{right side of their normal equations})$$

In this example there are three coefficients estimated by the normal equations, and so

$$\text{Reduction } \underset{3df}{SS} = b_1\left(\sum x_1 y\right) + b_2\left(\sum x_2 y\right) + b_3\left(\sum x_3 y\right)$$

$$= (0.53926)(6428.7858) + (0.47368)(2632.2143) + (0.23598)(3327.9286)$$

$$= 5498.9389$$

The residual df and sum of squares are obtained by subtraction. Thus the analysis becomes

Source of Variation	Degrees of Freedom	Sums of Squares	Mean Squares
Reduction due to X_1, X_2, and X_3	3	5498.9389	1832.9796
Residuals	24	475.7754	19.8240
Total	27	5974.7143	

To test the regression we compute

$$F_{3/24df} = \frac{1832.9696}{19.8240} = 92.46$$

which is significant at the 0.01 level.

Often we will want to test individual terms of the regression. In the previous example we might want to test the hypothesis that the true value of b_3 is zero. This would be equivalent to testing whether the viable X_3 makes any contribution to the prediction of Y. If we decide a b_3 may be equal to zero, we might rewrite the equation in terms of X_1 and X_2. Similarly, we could test the hypothesis that b_1 and b_3 are both equal to zero.

To test the contribution of any set of the independent variables in the presence of the remaining variables:

1. Fit all independent variables and compute the reduction and residual sums of squares.
2. Fit a new regression that includes only the variables not being tested. Compute the reduction due to this regression.
3. The reduction obtained in the first step minus the reduction in the second step is the gain due to the variables being tested.
4. The mean square for the gain (step 3) is tested against the mean square residual from the first step.

7.17.2.2 Coefficient of Multiple Determination

As a measure of how well the regression fits the data it is customary to compute the ratio of the reduction sum of squares to the total sum of squares. This ratio is symbolized by R^2 and is sometimes called the coefficient of determination:

$$R^2 = \frac{\text{Reduction } SS}{\text{Total } SS}$$

For the regression of Y on X_1, X_2, and X_3,

$$R^2 = \frac{5498.9389}{5974.7143} = 0.92$$

The R^2 value is usually referred to by saying that a certain percentage (92 in this case) of the variation in Y is associated with regression. The square root (R) of the ratio is the multiple correlation coefficient.

7.17.2.3 The c-Multipliers

Putting confidence limits on a multiple regression requires computation of the Gauss or c-multipliers. The c-multipliers are the elements of the inverse of the matrix of corrected sums of squares and products as they appear in the normal equations.

7.17.3 CURVILINEAR REGRESSIONS AND INTERACTIONS

7.17.3.1 Curves

Many forms of curvilinear relationships can be fitted by the regression methods that have been described in the previous sections. If the relationship between height and age is assumed to be hyperbolic so

$$\text{Height} = a + \frac{b}{\text{Age}}$$

then we could let Y = height and X_1 = 1/age and fit

$$Y = a + b_1 X_1$$

Similarly, if the relationship between Y and X is quadratic

$$Y = a + bX + cX^2$$

then we can let $X = X_1$ and $X^2 = X_2$ and fit

$$Y = a + b_1 X_1 + b_2 X_2$$

Functions such as

$$Y = aX^b$$

$$Y = a(b^x)$$

$$10^Y = aX^b$$

which are nonlinear in the coefficients can sometimes be made linear by a logarithmic transformation. The equation

$$Y = aX^b$$

would become

$$\log Y = \log a + b(\log X)$$

which could be fitted by

$$Y' = a + b_1 X_1$$

where $Y' = \log Y$, and $X_1 = \log X$.
     The second equation transforms to

$$\log Y = \log a + (\log b) X$$

The third becomes

$$Y = \log a + b(\log X)$$

Both can be fitted by the linear model.
     When making these transformations, the effect on the assumption of homogeneous variance must be considered. If Y has homogeneous variance, then log Y probably will not—and *vice versa*. Some curvilinear models cannot be fitted by the methods that have been described, including

$$Y = a + b^x$$
$$Y = a(X - b)^2$$
$$Y = a(X_1 - b)(X_2 - c)$$

Fitting these models requires more cumbersome procedures.

7.17.3.2   Interactions

Suppose that there is a simple linear relationship between Y and X_1. If the slope (b) of this relationship varies, depending on the level of some other independent variable (X_2), then X_1 and X_2 are said to interact. Such interactions can sometimes be handled by introducing interaction variables. To illustrate, suppose that we know that there is a linear relationship between Y and X_1:

$$Y = a + bX_1$$

Suppose further that we know or suspect that the slope (b) varies linearly with Z:

$$b = a' + b'Z$$

This implies the relationship

$$Y = a + (a' + b'Z)X_1$$

or

$$Y = a + a'X_1 + b'X_1Z$$

where $X_2 = X_1Z$, an interaction variable.

If the Y-intercept is also a linear function of Z, then

$$a = a'' + b''Z$$

and the form of relationship is

$$Y = a'' + b''Z + a'X_1 + b'X_1Z$$

7.17.4   GROUP REGRESSIONS

Linear regressions of Y on X were fitted for each of two groups:

Group A									Sum	Mean
Y	3	7	9	6	8	13	10	12	14	82	9.111
X	1	4	7	7	2	9	10	6	12	58	6.444

where

$$n = 9, \quad \sum Y^2 = 848, \quad \sum XY = 609, \quad \sum X^2 = 480, \quad \sum y^2 = 100.8889$$

$$\sum xy = 80.5556, \quad \sum x^2 = 106.2222, \quad \hat{Y} = 4.224 + 0.7584X$$

Residual $SS = 39.7980$, with 7 degrees of freedom

Group B												Sum	Mean
Y	4	6	12	2	8	7	0	5	9	2	11	3	10	79	6.077
X	4	9	14	6	9	12	2	7	5	5	11	2	13	99	7.616

where

$$n = 13, \quad \sum Y^2 = 653, \quad \sum XY = 753, \quad \sum X^2 = 951, \quad \sum y^2 = 172.9231$$

$$\sum xy = 151.3846, \quad \sum x^2 = 197.0796, \quad \hat{Y} = 0.228 + 0.7681X$$

Residual $SS = 56.6370$, with 11 degrees of freedom

Now, we might ask, are these really different regressions? Or could the data be combined to produce a single regression that would be applicable to both groups? If there is no significant difference between the mean square residuals for the two groups (this matter may be determined by Bartlett's test, see Section 7.14.3), the test described below helps to answer the question.

7.14.4.1 Testing for the Common Regressions

Simple linear regressions may differ either in their slope or in their level. When testing for common regressions the procedure is to test first for common slopes. If the slopes differ significantly, the regressions are different and no further testing is needed. If the slopes are not significantly different, the difference in level is tested. The analysis table is

Line	Group	df	Σy^2	Σxy	Σx^2	df	SS	MS
							Residuals	
1	A	8	100.8889	80.5556	106.2222	7	39.7980	—
2	B	12	172.9231	151.3846	97.0769	11	56.6370	—
3	Pooled residuals					18	96.4350	5.3575
4	Difference for testing common slopes					1	0.0067	0.0067
5	Common slope	20	273.8120	231.9402	303.2991	19	5.0759	96.4417
6	Difference for testing trends					1	80.1954	80.1954
7	Single regression	21	322.7727	213.0455	310.5909	20	176.6371	—

The first two lines in this table contain the basic data for the two groups. To the left are the total degrees of freedom for the groups (8 for A and 12 for B). In the center are the corrected sums of squares and products. The right side of the table gives the residual sum of squares and degrees of freedom. Since only simple linear regressions have been fitted, the residual degrees of freedom of each group are one less than the total degrees of freedom. The residual sum of squares is obtained by first computing the reduction sum of squares for each group.

$$\text{Reduction } SS = \frac{\left(\sum xy\right)^2}{\sum x^2}$$

This reduction is then subtracted from the total sum of squares (Σy^2) to give the residuals.

Line 3 is obtained by pooling the residual degrees of freedom and residual sums of squares for the groups. Dividing the pooled sum of squares by the pooled degrees of freedom gives the pooled mean square. The left side and center of line (we will skip line 4 for the moment) is obtained by pooling the total degrees of freedom and the corrected sums of squares and products for the groups. These are the values that are obtained under the assumption of no difference in the slopes of the group regressions. If the assumption is wrong, the residuals about this common slope regression will be considerably larger than the mean square residual about the separate regressions. The residual degrees of freedom and sum of squares are obtained by fitting a straight line to these pooled data. The residual degrees of freedom are, of course, one less than the total degrees of freedom. The residual sum of squares is, as usual,

$$\text{Reduction } SS = 273.8120 - \frac{(231.9402)^2}{303.2991} = 96.4417$$

Now, the difference between these residuals (line 4 = line 5 − line 3) provides a test of the hypothesis of common slopes. The error term for this test is the pooled mean square from line 3:

$$\text{Test of common slopes: } F_{1/13df} = \frac{0.0067}{5.3575}$$

The difference is not significant.

If the slopes differed significantly, the groups would have different regressions, and we would stop here. Because the slops did not differ, we now go on to test for a difference in the levels of the regression.

Line 7 is what we would have if we ignored the groups entirely, lumped all the original observations together, and fitted a single linear regression. The combined data are as follows:

$$n = (9 + 13) = 22, \text{ so the degrees of freedom} = 21$$

$$\sum Y = (82 + 79) = 161$$

$$\sum Y^2 = (848 + 653) = 1501$$

$$\sum y^2 = 1501 - \frac{(161)^2}{22} = 322.7727$$

$$\sum X = (58 + 99) = 157$$

$$\sum X^2 = (480 + 951) = 1431$$

$$\sum x^2 = 1431 - \frac{(157)^2}{22} = 310.5909$$

$$\sum XY = (609 + 753) = 1362$$

$$\sum xy = 1362 - \frac{(157)(161)}{22} = 213.0455$$

From this we obtain the residual values on the right side of line 7.

$$\text{Residual } SS = 322.7727 - \frac{(213.0455)^2}{310.5909} = 176.6371$$

If there is a real difference among the levels of the groups, the residuals about this single regression will be considerably larger than the mean square residual about the regression that assumed the same slopes but different levels. This difference (line 6 = line 7 − line 5) is tested against the residual mean square from line 5.

$$\text{Test of levels: } F_{1/18df} = \frac{80.1954}{5.0759} = 15.80$$

As the levels differ significantly, the groups do not have the same regressions.

The test is easily extended to cover several groups, though there may be a problem in finding which groups are likely to have separate regressions and which can be combined. The test can also be extended to multiple regressions.

7.17.5   ANALYSIS OF COVARIANCE IN A RANDOMIZED BLOCK DESIGN

A test was made of the effect of three soil treatments on the height growth of 2-year-old seedlings. Treatments were assigned at random to the three plots within each of 11 blocks. Each plot was made up of 50 seedlings. Average 5-year height growth was the criterion for evaluating treatments. Initial heights and 5-year growths, all in feet, were as follows:

Block	Treatment A Height	Treatment A Growth	Treatment B Height	Treatment B Growth	Treatment C Height	Treatment C Growth	Block Totals Height	Block Totals Growth
1	3.6	8.9	3.1	10.7	4.7	12.4	11.4	32.0
2	4.7	10.1	4.9	14.2	2.6	9.0	12.2	33.3
3	2.6	6.3	0.8	5.9	1.5	7.4	4.9	19.6
4	5.3	14.0	4.6	12.6	4.3	10.1	14.2	36.7
5	3.1	9.6	3.9	12.5	3.3	6.8	10.3	28.9
6	1.8	6.4	1.7	9.6	3.6	10.0	7.1	26.0
7	5.8	12.3	5.5	12.8	5.8	11.9	17.1	37.0
8	3.8	10.8	2.6	8.0	2.0	7.5	8.4	26.3
9	2.4	8.0	1.1	7.5	1.6	5.2	5.1	20.7
10	5.3	12.6	4.4	11.4	5.8	13.4	15.5	37.4
11	3.6	7.4	1.4	8.4	4.8	10.7	9.8	26.5
Sums	42.0	106.4	34.0	113.6	40.0	104.4	116.0	324.4
Means	3.82	9.67	3.09	10.33	3.64	9.49	3.52	9.83

The analysis of variance of growth is

Source of Variation	Degrees of Freedom	Sums of Squares	Mean Squares
Blocks	10	132.83	—
Treatment	2	4.26	2.130
Error	20	68.88	3.444
Total	32	205.97	—

$$F(\text{for testing treatments})_{2/20df} = \frac{2.130}{3.444}$$

which is not significant at the 0.05 level.

There is no evidence of a real difference in growth due to treatments. There is, however, reason to believe that, for young seedlings, growth is affected by initial height. A glance at the block totals seems to suggest that plots with greatest initial height had the greatest 5-year growth. The possibility that effects of treatment are being obscured by differences in initial heights raises the question of how the treatments would compare if adjusted for differences in initial heights.

If the relationship between height growth and initial height is linear and if the slope of the regression is the same for all treatments, the test of adjusted treatment means can be made by an analysis of covariance as described below. In this analysis, the growth will be labeled Y and initial height X.

Computationally the first step is to obtain total, block, treatment, and error sums of squares of X (SS_x) and sums of products of X and Y (SP_{xy}), just as has already been done for Y.

For X:

$$CT_x = \frac{(116.0)^2}{33} = 407.76$$

$$\text{Total } SS_x = \left(3.6^2 + \ldots + 4.8^2\right) - CT_x = 73.26$$

$$\text{Block } SS_x = \left(\frac{11.4^2 + \ldots 9.8^2}{3}\right) - CT_x = 54.31$$

$$\text{Treatment } SS_x = \left(\frac{42.0^2 + 34.0^2 + 40.0^2}{11}\right) - CT_x = 3.15$$

$$\text{Error } SS_x = \text{Total } SS_x - \text{Block } SS_x - \text{Treatment } SS_x = 15.80$$

For XY:

$$CT_{xy} = \frac{(116.0)(324.4)}{33} = 1140.32$$

$$\text{Total } SP_{xy} = (3.6)(8.9) + \ldots + (4.8)(10.7) - CT_{xy} = 103.99$$

$$\text{Block } SP_{xy} = \left(\frac{(11.4)(32.0) + \ldots + (9.8)(26.5)}{3}\right) - CT_{xy} = 82.71$$

$$\text{Treatment } SP_{xy} = \left(\frac{(42.0)(106.4) + (34.0)(113.6) + (40.0)(104.4)}{11}\right) - CT_{xy} = -3.30$$

$$\text{Error } SP_{xy} = \text{Total } SP_{xy} - \text{Block } SP_{xy} - \text{Treatment } SP_{xy} = 24.58$$

These computed terms are arranged in a manner similar to that for the test of group regressions (which is exactly what the covariance analysis is). One departure is that the total line is put at the top.

Source of Variation	df	SS_y	SP_{xy}	SS_x	df	SS	MS
						Residuals	
Total	32	205.97	103.99	73.26			
Blocks	10	132.83	82.71	54.31			
Treatment	2	4.26	−3.30	3.15			
Error	20	68.88	24.58	15.80	19	30.641	1.613

On the error line, the residual sum of squares after adjusting for a linear regression is

$$\text{Residual } SS = SS_y - \frac{\left(SP_{xy}\right)^2}{SS_x} = 68.88 - \frac{(24.58)^2}{15.80} = 30.641$$

This sum of squares has one degree of freedom less than the unadjusted sum of squares.

To test treatments we first pool the unadjusted degree of freedom and sums of squares and products for treatment and error. The residual terms for this pooled line are then computed just as they were for the error line:

Source of Variation	df	SS_y	SP_{xy}	SS_x	df	SS
						Residuals
Treatment plus error	22	73.14	21.28	18.95	21	49.244

Then to test for a difference among treatments after adjustment for the regression of growth on initial height, we compute the difference in residuals between the error and the treatment + error lines:

Source of Variation	Degrees of Freedom	Sums of Squares	Mean Squares
Difference for testing adjusted treatments	2	18.603	9.302

The mean square for the difference in residual is now tested against the residual mean square for error.

$$F_{2/19df} = \frac{9.302}{1.613} = 5.77$$

Thus, after adjustment, the difference in treatment means is found to be significant at the 0.05 level. It may also happen that differences that were significant before adjustment are not significant afterwards.

If the independent variable has been affected by treatments, interpretation of a covariance analysis requires careful thinking. The covariance adjustment may have the effect of removing the treatment differences that are being tested. On the other hand, it may be informative to know that treatments are or are not significantly different in spite of the covariance adjustment. The beginner who is uncertain of the interpretations would do well to select as covariates only those that have not been affected by treatments.

The covariance test may be made in a similar manner for any experimental design and, if desired (and justified), adjustment may be made for multiple or curvilinear regressions.

The entire analysis is usually presented in the following form:

Source	df	SS_y	SP_y	SS_x	Adjusted df	Adjusted SS	Adjusted MS
Total	32	205.97	103.99	73.26			
Blocks	10	132.83	82.71	54.31			
Treatment	2	4.26	−3.30	3.15			
Error	20	68.88	24.58	15.80	19	30.641	1.613
Treatment + error	22	73.14	21.28	18.95	21	49.244	—
Difference for testing adjusted treatment means					2	18.603	9.302

Unadjusted treatments: $F_{2/20dt} = \dfrac{2.130}{3.444}$, not significant

Adjusted treatments: $F_{2/19dt} = \dfrac{9.302}{1.613} = 5.77$, significant at 0.05 level

7.17.5.1 Adjusted Means

If we wish to know what the treatment means are after adjustment for regression, the equation is

$$\text{Adjusted } \bar{Y}_i = \bar{Y}_i - b\left(\bar{X}_i - \bar{X}\right)$$

where

$\bar{Y}_i$ = Unadjusted mean for treatment i.

b = Coefficient of the linear regression = $\dfrac{\text{Error } SP_{xy}}{\text{Error } SS_x}$

$\bar{X}_i$ = Mean of the independent variable for treatment i.

$\bar{X}$ = Mean of X for all treatments.

In the example, we had $\overline{X}_A = 3.82$, $\overline{X}_B = 3.09$, $\overline{X}_C = 3.64$, $\overline{X} = 3.52$, and

$$b = \frac{24.58}{15.80} = 1.56$$

So, the unadjusted and adjusted mean growths are

	Mean Growth	
Treatment	Unadjusted	Adjusted
A	9.67	9.20
B	10.33	11.00
C	9.49	9.30

7.17.5.2 Tests among Adjusted Means

In an earlier section we encountered methods of making further tests among the means. Ignoring the covariance adjustment, we could, for example, make an F test for pre-specified comparisons such as A + C vs. B, or A vs. C. Similar tests can also be made after adjustment for covariance, through they involve more labor. The F test will be illustrated for the comparison B vs. A + C after adjustment.

As might be suspected, to make the F test we must first compute sums of squares and products of X and Y for the specified comparison:

$$SS_y = \frac{\left[2\left(\sum Y_B\right) - \left(\sum Y_A + \sum Y_C\right)\right]^2}{(2^2 + 1^2 + 1^2)(11)} = \frac{\left[2(113.6) - (106.4 + 104.4)\right]^2}{66} = 4.08$$

$$SS_x = \frac{\left[2\left(\sum X_B\right) - \left(\sum X_A + \sum X_C\right)\right]^2}{(2^2 + 1^2 + 1^2)(11)} = \frac{\left[2(34.0) - (42.0 + 40.0)\right]^2}{66} = 2.97$$

$$SP_{xy} = \frac{\left[2\left(\sum X_B\right) - \left(\sum X_A + \sum X_C\right)\right]^2}{(2^2 + 1^2 + 1^2)(11)} = \frac{\left[2(34.0) - (42.0 + 40.0)\right]^2}{66} = -3.48$$

From this point on, the F test of A + B vs. C is made in exactly the same manner as the test of treatments in the covariance analysis.

					Residuals		
Source	df	SS_y	SP_{xy}	SS_x	df	SS	MS
2B – (A + C)	1	4.08	–3.48	2.97	—	—	—
Error	20	68.88	24.58	15.80	19	30.641	1.613
Sum	21	72.96	21.10	18.77	20	49.241	—
Difference for testing adjusted comparison					1	18.600	18.600

$F_{1/19df} = 11.531$, which is significant at the 0.01 level.

REFERENCES AND RECOMMENDED READING

Addelman, S. (1969). The generalized randomized block design. *American Statistician*, 23(4), 35–36.

Addelman, S. (1970). Variability of treatments and experimental units in the design and analysis of experiments. *Journal of American Statistical Association*, 65(331), 1095–1108.

Fisher, R. (1926). The arrangement of field experiments. *Journal of the Ministry of Agriculture of Great Britain*, 33, 503–513.

Freese, F. (1962). *Elementary Forest Sampling*. U.S. Department of Agriculture, Washington, DC.

Freese, F. (1967). *Elementary Statistical Methods for Foresters*, Handbook 317. U.S. Department of Agriculture, Washington, DC.

Hamburg, M. (1987). *Statistical Analysis for Decision Making*, 4th ed. Harcourt Brace Jovanovich, New York.

Mayr, E. (1970). *Populations, Species, and Evolution: An Abridgement of Animal Species and Evolution*. Belknap Press, Cambridge, MA.

Mayr, E. (2002). The biology of race and the concept of equality. *Daedalus*, 131, 89–94.

Wadsworth, H.M. (1990). *Handbook of Statistical Methods for Engineers and Scientists*. McGraw-Hill, New York.

8 Computation and Measurement of Risk

… the currents swirled about me; all your waves and breakers swept over me.

—Jonah 2:3

8.1 INTRODUCTION*

Environmental practitioners are concerned with a broad range of environmental and public health issues related to everyday activities and natural occurrences that lead to environmental degradation, such as the creation of wastes, emissions, and resource depletion. By another name, these are *risks*. Environmental risk managers protect human health and the environment by

- Controlling and preventing air pollution emissions
- Supporting Homeland Security programs by investigating ways to decontaminate buildings
- Reducing greenhouse gas (GHG) emissions through new technologies
- Identifying and reducing water quality and availability issues
- Protecting groundwater
- Providing technical support on below-ground pollution and ecosystem restoration
- Determining which oil spill dispersants are best
- Helping communities achieve growth goals, improving residents' quality of life, and enhancing financial and environmental sustainability
- Working on soil and sediment contamination issues
- Cleaning up waterways to make them usable for drinking, swimming, and fishing
- Providing choices for managing wastes
- Helping communities with their land use decisions
- Identifying consequences of technology changes and presenting sustainable alternatives
- Analyzing product processes or services and their adverse impacts and recommending greener choices
- Demonstrating practices that meet the needs of the present without compromising the ability of future generations to meet theirs
- Developing approaches and tools to monitor, treat, protect, and restore impaired waterways
- Improving drinking water and wastewater systems

Environmental practitioners accomplish these tasks by working with outside organizations to design, develop, and evaluate technologies and methods; by providing assistance to develop and apply environmental technologies; and by incorporating cost-effective techniques that lead to successful pollution prevention and control strategies.

---

* Material presented in this chapter is adapted from Spellman, F.R. and Stoudt, M.L., *The Handbook of Environmental Health*, Scarecrow Press, Lanham, MD, 2013; CDC, *Principles of Epidemiology in Public Health*, Centers for Disease Control and Prevention, Atlanta, GA, 2012.

Many of these pollution prevention and risk control strategies depend on mathematical computations and measures. Chapter 7 described measures of central location and spread, among other statistical functions, which are useful for summarizing continuous variables. However, many variables used by field environmental practitioners (e.g., epidemiologists) are categorical variables, some of which have only two categories—exposed/not exposed, test positive/test negative, case/control, and so on. Because many of the variables encountered in field environmental health are nominal-scale variables, frequency measures are used quite commonly in environmental health. These variables have to be summarized with frequency measures such as ratios, proportions, and rates. Incidence, prevalence, and mortality rates are three frequency measures that are used in public health to characterize the occurrence of health events in a population.

In this chapter, we calculate and interpret environmental health measures using ratio, proportion, incidence proportion (attack rate), incidence rate, prevalence, and mortality rate. Specifically, we will discuss frequency measures, morbidity frequency measures, mortality frequency measures, natality (birth) measures, measures of association, and measures of public health impact. Note that many of the examples presented in the following are commonly used in epidemiological practices; they are presented here to demonstrate the computation and measurement of risk used in all fields.

8.2 FREQUENCY MEASURES

A measure of central location provides a single value that summarizes an entire distribution of data. In contrast, a frequency measure characterizes only part of the distribution. Frequency measures compare one part of the distribution to another part of the distribution, or to the entire distribution. Common frequency measures are *ratios*, *proportions*, and *rates*. All three frequency measures have the same basic form:

$$\frac{\text{Numerator}}{\text{Denominator}} \times 10^n$$

Recall that

$10^0 = 1$ (anything raised to the 0 power equals 1).
$10^1 = 10$ (anything raised to the 1st power is the value itself).
$10^2 = 10 \times 10 = 100$.
$10^3 = 10 \times 10 \times 10 = 1000$.

So, the fraction of numerator/denominator can be multiplied by 1, 10, 100, 1000, and so on. This multiplier varies by measure and will be addressed in each section.

8.2.1 RATIO

Recall that we presented an introduction to ratios in Section 2.8 of this text. Here, we review the basics and then demonstrate the actual use of ratios in measuring risk in environmental health and public health practice. A ratio is the relative magnitude of two quantities or a comparison of any two values. It is calculated by dividing one interval- or ratio-scale variable by the other. The numerator and denominator need not be related; therefore, one could compare apples with oranges or apples with number of chemical spills. The method for calculating a ratio is

$$\frac{\text{Number or rate of events, items, persons, etc. in one group}}{\text{Number or rate of events, items, persons, etc. in another group}}$$

Two examples of ratios are

$$\frac{\text{Number of women in State A who died from heart disease in 2011}}{\text{Number of women in State A who died from cancer in 2011}}$$

$$\frac{\text{Number of women in State A who died from lung cancer in 2010}}{\text{Estimated revenue (in dollars) in State A from cigarette sales in 2010}}$$

After the numerator is divided by the denominator, the result is often expressed as the result "to one" or written as the result ":1."

Note that, in certain ratios, the numerator and denominator are different categories of the same variable, such as males and females, or persons 20 to 29 years and 30 to 39 years of age. In other ratios, the numerator and denominator are completely different variables, such as the number of environmental laboratories in a city and the number of manufacturing industries operating in that city.

■ **EXAMPLE 8.1**

Problem: Between 1971 and 1975, as part of the National Health and Nutrition Examination Survey (NHANES), 7381 persons ages 40 to 77 years were enrolled in a follow-up study (Kleinman et al., 1988). At the time of enrollment, each study participant was classified as having or not having diabetes. From 1982 to 1984, enrollees were documented as either having died or being still alive. The results are summarized as follows.

	Open Enrollment (1971–1975)	Dead at Follow-Up (1982–1984)
Diabetic men	189	100
Non-diabetic men	3151	811
Diabetic women	218	74
Non-diabetic women	3823	511

Of the men enrolled in the NHANES follow-up study, 3151 were non-diabetic and 189 were diabetic. Calculate the ratio of non-diabetic to diabetic men.

Solution:

$$\text{Ratio} = (3151/189) \times 1 = 16.7:1$$

8.2.1.1 Properties and Uses of Ratios

Ratios are common descriptive measures used in all fields. In environmental health and epidemiology, ratios are used as both descriptive measures and as analytic tools. As a descriptive measure, ratios can describe the male-to-female ratio of participants in a study, or the ratio of controls to cases (e.g., two controls per case). As an analytical tool, ratios can be calculated for the occurrence of illness, injury, or death between two groups. These ratio measures, including risk ratio (relative risk), rate ratio, and odds ratio, are described later in this chapter.

As noted previously, the numerators and denominators of a ratio can be related or unrelated. In other words, you are free to use a ratio to compare the number of males in a population with the number of females, or to compare the number of residents in a population with the number of hospitals or dollars spent on over-the-counter medicines. Usually, the values of both the numerator and denominator of a ratio are divided by the value of one or the other so that either the numerator or the denominator equals 1.0; thus, the ratio of non-diabetics to diabetics cited in Example 8.1 is more likely to be reported as 16.7:1 than 3151:189. Calculating ratios for different variables is demonstrated by Example 8.2A and Example 8.2B.

■ **EXAMPLE 8.2A**

Problem: A city of 4,000,000 persons has 500 clinics. Calculate the ratio of clinics per person.

Solution:

$$(500/4,000,000) \times 10^n = 0.000125 \text{ clinics per person}$$

To get a more easily understood result, we could set $10^n = 10^4 = 10,000$. Then the ratio becomes:

$$(0.000125) \times 10,000 = 1.25 \text{ clinics per 10,000 persons}$$

We could also divide each value by 1.25 and express this ratio as 1 clinic for every 8000 persons.

■ **EXAMPLE 8.2B**

Problem: Delaware's infant mortality rate in 2001 was 10.7 per 1000 live births (Arias et al., 2003). New Hampshire's infant mortality rate in 2001 was 3.8 per 1000 live births. Calculate the ratio of the infant mortality rate in Delaware to that in New Hampshire.

Solution:

$$(10.7/3.8) \times 1 = 2.8\text{:}1$$

Thus, Delaware's infant mortality rate was 2.8 times as high as New Hampshire's infant mortality rate in 2001.

8.2.1.2 A Common Environmental Health Ratio: Death-to-Case Ratio

Death-to-case ratio is the number of deaths attributed to a particular disease during a specified period divided by the number of new cases of that disease identified during the same period. It is used as a measure of the severity of illness. The death-to-case ratio for rabies is close to 1 (that is, almost everyone who develops rabies dies from it), whereas the death-to-case ratio for the common cold is close to 0. In the United States in 2002, for example, a total of 15,075 new cases of tuberculosis were reported (CDC, 2004). During the same year, 802 deaths were attributed to tuberculosis. The tuberculosis death-to-case ratio for 2002 can be calculated as 802/15,075. Dividing both numerator and denominator by the numerator yields 1 death per 18.8 new cases. Dividing both numerator and denominator by the denominator (and multiplying by $10^n = 100$) yields 5.3 deaths per 100 new cases. Both expressions are correct. Note that, presumably, many of those who died had initially contracted tuberculosis years earlier. Thus, many of the 802 in the numerator are not among the 15,075 in the denominator; therefore, the death-to-case ratio is a ratio, but not a proportion.

8.2.2 PROPORTION

A proportion is the comparison of a part to the whole. It is a type of ratio in which the numerator is included in the denominator. We might use a proportion to describe what fraction of clinic patients tested positive for HIV, or what percentage of the population is younger than 25 years of age. A proportion may be expressed as a decimal, a fraction, or a percentage. The method for calculating a proportion is

$$\frac{\text{Number of persons or events with a particular characteristic}}{\text{Total number of persons or events, of which the numerator is a subset}} \times 10^n$$

Two examples of proportions are

$$\frac{\text{Number of women in State A who died from heart disease in 2008}}{\text{Number of women in State A who died in 2008}}$$

$$\frac{\text{Number of women in State A who died from lung cancer in 2004}}{\text{Number of women in State A who died from cancer (all types) in 2004}}$$

For a proportion, 10^n is usually 100 (or $n = 2$) and is often expressed as a percentage. Example 8.3 and Example 8.4 illustrate how proportion is calculated.

■ EXAMPLE 8.3

Problem: Calculate the proportion of men in the NHANES follow-up study who were diabetics.

Solution:

Numerator = 189 diabetic men
Denominator = Total number of men = 189 + 3151 = 3340
Proportion = (189/3340) × 100 = 5.66%

■ EXAMPLE 8.4

Problem: Calculate the proportion of deaths among men.

Solution:

Numerator = deaths in men
      = 100 deaths in diabetic men + 811 deaths in non-diabetic men
      = 911 deaths in men

Notice that the numerator (911 deaths in men) is a subset of the denominator.

Denominator = all deaths
      = 911 deaths in men + 72 deaths in diabetic women + 511 deaths in non-diabetic women
      = 1494 deaths

Proportion = (911/1494) × 100 = 60.98%, or 61%

8.2.2.1 Properties and Uses of Proportions

Proportions are common descriptive measures used in all fields. In environmental health and epidemiology, proportions are used most often as descriptive measures. For example, one could calculate the proportion of persons enrolled in a study among all those eligible ("participation rate"), the proportion of children in a village vaccinated against measles, or the proportion of persons who developed illness among all passengers of a cruise ship. Proportions are also used to describe the amount of disease that can be attributed to a particular exposure. For example, on the basis of studies of smoking and lung cancer, public health officials have estimated that greater than 90% of the lung cancer cases that occur are attributable to cigarette smoking. In a proportion, the numerator must be included in the denominator. Thus, the number of apples divided by the number or oranges is not a proportion, but the number of apples divided by the total number of fruits of all kinds is a proportion. Remember, the numerator is always a subset of the denominator.

A proportion can be expressed as a fraction, a decimal, or a percentage. The statements "one fifth of the residents became ill" and "twenty percent of the residents because ill" are equivalent. Proportions can easily be converted to ratios. If the numerator is the number of women (179) who attended a clinic and the denominator is all the clinic attendees (341), the proportion of clinic attendees who were women is 179/341, or 53% (a little more than half). To convert to a ratio, subtract

DID YOU KNOW?

If the numerator and denominator of a ratio together make up an entire population, the ratio can be converted to a proportion by adding the numerator and denominator to form the denominator of the proportion.

the numerator from the denominator to get the number of clinic patients who were not women (i.e., the number of men): 341 − 179 = 162 men. Thus, the ratio of women to men could be calculated from the proportion as

$$\text{Ratio} = [179/(341 - 179)] \times 1 = 179/162 = 1.1 \text{ to } 1 \text{ female-to-male ratio}$$

8.2.2.2 A Specific Type of Environmental Health: Proportionate Mortality

Proportionate mortality is the proportion of deaths in a specified population during a period of time that are attributable to different causes. Each cause is expressed as a percentage of all deaths, and the sum of the causes adds up to 100%. These proportions are not rates because the denominator is all deaths, not the size of the population in which the deaths occurred. Table 8.1 lists the primary causes of death in the United States in 2003 for persons of all ages and for persons ages 25 to 44 years, by number of deaths, proportionate mortality, and rank. As illustrated in Table 8.1, the proportionate mortality for HIV was 0.5% among all age groups, and 5.3% among those ages 25 to 44 years. In other words, HIV infection accounted for 0.5% of all deaths, and 5.3% of the deaths among 25 to 44 year olds.

TABLE 8.1
Number, Proportionate Mortality, and Ranking of Deaths for Leading Causes of Death, All Ages and 25- to 44-Year Age Group—United States, 2003

	All Ages			Ages 25–44 Years		
	Number	Percentage	Rank	Number	Percentage	Rank
All causes	2,443,930	100.0	—	128,924	100.0	—
Diseases of heart	684,462	28.0	1	16,283	12.6	3
Malignant hepatitis	554,643	22.7	2	19,041	14.8	2
Cerebrovascular disease	157,803	6.5	3	3004	2.3	8
Chronic lower respiratory diseases	126,128	5.2	4	401	0.3	—[a]
Accidents (unintentional injuries)	105,695	4.3	5	27,844	21.6	1
Diabetes mellitus	73,965	3.0	6	2662	2.1	9
Influenza & pneumonia	64,847	2.6	7	1337	1.0	10
Alzheimer's disease	63,343	2.6	8	0	0.0	—[a]
Nephritis, nephritic syndrome, nephrosis	33,615	1.4	9	305	0.2	—[a]
Septicemia	34,243	1.4	10	328	0.2	—[a]
Intentional self-harm (suicide)	30,642	1.3	11	11,251	8.7	4
Chronic liver disease and cirrhosis	27,201	1.1	12	3288	2.6	7
Assault (homicide)	17,096	0.7	13	7367	5.7	5
HIV disease	13,544	0.5	—[a]	6879	5.3	6
All other	456,703	18.7	—	29,480	22.9	—

[a] Not among top-ranked causes.

Sources: Data from CDC (2005) and Hoyert et al. (2005).

8.2.3 RATE

In environmental health, a rate is a measure of the frequency with which an event occurs in a defined population over a specified period of time. Because rates put disease frequency in the perspective of the size of the population, rates are particularly useful for comparing disease frequency in different locations, at different times, or among different groups of persons with potentially different sized populations; that is, a rate is a measure of risk.

To a non-environmental health practitioner, rate means how fast something is happening or going. The speedometer of a car indicates the car's speed or rate of travel in miles or kilometers per hour. This rate is always reported per some unit of time. Some environmental practitioners restrict use of the term *rate* to similar measures that are expressed per unit of time. For these environmental practitioners, especially those practicing environmental health, a rate describes how quickly disease occurs in a population—for example, 70 new cases of breast cancer per 1000 women per year. This measure conveys a sense of the speed with which disease occurs in a population and seems to imply that this pattern has occurred and will continue to occur for the foreseeable future. This rate is an *incidence rate*, described later in Section 8.3.

Other environmental health professionals use the term more loosely, referring to proportions with case counts in the numerator and size of population in the denominator as rates. Thus, an *attack rate* is the proportion of the population that develops illness during an outbreak; for example, 20 of 130 persons developed diarrhea after attending a picnic. (An alternative and more accurate phrase for attack rate is *incidence proportion*.) A *prevalence rate* is the proportion of the population that has a health condition at a point in time; for example, 70 influenza patients being reported in a county in March 2005. A *case-fatality rate* is the proportion of persons with the disease who die from it—for example, one death due to meningitis among a county's population. All of these measures are proportions, and none is expressed per units of time; therefore, these measures are not considered "true" rates by some, although use of the terminology is widespread.

An example of rate is

$$\frac{\text{Number of women in State A who died from heart disease in 2009}}{\text{Estimated number of women living in State A on July 1, 2009}}$$

Table 8.2 summarizes some of the common environmental health measures as ratios, proportions, or rates.

TABLE 8.2
Environmental Health Measures Categorized as Ratio, Proportion, or Rate

Condition	Ratio	Proportion	Rate
Morbidity (disease)	Risk ratio (relative risk)	Attack rate (incidence proportion)	Person-time incidence rate
	Rate ratio	Secondary attack rate	
	Odds ratio	Point prevalence	
	Period prevalence	Attributable proportion	
Mortality (death)	Death-to-case ratio	Proportionate mortality	Crude mortality rate
			Case-fatality rate
			Cause-specific mortality rate
			Age-specific mortality rate
			Maternal mortality rate
			Infant mortality rate
Natality (birth)	—	—	Crude birth rate
			Crude fertility rate

TABLE 8.3

Frequently Used Measures of Morbidity

Measure	Numerator	Denominator
Incidence proportion (or attack rate or risk)	Number of new cases of disease during specified time interval	Population at start of time interval
Secondary attack rate	Number of new cases among contacts	Total number of contacts
Incidence rate (or person-time rate)	Number of new cases of disease during specified time interval	Summed person-years of observation or average population during time interval
Point prevalence	Number of current cases (new and preexisting) at a specified point in time	Population at the same specified point
Period prevalence	Number of current cases (new and preexisting) over a specified period of time	Average or mid-interval population

8.3 MORBIDITY FREQUENCY MEASURES

Morbidity has been defined as any departure, subjective or objective, from a state of physiological or psychological well-being. In practice, morbidity encompasses disease, injury, and disability. In addition (although for this book the term refers to the number of persons who are ill), it can also be used to describe the periods of illness that these persons experience, or the duration of these illnesses (Last, 2001). Measures of morbidity frequency characterize the number of persons in a population who become ill (incidence) or are ill at the given time (prevalence). Commonly used measures are listed in Table 8.3. *Incidence* refers to the occurrence of new cases of disease or injury in a population over a specified period of time. Although some environmental health practitioners use incidence to mean the number of new cases in a community, others use incidence to mean the number of new cases per unit of population. Two types of incidence are commonly used—*incidence proportion* and *incidence rate*.

8.3.1 INCIDENCE PROPORTION OR RISK

Incidence proportion is the proportion of an initially disease-free population that develops disease, becomes injured, or dies during a specified (usually limited) period of time. Synonyms include attack rate, risk, probability of getting diseases, and cumulative incidence. Incidence proportion is a proportion because the persons in the numerator, those who develop disease, are all included in the denominator (the entire population). Incidence proportion (risk) is calculated by

$$\frac{\text{Number of new cases of disease or injury during specified period}}{\text{Size of population at start of period}}$$

■ EXAMPLE 8.5

Problem: In the study of diabetics, 100 of the 189 diabetic men died during the 13-year follow-up period. Calculate the risk of death for these men.

Solution:

Numerator = 100 deaths among the diabetic men
Denominator = 189 diabetic men
$10^n = 10^2 = 100$

$$\text{Risk} = (100/189) \times 100 = 52.9\%$$

■ EXAMPLE 8.6

Problem: In an outbreak of gastroenteritis among attendees of a corporate picnic, 99 persons ate potato salad, and 30 of those developed gastroenteritis. Calculate the risk of illness among the persons who ate the potato salad.

Solution:

Numerator = 30 persons who ate potato salad and developed gastroenteritis
Denominator = 99 persons who ate potato salad
$10^n = 10^2 = 100$

$$\text{Risk (``food-specific attack rate'')} = (30/90) \times 100 = 0.303 \times 100 = 30.3\%$$

Two example fractions used in incidence proportion problems are

$$\frac{\text{Number of women in Framingham Study who died through last year from heart disease}}{\text{Number of women initially enrolled in Framingham Study}}$$

$$\frac{\text{Number of women in Framingham Study newly diagnosed with heart disease last year}}{\text{Number of women in Framingham Study without heart disease at beginning of same year}}$$

8.3.1.1 Properties and Uses of Incidence Proportions

Incidence proportion is a measure of the risk of diseases or the probability of developing a particular disease during a specified period. As a measure of incidence, it includes only new cases of disease in the numerator. The denominator is the number of persons in the population at the start of the observation period. Because all of the persons with new cases of disease (numerator) are also represented in the denominator, a risk is also a proportion. In the outbreak setting, the term *attack rate* is often used as a synonym for risk. It is the risk of getting the disease during a specified period, such as the duration of an outbreak. A variety of attack rates can be calculated. *Overall attack rate* is the total number of new cases divided by the total population. A *food-specific attack rate* is the number of persons who ate a specified food and became ill divided by the total number of persons who ate the food, as illustrated in the previous potato salad example. A *secondary attack rate* is sometimes calculated to document the difference between community transmission of illness vs. transmission of illness in a household, barracks, or other closed population. It is calculated as

$$\frac{\text{Number of cases among contacts of primary cases}}{\text{Total number of contacts}} \times 10^n$$

DID YOU KNOW?

The denominator of an incidence proportion is the number of persons at the start of the observation period. The denominator should be limited to the "population at risk" for developing the disease (i.e., persons who have the potential to get the disease and be included in the numerator). For example, if the numerator represents new cases of ovarian cancer, the denominator should be restricted to women, because men do not have ovaries. This is easily accomplished because census data by sex are readily available. In fact, ideally the denominator should be restricted to women with ovaries, excluding women who have had their ovaries removed surgically (often done in conjunction with a hysterectomy), but this is not usually practical. This is an example of field environmental health practitioners doing the best they can with the data they have.

Often, the total number of contacts in the denominator is calculated as the total population in the households of the primary cases minus the number of primary cases. For a secondary attack rate, 10^n usually is 100%. An example of calculating secondary attack rates is illustrated below.

■ EXAMPLE 8.7

Problem: Consider an outbreak of shigellosis in which 18 persons in 18 different households all became ill. If the population of the community was 1000, then the overall attack rate was 18/1000 × 100% = 1.8%. One incubation period later, 17 persons in the same households as these "primary" cases developed shigellosis. If the 18 households included 86 persons, calculate the secondary attack rate.

Solution:

$$\text{Secondary attack rate} = [17/(86 - 18)] \times 100\% = (17/68) \times 100\% = 25.0\%$$

8.3.2 Incidence Rate or Person-Time Rate

Incidence rate or person-time rate is a measure of incidence that incorporates time directly into the denominator. A person-time rate is generally calculated from a long-term cohort follow-up study, wherein enrollees are followed over time and the occurrence of new cases of disease is documented. Typically, each person is observed from an established starting time until one of four "end points" is reached: onset of disease, death, migration out of the study ("lost to follow-up"), or the end of the study. Similar to the incidence proportion, the numerator of the incidence rate is the number of new cases identified during the period of observation. However, the denominator differs. The denominator is the sum of the time each person was observed, totaled for all persons. This denominator represents the total time the population was at risk of and being watched for disease. Thus, the incidence rate is the ratio of the number of cases to the total time the population is at risk of disease:

$$\frac{\text{Number of new cases of disease or injury during specified period}}{\text{Time each person was observed, totaled for all persons}}$$

In a long-term follow-up study of morbidity, each study participant may be followed or observed for several years. One person followed for 5 years without developing disease is said to contribute 5 person-years of follow-up.

What about a person followed for one year before being lost to follow-up at year 2? Many researchers assume that persons lost to follow-up were, on average, disease-free for half the year, and thus contribute 1/2 year to the denominator. Therefore, the person followed for one year before being lost to follow-up contributes 1.5 person-years. The same assumption is made for participants diagnosed with the disease at the year 2 examination—some may have developed illness in month 1, and others in months 2 through 12. So, on average, they developed illness halfway through the year. As a result, persons diagnosed with the disease contribute 1/2 year of follow-up during the year of diagnosis.

DID YOU KNOW?

A cohort study is a systematic review of the management of patients with disease and their contacts. A "cohort" is a group of disease cases counted over a specific period of time, usually 3 months. Disease cases are reviewed for the patient's clinical status, the adequacy of the medication regimen, treatment adherence of completion, and the results of contact investigation.

The denominator of the person-time rate is the sum of all of the person-years for each study participant. So, someone lost to follow-up in year 3 and someone diagnosed with the disease in year 3 each contributes 2.5 years of disease-free follow-up to the denominator.

8.3.2.1 Properties and Uses of Incidence Rates

An incidence rate describes how quickly disease occurs in a population. It is based on person-time, so it has some advantages over an incidence proportion. Because person-time is calculated for each subject, it can accommodate persons coming into and leaving the study. As noted in the previous example, the denominator accounts for study participants who are lost to follow-up or who die during the study period. In addition, it allows enrollees to enter the study at different times. Person-time has one important drawback. Person-time assumes that the probability of disease during the study period is constant, so that 10 persons followed for one year equals one person followed for 10 years. Because the risk of many chronic diseases increases with age, this assumption is often not valid.

Long-term cohort studies of the type described here are not very common; however, environmental health practitioners far more commonly calculate incidence rates based on a numerator of cases observed or reported and a denominator based on the mid-year population. This type of incident rate turns out to be comparable to a person-time rate.

Finally, if you report the incidence rate of, say, the heart disease study as 2.5 per 1000 person-years, environmental scientists might understand, but most others will not. Person-time is environmental health jargon. To convert this jargon to something understandable, simply replace "person-years" with "persons per year." Reporting the results as 2.5 new cases of heart disease per 1000 persons per year sounds like English rather than jargon. It also conveys the sense of the incidence rate as a dynamic process, the speed at which new cases of disease are occurring in the population.

Two example fractions commonly used in determining incidence rate are

$$\frac{\text{Number of women in Framingham Study who}}{\text{Number of person-years contributed through last year}}$$
$$\frac{\text{died through last year from heart disease}}{\text{by women initally enrolled in Framingham Study}}$$

$$\frac{\text{Number of women in State A newly diagnosed with heart disease in 2011}}{\text{Estimated number of women living in State A on July 1, 2011}}$$

■ **EXAMPLE 8.8**

Problem: Investigators enrolled 2100 women in a study and followed them annually for 4 years to determine the incidence rate of heart disease. After 1 year, none had a new diagnosis of heart disease, but 100 had been lost to follow-up. After 2 years, one had a new diagnosis of heart disease and another 99 had been lost to follow-up. After 3 years, another 7 had new diagnoses of heart disease and 793 had been lost to follow-up. After 4 years, another 8 had new diagnoses with heart disease and 392 more had been lost to follow-up. The study results could also be described as follows: No heart disease was diagnosed at the first year. Heart disease was diagnosed in one woman at the second year, in seven women at the third year, and in eight women at the fourth year of follow-up. One hundred women were lost to follow-up by the first year, another 99 were lost to follow-up after 2 years, another 793 were lost to follow-up after 3 years, and another 392 women were lost to follow-up after 4 years, leaving 700 women who were followed for 4 years and remained disease free. Calculate the incidence rate of heart disease among this cohort. Assume that persons with new diagnoses of heart disease and those lost to follow-up were disease free for half the year and thus contribute 1/2 year to the denominator.

Solution:

Numerator = Number of new cases of heart disease

$$= 0 + 1 + 7 + 8 = 16$$

Denominator = Person-years of observation

$$= \left[2000 + (1/2 \times 100)\right] + \left[1900 + (1/2 \times 1) + (1/2 \times 99)\right] + \left[1100 + (1/2 \times 7) + (1/2 \times 793)\right]$$
$$+ \left[700 + (1/2 \times 8) + (1/2 \times 392)\right]$$
$$= 6400 \text{ person-years of follow-up}$$

or

Denominator = Person-years of observation

$$= (1 \times 1.5) + (7 \times 2.5) + (8 \times 3.5) + (100 \times 0.5) + (99 \times 1.5) + (793 \times 2.5)$$
$$+ (392 \times 3.5) + (700 \times 4.0)$$
$$= 6400 \text{ person-years of follow-up}$$

$$\text{Person-time rate} = \frac{\text{Number of new cases of disease or injury during specified period}}{\text{Time each person was observed, totaled for all persons}}$$
$$= 16 / 6400 = 0.0025 \text{ cases per person-year}$$
$$= 2.5 \text{ cases per 1000 person-years}$$

In contrast, the incidence proportion can be calculated as 16/2100 = 7.6 cases per 1000 population during the 4-year period, or an average of 1.9 cases per 1000 per year (7.6 divided by 4 years). The incidence proportion underestimates the true rate because it ignores persons lost to follow-up and assumes that they remained disease free for all 4 years.

■ EXAMPLE 8.9

Problem: A diabetes follow-up study included 218 diabetic women and 3823 non-diabetic women. By the end of the study, 72 of the diabetic women and 511 of the non-diabetic women had died. The diabetic women were observed for a total of 1862 person-years; the non-diabetic women were observed for a total of 36,653 person-years. Calculate the incidence rates of death for the diabetic and non-diabetic women.

Solution:

For diabetic women:

$$\text{Numerator} = 72$$
$$\text{Denominator} = 1862$$
$$\text{Person-time rate} = 72/1862$$
$$= 0.0386 \text{ deaths per person-year}$$
$$= 38.6 \text{ deaths per 1000 person-years}$$

For non-diabetic women:

$$\text{Numerator} = 511$$
$$\text{Denominator} = 36,653$$
$$\text{Person-time rate} = 511/36,653$$
$$= 0.0139 \text{ deaths per person-year}$$
$$= 13.9 \text{ deaths per 1000 person-years}$$

■ EXAMPLE 8.10

Problem: In 2003, 44,232 new cases of acquired immunodeficiency syndrome (AIDS) were reported in the United States (CDC, 2005). The estimated mid-year population of the United States in 2003 was approximately 290,809,777 (U.S. Census Bureau, 2006). Calculate the incidence rate of AIDS in 2003.

Solution:

$$\text{Numerator} = 44,232 \text{ new cases of AIDS}$$
$$\text{Denominator} = 290,809,777 \text{ estimated mid-year population}$$
$$10^n = 100,000$$
$$\text{Incidence rate} = (44,232/290,809,777) \times 100,000$$
$$= 15.21 \text{ new cases of AIDS per 100,000 population}$$

8.3.3 PREVALENCE

Prevalence, sometimes referred to as *prevalence rate*, is the proportion of persons in a population who have a particular disease or attribute at a specified point in time or over a specified period of time. Prevalence differs from incidence in that prevalence includes all cases, both new and preexisting, in the population at the specified time, whereas incidence is limited to new cases only. *Point prevalence* refers to the prevalence measured at a particular point in time. It is the proportion of persons with a particular disease or attribute on a particular date. *Period prevalence* refers to prevalence measured over an interval of time. It is the proportion of persons with a particular disease or attribute at any time during the interval.

- Method for calculating prevalence of disease:

$$\frac{\text{All new and pre-existing cases during a given time period}}{\text{Population during the same time period}} \times 10^n$$

- Method for calculating prevalence of an attribute:

$$\frac{\text{Persons having a particular attribute during a given time period}}{\text{Population during the same time period}} \times 10^n$$

The value of 10^n is usually 1 or 100 for common attributes. The value of 10^n might be 1000, 100,000, or even 1,000,000 for rare attributes and for most diseases.

Two examples of prevalence fractions are

$$\frac{\text{Number of women in town of Framingham who reported having heart disease in recent health survey}}{\text{Estimated number of women residents of Framingham during same period}}$$

$$\frac{\text{Estimated number of women smokers in State A according to 2004 Behavior Risk Factor Survey}}{\text{Estimated number of women living in State A on July 1, 2004}}$$

■ EXAMPLE 8.11

Problem: In a survey of 1150 women who gave birth in Maine in 2000, a total of 468 reported taking a multivitamin at least 4 times a week during the month before becoming pregnant (Williams et al., 2003). Calculate the prevalence of frequent multivitamin use in this group.

Solution:

Numerator = 468 multivitamin users
Denominator = 1150 women
Prevalence = (448/1150) × 100 = 0.407 × 100 = 40.7%

8.3.3.1 Properties and Uses of Prevalence

Prevalence and incidence are frequently confused. Prevalence refers to proportion of persons who *have* a condition at or during a particular time period, whereas incidence refers to the proportion or rate of persons who *develop* a condition during a particular time period. So, prevalence and incidence are similar, but prevalence includes new and preexisting cases, whereas incidence includes new cases only. The key difference is in their numerators:

- *Numerator of incidence* = New cases that occurred during a given time period.
- *Numerator of prevalence* = All cases present during a given time period.

The numerator of an incidence proportion or rate consists only of persons whose illness began during the specified interval. The numerator for prevalence includes all persons ill from a specified cause during the specified interval *regardless of when the illness began*. It includes not only new cases but also preexisting cases representing persons who remained ill during some portion of the specified interval.

Prevalence is based on both incidence and duration of illness. High prevalence of a disease within a population might reflect high incidence or prolonged survival without cure or both. Conversely, low prevalence might indicate low incidence, a rapidly fatal process, or rapid recovery. Prevalence rather than incidence is often measured for chronic diseases such as diabetes or osteoarthritis that have long duration and dates of onset that are difficult to pinpoint.

Figure 8.1 shows 10 new cases of illness over about 15 months in a population of 20 persons. Each horizontal line represents one person. The down arrow indicates the date of onset of illness. The solid line represents the duration of illness. The up arrow and the cross represent the date of recovery and date of death, respectively. Use Figure 8.1 to solve the problems presented in Example 8.12, Example 8.13, and Example 8.14.

■ EXAMPLE 8.12

Problem: Calculate the incidence rate from October 1, 2004, to September 30, 2005, using the midpoint population (population alive on April 1, 2005) as the denominator. Express the rate per 100 population.

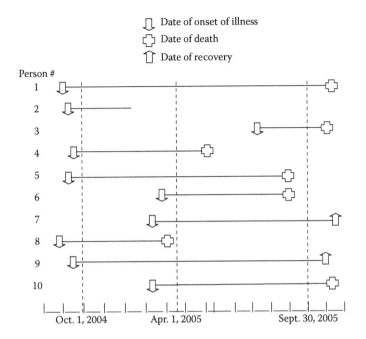

FIGURE 8.1 New cases of illness from October 1, 2004, to September 30, 2005. (From CDC, *Principles of Epidemiology in Public Health Practice*, 3rd ed., U.S. Centers for Disease Control and Prevention, Atlanta, GA, 2012.)

Solution:

Incidence rate numerator = Number of new cases between October 1 and September 30

= 4 (the other 6 all had onsets before October 1 and are not included)

Incidence rate denominator = April 1 population

= 18 (persons 2 and 8 died before April 1)

Incidence rate = $(4/18) \times 100$

= 22 new cases per 100 population

■ EXAMPLE 8.13

Problem: Calculate the point prevalence on April 1, 2005. Point prevalence is the number of persons ill on the date divided by the population on that date. On April 1, 7 persons (persons 1, 4, 5, 7, 9, and 10) were ill.

Solution:

$$\text{Point prevalence} = (7/18) \times 100 = 38.89\%$$

■ EXAMPLE 8.14

Problem: Calculate the period prevalence from October 1, 2004, to September 30, 2005. The numerator of period prevalence includes anyone who was ill any time during the period. In Figure 8.1, the first 10 persons were all ill at the same time during the period.

Solution:

$$\text{Period prevalence} = (10/20) \times 100 = 50.0\%$$

TABLE 8.4

Frequently Used Measures of Mortality

Measure	Numerator	Denominator	10^n
Crude death rate	Total numerator of deaths during a given time interval	Mid-interval population	1000 or 100,000
Cause-specific death rate	Number of deaths assigned to a specific cause during a given time interval	Mid-interval population	100,000
Proportionate mortality	Number of deaths assigned to a specific cause during a given time interval	Total number of deaths from all causes during the same time interval	100 or 1000
Death-to-case ratio interval	Number of deaths assigned to a specific cause during a given time interval	Number of new cases of same disease reported during the same time interval	100
Neonatal mortality rate	Number of deaths among children < 28 days of age during a given time interval	Number of live births during the same time interval	1000
Postneonatal mortality rate	Number of deaths among children 28–364 days of age during a given time interval	Number of live births during the same time interval	1000
Infant mortality rate	Number of deaths among children < 1 year of age during a given time interval	Number of live births during the same time interval	1000
Maternal mortality rate	Number of deaths assigned to pregnancy-related causes during a given time interval	Number of live births during the same time interval	100,000

8.4 MORTALITY FREQUENCY MEASURES

8.4.1 MORTALITY RATE

A *mortality rate* is a measure of the frequency of occurrence of death in a defined population during a specified interval. Morbidity and mortality measures are often the same mathematically; it's just a matter of what we choose to measure, illness or death. The formula for the mortality of a defined population, over a specified period of time, is

$$\frac{\text{Deaths occurring during given time period}}{\text{Size of the population among which the deaths occurred}} \times 10^n$$

When mortality rates are based on vital statistics (e.g., counts of death certificates), the denominator most commonly used is the size of the population at the middle of the time period. In the United States, values of 1000 and 100,000 are both used for 10^n for most types of mortality rates. Table 8.4 summarizes the formulas of frequently used mortality measures.

8.4.1.1 Crude Mortality Rate (Crude Death Rate)

The crude mortality rate is the mortality rate from all causes of death for a population. In the United States in 2003, a total of 2,419,921 deaths occurred. The estimated population was 290,809,777. The crude mortality rate in 2003 was, therefore, (2,419,921/290,809,777) × 100,000, or 832.1 deaths per 100,000 population (WISQARS, 2012).

8.4.1.2 Cause-Specific Mortality Rate

The cause-specific mortality rate is the mortality rate from a specified cause for a population. The numerator is the number of deaths attributed to a specific cause. The denominator remains the size of the population at the midpoint of the time period. The fraction is usually expressed per

100,000 population. In the United States in 2003, a total of 108,256 deaths were attributed to accidents (unintentional injuries), yielding a cause-specific mortality rate of 37.2 per 100,000 population (WISQARS, 2012).

8.4.1.3 Age-Specific Mortality Rate

An age-specific mortality rate is a mortality rate limited to a particular age group. The numerator is the number of deaths in that age group; the denominator is the number of persons in that age group in the population. In the United States in 2003, a total of 130,761 deaths occurred among persons ages 25 to 44 years, for an age-specific mortality rate of 153.0 per 100,000 25- to 44-year-olds (WISQARS, 2012). Some specific types of age-specific mortality rates are neonatal, postneonatal, and infant mortality rates, as described in the following sections.

8.4.1.4 Infant Mortality Rate

The infant mortality rate is perhaps the most commonly used measure for comparing health status among nations. It is calculated as follows:

$$\frac{\text{Number of deaths among children} < 1 \text{ year of age reported during a given time period}}{\text{Number of live births reported during the same time period}} \times 10^n$$

The infant mortality rate is generally calculated on an annual basis. It is a widely used measure of health status because it reflects the health of the mother and infant during pregnancy and the year thereafter. The health of the mother and infant, in turn, reflects a wide variety of factors, including access to prenatal care, prevalence of prenatal maternal health behaviors (such as alcohol or tobacco use and proper nutrition during pregnancy), postnatal care and behaviors (including childhood immunizations and proper nutrition), sanitation, and infection control.

Is the infant mortality rate a ratio? Yes. Is it a proportion? No, because some of the deaths in the numerator occurred among children born the previous year. Consider the infant mortality rate in 2003. That year, 28,025 infants died and 4,089,950 children were born, for an infant mortality rate of 6951 per 1000. Undoubtedly, some of the deaths in 2003 occurred among children born in 2002, but the denominator includes only children born in 2003.

Is the infant mortality rate truly a rate? No, because the denominator is not the size of the mid-year population of children less than 1 year of age in 2003. In fact, the age-specific death rate for children less than 1 year of age for 2003 was 694.7 per 100,000 (WISQARS, 2012). Obviously, the infant mortality rate and the age-specific death rate for infants are very similar (695.1 vs. 694.7 per 100,000) and close enough for most purposes. They are not exactly the same, however, because the estimated number of infants residing in the United States on July 1, 2003, was slightly larger than the number of children born in the United States in 2002, presumably because of immigration.

8.4.1.5 Neonatal Mortality Rate

The neonatal period covers birth up to but not including 28 days. The numerator of the neonatal mortality rate therefore is the number of deaths among children under 28 days of age during a given time period. The denominator of the neonatal mortality rate, like that of the infant mortality rate, is the number of live births reported during the same time period. The neonatal mortality rate is usually expressed per 1000 live births. In 2003, the neonatal mortality rate in the United States was 4.7 per 1000 live births (WISQARS, 2012).

8.4.1.6 Postneonatal Mortality Rate

The postneonatal period is defined as the period from 28 days of age up to but not including 1 year of age. The numerator of the postneonatal mortality rate therefore is the number of deaths among children from 28 days up to but not including 1 year of age during a given time period. The denominator is the number of live births reported during the same time period. The postneonatal mortality rate is usually expressed per 1000 live births. In 2003, the postneonatal mortality rate in the United States was 2.3 per 1000 live births (WISQARS, 2012).

8.4.1.7 Maternal Mortality Rate

The maternal mortality rate is really a ratio used to measure mortality associated with pregnancy. The numerator is the number of deaths during a given time period among women while pregnant or within 42 days of termination of pregnancy, irrespective of the duration and site of the pregnancy, from any cause related to or aggravated by the pregnancy or its management, but not from accidental or incidental causes. The denominator is the number of live births reported during the same time period. Maternal mortality rate is usually expressed per 100,000 live births. In 2003, the U.S. maternal mortality rate was 8.9 per 100,000 live births (WISQARS, 2012).

8.4.1.8 Sex-Specific Mortality Rate

A sex-specific mortality rate is a mortality rate on either males or females. Both numerator and denominator are limited to the one sex.

8.4.1.9 Race-Specific Mortality Rate

A race-specific mortality rate is a mortality rate related to a specified racial group. Both numerator and denominator are limited to the specified race.

8.4.1.10 Combinations of Specific Mortality Rates

Mortality rates can be further stratified by combinations of cause, age, sex, and/or race. For example, in 2002, the death rate from diseases of the heart among women ages 45 to 54 years was 50.6 per 100,000. The death rate from diseases of the heart among men in the same age group was 138.4 per 100,000, or more than 2.5 times as high as the comparable rate for women. These rates are a cause-, age-, and sex-specific rates, because they refer to one cause (diseases of the heart), one age group (45 to 54 years), and one sex (female or male).

8.4.1.11 Calculating Mortality Rates

Table 8.5 provides the number of deaths from all causes and from accidents (unintentional injuries) by age group in the United States in 2002. In the following, we present various rates and demonstrate how to calculate each using the data provided in Table 8.5.

- Unintentional-injury-specific mortality rate for the entire population (cause-specific mortality rate):

$$\text{Rate} = \frac{\text{Number of unintentional injury deaths in the entire population}}{\text{Estimated mid-year population}} \times 100,000$$

$$= \frac{106,742}{288,356,000} \times 100,000$$

$$= 37.0 \text{ unintentional-injury-related deaths per 100,000 population}$$

- All-cause mortality rate for 25- to 34-year-olds (age-specific mortality rate):

$$\text{Rate} = \frac{\text{Number of deaths from all causes among 25- to 34-year-olds}}{\text{Estimated mid-year poulation of 25- to 34-year-olds}} \times 100,000$$

$$= \frac{41,355}{39,928,000} \times 100,000$$

$$= 103.6 \text{ deaths per 100,000 25- to 34-year-olds}$$

TABLE 8.5

All-Case and Unintentional Injury Mortality and Estimated Population by Age Group for Both Sexes and for Males Alone—United States, 2002

Age Group (Years)	All Races, Both Sexes			All Races, Males		
	All Causes	Unintentional Injuries	Estimated Population (x 1000)	All Causes	Unintentional Injuries	Estimated Population (x 1000)
0–4	32,892	2587	19,597	18,523	1577	10,020
5–14	7150	2718	41,037	4198	1713	21,013
15–24	33,046	15,412	40,590	24,416	11,438	20,821
25–34	41,355	12,569	39,928	28,736	9635	20,203
35–44	91,140	16,710	44,917	57,593	12,012	22,367
45–54	172,385	14,675	40,084	107,722	10,492	19,676
55–64	253,342	8345	26,602	151,363	5781	12,784
65+	1,811,720	33,641	35,602	806,431	16,535	14,772
Not stated	357	85	0	282	74	0
Total	2,443,387	106,742	288,357	1,199,264	69,257	141,656

Source: Web-based Injury Statistics Query and Reporting System (WISQARS), http://www.cdc.gov/injury/wisqars.

- All-cause mortality among males (sex-specific mortality rate):

$$\text{Rate} = \frac{\text{Number of deaths from all causes among males}}{\text{Estimated mid-year population of males}} \times 100,000$$

$$= \frac{1,199,264}{141,656,000} \times 100,000$$

$$= 846.6 \text{ deaths per } 100,000 \text{ males}$$

- Unintentional-injury specific mortality among 25 to 34 year old males (cause-specific, age-specific, and sex-specific mortality rate):

$$\text{Rate} = \frac{\text{Number of unintentional injury deaths among 25- to 34-year-old males}}{\text{Estimated mid-year population of 25- to 34-year-old males}} \times 100,000$$

$$= \frac{9635}{20,203,000} \times 100,000$$

$$= 47.7 \text{ unintentional-injury-related deaths per } 100,000 \text{ 25- to 34-year-olds}$$

■ **EXAMPLE 8.15**

Problem: In 2001, a total of 15,555 homicide deaths occurred among males and 4753 homicide deaths occurred among females. The estimated 2001 mid-year populations for males and females were 139,813,000 and 144,984,000, respectively. Calculate the homicide-related death rates for males and for females.

Solution:

- Homicide-related death rate (males)

    = (Number of homicide deaths among males/male population) × 100,000
    = 15,555/139,813,000 × 100,000
    = 11.1 homicide deaths per 100,000 population among males

- Homicide-related death rate (females)

    = (Number of homicide deaths among females/female population) × 100,000
    = 4,753/144,984,000 × 100,000
    = 3.3 homicide deaths per 100,000 population among females

8.4.1.12 Age-Adjusted Mortality Rates

Mortality rates can be used to compare the rates in one area with the rates in another area, or to compare rates over time. However, because mortality rates obviously increase with age, a higher mortality rate among one population than among another might simply reflect the fact that the first population is older than the second. Consider that the mortality rates in 2002 for the states of Alaska and Florida were 472.2 and 1005.7 per 100,000, respectively (see Table 8.6). Should everyone from Florida move to Alaska to reduce their risk of death? No, the reason that Alaska's mortality rate is so much lower than Florida's is that Alaska's population is considerably younger. Indeed, for seven age groups, the age-specific mortality rates in Alaska are actually higher than Florida's.

TABLE 8.6
All-Cause Mortality by Age Group—Alaska and Florida, 2002

Age Group (years)	Alaska Population	Alaska Deaths	Alaska Death Rate (per 100,000)	Florida Population	Florida Deaths	Florida Death Rate (per 100,000)
<1	9938	55	553.4	205,579	1548	753.0
1–4	38,503	12	31.2	816,570	296	36.2
5–9	50,400	6	11.9	1,046,504	141	13.5
10–14	57,216	24	41.9	1,131,068	219	19.4
15–19	56,634	43	75.9	1,073,470	734	68.4
20–24	42,929	63	146.8	1,020,856	1146	112.3
25–34	84,112	120	142.7	2,090,312	2627	125.7
35–44	107,305	280	260.9	2,516,004	5993	238.2
45–54	103,039	427	414.4	2,225,957	10,730	482.0
55–64	52,543	480	913.5	1,694,574	16,137	952.3
65–74	24,096	502	2083.3	1,450,843	28,959	1996.0
65–84	11,784	645	5473.5	1,056,275	50,755	4805.1
85+	3117	373	11,966.0	359,056	48,486	13,503.7
Unknown	NA	0	NA	NA	43	NA
Total	3000	3030	472.2	16,687,068	167,814	1005.7
Age-adjusted rate			794.1			787.8

Source: Web-based Injury Statistics Query and Reporting System (WISQARS), http://www.cdc.gov/injury/wisqars.

To eliminate the distortion caused by different underlying age distributions in different populations, statistical techniques are used to adjust or standardize the rates among the populations to be compared. These techniques take a weighted average of the rate-specific mortality rates and eliminate the effect of different age distributions among the different populations. Mortality rates computed with these techniques are *age-adjusted* or *age-standardized mortality rates*. Alaska's 2002 age-adjusted mortality rate (794.1 per 100,000) was higher than Florida's (787.8 per 100,000), which is not surprising given that 7 of 13 age-specific mortality rates were higher in Alaska than Florida.

8.4.2 DEATH-TO-CASE RATIO

The death-to-case ratio is the number of deaths attributed to a particular disease during a specified time period divided by the number of new cases of that disease identified during the same time period. The death-to-case ratio is a ratio but not necessarily a proportion, because some of the deaths that are counted in the numerator might have occurred among persons who developed the disease in an earlier period and are therefore not counted in the denominator.

8.4.2.1 Method for Calculating Death-to-Case Ratio

$$\frac{\text{Number of deaths attributed to a particular disease during specified period}}{\text{Number of new cases of the disease identified during the specified period}} \times 10^n$$

■ EXAMPLE 8.16

Problem: Between 1940 and 1949, a total of 143,497 incident cases of diphtheria were reported. During the same decade, 11,228 deaths were attributed to diphtheria. Calculate the death-to-case ratio.
Solution:

$$\text{Death-to-case ratio} = 11,228/143,497 \times 1 = 0.0783$$

or

$$\text{Death-to-case ratio} = 11,228/143,497 \times 100 = 7.83 \text{ per } 100$$

■ EXAMPLE 8.17

Problem: The following table provides the number of reported cases of diphtheria and the number of diphtheria-associated deaths in the United States by decade. Calculate the death-to-case ratio by decade. Describe the data in the table, including your results (CDC, 1999, 2001, 2003).

Number of Cases and Deaths from Diphtheria by Decade—United States, 1940–1999

Decade	Number of New Cases	Number of Deaths	Death-to-Case Ratio ($\times 100$)
1940–1949	143,497	11,228	7.62
1950–1959	23,750	1710	————
1960–1969	3679	390	————
1970–1979	1956	90	————
1980–1989	27	3	————
1990–1999	22	5	————

Solution:

Number of Cases and Deaths from Diphtheria by Decade—United States, 1940–1999

Decade	Number of New Cases	Number of Deaths	Death-to-Case Ratio ($\times 100$)
1940–1949	143,497	11,228	7.62
1950–1959	23,750	1710	7.20
1960–1969	3679	390	10.60
1970–1979	1956	90	4.60
1980–1989	27	3	11.11
1990–1999	22	5	22.72

The number of new cases and deaths from diphtheria declined dramatically from the 1940s through the 1980s but remained roughly level at very low levels in the 1990s. The death-to-case ratio was actually higher in the 1980s and 1990s than in 1940s and 1950s. From these data one might conclude that the decline in death is a result of the decline in cases—that is, from prevention rather than from any improvement in the treatment of cases that do occur.

8.4.3 Case-Fatality Rate

The case-fatality rate is the proportion of persons with a particular condition (cases) who die from that condition. It is a measure of the severity of the condition. The formula is

$$\frac{\text{Number of cause-specific deaths among the incident cases}}{\text{Number of incident cases}} \times 10^n$$

The case-fatality rate is a proportion, so the numerator is restricted to deaths among people included in the denominator. The time periods for the numerator and the denominator do not need to be the same; the denominator could be cases of HIV/AIDS diagnosed during the calendar year 1990, and the numerator, deaths among those diagnosed with HIV in 1990, could be from 1990 to the present.

■ EXAMPLE 8.18

Problem: In an epidemic of hepatitis A traced to green onions from a restaurant, 555 cases were identified. Three of the case-patients died as a result of their infections. Calculate the case-fatality rate.

Solution:

$$\text{Case-fatality rate} = (3/555) \times 100 = 0.5\%$$

The concept behind the case-fatality rate and the death-to-case ratio is similar, but the formulations are different. The death-to-case ratio is simply the number of cause-specific deaths that occurred during a specified time divided by the number of new cases of that disease that occurred during the same time. The deaths included in the numerator of the death-to-case ratio are not restricted to the new cases in the denominator; in fact, the deaths included in the numerator are restricted to the cases in the denominator.

DID YOU KNOW?

The case-fatality rate is a proportion, not a true rate. As a result, some environmental practitioners prefer the term *case-fatality ratio*.

8.4.4   PROPORTIONATE MORTALITY

Proportionate mortality describes the proportion of deaths in a specified population over a period of time attributable to different causes. Each cause is expressed as a percentage of all deaths, and the sum of the causes must add to 100%. These proportions are not mortality rates, because the denominator is all deaths rather than the population in which the deaths occurred.

8.4.4.1   Method for Calculating Proportionate Mortality

For a specified population over a specified period,

$$\frac{\text{Deaths caused by a particular cause}}{\text{Deaths from all causes}} \times 100$$

The distribution of primary causes of death in the United States in 2003 for the entire population (all ages) and for persons ages 25 to 44 years are provided in Table 8.1. As illustrated in that table, accidents (unintentional injuries) accounted for 4.3% of all deaths but 21.6% of deaths among 25- to 44-year-olds (WISQARS, 2012).

Sometimes, particularly in occupational environmental health, proportionate mortality is used to compare deaths in a population of interest (say, a workplace) with the proportionate mortality in the broader population. This comparison of two proportionate mortalities is referred to as a *proportionate mortality ratio*, or PMR for short. A PMR greater than 1.0 indicates that a particular cause accounts for a greater proportion of deaths in the population of interest than might be expected; for example, construction workers may be more likely to die of injuries than the general population. PMRs can be misleading, though, because they are not based on mortality rates. A low cause-specific mortality rate in the population of interest can elevate the proportionate mortalities of all of the other causes, because they must add up to 100%. Those workers with a high injury-related proportionate mortality very likely have lower proportionate mortalities for chronic or disabling conditions that keep people out of the workforce. In other words, people who work are more likely to be healthier than the population as a whole—this is known as the *healthy worker effect*.

■ EXAMPLE 8.19

Problem: Using the data in Table 8.7, calculate the missing proportionate mortalities for persons ages 25-44 years for diseases of the heart and assaults (homicide).

Solution:

Proportionate mortality for diseases of the heart, 25- to 44-year-olds

$$= \frac{\text{Number of deaths from diseases of the heart}}{\text{Number of deaths from all causes}} \times 100$$

$$= (16,283/128,294) \times 100 = 12.6\%$$

Proportionate mortality for assaults (homicide), 25- to 44-year-olds

$$= \frac{\text{Number of deaths from assaults (homicides)}}{\text{Number of deaths from all causes}} \times 100$$

$$= (7367/128,294) \times 100 = 5.7\%$$

TABLE 8.7

Number, Proportion (Percentage), and Ranking of Deaths for Leading Causes of Death, All Ages and Ages 25 to 44 Years—United States, 2003

	All Ages			Ages 25 to 44 Years		
	Number	Percentage	Rank	Number	Percentage	Rank
All causes	2,443,930	100.0		128,924	100.0	
Diseases of heart	684,462	28.0	1	16,283		3
Malignant neoplasms	554,643	22.7	2	19,041	14.8	2
Cerebrovascular disease	157,803	6.5	3	3004	2.3	8
Chronic lower respiratory diseases	126,128	5.2	4	401	0.3	—[a]
Accidents (unintentional injuries)	105,695	4.3	5	27,844	21.6	1
Diabetes mellitus	73,965	3.0	6	2662	2.1	9
Influenza and pneumonia	64,847	2.6	7	1337	1.0	10
Alzheimer's disease	63,343	2.6	8	0	0.0	—[a]
Nephritis, nephritic syndrome	33,615	1.4	9	305	0.2	—[a]
Septicemia	34,243	1.4	10	328	0.2	—[a]
Intentional self-harm (suicide)	30,642	1.3	11	11,251	8.7	4
Chronic liver disease and cirrhosis	27,201	1.1	12	3288	2.6	7
Assault (homicide)	17,095	0.7	13	7367	—[a]	5
HIV disease	13,544	0.5	—[a]	6879	5.3	6
All other	456,703	18.7		29,480	22.9	

[a] Not among top-ranked causes.

Source: Data from CDC (2005) and Hoyert et al. (2005).

8.4.5 Years of Potential Life Lost

Years of potential life lost (YPLL) is one measure of the impact of premature mortality on a population. Additional measures incorporate disability and other measures of quality of life. YPLL is calculated as the sum of the differences between a predetermined end point and the ages of death for those who died before that end point. The two most commonly used end points are age 65 years and average life expectancy.

The use of YPLL is affected by this calculation, which implies a value system in which more weight is given to a death when it occurs at an earlier age. Thus, deaths at older ages are "devalued." However, the YPLL before age 65 (YPLL$_{65}$) places much more emphasis on deaths at early ages than does YPLL based on remaining life expectancy (YPLL$_{LE}$). In 2000, the remaining life expectancy was 21.6 years for a 60-year-old, 11.3 years for a 70-year-old, and 8.6 for an 80-year-old. YPLL$_{65}$ is based on the fewer than 30% of deaths that occur among persons younger than 65. In contrast, YPLL for life expectancy (YPLL$_{LE}$) is based on deaths among persons of all ages, so it more closely resembles crude mortality rates (Wise et al., 1988). YPLL rates can be used to compare YPLL among populations of different sizes. Because different populations may also have different age distributions, YPLL rates are usually age adjusted to eliminate the effect of differing age distributions.

8.4.5.1 Method for Calculating YPLL from a Line Listing

1. Decide on an end point (65 years, average life expectancy, or other).
2. Exclude records of all persons who died at or after the end point.

3. For each person who died before the end point, calculate the person's YPLL by subtracting the age at death from the end point:

$$YPLL_{individual} = End\ point - Age\ at\ death$$

4. Sum the individual YPLLs:

$$YPLL = \Sigma YPLL_{individual}$$

8.4.5.2 Method for Calculating YPLL from a Frequency

1. Ensure that age groups break at the identified end point (e.g., 65 years). Eliminate all age groups older than the endpoint.
2. For each age group younger than the end point, identify the midpoint of the age group, where midpoint is determined as

$$\frac{Age\ group's\ youngest\ age\ in\ years + oldest\ age + 1}{2}$$

3. For each age group younger than the end point, identify that age group's YPLL by subtracting the midpoint from the end point.
4. Calculate age-specific YPLL by multiplying the age group's YPLL times the number of persons in that age group.
5. Sum the age-specific YPLLs.

The *YPLL rate* represents years of potential life lost per 1000 population below the end-point age, such as 65 years. YPLL rates should be used to compare premature mortality in different populations, because YPLL does not take into account differences in population sizes. The formula for YPLL rate is as follows:

$$YPLL\ rate = \frac{Years\ of\ potential\ life\ lost}{Population\ under\ age\ 65\ years} \times 10^n$$

8.5   NATALITY (BIRTH) MEASURES

Natality measures are population-based measures of birth. These measures are used primarily by persons working in the field of maternal and child health. Table 8.8 includes some of the commonly used measures of natality.

TABLE 8.8
Frequently Used Measures of Natality

Measure	Numerator	Denominator	10^n
Crude birth rate	Number of live births during a specified time interval	Mid-interval population	1000
Crude fertility rate	Number of live births during a specified time interval	Number of women ages 15 to 44 at mid-interval	1000
Crude rate of natural increase	Number of live births minus number of deaths during a specified time interval	Mid-interval population	1000
Low-birth-weight ratio	Number of live births < 2500 grams during a specified time interval	Number of live births during the same time interval	100

8.6   MEASURES OF ASSOCIATION

The key to environmental health analysis is comparison. Occasionally we might observe an incidence rate among a population that seems high and wonder whether it is actually higher than what should be expected based on, say, the incidence rates in other communities. Or, we might observe that, among a group of case-patients in an outbreak, several report having eaten at a particular restaurant. Is the restaurant just a popular one, or have more case-patients eaten there than would be expected? The way to address that concern is by comparing the observed group with another group that represents the expected level.

A measure of association quantifies the relationship between exposure and disease among the two groups. Exposure is used loosely to mean not only exposure to foods, mosquitoes, a partner with a sexually transmissible disease, or a toxic waste dump, but also inherent characteristics of persons (for, example, age, race, sex), biological characteristics (immune status), acquired characteristics (marital status), activities (occupation, leisure activities), or conditions under which they live (socioeconomic status or access to medical care).

The measures of association described in the following section compare disease occurrence among one group with disease occurrence in another group. Examples of measure of association include risk ratio (relative risk), ratio, odds ratio, and proportionate mortality ratio.

8.6.1   RISK RATIO

A risk ratio (RR), also called *relative risk*, compares the risk of a health event (disease, injury, risk factor, or death) among one group with the risk among another group. It does so by dividing the risk (incidence proportion, attack rate) in group 1 by the risk (incidence proportion, attack rate) in group 2. The two groups are typically differentiated by such demographic factors as sex (e.g., males vs. females) or by exposure to a suspected risk factor (e.g., did or did not eat potato salad). Often, the group of primary interest is labeled the exposed group, and the comparison group is labeled the unexposed group.

8.6.1.1   Method for Calculating Risk Ratio

The formula for risk ratio (RR) is

$$\frac{\text{Risk of disease (incidence proportion, attack rate) in group of primary interest}}{\text{Risk of disease (incidence proportion, attack rate) in comparison group}}$$

A risk ratio of 1.0 indicates identical risk among the two groups. A risk ratio greater than 1.0 indicates an increased risk for the group in the numerator, usually the exposed group. A risk ratio less than 1.0 indicates a decreased risk for the exposed group, indicating that perhaps exposure actually protects against disease occurrence.

■ EXAMPLE 8.20

Problem: In an outbreak of tuberculosis among prison inmates in South Carolina in 1999, 28 of 157 inmates residing on the east wing of the dormitory developed tuberculosis, compared with 4 of 137 inmates residing on the west wing (McLaughlin et al., 2003). These data are summarized in the two-by-two (2×2) table (see Table 8.9A), so called because it has two rows for the exposure and two columns for the outcome. In this example, the exposure is the dormitory wing and the outcome is tuberculosis (see Table 8.9B). Calculate the risk ratio.

Solution: To calculate the risk ratio, first calculate the risk or attack rate for each group. Here are the formulas:

TABLE 8.9A

General Format and Notation for a Two-by-Two Table

	Ill	Well	Total
Exposed	a	b	$a + b = H_1$
Unexposed	c	d	$c + d = H_0$
Total	$a + c = V_1$	$b + d = V_0$	294

TABLE 8.9B

Incidence of *Mycobacterium Tuberculosis* Infection Among Congregated, HIV-Infected Prison Inmates by Dormitory Wing—South Caroline, 1999

	Developed Tuberculosis?		
	Yes	No	Total
East wing	$a = 28$	$b = 129$	$H_1 = 157$
West wing	$c = 4$	$d = 133$	$H_0 = 137$
Total	32	262	294

Source: Data from CDC (2012) and McLaughlin et al. (2003).

Attack rate for exposed = $a/(a + b)$
Attack rate for unexposed = $c/(c + d)$

For this example:

Risk of tuberculosis among east-wing residents = 28/157 = 0.178 = 17.8%
Risk of tuberculosis among west-wing residents = 4/137 = 0.029 = 2.9%

The risk ratio is simply the ratio of these two risks:

Risk ratio = 17.8/2.9 = 6.1

Thus, inmates who resided in the east wing of the dormitory were 6.1 times as likely to develop tuberculosis as those who resided in the west wing.

■ **EXAMPLE 8.21**

Problem: In an outbreak of *Varicella* (chickenpox) in Oregon in 2002, *Varicella* was diagnosed in 18 of 152 vaccinated children compared with 3 of 7 unvaccinated children. Using data contained in Table 8.10, calculate the risk ratio.

Solution:

Risk of *Varicella* among vaccinated children = 18/152 = 0.118 = 11.8%
Risk of *Varicella* among unvaccinated children = 2/7 = 0.429 = 42.9%

Risk ratio = 0.118/0.429 = 0.28

TABLE 8.10

Incidence of *Varicella* Among Schoolchildren in Nine Affected Classrooms—Oregon, 2002

	Varicella	Non-case	Total
Vaccinated	$a = 18$	$b = 134$	152
Unvaccinated	$c = 3$	$d = 4$	7
Total	21	138	159

Source: Data from CDC (2012) and Tugwell et al. (2004).

The risk ratio is less than 1.0, indicating a decreased risk or protective effect for the exposed (vaccinated) children. The risk ratio of 0.28 indicates that vaccinated children were only approximately one-fourth as likely (28%, actually) to develop *Varicella* as were unvaccinated children.

8.6.2 Rate Ratio

A rate ratio compares the incidence rates, person-time rates, or mortality rates of two groups. As with the risk ratio, the two groups are typically differentiated by demographic factors or by exposure to a suspected causative agent. The rate for the group of primary interest is divided by the rate for the comparison group:

$$\frac{\text{Rate for group of primary interest}}{\text{Rate for comparison group}}$$

The interpretation of the value of a rate is similar to that of the risk ratio. That is, a rate ratio of 1.0 indicates equal rates in the two groups, a rate ratio greater than 1.0 indicates an increased risk for the group in the numerator, and rate ratio less than 1.0 indicates a decreased risk for the group in the numerator.

■ EXAMPLE 8.22

Problem: Public health officials were called to investigate a perceived increase in visits to ships' infirmaries for acute respiratory illness (ARI) by passengers of cruise ships in Alaska in 1998 (Uyeki et al., 2003). The officials compared passenger visits to ship infirmaries for ARI during May–August 1998 with the same period in 1997. They recorded 11.6 visits for ARI per 1000 tourists per week in 1998, compared with 5.3 visits per 1000 tourists per week in 1997. Calculate the rate ratio.

Solution:

$$\text{Rate ratio} = 11.6/5.3 = 2.2$$

Passengers on cruise ships in Alaska during May–August 1998 were more than twice as likely to visit ships' infirmaries for ARI as were passengers in 1997. (*Note:* Of 58 viral isolates identified from nasal cultures from passengers, most were influenza A, making this the largest summertime influenza outbreak in North America.)

8.6.3 ODDS RATIO

An odds ratio (OR) is another measure of association that quantifies the relationship between an exposure with two categories and health outcomes. Referring to the four cells in Table 8.11, the odds ratio is calculated as

$$\text{Odds ratio} = \left(\frac{a}{b}\right)\left(\frac{c}{d}\right) = ad\,/\,bc$$

where
  a = Number of persons exposed and with disease.
  b = Number of persons exposed but without disease.
  c = Number of persons unexposed but with disease.
  d = Number of persons unexposed and without disease.
  $a + c$ = Total number of persons with disease (case-patients).
  $b + d$ = Total number of persons without disease (controls).

The odds ratio is sometimes called the *cross-product ratio* because the numerator is based on multiplying the value in cell a times the value in cell d, whereas the denominator is the product of cell b and cell c. A line from cell a to cell d (for the numerator) and another from cell b to cell c (for the denominator) creates an X, or cross, on the two-by-two table.

■ EXAMPLE 8.23

Problem: Use the data in Table 8.11 to calculate the risk and odds ratios.

Solution:

  1. Risk ratio
$$5.0/1.0 = 5.0$$

  2. Odds ratio
$$(100 \times 7920)/(1900 \times 80) = 5.2$$

Notice that the odds ratio of 5.2 is close to the risk ratio of 5.0. That is one of the attractive features of the odds ratio—when the health outcome is uncommon, the odds ratio provides a reasonable approximation of the risk ratio. Another attractive feature is that the odds ratio can be calculated with data from a case-control study, whereas neither a risk ratio nor a rate ratio can be calculated.

  The odds ratio is a *case-control study*, which is based on enrolling a group of persons with disease ("case-patients") and a comparable group without disease ("controls"). The number of persons in the control group is usually decided by the investigator. Often, the size of the population from

---

TABLE 8.11
Exposure and Disease in a Hypothetical Population of 10,000 Persons

	Disease	No Disease	Total	Risk
Exposed	$a = 100$	$b = 1900$	2000	5.0%
Not exposed	$c = 80$	$d = 7920$	8000	1.0%
Total	180	9820	10,000	—

which the case-patients came is not known. As a result, risks, rates, risk ratios, or rate ratios cannot be calculated from the typical case-control study. However, you can calculate an odds ratio and interpret it as an approximation of the risk ratio, particularly when the disease is uncommon in the population.

8.7   MEASURES OF PUBLIC HEALTH IMPACT

A measure of public health impact is used to place the association between an exposure and an outcome into a meaningful public health context. Whereas a measure of association quantifies the relationship between exposure and disease, and thus begins to provide insight into causal relationships, measures of public health impact reflect the burden that an exposure contributes to the frequency of disease in the population. Two measures of public health impact often used are the *attributable proportion* and *efficacy* or *effectiveness*.

8.7.1   ATTRIBUTABLE PROPORTION

The attributable proportion, also known as the *attributable risk percent*, is a measure of the public health impact of a causative factor. The calculation of the measure assumes that the occurrence of disease in the unexposed group represents the baseline or expected risk for the disease. It further assumes that if the risk of disease in the exposed group is higher than the risk in the unexposed group, the difference can be attributed to the exposure. Thus, the attributable proportion is the amount of disease in the exposed group attributable to the exposure. It represents the expected reduction in disease if the exposure could be removed (or never existed).

8.7.1.1   Method for Calculating Attributable Proportion

Attributable proportion is calculated as follows:

$$\frac{\text{Risk for exposed group} - \text{risk for unexposed group}}{\text{Risk for exposed group}} \times 100\%$$

Attributable proportion can be calculated for rates in the same way.

■ EXAMPLE 8.24

Problem: In another study of smoking and lung cancer, the lung cancer mortality rate among non-smokers was 0.07 per 1000 persons per year (Doll and Hill, 1950). The lung cancer mortality rate among persons who smoked 1 to 14 cigarettes per day was 0.57 lung cancer deaths per 1000 persons per year. Calculate the attributable proportion.

Solution:

$$\text{Attributable proportion} = [(0.57 - 0.07)/0.57] \times 100\% = 87.7\%$$

DID YOU KNOW?

Appropriate use of attributable proportion depends on a single risk factor being responsible for a condition. When multiple risk factors may interact (e.g., physical activity and age or health status), this measure may not be appropriate.

Given the proven causal relationship between cigarette smoking and lung cancer, and assuming that the groups were comparable in all other ways, once could say that about 88% of the lung cancer among smokers of 1 to 14 cigarettes per day might be attributable to their smoking. The remaining 12% of the lung cancer cases in this group would have occurred anyway.

8.7.2 VACCINE EFFICACY OR VACCINE EFFECTIVENESS

Vaccine efficacy and vaccine effectiveness measure the proportionate reduction in cases among vaccinated persons. Vaccine efficacy is used when a study is carried out under ideal conditions—for example, during a clinical trial. Vaccine effectiveness is used when a study is carried out under typical field (that is, less than perfectly controlled) conditions. Vaccine efficacy/effectiveness (VE) is measured by calculating the risk of disease among vaccinated and unvaccinated persons and determining the percentage reduction in risk of disease among vaccinated persons relative to unvaccinated persons. The greater the percentage reduction of illness in the vaccinated group, the greater the vaccine efficacy/effectiveness. The basic formula is written as

$$\frac{\text{Risk among unvaccinated group} - \text{risk among vaccinated group}}{\text{Risk among unvaccinated group}}$$

or

$$1 - \text{Risk ratio}$$

In the first formula, the numerator (risk among unvaccinated – risk among vaccinated) is sometimes called the *risk difference* or *excess risk*.

Vaccine efficacy/effectiveness is interpreted as the proportionate reduction in disease among the vaccinated group, so a VE of 90% indicates a 90% reduction in disease occurrence among the vaccinated group, or a 90% reduction from the number of cases you would expect if they have not been vaccinated.

■ EXAMPLE 8.25

Problem: Calculate the vaccine effectiveness from the *Varicella* data in Table 8.10.

Solution:

$$VE = (42.9 - 11.8)/42.9 = 31.1/42.9 = 72\%$$

Alternatively,

$$VE = 1 - RR = 1 - 0.28 = 72\%$$

So, the vaccinated group experienced 72% fewer *Varicella* cases than they would have if they had not been vaccinated.

REFERENCES AND RECOMMENDED READING

Arias, E., Anderson, R.N., Kung, H-F., Murphy, SI., and Kochanek, K.D. (2003). Deaths: final data for 2001. *National Vital Statistics Reports*, 52(3), 1–116.

CDC. (1999). Summary of notifiable diseases—United States, 1998. *MMWR*, 47(53), 1–93.

CDC. (2001). Summary of notifiable diseases—United States, 1999. *MMWR* 48(53), 1–104.

CDC. (2003). Summary of notifiable disease—United States, 2001. *MMWR*, 50(53), 1–108.

CDC. (2004). *Reported Tuberculosis in the United States, 2003.* U.S. Centers for Disease Control and Prevention, Atlanta, GA.

CDC. (2005). Summary of notifiable diseases—United States, 2003. *MMWR*, 2(54), 1–85.

CDC. (2012). *Principles of Epidemiology in Public Health Practice*, 3rd ed. U.S. Centers for Disease Control and Prevention, Atlanta, GA.

Doll, R. and Hill, A.B. (1950). Smoking and carcinoma of the lung. *British Medical Journal*, 1, 739–748.

Hoyert, D.L., Kung, H.C., and Smith, B.L. (2005). Deaths: preliminary data for 2003. *National Vital Statistics Reports*, 53(15), 1–48.

Kleinman, J.C., Donahue, R.P., Harris, M.I., Finucane, F.F., Madans, J.H., and Brock, D.B. (1988). Mortality among diabetics in a national sample. *American Journal of Epidemiology*, 128, 389–401.

Last, J.M. (2001). *A Dictionary of Epidemiology*, 4th ed. Oxford University Press, Oxford, U.K.

McLaughlin, S.I., Spradling, P., Drociuk, D., Ridzon, R., Pozsik, C.J., and Onorato, I. (2003). Extensive transmission of *Mycobacterium tuberculosis* among congregated, HIV-infected prison inmates in South Carolina, United States. *International Journal of Tuberculosis and Lung Disease*, 7, 665–672.

Tugwell, B.D., Lee, L.E., Gillette, H., Lorber, E.M., Hedberg, K., and Cieslak, P.R. (2004). Chickenpox outbreak in a highly vaccinated school population. *Pediatrics*, 113(3, Pt. 1), 455–459.

Uyeki, T.M., Zane, S.B., Bodnar, U.R., Fielding, K.L., Buxton, J.A., Miller, J.M. et al. (2003). Large summertime influenza A outbreak among tourists in Alaska and the Yukon Territory. *Clinical Infectious Diseases*, 36, 1095–1102.

U.S. Census Bureau. (2006). *Population Estimates*, http://www.census.gov/popest.

Williams, L.M., Morrow, B., and Lansky, A. (2003). Surveillance for selected maternal behaviors and experiences before, during, and after pregnancy: Pregnancy Risk Assessment Monitoring System (PRAMS). *MMWR Surveillance Summaries*, 52(SS-11), 1–14.

Wise, R.P., Livengood, J.R., Berkelman, R.L., and Goodman, R.A. (1988). Methodologic alternatives for measuring premature mortality. *American Journal of Preventive Medicine*, 4, 268–273.

WISQARS. (2012). *Web-based Injury Statistics Query and Reporting System*. U.S. Centers for Disease Control and Prevention, Atlanta, GA, http://www.cdc.gov./injury/wisqars.

9 Boolean Algebra

$$A \cdot (B + C) = (A \cdot B) + (A \cdot C)$$

$$A + B = B + A$$

$$A + (B \cdot C) = (A + B) \cdot (A + C)$$

$$A \cdot B = B \cdot A$$

For the [environmental health] practitioner, total hazard elimination is a noble goal. Yet, as long as the human element is present in the system, perfection will be impossible to attain. The recognition that hazards exist at different degrees of severity and accident causation leads to the concept of eliminating the most "important" hazards first. Thus an evaluation to determine the important of hazards is essential if there is to be any boundary whatsoever on the problem.

—**Brown (1976)**

9.1 WHY BOOLEAN ALGEBRA?[*]

Environmental health practitioners (along with just about everyone else) intuitively know that no rational person wants to cause pain or injury either to him- or herself or to their fellow humans. The continuing occurrence of environmental accidents (and other types of accidents) despite the obvious lack of intent indicates that there is a logical flaw in some part of the reasoning process. No doubt you have heard the old saying, "The best laid plans …," and so forth and so on. Simply, when an environmental accident occurs, a breakdown in the cause-and-effect reasoning process is apparent. In this chapter, the goal is to describe a basic procedure for evaluating the logical reasoning process (system safety analysis) commonly used by environmental practitioners involved primarily in occupational health and safety and industrial hygiene activities.

The methodology we briefly describe is based on the concepts of Boolean algebra. One might ask, "Why describe the basic concepts of Boolean algebra?" The simple answer is that environmental health practitioners are expected to use the tools of logic to mitigate hazardous situations; Boolean algebra is one of these tools. The compound answer is that environmental health practitioners are often expected to know the basics of Boolean algebra to score well on certification examinations for licensure, or they are taught Boolean algebra concepts in formalized training programs or on-the-job training. Moreover, Boolean algebra plays an instrumental role in probability theory and reliability and in various environmental engineering and environmental health studies.

Having evolved in the 1950s, Boolean algebra is a branch of mathematics (a variant of algebra) that was developed systematically, because of its applications to logic, by the English mathematician George Boole. Closely related are its applications to sets and probability. A *set* is any well-defined list or collection of objects (elements). Usually, sets are denoted by capital letters such as A, B, and C and their elements by the lower case letters such as e, f, and g. Boolean algebra also underlies the theory of relations. The most prominent use of Boolean algebra is in the design of electronic switching circuits used in digital computers (Marcus, 1967). Its original application, however, was in the

---

[*] The material presented in this chapter is adapted from Spellman, F.R. and Whiting, N.E., *The Handbook of Safety Engineering: Principles and Applications*, Government Institutes Press, Lanham, MD, 2010.

context of logical reasoning (Boole 1951, 2003). Common practice today is to use Boolean algebra to solve the logical portion of problems involving probability calculations (Boole 1952, 2003). All of these applications can be of some benefit to the safety practitioner in his or her goal of reducing accidents. Of particular interest is the use of Boolean techniques in fault-tree analysis (FTA), which is used in reliability analysis of engineering systems (Clemens and Simmons, 1986; Kolodner, 1971). Fault-tree analysis is not only an effective combination of probability and backward reasoning but also Boolean algebra. The probability of some high-level event is the result of a combination of lower level events. Estimating (or knowing) the probabilities of failure of events in the fault tree allows one to estimate the probability of success or failure.

Environmental practitioners and students of environmental disciplines are familiar with algebra; it is a required core subject. Algebra is a logical outgrowth of arithmetic, and many of the methods of arithmetic are used in algebra, although in modified, expanded, or original form. To a limited extent, the same relationship can be used in describing Boolean algebra. That is, many of the laws for Boolean variables are not much different than the laws of numeric algebra. This relationship is readily seen in the commutative law, distributive law, and less so in addition (identity and inverse variables) used for real algebra operations.

9.1.1 Fault-Tree Analysis

In environmental practice, inductive methods of analysis analyze the components of the system and postulate the effects of their failure on total system performance. Deductive methods of analysis move from the end event to try to determine the possible causes. They determine how a given end event *could* have happened. One widespread application of deductive systems to environmental health and safety analysis is fault-tree analysis, which postulates the possible failure of a system and then identifies component states that would contribute to the failure. It reasons backwards from the undesired event to identify all of the ways in which such an event could occur and, in doing so, identifies the contributory causes. The lowest levels of a fault tree involve individual components or processes and their failure modes. This level of analysis generally corresponds to the starting point in *failure mode and effect analysis* (FMEA), which is a system reliability analysis that is organized around the basic question "What if ...?"

Fault-tree analysis uses Boolean logic and algebra to represent and quantify the interactions between events. The primary Boolean operators are AND and OR gates. With an AND gate, the output of the gate—the event that is at the top of the symbol—occurs only if all of the conditions below the gate, and feeding into the gate, coexist. With the OR gate, the output event occurs if any one of the input events occurs.

When the probabilities of initial events or conditions are known, the probabilities of succeeding events can be determined through the application of Boolean algebra. For an AND gate, the probability of the output event is the intersection of the Boolean probabilities, or the product of the probabilities of the input events, or:

$$\text{Probability (output)} = (\text{Prob Input 1}) \times (\text{Prob Input 2}) \times (\text{Prob Input 3})$$

For an OR gate, the probability of the output event is the sum of the "union" of the Boolean probabilities, or the sum of the probabilities of the input events minus all of the products:

$$\text{Probabilty (output)} = (\text{Prob Input 1}) + (\text{Prob Input 2}) + (\text{Prob Input 3})$$
$$- \begin{bmatrix} (\text{Prob Input 1}) \times (\text{Prob Input 2}) \\ + (\text{Prob Input 2}) \times (\text{Prob Input 3}) \\ + (\text{Prob Input 1}) \times (\text{Prob Input 1}) \times (\text{Prob Input 3}) \\ + (\text{Prob Input 1}) \times (\text{Prob Input 2}) \times (\text{Prob Input 3}) \end{bmatrix}$$

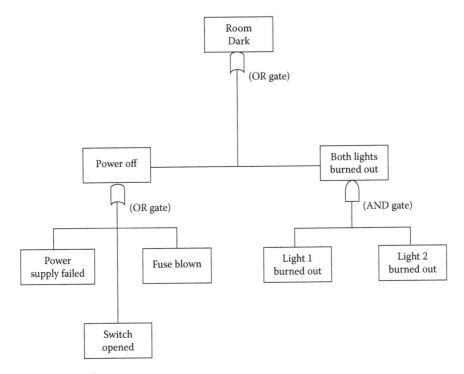

FIGURE 9.1   Fault tree analysis of light bulbs.

Where the probabilities of the input events are small (less than 0.1, for example), the probability of the output event for an OR gate can be estimated by the sum of the probabilities of the input events, or

$$\text{Probability (output)} = (\text{Prob Input 1}) + (\text{Prob Input 2}) + (\text{Prob Input 3})$$

Figure 9.1 illustrates a simple FTA. The basic events in Figure 9.1 are represented as rectangles rather than circles because they could possibly be developed (reduced) further. Assume, for example, that in Figure 9.1 the following probabilities exist:

Probability of power supply failing = 0.0010
Probability of switch open = 0.0030
Probability of blown fuse = 0.0020
Probability of light 1 out = 0.0300
Probability of light 2 out = 0.0400

Because the power will be off if the power supply fails or the switch is open or the fuse is blown, the probability of the power being off is the sum of the probabilities of the power supply failing, the switch being open, and the fuse being blown, or 0.0010 + 0.0030 + 0.0020 = 0.0060. Both lights will be burned out if light 1 is out and light 2 is out. The probability of both lights being out is 0.0300 × 0.0400, or 0.0012. The probability of the room being dark because the power is off or both lights are burned out is 0.0060 + 0.0012, or 0.0072.

The concepts of *cut sets* and *path sets* are useful in the analysis of fault trees. A cut set is any group of contributing elements that if *all* occur will *cause* the end event to occur. A path set is any grouped of contributing elements that if *none* occurs will *prevent* the occurrence of the end event. For the example in Figure 9.1, the end event will occur if

- The power supply fails.
- The switch is open.
- The fuse is blown.
- Light 1 and light 2 are burned out.

Each of the four sets above represents a cut set, because, if all of the events in any of the sets occur, then the end event (room being dark) will occur. For the example in Figure 9.1, the end event will not occur if no events in either of the following sets occur:

- The power supply fails, the switch is open, the fuse is blown, and light 1 is burned out, or
- The power supply fails, the switch is open, the fuse is blown, and light 2 is burned out.

Each of the two sets above represents a path set, because, if none of the events in either of the sets occurs, then the end event (room being dark) cannot occur.

9.1.2 Key Terms

Boolean variable is usually represented by a capital letter, representing a distinct event or fact.
A, B, and C are Boolean variables.
(+) is the logic OR operator.
(•) is the logic AND operator.

9.2 TECHNICAL OVERVIEW

Boolean algebra variables are usually depicted by capital letters representing distinct events or facts. For example, we may let A represent the event that the belt on a certain machine pulley system breaks. If this occurs, we say that $A = T$ or A is true. If the event fails to occur, we say $A = F$ or A is false. Of course, there must be some finite time during which the system is under consideration, and there is a probability associated with event A, although it is often unknown.

The most obvious way to simplify Boolean expressions is to manipulate them in the same way as normal algebraic expressions are manipulated. The manipulations, expressed as true or false (i.e., occurrence or nonoccurrence), are two modes of a Boolean algebra variable. These modes can be formed into functions, as a combination of Boolean algebra variables. With regard to logic relations in digital forms, a set of rules for symbolic manipulation is needed in order to solve for the unknowns.

A set of rules formulated by the originator of Boolean expressions, George Boole, described certain propositions whose outcomes correspond to the true or false relationship described above. Again, with regard to digital logic, these rules are used to describe circuits or possibilities whose state can be either 1 (true) or 0 (false). In order to fully understand this, the AND, OR, and NOT operators should be appreciated. A number of rules can be derived from these relations, as Table 9.1 demonstrates.

Table 9.2 shows the basic Boolean laws we are concerned with in this text. Note that every law has two expressions, (a) or (b). This is known as *duality*. These are obtained by changing every AND (•) to OR (+), every OR (+) to AND (•), and all 1's to 0's and *vice versa*. In most cases, it has become conventional to drop the AND symbol (•); that is, $A • B$ is simply written as AB.

9.2.1 Commutative Law

The *commutative law* for Boolean algebra is not much different than the commutative law for numeric algebra. For both the AND or OR logic operations, the order in which a variable is presented will not affect the outcome. So the first equation, $A + B = B + A$, may be read as "A OR B equals B OR A." It follows that $AB + BA$ also reads as "A AND B equals B AND A."

TABLE 9.1	
Boolean Postulates	
Postulate 1	$X = 0$ or $X = 1$
Postulate 2	$0 \cdot 0 = 0$
Postulate 3	$1 + 1 = 1$
Postulate 4	$0 + 0 = 0$
Postulate 5	$1 \cdot 1 = 1$
Postulate 6	$1 \cdot 0 = 0 \cdot 1 = 0$
Postulate 7	$1 + 0 = 0 + 1 = 1$

TABLE 9.2	
Boolean Laws	
Commutative law	$A + B = B + A$
	$AB = BA$
Distributive law	$A(B + C) = AB + AC$
	$A + (BC) = (A + B)(A + C)$
Identity and inverse laws	$A + A = A$
	$AA = A$
	$A + \text{not }(A) = 1$
	$A \cdot \text{not }(A) = 0$

Note: In algebra, the commutative laws for addition and multiplication are similar to the commutative law for Boolean variable. The *commutative law for addition* (algebra) states that the sum of two quantities is the same in whatever order they are added. The *commutative law for multiplication* (algebra) states that the product of two quantities is the same whatever the order of multiplication.

9.2.2 DISTRIBUTIVE LAW

The equation $A(B + C) = (AB) + (AC)$ is the first distributive law for Boolean variables, which, again, is not that different from the distributive law for numeric algebra. This simply states that A AND with the results of B OR C is the same as the results of A AND B and A AND C undergoing OR. On the other hand, the equation $A + (BC) = (A + B)(A + C)$ is the second distributive law for Boolean logic, which is not so logical from real algebra. It states that A OR with the results of B AND C is the same as the results of A OR B and A OR C undergoing AND.

Note: In algebra, the distributive law states that the product of an expression of two or more terms multiplied by a single factor is equal to the sum of the products of each term of the expression multiplied by the single factor.

9.2.3 IDENTITY AND INVERSE VARIABLES

Equations $A + A = A$, $AA = A$, $A + \text{not }(A) = 1$, and $A \cdot \text{not }(A) = 0$ are statements for the existence of an identity and inverse element for a given Boolean variable. Similar to addition for real algebra, the identity for the OR operator is 0. For the AND operator it is 1, similar to multiplication in real algebra.

Note: Because the *bit* of a Boolean variable (similar to the digit of a real number variable) is either 0 or 1 by definition, then the inverse is found by simply switching all the bits to the other Boolean value (e.g., 0 becomes 1 and 1 becomes 0). Unlike numeric algebra the inverse of Boolean variable AND with the Boolean variable is 0. If the Boolean variable undergoes the OR operation with its inverse then the result is 1. Also, if a Boolean variable undergoes either the OR or AND operation with itself, the result is simply the original Boolean variable.

9.3 BOOLEAN SYNTHESIS

To this point we have focused briefly on those Boolean laws and variables important to the environmental practitioner. There is, however, much more to Boolean algebra, but the additional concepts are beyond the scope of this text. Notwithstanding the brevity of the overview of Boolean principles provided to this point, we can now provide a basic illustration of how these Boolean principles can be applied in the real world of environmental safety and health.

■ EXAMPLE 9.1*

The following facts are to be considered in this example:

1. The time was 8:40 a.m. on a Sunday.
2. A single car, occupied by the driver only, struck the concrete uprights of a bridge pier.
3. The driver was pronounced dead on arrival at the hospital 15 minutes after the crash. He suffered a compound comminuted fracture of the skull after being thrown from the vehicle and his head struck a concrete pillar.
4. The roadway was wet, and it was possibly raining hard during the accident; it was daylight.
5. The roadway was a divided, limited-access highway with three lanes each way.
6. The accident took place 200 feet before an exit ramp to the right. The uprights that were struck were off the road to the right.
7. Traffic controls consisted of speed-limit signs (45 mph), broken white-lane-divider lines, a solid white line dividing the roadway from the 10-foot shoulder, and exit signs.
8. The vehicle was a 1964 Pontiac Catalina, recently inspected, that had traveled 72,843 miles. Mismatched tire sizes and tread types may have contributed to the accident.
9. At impact the left front door detached as a result of broken hinges. The driver was thrown from the vehicle. Lap-belts were not being used at the time of the accident. The car spun around 180 degrees clockwise.
10. The driver, 19 years of age, was alleged to suffer blackouts and nystagmus. He was familiar with the area. Alcohol tests proved negative.
11. According to a witness, the vehicle was traveling at 90 to 100 mph when the vehicle veered to the right beneath the overpass and struck one of the concrete uprights.
12. After impact the left rear tire was seen to be deflated.

The pertinent facts for Boolean analysis are as follows (with the applicable item number in parentheses):

- The wet roadway contributed to the loss of control (4).
- It was raining hard, with limited visibility (4).
- Bridge uprights and exit ramp caused confusion (6).
- Improper traffic controls were in place (7).
- Mismatched tires caused loss of control (8).
- Door hinges were weak (9).
- Lap belts were not installed (9).
- Shoulder belts were not installed.
- Driver illness caused the accident (10).
- Speed contributed to the loss of control (11).
- The left rear tire blew out and caused the accident (12).
- There were no guardrails to prevent direct impact with the bridge piers.

Although the Boolean operators are stated as facts, there is no implication that they are all true. In fact, the following "contradictions" are significant.

- Not (C): Driver was familiar with the area, thus no confusion.
- Not (D): Traffic controls were standard for this type of highway and speed limit (45 mph) and were adequate.

---
* This example is from USDOT, *Summary of Multidisciplinary Accident Investigation Reports*, 2(4), 143, 1998.

- Not (E): Mismatched tires would only cause loss of control during sudden braking; no evidence of this was available.
- Not (K): Deflation of the left rear tire would have caused a pull to the left, which was not observed.

Also, if the tire blew out, this would make the effect of mismatched tires insignificant, further strengthening the probability of not (E). The deflation of the left rear tire could have been an effect of the car spinning around after impact.

The objective now is to synthesize a Boolean expression for the accident in terms of each contributing factor. First consider the hazard caused by the weather (W), where

$$W = A + B$$

A second hazard could have been caused by the roadway (R), where

$$R = C + D$$

A third hazard could have been caused by the vehicle (V), where

$$V = E + K$$

A fourth factor will be charged to shortcomings of the driver. These could be in terms of errors in judgment, illness, or intentional violation of the law. Calling this factor P,

$$P = I + J$$

Finally, there are factors that did not cause the accident but did add to the severity. Calling these severity factors S,

$$S = F + G + H + L$$

In the absence of other evidence, it will be assumed that the concurrence of events W, R, V, P, and S led to the particular accident. Thus, letting X be the occurrence of an accident of this type and severity,

$$X = WRVPS = (A + B)(C + D)(E + K)(I + J)(F + G + H + L)$$

Note: This Boolean expression is in its simplest form and would be best left as a product of sums.

Before leaving this example, consider the reduction that would result if only the presence of an accident was considered, and not total event occurrence X. Letting Y be the accident only,

$$Y = WRVP$$

Most importantly, those who are to perform modifications on the roadway, vehicles, or drivers might restrict their considerations to those pertinent subexpressions.

REFERENCES AND RECOMMENDED READING

Boole, G. (1854/2003). *An Investigation of the Laws of Thought.* Prometheus Books, New York.
Boole, G. (1847/1951). *The Mathematical Analysis of Logic.* Basil Blackwell, Oxford, U.K.
Boole, G. (1952). *Studies in Logic and Probability.* Open Court Publishing, LaSalle, IL.

Brown, D.B. (1976). *Systems Analysis and Design for Safety.* Prentice-Hall, Englewood Cliffs, NJ.

Clemens, P. and Simmons, R. (1986). *System Safety and Risk Management.* U.S. Department of Health and Human Services, Cincinnati, OH.

Kolodner, H.H. (1971). *The Fault Tree Techniques of System Safety Analysis as Applied to the Occupational Safety Situation.* American Society of Safety Engineers, Park Ridge, IL.

Marcus, M.P. (1967). *Switching Circuits for Engineers.* Prentice-Hall, Englewood Cliffs, NJ.

Section III

Economics

Environmental activities of research and analysis are not an end in themselves, but are a means for satisfying human wants. Thus, environmental practice has several aspects. One aspect concerns itself with the materials and forces of nature. Because of this, environmental practice must be closely associated with economics.

10 Environmental Economics

$$A = P\left(\frac{i(1+i)^n}{(1+i)^n-1}\right) \qquad P = A\left(\frac{(1+i)^n-1}{i(1+i)^n}\right)$$

The NBER Environmental Economics Working Group undertakes theoretical or empirical studies of the economic effects of national or local environmental policies around the world.... Particular issues include the costs and benefits of alternative environmental policies to deal with air pollution, water quality, toxic substances, solids waste, and global warming.

—National Bureau of Economic Research (2013)

10.1 ENVIRONMENTAL PRACTICE AND ECONOMICS[*]

Environmental practitioners don't necessarily have to be economists, but they should have a basic understanding of various economic principles. This makes sense when we consider that most environmental decisions are based on economic considerations—a situation that is unlikely to change in the years ahead. Moreover, it is also important to consider that maintaining or sustaining our environment along with preventing environmental harm and correcting harmful situations cannot be achieved without incurring some cost. Unfortunately, even the most conscientious environmental practitioner often forgets or overlooks the financial implications when designing measures to minimize environmental impact or when formulating corrective actions for mitigation; unfortunately, by doing so, the plan may not get funded or may be underfunded.

In addition, the current trend in environmental practice (basically the mantra these days) is that a number of long-term economic, social, and environmental trends—Elkington's (1997) so-called *triple bottom line* (people, planet, profit)—are evolving around us. Many of these long-term trends are developing because of us and specifically for us or simply to sustain us. They all follow a general course and can be described by the jargon of the day; that is, they can be referred to by specific buzzwords in common usage today. We frequently hear such buzzwords in general conversation (especially in abbreviated texting form)—buzzwords such as empowerment, outside the box, streamline, wellness, synergy, generation X, face time, exit strategy, clear goal, and so on and so forth.

A popular buzzword that environmentalists and others are concerned with is *sustainability*, a term often used in business. However, in environmental practice, sustainability is much more than a buzzword; it is a way of life (or should be). The numerous definitions of sustainability are usually either overwhelming or too vague. For our purposes, we can come up with a long definition and short definition of sustainability. The long definition is *ensuring that environmental practices and operations occur indefinitely without negative impact*. The short definition is *the capacity of our environment to endure*. Note, however, that sustainability in environmental operations can be characterized in broader terms than these simple definitions. Under the triple bottom line scenario, the social, environmental, and economic aspects of sustainability can better define today's and tomorrow's needs.

---

[*] The material presented in this chapter is adapted from Spellman, F.R. and Whiting, N.E., *The Handbook of Safety Engineering: Principles and Applications*, Government Institutes Press, Lanham, MD, 2010.

Returning to the focus of this chapter and this book, the person tasked with protecting people, property, and the environment through well-founded and well-grounded environmental practices often feels that nothing is more important than accomplishing this goal. When told that plans to do just that must be justified through cost–benefit analysis and must add value to the business or enterprise at hand, environmental professionals sometimes balk at the notion that anyone can, could, or should put a price on life or the environment. The fact is, in the real world, we are required do this every day—environmental professionals must justify their existence within the organization. Although environmental professionals feel that environmental compliance is the *sine qua non* of any business success, the average business manager views environmental concerns as being costly measures that do not add to the bottom line.

Those of us who have worked in the environmental profession for any length of time are accustomed to this type of dysfunctional thinking. It must be pointed out, however, that we also learn (sooner rather than later) that most of us work in the real world, where we have to deal with or within the constraints of an economic bottom line. It does not take a rocket science mentality to understand the implausibility of recommending a very costly fix for a current or potential environmental hazard or situation when, at the same time, bringing about such a fix would bankrupt the company. The implementation of environmental compliance and remediation practices must be tempered not only by common sense but also by the economic bottom line.

With the understanding that cost–benefit analysis has its place in the environmental professions, this chapter presents a few of the economic principles that environmental practitioners should be familiar with. That is, we present mathematical techniques and provide practical advice for evaluating decisions during the design and preparation of environmental practices and procedures. These procedures support the selection and justification of design alternatives, operating policies, and capital expenditures. Thus, what follows is a brief introduction to economic equations and formulas commonly used in the environmental profession. Keep in mind that, historically, many of the math operations presented here may be encountered again during professional certification examinations.

10.1.1 Key Terms

A is an end-of-period cash receipt or disbursement in a uniform series, continuing for *n* periods, where the entire series is equivalent to *P* or *F* at interest rate *i*.

F is a future sum of money that is an amount, *n* interest periods from the present, that is equivalent to *P* with interest rate *i*.

i is the interest rate per interest period; in the equations, the interest rate is stated as a decimal (e.g., 6% interest is 0.06).

n is the number of interest periods.

P is a present sum of money.

Sinking fund is a separate fund into which one makes a uniform series of money deposits (*A*) with the goal of accumulating some desired future sum (*F*) at a given future point in time.

10.2 CAPITAL-RECOVERY FACTOR (EQUAL-PAYMENT SERIES)

Annual amounts of money to be received or paid are the equivalent of either a single amount in the future or a single amount in the present, when the annual amounts are compounded over a period of years at a given interest rate (*i*). The value of the annual amounts can be calculated from a single present amount (*P*) or a single future amount (*F*). We can use the *capital-recovery factor* (sometimes called the *uniform series capital-recovery factor* or *annual payment from a present value*) to determine the annual payments (*A*) from an investment. This is accomplished using Equation 10.1. The equation is based on present value (*P*), the interest rate (*i*) at which that present value is invested, and the period (term) over which it is invested (*n*).

$$A = P\left(\frac{i(1+i)^n}{(1+i)^n - 1}\right) \tag{10.1}$$

where
  A = Annual investment or payment ($).
  P = Present value ($).
  i = Interest rate (%).
  n = Number of years.

■ **EXAMPLE 10.1**

Problem: How much will an investment of $5000 yield annually over 8 years at an interest rate of 5%?

Solution:

$$A = \$5000\left(\frac{0.05(1+0.05)^8}{(1+0.05)^8 - 1}\right) = 0.1547$$

$$= \$5000(0.1547)$$

$$= \$773.50$$

$5000 invested at 5% for 8 years will yield an annual payment of $773.50.

Note: The higher the interest rate (i) earned by the investment, the higher the annual amount will be, because annual amounts compound at a higher rate. On the other hand, the longer the term of the investment (n), the lower the annual amount will be, because there are more annual payments being made that compound for a longer time.

10.3   UNIFORM SERIES PRESENT WORTH (VALUE) FACTOR

The present worth of an amount of money is the equivalent of either a single amount in the future (the future amount) of a series of amounts to be received or paid annually over a period of years as compounded at an interest rate over a period of years. Stated differently, what must the investment be now so a future series of money can be received? The present worth can be calculated from a single future amount (F) or an annual amount (A). Here, the present worth (P) of a series of equal annual amounts (A) can be calculated by using Equation 10.2, which compounds the interest (%) at which the annual amounts are invested over the term of the investment in years (n):

$$P = A\left(\frac{(1+i)^n - 1}{i(1+i)^n}\right) \tag{10.2}$$

■ **EXAMPLE 10.2**

Problem: The present worth of a series of eight equal annual payments of $154.72 at an interest rate of 5% compounded annually will be what?

Solution:

$$P = \$154.70\left(\frac{(1+0.05)^8 - 1}{0.05(1+0.05)^8}\right)$$

$$= \$154.72(6.4632)$$

$$= \$1000$$

10.4 FUTURE VALUE

The future value (or uniform series compound amount factor) of an amount of money is the equivalent of either a single amount today (the present amount) or a series of amounts to be received or paid annually over a period of years as compounded at an interest rate over a period of years. The future value can be calculated from either a single present amount (P), or an annual amount (A). The future (F) value of a series of equal annual amounts can be calculated by using Equation 10.3. The equation compounds the interest (i) at which the annual amounts (A) are invested over the term of the investment in years (n).

$$F = A\left(\frac{(1+i)^n - 1}{i}\right)$$
(10.3)

■ **EXAMPLE 10.3**

Problem: A woman deposits \$500 in a bank at the end of each year for 5 years. The bank pays 5% interest, compounded annually. At the end of 5 years, immediately following her fifth deposit, how much will she have in her account?

Solution: Given that A = \$500, i = 0.05, and n = 5,

$$F = A\left(\frac{(1+i)^n - 1}{i}\right)$$

$$= \$500\left(\frac{(1+0.05)^5 - 1}{0.05}\right)$$

$$= \$500(5.526)$$

$$= \$2763$$

She will have \$2763 in her account following the fifth deposit.

Note: The higher the interest rate (i) earned by the investment, the higher the future value will be because the investment compounds at a higher rate. The longer the term of the investment (n), the higher the future value will be because there are more annual payments being made that compound for a longer time.

10.5 ANNUAL PAYMENT (UNIFORM SERIES SINKING FUND)

Annual amounts of money to be received or paid are the equivalent of either a single amount in the future or a single amount in the present, when the annual amounts are compounded over a period of years at a given interest rate. The value of the annual amounts can be calculated from a single present amount (P), or a single future amount (F). The value of the annual amount can be calculated from a single present amount (P) or a single future amount (F). The annual payments into an investment can be calculated by using Equation 10.4:

$$A = F\left(\frac{i}{(1+i)^n - 1}\right)$$
(10.4)

■ **EXAMPLE 10.4**

Problem: A man read that in the western United States a 10-acre parcel of land could be purchased for $1000 cash. The man decided to save a uniform amount at the end of each month so that he would have the required $1000 at the end of one year. The local bank pays 1/2% (0.005) interest, compounded monthly. How much would the man have to deposit each month?

Solution: Given that $F = \$1000$, $i = 0.05$, and $n = 12$,

$$A = F\left(\frac{i}{(1+i)^n - 1}\right)$$

$$= \$1000\left(\frac{0.005}{(1+0.005)^{12} - 1}\right)$$

$$= \$1000(0.0811)$$

$$= \$81.10$$

The man would have to deposit $81.10 each month.

Note: The higher the interest rate (i) earned by the investment, the lower the annual amount will be, because the annual amounts can compound at a lower rate to reach the same future amount. The longer the term of the investment (n), the higher the annual amount will be, because there are more annual payments being made that compound for a longer time.

10.6   PRESENT VALUE OF FUTURE DOLLAR AMOUNT

The present value of an amount of money is the equivalent of either a single amount in the future (the future amount) or a period of years as compounded at an interest rate over a period of years. The present value can be calculated from a single amount (F) or an annual amount (A). The present value (P) of a future dollar amount (F) can be calculated by using Equation 10.5. The equation compounds the interest in percent (i) at which the present value (P) is invested over the term of the investment in years (n).

$$P = F(1+i)^{-n} \tag{10.5}$$

■ **EXAMPLE 10.5**

Problem: What is the present value of $6000 to be received in 5 years if it is invested at 6%?

Solution:

$$P = F(1+i)^{-n}$$

$$= \$6000(1+0.06)^{-5} = \$6000\left(\frac{1}{0.06}\right)^5$$

$$= \$6000(0.747)$$

$$= \$4482$$

The present value of $6000 to be received in 5 years, if it is invested at 6%, is $4482.

Note: The higher the interest rate (i) earned by the investment, the lower the present value will be because the investment compounds at a higher rate. The longer the term of the investment (n), the lower the present value will be because the investment compounds over a longer time.

10.7 FUTURE VALUE OF A PRESENT AMOUNT

The future value of an amount of money is the equivalent of either a single amount today (the present amount) or series of amounts to be received or paid annually over a period of years as compounded at an interest rate over a period of years. The future value can be calculated from either a single present amount (P) or an annual amount (A). The future value (F) of a present dollar amount can be calculated by using Equation 10.6. The equation compounds the interest in dollars (i) at which the present value (P) is invested over the term of the investment in years (n).

$$F = P(1+i)^n \tag{10.6}$$

■ EXAMPLE 10.6

Problem: If $6000 is invested for 5 years at 6% interest per year, what will the investment be worth in 5 years?

Solution:

$$F = P(1+i)^n$$

$$= \$6000(1+0.06)^5$$

$$= \$6000(1.06)^5$$

$$= \$6000(1.338)$$

$$= \$8029$$

The $6000 investment will be worth $8029 in 5 years.

Note: The higher the interest rate (i) earned by the investment, the higher the future value will be because the investment compounds at a higher rate. The longer the term of the investment (n), the higher the future value will be because the investment compounds for a longer time.

REFERENCES AND RECOMMENDED READING

Blank, L.T. and Tarquin, A.J. (1997). *Engineering Economy.* McGraw-Hill, New York.

Elkington, J. (1997). *Cannibals with Forks.* Capstone Publishing, Oxford, U.K.

National Bureau of Economic Research. (2013). *NBER Working Group Descriptions: Environmental Economics*, http://www.nber.org/programs/eee/ee_oldworkinggroup_directory/ee.html.

Thuesen, H.G., Fabrycky, W.J., and Thuesen, G.J. (1971). *Engineering Economy*, 4th ed. Prentice-Hall, Englewood Cliffs, NJ.

Section IV

Engineering Fundamentals

Thank God men cannot fly, and lay waste the sky as well as the earth.

—**Henry David Thoreau**

11 Fundamental Engineering Concepts

$$N = W\cos\emptyset$$

The environment is everything that is not me.

—Albert Einstein

A life without adventure is likely to be unsatisfying, but a life in which adventure is allowed to take whatever form it will, is likely to be short.

—Bertrand Russell

11.1 INTRODUCTION

Education can only go so far in preparing the environmental practitioner for on-the-job performance. A person who wishes to become an environmental practitioner is greatly assisted by two personal characteristics. First, a well-rounded, broad development of experience in many areas is required and produces the classic generalist. Second, although environmental practitioners cannot possibly attain great depth in all areas, they must have the desire and the aptitude to do so. They must be interested in—and well informed about—many widely differing fields of study. The necessity for this in the environmental application is readily apparent. Why? Simply because the range of problems encountered is so immense that a narrow education will not suffice; environmental practitioners must handle situations that call upon skills as widely diverse as the ability to solve psychological, sociological, and economic problems (remember the triple bottom line), along with the ability to perform calculations required in fundamental engineering, mechanics, and the structural–construction–maintenance–environmental interface. The would-be practicing environmental professional can come from just about any background, and a narrow education does not preclude students and others from broadening their education later; however, quite often those who are very specialized and have a very narrow focus lack appreciation for other disciplines, as well as the adaptability necessary for environmental practice.

Why do we say this? What makes us so certain that the practicing environmental professional is better off if he or she is a generalist with a very wide-range of knowledge vs. being a narrowly focused expert? Two things convince us: our more than 50 years of combined personal experience in environmental practice and knowing the situations that the environmental professional may be called upon to investigate and analyze. Consider the following short list suggested by Parkhurst (2006):

- Acorn production, germination, and mating patterns
- Air pollution toxicity
- Algal growth
- Animal heat loss
- Bird flight speed
- Chemical mixture
- Coliform bacteria in water
- Copper sulfate dispersal
- Dissolved oxygen

- Doubling time
- Insect pest dispersal
- Mountain streamflow
- PCB dumping
- Wastewater treatment
- Water treatment
- Soil remediation
- Public health
- Water volume
- Thermal radiation exchange

Though individual learning style is important in choosing a career as an environmental professional, again we stress that generalized education is the key ingredient in the mix that produces the fully educated environmental professional. Based on the list above, it can be seen that, along with an education in the basic and applied sciences of mathematics, natural science, and behavioral science (which are applied to the solution of technological, biological, and behavioral problems), an education in engineering and technology is a must. Topics such as applied mechanics, properties of materials, electrical circuits and machines, fire science hydraulics, principles of engineering design, and computer science fall into this category.

In this chapter, we concentrate on applied mechanics and, in particular, forces and the resolution of forces. Why? Because many environmental incidents or accidents and subsequent damage to the environment and injuries to people and animals are caused by forces being of too great a magnitude for a particular machine, material, or structure. To inspect systems, devices, or products to ensure their safety, environmental professionals must account for the forces that act or might act on them. Environmental professionals must also account for forces from objects that may act on the human body (an area of focus that is often overlooked).

Important areas that are part of or that interface with applied mechanics are the properties of materials and engineering design considerations. (Fire science hydraulics and electrical circuits, as well as safety concerns associated with both of these important areas are addressed separately later in this text.) We do not discuss all of the environmental engineering aspects related to these areas in this text—this is not an engineering text. Instead, we look at a few fundamental engineering concepts and their applications to environmental professions.

11.2 RESOLUTION OF FORCES

In the environmental and occupational health aspects of environmental engineering, we tend to focus our attention on those forces that are likely to cause failure or damage to some device or system, resulting in an occurrence that is likely to produce secondary and tertiary damage to other devices or systems and harm to individuals. Typically, large forces are more likely to cause failure or damage than small ones. Environmental engineers must understand force and how a force acts on a body, particularly (1) the direction of force, (2) point of application (location) of force, (3) the area over which force acts, (4) the distribution or concentration of forces that act on bodies, and (5) how essential these elements are in evaluating the strength of materials. For example, a 40-lb force applied to the edge of a sheet of plastic and parallel to it probably will not break it. If a sledgehammer strikes the center of the sheet with the same force, the plastic will probably break. A sheet metal panel of the same size undergoing the same force will not break.

Practice tells us that different materials have different strength properties. Striking a plastic panel will probably cause it to break, whereas striking a sheet metal panel will cause a dent. The strength of a material and its ability to deform are directly related to the force applied. Important physical, mechanical, and other properties of materials include the following:

- Crystal structure
- Strength
- Melting point
- Density
- Hardness
- Brittleness
- Ductility
- Modulus of elasticity
- Wear properties
- Coefficient of expansion
- Contraction
- Conductivity
- Shape
- Exposure to environmental conditions
- Exposure to chemicals
- Fracture toughness

Note: All of these properties can vary, depending on whether the force is crushing, corroding, cutting, pulling, or twisting.

The forces an object can encounter are often different from the forces that an object can withstand. An object may be designed to withstand only minimal force before it fails (a toy doll may be designed of very soft, pliable materials or designed to break or give way in certain places when a child falls on it, thus preventing injury). Other devices may be designed to withstand the greatest possible load and shock (e.g., a building constructed to withstand an earthquake).

When working with any material that will go in an area with a concern for safety, a safety factor (SF) is often introduced. As defined by ASSE (1988), a safety factor is the ratio allowed for in design between the ultimate breaking strength of a member, material, structure, or equipment and the actual working stress or safe permissible load placed on it during ordinary use. Simply put, including a factor of safety—into the design of a machine, for example—makes an allowance for many unknowns (inaccurate estimates of real loads or irregularities in materials, for example) related to the materials used to make the machine, related to the machine's assembly, and related to the use of the machine. Safety factors can be determined in several ways. One of the most commonly used ways is

$$SF = \frac{\text{Failure} - \text{Producing load}}{\text{Allowable stress}} \tag{11.1}$$

Forces on a material or object are classified by the way they act on the material. For example, a force that pulls a material apart is called the *tensile force*. Forces that squeeze a material or object are called *compression forces*. *Shear forces* cut a material or object. Forces that twist a material or object are called *torsional forces*. Forces that cause a material or object to bend are called *bending forces*. A *bearing force* occurs when one material or object presses against or bears on another material or body.

So, what is force? *Force* is typically defined as any influence that tends to change the state of rest or the uniform motion in a straight line of a body. The action of an unbalanced or resultant force results in the acceleration of a body in the direction of action of the force, or it may (if the body is unable to move freely) result in its deformation. Force is a vector quantity, possessing both magnitude and direction (see Figure 11.1); its SI unit is the newton (equal to 3.6 ounces, or 0.225 lb).

According to Newton's second law of motion, the magnitude of a resultant force is equal to the rate of change of momentum of the body on which it acts. The unit of force is the pound force in the English or engineering system and is the newton in the SI system. The pound force is defined as the

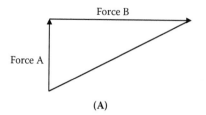

(A)

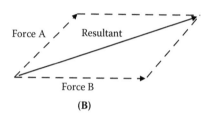

(B)

FIGURE 11.1   Force vector quantities.

force required to give 1 slug of mass an acceleration of 1 foot per second per second (ft/s²). The new-ton is defined as the force required to give 1 kilogram of mass acceleration of 1 meter per second per second. A slug is the mass of 32.2 standard pounds. Since the mass of an object is commonly expressed as its weight in pounds, we can make the conversion to slugs by dividing this weight by the gravitational constant, $g_c = 32.2$ ft/s², for use in the following formula:

$$F = m \times a \times SF \tag{11.2}$$

■ EXAMPLE 11.1

Problem: How much force must a seat belt be capable of withstanding to safely restrain a 180-lb woman when her car comes to a sudden (i.e., 1-second) stop if initially traveling at 60 mph? (Assume a safety factor of 4).

Solution:

$$m = (180 \text{ lb})/(32.3 \text{ ft/s}^2) = 5.59 \text{ slugs}$$

$$a = V/T = [(60 \text{ mph})(5280 \text{ ft/1 m})(1 \text{ hr/3600 s})]/1 \text{ sec} = 88 \text{ ft/s}^2$$

$$F = m \times a \times SF = 5.59 \times 88 \text{ ft/s}^2 \times 4 = 1968 \text{ lb}$$

Another important relationship where force is a key player is the concept of *work*. Work is the product of the force and the effective displacement of its application point. The equation for calculating force is

$$W = F \times s \tag{11.3}$$

where
    W = Work in foot-pounds (ft-lb).
    F = Force in pounds (lb).
    s = Distance in feet (ft).

If a force is applied to an object and no movement occurs, no effective work is done. The energy possessed by a body determines the amount of work it can do. Newton's third law states that for every action there is an equal and opposite reaction.

■ EXAMPLE 11.2

Problem: A system of pulleys having a mechanical advantage of 5 is used to move a 1-ton weight up an inclined plane. The plane has an angle of 45° from horizontal. The weight has to be moved a vertical distance of 30 ft. Ignoring friction, how much work is required to be done to move the weight up to the top (vertical lift work)?

Solution:

$$W = F \times s$$

$$W = 2000 \text{ lb} \times 30 \text{ ft} = 60,000 \text{ ft-lb}$$

With regard to the safety aspects of environmental engineering, a key relationship between force F and a body on which it acts is

$$F = s \times A \tag{11.4}$$

where
  s = Force or stress per unit area (e.g., pounds per square inch).
  A = Area (square inches, square feet, etc.) over which a force acts.

Frequently, two or more forces act together to produce the effect of a single force, called a *resultant*. This resolution of forces can be explained by either the triangle law or the parallelogram law. The *triangle law* provides that if two concurrent forces are laid out vectorially with the beginning of the second force at the end of the first, the vector connecting the beginning and the end of the forces represents the resultant of the two forces (see Figure 11.1A). The *parallelogram law* provides that if two concurrent forces are laid out vectorially, with either forces pointing toward or both away from their point of intersection, a parallelogram represents the resultant of the force. The concurrent forces must have both direction and magnitude if their resultant is to be determined (see Figure 11.1B). If the individual forces are known or if one of the individual forces and the resultant are known, the resultant force may be simply calculated by either the trigonometric method (sines, cosines, and tangents) or the graphic method (which involves laying out the known force, or forces, at an exact scale and in the exact directions in either a parallelogram or triangle and then measuring the unknown to the same scale).

11.3 SLINGS

Slings must be used in accordance with recommendations of the sling manufacturer (PNNL, 2013). Slings manufactured from conventional three-strand natural or synthetic fiber rope are not recommended for use in lifting service. Natural or synthetic fiber rope slings must be used only if other sling types are not suitable for the unique application. For natural or synthetic rope slings, the

DID YOU KNOW?

The stress a material can withstand is a function of the material and the type of loading.

requirements of ASME B30.9, and OSHA 1910.184(h) must be followed. All types of slings must have, as a minimum, the rated capacity clearly and permanently marked on each sling. Each sling must receive a documented inspection at least annually, more frequently if recommended by the manufacturer or made necessary by service conditions.

Note: Slings are commonly used between cranes, derricks, or hoists and the load, so the load may be lifted and moved to a desired location. For the safety engineer, the properties and limitations of the sling, the type and condition of material being lifted, the weight and shape of the object being lifted, the angle of the lifting sling to the load being lifted, and the environment in which the lift is to be made are all important considerations to be evaluated—before the transfer of material can take place safely.

Let's take a look at a few example problems involving forces that the environmental engineer might be called upon to calculate. In our examples, we use lifting slings under different conditions of loading.

■ EXAMPLE 11.3

Problem: Let us assume a load of 2000 lb supported by a two-leg sling; the legs of the sling make an angle of 60° with the load. What force is exerted on each leg of the sling?

Solution: When solving this type of problem, always draw a rough diagram as shown in Figure 11.2. A resolution of forces provides the answer. We will use the trigonometric method to solve this problem, but remember that it may also be solved using the graphic method. Using the trigonometric method with the parallelogram law, the problem could be solved as described below. Again, make a drawing to show a resolution of forces similar to that shown in Figure 11.3.

We could consider the load (2000 lb) as being concentrated and acting vertically, which can be indicated by a vertical line. The legs of the slings are at a 60° angle, which can be shown as **ab** and **ac**. The parallelogram can now be constructed by drawing lines parallel to **ab** and **ac**, intersecting at **d**. The point where **cb** and **ad** intersect can be indicated as **e**. The force on each leg of the sling (**ab**, for example) is the resultant of two forces, one acting vertically (**ae**), the other horizontally (**be**), as shown in the force diagram. Force **ae** is equal to one-half of **ad** (the total force acting vertically, 2000 lb), so **ae** = 1000. This value remains constant regardless of the angle **ab** makes with **bd,** because as the angle increases or decreases, **ae** also increases or decreases. But **ae** is always **ad**/2. The force **ab** can be calculated by trigonometry using the right triangle **abe:**

$$\text{Sine of an angle} = \frac{\text{Opposite side}}{\text{Hypotenuse}}$$

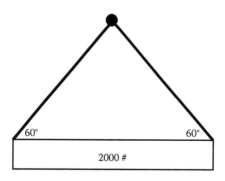

FIGURE 11.2   Illustration for Example 11.3.

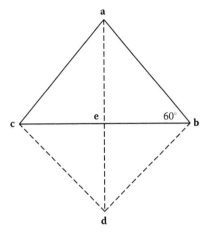

FIGURE 11.3   Illustration for Example 11.3.

therefore,

$$\sin 60° = \frac{ae}{ab}$$

transposing,

$$ab = \frac{ae}{\sin 60°}$$

substituting known values,

$$ab = \frac{1000}{0.866} = 1155$$

The total weight on each leg of the sling at a 60° angle from the load is 1155 lb. Note that the weight is more than half the load, because the load is made up of two forces—one acting vertically, the other horizontally. An important point to remember is that the smaller the angle, the greater the load (force) on the sling. For example, at a 15° angle, the force on each leg of a 2000-lb load increases to 3864 lb.

Note: Sling angles less than 30° *not* recommended.

Let's take a look at what the force would be on each leg of a 2000-lb load at various angles that are common for lifting slings (Figure 11.4), and work a couple of example problems.

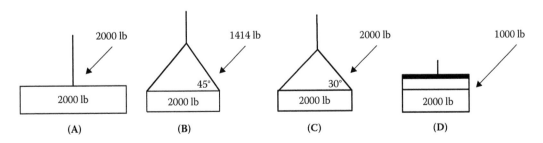

FIGURE 11.4   Sling angle and lead examples.

■ **EXAMPLE 11.4**

Problem: We have a 3000-lb load to be lifted with a two-leg sling whose legs are at a 30° angle from the load. The load (force) on each leg of the sling is

Solution:

$$\sin A = \frac{a}{c}$$

$$\sin 30 = 0.500$$

$$a = \frac{3000 \text{ lb}}{2} = 1500$$

$$c = \frac{a}{\sin A} = \frac{1500}{0.5} = 3000$$

■ **EXAMPLE 11.5**

Problem: Given a two-rope sling supporting 10,000 lb, what is the load (force) on the left sling? Sling angle to load is 60°.

Solution:

$$\sin A = \frac{60}{0.866}$$

$$a = \frac{10,000 \text{ lb}}{2} = 1500$$

$$c = \frac{a}{\sin A} = \frac{5000}{0.866} = 5774$$

11.3.1 RATED SLING LOADS

In the preceding section we demonstrated simple math operations used to determine the rated sling load that a particular sling can safely bear. In the field, on the job, knowing how to use simple math to make sling angle-load determinations is important. It is also important to point out, however, that many tables showing rated loads on slings are available. Table 11.1, for example, shows rated loads for alloy steel chain slings.

11.4 INCLINED PLANE

Another common problem encountered by environmental engineers involved in the resolution of forces occurs in material handling operations in moving a load (a cart, for example) up or down an *inclined plane* (a ramp or tilted surface, in our example). The safety implications in this type of work activity should be obvious. Objects are known to accelerate down inclined planes because of an unbalanced force (anytime we deal with unbalanced forces, safety issues are present and must be addressed). To understand this type of motion, it is important to analyze the forces acting upon an object on an inclined plane. Figure 11.5 depicts the two forces acting upon a load positioned on an inclined plane (assuming no friction). As shown in Figure 11.5, there are always at least two forces acting upon any load that is positioned on an inclined plane—the force of gravity (also known as weight) and the normal (perpendicular) force. The force of gravity acts in a downward direction; yet, the normal force acts in a direction perpendicular to the surface. Let's take a look at a typical example of how to determine the force needed to pull a fully loaded cart up a ramp (an inclined plane).

TABLE 11.1
Alloy Steel Chain Sling Load Angle Factors[a]

Horizontal Sling Angle	Load Angle Factor
90°	1.000
85°	1.004
80°	1.015
75°	1.035
70°	1.064
65°	1.104
60°	1.155
55°	1.221
50°	1.305
45°	1.414
40°	1.555
35°	1.742
30°	2.000
25°	2.364[b]
20°	2.924[b]
15°	3.861[b]
10°	5.747[b]
5°	11.490[b]

Source: PNNL, *PNNL Hoisting and Rigging Manual*, Pacific Northwest National Laboratory, Richland, WA, 2013 (http://www.pnl.gov/contracts/hoist_rigging/slings.asp).

[a] Tension in each sling leg = Load/2 = Load angle factor.
[b] Not recommended.

■ EXAMPLE 11.6

Problem: We assume that a fully loaded cart weighing 400 lb is to be pulled up a ramp that has a 5-ft rise for each 12 ft, measured along the horizontal direction (make a rough drawing; see Figure 11.6). What force is required to pull it up the ramp?

Note: For illustrative purposes, we assume no friction. Without friction, of course, the work done in moving the cart in a horizontal direction would be zero; once the cart was started, it would move with constant velocity—the only work required is that necessary to get it started. However, a force equal to J is necessary to pull the cart up the ramp or to maintain the car at

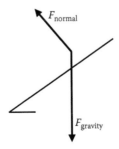

FIGURE 11.5 Forces acting on an inclined plane.

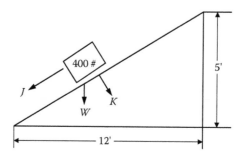

FIGURE 11.6   Illustration for Example 11.6.

rest (in equilibrium). As the angle (slope) of the ramp is increased, greater force is required to move it, because the load is being raised as it moves along the ramp, thus doing work. Remember that this is not the case when the cart is moved along a horizontal plane without friction; however, in actual practice friction can never be ignored, and some work is accomplished in moving the cart.

Solution: To determine the actual force involved, we can again use a resolution of forces. The first step is to determine the angle of the ramp. This can be calculated by the formula:

$$\text{Tangent (angle of ramp)} = \frac{\text{Opposite side}}{\text{Adjacent side}} = \frac{5}{12} = 0.42$$

and arctan 0.42 = 22.8°.

Now we need to draw a force parallelogram (see Figure 11.7) and apply the trigonometric method. The weight of the cart (*W*) (shown as force acting vertically) can be resolved into two components: force *J* parallel to the ramp and force *K* perpendicular to the ramp. Component *K*, being perpendicular to the inclined ramp, does not hinder movement up the ramp. Component *J* represents a force that would accelerate the cart down the ramp. To pull the cart up the ramp, a force equal to or greater than *J* is necessary.

Applying the trigonometric method, angle *WOK* is the same as the angle of the ramp.

$$OJ = WK + OW = 400 \text{ lb}$$

$$\text{Sine of angle } WOK \ (22.8°) = \frac{\text{Opposite side } (WK)}{\text{Adjacent side } (OW)}$$

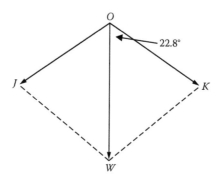

FIGURE 11.7   Force parallelogram.

Transposing,

$$WK = OW \times \sin 22.8° = 400 \times 0.388 = 155.2$$

Thus, a force of 155.2 lb is necessary to pull the cart up the 22.8° angle of the ramp (friction ignored). Note that the total amount of work is the same, whether the cart is lifted vertically (400 lb × 5 ft = 2000 ft-lb), or pulled up the ramp (155.2 lb × 13 ft = 2000 ft-lb). The advantage gained in using a ramp instead of a vertical lift is that less force is required—but through a greater distance.

11.5 PROPERTIES OF MATERIALS

When we speak of the properties of materials or a material's properties, what are we referring to and why should we concern ourselves with this topic? The best way to answer this question is to use an example where the environmental engineer, working with design engineers in a preliminary design conference, might typically be exposed to engineering design data, parameters, and specifications related to the properties of a particular construction material to be used in the fabrication of, for example, a large mezzanine in a warehouse. In constructing this particular mezzanine, consideration must be given to the fact that it will be used to store large, heavy equipment components. The demands placed on the finished mezzanine create the need for the mezzanine to be built using materials that can safely support a heavy load.

For illustration, let's say that the design engineers plan to use an aluminum alloy (structural, No. 17ST). Before they decide upon using No. 17ST and determining the required quantity required to build the mezzanine, they must examine its mechanical properties to ensure that it will be able to handle the intended load (they will also factor in, many times over, for safety, the use of a material that will handle a load much greater than expected). Using a table on the mechanical properties of engineering materials in Urquhart's *Civil Engineering Handbook*, they found the following information for No. 17ST:

1. Ultimate strength (defined as the ultimate strength in compression for ductile materials, usually taken as the yield point), including tension, 58,000 psi; compression strength, 35,000 psi; shear strength, 35,000 psi
2. Yield point tension, 35,000 psi
3. Modulus of elasticity, tension or compression, 10,000,000 psi
4. Modulus of elasticity, shear, 3,750,000 psi
5. Weight, 0.10 lb/in.^{3}

Is this information important to the environmental engineer? Well, in a specific sense, no—not exactly; however, in a general sense, yes. What is important to the environmental engineer is that procedures such as the one just described actually occur; that is, professional engineers do take the time to determine the correct materials to use in constructing, for example, a mezzanine. Also, when exposed to this type of information, the environmental engineer must know enough about the language used to know what the design engineers are talking about—and to understand its significance. A materials property is an intensive, often quantitative property of a material. Thus, it is important for environmental engineers to understand the material property descriptor and also the units used as metrics of value. In short, we must know and understand the meaning of nomenclature of the materials to be used and their limits. This is important in order to compare the benefits of one material vs. another when selecting the appropriate materials to be used. Remember Voltaire: "If you wish to converse with me, define your terms."

Let's take a look at a few other engineering terms and their definitions, so that we will be able to converse. Keep in mind that many of these definitions are defined exactly and precisely in mathematical formulas and computation. These exact and precise mathematical definitions are beyond the

FIGURE 11.8 (A) Stress is measured in terms of the applied load over the area. (B) Strain is expressed in terms of amount per square inch.

scope of this text, but we do include a few here that are pertinent to the environmental profession. Many of the engineering terms defined below are from Heisler (1998), Tapley (1990), and Giachino and Weeks (1985), all of which we highly recommend and should be standard reference texts for environmental engineers.

Mechanics—Branch of science that deals with forces and motion.

Rupture—The ultimate failure of tough ductile materials loaded in tension.

Buckling—Failure mode characterized by the sudden failure of a structural member subjected to high compressive stresses, where the actual compressive stress at the point of failure is less than the ultimate compressive stresses that the material is capable of withstanding.

Stress—The internal resistance a material offers to being deformed; it is measured in terms of the applied load over the area (see Figure 11.8A)

Strain—Deformation that results from a stress; it is expressed in terms of the amount of deformation per square inch (see Figure 11.8B).

Corrosion—The breaking down of essential properties in a material due to chemical reactions with its surroundings.

Intensity of stress—Stress per unit area, usually expressed in pounds per square inch; it is the result of a force of *P* pounds producing tension, compression, or shear on an area of *A* square inches, over which it is uniformly distributed. The simple term *stress* is normally used to indicate intensity of stress.

Creep—The tendency of a solid material to slowly move or deform permanently under the influence of stresses.

Ultimate stress—The greatest stress that can be produced in a body before rupture occurs.

Allowable stress or working stress—The intensity of stress that the material of a structure or a machine is designed to resist.

Elastic limit—The maximum intensity of stress to which a material may be subjected and return to its original shape upon the removal of stress (see Figure 11.9).

Yield point—The intensity of stress beyond which the change in length increases rapidly with little (if any) increase in stress.

Fracture—Local separation of an object or material into two or more pieces under the action of stress.

FIGURE 11.9 Elasticity and elastic limit; a metal has the ability to return to its original shape after being elongated or distorted, unless it reaches its maximum stress point.

DID YOU KNOW?

Cork has a Poisson ratio of practically zero. This is why cork is used as a stopper in wine bottles. As the cork is inserted into the bottle, the upper part that is not yet inserted will not expand as the lower part is compressed.

Modulus of elasticity—Ratio of stress to strain, for stresses below the elastic limit. By checking the modulus of elasticity, the comparative stiffness of different materials can readily be ascertained. Rigidity and stiffness are very important considerations for many machine and structural applications.

Poisson's ratio—Ratio of the relative change of diameter of a bar to its unit change of length under an axial load that does not stress it beyond the elastic limit. Stated differently, Poisson's ratio (υ) is the negative ratio of transverse to axial strain. When a sample object is stretched (or squeezed) to an extension (or contraction) in the direction of the applied load, it corresponds to a contraction (or extension) in a direction perpendicular to the applied load (Gercek, 2007).

$$\upsilon = -\frac{d\varepsilon_{trans}}{d\varepsilon_{axial}} - = \frac{d\varepsilon_y}{d\varepsilon_x} - = \frac{d\varepsilon_z}{d\varepsilon_x}$$

where

υ = Poisson's ratio.

ε_{trans} = Transverse strain, which is negative for axial tension (stretching) and positive for axial compression.

ε_{axial} = Axial strain, which is positive for axial tension and negative for axial compression.

Note: Later, in our discussion of hydraulics, we talk about pressurized piping systems. When the air or liquid in a pipe is highly pressurized it exerts a uniform force on the inside of the pipe, resulting in a radial stress within the pipe material. Due to Poisson's effect, this radial stress will cause the pipe to slightly increase in diameter and decrease in length. The decrease in length, in particular, can have a noticeable effect upon the pipe joints, as the effect will accumulate for each section of pipe joined in series. Obviously, a restrained joint may be pulled apart or otherwise prone to failure. This could result in a catastrophic spill of piping contents onto nearby personnel and into the environment.

Tensile strength—Resistance to forces acting to pull the metal apart, a very important factor in the evaluation of a metal (see Figure 11.10).

Compressive strength—Ability of a material to resist being crushed (see Figure 11.11).

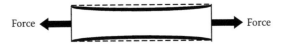

FIGURE 11.10 A metal with tensile strength resists pulling forces.

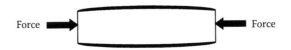

FIGURE 11.11 Compressive strength is the ability of a metal to resist crushing forces.

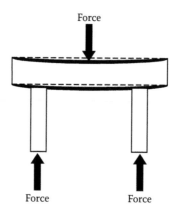

FIGURE 11.12 Bending strength (stress) is a combination of tensile strength and compressive strength.

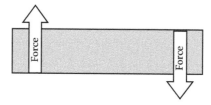

FIGURE 11.13 Torsional strength is the ability of a metal to withstand twisting forces.

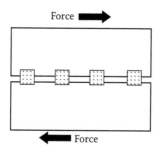

FIGURE 11.14 Sheer strength determines how well a member can withstand two equal forces acting in opposite directions.

Bending strength—Resistance to forces causing a member to bend or deflect in the direction in which the load is applied; it is actually a combination of tensile and compressive stresses (see Figure 11.12).

Torsional strength—Resistance to forces causing a member to twist (see Figure 11.13).

Shear strength—Resistance to two equal forces acting in opposite directions (see Figure 11.14).

Fatigue strength—Resistance to various kinds of rapidly alternating stresses.

Impact strength—Resistance to loads that are applied suddenly and often at high velocity.

Ductility—Ability of a metal to stretch, bend, or twist without breaking or cracking.

Hardness—Property in steel that resists indentation or penetration.

Brittleness—A condition whereby a metal will easily fracture under low stress.

Toughness—May be considered as strength, together with ductility. A tough material can absorb large amounts of energy without breaking.

Malleability—Ability of a metal to be deformed by compression forces without developing defects, such as forces encountered in rolling, pressing, or forging.

Mechanical overload—Failure or fracture of a product or component in a single event.

11.5.1 FRICTION

Earlier, in discussing the principle of the inclined plane, we ignored the effect of friction. In actual use, friction cannot be ignored and you must have some understanding of its characteristics and applications. Friction results when an object on the verge of sliding, rotating, rolling, or spinning, or in the process of any of these, is in contact with another body. Friction allows us to walk, ski, drive vehicles, and power machines, among other things. Whenever one object slides over another, frictional forces opposing the motion are developed between them. Friction force is the force tangent to the contact surface that resists motion. If motion occurs, the resistance is due to kinetic friction, which is normally lower than the value for static friction. Contrary to common perception, the degree of smoothness of a surface area is not responsible for these frictional forces; instead, the molecular structure of the materials is responsible. The coefficient of friction (M) (which differs among different materials) is the ratio of the frictional force (F) to the normal force (N) between two bodies.

$$M = \frac{F}{N} \tag{11.5}$$

For dry surfaces, the coefficient of friction remains constant, even if the weight of an object (i.e., force N) is changed. The force of friction (F) required to move the block changes proportionally. Note that the coefficient of friction is independent of the area of contact, which means that pushing a brick across the floor requires the same amount of force whether it is on end, on edge, or flat. The coefficient of friction is useful in determining the force necessary to do a certain amount of work. Temperature changes only slightly affect friction. Friction causes wear. To overcome this wear problem, lubricants are used to reduce friction.

■ **EXAMPLE 11.7**

Problem: How much force is required to more a 300-lb box if the static coefficient of friction between it and the horizontal surface upon which it is resting is 0.66?

Solution:

$$F = (0.66)(300 \text{ lb}) = 396 \text{ lb}$$

■ **EXAMPLE 11.8**

Problem: A 200-lb box is placed on a plane inclined at a 30° angle from the horizontal. The plane has a static coefficient of friction of 0.66. What is the minimum push or pull force required to move the box down the plane?

Solution:

$$F = (0.66)(200 \text{ lb} \times \cos 30) = (0.66)(200 \text{ lb} \times 0.866) = (0.66)(173.2 \text{ lb}) = 114 \text{ lb}$$

Note, however that a component of the weight of the box is causing a force (due to gravity) to already be acting in the downward direction of incline. This force is equal to $W\sin\theta$, or $200 \times 0.50 = 50$ lb. Therefore, the only additional push/push force needed to get the box moving is

$$114 \text{ lb} - 50 \text{ lb} = 64 \text{ lb}$$

11.5.2 Specific Gravity

Specific gravity is the ratio of the weight of a liquid or solid substance to the weight of an equal volume of water, a number that can be determined by dividing the weight of a body by the weight of an equal volume of water. Because the weight of any body per unit of volume is its density, then

$$\text{Specific gravity} = \frac{\text{Density of body}}{\text{Density of water}} \qquad (11.6)$$

■ EXAMPLE 11.9

Problem: The density of a particular material is 0.24 lb/in.^{3}, and the density of water is 0.0361 lb/in.^{3}. What is the specific gravity?

Solution:

$$\text{Specific gravity} = \frac{\text{Density of body}}{\text{Density of water}} = \frac{0.24}{0.0361} = 6.6$$

The material is 6.6 times as heavy as water. This ratio does not change, regardless of the units that may be used, which is an advantage for two reasons: (1) the ratio will always be the same for the same material, and (2) specific gravity is less confusing than the concept of density, which changes as the units change.

11.5.3 Force, Mass, and Acceleration

Let's review what we learned earlier about force, mass, and acceleration. According to Newton's second law of motion:

> The acceleration produced by unbalanced force acting on a mass is directly proportional to the unbalanced force, in the direction of the unbalanced force, and inversely proportional to the total mass being accelerated by the unbalanced force.

If we express Newton's second law mathematically, it is greatly simplified and becomes

$$F = ma \qquad (11.7)$$

where F is force, m is mass, and a is acceleration; 1 Newton = 1 kg $\times$ m/s^2. This equation is extremely important in physics and engineering. It simply relates acceleration to force and mass. Acceleration is defined as the change in velocity divided by the time taken. This definition tells us how to measure acceleration. $F = ma$ tells us what causes the acceleration—an unbalanced force. Mass may be defined as the quotient obtained by dividing the weight of a body by the acceleration caused by gravity. Because gravity is always present, we can, for practical purposes, think of mass in terms of weight, making the necessary allowance for gravitational acceleration.

11.5.4 Centrifugal and Centripetal Forces

Two terms with which the environmental professional should be familiar are *centrifugal force* and *centripetal force*. Centrifugal force is a concept based on an apparent (but not real) force. It may be regarded as a force that acts radially outward from a spinning or orbiting object (a ball tied to a string whirling about), thus balancing a real force, the centripetal force (the force that acts radially inwards). This concept is important in environmental engineering, because many of the machines encountered on the job may involve rapidly revolving wheels or flywheels. If the wheel is revolving fast enough, and if the molecular structure of the wheel is not strong enough to overcome the centrifugal force, it may fracture and pieces of the wheel would fly off tangent to the arc described by the wheel. The safety implications are obvious. Any worker using such a device, or near it may be severely injured when the rotating member ruptures. This is what happens when a grinding wheel on a pedestal grinder "bursts." Rim speed determines the centrifugal force, and rim speed involves both the speed (rpm) of the wheel and the diameter of the wheel.

11.5.5 STRESS AND STRAIN

In materials, stress is a measure of the deforming force applied to a body. Strain (which is often erroneously used as a synonym for stress) is really the resulting change in its shape (deformation). For perfectly elastic material, stress is proportional to strain. This relationship is explained by Hooke's law, which states that the deformation of a body is proportional to the magnitude of the deforming force, provided that the body's elastic limit is not exceeded. If the elastic limit is not reached, the body will return to its original size once the force is removed. For example, if a spring is stretched 2 cm by a weight of 1 N, it will be stretched 4 cm by a weight of 2 N, and so on; however, once the load exceeds the elastic limit for the spring, Hooke's law will no longer be obeyed, and each successive increase in weight will result in a greater extension until the spring finally breaks.

Stress forces are categorized in three ways:

1. Tension (or tensile stress), in which equal and opposite forces that act away from each other are applied to a body; tends to elongate a body.
2. Compression stress, in which equal and opposite forces that act toward each other are applied to a body; tends to shorten a body.
3. Shear stress, in which equal and opposite forces that do not act along the same line of action or plane are applied to a body; tends to change the shape of a body without changing its volume.

11.6 MATERIALS AND PRINCIPLES OF MECHANICS

To be able to recognize hazards and to select and implement appropriate controls, environmental engineers must have a good understanding of the properties of materials and principles of mechanics. In this section, we start with the properties of materials, and then cover the wide spectrum that is comprised of mechanics and soil mechanics. Our intent is to clearly illustrate the wide scope of knowledge required in areas germane to the properties of materials and the principles of mechanics, as well as those topics on the periphery, all of which are blended in the mix—the safety knowledge mix that helps to produce the well-rounded, knowledgeable environmental engineer. Mechanics is the cornerstone of physics and is of primary interest because the world is full of many kinds of motions that are often used for practical purposes such as the motions of falling objects, of cars, of boats, of planes, of rolling wheels, of flowing liquids, of moving air masses in meteorology, and so forth (Reif, 1996). In this section, we discuss the mechanical principles of statics, dynamics, moments, beams, columns, floors, industrial noise, and radiation. The field environmental engineer should have at least some familiarity with all of these. (Note that the environmental engineer whose function is to verify design specifications, with safety in mind, should have more than just a passing familiarity with these topics.)

11.6.1 STATICS

Statics is the branch of mechanics concerned with the behavior of bodies at rest and forces in equilibrium and is distinguished from dynamics (concerned with the behavior of bodies in motion). Forces acting on statics do not create motion. Static applications are bolts, welds, rivets, load-carrying components (ropes and chains), and other structural elements. A common example of a static situation is shown in the bolt-and-plate assembly. The bolt is loaded in tension and holds two elements together. One force acting on it is the load on the lower element (160-lb load plus 15 lb of suspending elements). Another force is that caused by the tightened nut (25 lb). The total effective load on the bolt is 200 lb (160 + 15 + 25). The plate will fail in shear if the head of the bolt pulls through the plate.

11.6.2 DYNAMICS

Dynamics (kinetics in mechanics) is the mathematical and physical study of the behavior of bodies under the action of forces that produce changes of motion in them. In dynamics, certain properties are important: displacement, velocity, acceleration, momentum, kinetic energy, potential energy, work, and power. Environmental engineers work with these properties to determine, for example, if rotating equipment will fly apart and cause injury to workers or to determine the distance required to stop a vehicle in motion.

11.6.3 HYDRAULICS AND PNEUMATICS—FLUID MECHANICS

Hydraulics (liquids only) and pneumatics (gases only) make up the study of fluid mechanics, which in turn is the study of forces acting on fluids (both liquids and gases). Environmental engineers encounter many fluid mechanics problems and applications of fluid mechanics. In particular, environmental engineers working in chemical industries, or in or around processes using or producing chemicals, need to have an understanding of flowing liquids or gases to be able to predict and control their behavior.

11.6.4 WELDS

Welding is a method of joining metals to achieve a more efficient use of the materials and faster fabrication and erection. Welding also permits the designer to develop and use new and aesthetically appealing designs, and it saves weight because connecting plates are not needed and allowances need not be made for reduced load-carrying ability due to holes for rivets, bolts, and so on (Heisler, 1998). Simply put, the welding process joins two pieces of metal together by establishing a metallurgical bond between them. Most processes use a fusion technique; the two most widely used are arc welding and gas welding. In the welding process, where two pieces of metal are joined together, the mechanical properties of metals are important, of course. The mechanical properties of metals primarily determine how materials behave under applied loads—in other words, how strong a metal is when it comes in contact with one or more forces. The important point is that if you apply knowledge about the strength properties of a metal, you can build a structure that is both safe and sound. The welder must know the strength of his weld as compared with the base metal to produce a weldment that is strong enough to do the job. Thus, the welder is just as concerned with the mechanical properties of metals as is the engineer.

11.6.5 MOMENT

Moment is a synonym for *torque*; however, in many engineering applications, the two terms are not interchangeable. Torque is usually used to describe a rotational force down a shaft (e.g., turning of a pump shaft), whereas moment is more often used to describe a bending force on a beam. Moment is the product of the force magnitude (F) and the distance from the point to its action line. The perpendicular distance (d) is called the arm of the force, and the point is the origin or center of the moment. The product is the measure of the tendency of the force to cause rotation (e.g., bending, twisting). The unit of measurement for moment is a combination of the names of the force and distance units, such as pound-foot (lb-ft), to distinguish it from the unit for work or energy, the foot-pound (ft-lb).

$$F_1 d_1 = F_2 d_2$$

$$\sum M_0 = 0 \tag{11.8}$$

11.6.6 BEAMS AND COLUMNS

A *beam* is a structural member whose length is large compared to its transverse dimensions and is subjected to forces acting transverse to its longitudinal axis (Tapley, 1990). Environmental engineers are concerned not only with structural members (e.g., beams, flooring support members) but also with columns. *Columns* are structural members with an unsupported length 10 times greater than the smallest lateral dimension, and they are loaded in compression. When a column is subjected to small compressive loads, the column axially shortens. If continually larger loads are applied, a load is reached at which the column suddenly bows out sideways. This load is referred to as the column's critical or buckling load. These sideways deformations are normally too large to be acceptable; consequently, the column is considered to have failed. For slender columns, the axial stress corresponding to the critical load is generally below the yield strength of the material. Because the stresses in the column just prior to buckling are within the elastic range, the failure is referred to as *elastic buckling.* The term *elastic stability* is commonly used to designate the study of elastic buckling problems. For short columns, yielding or rupture of the column may govern failure while it is still axially straight. Failure of short columns may also be caused by inelastic buckling; that is, large sideways deformation that occur when the nominal axial stress is greater than the yield strength.

Beams, floors, and columns are all critical elements for safe loading. As an example, in late October of 2003, the top five floors of a parking garage under construction in Atlantic City, NJ, collapsed while workmen were pouring concrete on the structure's top floor, killing four workers and critically injuring six others. OSHA's investigation of the collapse involved close examination of the blueprints for the garage and evaluation of the cure rate for the concrete. No matter who or what element of construction, design, or engineering was ultimately to blame in this collapse, obviously the load limits were exceeded for that moment. Perhaps a longer concrete curing interval would have prevented the collapse. Regardless of the cause, however, the cost in lives and dollars lost was too high.

Environmental engineers are primarily interested in beams because the load on a beam induces stresses in the material that could be dangerous. The structural aspect of beams is most important to the environmental engineer, because the strength of the beam material and the kind of loading determine the size of load that it can safely carry. For example, in the construction of the storage mezzanine discussed earlier, and in the construction of other load-bearing structures, the beams used to support the load are an important (critical) consideration.

Refer to Figure 11.15. The neutral axis is the plane that undergoes no change in length from bending and along which the direct stress is zero. The fibers on one side of the neutral axis are stressed in tension, and on the other side in compression, and the intensities of these stresses in homogeneous beams are directly proportional to the distances of the fibers from the neutral axis (Heisler, 1998).

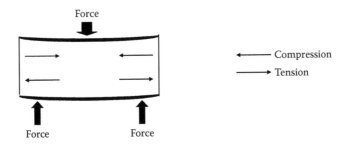

FIGURE 11.15   Distribution of stress in a beam cross-section during bending.

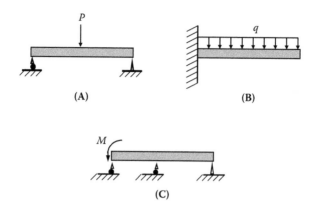

$$(A) \qquad\qquad (B)$$

$$(C)$$

FIGURE 11.16 Classification of beams: (A) simple beam, concentrated load; (B) cantilever beam, distributed load; and (C) continuous beam, concentrated moment.

When determining the load that can be carried, two properties of the beam are important: the moment of inertia (I) and the section modulus (Z). The moment of inertia (I) is the sum of differential areas multiplied by the square of the distance from a reference plane (usually the neutral axis) to each differential area. Note that the strength of a beam increases rapidly as its cross section is moved farther from the neutral axis because the distance is squared. This is why a rectangular beam is much stronger when it is loaded along its thin dimension than along its flat dimension. The section modulus (Z) is the moment of inertia divided by the distance from the neutral axis to the outside of the beam cross-section.

Of special interest to environmental engineers is the type of beams and allowable loads on beams. Allowable loads differ from maximum loads that produce failure by some appropriate factor of safety (Brauer, 1994). The following types of beams are approximations of actual beams used in practice (Tapley 1990):

- *Simple beams* are supported beams that have a roller support at one end and pin support at the other. The ends of a simple beam cannot support a bending moment but can support upward and downward vertical loads. Stated differently, the ends are free to rotate but cannot translate in the vertical direction. The end with the roller support is free to translate in the axial direction (see Figure 11.16A for concentrated load).
- *Cantilever beams* are beams rigidly supported on only one end. The beam carries the load to the support, where it is resisted by moment and shear stress (see Figure 11.16B for distributed load).
- *Continuously supported beams* rest on more than two supports (see Figure 11.16C for concentrated moment).
- *Fixed beams* are rigidly fixed at both ends.
- *Restrained beams* are rigidly fixed at one end and simply supported at the other.
- *Overhanging beams* project beyond one or both ends of their supports.

11.6.7 BENDING MOMENT

A bending moment (internal torque) exists in a structural element when a moment is applied to the element so that the element bends. In engineering design (and safety engineering), it is important to determine the points along a beam where the bending moment (and shear force) is maximum since it is at these points that the bending stresses reach their maximum values. When a beam is in equilibrium, the sum of all moments about a particular point is zero. To determine the maximum bending moment (uniform loading), we use the following equation:

$$M = WL^2/8 \qquad (11.9)$$

where
  M = Maximum bending moment (ft-lb).
  W = Uniform weight loading per foot of the beam (lb/ft).
  L = Length of the beam (ft).

■ EXAMPLE 11.10

Problem: An 11-ft-long beam supported at two points, one at each end, is uniformly loaded at a rate of 250 lb/ft. What is the maximum bending moment in this bean?

Solution:

$$M = WL^2/8 = [(250 \text{ lb/ft})(11 \text{ ft})^2]/8 = 3781 \text{ ft-lb}$$

To determine the bending moment (concentrated load at center) the following equation can be used:

$$M = PL/4 \qquad (11.10)$$

where
  M = Maximum bending moment (ft-lb).
  P = Concentrated load applied at the center of the beam (lb).
  L = Length of the beam (ft).

■ EXAMPLE 11.11

Problem: A 10-ton hoist is suspended at the mid-point of a 10-ft-long beam that is supported at each end. What is the maximum bending moment in this beam? (Neglect the weight of the beam).

Solution:

$$M = PL/4 = [(20,000 \text{ lb})(10 \text{ ft})]/4 = 50,000 \text{ ft-lb}$$

To determine the bending moment (concentrated load off-center) the following equation can be used:

$$M = Pab/L \qquad (11.11)$$

where
  M = Maximum bending moment (ft-lb).
  P = Concentrated load on the beam (lb).
  a = Distance from the left support of the beam (ft).
  b = Distance from the right support of the beam (ft).
  L = Length of the beam (ft).

■ EXAMPLE 11.12

Problem: A 10-ft-long beam supported at each end is loaded with a weight of 1400 lb at a point 2 ft left of its center. What is the maximum bending moment in this beam? (Neglect weight of the beam.)

Solution:

$$M = Pab/L = [(1400 \text{ lb})(2 \text{ ft})(8 \text{ ft})]/10 \text{ ft} = 2240 \text{ ft-lb}$$

11.6.8 RADIUS OF CURVATURE (BEAM)

The radius of curvature provides one of several measures of the deformation of a beam. The radius of curvature of a beam experiencing a bending moment (inches, feet, meters) can be determined by using the following equation:

$$p = EI/M \tag{11.12}$$

where

    p = Radius of curvature (in., ft, m).
    E = Modulus of elasticity of the beam material (psi, KPa).
    I = Cross-sectional area moment of inertia about its centroid (m^4, $in.^4$, ft^4).
    M = Moment or torque applied at the cross-section (ft-lb).

11.6.9 COLUMN STRESS

Stress is defined as force per unit area and is expressed in pounds per square inch (psi). The amount of stress is a real indicator of how severely the member is loaded. *Tensile stress* occurs when a member is in tension (force acts to stretch it); *compressive stress* occurs when a member is in compression (force acts to shorten or flatten it); and, *shear stress* occurs when a member is in shear (force acts to cause one part of the material to slide over another part). To determine the stress (loading) of a column, we use the following equation:

$$\sigma = P/A \tag{11.13}$$

where

    σ = Stress ($lb/in.^2$ or psi).
    P = Loading on the column (lb).
    A = Cross-sectional area of the member ($in.^2$).

■ EXAMPLE 11.13

Problem: A column 4 in. in diameter supports a load of 6000 lb. What is the stress on it?

Solution:

$$A = \pi r^2 = (3.14)(2 \text{ in.})^2 = 12.56 \text{ in.}^2$$

$$\sigma = P/A = (6000 \text{ lb})/12.56 \text{ in.}^2 = 478 \text{ psi}$$

11.6.10 BEAM FLEXURE

In mechanics, flexure (also known as *bending*) characterizes the behavior of a structural element (beam) subjected to an external load applied perpendicular to the axis of the element. A closet rod sagging under the weight of clothes on clothes hangers is an example of a beam experiencing flexure (or bending). The equation used to calculate flexure is known as the *flexure formula* (see below), which shows the relationship between the maximum bending stress (σ) and the maximum bending moment (M):

$$\sigma = MC/L \tag{11.14}$$

where

    σ = Maximum stress ($lb/in.^2$ or psi).
    M = Moment (in.-lb).
    C/L = Inverse of the section modulus ($in.^3$) (taken from an applicable table).

■ EXAMPLE 11.14

Problem: A rectangular beam that is 12 ft long by 6 in. high by 3 in. wide has a 2-ton concentrated load applied at its center. What is the maximum bending stress on this beam?

Solution:

$$M = PL/4 = [(4000 \text{ lb})(12 \text{ ft.} \times 12 \text{ in./ft})]/4 = 144{,}000 \text{ in.-lb}$$

$$C/L = 6/bD^2 \text{ (from applicable table)} = 6/[(3 \text{ in.})(6 \text{ in.})^2] = 0.05555 \text{ in.}^3$$

$$\sigma = MC/L = (144{,}000 \text{ in.-lb})(0.05555 \text{ in.}^3) = 7999 \text{ psi.}$$

11.6.11   REACTION FORCE (BEAMS)

11.6.11.1   Uniform Loading

Unknown forces exerted upward by the supports at each end of a beam to hold it up are called *reactions*. Known forces that act on beams are called *loads*. Reactions balance the load to keep the beam in a state of equilibrium. To determine the reactions on a beam's supports we use the following equation:

$$R_L = R_R = WL/2 \tag{11.15}$$

where
R_L = Left reaction (lb).
R_R = Right reaction (lb).
W = Uniform weight loading per foot of the bean (lb/ft).
L = Length of the beam (ft).

■ EXAMPLE 11.15

Problem: A 10-ft-long beam supported at two points, one at each end, is uniformly loaded at a rate of 200 lb/ft. What are the reactions on the supports?

Solution:

$$R_L = R_R = WL/2 = [(200 \text{ lb/ft})(10 \text{ ft})]/2 = 1000 \text{ lb}$$

11.6.11.2   Concentrated Load at Center

To calculate the reactions R_L and R_R in simple beams (those beams supported by two points, one at each end) of any length that are loaded at their center with a concentrated load, we use the following equation.

$$R_L = R_R = P/2 \tag{11.16}$$

where
R_L = Left reaction (lb).
R_R = Right reaction (lb).
P = Concentrated load at the center of the beam (lb/ft).

■ EXAMPLE 11.16

Problem: A 10-ton hoist is suspended at the midpoint of a beam that is supported at each end. What would the design basis need to be for these supports if we specified a safety factor of 4?

Solution:

$$10 \text{ tons} = 20{,}000 \text{ lb}$$
$$R_L = R_R = P/2 = (20{,}000 \text{ lb})/2 = 10{,}000 \text{ lb}$$

Applying a safety factor of 4 requires that each support be designed to handle at least 32,000 lb, or 16 tons of force.

11.6.11.3 Load at Any Point

The equation below is used to calculate the reactions R_L and R_R in simple beams of length L that are loaded off-center with a weight of P:

$$R_L = Pa/L \quad R_R = Pb/L \tag{11.17}$$

where
  R_L = Left reaction (lb).
  R_R = Right reaction (lb).
  P = Concentrated load on the beam (lb).
  a = Distance from the left support of the beam (ft).
  b = Distance from the right support of the beam (ft).
  L = Length of the beam (ft).

■ **EXAMPLE 11.17**

Problem: A 10-ft-long beam supported at each end is loaded with a weight of 1200 lb at a point 2 ft left of its center. What are the reactions on the supports? (Neglect weight of the beam.)

Solution:

$$R_L = Pa/L = [(1200 \text{ lb})(2 \text{ ft})]/10 \text{ ft} = 240 \text{ lb}$$

$$R_R = Pb/L = [(1200 \text{ lb})(8 \text{ ft})]/2 \text{ ft} = 960 \text{ lb}$$

11.6.12 BUCKLING STRESS (WOOD COLUMNS)

Columns that support a load have a tendency to bend. If the bending becomes too great, the column will become unstable and fail by buckling under the pressure from the load. The equation below is for columns made of wood, with intermediate length:

$$P/A = \sigma[L - 1/3(L/KD)]^4 \tag{11.18}$$

where
  P = Maximum acceptable load applied to the column (lb, N).
  A = Cross-sectional area of the column (in.^{2}, m^2).
  σ = Allowable unit compressive stress parallel to the grain (psi, kPa).
  L = Free unsupported length of the column (in., m).
  K = Minimum value of l/D for which Euler column mechanics may be used.
  D = Dimension (width) of the column in the expected direction of buckling (in., m).

11.6.13 FLOORS

Environmental engineers commonly spend a considerable amount of time and attention on ensuring the proper maintenance of floors and flooring in general, in ways that range from the mundane (housekeeping) to the structurally essential (calculating floor load). Housekeeping is always a focal point; passageways, storerooms, and service rooms need to be scrutinized daily to ensure that they are kept clean and orderly, in a sanitary condition, and free from fire hazards. Typically the responsibility for housekeeping (in most places of employment) falls to the supervisor and to the employees themselves; however, safety engineers cannot avoid the responsibility of ensuring that workplace housekeeping is kept to the highest standards. Housekeeping also includes maintaining the floors of every workroom in a clean and dry condition. Where wet processes are used, drainage must be maintained, and platforms, mats, or other dry standing places must be provided where practicable. All floors must be kept free from protruding nails, splinters, holes, or loose boards.

Along with housekeeping, safety engineers are concerned with floor load protection. In fact, one of the environmental engineer's commonly requested services is to determine the safe loads on a floor. To determine the safe floor load for any floor, safety engineers have to take into consideration two load components: (1) *dead loads* (the weight of the building and its components) and (2) *live loads* (loads placed on the floor). The environmental engineer's main safety concern is to ensure that no load exceeds that for which a floor (or roof) is approved by the design engineer and/or approving official. Environmental engineers should ensure that the approved floor rating is properly posted in a conspicuous place in each space to which it relates.

11.7 PRINCIPLES OF ELECTRICITY*

Why does the environmental professional need to possess an understanding of the fundamentals of electricity? Good question. The simple answer: Again, the environmental professional must be a generalist, possessing a sampling of knowledge in a wide-range of topics and specialty areas. The compound answer: Beyond knowing that sticking one's finger into a live light socket may scare and shock the bejesus out of him or her, or may turn him or her into a crispy critter, or just kill the recipient outright, the environmental impact of electricity generation is significant because modern society uses large amounts of electrical power. This power is normally generated at power plants that convert some other kind of energy (e.g., fossil fuels, nuclear power, hydroelectric power, tidal power, biomass, wind power, geothermal power, solar power) into electrical power. Each system has advantages and disadvantages, but many of them pose environmental concerns. Thus, because of the environmental impact of electricity generation and use, environmental practitioners should be well-grounded in the basics of electricity. This includes an understanding and proper use of simple electrical calculations.

We begin our discussion by asking what is *electricity*? Water and wastewater operators generally have little difficulty in recognizing electrical equipment. Electrical equipment is everywhere and is easy to spot; for example, typical plant sites are outfitted with equipment to

- Generate electricity (a generator or emergency generator)
- Store electricity (batteries)
- Change electricity from one form to another (transformers)
- Transport or transmit and distribute electricity throughout the plant site (wiring distribution systems)
- Measure electricity (meters)
- Convert electricity into other forms of energy (mechanical energy, heat energy, light energy, chemical energy, or radio energy)
- Protect other electrical equipment (fuses, circuit breakers, or relays)
- Operate and control other electrical equipment (motor controllers)
- Convert some condition or occurrence into an electric signal (sensors)
- Convert some measured variable to a representative electrical signal (transducers or transmitters).

---

* This section is adapted from Spellman, F.R., *Handbook of Water and Wastewater Treatment Plant Operations*, 3rd ed., CRC Press, Boca Raton, FL, 2013.

Recognizing electrical equipment is easy because we use so much of it. If we ask typical operators where such equipment is located in their plant site, they know, because they probably operate these devices or their ancillaries. If we asked these same operators what a particular electrical device does, they could probably tell us. If we were to ask if their plant electrical equipment was important to plant operations, the chorus would be a resounding "absolutely."

Here is another question that does not always result in such an assured answer. If we were to ask these same operators to explain to us in very basic terms how electricity works to make their plant equipment operate, the answers we would receive probably would be varied, jumbled, disjointed—and probably not all that accurate. Even on a more basic level, how many operators would be able to accurately answer the question: What is electricity?

Probably very few. Why do so many workplace employees know so little about electricity? Part of the answer resides in the fact that employees, other than electricians, are expected to know their jobs and not much more.

11.7.1 NATURE OF ELECTRICITY

The word "electricity" is derived from the Greek word *electron* for "amber," which is a translucent (semitransparent) yellowish fossilized mineral resin. The ancient Greeks used the words "electric force" to refer to the mysterious forces of attraction and repulsion exhibited by amber when it was rubbed with a cloth. They did not understand the nature of this force and could not answer the question: "What is electricity?" The fact is this question remains unanswered. Today, we often attempt to answer this question by describing the effect and not the force; that is, the standard answer given is that electricity is "the force that moves electrons," which is about the same as defining a sail as "the force that moves a sailboat."

At the present time, little more is known than the ancient Greeks knew about the fundamental nature of electricity, but we have made tremendous strides in harnessing and using it. As with many other unknown (or unexplainable) phenomena, elaborate theories concerning the nature and behavior of electricity have been advanced and have gained wide acceptance because of their apparent truth—and because they work.

Scientists have determined that electricity seems to behave in a constant and predictable manner in given situations or when subjected to given conditions. Scientists such as Faraday, Ohm, Lenz, and Kirchhoff have described the predictable characteristics of electricity and electric current in the form of certain rules. These rules are often referred to as *laws*. Thus, although electricity itself has never been clearly defined, its predictable nature and ease of use have made it one of the most widely used power sources in modern times.

The bottom line: We can still learn about electricity by studying the rules, or laws, applying to the behavior of electricity and by understanding the methods of producing, controlling, and using it. Thus, this learning about electricity can be accomplished without ever having determined its fundamental identity.

You are probably scratching your head in puzzlement, and a question almost certainly running through your brain at this exact moment is "This is a text about basic electricity and the authors can't even explain what electricity is?" That is correct. We cannot. The point is, no one can definitively define electricity. Electricity is one of those subject areas where the old saying "We don't know what we don't know about it" fits perfectly.

Again, a few theories about electricity have so far stood the test of extensive analysis and much time (relatively speaking, of course). One of the oldest and most generally accepted theories concerning electric current flow (or electricity) is known as the *electron theory*, which states that electricity or current flow is the result of the flow of free electrons in a conductor. Thus, electricity is the flow of free electrons or simply electron flow. In this text, this is how we define electricity; that is, *electricity is the flow of free electrons.*

COULOMB'S LAW

Simply put, Coulomb's law points out that the amount of attracting or repelling force that acts between two electrically charged bodies in free space depends on two things:

- Their charges
- The distance between them

Specifically, Coulomb's law states: *Charged bodies attract or repel each other with a force that is directly proportional to the product of their charges and is inversely proportional to the square of the distance between them.*

Note: The magnitude of electric charge a body possesses is determined by the number of electrons compared with the number of protons within the body. The symbol for the magnitude of electric charge is Q, expressed in units of coulombs (C). A charge of 1 positive coulomb means a body contains a charge of 6.25×10^{18}. A charge of 1 negative coulomb $(-Q)$ means a body contains a charge of 6.25×10^{18} more electrons than protons.

11.7.2 SIMPLE ELECTRICAL CIRCUIT

An electrical circuit includes the following components (Figure 11.17):

- An energy source—a source of electromotive force (emf), or voltage, such as a battery or generator
- A conductor (wire)
- A load
- A means of control

The energy source could be a battery, as shown in Figure 11.17, or some other means for producing a voltage. The load that dissipates the energy could be a lamp, a resistor, or some other device (or devices) that does useful work, such as an electric toaster, a power drill, a radio, or a soldering iron. Conductors are wires that offer low resistance to current; they connect all the loads in the circuit to the voltage source. No electrical device dissipates energy unless current flows through it. Because conductors, or wires, are not perfect conductors, they heat up (dissipate energy), so they are actually

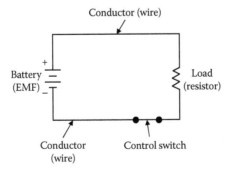

FIGURE 11.17   Simple closed circuit.

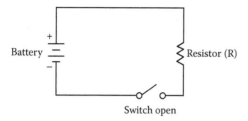

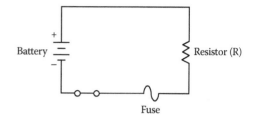

FIGURE 11.18   Open circuit.   **FIGURE 11.19**   Simple fused circuit.

part of the load. For simplicity, however, we usually think of the connecting wiring as having no resistance, as it would be tedious to assign a very low resistance value to the wires every time we wanted to solve a problem. Control devices might be switches, variable resistors, circuit breakers, fuses, or relays.

A complete pathway for current flow, or closed circuit (Figure 11.17), is an unbroken path for current from the emf, through a load, and back to the source. A circuit is open (see Figure 11.18) if a break in the circuit (e.g., open switch) does not provide a complete path for current.

Key Point: Current flows from the negative (–) terminal of the battery, shown in Figures 11.17 and 11.18, through the load to the positive (+) battery terminal and then continues through the battery from the positive (+) terminal to the negative (–) terminal. As long as this pathway is unbroken, it is a closed circuit and current will flow; however, if the path is broken at *any* point, it becomes an open circuit and no current flows.

To protect a circuit, a fuse is placed directly in the circuit (see Figure 11.19). A fuse will open the circuit whenever a dangerously large current begins to flow (i.e., when a short circuit condition occurs caused by an accidental connection between two points in a circuit offering very little resistance). A fuse will permit currents smaller than the fuse value to flow but will melt and therefore break or open the circuit if a larger current flows.

11.7.2.1 Schematic Representation

The simple circuits shown in Figures 11.17, 11.18, and 11.19 are displayed in schematic form. A *schematic diagram* (usually shortened to "schematic") is a simplified drawing that represents the electrical, not the physical, situation in a circuit. The symbols used in schematic diagrams are the electrician's shorthand; they make the diagrams easier to draw and easier to understand. Consider the symbol used to represent a battery power supply (see Figure 11.20). The symbol is rather simple and straightforward but is very important; for example, by convention, the shorter line in the symbol for a battery represents the negative terminal. It is important to remember this, because it is sometimes necessary to note the direction of current flow, which is from negative to positive, when examining a schematic. The battery symbol shown in Figure 11.20 has a single cell, so only one short and one long line are used. The number of lines used to represent a battery vary (and they are not necessarily equivalent to the number of cells), but they are always in pairs, with long and short lines alternating. In the circuit shown in Figure 11.19, the current would flow in a *counterclockwise* direction—that is, opposite the direction that the hands of a clock move. If the long and short lines of the battery symbol (symbol shown in Figure 11.20) were reversed, the current in the circuit shown in Figure 11.19 would flow *clockwise*—that is, in the direction that the hands of a clock move.

FIGURE 11.20   Schematic symbol for a battery.

Note: In studies of electricity and electronics, many circuits are analyzed that consist mainly of specially designed resistive components. As previously stated, these components are called *resistors*. Throughout our remaining analysis of the basic circuit, the resistive component will be a physical resistor; however, the resistive component could be any one of several electrical devices.

Keep in mind that the simple circuits shown in the figures to this point illustrate only a few of the many symbols used in schematics to represent circuit components. Other symbols will be introduced, as we need them. It is also important to keep in mind that a closed loop of wire (conductor) is not necessarily a circuit. A source of voltage must be included to make it an electric circuit. In any electric circuit where electrons move around a closed loop, current, voltage, and resistance are present. The physical pathway for current flow is actually the circuit. By knowing any two of the three quantities, such as voltage and current, the third (resistance) may be determined. This is done mathematically using *Ohm's law*, which is the foundation on which electrical theory is based.

11.7.3 OHM'S LAW

Simply put, Ohm's law defines the relationship between current, voltage, and resistance in electric circuits. Ohm's law can be expressed mathematically in three ways:

1. The *current* (*I*) in a circuit is equal to the voltage applied to the circuit divided by the resistance of the circuit. Stated another way, the current in a circuit is *directly* proportional to the applied voltage and *inversely* proportional to the circuit resistance. Ohm's law may be expressed as

$$I = \frac{E}{R} \qquad (11.19)$$

where
    I = Current in amps.
    E = Voltage in volts.
    R = Resistance in ohms.

2. The *resistance* (*R*) of a circuit is equal to the voltage applied to the circuit divided by the current in the circuit:

$$R = \frac{E}{I} \qquad (11.20)$$

3. The applied *voltage* (*E*) to a circuit is equal to the product of the current and the resistance of the circuit:

$$E = I \times R = IR \qquad (11.21)$$

If any two of the quantities in Equations 11.19 through 11.21 are known, the third may be easily found. Let us look at an example.

■ EXAMPLE 11.18

Problem: Figure 11.21 shows a circuit containing a resistance (*R*) of 6 ohms and a source of voltage (*E*) of 3 volts. How much current (*I*) flows in the circuit?

Solution:

$$I = \frac{E}{R} = \frac{3}{6} = 0.5 \text{ amp}$$

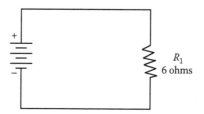

FIGURE 11.21   Determining current in a simple circuit.

To observe the effect of source voltage on circuit current, we use the circuit shown in Figure 11.21 but double the voltage to 6 volts. Notice that, as the source of voltage doubles, the circuit current also doubles.

■ **EXAMPLE 11.19**

Problem: Given that $E = 6$ volts and $R = 6$ ohms, what is I?

Solution:

$$I = \frac{E}{R} = \frac{6}{6} = 1 \text{ amp}$$

Key Point: Circuit current is directly proportional to applied voltage and will change by the same factor that the voltage changes.

To verify that current is inversely proportional to resistance, assume that the resistor in Figure 11.21 has a value of 12 ohms.

■ **EXAMPLE 11.20**

Problem: Given that $E = 3$ volts and $R = 12$ ohms, what is I?

Solution:

$$I = \frac{E}{R} = \frac{3}{12} = 0.25 \text{ amps}$$

Comparing the current of 0.25 amp for the 12-ohm resistor to the 0.5-amp current obtained with the 6-ohm resistor shows that doubling the resistance will reduce the current to one half the original value. The point here is that *circuit current is inversely proportional to the circuit resistance.*

Recall that, if we know any two quantities (E, I, or R), we can calculate the third. In many circuit applications, current is known and either the voltage or the resistance will be the unknown quantity. To solve a problem in which current (I) and resistance (R) are known, for example, the basic formula for Ohm's law must be transposed to solve for E. The Ohm's law equations can be memorized and practiced effectively by using an Ohm's law circle (see Figure 11.22). To find the equation for E, I, or R when two quantities are known, cover the unknown third quantity with your finger, ruler, or piece of paper as shown in Figure 11.23.

■ **EXAMPLE 11.21**

Problem: An electric light bulb draws 0.5 A when operating on a 120-V DC circuit. What is the resistance of the bulb?

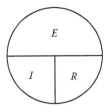

FIGURE 11.22   Ohm's law circle.

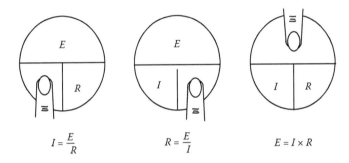

$$I = \frac{E}{R} \qquad\qquad R = \frac{E}{I} \qquad\qquad E = I \times R$$

FIGURE 11.23   Putting the Ohm's law circle to work.

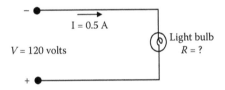

FIGURE 11.24   Simple circuit.

Solution: The first step in solving a circuit problem is to sketch a schematic diagram of the circuit itself, labeling each of the parts and showing the known values (see Figure 11.24). Because I and E are known, we can use Equation 11.20 to solve for R:

$$R = \frac{E}{I} = \frac{120}{0.5} = 240 \text{ ohms}$$

11.7.4   ELECTRICAL POWER

Power, whether electrical or mechanical, pertains to the rate at which work is being done, so the power consumption in a plant is related to current flow. A large electric motor or air dryer consumes more power (and draws more current) in a given length of time than, for example, an indicating light on a motor controller. Work is done whenever a force causes motion. If a mechanical force is used to lift or move a weight, work is done; however, force exerted *without* causing motion, such as the force of a compressed spring acting between two fixed objects, does not constitute work.

Note: Power is the rate at which work is done.

11.7.4.1   Electrical Power Calculations

The electric power (P) used in any part of a circuit is equal to the voltage (E) across that part of the circuit multiplied by the current (I) in that part. In equation form:

$$P = E \times I \tag{11.22}$$

where
  P = Power (watts, W).
  E = Voltage (volts, V).
  I = Current (amps, A).

If we know the current (I) and the resistance (R) but not the voltage, we can find the power (P) by using Ohm's law for voltage, so by substituting Equation 11.21:

$$E = I \times R = IR$$

into Equation 11.22, we obtain:

$$P = IR \times I = I^2 R \tag{11.23}$$

In the same manner, if we know the voltage and the resistance but not the current, we can find the P by using Ohm's law for current, so by substituting Equation 11.19:

$$I = \frac{E}{R}$$

into Equation 11.22, we obtain:

$$P = E \times \frac{E}{R} = \frac{E^2}{R} \tag{11.24}$$

Note: If we know any two quantities, we can calculate the third.

■ EXAMPLE 11.22

Problem: The current through a 200-Ω resistor to be used in a circuit is 0.25 A. Find the power rating of the resistor.

Solution: Because the current (I) and resistance (R) are known, use Equation 11.23 to find P:

$$P = I^2 \times R = (0.25)^2 \times 200 = 0.0625 \times 200 = 12.5 \text{ W}$$

Note: The power rating of any resistor used in a circuit should be twice the wattage calculated by the power equation to prevent the resistor from burning out; thus, the resistor used in Example 11.22 should have a power rating of 25 W.

■ EXAMPLE 11.23

Problem: How many kilowatts of power are delivered to a circuit by a 220-V generator that supplies 30 A to the circuit?

Solution: Because the voltage (E) and current (I) are given, use Equation 11.22 to find P:

$$P = E \times I = 220 \times 30 = 6600 \text{ W} = 6.6 \text{ kW}$$

■ EXAMPLE 11.24

Problem: If the voltage across a 30,000-Ω resistor is 450 V, what is the power dissipated in the resistor?

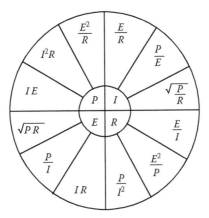

FIGURE 11.25 Ohm's law circle—summary of the basic formulas.

Solution: Because the resistance (R) and voltage (E) are known, use Equation 11.24 to find P:

$$P = \frac{E^2}{R} = \frac{(450)^2}{30,000} = \frac{202,500}{30,000} = 6.75 \text{ W}$$

In this section, P was expressed in terms of various pairs of the other three basic quantities E, I, and R. In practice, we should be able to express any one of the three basic quantities, as well as P, in terms of any two of the others. Figure 11.25 is a summary of the 12 basic formulas we should know. The four quantities E, I, R, and P are at the center of the figure. Adjacent to each quantity are three segments. Note that in each segment the basic quantity is expressed in terms of two other basic quantities, and no two segments are alike.

11.7.5 ELECTRICAL ENERGY (KILOWATT-HOURS)

Energy (the mechanical definition) is defined as the ability to do work (energy and time are essentially the same and are expressed in identical units). Energy is expended when work is done, because it takes energy to maintain a force when that force acts through a distance. The total energy expended to do a certain amount of work is equal to the working force multiplied by the distance through which the force moves to do the work. In electricity, total energy expended is equal to the *rate* at which work is done, multiplied by the length of time the rate is measured. Essentially, energy (W) is equal to power (P) times time (t).

The kilowatt-hour (kWh) is a unit commonly used for large amounts of electric energy or work. The amount of kilowatt-hours is calculated as the product of the power in kilowatts (kW) and the time in hours (hr) during which the power is used:

$$\text{kWh} = \text{kW} \times \text{hr} \tag{11.25}$$

■ **EXAMPLE 11.25**

Problem: How much energy is delivered in 4 hours by a generator supplying 12 kW?

Solution:

$$\text{kWh} = \text{kW} \times \text{hr} = 12 \times 4 = 48 \text{ kWh}$$

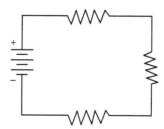

FIGURE 11.26   Series circuit.

11.7.6   SERIES DC CIRCUIT CHARACTERISTICS

As previously mentioned, an electric circuit is made up of a voltage source, the necessary connecting conductors, and the effective load. If the circuit is arranged so the electrons have only *one* possible path, the circuit is a *series circuit*. A series circuit, then, is defined as a circuit that contains only one path for current flow. Figure 11.26 shows a series circuit having several loads (resistors).

Key Point: A series circuit is a circuit having only one path for the current to flow along.

11.7.6.1   Series Circuit Resistance

To follow its electrical path, the current in a series circuit must flow through resistors inserted in the circuit (see Figure 11.27); thus, each additional resistor offers added resistance. In a series circuit, the *total circuit resistance* (R_T) is equal to the sum of the individual resistances, or

$$R_T = R_1 + R_2 + R_3 + \dots + R_n \tag{11.26}$$

where

R_T = Total resistance (Ω).
R_1, R_2, R_3 = Resistance in series (Ω).
R_n = Any number of additional resistors in the series.

■ **EXAMPLE 11.26**

Problem: Three resistors of 10 ohms, 12 ohms, and 25 ohms are connected in series across a battery whose emf is 110 volts (Figure 11.27). What is the total resistance?

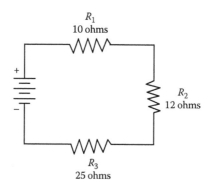

FIGURE 11.27   Solving for total resistance in a series circuit.

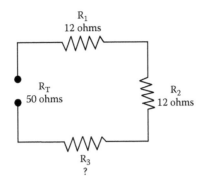

FIGURE 11.28   Calculating the value of one resistance in a series circuit.

Solution:

Given:

$R_1 = 10$ ohms
$R_2 = 12$ ohms
$R_3 = 25$ ohms

$$R_T = R_1 + R_2 + R_3$$

$$R_T = 10 + 12 + 25 = 47 \, \Omega$$

Equation 11.26 can be transposed to solve for the value of an unknown resistance; for example, transposition can be used in some circuit applications where the total resistance is known but the value of a circuit resistor has to be determined.

■ **EXAMPLE 11.27**

Problem: The total resistance of a circuit containing three resistors is 50 ohms (see Figure 11.28). Two of the circuit resistors are 12 ohms each. Calculate the value of the third resistor (R_3).

Solution:

Given:

$R_T = 50$ ohms
$R_1 = 12$ ohms
$R_2 = 12$ ohms

$$R_T = R_1 + R_2 + R_3$$

$$R_3 = R_T - R_1 - R_2$$

$$R_3 = 50 - 12 - 12 = 26 \, \Omega$$

Note: When resistances are connected in series, the total resistance in the circuit is equal to the sum of the resistances of all the parts of the circuit.

11.7.6.2   Series Circuit Current

Because there is but one path for current in a series circuit, the same current (I) must flow through each part of the circuit. Thus, to determine the current throughout a series circuit, only the current through one of the parts must be known. The fact that the same current flows through each part of a series circuit can be verified by inserting ammeters into the circuit at various points as shown in Figure 11.29. As indicated in Figure 11.29, each meter indicates the same value of current.

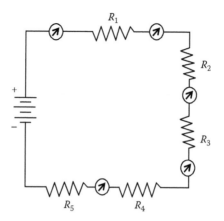

FIGURE 11.29   Current in a series circuit.

Note: In a series circuit, the same current flows in every part of the circuit. Do *not* add the currents in each part of the circuit to obtain *I*.

11.7.6.3   Series Circuit Voltage

The *voltage drop* across the resistor in the basic circuit is the total voltage across the circuit and is equal to the applied voltage. The total voltage across a series circuit is also equal to the applied voltage but consists of the sum of two or more individual voltage drops. This statement can be proven by an examination of the circuit shown in Figure 11.29. In this circuit, a source potential (E_T) of 30 volts is impressed across a series circuit consisting of two 6-ohm resistors. The total resistance (Figure 11.30) of the circuit is equal to the sum of the two individual resistances, or 12 ohms. Using Ohm's law, the circuit current may be calculated as follows:

$$I = \frac{E_T}{R_T} = \frac{30}{12} = 2.5 \text{ amps}$$

Because we know that the value of the resistors is 6 ohms each, and the current through the resistors is 2.5 amp, we can calculate the voltage drops across the resistors. The voltage (E_1) across R_1 is, therefore,

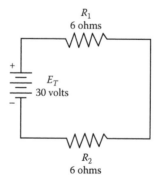

FIGURE 11.30   Calculating total resistance in a series circuit.

$$E_1 = I \times R_1$$

$$E_1 = 2.5 \text{ amps} \times 6 \text{ ohms}$$

$$E_1 = 15 \text{ volts}$$

Because R_2 is the same ohmic value as R_1 and carries the same current, the voltage drop across R_2 is also equal to 15 volts. Adding these two 15-volt drops together gives a total drop of 30 volts, exactly equal to the applied voltage. For a series circuit then,

$$E_T = E_1 + E_2 + E_3 \dots E_n \qquad (11.27)$$

where
  E_T = Total voltage (V).
  E_1 = Voltage across resistance R_1 (V).
  E_2 = Voltage across resistance R_2 (V).
  E_3 = Voltage across resistance R_3 (V).

■ **EXAMPLE 11.28**

Problem: A series circuit consists of three resistors having values of 10 ohms, 20 ohms, and 40 ohms, respectively. Find the applied voltage if the current through the 20-ohm resistor is 2.5 amp.

Solution: To solve this problem, first draw a circuit diagram and label it as shown in Figure 11.31.

  Given:
    R_1 = 10 ohms
    R_2 = 20 ohms
    R_3 = 40 ohms
    I = 2.5 amps

  Because the circuit involved is a series circuit, the same 2.5 amp of current flows through each resistor. Using Ohm's law, the voltage drops across each of the three resistors can be calculated:

$$E_1 = 25 \text{ volts}$$

$$E_2 = 50 \text{ volts}$$

$$E_3 = 100 \text{ volts}$$

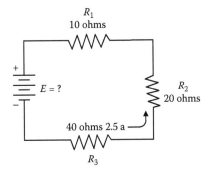

FIGURE 11.31 Solving for applied voltage in a series circuit.

When the individual drops are known, they can be added to find the total or applied voltage by using Equation 11.27:

$$E_T = E_1 + E_2 + E_3$$

$$E_T = 25 \text{ V} + 50 \text{ V} + 100 \text{ V}$$

$$E_T = 175 \text{ V}$$

Key Point 1: The total voltage (E_T) across a series circuit is equal to the sum of the voltages across each resistance of the circuit.

Key Point 2: The voltage drops that occur in a series circuit are in direct proportions to the resistance across which they appear. This is the result of having the same current flow through each resistor. Thus, the larger the resistor, the larger will be the voltage drop across it.

11.7.6.4 Series Circuit Power

Each resistor in a series circuit consumes *power*. This power is dissipated in the form of heat. Because this power must come from the source, the total power must be equal in amount to the power consumed by the circuit resistances. In a series circuit, the total power is equal to the sum of the powers dissipated by the individual resistors. Total power (P_T) is thus equal to

$$P_T = P_1 + P_2 + P_3 + \dots + P_n \tag{11.28}$$

where
P_T = Total power (W).
P_1 = Power used in first part (W).
P_2 = Power used in second part (W).
P_3 = Power used in third part (W).
P_n = Power used in nth part (W).

■ EXAMPLE 11.29

Problem: A series circuit consists of three resistors having values of 5 ohms, 15 ohms, and 20 ohms, respectively. Find the total power dissipation when 120 volts is applied to the circuit (see Figure 11.32).

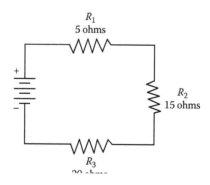

FIGURE 11.32 Solving for total power in a series circuit.

Solution:

Given:

$R_1 = 5$ ohms
$R_2 = 15$ ohms
$R_3 = 20$ ohms
$E = 120$ volts

The total resistance is found first:

$$R_T = R_1 + R_2 + R_3$$

$$R_T = 5 + 15 + 20 = 40 \text{ ohms}$$

Using total resistance and the applied voltage, we can calculate the circuit current:

$$I = \frac{E_T}{R_T} = \frac{120}{40} = 3 \text{ amps}$$

Using the power formula, we can calculate the individual power dissipations:

For resistor R_1:

$$P_1 = I^2 \times R_1$$

$$P_1 = (3)^2 \times 5 = 45 \text{ watts}$$

For resistor R_2:

$$P_2 = I^2 \times R_2$$

$$P_2 = (3)^2 \times 15 = 135 \text{ watts}$$

For resistor R_3:

$$P_3 = I^2 \times R_3$$

$$P_3 = (3)^2 \times 20 = 180 \text{ watts}$$

To obtain total power:

$$P_T = P_1 + P_2 + P_3$$

$$P_T = 45 + 135 + 180 = 360 \text{ watts}$$

To check our answer, the total power delivered by the source can be calculated:

$$P = E \times I$$

$$P = 120 \text{ volts} \times 3 \text{ amps} = 360 \text{ watts}$$

Thus, the total power is equal to the sum of the individual power dissipations.

Note: We found that Ohm's law can be used for total values in a series circuit as well as for individual parts of the circuit. Similarly, the formula for power may be used for total values:

$$P_T = E_T \times I \tag{11.29}$$

11.7.6.5 General Series Circuit Analysis

Now that we have discussed the pieces involved in putting together the puzzle for solving series circuit analysis, we can move on to the next step in the process: solving series circuit analysis in total.

■ EXAMPLE 11.30

Problem: Three resistors of 20 ohms, 20 ohms, and 30 ohms are connected across a battery supply rated at 100-volt terminal voltage. Completely solve the circuit shown in Figure 11.33.

Note: To solve the circuit, the total resistance must be found first, then the circuit current can be calculated. When the current is known, the voltage drops and power dissipations can be calculated.

Solution: The total resistance is

$$R_T = R_1 + R_2 + R_3$$

$$R_T = 20 \text{ ohms} + 20 \text{ ohms} + 30 \text{ ohms} = 70 \text{ ohms}$$

By Ohm's law, the current is

$$I = \frac{E}{R_T} = \frac{100}{70} = 1.43 \text{ amps}$$

The voltage (E_1) across R_1 is

$$E_1 = I \times R_1$$

$$E_1 = 1.43 \text{ amps} \times 20 \text{ ohms} = 28.6 \text{ volts}$$

The voltage (E_2) across R_2 is

$$E_2 = I \times R_2$$

$$E_2 = 1.43 \text{ amps} \times 20 \text{ ohms} = 28.6 \text{ volts}$$

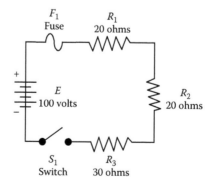

FIGURE 11.33   Solving for various values in a series circuit.

The voltage (E_3) across R_3 is

$$E_3 = I \times R_3$$

$$E_3 = 1.43 \text{ amps} \times 30 \text{ ohms} = 42.9 \text{ volts}$$

The power dissipated by R_1 is

$$P_1 = E_1 \times I$$

$$P_1 = 28.6 \text{ volts} \times 1.43 \text{ amps} = 40.9 \text{ watts}$$

The power dissipated by R_2 is

$$P_2 = E_2 \times I$$

$$P_2 = 28.6 \text{ volts} \times 1.43 \text{ amps} = 40.9 \text{ watts}$$

The power dissipated by R_3 is

$$P_3 = E_3 \times I$$

$$P_3 = 42.9 \text{ volts} \times 1.43 \text{ amps} = 61.3 \text{ watts}$$

The total power dissipated is

$$P_T = E_T \times I$$

$$P_T = 100 \text{ volts} \times 1.43 \text{ amps} = 143 \text{ watts}$$

Note: Keep in mind when applying Ohm's law to a series circuit to consider whether the values used are component values or total values. When the information available allows the use of Ohm's law to find total resistance, total voltage, and total current, then total values must be inserted into the formula.

To find total resistance:

$$R_T = \frac{E_T}{I_T}$$

To find total voltage:

$$E_T = I_T \times R_T$$

To find total current:

$$I_T = \frac{E_T}{R_T}$$

11.7.6.6 Kirchhoff's Voltage Law

Kirchhoff's voltage law states that the voltage applied to a closed circuit equals the sum of the voltage drops in that circuit. It should be obvious that this fact was used in the study of series circuits to this point. It was expressed as follows:

Voltage applied = Sum of voltage drops

$$E_A = E_1 + E_2 + E_3$$

where E_A is the voltage applied, and E_1, E_2, and E_3 are voltage drops.

Another way of stating Kirchhoff's law is that the algebraic sum of the instantaneous emf values and voltage drops around any closed circuit is zero. Through the use of Kirchhoff's law, circuit problems can be solved that would be difficult and often impossible with only knowledge of Ohm's law. When Kirchhoff's law is properly applied, an equation can be set up for a closed loop and the unknown circuit values may be calculated.

11.7.6.7   Polarity of Voltage Drops

When a voltage drop occurs across a resistance, one end must be more positive or more negative than the other end. The polarity of the voltage drop is determined by the direction of current flow. In the circuit shown in Figure 11.34, the current is seen to be flowing in a counterclockwise direction due to the arrangement of the battery source (E). Notice that the end of resistor R_1 into which the current flows is marked negative (–). The end of R_1 at which the current leaves is marked positive (+). These polarity markings are used to show that the end of R_1 into which the current flows is at a higher negative potential than is the end of the resistor from which the current leaves. Point A is thus more negative than point B.

Point C, which is at the same potential as point B, is labeled negative. This indicates that point C, although positive with respect to point A, is more negative than point D. To say that a point is positive (or negative), without stating what it is positive (or negative) with respect to, has no meaning.

Kirchhoff's voltage law can be written as an equation as shown below:

$$E_a + E_b + E_c + \ldots E_n = 0 \tag{11.30}$$

where E_a, E_b, etc. are the voltage drops and emf values around any closed circuit loop.

■ EXAMPLE 11.31

Problem: Three resistors are connected across a 60-volt source. What is the voltage across the third resistor if the voltage drops across the first two resistors are 10 volts and 20 volts?

Solution: First, draw a diagram like the one shown in Figure 11.35. Next, assume a direction of current as shown. Using this current, place the polarity markings at each end of each resistor and on the terminals of the source. Starting at point A, trace around the circuit in the direction of current flow,

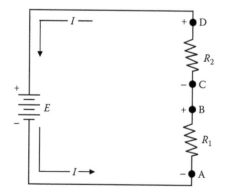

FIGURE 11.34   Polarity of voltage drops.

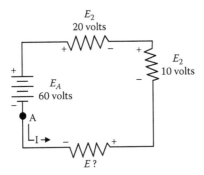

FIGURE 11.35 Determining unknown voltage in a series circuit.

recording the voltage and polarity of each component. Starting at point A, these voltages would be as follows:

Basic formula:

$$E_a + E_b + E_c \ldots E_n = 0$$

From the circuit:

$$(+E_?) + (+E_2) + (+E_3) - (E_A) = 0$$

Substituting values from the circuit, we obtain

$$E_? + 10 + 20 - 60 = 0$$

$$E_? - 30 = 0$$

$$E_? = 30 \text{ volts}$$

Thus, the unknown voltage $(E_?)$ is found to be 30 volts.

Using the same idea as above, a problem can be solved in which the current is the unknown quantity.

11.7.6.8 Series Aiding and Opposing Sources

Sources of voltage that cause current to flow in the same direction are considered to be *series aiding*, and their voltages add. Sources of voltage that would tend to force current in opposite directions are said to be *series opposing*, and the effective source voltage is the difference between the opposing voltages. When two opposing sources are inserted into a circuit, current flow would be in a direction determined by the larger source. Examples of series aiding and opposing sources are shown in Figure 11.36.

11.7.6.9 Kirchhoff's Law and Multiple Source Solutions

Kirchhoff's law can be used to solve multiple source circuit problems. When applying this method, the exact same procedure is used for multiple source circuits as was used for single source circuits. This is demonstrated by the following example.

■ **EXAMPLE 11.32**

Problem: Find the amount of current in the circuit shown in Figure 11.37.

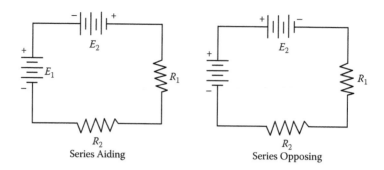

FIGURE 11.36 Series aiding and opposing sources.

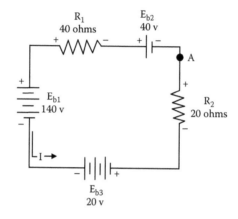

FIGURE 11.37 Solving for circuit current in a multiple-source circuit.

Solution: Start at point A.

Basic equation:

$$E_a + E_b + E_c + \dots + E_n = 0$$

From the circuit:

$$E_{b2} + E_1 - E_{b1} + E_{b3} + E_2 = 0$$

$$40 + 40I - 140 + 20 + 20I = 0$$

Combining like terms, we obtain

$$60I - 80 = 0$$

$$60I = 80$$

$$I = 1.33 \text{ amps}$$

11.7.7 PARALLEL DC CIRCUITS

The principles we applied to solving simple series circuit calculations for determining the reactions of such quantities as voltage, current, and resistance can be used in parallel and series–parallel circuits.

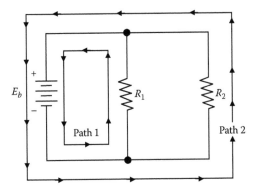

FIGURE 11.38 Basic parallel circuit.

11.7.7.1 Parallel Circuit Characteristics

A parallel circuit is defined as a circuit having two or more components connected across the same voltage source (see Figure 11.38). Recall that a series circuit has only one path for current flow. As additional loads (resistors, etc.) are added to the circuit, the total resistance increases and the total current decreases. This is *not* the case in a parallel circuit. In a parallel circuit, each load (or branch) is connected directly across the voltage source. In Figure 11.38, commencing at the voltage source (E_b) and tracing counterclockwise around the circuit, two complete and separate paths can be identified in which current can flow. One path is traced from the source through resistance R_1 and back to the source, the other from the source through resistance R_2 and back to the source.

11.7.7.2 Voltage in Parallel Circuits

Recall that in a series circuit the source voltage divides proportionately across each resistor in the circuit. In a parallel circuit (see Figure 11.38), the same voltage is present across all of the resistors of a parallel group. This voltage is equal to the applied voltage (E_b) and can be expressed in equation form as

$$E_b = E_{R1} = E_{R2} = E_{Rn} \tag{11.31}$$

We can verify Equation 11.31 by taking voltage measurements across the resistors of a parallel circuit, as illustrated in Figure 11.39. Notice that each voltmeter indicates the same amount of voltage; that is, the voltage across each resistor is the same as the applied voltage.

Note: In a parallel circuit, the voltage remains the same throughout the circuit.

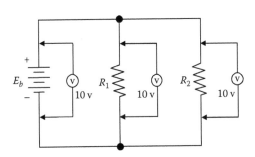

FIGURE 11.39 Voltage comparison in a parallel circuit.

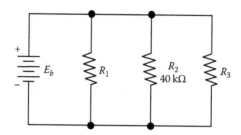

FIGURE 11.40   Illustration for Example 11.33.

■ **EXAMPLE 11.33**

Problem: Assume that the current through a resistor of a parallel circuit is known to be 4 milliamperes (mA) and the value of the resistor is 40,000 ohms. Determine the potential (voltage) across the resistor. The circuit is shown in Figure 11.40.

Solution:

Given:

$$R_2 = 40 \text{ kW}$$
$$I_{R2} = 4 \text{ mA}$$

Find E_{R2} and E_b. Select the appropriate equation:

$$E = I \times R$$

Substitute known values:

$$E_{R2} = I_{R2} \times R_2$$

$$E_{R2} = 4 \text{ mA} \times 40,000 \text{ ohms}$$

Using power of tens,

$$E_{R2} = (4 \times 10^{-3}) \times (40 \times 10^3)$$

$$E_{R2} = 4.0 \times 40$$

we obtain

$$E_{R2} = 160 \text{ V}$$

Therefore,

$$E_b = 160 \text{ V}$$

11.7.7.3   Current in Parallel Circuits

In a series circuit, a single current flows. Its value is determined in part by the total resistance of the circuit; however, the source current in a parallel circuit divides among the available paths in relation to the value of the resistors in the circuit. Ohm's law remains unchanged. For a given voltage, current varies inversely with resistance.

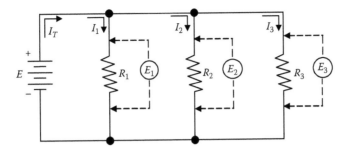

FIGURE 11.41 Parallel circuit.

Note: Ohm's law states that the *current in a circuit is inversely proportional to the circuit resistance*. This fact, important as a basic building block of electrical theory, is also important in the following explanation of current flow in parallel circuits.

The behavior of current in a parallel circuit is best illustrated by example (see Figure 11.41). The resistors R_1, R_2, and R_3 are in parallel with each other and with the battery. Each parallel path is then a branch with its own individual current. When the total current (I_T) leaves the voltage source (E), part I_1 of current I_T will flow through R_1, part I_2 will flow through R_2, and I_3 through R_3. The branch currents I_1, I_2, and I_3 can be different; however, if a voltmeter (used for measuring the voltage of a circuit) is connected across R_1, R_2, and R_3, then the respective voltages E_1, E_2, and E_3 will be equal. Therefore,

$$E = E_1 = E_2 = E_3 \tag{11.32}$$

The total current I_T is equal to the sum of all branch currents:

$$I_T = I_1 = I_2 = I_3 \tag{11.33}$$

This formula applies for any number of parallel branches, whether the resistances are equal or unequal.

By Ohm's law, each branch current equals the applied voltage divided by the resistance between the two points where the voltage is applied. Hence, for each branch we have the following equations:

$$\text{Branch 1:} \quad I_1 = \frac{E_1}{R_1} = \frac{V}{R_1}$$

$$\text{Branch 2:} \quad I_2 = \frac{E_2}{R_2} = \frac{V}{R_2} \tag{11.34}$$

$$\text{Branch 3:} \quad I_3 = \frac{E_3}{R_3} = \frac{V}{R_3}$$

With the same applied voltage, any branch that has less resistance allows more current through it than a branch with higher resistance.

■ EXAMPLE 11.34

Problem: Two resistors, each drawing 2 amps, and a third resistor drawing 1 amp are connected in parallel across a 100-V line (see Figure 11.42). What is the total current?

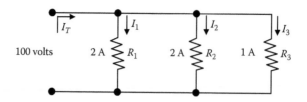

FIGURE 11.42   Illustration for Example 11.34.

Solution: The formula for total current is

$$I_T = I_1 = I_2 = I_3$$

Thus,

$$I_T = 2 + 2 + 1 = 5 \text{ amps}$$

The total current, then, is 5 amps.

■ EXAMPLE 11.35

Problem: Two branches, R_1 and R_2, across a 100-V power line draw a total line current of 20 amps (Figure 11.43). Branch R_1 takes 10 amps. What is the current (I_2) in branch R_2?

Solution: Beginning with Equation 11.33, transpose to find I_2 and then substitute given values:

$$I_T = I_1 + I_2$$

$$I_2 = I_T - I_1$$

$$I_2 = 20 - 10 = 10 \text{ amps}$$

The current in branch R_2, then, is 10 amps.

■ EXAMPLE 11.36

Problem: A parallel circuit consists of two 15-ohm and one 12-ohm resistors across a 120-V line (see Figure 11.44). What current will flow in each branch of the circuit and what is the total current drawn by all the resistors?

Solution: There is 120-V potential across each resistor. Using Equation 11.34, apply Ohm's law to each resistor:

$$I_1 = \frac{V}{R_1} = \frac{120}{15} = 8 \text{ amps}, \quad I_2 = \frac{V}{R_2} = \frac{120}{15} = 8 \text{ amps}, \quad I_3 = \frac{V}{R_3} = \frac{120}{12} = 10 \text{ amps}$$

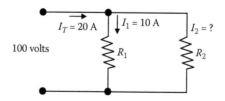

FIGURE 11.43   Illustration for Example 11.35.

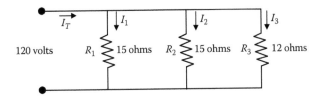

FIGURE 11.44 Illustration for Example 11.36.

Now find total current using Equation 11.34:

$$I_T = I_1 + I_2 + I_3$$

$$I_T = 8 + 8 + 10 = 26 \text{ amps}$$

11.7.7.4 Parallel Circuits and Kirchhoff's Current Law

The division of current in a parallel network follows a definite pattern. This pattern is described by *Kirchhoff's current law*, which is stated as follows:

The algebraic sum of the currents entering and leaving any junction of conductors is equal to zero.

This can be stated mathematically as

$$I_a + I_b + \dots + I_n = 0 \qquad (11.35)$$

where I_a, I_b, ..., I_n are the currents entering and leaving the junction. Currents entering the junction are assumed to be positive, and currents leaving the junction are considered negative. When solving a problem using Equation 11.35, the currents must be placed in the equation with the proper polarity.

■ EXAMPLE 11.37

Problem: Solve for the value of I_3 in Figure 11.45.

Solution: First, give the currents the appropriate signs:

$I_1 = +10$ amps
$I_2 = -3$ amps
$I_3 = ?$ amps
$I_4 = -5$ amps

Then, place these currents into Equation 11.35:

$$I_a + I_b + \dots + I_n = 0$$

with the proper signs as follows:

$$I_1 + I_2 + I_3 + I_4 = 0$$

$$(+10) + (-3) + (I_3) + (-5) = 0$$

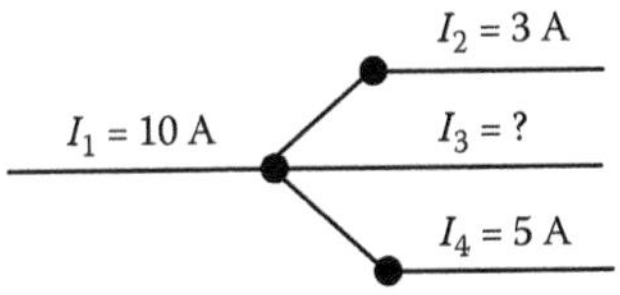

$$I_2 = 3 \text{ A}$$
$$I_1 = 10 \text{ A} \qquad I_3 = ?$$
$$I_4 = 5 \text{ A}$$

FIGURE 11.45 Illustration for Example 11.37.

Combining like terms, we obtain:

$$I_3 + 2 = 0$$

$$I_3 = -2 \text{ amps}$$

Thus, I_3 has a value of 2 amps, and the negative sign shows it to be a current leaving the junction.

11.7.7.5 Parallel Circuit Resistance

Unlike series circuits, where total resistance (R_T) is the sum of the individual resistances, in a parallel circuit the total resistance is *not* the sum of the individual resistances. In a parallel circuit, we can use Ohm's law to find total resistance:

$$R = \frac{E}{I}$$

or

$$R_T = \frac{E_S}{I_T}$$

where R_T is the total resistance of all of the parallel branches across the voltage source E_S, and I_T is the sum of all the branch currents.

■ EXAMPLE 11.38

Problem: Given that $E_S = 120$ volts and $I_T = 26$ amps, what is the total resistance of the circuit shown in Figure 11.46?

Solution: In Figure 11.46, the line voltage is 120 volts and the total line current is 26 amps; therefore,

$$R_T = \frac{E_S}{I_T} = \frac{120}{26} = 4.62 \text{ ohms}$$

Note: Notice that R_T is smaller than any of the three resistances in Figure 11.46. This fact may surprise you; it may seem strange that the total circuit resistance is *less* than that of the smallest resistor (R_3, 12 ohms). If we refer back to the water analogy we have used previously, it makes sense. Consider water pressure and water pipes, and assume that we can keep the water pressure constant. A small pipe offers more resistance to the flow of water than a larger pipe, but if we add another pipe in parallel, one of even smaller diameter, the total resistance to water flow is decreased. In an electrical circuit, even a larger resistor in another parallel branch provides an additional path for current flow, so the total resistance is less. Remember, if we add one more branch to a parallel circuit, the total resistance decreases and the total current increases.

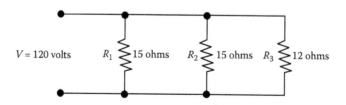

FIGURE 11.46 Illustration for Example 11.38.

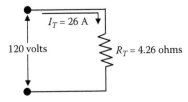

FIGURE 11.47   Circuit equivalent to that of Figure 11.46.

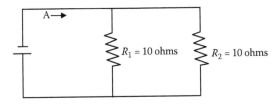

FIGURE 11.48   Two equal resistors connected in parallel.

Back to Example 11.38 and Figure 11.46. What we essentially demonstrated in working this particular problem is that the total load connected to the 120-V line is the same as the single equivalent resistance of 4.62 ohms connected across the line. It is probably more accurate to call this total resistance the *equivalent resistance*, but by convention R_T (total resistance) is generally used, although they are often used interchangeably. The equivalent resistance is illustrated in the equivalent circuit shown in Figure 11.47. Other methods are used to determine the equivalent resistance of parallel circuits. The most appropriate method for a particular circuit depends on the number and value of the resistors; for example, consider the parallel circuit shown in Figure 11.48. For this circuit, the following simple equation is used:

$$R_{eq} = \frac{R}{N} \tag{11.36}$$

where
  R_{eq} = Equivalent parallel resistance.
  R = Ohmic value of one resistor.
  N = Number of resistors.

Thus,

$$R_{eq} = \frac{10 \text{ ohms}}{2} = 5 \text{ ohms}$$

Note: Equation 11.36 is valid for any number of equal value parallel resistors.

Key Point: When two equal value resistors are connected in parallel, they present a total resistance equivalent to a single resistor of one half the value of either of the original resistors.

■ EXAMPLE 11.39

Problem: Five 50-ohm resistors are connected in parallel. What is the equivalent circuit resistance?

Solution: Using Equation 11.36:

$$R_{eq} = \frac{R}{N} = \frac{50}{5} = 10 \text{ ohms}$$

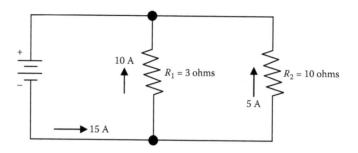

FIGURE 11.49 Illustration for Example 11.40.

What about parallel circuits containing resistance of unequal value? How is equivalent resistance determined? Example 11.40 demonstrates how this is accomplished.

■ EXAMPLE 11.40

Problem: Refer to Figure 11.49.

Solution:

Given:

$R_1 = 3$ ohms
$R_2 = 10$ ohms
$E_a = 30$ volts

We know that

$I_1 = 10$ amps
$I_2 = 5$ amps
$I_t = 15$ amps

and can now determine R_{eq}:

$$R_{eq} = \frac{E_a}{I_t} = \frac{30}{15} = 2 \text{ ohms}$$

Key Point: In Example 11.40, the equivalent resistance of 2 ohms is less than the value of either branch resistor. Remember, in parallel circuits the equivalent resistance will always be smaller than the resistance of any branch.

11.7.7.6 Reciprocal Method

When circuits are encountered in which resistors of unequal value are connected in parallel, the equivalent resistance may be computed by using the *reciprocal method*.

Note: A *reciprocal* is an inverted fraction; the reciprocal of the fraction 3/4, for example, is 4/3. We consider a whole number to be a fraction with 1 as the denominator, so the reciprocal of a whole number is that number divided into 1; for example, the reciprocal of R_T is $1/R_T$.

The equivalent resistance in parallel is given by the following formula:

$$\frac{1}{R_T} = \frac{1}{R_1} + \frac{1}{R_2} + \frac{1}{R_3} + \cdots + \frac{1}{R_n} \tag{11.37}$$

where R_T is the total resistance in parallel, and R_1, R_2, R_3, and R_n are the branch resistances.

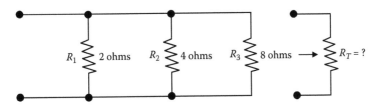

FIGURE 11.50   Illustration for Example 11.41.

■ **EXAMPLE 11.41**

Problem: Find the total resistance of a 2-ohm, a 4-ohm, and an 8-ohm resistor in parallel (Figure 11.50).

Solution: Write the formula for the three resistors in parallel:

$$\frac{1}{R_T} = \frac{1}{R_1} + \frac{1}{R_2} + \frac{1}{R_3}$$

Substitute the resistance values:

$$\frac{1}{R_T} = \frac{1}{2} + \frac{1}{4} + \frac{1}{8}$$

Add the fractions:

$$\frac{1}{R_T} = \frac{4}{8} + \frac{2}{8} + \frac{1}{8} = \frac{7}{8}$$

Invert both sides of the equation to solve for R_T:

$$\frac{1}{R_T} = \frac{8}{7} = 1.14 \text{ ohms}$$

Note: When resistances are connected in parallel, the total resistance is always less than the smallest resistance of any single branch.

11.7.7.7   Product Over the Sum Method

When any two unequal resistors are in parallel, it is often easier to calculate the total resistance by multiplying the two resistances and then dividing the product by the sum of the resistances:

$$R_T = \frac{R_1 \times R_2}{R_1 + R_2} \tag{11.38}$$

where R_T is the total resistance in parallel, and R_1 and R_2 are the two resistors in parallel.

■ **EXAMPLE 11.42**

Problem: What is the equivalent resistance of a 20-ohm and a 30-ohm resistor connected in parallel?

Solution:

   Given:

   $R_1 = 20$ ohms
   $R_2 = 30$ ohms

$$R_T = \frac{R_1 \times R_2}{R_1 + R_2} = \frac{20 \times 30}{20 + 30} = 12 \text{ ohms}$$

11.7.7.8 Power in Parallel Circuits

As in the series circuit, the total *power* consumed in a parallel circuit is equal to the sum of the power consumed in the individual resistors.

Note: Because power dissipation in resistors consists of a heat loss, power dissipations are additive regardless of how the resistors are connected in the circuit.

$$P_T = P_1 + P_2 + P_3 + \ldots + P_n \tag{11.39}$$

where P_T is the total power, and P_1, P_2, P_3, ... P_n are the branch powers.

Total power can also be calculated by the following equation:

$$P_T = E \times I_T \tag{11.40}$$

where P_T is the total power, E is the voltage source across all parallel branches, and I_T is the total current. The power dissipated in each branch is equal to $E \times I$ and equal to V^2/R.

Note: In both parallel and series arrangements, the sum of the individual values of power dissipated in the circuit equals the total power generated by the source. The circuit arrangements cannot change the fact that all power in the circuit comes from the source.

11.7.8 SERIES–PARALLEL CIRCUITS

So far, we have discussed series and parallel DC circuits; however, operators will seldom encounter a circuit that consists solely of either type of circuit. Most circuits consist of both series and parallel elements. A circuit of this type is referred to as a *series–parallel circuit* or as a *combination circuit*. Analyzing a series–parallel (combination) circuit is simply a matter of applying the laws and rules discussed up to this point.

11.7.8.1 Solving a Series–Parallel Circuit

At least three resistors are required to form a series–parallel circuit: two parallel resistors connected in series with at least one other resistor. In a circuit of this type, the current (I_T) divides after it flows through R_1; part of it flows through R_2, and part flows through R_3. Then, the current joins at the junction of the two resistors and flows back to the positive terminal of the voltage source (E) and through the voltage source to the positive terminal.

When solving for values in a series–parallel circuit (current, voltage, and resistance), follow the rules that apply to a series circuit for the series part of the circuit, and follow the rules that apply to a parallel circuit for the parallel part of the circuit. Solving series–parallel circuits is simplified if all parallel and series groups are first reduced to single equivalent resistances and the circuits are redrawn in simplified form. Recall that the redrawn circuit is called an *equivalent circuit*.

Note: No general formulas are available for solving series–parallel circuits because so many different forms of these circuits exist.

Note: The total current in the series–parallel circuit depends on the effective resistance of the parallel portion and on the other resistances.

11.7.9 CONDUCTORS

Earlier we mentioned that electric current moves easily through some materials but with greater difficulty through others. Three good electrical conductors are copper, silver, and aluminum (generally, we can say that most metals are good conductors). Today, copper is the material of choice for electrical conductors. Under special conditions, certain gases are also used as conductors; for example, neon gas, mercury vapor, and sodium vapor are used in various kinds of lamps.

The function of the wire conductor is to connect a source of applied voltage to a load resistance with a minimum *IR* voltage drop in the conductor so most of the applied voltage can produce current in the load resistance. Ideally, a conductor must have a very low resistance; a typical value for a conductor such as copper is less than 1 ohm per 10 feet.

Because all electrical circuits utilize conductors of one type or another, in this section we discuss the basic features and electrical characteristics of the most common types of conductors. Moreover, because conductor splices and connections (and insulation of such connections) are also an essential part of any electric circuit, they are also discussed.

11.7.9.1 Unit Size of Conductors

A standard (or unit size) of a conductor has been established to compare the resistance and size of one conductor with another. The unit of linear measurement used (with regard to the diameter of a piece of wire) is the *mil* (0.001 of an inch). A convenient unit of wire length is the foot. Thus, the standard unit of size in most cases is the mil-foot; that is, a wire will have unit size if it has diameter of 1 mil and a length of 1 foot. The resistance in ohms of a unit conductor or a given substance is called the *resistivity* (or *specific resistance*) of the substance. As a further convenience, gauge numbers are also used to compare the diameter of wires. The Browne and Sharpe (B&S) gauge was used in the past; now the most commonly used gauge is the American Wire Gauge (AWG).

11.7.9.2 Square Mil

Figure 11.51 shows a *square mil*, which is a convenient unit of cross-sectional area for square or rectangular conductors. As shown in Figure 11.51, a square mil is the area of a square, the sides of which are 1 mil. To obtain the cross-sectional area in square mils of a square conductor, square one side measured in mils. To obtain the cross-sectional area in square mils of a rectangular conductor, multiply the length of one side by that of the other, each length being expressed in mils.

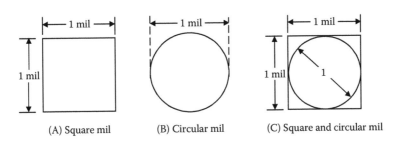

(A) Square mil   (B) Circular mil   (C) Square and circular mil

FIGURE 11.51   (A) Square mil; (B) circular mil; (C) comparison of circular to square mil.

■ **EXAMPLE 11.43**

Problem: Find the cross-sectional area of a large rectangular conductor 5/8 inch thick and 5 inches wide.

Solution: The thickness may be expressed in mils as $0.625 \times 1000 = 625$ mils and the width as $5 \times 1000 = 5000$ mils. The cross-sectional area is 625×5000, or 3,125,000 square mils.

11.7.9.3 Circular Mil

The *circular mil* is the standard unit of wire cross-sectional area used in most wire tables. To avoid the use of decimals (because most wires used to conduct electricity may be only a small fraction of an inch), it is convenient to express these diameters in mils. As an example, the diameter of a wire is expressed as 25 mils instead of 0.025 inch. A circular mil is the area of a circle having a diameter of 1 mil, as shown in Figure 11.51B. The area in circular mils of a round conductor is obtained by squaring the diameter measured in mils. Thus, a wire having a diameter of 25 mils has an area of $(25)^2$ or 625 circular mils. By way of comparison, the basic formula for the area of a circle is

$$A = \pi r^2 \tag{11.41}$$

In this example, the area in square inches is

$$A = \pi r^2 = 3.14 \times (0.0125)^2 = 0.00049 \text{ in.}^2$$

If D is the diameter of a wire in mils, the area in square mils can be determined using

$$A = \pi \times (D/2)^2 \tag{11.42}$$

which translates to

$$A = 3.14/4 \times D^2 = 0.785 \times D^2 \text{ square mils}$$

Thus, a wire 1 mil in diameter has an area of

$$A = 0.785 \times (1)^2 = 0.785 \text{ square mils}$$

which is equivalent to 1 circular mil. The cross-sectional area of a wire in circular mils is therefore determined as

$$A = \frac{0.785D^2}{0.785} = D^2 \text{ circular mils}$$

where D is the diameter in mils; therefore, the constant $\pi/4$ is eliminated from the calculation.

It should be noted that in comparing square and round conductors that the circular mil is a smaller unit of area than the square mil; therefore, there are more circular mils than square mils in any given area. The comparison is shown in Figure 11.51C. The area of a circular mil is equal to 0.785 of a square mil.

Note: To determine the circular mil area when the square mil area is given, divide the area in square mils by 0.785. Conversely, to determine the square mil area when the circular mil area is given, multiply the area in circular mils by 0.785.

■ EXAMPLE 11.44

Problem: A No. 12 wire has a diameter of 80.81 mils. What are (1) its area in circular mils and (2) its area in square mils?

Solution:
  1. $A = D^2 = (80.81)^2 = 6530$ circular mils
  2. $A = 0.785 \times 6530 = 5126$ square mils

■ EXAMPLE 11.45

Problem: A rectangular conductor is 1.5 inches wide and 0.25 inch thick. (1) What is its area in square mils? (2) What size of round conductor in circular mils is necessary to carry the same current as the rectangular bar?

Solution:
  1. 1.5 in. = 1.5 × 1000 = 1500 mils, and 0.25 in. = 0.25 × 1000 = 250 mils; thus,

$$A = 1500 \times 250 = 375,000 \text{ square mils}$$

  2. To carry the same current, the cross-sectional area of the rectangular bar and the cross-sectional area of the round conductor must be equal. There are more circular mils than square mils in this area; therefore,

$$A = \frac{375,000}{0.785} = 477,700 \text{ circular mils}$$

Note: Many electric cables are composed of stranded wires. The strands are usually single wires twisted together in sufficient numbers to make up the necessary cross-sectional area of the cable. The total area in circular mils is determined by multiplying the area of one strand in circular mils by the number of strands in the cable.

11.7.9.4 Circular-Mil-Foot

As shown in Figure 11.52, a *circular-mil-foot* is actually a unit of volume. More specifically, it is a unit conductor 1 foot in length and having a cross-sectional area of 1 circular mil. The circular-mil-foot is useful in making comparisons between wires that are made of different metals because it is considered a unit conductor; for example, a basis of comparison of the resistivity of various substances may be the resistance of a circular-mil-foot of each of the substances.

Note: It is sometimes more convenient to employ a different unit of volume when working with certain substances. Accordingly, unit volume may also be taken as the centimeter cube. The inch cube may also be used. The unit of volume employed is given in tables of specific resistances.

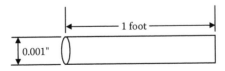

FIGURE 11.52 Circular-mil-foot.

11.7.9.5 Resistivity

All materials differ in their atomic structure and therefore in their ability to resist the flow of an electric current. The measure of the ability of a specific material to resist the flow of electricity is called its *resistivity* or *specific resistance*—the resistance in ohms offered by unit volume (the circular-mil-foot) of a substance to the flow of electric current. Resistivity is the reciprocal of conductivity (i.e., the ease with which current flows in a conductor). A substance that has a high resistivity will have a low conductivity, and *vice versa*.

The resistance of a given length for any conductor depends on the resistivity of the material, the length of the wire, and the cross-sectional area of the wire according to the following equation:

$$R = \rho \frac{L}{A}$$

(11.43)

where

R = Resistance of the conductor (ohms).
ρ = Specific resistance or resistivity (circular mil ohm/ft).
L = Length of the wire (ft).
A = Cross-sectional area of the wire (circular mil).

The factor ρ (Greek letter rho) permits different materials to be compared for resistance according to their nature without regard to different lengths or areas. Higher values of ρ mean more resistance.

Key Point: The resistivity of a substance is the resistance of a unit volume of that substance.

11.7.10 Magnetic Units

The law of current flow in the electric circuit is similar to the law for the establishing of flux in the magnetic circuit. The *magnetic flux* (ϕ, phi) is similar to current in the Ohm's law formula and is the total number of lines of force existing in the magnetic circuit. The Maxwell (Ma) is the unit of flux; that is, 1 line of force is equal to 1 Maxwell.

Note: The Maxwell is often referred to as simply a *line of force*, *line of induction*, or *line*.

The *strength* of a magnetic field in a coil of wire depends on how much current flows in the turns of the coil—the more current, the stronger the magnetic field. In addition, the more turns, the more concentrated are the lines of force. The *force* that produces the flux in the magnetic circuit (comparable to electromotive force in Ohm's law) is known as *magnetomotive force* (mmf). The practical unit of magnetomotive force is the ampere-turn (At). In equation form,

$$F \text{ (ampere-turns)} = N \times I$$

(11.44)

where

F = Magnetomotive force (At).
N = Number of turns.
I = Current (A).

■ EXAMPLE 11.46

Problem: Calculate the ampere-turns for a coil with 2000 turns and a 5-Ma current.

Solution: Use Equation 11.44 and substitute N = 2000 and $I = 5 \times 10^{-3}$ A:

$$N \times I = 2000 \times (5 \times 10^{-3}) = 10 \text{ At}$$

The unit of *intensity* of magnetizing force per unit of length is designated as H and is sometimes expressed as Gilberts per centimeter of length. Expressed as an equation:

$$H = \frac{N \times I}{L} \qquad (11.45)$$

where
  H = Magnetic field intensity (ampere-turns per meter, At/m).
  N = Number of turns.
  I = Current (A).
  L = Length between poles of the coil (m).

Note: Equation 11.45 is for a solenoid. H is the intensity of an air core. For an iron core, H is the intensity through the entire core, and L is the length or distance between poles of the iron core.

11.7.11 AC Theory

Because voltage is induced in a conductor when lines of force are cut, the amount of the induced emf depends on the number of lines cut in a unit time. To induce an emf of 1 volt, a conductor must cut 100,000,000 lines of force per second. To obtain this great number of cuttings, the conductor is formed into a loop and rotated on an axis at great speed (see Figure 11.53). The two sides of the loop become individual conductors in series, each side of the loop cutting lines of force and inducing twice the voltage that a single conductor would induce. In commercial generators, the number of cuttings and the resulting emf are increased by: (1) increasing the number of lines of force by using more magnets or stronger electromagnets, (2) using more conductors or loops, and (3) rotating the loops faster.

How an alternating-current (AC) generator operates to produce an AC voltage and current is a basic concept taught in elementary and middle school science classes. Of course, we accept technological advances as commonplace today. We surf the Internet, watch cable television, use our cell phones, take space flight as a given—and consider production of the electricity that makes all these technologies possible as our right. These technologies are bottom shelf to us today; they are available to us so we use them.

In the groundbreaking years of electric technology development, the geniuses of the science of electricity (including George Simon Ohm) achieved their technological breakthroughs in faltering steps. We tend to forget that these first faltering steps were achieved using crude and, for the

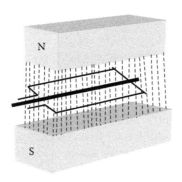

FIGURE 11.53 Loop rotating in a magnetic field produces an AC voltage.

most part, homemade apparatus. Indeed, the innovators of electricity had to fabricate nearly all of the laboratory equipment used in their experiments. At the time, the only convenient source of electrical energy available to these early scientists was the voltaic cell, invented some years earlier. Because cells and batteries were the only sources of power available, some of the early electrical devices were designed to operate from direct current (DC). For this reason, initially direct current was used extensively; however, when the use of electricity became widespread, certain disadvantages in the use of direct current became apparent. In a DC system, the supply voltage must be generated at the level required by the load. To operate a 240-volt lamp, for example, the generator must deliver 240 volts. A 120-volt lamp could not be operated from this generator by any convenient means. A resistor could be placed in series with the 120-volt lamp to drop the extra 120 volts, but the resistor would waste an amount of power equal to that consumed by the lamp.

Another disadvantage of DC systems is the large amount of power lost due to the resistance of the transmission wires used to carry current from the generating station to the consumer. This loss could be greatly reduced by operating the transmission line at very high voltage and low current. This is not a practical solution in a DC system, however, because the load would also have to operate at high voltage. Because of the difficulties encountered with direct current, practically all modern power distribution systems use alternating current, including water/wastewater treatment plants.

Unlike DC voltage, AC voltage can be stepped up or down by a device called a *transformer*. Transformers permit the transmission lines to be operated at high voltage and low current for maximum efficiency. At the consumer end, the voltage is stepped down to whatever value the load requires by using a transformer. Due to its inherent advantages and versatility, alternating current has replaced direct current in all but a few commercial power distribution systems.

11.7.11.1 Basic AC Generator

As shown in Figure 11.53, an AC voltage and current can be produced when a conductor loop rotates through a magnetic field and cuts lines of force to generate an induced AC voltage across its terminals. This describes the basic principle of operation of an alternating current generator, or alternator. An alternator converts mechanical energy into electrical energy. It does this by utilizing the principle of electromagnetic induction. The basic components of an alternator are an armature, about which many turns of conductor are wound and which rotates in a magnetic field, and some means of delivering the resulting alternating current to an external circuit.

11.7.11.2 Cycle

An AC voltage is one that continually changes in magnitude and periodically reverses in polarity (see Figure 11.54). The zero axis is a horizontal line across the center. The vertical variations on the voltage wave show the changes in magnitude. The voltages above the horizontal axis have positive (+) polarity, and voltages below the horizontal axis have negative (–) polarity.

Figure 11.54 shows a suspended loop of wire (conductor or armature) being rotated (moved) in a counterclockwise direction through the magnetic field between the poles of a permanent magnet. For ease of explanation, the loop has been divided into a thick and thin half. Notice that in part A the thick half is moving along (parallel to) the lines of force; consequently, it is cutting none of these lines. The same is true of the thin half, moving in the opposite direction. Because the conductors are not cutting any lines of force, no emf is induced. As the loop rotates toward the position shown in part B, it cuts more and more lines of force per second because it is cutting more directly across the field (lines of force) as it approaches the position shown in part B. At position B, the induced voltage is greatest because the conductor is cutting directly across the field.

As the loop continues to be rotated toward the position shown in part C, it cuts fewer and fewer lines of force per second. The induced voltage decreases from its peak value. Eventually, the loop is once again moving in a plane parallel to the magnetic field, and no voltage (zero voltage) is induced.

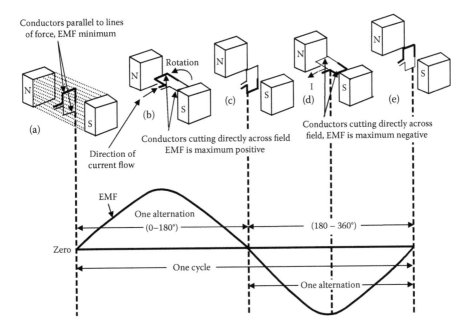

FIGURE 11.54 Basic AC sine wave and AC generator.

The loop has now been rotated through half a circle (one alternation, or 180°). The sine curve shown in the lower part of Figure 11.54 shows the induced voltage at every instant of rotation of the loop. Notice that this curve contains 360°, or two alternations. Two alternations represent one complete circle of rotation.

Note: Two complete alternations in a period is called a *cycle*.

In Figure 11.54, if the loop is rotated at a steady rate and if the strength of the magnetic field is uniform, the number of cycles per second (cps), or hertz, and the voltage will remain at fixed values. Continuous rotation will produce a series of sine-wave voltage cycles, or, in other words, an AC voltage. In this way, mechanical energy is converted into electrical energy.

11.7.11.3 Frequency, Period, and Wavelength

The *frequency* of an alternating voltage or current is the number of complete cycles occurring in each second of time. It is indicated by the symbol f and is expressed in hertz (Hz). One cycle per second equals 1 hertz; thus, 60 cycles per second (cps) equals 60 Hz. A frequency of 2 Hz (Figure 11.55A) is twice the frequency of 1 Hz (Figure 11.55B). The amount of time for the completion of 1 cycle is the *period*. It is indicated by the symbol T for time and is expressed in seconds (sec). Frequency and period are reciprocals of each other:

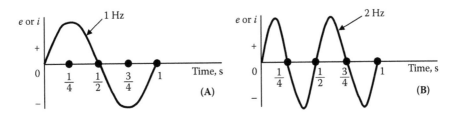

FIGURE 11.55 Comparison of frequencies.

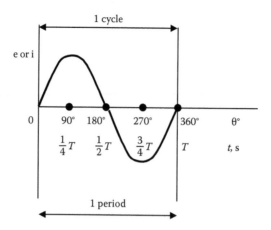

FIGURE 11.56   Relationship between electrical degrees and time.

$$f = \frac{1}{T} \tag{11.46}$$

$$T = \frac{1}{f} \tag{11.47}$$

Note: The higher the frequency, the shorter the period.

The angle of 360° represents the time for 1 cycle, or the period T; therefore, we can show the horizontal axis of the sine wave in units of either electrical degrees or seconds (see Figure 11.56).

The *wavelength* is the length of one complete wave or cycle. It depends on the frequency of the periodic variation and its velocity of transmission. It is indicated by the symbol λ (lambda). Expressed as a formula:

$$\lambda = \frac{\text{Velocity}}{\text{Frequency}} \tag{11.48}$$

11.7.11.4 Characteristic Values of AC Voltage and Current

Because an AC sine wave voltage or current has many instantaneous values throughout the cycle, it is convenient to specify magnitudes for comparing one wave with another. The peak, average, or root-mean-square (RMS) value can be specified (see Figure 11.57). These values apply to current or voltage.

11.7.11.5 Peak Amplitude

One of the most frequently measured characteristics of a sine wave is its amplitude. Unlike DC measurement, the amount of alternating current or voltage present in a circuit can be measured in various ways. In one method of measurement, the maximum amplitude of either the positive or the negative alternation is measured. The value of current or voltage obtained is called the *peak voltage* or the *peak current*. To measure the peak value of current or voltage, an oscilloscope must be used. The peak value is illustrated in Figure 11.57.

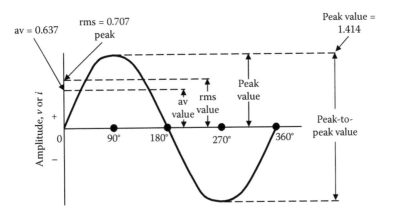

FIGURE 11.57 Amplitude values for AC sine wave.

11.7.11.6 Peak-to-Peak Amplitude

A second method of indicating the amplitude of a sine wave consists of determining the total voltage or current between the positive and negative peaks. This value of current or voltage is the *peak-to-peak value* (see Figure 11.57). Because both alternations of a pure sine wave are identical, the peak-to-peak value is twice the peak value. Peak-to-peak voltage is usually measured with an oscilloscope, although some voltmeters have a special scale calibrated in peak-to-peak volts.

11.7.11.7 Instantaneous Amplitude

The *instantaneous value* of a sine wave of voltage for any angle of rotation is expressed by the following formula:

$$e = E_m \times \sin\theta \tag{11.49}$$

where

e = Instantaneous voltage.
E_m = Maximum or peak voltage.
$\sin\theta$ = Sine of angle at which e is desired.

Similarly the equation for the instantaneous value of a sine wave of current is

$$i = I_m \times \sin\theta \tag{11.50}$$

where

i = Instantaneous current.
I_m = Maximum or peak current.
$\sin\theta$ = Sine of the angle at which i is desired.

Note: The instantaneous value of voltage constantly changes as the armature of an alternator moves through a complete rotation. Because current varies directly with voltage, according to Ohm's law, the instantaneous changes in current also result in a sine wave whose positive and negative peaks and intermediate values can be plotted exactly as we plotted the voltage sine wave. Because instantaneous values are not useful in solving most AC problems, an effective value is used instead.

11.7.21.8   Effective or RMS Value

The *effective value* of an AC voltage or current of sine waveform is defined in terms of an equivalent heating effect of a direct current. Heating effect is independent of the direction of current flow.

Key Point: Because all instantaneous values of induced voltage are somewhere between 0 and E_m (maximum or peak voltage), the effective value of a sine wave voltage or current must be greater than 0 and less than E_m.

The AC of a sine waveform having a maximum value of 14.14 amps produces the same amount of heat in a circuit having a resistance of 1 ohm as a direct current of 10 amps. For this reason, we can work out a constant value for converting any peak value to a corresponding effective value. In the simple equation below, x represents this constant (solve for x to three decimal places):

$$14.14x = 10$$

$$x = 0.707$$

The effective value is also called the *root-mean-square* (RMS) value because it is the square root of the average of the squared values between zero and maximum. The effective value of an AC current is stated in terms of an equivalent DC current. The phenomenon used for standard comparisons is the heating effect of the current.

Note: Anytime an AC voltage or current is stated without any qualifications, it is assumed to be an effective value.

In many instances, it is necessary to convert from effective to peak or *vice versa* using a standard equation. Figure 11.57 shows that the peak value of a sine wave is 1.414 times the effective value; therefore, the equation we use is

$$E_m = E \times 1.414 \qquad\qquad (11.51)$$

where
  E_m = Maximum or peak voltage.
  E = Effective or RMS voltage.

and

$$I_m = I \times 1.414 \qquad\qquad (11.52)$$

where
  I_m = Maximum or peak current.
  I = Effective or RMS current.

Occasionally, it is necessary to convert a peak value of current or voltage to an effective value. This is accomplished by using the following equations:

$$E = E_m \times 0.707 \qquad\qquad (11.53)$$

where
  E = Effective voltage.
  E_m = Maximum or peak voltage.

$$I = I_m \times 0.707 \qquad\qquad (11.54)$$

where
  I = Effective current.
  I_m = Maximum or peak current.

TABLE 11.2
AC Sine Wave Conversion Table

Multiply the Value:	by	to Obtain:
Peak	2	Peak-to-peak
Peak-to-peak	0.5	Peak
Peak	0.637	Average
Average	1.637	Peak
Peak	0.707	RMS (effective)
RMS (effective)	1.414	Peak
Average	1.110	RMS (effective)
RMS (effective)	0.901	Average

11.7.11.9 Average Value

Because the positive alternation is identical to the negative alternation, the *average value* of a complete cycle of a sine wave is zero. In certain types of circuits however, it is necessary to compute the average value of one alternation. Figure 11.57 shows that the average value of a sine wave is 0.637 x peak value:

$$\text{Average value} = 0.637 \times \text{peak value} \tag{11.55}$$

or

$$E_{avg} = E_m \times 0.637$$

where
  E_{avg} = Average voltage of one alternation.
  E_m = Maximum or peak voltage.

Similarly,

$$I_{avg} = I_m \times 0.637 \tag{11.56}$$

where
  I_{avg} = Average current in one alternation.
  I_m = Maximum or peak current.

Table 11.2 lists the various values of sine wave amplitude used for the conversion of AC sine wave voltage and current.

11.7.11.10 Resistance in AC Circuits

If a sine wave of voltage is applied to a resistance, the resulting current will also be a sine wave. This follows Ohm's law, which states that the current is directly proportional to the applied voltage. Figure 11.58 shows a sine wave of voltage and the resulting sine wave of current superimposed on the same time axis. Notice that as the voltage increases in a positive direction the current increases along with it. When the voltage reverses direction, the current reverses direction. At all times, the voltage and current pass through the same relative parts of their respective cycles at the same time. When two waves, such as those shown in Figure 11.58, are precisely in step with one another, they are said to be *in phase*. To be in phase, the two waves reach their maximum and minimum points at the same time and in the same direction. In some circuits, several sine waves can be in phase with each other; thus, it is possible to have two or more voltage drops in phase with each other and in phase with the circuit current.

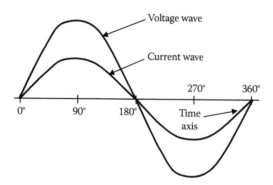

FIGURE 11.58   Voltage and current waves in phase.

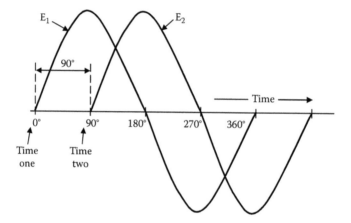

FIGURE 11.59   Voltage waves 90° out of phase.

Note: It is important to remember that Ohm's law for DC circuits is applicable to AC circuits with resistance only.

Voltage waves are not always in phase. Figure 11.59 shows a voltage wave (E_1) considered to start at 0° (time 1). As voltage wave E_1 reaches its positive peak, a second voltage wave (E_2) begins to rise (time 2). Because these waves do not pass through their maximum and minimum points at the same instant of time, a *phase difference* exists between the two waves. The two waves are said to be *out of phase*. For the two waves in Figure 11.59, this phase difference is 90°.

11.7.11.11   Phase Relationships

In the preceding section, we discussed the important concepts of *in phase* and *phase difference*. Another important phase concept is *phase angle*. The phase angle between two waveforms of the same frequency is the angular difference at a given instant of time. As an example, the phase angle between waves B and A (see Figure 11.60) is 90°. Take the instant of time at 90°. The horizontal axis is shown in angular units of time. Wave B begins at maximum value and reduces to 0 value at 90°, whereas wave A begins at 0 and increases to maximum value at 90°. Wave B reaches its maximum value 90° ahead of wave A, so wave B leads wave A by 90° (and wave A lags wave B by 90°). This 90° phase angle between waves B and A is maintained throughout the complete cycle and all successive cycles. At any instant of time, wave B has the value that wave A will have 90° later. Wave B is a cosine wave because it is displaced 90° from wave A, which is a sine wave.

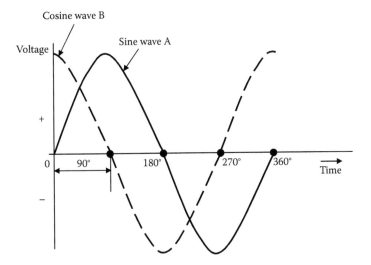

FIGURE 11.60 Wave B leads wave A by a phase angle of 90°.

Key Point: The amount by which one wave leads or lags another is measured in degrees.

To compare phase angles or phases of alternating voltages or currents, it is more convenient to use vector diagrams corresponding to the voltage and current waveforms. A *vector* is a straight line used to denote the magnitude and direction of a given quantity. The length of the line drawn to scale denotes magnitude, and the direction is indicated by the arrow at one end of the line, together with the angle that the vector makes with a horizontal reference vector.

Note: In electricity, because different directions really represent time expressed as a phase relationship, an electrical vector is called a *phasor*. In an AC circuit containing only resistance,
the voltage and current occur at the same time, or are in phase. To indicate this condition by
means of phasors, all that is necessary is to draw the phasors for the voltage and current in
the same direction. The length of the phasor indicates the value of each.

A vector, or phasor, diagram is shown in Figure 11.61, where vector V_B is vertical to show the phase angle of 90° with respect to vector V_A, which is the reference. Because lead angles are shown in the counterclockwise direction from the reference vector, V_B leads V_A by 90°.

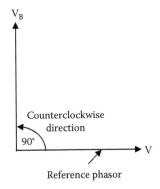

FIGURE 11.61 Phasor diagram.

11.7.12 Inductance

To this point, we have learned the following key points about magnetic fields:

- A field of force exists around a wire carrying a current.
- This field has the form of concentric circles around the wire in planes perpendicular to the wire, with the wire at the center of the circles.
- The strength of the field depends on the current. Large currents produce large fields; small currents produce small fields.
- When lines of force cut across a conductor, a voltage is induced in the conductor.

Moreover, we have studied circuits that have been resistive (i.e., resistors presented the only opposition to current flow). Two other phenomena—*inductance* and *capacitance*—exist in DC circuits to some extent, but they are major players in AC circuits. Both inductance and capacitance present a kind of opposition to current flow that is called *reactance*. (Other than this brief introduction to capacitance and reactance, we do not discuss these two electrical properties in detail in this text; instead, our focus is on the basics, covering only those electrical properties important to water/wastewater operators.)

Inductance is the characteristic of an electrical circuit that makes itself evident by opposing the starting, stopping, or changing of current flow. A simple analogy can be used to explain inductance. We are all familiar with how difficult it is to push a heavy load (such as a cart full of heavy items). It takes more work to start the load moving than it does to keep it moving. This is because the load possesses the property of *inertia*. Inertia is the characteristic of mass that opposes a change in velocity; it can hinder us in some ways and help us in others. Inductance exhibits the same effect on current in an electric circuit as inertia does on velocity of a mechanical object. The effects of inductance are sometimes desirable and sometimes undesirable.

Note: Simply put, inductance is the characteristic of an electrical conductor that opposes a change in current flow.

Inductance is the property of an electric circuit that opposes any change in the current passing through that circuit, so, if the current increases, a self-induced voltage opposes this change and delays the increase. On the other hand, if the current decreases, a self-induced voltage tends to aid (or prolong) the current flow, delaying the decrease. Thus, current can neither increase nor decrease as quickly in an inductive circuit as it can in a purely resistive circuit. In AC circuits, this effect becomes very important because it affects the phase relationships between voltage and current. Earlier, we learned that voltages (or currents) could be out of phase if they are induced in separate armatures of an alternator. In that case, the voltage and current generated by each armature were in phase. When inductance is a factor in a circuit, the voltage and current generated by the same armature are out of phase. We will examine these phase relationships later. Our objective in this chapter is to understand the nature and effects of inductance in an electric circuit.

The unit for measuring inductance (*L*) is the *henry* (named for the American physicist Joseph Henry), which is abbreviated as h. Figure 11.62 shows the schematic symbol for an inductor. An inductor has an inductance of 1 henry if an emf of 1 volt is induced in the inductor when the current through the inductor is changing at the rate of 1 ampere per second. The relationships among the induced voltage, inductance, and rate of change of current with respect to time can be stated mathematically as

$$E = L\frac{I}{t} \qquad (11.57)$$

where
  E = Induced emf (volts).
  L = Inductance (henrys).
  ΔI = Change in amperes occurring in Δt seconds, where Δ (delta) means "a change in."

FIGURE 11.62 Schematic symbol for an inductor.

The henry is a large unit of inductance and is used with relatively large inductors. The unit employed with small inductors is the millihenry (mh). For still smaller inductors, the unit of inductance is the microhenry (μh).

10.7.12.1 Self-Inductance

As previously explained, current flow in a conductor always produces a magnetic field surrounding, or linking with, the conductor. When the current changes, the magnetic field changes, and an emf is induced in the conductor. This emf is referred to as a *self-induced emf* because it is induced in the conductor carrying the current.

Note: Even a perfectly straight length of conductor has some inductance.

The direction of the induced emf has a definite relation to the direction in which the field that induces the emf varies. When the current in a circuit is increasing, the flux linking with the circuit is increasing. This flux cuts across the conductor and induces an emf in the conductor in such a direction as to oppose the increase in current and flux. This emf is sometimes referred to as *counterelectromotive force* (cemf). The two terms are used synonymously throughout this manual. Likewise, when the current is decreasing, an emf is induced in the opposite direction and opposes the decrease in current.

Note: The effects just described are summarized by *Lenz's law*, which states that the induced emf in any circuit is always in a direction opposed to the effect that produced it.

Shaping a conductor so the electromagnetic field around each portion of the conductor cuts across some other portion of the same conductor increases inductance, as shown in its simplest form in Figure 11.63A. The conductor is looped so two portions of the conductor lie adjacent and parallel to one another. These portions are labeled conductor 1 and conductor 2. When the switch is closed, electron flow through the conductor establishes a typical concentric field around all portions of the conductor. The field is shown in a single plane (for simplicity) that is perpendicular to both conductors. Although the field originates simultaneously in both conductors, it is considered as originating in conductor 1, and its effect on conductor 2 will be noted. With increasing current, the field expands outward, cutting across a portion of conductor 2. The dashed arrow shows the resultant induced emf in conductor 2. Note that it is in opposition to the battery current and voltage, according to Lenz's law. In Figure 11.63B, the same section of conductor 2 is shown but with the switch open and the flux collapsing.

Note: In Figure 11.63, the important point to note is that the voltage of self-induction opposes both changes in current. It delays the initial buildup of current by opposing the battery voltage and delays the breakdown of current by exerting an induced voltage in the same direction in which the battery voltage acted.

Four major factors affect the self-inductance of a conductor, or circuit:

1. *Number of turns*—Inductance depends on the number of wire turns. Wind more turns to increase inductance; take turns off to decrease the inductance. Figure 11.64 compares the inductance of two coils made with different numbers of turns.
2. *Spacing between turns*—Inductance depends on the spacing between turns, or the length of the inductor. Figure 11.65 shows two inductors with the same number of turns. The turns of the first inductor have a wide spacing. The turns of the second inductor are close

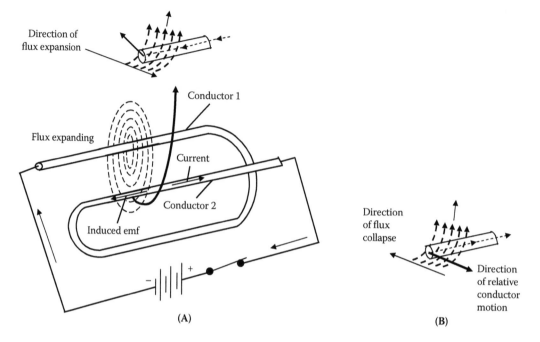

FIGURE 11.63   Self-inductance.

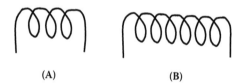

FIGURE 11.64   (A) Few turns, low impedance; (B) more turns, higher inductance.

FIGURE 11.65   (A) Wide spacing between turns, low inductance; (B) close spacing between turns, higher inductance.

together. The second coil, though shorter, has a larger inductance value because of its close spacing between turns.

3. *Coil diameter*—Coil diameter, or cross-sectional area, is highlighted in Figure 11.66. The larger diameter inductor has more inductance. Both coils shown have the same number of turns, and the spacing between turns is the same. The first inductor has a small diameter, and the second one has a larger diameter. The second inductor has more inductance than the first one.

4. *Type of core material*—Permeability is a measure of how easily a magnetic field goes through a material. Permeability also tells us how much stronger the magnetic field will be with the material inside the coil.

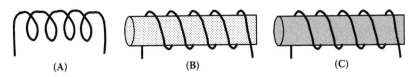

FIGURE 11.66 (A) Small diameter, low inductance; (B) larger diameter, higher inductance.

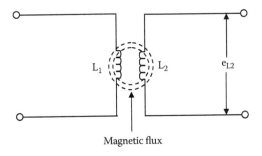

FIGURE 11.67 (A) Air core, low inductance; (B) powdered iron core, higher inductance; (C) soft iron core, highest inductance.

Figure 11.67 shows three identical coils. One has an air core, one has a powdered iron core in the center, and the other has a soft iron core. This figure illustrates the effects of core material on inductance. The inductance of a coil is affected by the magnitude of current when the core is a magnetic material. When the core is air, the inductance is independent of the current.

Note: The inductance of a coil increases very rapidly as the number of turns is increased. It also increases as the coil is made shorter, the cross-sectional area is made larger, or the permeability of the core is increased.

11.7.12.2 Mutual Inductance

When the current in a conductor or coil changes, the varying flux can cut across any other conductor or coil located nearby, thus inducing voltages in both. A varying current in L_1, therefore, induces voltage across L_1 and across L_2 (Figure 11.68; see Figure 11.69 for the schematic symbol for two coils with mutual inductance). When the induced voltage e_{L2} produces current in L_2, its varying magnetic field induces voltage in L_1; hence, the two coils L_1 and L_2 have *mutual inductance* because current change in one coil can induce voltage in the other. The unit of mutual inductance is the henry, and the symbol is L_M. Two coils have L_M of 1 henry when a current change of 1 A/sec in one coil induces 1 E in the other coil. Factors affecting the mutual inductance of two adjacent coils include the following:

- Physical dimensions of the two coils
- Number of turns in each coil
- Distance between the two coils
- Relative positions of the axes of the two coils
- Permeability of the cores

FIGURE 11.68 Mutual inductance between L_1 and L_2.

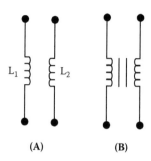

FIGURE 11.69 (A) Schematic symbol for two coils (air core) with mutual inductance; (B) two coils (iron core) with mutual inductance.

Key Point: The amount of mutual inductance depends on the relative position of the two coils. If the coils are separated a considerable distance, the amount of flux common to both coils is small and the mutual inductance is low. Conversely, if the coils are close together so nearly all the flow of one coil links the turns of the other, mutual inductance is high. The mutual inductance can be increased greatly by mounting the coils on a common iron core.

11.7.12.3 Calculation of Total Inductance

(In the study of advanced electrical theory, it is necessary to know the effect of mutual inductance when solving for total inductance in both series and parallel circuits. For our purposes here, however, we will not attempt to make these calculations; instead, we discuss the basic total inductance calculations with which the maintenance operator should be familiar.)

If inductors in series are located far enough apart, or well shielded to make the effects of mutual inductance negligible, the total inductance is calculated in the same manner as for resistances in series; we merely add them:

$$L_t = L_1 + L_2 + L_3 + \ldots \qquad (11.58)$$

■ EXAMPLE 11.47

Problem: If a series circuit contains three inductors with values of 40 μh, 50 μh, and 20 μh, what is the total inductance?

Solution:

$$L_t = 40 \text{ μh} + 50 \text{ μh} + 20 \text{ μh} = 110 \text{ μh}$$

In a parallel circuit containing inductors (without mutual inductance), the total inductance is calculated in the same manner as for resistances in parallel:

$$\frac{1}{L_t} = \frac{1}{L_1} + \frac{1}{L_2} + \frac{1}{L_3} + \cdots \qquad (11.59)$$

■ EXAMPLE 11.48

Problem: A circuit contains three totally shielded inductors in parallel. The values of the three inductances are 4 mh, 5 mh, and 10 mh. What is the total inductance?

Solution:

$$\frac{1}{L_t} = \frac{1}{4} + \frac{1}{5} + \frac{1}{10} = 0.25 + 0.2 + 0.1 = 0.55$$

$$L_t = \frac{1}{0.55} = 1.8 \text{ mh}$$

REFERENCES AND RECOMMENDED READING

AISC. (1980). *AISC Manual of Steel Construction*, 8th ed. American Institute of Steel Construction, Chicago, IL.

ASCE. (2005). *Minimum Design Loads for Buildings and Other Structures*, ASCE/SEI 7-05. New York: American Society of Civil Engineers, New York.

ASSE. (1988). *The Dictionary of Terms Used in the Safety Profession*, 3rd ed. American Society of Safety Engineers, Des Plaines, IL.

ASSE. (1991). *CSP Refresher Guide*. American Society of Safety Engineers, Des Plaines, IL.

Bennett, C.S. (1967). *College Physics*. Harper and Row, New York.

Brauer, R.L. (1994). *Safety and Health for Engineers*. Van Nostrand Reinhold, New York.

Gercek, H. (2007). Poisson's ratio values for rocks. *International Journal of Rock Mechanics and Mining Sciences*, 44(1), 1–13.

Giachino, J.W. and Weeks, W. (1985). *Welding Skills*. American Technical Publishers, Homewood, IL.

Heisler, S.I. (1998). *Wiley Engineer's Desk Reference*, 2nd ed. John Wiley & Sons, New York.

Kirschner, M.W., Marincola, E., and Olmsted Teisberg, E. (1994). The role of biomedical research in health care reform. *Science*, 266(5182), 49–51.

Levy, M. and Salvadori, M. (1992). *Why Buildings Fall Down: How Structures Fail*. W.W. Norton, New York.

Merritt, F.S., Ed. (1983). *Standard Handbook for Civil Engineers*, 3rd ed. McGraw-Hill, New York, 1983.

Parkhurst, D.F. (2006). *Introduction to Applied Mathematics for Environmental Science*. Springer, New York.

PNNL. (2013). *PNNL Hoisting and Rigging Manual*. Pacific Northwest National Laboratory, Richland, WA, http://www.pnl.gov/contracts/hoist_rigging/slings.asp.

Reif, F. (1996). *Understanding Basic Mechanics*. John Wiley & Sons, New York.

Spellman, F.R. and Drinan, J. (2001). *Electricity*. CRC Press, Boca Raton, FL.

Tapley, B., Ed. (1990). *Eshbach's Handbook of Engineering Fundamentals*, 4th ed. John Wiley & Sons, New York.

Urquhart, L. (1959). *Civil Engineering Handbook*, 4th ed. McGraw-Hill, New York.

Section V

Soil Mechanics

Nature, to be commanded, must be obeyed.

—Francis Bacon (1561–1626)

12 Fundamental Soil Mechanics

When dealing with nature's natural building material, soil, the environmental engineer (including any engineer or responsible person in charge ... of anything) should keep the following statement in mind:

> Observe always that everything is the result of change, and get used to thinking that there is nothing Nature loves so well as to change existing forms and to make new ones like them.
>
> **—Marcus Aurelius** (*Meditations*)

12.1 INTRODUCTION*

If a man from today were transported back in time to a particular location, he would instantly recognize the massive structure before him, even though he might be taken aback at what he was seeing: a youthful mountain range with considerable mass, steep sides, and a height that certainly reached beyond any cloud. He would instantly relate to one particular peak—the tallest, most massive one. The polyhedron-shaped object, with its polygonal base and triangular faces culminating in a single sharp-tipped apex, would look familiar—comparable in shape, but larger in size, to the largest of the Great Egyptian Pyramids, although the Pyramids were originally covered in a sheet of limestone, not the thick, perpetual sheet of solid ice and snow covering this mountain peak.

If that same man were to walk this same site today, if he knew what had once stood upon this site, the changes would be obvious and startling—and entirely relative to time. Otherwise, he wouldn't give a second thought while walking across its remnants and through the vegetation growing from its pulverized and amended remains. Over 300 million years ago, the pyramid-shaped mountain peak stood in full, unchallenged splendor above the clouds, wrapped in a cloak of ice, a mighty fortress of stone, seemingly vulnerable to nothing, standing tallest of all—higher than any mountain ever stood—or will ever stand—on Earth.

And so it stood, for millions upon millions of passings of the Earth around the sun. Born when Mother Earth took a deep breath, the pyramid-shaped peak stood tall and undisturbed until millions of years later, when Mother Earth stretched. Today we would call this stretch a massive earthquake, although humans have never witnessed one of such magnitude. Rather than registering on the Richter scale, it would have destroyed it.

When this massive earthquake shattered the Earth's surface, nothing we would call intelligent life lived on Earth—and it's a good thing. During this massive upheaval, the peak shook to its very foundations, and after the initial shockwave and hundreds of aftershocks, the solid granite structure was fractured. This immense fracture was so massive that each aftershock widened it and loosened the base foundation of the pyramid-shaped peak itself. Only 10,000 years later (a

---

* Introduction adapted from Spellman, F.R. and Stoudt, M.L., *Environmental Science*, Scarecrow Press, Lanham, MD, 2013.

few seconds relative to geologic time), the fracture's effects totally altered the shape of the peak forever. During a horrendous windstorm, one of an intensity known only in Earth's earliest days, a sharp tremor (emanating from deep within the Earth and shooting up the spine of the mountain itself, up to the very peak) widened the gaping wound still more.

Decades of continued tremors and terrible windstorms passed (no present-day structure could withstand such a blasting), and, finally, the highest peak of that time, of all time, fell. It broke off completely at its base, and, following the laws of gravity (as effective and powerful a force then as today, of course), it tumbled from its pinnacle position and fell more than 20,000 feet, straight down. It collided with the expanding base of the mountain range, the earth-shattering impact destroying several thousand acres. What remained intact finally came to rest on a precipitous ledge, 15,000 feet in elevation. The pyramid-shaped peak, much smaller now, sat precariously perched on the precipitous ledge for about 5 million years.

Nothing, absolutely nothing, is safe from time. The most inexorable natural law is that of entropy. Time and entropy mean change and decay—harsh, sometimes brutal, but always inevitable. The bruised, scarred, truncated, but still massive rock form, once a majestic peak, was now a victim of Nature's way. Nature, with its chief ally, time, at its side works to degrade anything and everything that has substance and form. For better or for worse, in doing so Nature is ruthless, sometimes brutal, and always inevitable—but never without purpose.

While resting on the ledge, the giant rock, over the course of that 5 million years, was exposed to constantly changing conditions. For several thousand years, Earth's climate was unusually warm—almost tropical—everywhere. Throughout this warm era, the rock was not covered with ice and snow, but instead baked in intense heat, steamed in hot rain, and endured gritty, violent windstorms that arose and released their abrasive fury, sculpting the rock's surface each day for more than 10,000 years.

Then came a pause in the endless windstorms and upheavals of the young planet, a span of time when the weather wasn't furnace-hot or arctic-cold, but moderate. The rock was still exposed to sunlight but at lower temperatures, to rainfall at increased levels, and to fewer windstorms of increased fury. The climate remained so for some years—then the cycle repeated itself—arctic cold, moderately warm, furnace hot—and repeated itself and again.

During the last of these cycles, the rock, considerably affected by physical and chemical exposure, was reduced in size even more. Considerably smaller now than when it landed on the ledge, and a mere pebble compared to its former size, it fell again, 8000 feet to the base of the mountain range, coming to rest on a bed of talus. Reduced in size still more, it remained on its sloping talus bed for many more thousands of years.

Somewhere around 15,000 BC, the rock form, continuously exposed to chemical and mechanical weathering, its physical structure weakened by its long-ago falls, fractured, split, and broke into ever-decreasing-sized rocks, until the largest intact fragment left from the original rock was no bigger than a four-bedroom house. But change did not stop, and neither did time, rolling on until about the time when the Egyptians were building their pyramids. By now, the rock had been reduced, by this long, slow decaying process, to roughly ten feet square.

Over the next thousand years, the rock continued to decrease in size, wearing, crumbling, flaking away, surrounded by fragments of its former self, until it was about the size of a beach ball. Covered with moss and lichen, a web of fissures, tiny crevices and fractures were now woven through the entire mass. Over the next thousand or so years, via *bare rock succession*, what had once been the mother of all mountain peaks, the highest point on Earth, had been reduced to nothing more than a handful of soil.

How did this happen? What is bare rock succession? If a layer of soil is completely stripped off land by natural means (e.g., water, wind), by anthropogenic means (tillage plus erosion), or by cataclysmic occurrence (a massive landslide or earthquake), only after many years can a soil-denuded area return to something approaching its original state, or can a bare rock be converted to soil.

But, given enough time—perhaps a millennium—the scars heal over, and a new, virgin layer of soil forms where only bare rock once existed. The series of events taking place in this restoration process is known as bare rock succession. It is indeed a true "succession"—with identifiable stages. Each stage in the pattern dooms the existing community as it succeeds the state that existed before.

Bare rock, however it is laid open to view, is exposed to the atmosphere. The geologic processes that cause weathering begin breaking down the surface into smaller and smaller fragments. Many forms of weathering exist, and all effectively reduce the bare rock surface to smaller particles or chemicals in solution. Lichens appear to cover the bare rock first. These hardy plants grow on the rock itself. They produce weak acids that assist in the slow weathering of the rock surface. The lichens also trap wind-carried soil particles, which eventually produce a very thin soil layer—a change in environmental conditions that gives rise to the next stage in bare rock succession.

Mosses replace lichens, growing in the meager soil the lichens and weathering provide. They produce a larger growing area and trap even more soil particles, providing a more moist bare rock surface. The combination of more soil and moisture establishes abiotic conditions that favor the next succession stage. Now the seeds of herbaceous plants invade what was once bare rock. Grasses and other flowering plants take hold. Organic matter provided by the dead plant tissue is added to the thin soil, while the rock still weathers from below. More and more organisms join the community as it becomes larger and more complex.

By this time, the plant and animal community is fairly complicated. The next major invasion is by weedy shrubs that can survive in the amount of soil and moisture present. As time passes, the process of building soil speeds up as more and more plants and animals invade the area. Soon trees take root and forest succession is evident. Many years are required, of course, before a climax forest will grow here, but the scene is set for that to occur (Tomera, 1989).

Today, only the remnants of the former, incomparable pyramid-shaped peak are left in the form of soil—soil packed full of organic humus, soil that looks like mud when wet and that, when dry, most people would think was just a handful of dirt.

12.2 SOIL: WHAT IS IT?

In any discussion about soil (after air and water, the third environmental medium), we must initially define exactly what soil is and explain why soil is so important to us. Having said the obvious, we must also clear up a major misconception about soil. As the chapter's introduction indicates, people often confuse soil with dirt. Soil is not dirt. Dirt is misplaced soil—soil where we don't want it, contaminating our hands or clothes, tracked in on the floor. Dirt we try to clean up and keep out of our environment.

But *soil* is special—mysterious, critical to our survival, and, whether we realize it or not, essential to our existence. We have relegated soil to an ignoble position. We commonly degrade it—we consider only feces to be a worse substance, but soil deserves better. Before moving on, let's take another look at that handful of "dirt" that our modern man is holding after the mountain peak was crafted into soil by the sure hand of Nature over millions and millions of years.

What is soil, really? Perhaps no word causes more confusion in communications among various groups of laypersons and professionals—environmental scientists, environmental engineers, specialized groups of earth scientists, and engineers in general—than the word "soil." Why? From the professional's perspective, the problem lies in the reasons why different groups study soils.

Pedologists (soil scientists) are interested in soils as a medium for plant growth. Representing a corresponding branch of engineering soils specialists, *soil engineers* look at soil as a medium that can be excavated with tools. A *geologist*'s view of soil falls somewhere between that of pedologists and soil engineers—they are interested in soils and the weathering processes as past indicators of climatic conditions and in relation to the geologic formation of useful materials ranging from clay deposits to metallic ores.

To clear up this confusion, let's view that handful of soil from a different—but much more basic and revealing—perspective. Consider the following descriptions of soil to better understand what soil is and why it is critically important to us all:

1. A handful of soil is alive, a delicate living organism—as lively as an army of migrating caribou and as fascinating as a flock of egrets. Literally teeming with life of incomparable forms, soil deserves to be classified as an independent ecosystem or, more correctly stated, as many ecosystems.
2. When we reach down and pick up an handful of soil, exposing the stark bedrock surface, it should remind us, maybe startle some of us, that without its thin living soil layer Earth is a planet as lifeless as our own moon.

If you still prefer to call soil dirt, that's okay. Maybe you view dirt in the same way as E.L. Konigsburg's character Ethan does:

The way I see it, the difference between farmers and suburbanites is the difference in the way we feel about dirt. To them, the earth is something to be respected and preserved, but dirt gets no respect. A farmer likes dirt. Suburbanites like to get rid of it. Dirt is the working layer of the earth, and dealing with dirt is as much a part of farm life as dealing with manure: neither is user-friendly, but both are necessary (Konigsburg, 1996, p. 64).

12.3 SOIL BASICS

Soil is the layer of bonded particles of sand, silt, and clay that covers the land surface of the Earth. Most soils develop multiple layers. The topmost layer (*topsoil*) is the layer in which plants grow. This topmost layer is actually an ecosystem composed of both biotic and abiotic components—inorganic chemicals, air, water, decaying organic material that provides vital nutrients for plant photosynthesis, and living organisms. Below the topmost layer (usually no more than a meter in thickness), is the *subsoil*, which is much less productive, partly because it contains much less organic matter. Below that is the *parent material*, the bedrock or other geologic material from which the soil is ultimately formed. The general rule of thumb is that it takes about 30 years to form one inch of topsoil from subsoil; it takes much longer than that for subsoil to be formed from parent material, the length of time depending on the nature of the underlying matter (Franck and Brownstone, 1992).

12.3.1 SOIL FOR CONSTRUCTION

By the time most students reach the third or fourth year of elementary school, they are familiar with the Leaning Tower of Pisa, and many are also familiar with Galileo's experiments with gravity and the speed of falling objects dropped from the top of the tower. This 12th-century bell tower has been a curiosity for literally millions of people from the time it was first built to the present. Eight stories high and 180 feet tall, with a base diameter of 52 feet, the tower began to lean by the time the third story was completed, and leans about 1/25 inch further each year.

How many people know why the tower is leaning in the first place—and who would be more than ordinarily curious about why the Leaning Tower leans? If you are a soil scientist or an engineer, this question has real significance and requires an answer. In fact, the Leaning Tower of Pisa should never have acquired the distinction of being a leaning tower in the first place. The problem is that the Leaning Tower of Pisa rests on a non-uniform consolidation of clay, and the ongoing process of leaning may eventually lead to failure of the building.

As you might have guessed, the mechanics of why the Leaning Tower of Pisa leans is what this section is all about. More specifically, it is about the mechanics and physics of the soil—important factors in making the determination as to whether a particular building site is viable for building. Simply put, these two factors are essential in answering the question "Will the soils present support buildings?"

12.4 SOIL CHARACTERISTICS

When we refer to the characteristics of soils, we are referring to the mechanical characteristics, physical factors important to environmental engineers. Environmental engineers focus on the characteristics of the soil related to its suitability as a construction material and its ability to be excavated. Simply put, the environmental engineer must understand the response of a particular volume of soil to internal and external mechanical forces. Obviously, it is important to be able to determine the soil's ability to withstand the load applied by structures of various types and its ability to remain stable when excavated. From a purely engineering point of view, soil is any surficial (near the surface) material that is unconsolidated enough to be excavated with tools (from bulldozers to shovels). The engineer takes into consideration both the advantages and disadvantages of using soil for engineering purposes. The obvious key advantage of using soil for engineering is that there is (in many places) no shortage of it—it may already be on the construction site, thus avoiding the expense of hauling it from afar. Another advantage of using soil for construction is its ease of manipulation; it may be easily shaped into almost any desired form. Soil also allows for the passage of moisture, or, as needed, it can be made impermeable.

The environmental engineer looks at both the advantages and disadvantages of using soil for construction projects. The most obvious disadvantage of using soil is its variability from place to place and from time to time. Soil is not a uniform material for which reliable data related to strength can be compiled or computed. Cycles of wetting and drying and freezing and thawing affect the engineering properties of soil. A particular soil may be suitable for one purpose but not for another. Stamford clay in Texas, for example, is rated as "very good" for sealing of farm ponds but "very poor" for use as base for roads and buildings (Buol et al., 1980).

To determine whether a particular soil is suitable for use as a base for roads or buildings, the environmental engineer studies soil survey maps and reports. The environmental engineer also checks with soil scientists and other engineers familiar with the region and the soil types of that region. Any good engineer will also want to conduct field sampling to ensure that the soil product he or she will be working with possesses the soil characteristics required for its intended purpose.

Important characteristics of soils for environmental engineering purposes include the following:

- Soil texture
- Kinds of clay present
- Depth to bedrock
- Soil density
- Erodibility
- Corrosivity
- Surface geology
- Plasticity
- Content of organic matter
- Salinity
- Depth to seasonal water table

The environmental engineer will also want to know the soil's density, space–volume and weight–volume relationships, stress and strain, slope stability, and compaction. Because these concepts are of paramount importance to the engineer, these concepts are discussed in the following sections.

12.4.1 SOIL WEIGHT–VOLUME AND SPACE–VOLUME RELATIONSHIPS

All natural soil consists of at least three primary components or phases: solid particles (minerals), water, and air (void spaces between the solid particles). The physical relationships (for soils in particular) between these phases must be examined. The volume of the soil mass is the sum of the volumes of three components, or

$$V_T = V_a + V_w + V_s \tag{12.1}$$

where

V_T = Total volume.
V_a = Air volume.
V_w = Water volume.
V_s = Solids volume.

The volume of the voids is the sum of V_a and V_w. However, because the weighing of air in the soil voids would be done within the Earth's atmosphere as with other weighings, the weight of the solids is determined on a different basis. We consider the weight of air in the soil to be zero and the total weight is expressed as the sum of the weights of the soil solids and the water:

$$W_T = W_s + W_w \tag{12.2}$$

where

W_T = Total weight.
W_s = Solids weight.
W_w = Water weight.

The relationship between weight and volume can be expressed as

$$W_m = V_m G_{m-w} \tag{12.3}$$

where

W_m = Weight of the material (solid, liquid, or gas).
V_m = Volume of the material.
G_m = Specific gravity of the material (dimensionless).
w = Unit weight of water.

With the relationships described above, a few useful problems can be solved. When an environmental engineer determines that, within a given soil, the proportions of the three major components need to be mechanically adjusted, this can be accomplished by reorienting the mineral grains by compaction or tilling. The environmental engineer may want to blend soil types to alter the proportions, such as increasing or decreasing the percentage of void space.

How do we go about doing this? Relationships between volumes of soil and voids are described by the void ratio (e) and porosity (η). To accomplish this, we must first determine the void ratio (the ratio of the void volume to the volume of solids):

$$e = \frac{V_v}{V_s} \tag{12.4}$$

We must also determine the ratio of the volume of void spaces to the total volume. This can be accomplished by determining the porosity (η) of the soil, which is the ratio of void volume to total volume. Porosity is usually expressed as a percentage:

$$\eta = V_v/V_T \times 100\% \tag{12.5}$$

where

V_v = Void space volume.
V_T = Total volume.

Two additional relationships, moisture content (w) and degree of saturation (S), relate the water content of the soil and the volume of the water in the void space to the total void volume:

$$w = \frac{W_w}{W_s} \times 100\%$$

(12.6)

and

$$S = \frac{V_w}{V_v} \times 100\%$$

(12.7)

12.4.2 SPECIFIC GRAVITY

The specific gravity of a substance is the ratio of the unit weight of that substance to the unit weight of water at 20°C. This is expressed in equation form as

$$SG = \frac{W_{substance}}{W_{water}}$$

(12.8)

The specific gravity of a soil's solids is abbreviated G_S. This value is the ratio of the unit weight of the soil solids condensed into a solid mass to the unit weight of water. The equation for specific gravity for soil's solids is

$$G_s = \frac{W_s/V_s}{W_{water}}$$

(12.9)

where
G_S = Soil solids.
W_S = Weight of a soil mass after drying.
V_S = Solids volume.

A laboratory test may be performed to determine the specific gravity of the soil solids; however, in some field situations, such data may not be available and an estimate must necessarily be used. The specific gravity of a soil depends on the mineralogy of the soil grains. Most oils are a blend of several basic minerals such as quartz, feldspar, hornblende, biotite, calcite, and others. A determination of the constituents of a soil is helpful in estimating a value for the soil's specific gravity. Table 12.1 provides the specific gravity values for some of the more important soil minerals.

TABLE 12.1
Specific Gravity of Select Soil Minerals

Mineral	Specific Gravity	Mineral	Specific Gravity
Montmorillonite	2.65–2.8	Dolomite	2.87
Kaolinite	2.6	Hornblende	3.2–3.5
Illite	2.8	Magnetite	5.17
Chlorite	2.60–3.00	Quartz	2.66
Calcite	2.72	Biotite	3.0–3.1

DID YOU KNOW?

Poisson's effect (see Chapter 11) is also applicable in the realm of structural geology. Rocks, like most materials, are subject to Poisson's effect while under stress. On a geological time scale, excessive erosion or sedimentation of Earth's crust can either create or remove large vertical stresses upon the underlying rock. This rock will expand or contract in the vertical direction as a direct result of the applied stress, and it will also deform in the horizontal direction as a result of Poisson's effect. This change in strain in the horizontal direction can affect or form joints and dormant stresses in the rock (Engelder, 2013).

Many sands and gravels are composed primarily of quartz. A value of 2.66 is commonly assumed for the specific gravity for these soils. Specific gravities of sands and gravels derived from granites or limestones might be higher. Soils with a high percentage of silt-size particles generally have a specific gravity of about 2.68, because quartz is usually a major constituent, and small additional amounts of clay minerals slightly increase the value. Clay soils may have specific gravity values ranging from about 2.60 to 2.80. An average value of 2.7 is commonly assumed. Soils that contain a large amount of micaceous flakes and soils with significant amounts of hematite or magnetite may have quite high specific gravities, ranging from 2.75 to 3.3. Test data are usually required to accurately determine specific gravities for these unusual soils.

12.5   SOIL PARTICLE CHARACTERISTICS

The size and shape of particles in the soil, as well as density and other characteristics, relate to sheer strength, compressibility, and other aspects of soil behavior. Engineers use these index properties to form engineering classifications of soil. Simple classification tests are used to measure index properties (see Table 12.2) in the lab or the field. From Table 12.2, we see that an important division of soils (from the engineering point of view) is the separation of the cohesive (fine-grained) from the incohesive (coarse-grained) soils. Let's take a closer look at these two important terms.

TABLE 12.2
Index Property of Soils

Soil Type	Index Property
Cohesive (fine-grained)	Water content
	Sensitivity
	Type and amount of clay
	Consistency
	Atterberg limits
Incohesive (coarse-grained)	Relative density
	In-place density
	Particle-size distribution
	Clay content
	Shape of particles

Source: Adaptation from Kehew, A.E., *Geology for Engineers and Environmental Scientists*, 2nd ed., Prentice-Hall, Englewood Cliffs, NJ, 1995, p. 284.

Cohesion indicates the tendency of soil particles to stick together. Cohesive soils contain silt and clay. The clay and water content makes these soils cohesive through the attractive forces between individual clay and water particles. The influence of the clay particles makes the index properties of cohesive soils somewhat more complicated than the index properties of cohesionless soils. The resistance of a soil at various moisture contents to mechanical stresses or manipulations depends on the soil's *consistency*, the arrangement of clay particles, and is the most important characteristic of cohesive soils.

Another important index property of cohesive soils is *sensitivity*. Simply defined, sensitivity is the ratio of unconfined compressive strength in the undisturbed state to strength in the remolded state (see Equation 12.10). Soils with high sensitivity are highly unstable.

$$\text{Sensitivity} = \frac{\text{Strength in undisturbed condition}}{\text{Strength in remolded condition}} \tag{12.10}$$

Soil water content is an important factor that influences the behavior of the soil. The water content values of soil are known as the *Atterburg limits*, a collective designation of so-called limits of consistency of fine-grained soils which are determined with simple laboratory tests. They are usually presented as the *liquid limit* (LL), *plastic limit* (PL), and *shrinkage limit* (SL). The plastic limit is the water level at which soil begins to be malleable in a semisolid state, but molded pieces crumble easily when a little pressure is applied. When the volume of the soil becomes nearly constant (solid) with further decreases in water content, the soil has reached the shrinkage limit. The liquid limit is the water content at which the soil–water mixture changes from a liquid to a semifluid (or plastic) state and tends to flow when jolted. Obviously, an engineer charged with building a highway or building would not want to choose a soil for the foundation that tends to flow when wet. The difference between the liquid limit and the plasticity limit is the range of water content over which the soil is plastic and is called the *plasticity index*. Soils with the highest plasticity indices are unstable in bearing loads.

Several systems for classifying the stability of soil materials have been devised, but the best known (and probably the most useful) system is called the Unified System of Classification. This classification gives each soil type (14 classes) a two-letter designation, primarily based on particle-size distribution, liquid limit, and plasticity index.

Cohesionless coarse-grained soils behave much differently than cohesive soils and are based on (from index properties) the size and distribution of particles in the soil. Other index properties (particle shape, in-place density, and relative density, for example) are important in describing cohesionless soils, because they relate to how closely particles can be packed together.

12.6 SOIL STRESS AND STRAIN

If you are familiar with water pressure and its effect as you go deeper into the water (as when diving deep into a lake), it should not surprise you that the same concept applies to soil and pressure. Like water, pressure within the soil increases as the depth increases. A soil, for example, that has a unit weight of 75 lb/ft³ exerts a pressure of 75 psi at a 1-foot depth and 225 psi at 3 feet, etc. As you might expect, as the pressure on a soil unit increases, the soil particles reorient themselves structurally to support the cumulative load. This consideration is important, because the elasticity of the soil sample retrieved from beneath the load may not be truly representative once it is delivered to the surface. The importance of taking representative samples cannot be overstated. The response of a soil to pressure (stress) is similar to what occurs when a load is applied to a solid object; the stress is transmitted throughout the material. The load subjects the material to pressure, which equals the amount of load, divided by the surface area of the external face of the object over which it is applied. The response to this pressure or stress is called *displacement* or *strain*. *Stress* (like pressure), at any point within the object, can be defined as force per unit area.

12.7 SOIL COMPRESSIBILITY

When a vertical load such as a building or material stockpile is placed above a soil layer, some settlement can be expected. Settlement is the vertical subsidence of the building (or load) as the soil is compressed. Compressibility refers to the tendency of soil to decrease in volume under load. This compressibility is most significant in clay soils because of the inherent high porosity. Although the mechanics of compressibility and settlement are quite complex and beyond the scope of this text, the reader should know something about the actual evaluation process for these properties, which is accomplished in the consolidation test. This test subjects a soil sample to an increasing load. The change in thickness is measured after the application of each load increment.

12.8 SOIL COMPACTION

The goal of compaction is to reduce void ratio and thus increase the soil density, which, in turn, increases the shear strength. This is accomplished by working the soil to reorient the soil grains into a more compact state. If water content is within a limited range (sufficient enough to lubricate particle movement), efficient compaction can be obtained. The most effective compaction occurs when the soil placement layer (commonly called *lift*) is approximately 8 inches. At this depth, the most energy is transmitted throughout the lift. Note that more energy must be dispersed, and the effort required to accomplish maximum density is greatly increased when the lift is greater than 10 inches in thickness. For cohesive soils, compaction is best accomplished by blending or kneading the soil using sheepsfoot rollers and pneumatic tire rollers. These devices work to turn the soil into a denser state. To check the effectiveness of the compactive effort, the in-place dry density of the soil (weight of solids per unit volume) is tested by comparing the dry density of field-compacted soil to a standard prepared in an environmental laboratory. Such a test allows a percent compaction comparison to be made.

12.9 SOIL FAILURE

Construction, environmental, and design engineers must be concerned with soil structural implications involved with natural processes (such as frost heave, which could damage a septic system) and changes applied to soils during remediation efforts (e.g., when excavating to mitigate a hazardous materials spill in soil). Soil failure occurs whenever it cannot support a load. Failure of an overloaded foundation, collapse of the sides of an excavation, or slope failure on the sides of a dike, hill, or similar feature is termed *structural failure*. The type of soil structural failure that probably occurs more frequently than any other is slope failure (commonly known as a *cave-in*). A Bureau of Labor Statistics review of on-the-job mishaps found that cave-ins occur in construction excavations more frequently than you might think, considering the obvious dangers inherent in excavation.

What is an excavation? How deep does an excavation have to be to be considered dangerous? The answers to these questions could save your life or help you protect others when you become an engineer involved with safety. An excavation is any manmade cut, cavity, trench, or depression in the Earth's surface formed by earth removal. This can include excavations for anything from a remediation dig to sewer line installation. No excavation activity should be accomplished without keeping personnel safety in mind. Any time soil is excavated, care and caution are advised. As a rule of thumb (and as law under 29 CFR 1926.650–652), the Occupational Safety and Health Administration (OSHA) requires trench protection in any excavation 5 feet or more in depth. Before digging begins, proper precautions must be taken. The responsible party in charge (the competent person, according to OSHA) must

- Contact utility companies to ensure that underground installations are identified and located.
- Ensure that underground installations are protected, supported, or removed as necessary to safeguard workers.
- Remove or secure any surface obstacles (trees, rocks, and sidewalks, for example) that may create a hazard for workers.
- Classify the type of soil and rock deposits at the site as stable rock, type A, type B, or type C soil. One visual and at least one manual analysis must be performed to make the soil classification.

Let's take a closer look at the requirement to classify the type of soil to be excavated. Before an excavation can be accomplished, the soil type must be determined. The soil must be classified as stable rock, type A, type B, or type C soil. Remember, commonly you will find a combination of soil types at an excavation site. In this case, soil classification is used to determine the need for a protective system. Following is a description of the various soil classifications:

- *Stable rock* is a natural solid mineral material that can be excavated with vertical sides. Stable rock will remain intact while exposed, but keep in mind that, even though solid rock is generally stable, it may become very unstable when excavated (in practice you never work in this kind of rock).
- *Type A soil*, the most stable soil, includes clay, silty clay, sandy clay, clay loam, and sometimes silty clay loam and sandy clay loam.
- *Type B soil*, moderately stable, includes silt, silt loam, sandy loam, and sometimes silty clay loam and sand clay loam.
- *Type C soil*, the least stable, includes granular soils such as gravel, sand, loamy sand, submerged soil, soil from which water is freely seeping, and submerged rock that is not stable.

To test and classify soil for excavation, both visual and manual tests should be conducted. Visual soil testing looks at soil particle size and type. Of course, a mixture of soils will be visible. If the soil clumps when dug it could be clay or silt. Type B or C soil can sometimes be identified by the presence of cracks in walls and spalling (breaks up into chips or fragments). If you notice layered systems with adjacent hazardous areas—buildings, roads, and vibrating machinery—a professional engineer may be required for classification. Standing water or water seeping through trench walls automatically classifies the soil as type C.

Manual soil testing is required before a protective system (e.g., shoring or shoring box) is selected. A sample taken from soil dug out into a spoil pile should be tested as soon as possible to preserve its natural moisture. Soil can be tested either onsite or offsite. Manual soil tests include a sedimentation test, wet shaking test, thread test, and ribbon test.

A sedimentation test determines how much silt and clay are in sandy soil. Saturated sandy soil is placed in a straight-sided jar with about 5 inches of water. After the sample is thoroughly mixed (by shaking it) and allowed to settle, the percentage of sand is visible. A sample containing 80% sand, for example, will be classified as type C.

The wet shaking test is another way to determine the amount of sand vs. clay and silt in a soil sample. This test is accomplished by shaking a saturated sample by hand to gauge soil permeability based on the following facts: (1) shaken clay resists water movement through it, and (2) water flows freely through sand and less freely through silt.

The thread test is used to determine cohesion (remember, cohesion relates to stability—how well the grains hold together). After a representative soil sample is taken, it is rolled between the palms of the hands to about 1/8-inch diameter and several inches in length (any child who has played in

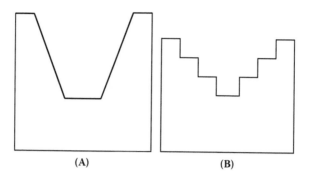

FIGURE 12.1   Sloped excavation.

dirt has accomplished this at one time or another—nobody said soil science has to be boring). The rolled piece is placed on a flat surface and then picked up. If a sample holds together for 2 inches, it is considered cohesive.

The ribbon test is used as a backup for the thread test. It also determines cohesion. A representative soil sample is rolled out (using the palms of your hands) to a 3/4-inch diameter and several inches in length. The sample is then squeezed between the thumb and forefinger into a flat unbroken ribbon 1/8 to 1/4 inch thick that is allowed to fall freely over the fingers. If the ribbon does not break off before several inches are squeezed out, the soil is considered cohesive.

Once soil has been properly classified, the correct protective system can be chosen. This choice is based on both soil classification and site restrictions. The two main types of protective systems are (1) *sloping* or *benching*, and (2) *shoring* or *shielding*. Sloping and benching are excavation protective measures that cut the walls of an excavation back at an angle to its floor. Examples of a sloped or angled cut and a bench system with one or more steps carved into the soil are shown in Figure 12.1. The angle used for sloping or benching is a ratio based on soil classification and site restrictions. In both systems, the flatter the angle, the greater the protection for workers. Reasonably safe side slopes for each of these soil types are presented in Table 12.3.

Shoring and shielding are two protective measures that add support to an existing excavation; they are generally used in excavations with vertical sides but can be used with sloped or benched soil. Shoring is a system designed to prevent cave-ins by supporting walls with vertical shores called *uprights* or *sheeting*. Wales are horizontal members along the sides of a shoring structure. Cross braces are supports placed horizontally between trench walls. Shielding is a system that employs a trench box or trench shield. These can be premanufactured or built onsite under the supervision of a licensed engineer. Shields are usually portable steel structures placed in the trench by heavy equipment. For deep excavations, trench boxes can be stacked and attached to each other with stacking lugs.

TABLE 12.3
Maximum Safe Side Slopes in Excavations

Soil Type	Side Slope (Vertical to Horizontal)	Side Slope (Degrees from Horizontal)
A	75:1	53°
B	1:1	45°
C	1.5:1	34°

Source: OSHA Excavation Standard 29 CFR 1926.650–652.

12.10 SOIL PHYSICS

Soil is a dynamic, heterogeneous body that is non-isotropic; that is, it does not have the same properties in all directions. As you might expect, because of these properties various physical processes are active in soil at all times. This important point was made clear by Winegardner (1996, p. 63): "All of the factors acting on a particular soil, in an established environment, at a specified time, are working from some state of imbalance to achieve a balance."

Most soil specialists have little difficulty in understanding why soils are very important to the existence of life on Earth. They know, for example, that soil is necessary (in a very direct sense) to sustain plant life and thus other life forms that depend on plants, and they know that soil functions to store and conduct water, serves a critical purpose in soil engineering involved with construction, and acts as a sink and purifying medium for waste disposal systems.

The environmental practitioner involved with soil management activities must be well versed in the physical properties of soil. Specifically, he or she must have an understanding of those physical processes that are active in soil. These factors include physical interactions related to soil water, soil grains, organic matter, soil gases, and soil temperature. To gain this knowledge, the environmental engineer must have training in basic geology, soil science, and engineering construction.

12.11 STRUCTURAL FAILURE

To this point, we have reviewed the basics involved with applied mechanics. This information is important to the environmental engineer because, without such knowledge, properly understanding the construction, function, and operation of workplace machines, equipment, and structures would be difficult. More importantly, a basic knowledge of applied mechanics also enables the environmental engineer to understand that systems fail, what causes them to fail, and (more significantly) how to prevent such failures. Many types of failures are possible, and failures occur for many reasons. Structural failures are important to the environmental engineer because when such failures occur they typically cause damage and injuries (or worse) to workers and others. Structural failures can be caused by any of the following: design errors, faulty material, physical damage, overloading, poor workmanship, and poor maintenance and inspection practices.

Design errors are not uncommon. They are usually the result of incorrect or poorly made assumptions. For example, the design engineer might assume some load or a maximum load for a design; however, the actual load may be much different in varying conditions. The 1981 Hyatt Regency Skywalk disaster (with 114 fatalities and over 200 injured) was caused by underestimating load, a badly designed and altered structural element design, and the failure to properly calculate and check the numbers (see Case Study 6.1).

Faulty materials cause structural failures for two main reasons: lack of uniformity of material and changes in properties (material strength, ductility, brittleness, and toughness) over time. Physical damage caused by usage, abuse, or unplanned events (natural or human-generated) are another cause of structural failure.

Overloading and inadequate support are common causes of structural failure. A particular structure might have originally been designed to house an office complex and then later reconfigured for use as a machine shop. The added weight of machines and ancillaries is a change in the use environment that may overload the original structure and eventually lead to failure because of inadequate structural support.

Poor and faulty workmanship is a factor that must always be considered when studying structural failures. Improper assembly and maintenance of devices, machines, and structures have certainly caused their share of failures and collapses. In fact, one of the major reasons many companies have employed quality control procedures in their manufacturing and construction activities is to guard against poor and faulty workmanship.

Poor maintenance, use, and inspection also play an important role in structural failures. Obviously, exposures to various conditions during their use changes structures. Improper maintenance can affect a structure's useful life. Improper use can have the same affect. Inspection is important to ensure that maintenance and use are providing effective care and to guard against unexpected failure.

Case Study 12.1. The Hyatt Regency Hotel Disaster[*]

On July 17, 1981, at 7:05 pm, the Kansas City, Missouri, Hyatt Regency Hotel Atrium held over 1600 people. At that moment, two suspended structurally connected skywalks connecting two towers on the second and fourth floor levels failed and fell, crushing the heavily occupied restaurant bar beneath them.

The death toll for this structural failure (the worst in U.S. history) was 114. Over 200 more people were injured, many permanently disabled. Plaintiff's claims amounted to over $3 billion. The hotel owner, Donald Hall, with an admirable sense of duty and decency, settled more than 90% of the claims.

Immediate theories on what caused this structural failure ranged from continued resonance to faulty materials to poor workmanship. A National Bureau of Standards investigation finally discovered the most probable cause—and their findings laid the blame squarely at the feet of the structural engineers. The cause? A design change submitted by the contractor and approved by the design engineers and architect. In the original designs, both skywalks were supported by nuts at both the second- and fourth-floor levels that were threaded onto single continuous hanger rods and spaced at regular intervals. Because of the single-rod design, while the roof trusses held the load of both, the welded box beams that supported the load of each skywalk were independent of the other.

The change? The contractor's design change shortened the hanger rods, added an extra hole to the fourth floor beams, and hung the second floor walkway by an independent second set of hanger rods from those box beam connections, thus putting the entire load of both walkways on the fourth floor walkway. This was not the only factor involved, however. In determining the dead load of the structure, the investigators discovered an 8% higher load than originally computed, because of changes and additions to decking and flooring materials. The live load, however, was well within the limits. The obviously weak element in the design was the fourth-floor box beams, so the investigators tested both new duplicates and undamaged beams and hanger rods from the Hyatt Regency Atrium.

The National Bureau of Standards report stated that the skywalks were underdesigned and that the design lacked redundancy. Six important points summarize the report:

- The collapse occurred under loads substantially less than those specified by the Kansas City Building Code.
- All the fourth-floor box beam–hanger connections were candidates for the initiation of walkway collapse.
- The box beam–hanger rod connections, the fourth-floor to ceiling hanger rods, and the third-floor-walkway hanger rods did not satisfy the design provisions of the Kansas City Building Code.
- The box beam–hanger to rod connections under the original hanger rod detail (continuous rod) would not have satisfied the Kansas City Building Code.
- Neither the quality of the workmanship nor the materials used in the walkway system played a significant role in initiating the collapse.

---

[*] Based on Levy, M. and Salvador, M., *Why Buildings Fall Down: How Structures Fail.* W.W. Norton, New York, 2002.

The Bureau's findings also made clear that, although the original design would not have met the Kansas City Building Code either, the original design might not have failed under the minor load present that day. In the inevitable court case that followed, the licenses of the principal and the project manager of the firm responsible for the design were revoked. The attorney for the Missouri State licensing board said, "It wasn't a matter of doing something wrong, they just never did it at all. Nobody ever did any calculations to figure out whether or not the particular connection that held the skywalks up would work. It got built without anybody ever figuring out if it would be strong enough."

REFERENCES AND RECOMMENDED READING

Andrews, Jr., J.S. (1992). The cleanup of Kuwait, in Kostecki, P.T. and Calabrese, E.J., Eds., *Hydrocarbon Contaminated Soils and Groundwater*, Vol. II. Lewis Publishers, Chelsea, MI.

API. (1980). *Landfarming: An Effective and Safe Way to Treat/Dispose of Oily Refinery Wastes*. Solid Waste Management Committee, American Petroleum Institute, Washington, DC.

Blackman, Jr., W.C. (1993). *Basic Hazardous Waste Management*. Lewis Publishers, Chelsea, MI.

Bossert, I. and Bartha, R. (1984). The fate of petroleum in soil ecosystems, in Atlas, R.M., Ed., *Petroleum Microbiology*. Macmillan, New York.

Brady, N.C. and Weil, R.R. (1996). *The Nature and Properties of Soils*, 11th ed. Prentice-Hall, New York.

Buol, S.W., Hole, F.E., and McCracken, R.J. (1980). *Soil Genesis and Classification*. Iowa State University, Ames.

Ehrhardt, R.F., Stapleton, P.J., Fry, R.L., and Stocker, D.J. (1986). *How Clean Is Clean? Cleanup Standards for Groundwater and Soil*. Edison Electric Institute, Washington, DC.

Engelder, T. (2013). *Poisson Effect*. Department of Geosciences, Penn State University, University Park, PA.

EPRI–EEI. (1988). *Remedial Technologies for Leaking Underground Storage Tanks*. Lewis Publishers, Chelsea, MI.

Franck, I. and Brownstone, D. (1992). *The Green Encyclopedia*. Prentice-Hall, New York.

Grady, Jr., C.P. (1985). *Biodegradation: Its Measurement and Microbiological Basis, Biotechnology and Bioengineering*, Vol. 27. John Wiley & Sons, New York.

Life Systems, Inc. (1985). *Toxicology Handbook*. U.S. Environmental Protection Agency, Washington, DC.

Jury, W.A. (1986). *Guidebook for Field Testing Soil Fate and Transport Models*, Final Report. U.S. Environmental Protection Agency, Washington, DC.

Kehew, A.E. (1995). *Geology for Engineers and Environmental Scientists*, 2nd ed. Prentice-Hall, Englewood Cliffs, NJ.

Konigsburg, E.L. (1996). *The View From Saturday*. Scholastic Books, New York.

Levy, M. and Salvador, M. (2002). *Why Buildings Fall Down: How Structures Fail*. W.W. Norton, New York.

MacDonald, J.A. (1997). Hard times for innovation cleanup technology. *Environmental Science & Technology*, 31(12), 560–563.

Mansdorf, S.Z. (1993). *Complete Manual of Industrial Safety*. Prentice-Hall, Englewood Cliffs, NJ.

Mehta, P.K. (1983). Pozzolanic and cementitious by-products as minor admixtures for concrete—a critical review, in Malhotra, V.M., Ed., *Fly Ash, Silica Fume, Slag, and Other Mineral By-Products in Concrete*. Vol. 1. American Concrete Institute, Farmington Hills, MI.

National Research Council. (1997). *Innovations in Groundwater and Soil Cleanup: From Concept to Commercialization*. National Academies Press, Washington, DC.

Testa, S.M. (1997). *The Reuse and Recycling of Contaminated Soil*. Lewis Publishers, Boca Raton, FL.

Tomera, A.N. (1989). *Understanding Basic Ecological Concepts*, J. Weston Walch, Publisher, Portland, ME.

Tucker, R.K. (1989). Problems dealing with petroleum contaminated soils: a New Jersey perspective, in Kostecki, P.T. and Calabrese, E.J., Eds., *Petroleum Contaminated Soils*, Vol. I. Lewis Publishers, Chelsea, MI.

USEPA. (1984). *Review of In-Place Treatment Techniques for Contaminated Surface Soils*. Vol. 1. *Technical Evaluation*, EPA/540/2-84-003. U.S. Environmental Protection Agency, Washington, DC.

USEPA. (1985). *Remedial Action at Waste Disposal Sites (Revised)*. U.S. Environmental Protection Agency, Washington, DC.

Wilson, J.T., Leach, L.E., Benson, M., and Jones, J.N. (1986). *In situ* biorestoration as a ground water remediation technique. *Ground Water Monitoring Review*, 6(4), 56–64.

Winegardner, D. C. (1996). *An Introduction to Soils for Environmental Professionals.* Lewis Publishers, Boca Raton, FL.

Woodward, H.P. (1936). Natural Bridge and Natural Tunnel, Virginia. *Journal of Geology,* 44(5), 604–616.

World Resources Institute. (1992). *World Resources 1992–93,* Oxford University Press, New York.

Section VI

Biomass Basic Computations

God has cared for these trees, saved them from drought, disease, avalanches, and a thousand tempests and floods. But he cannot save them from fools—only Uncle Sam can do that.

—John Muir (1897)

13 Forest-Based Biomass Basic Computations

> Wood is still the largest biomass energy resource today.
>
> —**NREL (2010)**

13.1 INTRODUCTION*

Currently, forest and grassland in the United States totals 747 million acres. Approximately 554 million acres are owned and managed by private land owners. The remaining forest acreage, 193 million acres, which is an area equivalent to the size of Texas, is managed by the U.S. Forest Service. Established in 1905, the U.S. Forest Service (USFS) is an agency of the U.S. Department of Agriculture. Whether they are privately or publicly owned, managing, maintaining, and sustaining U.S. forests and grasslands occupy a large and growing segment of work performed by professional environmental practitioners.

Most people understand that forest management activities include providing wood (obviously, the most famous product from the forest) and lesser known but still important non-timber produce, such as medicinal plants, honey, fruits, and bushment, but many are not really aware of less visible but still important reasons for maintaining forests:

- Forests deliver all kind of ecosystem services. Forests play an important role in the global and local water cycle. Forests attract rain water, purify water, and regulate water flows. In some areas, relief trees protect against erosion.
- Forests influence local climate. Depending on latitude, forests influence the temperature in a region. On a global level, forests stabilize climate by regulating energy and water cycles.
- Forests have cultural, religious, and spiritual significance; for example, sacred forests are often left untouched or are protected.
- Forests are extremely important in maintaining biodiversity.
- Forests provide feedstock (biomass) for renewable energy production.

The U.S. Forest Service's mission is to sustain the health, diversity, and productivity of the nation's forests and grasslands to meet the needs of present and future generations, including monitoring, protecting, and ensuring biomass feedstock supply for bioenergy production. The Forest Service professional is not a Paul Bunyan wielding a gigantic ax nor a steel-helmeted lumberjack chain-sawing immense portions of northern forest lands. Quite to the contrary, environmental professionals involved in forestry work, whether public or private, are highly educated practitioners who typically serve as park rangers, historians, forestry aides and technicians, forest firefighters, fishing guides, surveyors, researchers, and hydrologists, among many other potential functions. Moreover, because of current cultural and socioeconomic conditions, Forest Service professionals are also called upon to protect forest users, forest resources, USFS employees, and public property from criminals.

---

* Material presented in this chapter is adapted from Spellman, F.R., *Forest-Based Biomass Energy: Concepts and Applications*, CRC Press, Boca Raton, FL, 2012.

13.1.1 Forestry Service and Mathematics: The Interface

As mentioned throughout this text, environmental practitioners must be well versed in mathematical operations. There is no clearer or more fundamental need for mathematical skills than that of the professional forester who must perform all the duties listed to this point plus surveying and sampling forest biomass, especially with regard to wood-to-energy processes. Why do we sample forest biomass instead of just counting the trees, adding them up, and computing the total woody biomass contained? Freese (1976) observed that partial knowledge is a normal state of doing business in many professions; the same can be said for the practice of forestry. The complete census is rare, and the sample is commonplace. A forester must advertise timber sales with estimated volume, estimated grade yield and value, estimated cost, and estimated risk. The nurseryperson sows seed whose germination is estimated from a tiny fraction of the seedlot, and at harvest he or she estimates the seedling crop with sample counts in the nursery beds. Enterprising pulp companies, seeking a source of raw material in sawmill residue, may estimate the potential tonnage of chippable material by multiplying reported production by a set of conversion factors obtained at a few representative sawmills.

On the surface, and in many cases, it would seem better to measure and not to sample; however, there are several good reasons why sampling is often preferred. In the first place, complete measurement or enumeration may be impossible. That is, not all units in the population can be identified. For example, how does one accurately count each branch or twig on a tree? How do we test the quality of every drop of water in a reservoir? How do we weigh every fish in a stream or count all seedlings in a 1000-bed nursery, enumerate all the egg masses in a turpentine beetle infestation, or measure the diameter and height all merchantable trees in a 20,000-acre forest? Moreover, the nurseryperson might be somewhat better informed if he or she knew the germinative capacity of all the seed to be sown, but the destructive nature of the germination test precludes testing every seed. For identical reasons, it is impossible to conduct tests that are destructive on every chainsaw without destroying every chainsaw. Likewise, it is impossible to measure the bending strength of all the timbers to be used in a bridge, the tearing strength of all the paper to be put into a book, or the grade of all the boards to be produced for a timber sale. If the tests were permitted, no seedlings would be produced, no bridges would be built, no books printed, and no stumpage sold. Clearly, where testing is destructive, some sort of sampling is inescapable. Obviously, the enormity of the counting task or the destructive effects of testing demand some sort of sampling procedure.

Sampling will frequently provide the essential information at a far lower cost and in less time than a complete enumeration. Surveying 100% of the lumber market is not going to provide information that is very useful to a seller if it takes 11 months to complete the job. In addition, it is often the case that sampling information may at times be more reliable than that obtained by a 100% inventory. There are several reasons why this might be true. With fewer observations to be made, measurement of the units in the sample can be and is more likely to be made with greater care. Moreover, a portion of the savings resulting from sampling could be used to buy better instruments and to employ or train higher caliber personnel. It is not difficult to see that good measurements on 5% of the units in a population could provide more reliable information than sloppy measurements on 100% of the units.

The bottom line to making sampling effective and accurate is obtaining reliable data from the population sampled and making correct inferences about that population. The quality of the sampling depends on such factors as the rule by which the sample was drawn, the care exercised in measurement, and the degree to which bias was avoided (Avery and Burkhart, 2002). The triple bottom line in sampling forest biomass comes down to how the data are drawn and measured and made bias free, thus leading us to the final step of performing mathematical operations to produce representative results that can be analyzed to obtain the ultimate bottom line: the object of our work.

13.1.1.1 Necessary Terms and Concepts

Inventory—The systematic acquisition and analysis of information necessary to describe, characterize, or quantify vegetation. As might be expected, data for many different vegetation attributes can be collected. Inventories can be used not only for mapping and describing ecological sites but also for determining ecological status, assessing the distribution and abundance of species, and establishing baseline data for monitoring studies.

Population—A population (used here in the structural, not biological, sense) is a complete collection of objects (usually called *units*) about which one wishes to make statistical inferences. Population units can be individual plants, points, plots, quadrats, or transects.

Sample—A set of units selected from a population used to estimate something about the population (a procedure that statisticians refer to as *making inferences* about the population). In order to properly make inferences about a population, the units must be selected using some random procedure. The units selected are called *sampling units*.

Sampling—A means by which inferences about a plant community can be made based on information from an examination of a small proportion of that community. The most complete way to determine the characteristics of a population is to conduct a complete enumeration (or census). In a census, each individual unit in the population is sampled to provide the data for the aggregate. This process is both time consuming and costly, and it may also result in inaccurate values when individual sampling units are difficult to identify; therefore, the best way to collect vegetation data is to sample a small subset of the population. If the population is uniform, sampling can be conducted anywhere in the population, but most vegetation populations are not uniform. It is important that data be collected that can ensure that the sample represents the entire population. Sample design is an important consideration in collected representative data.

Sampling unit—One of a set of objects in a sample that is drawn to make inferences about a population of those same objects. A collection of sampling units is a sample. Sampling units can be individual plants, points, plot, quadrats, or transects.

Shrub characterization—This topic is addressed here because it is not covered in most of the techniques in this text. Shrub characterization is the collection of data on the shrub and tree component of a vegetation community. Attributes that could be important for shrub characterization are height, volume, foliage density, crown diameter, form class, age class, and total number of plants by species (density). Another important feature of shrub characterization is the collection of data on a vertical as well as horizontal plane. Canopy layering is almost as important. The occurrence of individual species and the extent of canopy cover of each species are recorded in layers. The number of layers chosen should represent the herbaceous layer, the shrub layer, and the tree layers, though additional layers can be added if needed.

Trend—Refers to the direction of change. Vegetation data are collected at different points in time on the same site, and the results are then compared to detect a change. Trend can be described as *moving toward meeting objectives*, *moving away from meeting objectives*, *not apparent*, or *state*. Trend data are important in determining the effectiveness of on-the-ground management actions. Trend data indicate whether rangeland is moving toward or away from specific objectives. The trend of a rangeland area may be judged by noting changes in vegetation attributes such as species composition, density, cover, production, and frequency. Trend data, along with actual use, authorized use, estimated use, utilization, climate, and other relevant data, are considered when evaluating activity plans.

13.2 FOREST BIOMASS COMPUTATIONS AND STATISTICS

Earlier, in Chapter 7, we provided a basic, fundamental review of statistics and a variety of example statistical operations, all for a good purpose. Simply put, you cannot expect to work even at the fringes or margin of any area of interest and specialization in the environmental profession without a strong foundation in statistical methods. This is certainly the case in activities related to forest and grassland management. Although we do not restate the statistical methods presented earlier, following is a list of some of the important statistical methods routinely used by environmental professionals in forest and grassland management:

- Simple random sampling
- Sample selection
- Standard errors
- Sampling and replacement
- Confidence limits for large and small samples
- Size of sample
- Variance
- Stratified random sampling
- Estimates
- Proportional and optimum allocation
- Regression estimation
- Family of regression estimators
- Ratio estimation
- Mean-of-ratios estimator
- Double sampling
- Pace frequency
- Point intercept
- Double-weight sampling

13.3 SAMPLE FOREST-BASED BIOMASS COMPUTATIONS

Wise utilization of the forest resource relates to awareness of its value (Ince, 1979). The amount of heat energy that can be recovered from wood or bark determines its fuel value. The amount of recoverable heat energy varies with moisture content and chemical composition. Recoverable heat energy varies among tree species and even within a species. In this chapter, we provide a summary of information that may be used to estimate recoverable heat energy in wood or bark fuel, biomass weight considerations, biomass weight examples, and methods for estimating stand-level biomass.

13.3.1 LOWER AND HIGHER HEATING VALUES

The lower heating value (LHV) of a fuel is defined as the amount of heat released by combusting a specified quantity (initially at 25°C) and returning the temperature of the combustion products to 150°C; it is assumed that the latent heat of vaporization of water in the reaction produces is not recovered. The LHV is the useful calorific value in boiler combustion plants and is frequently used in Europe. The *higher heating value* (HHV), the gross calorific value or gross energy, of a fuel is defined as the amount of heat released by a specified quantity (initially at 25°C) once it is combusted and the products have returned to a temperature of 25°C, which takes into account the latent heat of vaporization of water in the combustion products. The HHV is derived only under laboratory conditions and is frequently used in the United States for solid fuels. Heating values in units of MJ/kg are converted from heating values in units of Btu/lb. For solid fuels, the heating values in units of BTU/lb are converted from the heating values in units of Btu/ton. The lower and higher heating

TABLE 13.1
Lower and Higher Heating Values of Solid Fuels

Fuel	Lower Heating Value (LHV)			Higher Heating Value (HHV)		
	Btu/ton	Btu/lb	MJ/kg	Btu/ton	Btu/lb	MJ/kg
Farmed trees	16,811,000	8406	19,551	17,703,170	8852	20,589
Forest residue	13,243,490	6622	15,402	14,164,160	7082	16,473

Source: Transportation Fuel Cycle Analysis Model, GREET 1.8b, Argonne National Laboratory, Argonne, IL, 2008.

values of solid forest-based biomass fuels are listed in Table 13.1, and Table 13.2 illustrates the variation in reported heat content values (on a dry weight basis) in the U.S. and European literature based on values in the ECN Phyllis database (http://www.ecn.nl/phyllis/), the U.S. Department of Energy Biomass Feedstock Composition and Property database (http://www1.eere.energy.gov/biomass/feedstock_databases.html), and selected literature sources.

13.3.2 EFFECT OF FUEL MOISTURE ON WOOD HEAT CONTENT

Because recently harvested wood fuels usually contain 30 to 55% water, it is useful to understand the effect of moisture content on the heating value of wood fuels. Table 13.3 shows the effect of percent moisture content (MC) on the higher heating value as-fired (HHV-AF) of a wood sample starting at 8500 Btu/lb (oven-dry). Fuel moisture content is usually reported as the wet weight basis moisture content. Moisture content expressed on a wet weight basis (also called "green" or "as-fired" moisture content) is the decimal fraction of the fuel that consists of water; for example, a pound of wet wood fuel at 50% moisture content contains 0.50 pound of water and 0.50 pound of wood. Note that the

TABLE 13.2
Heat Content Ranges for Forest Biomass Fuels (Dry Weight Basis)

Fuel	English		Metric			
	Higher Heating Value		Higher Heating Value		Lower Heating Value	
	Btu/lb	MBtu/ton	kJ/kg	MJ/kg	kJ/kg	MJ/kg
Woody crops						
Black locust	8409–8582	16.8–17.2	19,547–19,948	19.5–19.9	18,464	18.5
Eucalyptus	8174–8432	16.3–16.9	19,000–19,599	19.0–19.6	17,963	18.0
Hybrid poplar	8183–8491	16.4–17.0	19,022–19,737	19.0–19.7	17,700	17.7
Willow	7983–8497	16.0–17.0	18,556–19,750	18.6–19.7	16,734–18,419	16.7–18.4
Forest residues						
Hardwood wood	8017–8920	16.0–17.5	18,635–20,734	18.6–20.7	—	—
Softwood wood	8000–9120	16.0–18.24	18,595–21,119	18.6–21.1	17,514–20,768	17.5–20.8

Sources: http://www1.eere.energy.gov/biomass/feedstock_databases.html; Bushnell, D., *Biomass Fuel Characterization: Testing and Evaluating the Combustion Characteristics of Selected Biomass Fuels,* Bonneville Power Administration, Portland, OR, 1989; Jenkins, B., *Properties of Biomass, Appendix to Biomass Energy Fundamentals,* EPRI Report TR-102107, Electric Power Research Institute, Palo Alto, CA, 1993; Jenkins, B.L. et al., *Fuel Processing Technology,* 54, 17–46, 1998; Tillman, D., *Wood as an Energy Resource,* Academic Press, New York, 1978.

TABLE 13.3

Effect of Fuel Moisture on Wood Heat Content

	Moisture Content (MC) Wet Basis (%)										
	0	**15**	**20**	**25**	**30**	**35**	**40**	**45**	**50**	**55**	**60**
Higher heating value as fired (HHV-AF) (Btu/lb)	8500	7275	6800	6375	5950	5525	5100	4575	4250	3825	3400

Sources: Borman, G.L. and Ragland, K.W., *Combustion Engineering*, McGraw-Hill, New York, 1998; Kluender, R.A., *The Forester's Wood Energy Handbook*, Publ. No. 80-A-12, American Pulpwood Association, Washington, D.C., 1980; Maker, T.M., *Wood-Chip Heating Systems: A Guide for Institutional and Commercial Biomass Installations*, Biomass Energy Resource Center, Montpelier, VT, 2004.

wet weight basis differs from the total dry weight basis method of expressing moisture content which is more commonly used for describing moisture content of finished wood products. The dry weight basis is the ratio of the weight of water in wood to the oven-dry weight of the wood. The formulas used require that moisture content be expressed on the wet weight basis (Ince, 1979).

13.3.2.1 Moisture Contents (MC) Wet and Dry Weight Basis Calculations

Moisture content (MC) on a wet or dry weight basis is calculated as follows:

$$\text{MC (dry basis)} = 100 \times \frac{(\text{Wet weight} - \text{dry weight})}{\text{Dry weight}} \tag{13.1}$$

$$\text{MC (wet basis)} = 100 \times \frac{(\text{Wet weight} - \text{dry weight})}{\text{Wet weight}} \tag{13.2}$$

To convert MC (wet basis) to MC (dry basis):

$$\text{MC (dry basis)} = \frac{100 \times \text{MC (wet basis)}}{100 - \text{MC (wet basis)}} \tag{13.3}$$

To convert MC (dry basis) to MC (wet basis):

$$\text{MC (wet basis)} = \frac{100 \times \text{MC (dry basis)}}{100 + \text{MC (dry basis)}} \tag{13.4}$$

Some sources report the heat contents of fuels "as-delivered" rather than at 0% moisture for practical reasons. Because most wood fuels have bone-dry (oven-dry) heat contents in the range of 7600 to 9600 Btu/lb (15,200,000 to 19,200,000 Btu/ton or 18 to 22 GJ/Mg), lower values will always mean that some moisture is included in the delivered fuel.

13.3.3 FORESTRY VOLUME UNIT TO BIOMASS WEIGHT CONSIDERATIONS

Biomass is frequently estimated from forestry inventory merchantable-volume data, particularly for purposes of comparing regional and national estimates of aboveground biomass and carbon levels. Making such estimations can be done several ways, but it always involves the use of conversion factors or biomass expansion factors (or both combined). Figure 13.1 defines what is included in each

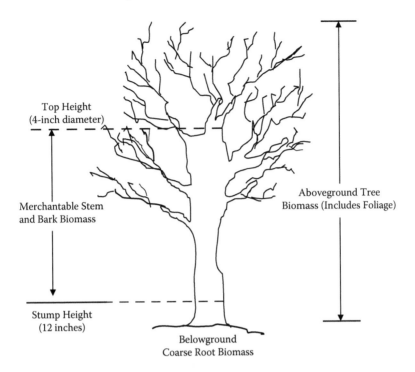

Top Height
(4-inch diameter)

Merchantable Stem
and Bark Biomass

Aboveground Tree
Biomass (Includes Foliage)

Stump Height
(12 inches)

Belowground
Coarse Root Biomass

FIGURE 13.1 U.S. forest inventory data. (Adapted from Jenkins, J.C. et al., *A Comprehensive Database of Diameter-Based Biomass Regressions for North American Tree Species*, Gen. Tech. Rep. NE-319, U.S. Department of Agriculture, Forest Service, Northeastern Research Station, Newtown Square, PA, 2004.)

category of volume or biomass units, and Figure 13.2 illustrates indirect methods of large-scale biomass estimation. Total volume or biomass includes stem, bark, stump, branches, and foliage, especially if evergreen trees are being measured. When estimating biomass available for bioenergy, the foliage is not included, and the stump may or may not be appropriate to include depending on whether harvest occurs at ground lever or higher. Both conversion and expansion factors can be used together to translate directly between merchantable volumes per unit area and total biomass per unit area, as demonstrated by the simple volume to weight conversion process shown below.

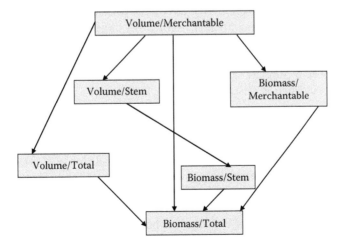

Volume/Merchantable

Volume/Stem

Biomass/
Merchantable

Volume/Total

Biomass/Stem

Biomass/Total

FIGURE 13.2 Indirect methods of large-scale biomass estimation. (Adapted from Somogyi, Z. et al., *European Journal of Forest Research*, 126(2), 197–207, 2007.)

13.3.3.1 Estimation of Biomass Weights from Forestry Volume Data

Equation 13.5 is used to estimate merchantable biomass from merchantable volume assuming that the specific gravity and moisture content are known and the specific gravity basis corresponds to the moisture content of the volume involved (Briggs, 1994). Specific gravity (SG) is a critical element of the volume to biomass estimation equation. The SG content should correspond to the moisture content of the volume involved. SG varies considerably from species to species, differs for wood and bark, and is closely related to the moisture content as explained in graphs and tables in Briggs (1994). The wood specific gravity of species can be found in several references, although the moisture content basis is not generally given. Briggs (1994) suggested using a moisture content of 12% as the standard upon which many wood properties measurements should be based.

$$\text{Weight} = \text{Volume} \times \text{Specific Gravity} \times \text{Density of } H_2O \times (1 + MC_{od}/100) \quad (13.5)$$

where the volume is expressed in cubic feet or cubic meters, the density of H_2O is 62.4 lb/ft^3 or 1000 kg/m^3, and MC_{od} is the oven-dry moisture content.

■ EXAMPLE 13.1

Problem: What is the weight of fiber in a 44-ft^3 oven-dry log with a specific gravity of 0.40?

Solution:

$$\text{Weight} = 44 \text{ ft}^3 \times 0.40 \times 62.4 \text{ lb ft}^3 \times (1 + 0/100) = 1098 \text{ lb, or } 9.549 \text{ dry ton}$$

13.3.3.2 Biomass Expansion Factors

Schroeder et al. (1997) described methods for estimating total aboveground dry biomass per unit area from growing stock volume data in the U.S. Forest Service Forest Inventory and Analysis (FIA) database. The growing stock volume data are limited to trees with diameters greater than or equal to 12.7 cm. It is highly recommended that the paper be studied for details of how the biomass expansion factors (BEFs) for oak–hickory and beech–birch were developed.

13.3.4 STAND-LEVEL BIOMASS ESTIMATION

At the individual field or stand level, biomass estimation is relatively straightforward, especially if it is being done for plantation-grown trees that are relatively uniform in size and other characteristics. The procedure involves first developing a biomass equation that predicts individual tree biomass as a function of diameter at breast height (DBH) or of DBH plus height. Second, the equation parameters (DBH and height) must be measured on a sufficiently large sample size to minimize variation around the mean values. Finally, the mean individual tree weight results are scaled to the area of interest based on percent survival or density information (trees per acre or hectare). Regression estimates are developed by directly sampling and weighing enough trees to cover the range of sizes being included in the estimation. They often take the form of

$$\ln Y = - \text{Factor } 1 + \text{Factor } 2 \times \ln X \quad (13.6)$$

where Y is weight in kilograms, and X is DBH or DBH^2 + height/100.

Regression equations can be found for many species in a wide range of literature. Examples for trees common to the Pacific Northwest are provided in Briggs (1994). The equations will differ depending on whether foliage or live branches are included, so care must be taken in interpreting the biomass data. For plantation trees grown on cropland or marginal cropland, it is usually assumed that tops and branches are included in the equations but foliage is not. For trees harvested from forests on lower quality land, it is usually recommended that tops and branches should not be removed (Pennsylvania DCNR, 2007) to maintain nutrient status and reduce erosion potential, thus biomass equations should assume regressions based on the stem weight only.

13.3.5 Biomass Equations*

The intensity of forest utilization has increased in recent years because of whole-tree harvesting and the use of wood for energy. Actually, estimating tree biomass (weight) based on parameters that are easily measured in the field is becoming a fundamental task in forestry and forest-based biomass technology. Traditionally, cubic-foot or board-foot volume of merchantable products, such as sawlogs or pulpwood, adequately described forest stands; however, the intensity of forest utilization has increased in recent years because of whole-tree harvesting and the use of wood for energy. All aboveground branches, leaves, bark, small trees, and trees of poor form or vigor are now commonly included in the harvested product and are listed as biomass of whole trees (WT) or individual components. With this increasing emphasis on complete tree utilization and use of wood as a source of energy, tables and equations have been developed to show the whole tree biomass as weights of total trees and their components.

Numerous equations for estimating tree biomass from dry weight in kilograms and DBH (tree diameter at breast height, in centimeters) have been developed by researchers based on local and tree species. For example, Landis and Mogren (1975) developed an equation for estimating the biomass of individual Engelmann spruce trees employing the following model:

$$Y = b_0 + b_1 \times \mathrm{DBH}^2 \tag{13.7}$$

where Y is tree component dry weight in kilograms, b_0 and b_1 are regression coefficients, and DBH is the tree diameter at breast height in centimeters.

Similar sets of equations have been developed for other species and locations. Regression equations are used to estimate tree biomass in both forestry and ecosystem studies. Examples of many common equations used in the northeastern United States are presented below. These equations are typically developed in the following way: Samples of major tree species are chosen for study, selected dimensions of each tree are recorded, the tree is felled and weighed either whole or in pieces, and subsamples are oven-dried and weighed again to determine tree moisture content. (Tree green weights are converted to dry weights by using moisture content values.) Because biomass is related to tree dimensions, regression analysis is used to estimate the constants or regression coefficients required for the actual calculation of biomass. The resultant regression equations may be used to estimate the biomass, by species, of all trees for which dimensional data are available. The equations that follow are of several different forms; they can be used to predict biomass from DBH or DBH and height. The most common forms used are allometric, exponential, and quadratic.

13.3.5.1 Key to Abbreviations Used in Example Biomass Equations

Br	Branch biomass
DBH	Diameter at breast height (1.37 m) measured in inches (in), millimeters (mm), or centimeters (cm)
DdBr	Dead branch biomass
ht	Tree height
Lf	Leaf biomass
Lf + Tw	Leaf and twig biomass
ln	Natural logarithm to the base e
log	Logarithm to the base 10
Rt	Root and stump biomass
St	Stem biomass
St + Br	Steam and branch biomass but not foliage

---

* Based on material contained in Tritton, L.M. and Hornbeck, J.W., *Biomass Equation for Major Tree Species of the Northeast*, U.S. Department of Agriculture, Washington, D.C., 1982.

Tw   Twig biomass
weight   Weight measured in pounds (lb), grams (gm), or kilograms (kg)
WT   Whole tree biomass (all above-ground components, including leaves, branches, stem)

13.3.5.2 Tree Species/Biomass Example Equations

- Balsam fir (*Abies balsams*) (Young et al., 1980)
  WT: ln (weight) = 0.5958 + 2.4017 × ln (DBH)
- Red maple (*Acer rubrum*) (Young et al., 1980)
  WT: ln (weight) = 0.9392 + 2.3804 × ln (DBH)
- Sugar maple (*Acer saccharum*) (Whittaker et al., 1974)
  St: log (weight) = 2.0877 + 2.3718 × log (DBH)
  Br: log (weight) = 0.6266 + 2.9740 × log (DBH)
  DdBr: log (weight) = 0.0444 + 2.2803 × log (DBH)
  Lf + Tw: log (weight) = 1.0975 + 1.9329 × log (DBH)
- Yellow birch (*Betula alleghaniensis* Britt.) (Ribe, 1973)
  Lf: log (weight) = 1.9962 + 1.9683 × log (DBH)
  Br: log (weight) = 2.5345 + 1.6179 × log (DBH)
  St: log (weight) = 2.9670 + 2.5330 × log (DBH)
- Black birch (*Betula lenta*) (Brenneman et al., 1978)
  WT: weight = 1.6542 × DBH$^{2.6606}$
- Paper birch (*Betula papyrifera* Marsh) (Kinerson and Bartholomew, 1977)
  St: ln (weight) = 3.720 + 2.877 × ln (DBH)
  Br: ln (weight) = −1.351 + 4.368 × ln (DBH)
- Gray birch (*Betula populifolia* Marsh) (Young et al., 1980)
  Wt: ln (weight) = 1.0931 + 2.3146 × ln (DBH)
- Hickory (*Carya* spp.) (Wiant et al., 1977)
  St + Br: weight = 1.93378 × DBH$^{2.62090}$
- Beech (*Fagus grandifolia* Ehrh.) (Ribe, 1973)
  Lf: log (weight) = 2.0660 + 1.8089 × log (DBH)
  Br: log (weight) = 2.5983 + 1.5402 × log (DBH)
  St: log (weight) = 3.0692 + 2.4868 × log (DBH)
- White ash (*Fraxinus americana*) (Brenneman et al., 1978)
  WT: weight = 2.3626 × DBH$^{2.4798}$
- Aspen (*Populus* spp.) (MacLean and Wein, 1976)
  WT: log (weight) = −0.7891 + 2.0673 × log (DBH)
- Spruce (*Picea* spp.) (MacLean and Wein, 1976)
  WT: log (weight) = −0.2112 + 1.5639 × log (DBH)
- Red pine (*Pinus resinosa* Ait.) (Dunlap and Shipman, 1967)
  St: weight = −113.954 + 35.265 × (DBH)
- White pine (*Pinus strobus*) (Swank and Schreuder, 1974)
  Lf: ln (weight) = 3.051 + 2.1354 × ln (DBH)
  Br: ln (weight) = 3.158 + e2.5328 × ln (DBH)
  St: ln (weight) = −2.788 + 2.1338 × ln (DBH)
- Yellow poplar (*Liriodendron tulipifera*) (Hitchcock, 1978)
  St + Br: log (weight) = 1.9167 + 0.7993 × log (DBH^2 × ht)
- Pin cherry (*Prunus pensylvanica*) (Young et al., 1980)
  WT: ln (weight) = 0.9758 + 2.1948 × ln (DBH)
- Black cherry (*Prunus serotina* Ehrh.) (Wiant et al., 1979)
  St + Br: weight = 0.12968 (DBH^2 × ht)$^{0.97028}$
- White oak (*Quercus alba*) (Reiners, 1972)
  Lf: log (weight) = 2.1426 + 1.6684 × log (DBH)

- Scarlet oak (*Quercus coccinea*) (Clark and Schroeder, 1977)
  St + Br: weight = $0.12161 \times (DBH^2 \times ht)^{1.00031}$
- Chestnut oak (*Quercus prinus*) (Wiant et al., 1979)
  St + Br: weight = $0.06834 \times (DBH^2 \times ht)^{1.06370}$
- Northern red oak (*Quercus rubra*) (Clark and Schroeder, 1977)
  WT: weight = $0.10987 \times (DBH^2 \times ht)^{1.00197}$
- Black oak (*Quercus velutina*) (Bridge, 1979)
  WT: $\ln (weight) = -0.34052 + 2.65803 \times \ln (DBH)$
- Hemlock (*Tsuga canadensis*) (Young et al., 1980)
  WT: $\ln (weight) = 0.6803 + 2.3617 \times \ln (DBH)$
- General hardwoods (Monk et al., 1970)
  WT: $\log (weight) = 1.9757 + 2.5371 \times \log (DBH)$
- General softwoods (Monteith, 1979)
  WT: weight = $4.5966 - (0.2364 \times DBH) + (0.00411 \times DBH^2)$

13.4 TIMBER SCALING AND LOG RULES

With regard to scaling and log rules associated with timber, the measurement of timber to be harvested (the cruise), timber cut and removed from the forest (scaling), timber not recovered in the harvesting process (waste processing), and the use of formulas or tables to estimate net yield for logs (log rule) are the basis of forest-based biomass harvesting operations.

13.4.1 THEORY OF SCALING[*]

Scaling is the determination (measuring) of the gross and net volume of logs by the customary commercial units for the product involved; volume may be expressed in terms of board feet, cords, cubic feet, cubic meter, linear feet, or number of pieces. The *cubic foot* is an amount of wood equivalent to a solid cube that measures $12 \times 12 \times 12$ inches and contains 1728 cubic inches. The *cubic meter*, used in countries that have adopted the metric system, contains 35.3 cubic feet. The *board foot* is a plank 1 inch thick and 12 inches square; it contains 144 cubic inches of wood. Scaling is not guessing; it is an art founded on applying specific rules in a consistent manner based on experienced judgment as to how serious certain external indicators of defect are in a specific locality.

The measuring standard used in scaling logs, called a *log rule*, is a table intended to show amounts of lumber that may be sawed from logs of different sizes under assumed conditions. At best, a log rule can only approximate salable manufactured volume because of constant changes in markets, machinery, manufacturing practices, and even the varying skills of individual sawyers. Thus, a log rule is an imaginary measure. Its application must not be varied according to the mill in which logs are sawed. The scaled volume of logs must be independent of variations in manufacture. The difference between the volume of log scale and the actual volume of lumber sawed from the same logs is called *overrun* if the volume tally exceeds log scale, or *underrun* if it is less.

There will generally be an overrun or an underrun when logs are scaled by a particular rule in a given locality and sawed by a mill. Basic assumptions in the log rules and assumptions in utilization practices cause overrun to vary with the size of the average log. Experience proves that this is true even for the International 1/4-Inch rule, although not to the same degree as for the Scribner Decimal C rule (both rules are discussed further below). This fact does not change scaling practices. Overrun

---

[*] This section is based on USDA, *National Forest Log Scaling Handbook*, FSH 2409.11, U.S. Department of Agriculture, Forest Service, Washington, DC, 1941; Freese, F., *A Collection of Log Rules*, General Technical Report FPL-01, U.S. Department of Agriculture, Forest Service, Forest Products Laboratory, Madison, WI, 1973; USDA, *National Forest Log Scaling Handbook*, FSH 2409.11, U.S. Department of Agriculture, Forest Service, Washington, DC, 2006.

DID YOU KNOW?

Tree cross-sections rarely form true circles, but they are normally presumed to be circular for purposes of computing cross-sectional areas.

DID YOU KNOW?

When measuring the cross-sectional area of a log, diameter instead of the radius of the log is measured; thus, this area in square inches may be determined by

$$\text{Area in square inches} = \pi D^2/4$$

(or underrun) is estimated in the process of appraising National Forest timber for sale and presumably by purchasers in determining what prices they will bid. Overrun or underrun is not considered in log scaling, even though it is very important to any mill.

As a general rule, timber is appraised, sold, and measured by customary commercial unit for the products involved. Standard practice is to scale saw timber by a board-foot log rule scale, mining timbers by the piece or linear foot, telephone poles by the linear foot or the piece of stated length, piling by the linear foot, pulpwood by the solid cubic foot or cord, and fuelwood, shingle bolts, and similar material by the cord. Other units may be used when better adapted to local trade customs or local situations.

13.4.2 AUTHORIZED LOG RULES

The Scribner Decimal C rule, the International 1/4-Inch rule, and the Smalian Cubic Volume rule are used by the U.S. Forest Service for uniform scaling of sawtimber. With the exception of the Smalian cubic volume rule, all specific rules are board-foot rules. Each board-foot rule is represented by a table showing an arbitrary estimate of the amount of lumber a log of given length and diameter can produce. Inasmuch as the tables for each rule have a different base, the scale of identical logs will differ according to the rule used.

13.4.2.1 Scribner Decimal C Rule

The Scribner Decimal C rule, developed by J.M. Scribner around 1846, is a standard rule for U.S. Forest service saw log scaling. This rule was derived from diagrams of 1-in. boards drawn to scale within cylinders of various sizes (see Figure 13.3). The rule rounds contents to the nearer 10 board feet; for example, logs that, according to the Scribner rule, have volumes between 136 and 145 board feet are rounded to 140 board feet and shown as 14.

13.4.2.2 International 1/4-Inch Rule

Developed in 1906, the International 1/4-Inch rule is based on a reasonably accurate mathematical formula that probably gives a closer lumber-volume estimate than other log rules in common use. This rule measures logs to the nearest 5 board feet. As the name implies, it allows for a saw kerf

DID YOU KNOW?

The Scribner rule gives a relatively high overrun (up to 30%) for logs under 14 inches. Above 14 inches, the overrun gradually decreases and flattens out around 28 inches to about 3 to 5%.

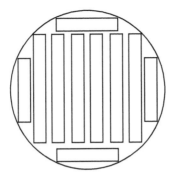

FIGURE 13.3 Diagram showing the number of 1-inch boards that can be cut from a specific log.

(width of the saw cut) of 1/4 inch. It is a rule based on a formula applied to each 4-foot section of the log. For practical purposes, the scaling cylinder becomes a part of a cone (a frustum) with a taper of 2 in. over 16 feet. This rule generally results in a log scale relatively close to lumber tally when logs are sawed in a reasonably efficient mill.

13.4.2.3 Smalian Cubic Volume Rule

The Smalian Cubic Volume rule requires measurement of the two inside bark diameters and the length. It can be shown generally in the form:

$$V = \frac{A+a}{2} \times L \tag{13.8}$$

where
V = Volume in cubic feet (ft^3).
A = Large-end cross-section area (ft^2).
a = Small-end cross-section area (ft^2).
L = Log length (ft).

13.4.2.4 Log Rule Development

A log rule is a table or formula showing the estimated net yield for logs of a given diameter and length. Ordinarily, the yield is expressed in terms of board feet of finished lumber, although a few rules give the cubic volume of the log or some fraction of it. As noted earlier, the board foot is equivalent to a plank 1 inch thick by 12 inches (1 foot) square; it contains 144 cubic inches of wood. Although the board foot has been a useful and fairly definitive standard for the measure of sawed lumber, it is an ambiguous and inconsistent unit for log scaling. Built into each log rule are allowances for losses due to such things as slabs, saw kerf, edgings, and shrinkage. The volume commonly used for determining the board-foot content of saw lumber is

$$\text{Board feet} = \frac{\text{Thickness (in.)} \times \text{Width (in.)} \times \text{Length (ft)}}{12} \tag{13.9}$$

At first glance and with the use of a simple equation (Equation 13.9), it would seem to be a relatively simple matter to devise such a rule and having done so that should be the end of the problem. But, it would seem so only to those who are unfamiliar with the great variations in the dimensions of lumber that may be produced from a log, with variations in the equipment used in producing this lumber and the skills of various operators, and, finally, with variations in the logs. All of these have an effect on the portion of the total log volume that ends up as usable lumber and the portion that becomes milling residue.

DID YOU KNOW?

By keeping careful lumber tallies of boards cut from various sized logs, any sawmill may construct its own empirical log rule. Such rules may provide excellent indicators of log volume at the particular sawmills where they are compiled.

Because no industrial lumber organization or government agency has control over the measurement of logs, districts and individual buyers have devised their own rules to fit a particular set of operating conditions. Thus, in the United States and Canada there are over 95 recognized rules bearing about 185 names. In addition, there are numerous location variations in the application of any given rule.

Three methods are employed to develop a new log rule. The most obvious is to record the volume of lumber produced from straight, defect-free logs of given diameters and lengths and accumulate such data until all sizes of logs have been covered. These "mill scale" or "mill tally" rules have the virtue of requiring no assumptions and of being perfectly adapted to all the conditions prevailing when the data were obtained. Their disadvantage, aside from the amount of recordkeeping required, is that they may have been produced in such a restricted set of conditions that the values are not applicable anywhere else.

The second method is to prescribe all of the pertinent conditions (e.g., allowance for saw kerf and shrinkage, thickness and minimum width and length of boards, taper assumptions) and then to draw diagrams in circles of various sizes, representing the sawing pattern on the small end of a log. These "diagram rules," of which the Scribner is an example, will be good or bad, depending on how well the sawmilling situation fits the assumptions used in producing the diagrams.

The third basic procedure is to start with the formula for some assumed geometric solid and then make adjustments to allow for losses to saw kerf, edgings, and so forth. These are referred to as "formula rules" and as is the case for any type of rule their applicability will depend on how well the facts fit all of the assumptions

The development of a rule may involve more than one of these procedures; thus, the step-like progression of values in a mill tally or diagram rule may be smoothed out by fitting a regression equation. Or, the allowance to be used for slabs and edgings in a formula rule may be estimated from mill tally data. Finally, there are the "combination" rules such as the Doyle–Scribner, which uses values from the Doyle rule for small logs and from the Scribner rule for large logs. The aim, of course, is to take advantage of either the best or the worst features of the different rules.

13.4.2.5 Basal Area Measurements

The cross-sectional area at tree stem breast height is called *basal area*. This is important because tree-stem measurements are often converted to cross-sectional areas. To compute tree basal area, one commonly assumes that the tree stem is circular in cross-section at breast height. Thus, the formula for calculating basal area in square feet (where the DBH is measured in inches) is

$$\text{Basal area (ft}^2) = \frac{\pi(\text{DBH})^2}{4 \times 144} = 0.005454 \times (\text{DBH})^2 \tag{13.10}$$

If metric units are used, basal area should be expressed in square meters (m²), and DBH is measured in centimeters:

$$\text{Basal area (m}^2) = \frac{\pi(\text{DBH})^2}{4 \times 10,000} = 0.00007854 \times (\text{DBH})^2 \tag{13.11}$$

■ **EXAMPLE 13.2**

Problem: Calculate the basal area (in ft²) for a tree measuring 6 in. DBH.

Solution:

$$\text{Basal area} = \frac{\pi(\text{DBH})^2}{4 \times 144} = \frac{3.14 \times 6^2}{4 \times 144} = \frac{113.04}{576} = 0.19625 \text{ ft}^2$$

13.4.3 WOOD DENSITY AND WEIGHT RATIOS

The weight (lb) per cubic foot of any tree species may be computed by using the moisture content and specific gravity (based on oven-dry weight and green volume):

$$\text{Density} = \text{Specific gravity} \times 62.4 \left(1 + \frac{\% \text{ Moisture content}}{100}\right) \tag{13.12}$$

■ **EXAMPLE 13.3**

Problem: Determine the volume (ft³) contained in the following wood weight, given that the weight, specific gravity, and moisture content of the wood, respectively, are 15,530 lb, 0.53, 100%.

Solution:

$$\text{Density} = \text{Specific gravity} \times 62.4 \left(1 + \frac{\% \text{ Moisture content}}{100}\right)$$

$$= 0.53 \times 62.4 \left(1 + \frac{100}{100}\right) = 33.07 \times 2 = 66.1 \text{ lb/ft}^3$$

$$\textit{Volume} = \frac{15,530 \text{ lb}}{66.1 \text{ lb/ft}^3} = 234.9 \text{ ft}^3$$

REFERENCES AND RECOMMENDED READING

Avery, T.E. and Burkhart, H.E. (2002). *Forest Measurements*, 5th ed. McGraw-Hill, New York.

Brenneman, B.B., Frederick, D.J., Gardner, W.E., Schoenhofen, L.H., and Marsh, P.L. (1978). Biomass of species and stands of West Virginia hardwoods, in Pope, P.E., Ed., *Proceedings of Central Hardwood Forest Conference II*, November 14–16, 1978, Purdue University, West Lafayette, IN.

Bonhan, C.D. (1989). *Measurements for Terrestrial Vegetation*. John Wiley & Sons, New York.

Bridge, J.A. (1979). Fuelwood Production of Mixed Hardwoods on Mesic Sites in Rhode Island, master's thesis, University of Rhode Island, Kingston.

Briggs, D. (1994). *Forest Products Measurements and Conversion Factors: With Special Emphasis on the U.S. Pacific Northwest*. College of Forest Resources University of Washington, Seattle, Chapter 1.

Cain, S.A. and De O. Castro, G.M. (1959). *Manual of Vegetation Analysis*. Harper & Brothers, New York.

Clark III, A. and Schroeder, J.G. (1977). *Biomass of Yellow Poplar in Natural Stands in Western North Carolina*, Paper SE-165. U.S. Department of Agriculture Forest Service, Washington, DC.

Clark III, A., Phillips, D.R., and Hitchcock, H.C. (1980). *Predicted Weights and Volumes of Scarlet Oak Trees on the Tennessee Cumberland Plateau*, Paper SE-214. U.S. Department of Agriculture Forest Service, Washington, DC.

Daubenmire, R.E. (1968). *Plant Communities: A Textbook of Plant Synecology*. HarperCollins, New York.

Dunlap, W.H. and Shipman, R.D. (1967). *Density and Weight Prediction of Standing White Oak, Red Maple, and Red Pine*, Research Brief. Pennsylvania State University, University Park, PA, pp. 66–69.

Elzinga, C.I., Salzer, D.W., and Willoughby, J.W. (1998). *Measuring and Monitoring Plant Populations*, Technical Reference 1730-1. U.S. Department of Interior, Bureau of Land Management, Denver, CO.

Hitchcock III, H.C. (1978). Aboveground tree weight equations for hardwood seedlings and saplings. *TAPPI Journal*, 61(10), 119–120.

Ince, P.J. (1979). *How to Estimate Recoverable Heat Energy in Wood or Bark Fuels*, General Technical Report FPL-GTR-29. U.S. Department of Agriculture, Forest Service, Forest Products Laboratory, Madison, WI.

Kinerson, R.S. and Bartholomew, I. (1977). *Biomass Estimation Equations and Nutrient Composition of White Pine, White Birch, Red Maple, and Red Oak in New Hampshire*, Research Report No. 62. New Hampshire Agricultural Experiment Station, University of New Hampshire, Durham.

Landis, T.D. and Mogren, E.W. (1975). Tree strata biomass of subalpine spruce–fir stands in southwestern Colorado. *Forest Science*, 21(1), 9–14.

MacLean, D.A. and Wein, R.W. (1976). Biomass of jack pine and mixed hardwood stands in northeastern New Brunswick. *Canadian Journal of Forest Research*, 6(4), 441–447.

Monk, C.D., Child, G.I., and Nicholson, S.A. (1970). Biomass, litter and leaf surface area estimates of an oak-hickory forest. *Oikos*, 21:138-141.

Monteith, D.B. (1979). *Whole-Tree Weight Table for New York*, AFRI Research Report 40. University of New York, Syracuse.

NREL. (2010). *Learning about Renewable Energy*. National Renewable Energy Laboratory, Golden, CO (http://www.nrel.gov/learning/).

Pennsylvania DCNR. (2007). *Guidance on Harvesting Woody Biomass for Energy in Pennsylvania*. Pennsylvania Department of Conservation and Natural Resources, Harrisburg (www.dcnr.state.pa.us/ PA_Biomass_guidance_final.pdf).

Reiners, W.A. (1972). Structure and energetics of three Minnesota forests. *Ecological Monographs*, 42(1), 71–94.

Ribe, J.H. (1973). *Puckerbrush Weight Tables*. University of Maine, Orono.

Schroeder, P., Brown, S., Mo, J., Birdsey, R., and Cieszewski, C. (1997). Biomass estimation of temperate broadleaf forests of the U.S. using forest inventory data. *Forest Science*, 43:424–434.

Swank, W.T. and Schreuder, H.T. (1974). Comparison of three methods of estimating surface area and biomass for a forest of young eastern white pine. *Forest Science*, 20, 91–100.

USDA. (1979). *How to Estimate Recoverable Heat Energy in Wood or Bark Fuels*. U.S. Department of Agriculture, Washington, DC.

Whittaker, R.H., Bormann, F.H., Likens, G.E., and Siccama, T.G. (1974). The Hubbard book ecosystem study: forest biomass and production. *Ecological Monographs*, 4, 233–254.

Wiant, Jr., H.V., Sheetz, C.E., Colaninno, A., DeMoss, J.C., and Castaneda, F. (1977). *Tables and Procedures for Estimating Weights of Some Appalachian Hardwoods*. West Virginia University, Agricultural and Forestry Experiment Station, Morgantown.

Wiant, Jr., H.V., Castaneda F., Sheetz, C.E., Colaninno, A., and DeMoss, J.C. (1979). Equations for predicting weights of some Appalachian hardwoods. *West Virginia Forestry Notes*, 7, 21–28.

Young, H.E., Ribe, J.H., and Wainwright, K. (1980). *Weight Tables for Tree and Shrub Species in Maine*. University of Maine, Life Sciences and Agriculture Experiment Station, Orono.

Section VII

Fundamental Science Computations

All is connected … no one thing can change by itself.

—Paul Hawken

14 Fundamental Chemistry and Hydraulics

As future working environmental professionals, it is not sufficient to understand the causes and effects of environmental problems in qualitative terms only. The environmental professional must also be able to express the perceived problem and its potential solution in quantitative terms. To do this, the environmental professional must be able to draw on the basic sciences such as chemistry and hydrology and others to predict the fate of pollutants in the environment and to design effective mitigation measures and treatment systems to reduce impacts. In this chapter, we discuss fundamental chemistry and basic hydraulics for environmental professionals.

14.1 FUNDAMENTAL CHEMISTRY

The chemists are a strange class of mortals, impelled by an almost insane impulse to seek their pleasure among smoke and vapor, soot and flame, poisons and poverty; yet among all these evils I seem to live so sweetly that I may die if I would change places with the Persian King.

—**Johann Joachim Becher (1635–1682)**

All matter on Earth consists of chemicals. This simplified definition may shock those that think chemistry is what happens between men and women. Chemistry is much more; it is the science of materials that make up the physical world. Chemistry is so complex that no one person could expect to master all aspects of such a vast field, so it has been found convenient to divide the subject into specialty areas, such as the following:

- *Organic chemists* study compounds of carbon. Atoms of this element can form stable chains and rings, giving rise to very large numbers of natural and synthetic compounds.
- *Inorganic chemists* are interested in all elements, but particularly in metals, and are often involved in the preparation of new catalysts.
- *Biochemists* concern themselves with the chemistry of the living world.
- *Physical chemists* study the structures of materials and the rates and energies of chemical reactions.
- *Theoretical chemists*, with the use of mathematics and computational techniques, derive unifying concepts to explain chemical behavior.
- *Analytical chemists* develop test procedures to determine the identity, composition, and purity of chemicals and materials. New analytical procedures often reveal the presence of previously unknown compounds.

Why should we care about chemistry? Isn't it enough to know that we don't want unnecessary chemicals in or on our food or harmful chemicals in our air, water, or soil? Chemicals are everywhere in our environment. The vast majority of these chemicals are natural. The chemist often copies from nature to create new substances that are often superior to and cheaper than natural materials. It is human nature to make nature serve us. Without chemistry (and the other sciences), we are at nature's mercy. To control nature, we must learn its laws and then use them.

TABLE 14.1

Water Properties (Temperature, Specific Weight, Density)

Temperature (°F)	Specific Weight (lb/ft³)	Density (slugs/ft³)	Temperature (°F)	Specific Weight (lb/ft³)	Density (slugs/ft³)
32	62.4	1.94	130	61.5	1.91
40	62.4	1.94	140	61.4	1.91
50	62.4	1.94	150	61.2	1.90
60	62.4	1.94	160	61.0	1.90
70	62.3	1.94	170	60.8	1.89
80	62.2	1.93	180	60.6	1.88
90	62.1	1.93	190	60.4	1.88
100	62.0	1.93	200	60.1	1.87
110	61.9	1.92	210	59.8	1.86
120	61.7	1.92			

Source: Spellman, F.R., *Handbook of Water and Wastewater Treatment Plant Operations*, 3rd ed., Lewis Publishers, Boca Raton, FL, 2013.

Environmental practitioners must also learn the laws of chemistry and to use them, as they must. But they must know even more. Environmental practitioners must know the ramifications of chemistry when it is out of control. Chemistry properly used can perform miracles. Out of control, chemicals and their effects can be devastating. In fact, many of the current environmental regulations dealing with chemical safety and emergency response for chemical spills resulted because of catastrophic events involving chemicals.

14.1.1 DENSITY AND SPECIFIC GRAVITY

When we say that iron is heavier than aluminum, we are saying that iron has a greater density than aluminum. In practice, what we are really saying is that a given volume of iron is heavier than the same volume of aluminum. Density (p) is the mass (weight) per unit volume of a substance at a particular temperature, though it generally varies with temperature. The weight may be expressed in terms of pounds, ounces, grams, kilograms, etc. The volume may be liters, milliliters, gallons, cubic feet, etc. Table 14.1 shows the relationship between temperature, specific weight and density of fresh water.

Suppose we had a tub of lard and a large box of crackers, each having a mass of 600 grams. The density of the crackers would be much less than the density of the lard because the crackers occupy a much larger volume than the lard occupies. The density of an object can be calculated by using the following formula:

$$\text{Density} = \frac{\text{Mass}}{\text{Volume}} \tag{14.1}$$

In water/wastewater operations, perhaps the most common measures of density are pounds per cubic foot (lb/ft³) and pounds per gallon (lb/gal):

- 1 cubic foot (ft³) of water weighs 62.4 lb, so its density is 62.4 lb/ft³.
- 1 gallon of water weighs 8.34 lb, so its density is 8.34 lb/gal.

The density of a dry material, such as cereal, lime, soda, or sand, is usually expressed in pounds per cubic foot. The densities of plain and reinforced concrete are 144 and 150 lb/ft³, respectively. The density of a liquid, such as liquid alum, liquid chlorine, or water, can be expressed either as pounds per cubic foot or as pounds per gallon. The density of a gas, such as chlorine gas, methane, carbon dioxide, or air, is usually expressed in pounds per cubic foot.

As shown in Table 14.1, the density of a substance such as water changes slightly as the temperature of the substance changes. This occurs because substances usually increase in volume as they become warmer. Because of this expansion with warming, the same weight is spread over a larger volume, so the density is lower when a substance is warm than when it is cold.

Specific gravity is defined as the weight (or density) of a substance compared to the weight (or density) of an equal volume of water. (The specific gravity of water is 1.) This relationship is easily seen when a cubic foot of water, which weighs 62.4 lb, is compared to a cubic foot of aluminum, which weights 178 lb. Aluminum is 2.7 times as heavy as water.

Finding the specific gravity of a piece of metal is not difficult. All we have to do is to weigh the metal in air, then weigh it under water. Its loss of weight is the weight of an equal volume of water. To find the specific gravity, divide the weight of the metal by its loss of weight in water:

$$\text{Specific gravity} = \frac{\text{Weight of substance}}{\text{Weight of equal volume of water}} \qquad (14.2)$$

■ EXAMPLE 14.1

Problem: Suppose a piece of metal weighs 150 lb in air and 85 lb under water. What is the specific gravity?

Solution: First, 150 lb – 85 lb = 65 lb loss of weight in water; thus,

$$\text{Specific gravity} = \frac{150}{65} = 2.3$$

Note: In a calculation of specific gravity, it is *essential* that the densities be expressed in the same units.

The specific gravity of water is 1, which is the standard, the reference against which all other liquid or solid substances are compared. Specifically, any object that has a specific gravity greater than 1 (e.g., rocks, steel, iron, grit, floc, sludge) will sink in water. Substances with a specific gravity of less than 1 (e.g., wood, scum, gasoline) will float. Considering the total weight and volume of a ship, its specific gravity is less than 1; therefore, it can float.

The most common use of specific gravity in water/wastewater treatment operations is in gallon-to-pound conversions. In many cases, the liquids being handled have a specific gravity of 1 or very nearly 1 (between 0.98 and 1.02), so 1 may be used in the calculations without introducing significant error. However, in calculations involving a liquid with a specific gravity of less than 0.98 or greater than 1.02, the conversions from gallons to pounds must consider the exact specific gravity. This technique is illustrated in the following example.

■ EXAMPLE 14.2

Problem: A basin holds 1455 gal of a certain liquid. If the specific gravity of the liquid is 0.94, how many pounds of liquid are in the basin?

Solution: Normally, for a conversion from gallons to pounds, we would use the factor 8.34 lb/gal (the density of water) if the specific gravity of the substance is between 0.98 and 1.02. However, in this instance the substance has a specific gravity outside this range, so the 8.34 factor must be adjusted.

1. Multiply 8.34 lb/gal by the specific gravity to obtain the adjusted factor:

$$8.34 \text{ lb/gal} \times 0.94 = 7.84 \text{ lb/gal (rounded)}$$

2. Then convert 1455 gal to pounds using the corrected factor:

$$1455 \text{ gal} \times 7.84 \text{ lb/gal} = 11,407 \text{ lb (rounded)}$$

■ EXAMPLE 14.3

Problem: The specific gravity of a liquid substance is 0.96 at 64°F. What is the weight of 1 gal of the substance?

Solution:

$$\text{Weight} = \text{Specific gravity} \times \text{weight of water} = 0.96 \times 8.34 \text{ lb/gal} = 8.01 \text{ lb}$$

■ EXAMPLE 14.4

Problem: A liquid has a specific gravity of 1.15. How many pounds is 66 gal of the liquid?

Solution:

$$\text{Weight} = 66 \text{ gal} \times 8.34 \text{ lb/gal} \times 1.15 = 633 \text{ lb}$$

■ EXAMPLE 14.5

Problem: If a solid in water has a specific gravity of 1.30, what percent heavier is it than water?

Solution:

$$\text{Percent heavier} = \frac{(\text{Specific gravity of solid}) - (\text{specific gravity of water})}{\text{Specific gravity of water}} \times 100$$

$$= \frac{1.30 - 1.0}{1.0} \times 100 = 30\%$$

14.1.2 WATER CHEMISTRY FUNDAMENTALS

Whenever we add a chemical substance to another chemical substance, such as adding sugar to tea or hypochlorite to water to make it safe to drink, we are performing the work of chemists. We are working as "chemists" because we are working with chemical substances, and how they react is important to us. Environmental practitioners involved with water treatment operations, for example, may be required to determine the amount of chemicals or chemical compounds to add (dosing) to various unit processes. Table 14.2 lists some of the chemicals and their common applications in water treatment operations.

14.1.2.1 Water Molecules

Just about everyone knows that water is a chemical compound of two simple and abundant elements: H_2O. Yet, scientists continue to argue the merits of rival theories on the structure of water. The fact is that we still know little about water; for example, we don't know how water works. The reality is that water is very complex and has many unique properties that are essential to life and determine

TABLE 14.2
Chemicals and Chemical Compounds Used in Water Treatment

Name	Common Application	Name	Common Application
Activated carbon	Taste and odor control	Aluminum sulfate	Coagulation
Ammonia	Chloramine disinfection	Ammonium sulfate	Coagulation
Calcium hydroxide	Softening	Calcium hypochlorite	Disinfection
Calcium oxide	Softening	Carbon dioxide	Recarbonation
Copper sulfate	Algae control	Ferric chloride	Coagulation
Ferric sulfate	Coagulation	Magnesium hydroxide	Defluoridation
Oxygen	Aeration	Potassium permanganate	Oxidation
Sodium aluminate	Coagulation	Sodium bicarbonate	pH adjustment
Sodium carbonate	Softening	Sodium chloride	Ion exchanger regeneration
Sodium fluoride	Fluoridation	Sodium fluosilicate	Fluoridation
Sodium hexametaphosphate	Corrosion control	Sodium hydroxide	pH adjustment
Sodium hypochlorite	Disinfection	Sodium silicate	Coagulation aid
Sodium thiosulfate	Dechlorination	Sulfur dioxide	Dechlorination
Sulfuric acid	pH adjustment		

its environmental chemical behavior. In a water molecule, the two hydrogen atoms *always* come to rest at an angle of approximately 105° from each other. The hydrogens tend to be positively charged and the oxygen tends to be negatively charged. This arrangement gives the water molecule an electrical polarity; that is, one end is positively charged and one end is negatively charged. This 105° relationship makes water lopsided, peculiar, and eccentric; it breaks all the rules (Figure 14.1). In the laboratory, pure water contains no impurities, but in nature water contains a lot of materials besides water. This is an important consideration for the environmental professional tasked with maintaining the purest or cleanest water possible. Water is often called the *universal solvent*, a fitting description when you consider that, given enough time, water will dissolve anything and everything on Earth.

14.1.2.2 Water Solutions

A *solution* is a condition in which one or more substances are uniformly and evenly mixed or dissolved. In other words, a solution is a homogeneous mixture of two or more substances. Solutions can be solids, liquids, or gases, such as drinking water, seawater, air, etc. Here, we focus primarily on liquid solutions. A solution has two components: a *solvent* and a *solute* (see Figure 14.2). The solvent is the component that does the dissolving. Typically, the solvent is the species present in the greater quantity. The solute is the component that is dissolved. When water dissolves substances, it creates solutions with many impurities. Generally, a solution is transparent and not cloudy. Because

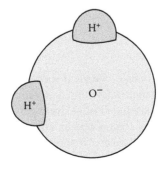

FIGURE 14.1 A molecule of water.

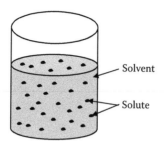

FIGURE 14.2   Solution with two components: solvent and solute. (From Spellman, F.R., *Handbook of Water and Wastewater Treatment Plant Operations*, 3rd ed., Lewis Publishers, Boca Raton, FL, 2013.)

usually water is colorless, the light necessary for photosynthesis can travel to considerable depths; however, a solution may be colored when the solute remains uniformly distributed throughout the solution and does not settle with time. When molecules dissolve in water, the atoms making up the molecules come apart, or dissociate, in the water. This dissociation in water is called *ionization*. When the atoms in the molecules come apart, they do so as charged atoms (both negatively and positively charged) which are called *ions*. The positively charged ions are called *cations* and the negatively charged ions are called *anions*. A good example of the ionization process is when calcium carbonate ionizes:

$$CaCO_3 \leftrightarrow Ca^{2+} + CO_3^{2-}$$

| Calcium carbonate | Calcium ion (cation) | Carbonate ion (anion) |

Another good example is the ionization that occurs when table salt (sodium chloride) dissolves in water:

$$NaCl \leftrightarrow Na^+ + Cl^-$$

| Sodium chloride | Sodium ion (cation) | Chloride ion (anion) |

Some of the common ions found in water and their symbols are provided below:

Hydrogen	H^+
Sodium	Na^+
Potassium	K^+
Chloride	Cl^-
Bromide	Br^-
Iodide	I^-
Bicarbonate	HCO^{3-}

Solutions serve as a vehicle to (1) allow chemical species to come into close proximity so that they can react; (2) provide a uniform matrix for solid materials, such as paints, inks, and other coatings so they can be applied to surfaces; and (3) dissolve oil and grease so they can be rinsed away. Water dissolves polar substances better than nonpolar substances. Polar substances (mineral acids, bases, and salts) are easily dissolved in water. Nonpolar substances (oils, fats, and many organic compounds) do not dissolve as easily in water.

14.1.2.3 Concentrations

Because the properties of a solution depend largely on the relative amounts of solvent and solute, the concentrations of each must be specified.

Note: Chemists use both relative terms, such as saturated and unsaturated, as well as more exact concentration terms, such as weight percentages, molarity, and normality.

Although polar substances dissolve better than nonpolar substances in water, polar substances dissolve in water only to a point; that is, only so much solute will dissolve at a given temperature. When that limit is reached, the resulting solution is saturated. At this point, the solution is in equilibrium—no more solute can be dissolved. A liquid/solids solution is supersaturated when the solvent actually dissolves more than an equilibrium concentration of solute (usually when heated).

Specifying the relative amounts of solvent and solute, or specifying the amount of one component relative to the whole, usually gives the exact concentrations of solution. Solution concentrations are sometimes specified as weight percentages:

$$\% \text{ of Solute} = \frac{\text{Mass of solute}}{\text{Total mass of solute}} \times 100 \tag{14.3}$$

To understand the concepts of molarity, molality, and normality, we must first understand the concept of a mole. The mole is defined as the amount of a substance that contains exactly the same number of items (i.e., atoms, molecules, or ions) as 12 g of carbon-12. By experiment, Avogadro determined this number to be 6.02×10^{23} (to three significant figures). If 1 mole of carbon atoms equals 12 g, for example, what is the mass of 1 mole of hydrogen atoms? Note that carbon is 12 times as heavy as hydrogen; therefore, we need only 1/12 the weight of hydrogen to equal the same number of atoms of carbon.

Note: One mole of hydrogen equals 1 gram.

By the same principle:

- One mole of CO_2 = 12 + 2(16) = 44 g
- One mole of Cl^- = 35.5 g
- One mole of Ra = 226 g

In other words, we can calculate the mass of a mole if we know the formula of the substance.

Molarity (*M*) is defined as the number of moles of solute per liter of solution. The volume of a solution is easier to measure in the lab than its mass:

$$M = \frac{\text{Number of moles of solute}}{\text{Number of liters of solution}} \tag{14.4}$$

Molality (*m*) is defined as the number of moles of solute per kilogram of solvent:

$$m = \frac{\text{Number of moles of solute}}{\text{Number of kilograms of solution}} \tag{14.5}$$

Note: Molality is not as frequently used as molarity, except in theoretical calculations.

Especially for acids and bases, the normality (N) rather than the molarity of a solution is often reported. Normality is the number of equivalents of solute per liter of solution (1 equivalent of a substance reacts with 1 equivalent of another substance):

$$N = \frac{\text{Number of equivalents of solute}}{\text{Number of liters of solution}} \qquad (14.6)$$

In acid/base terms, an *equivalent* (or gram equivalent weight) is the amount that will react with one mole of H^+ or OH^-; for example,

- One mole of HCl will generate 1 mole of H^+; therefore, 1 mole HCl = 1 equivalent.
- One mole of $Mg(OH)_2$ will generate 2 moles of OH^-; therefore, 1 mole of $Mg(OH)_2$ = 2 equivalents.

$$HCl \Rightarrow H^+ + Cl^-$$

$$Mg(OH)^{+2} \Rightarrow Mg^{+2} + 2OH^-$$

By the same principle:

- A 1-M solution of H_3PO_4 is 3 N.
- A 2-N solution of H_2SO_4 is 1 M.
- A 0.5-N solution of NaOH is 0.5 M.
- A 2-M solution of HNO_3 is 2 N.

Chemists titrate acid/base solutions to determine their normality. An endpoint indicator is used to identify the point at which the titrated solution is neutralized.

Key Point: If it takes 100 mL of 1-N HCl to neutralize 100 mL of NaOH, then the NaOH solution must also be 1 N.

14.1.2.4 Predicting Solubility

Predicting solubility is difficult, but there are a few general rules of thumb, such as "like dissolves like."

- *Liquid–liquid solubility*—Liquids with similar structure and hence similar intermolecular forces will be completely miscible. For example, we would correctly predict that methanol and water are completely soluble in any proportion.
- *Liquid–solid solubility*—Solids *always* have limited solubilities in liquids, in general because of the difference in magnitude of their intermolecular forces. Therefore, the closer the temperature is to its melting point, the better the match between a solid and a liquid.

Key Point: At a given temperature, lower melting solids are more soluble than higher melting solids. Structure is also important; for example, nonpolar solids are more soluble in nonpolar solvents.

- *Liquid–gas solubility*—As with solids, the more similar the intermolecular forces, the higher the solubility. Therefore, the closer the match between the temperature of the solvent and the boiling point of the gas, the higher the solubility. When water is the solvent, an additional *hydration* factor promotes solubility of charged species. Other factors that can significantly affect solubility are temperature and pressure. In general, raising the temperature typically increases the solubility of solids in liquids.

Key Point: Dissolving a solid in a liquid is usually an endothermic process (i.e., heat is absorbed), so raising the temperature will fuel this process. In contrast, dissolving a gas in a liquid is usually an exothermic process (i.e., it evolves heat), so lowering the temperature generally increases the solubility of gases in liquids.

Note: Thermal pollution is a problem because of the decreased solubility of O_2 in water at higher temperatures.

Pressure has only an appreciable effect on the solubility of gases in liquids. For example, carbonated beverages such as soda water are typically bottled at significantly higher atmospheres. When the beverage is opened, the decrease in the pressure above the liquid causes the gas to bubble out of solution. When shaving cream is used, dissolved gas comes out of solution, bringing the liquid with it as foam.

14.1.2.5 Colligative Properties

Properties of a solution that depend on the concentrations of the solute species rather than their identity include the following:

- Lowering vapor pressure
- Raising boiling point
- Decreasing freezing point
- Osmotic pressure

True colligative properties are directly proportional to the concentration of the solute but entirely independent of its identity.

- *Lowering of vapor pressure*—With all other conditions identical, the vapor pressure of water above the pure liquid is higher than that above sugar water. The vapor pressure above a 0.2-*M* sugar solution is the same as that above a 0.2-*M* urea solution. The lowering of vapor pressure above a 0.4-*M* sugar solution is twice as great as that above a 0.2-*M* sugar solution. Solutes lower vapor pressure because they lower the concentration of solvent molecules. To remain in equilibrium, the solvent vapor concentration must decrease (hence the vapor pressure decreases).
- *Raising the boiling point*—A solution containing a nonvolatile solute boils at a higher temperature than the pure solvent. The increase in boiling point is directly proportional to the increase in solute concentration in dilute solutions. This phenomenon is explained by the lowering of vapor pressure already described.
- *Decreasing the freezing point*—At low solute concentrations, solutions generally freeze or melt at lower temperatures than for the pure solvent.

Key Point: The presence of dissolved "foreign bodies" tends to interfere with freezing; therefore, solutions can only be frozen at temperatures below that of the pure solvent.

Key Point: We add antifreeze to the water in our radiators to both lower its freezing point and increase its boiling point.

- *Osmotic pressure*—Water moves spontaneously from an area of high vapor pressure to an area of low vapor pressure. If allowed to continue, in the end all of the water would move to the solution. A similar process will occur when pure water is separated from a concentrated solution by a semipermeable membrane (i.e., it only allows the passage of water molecules). The osmotic pressure is the pressure that is just adequate to prevent osmosis.

TABLE 14.3

Types of Colloids

Name	Dispersing Medium	Dispersed Phase
Solid sol	Solid	Solid
Gel	Solid	Liquid
Solid form	Solid	Gas
Sol	Liquid	Solid
Emulsion	Liquid	Liquid
Foam	Liquid	Gas
Solid aerosol	Gas	Solid
Aerosol	Liquid	Aerosol

Source: Adapted from Davies, P., *Types of Colloids*, University of Bristol, 2013, http://www.chm.bris. ac.uk/webprojects2002/pdavies/types.html.

In dilute solutions, the osmotic pressure is directly proportional to the solute concentration and is independent of its identity. The properties of electrolyte solutions follow the same trends as nonelectrolyte solutions but are also dependent on both the *nature* of the electrolyte as well as its concentration.

14.1.2.6 Colloids/Emulsions

A solution (e.g., seawater) is a homogeneous mixture of two or more substances. A suspension (e.g., sand and water) is a brief commingling of solvent and undissolved particles. A colloidal suspension is a commingling of particles not visible to the naked eye but larger than individual molecules.

Note: Colloidal particles do not settle out by gravity alone.

Colloidal suspensions can consist of

- Hydrophilic solutions of macromolecules (proteins, for example) that spontaneously form in water
- Hydrophobic suspensions, which gain stability from their repulsive electrical charges
- Micelles, special colloids having charged hydrophilic heads and long hydrophobic tails

Colloids are usually classified according to the original states of their constituent parts (see Table 14.3). The stability of colloids can be primarily attributed to *hydration* and *surface charge*, both of which help to prevent contact and subsequent coagulation.

Note: In many cases, water-based emulsions have been used to replace organic solvents (e.g., paints, inks), even though the compounds are not readily soluble in water.

In wastewater treatment, the elimination of colloidal species and emulsions is achieved by various means:

- Agitation
- Heat
- Acidification
- Coagulation (adding ions)
- Flocculation (adding bridging groups)

14.1.2.7 Water Constituents

Natural water can contain a number of substances known as impurities or constituents. When a particular constituent can affect the good health of the water user, it is called a *contaminant* or *pollutant*. It is these contaminants that the environmental practitioner works to keep from entering or to remove from the water supply.

14.1.2.7.1 Solids

Other than gases, all contaminants of water contribute to the solids content. Natural water carries many dissolved and undissolved solids. The undissolved solids are nonpolar substances and consist of relatively large particles of materials, such as silt, that will not dissolve. Classified by their size and state, by their chemical characteristics, and by their size distribution, solids can be dispersed in water in both suspended and dissolved forms. The size classifications for solids in water are

- Suspended solids
- Settleable solids
- Colloidal solids
- Dissolved solids

Total solids are the suspended and dissolved solids that remain behind when the water is removed by evaporation. Solids are also characterized as being *volatile* or *nonvolatile*.

Key Point: Though not technically accurate from a chemical point of view because some finely suspended material can actually pass through the filter, suspended solids are defined as those that can be filtered out in the suspended solids laboratory test. The material that passes through the filter is defined as dissolved solids. Colloidal solids are extremely fine suspended solids (particles) less than 1 micron in diameter; they are so small (though they still can make water cloudy) that they will not settle even if allowed to sit quietly for days or weeks.

14.1.2.7.2 Turbidity

Simply, turbidity refers to how clear the water is. The clarity of water is one of the first characteristics people notice. Turbidity in water is caused by the presence of suspended matter, which results in the scattering and absorption of light rays. The greater the amount of *total suspended solids* (TSS) in the water, the murkier it appears and the higher the measured turbidity. Thus, in plain English, turbidity is a measure of the light-transmitting properties of water. Natural water that is very clear (low turbidity) allows us to see images at considerable depths. High turbidity water, on the other hand, appears cloudy. Keep in mind that water of low turbidity is not necessarily without dissolved solids. Dissolved solids do not cause light to be scattered or absorbed; thus, the water looks clear. High turbidity causes problems for the waterworks operator, as components that cause high turbidity can cause taste and odor problems and will reduce the effectiveness of disinfection.

14.1.2.7.3 Color

Water can be colored, but often the color of water can be deceiving. In the first place, color is considered an aesthetic quality of water with no direct health impact. Second, many of the colors associated with water are not true colors but the result of colloidal suspension and are referred to as the *apparent color*. This apparent color can often be attributed to iron and to dissolved tannin extracted from decaying plant material. *True color* is the result of dissolved chemicals (most often organics) that cannot be seen.

TABLE 14.4

Common Metals Found in Water

Metal	Health Hazard
Barium	Circulatory system effects and increased blood pressure
Cadmium	Concentration in the liver, kidneys, pancreas, and thyroid
Copper	Nervous system damage and kidney effects; toxic to humans
Lead	Same as copper
Mercury	Central nervous system (CNS) disorders
Nickel	CNS disorders
Selenium	CNS disorders
Silver	Gray skin
Zinc	Taste effects; not a health hazard

Source: Spellman, F.R., *Handbook of Water and Wastewater Treatment Plant Operations*, 3rd ed., Lewis Publishers, Boca Raton, FL, 2013.

14.1.2.7.4 Dissolved Oxygen (DO)

Gases, including oxygen, carbon dioxide, hydrogen sulfide, and nitrogen, can be dissolved in water. Gases dissolved in water are important. For example, carbon dioxide is important because of the role it plays in pH and alkalinity. Carbon dioxide is released into the water by microorganisms and is consumed by aquatic plants. Dissolved oxygen (DO) in water is of the most importance to water-works operators because it is an indicator of water quality. We stated earlier that solutions could become saturated with solute. This is also the case with water and oxygen. The amount of oxygen that can be dissolved at saturation depends upon temperature of the water. However, in the case of oxygen, the effect is just the opposite of other solutes. The higher the temperature is, the lower the saturation level; the lower the temperature is, the higher the saturation level.

14.1.2.7.5 Metals

Metals are common constituents or impurities often carried by water. At normal levels, most metals are not harmful; however, a few metals can cause taste and odor problems in drinking water. Some metals may be toxic to humans, animals, and microorganisms. Most metals enter water as part of compounds that ionize to release the metal as positive ions. Table 14.4 lists some metals commonly found in water and their potential health hazards.

14.1.2.7.6 Organic Matter

Organic matter or organic compounds are those that contain the element carbon and are derived from material that was once alive (i.e., plants and animals):

- Fats
- Dyes
- Soaps
- Rubber product
- Wood
- Fuels
- Cotton
- Proteins
- Carbohydrates

Organic compounds in water are usually large, nonpolar molecules that do not dissolve well in water. They often provide large amounts of energy to animals and microorganisms.

TABLE 14.5
Relative Strengths of Acids in Water

Acid	Formula
Perchloric acid	$HClO_4$
Sulfuric acid	$H2SO_4$
Hydrochloric acid	HCl
Nitric acid	HNO_3
Phosphoric acid	H_3PO_4
Nitrous acid	HNO_2
Hydrofluoric acid	HF
Acetic acid	CH_3COOH
Carbonic acid	H_2CO_3
Hydrocyanic acid	HCN
Boric acid	H_3BO_3

Source: Spellman, F.R., *Handbook of Water and Wastewater Treatment Plant Operations*, 3rd ed., Lewis Publishers, Boca Raton, FL, 2013.

14.1.2.7.7 Inorganic Matter

Inorganic matter or inorganic compounds are carbon free, not derived from living matter, and easily dissolved in water; they are of mineral origin. The inorganics include acids, bases, oxides, salts, etc. Several inorganic components are important in establishing and controlling water quality.

14.1.2.7.8 Acids

An acid is a substance that produces hydrogen ions (H^+) when dissolved in water. Hydrogen ions are hydrogen atoms that have been stripped of their electrons. A single hydrogen ion is nothing more than the nucleus of a hydrogen atom. Lemon juice, vinegar, and sour milk are acidic or contain acid. The common acids used in treating water are hydrochloric acid (HCl), sulfuric acid (H_2SO_4), nitric acid (HNO_3), and carbonic acid (H_2CO_3). Note that in each of these acids hydrogen (H) is one of the elements. The relative strengths of acids in water, listed in descending order of strength, are shown in Table 14.5.

14.1.2.7.9 Bases

A base is a substance that produces hydroxide ions (OH^-) when dissolved in water. Lye or common soap (bitter things) contains bases. Bases used in waterworks operations are calcium hydroxide ($Ca(OH)_2$), sodium hydroxide ($NaOH$), and potassium hydroxide (KOH). Note that the hydroxyl group (OH) is found in all bases. Certain bases also contain metallic substances, such as sodium (Na), calcium (Ca), magnesium (Mg), and potassium (K). These bases contain the elements that produce the alkalinity in water.

14.1.2.7.10 Salts

When acids and bases chemically interact, they neutralize each other. The compound other than water that forms from the neutralization of acids and bases is a salt. Salts constitute, by far, the largest groups of inorganic compounds. A common salt used in waterworks operations, copper sulfate, is used to kill algae in water.

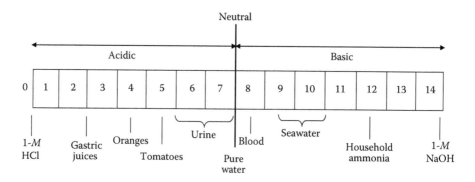

FIGURE 14.3 pH of selected liquids. (From Spellman, F.R., *Handbook of Water and Wastewater Treatment Plant Operations*, 3rd ed., Lewis Publishers, Boca Raton, FL, 2013.)

14.1.2.7.11   pH

pH is a measure of the hydrogen ion (H^+) concentration. Solutions range from very acidic (having a high concentration of H^+ ions) to very basic (having a high concentration of OH^- ions). The pH scale ranges from 0 to 14, with 7 being the neutral value (see Figure 14.3).

The pH of water is important to the chemical reactions that take place within water, and pH values that are too high or low can inhibit growth of microorganisms. High pH values are considered basic, and low pH values are considered acidic. Stated another way, low pH values indicate a high level of H^+ concentration, while high pH values indicate a low H^+ concentration. Because of this inverse logarithmic relationship, there is a tenfold difference in H^+ concentration. The pH is the logarithm of the reciprocal of the molar concentration of the hydrogen ion. In mathematical form,

$$pH = \log \frac{1}{H^+} \qquad (14.7)$$

Natural water varies in pH depending on its source. Pure water has a neutral pH, with an equal number of H^+ and OH^-. Adding an acid to water causes additional positive ions to be released so the H^+ ion concentration goes up and the pH value goes down:

$$HCl \rightarrow H^+ + Cl^-$$

Changing the hydrogen ion activity in solution can shift the chemical equilibrium of water. Thus, pH adjustment is used to optimize coagulation, softening, and disinfection reactions and for corrosion control. To control water coagulation and corrosion, it is necessary for the waterworks operator to test for the hydrogen ion concentration of the water to obtain its pH. In coagulation tests, as more alum (acid) is added, the pH value is lowered. If more lime (alkali, base) is added, the pH value is raised. This relationship is important—and if good floc is formed, the pH should then be determined and maintained at that pH value until there is a change in the new water.

■ **EXAMPLE 14.6**

Problem: A 0.1-*M* solution of acetic acid has a hydrogen ion concentration of 1.3×10^{-3} *M*. Find the pH of the solution.

Solution:

$$pH = \log \frac{1}{H^+} = \log \frac{1}{1.3 \times 10^{-3}} = \log \left(\frac{10^3}{1.3} \right) \log 10^3 - \log 1.3 = 3 - \log 1.3 = 3 - 0.11 = 2.89$$

14.1.2.7.12 Alkalinity

Alkalinity is the capacity of water to accept protons (positively charged particles); it can also be defined as a measure of the ability of the water to neutralize an acid. Stated in even simpler terms: Alkalinity is a measure of water's capacity to absorb hydrogen ions without significant pH change (i.e., capacity to neutralize acids). Bicarbonates, carbonates, and hydrogen cause alkalinity compounds in a raw or treated water supply. Bicarbonates are the major components, due to the action of carbon dioxide on basic materials of soil; borates, silicates, and phosphates may be minor components. Alkalinity of raw water may also contain salts formed from organic acids, such as humic acid. Alkalinity in water acts as a buffer that tends to stabilize and prevent fluctuations in pH. It is usually beneficial to have significant alkalinity in water because it would tend to prevent quick changes in pH. Quick changes in pH interfere with the effectiveness of common water treatment processes. Low alkalinity also contributes to corrosive tendencies of water. When alkalinity is below 80 mg/L, it is considered low.

14.1.2.7.13 Hardness

Hardness may be considered a physical or chemical parameter of water. It represents the total concentration of calcium and magnesium ions, reported as calcium carbonate. Hardness causes soaps and detergents to be less effective and contributes to scale formation in pipes and boilers. Hardness is not considered a health hazard; however, lime precipitation or ion exchange must often soften water that contains hardness. Low hardness contributes to the corrosive tendencies of water. Hardness and alkalinity often occur together because some compounds can contribute both alkalinity and hardness ions. Hardness is generally classified as shown in Table 14.6.

■ EXAMPLE 14.7

Problem: Find the molarity and normality of a solution if 22.4 g of Na_2CO_3 are dissolved in water and the solution is made up to 500 mL.

Solution: The molecular weight of Na_2CO_3 is 106. Molarity is number of moles/volume in liters. Number of moles is actual weight/molecular weight. The net positive valence of Na_2CO_3 is $1 \times 2 = 2$. The equivalent weight of Na_2CO_3 is $106/2 = 53$. So,

$$\text{Molarity} = \frac{\text{Actual weight/molecular wt.}}{\text{Volume (L)}} = \frac{\text{Actual weight}}{\text{Molecular wt.} \times \text{volume (L)}} = \frac{22.4}{106 \times 0.500} = 0.42\ M$$

$$\text{Normality} = \frac{\text{Actual weight}}{\text{Equivalent weight} \times \text{volume (L)}} = \frac{22.4}{53 \times 0.500} = 0.85\ N$$

TABLE 14.6
Water Hardness

Classification	mg/L $CaCo_3$
Soft	0–75
Moderately hard	75–150
Hard	150–300
Very hard	Over 300

Source: Spellman, F.R., *Handbook of Water and Wastewater Treatment Plant Operations*, 3rd ed., Lewis Publishers, Boca Raton, FL, 2013.

■ **EXAMPLE 14.8**

Problem: Find the molarity of 0.03 moles of NaOH in 90 mL of solution.

Solution:

$$\frac{0.03}{0.090} = 0.33 \ M$$

■ **EXAMPLE 14.9**

Problem: Find the normality of 0.3 moles $CaCl_2$ in 300 mL of solution.

Solution:

$$\frac{2 \times 0.3}{0.300} = 2.0 \ N$$

■ **EXAMPLE 14.10**

Problem: Find the molality of 5.0 g of NaOH in 500 g of water.

Solution:

$$\frac{5.0}{50 \times 0.500} = 0.2 \ m$$

14.1.2.8 Simple Solutions and Dilutions

A simple dilution is one in which a unit volume of a liquid material of interest is combined with an appropriate volume of a solvent liquid to achieve the desired concentration. The dilution factor is the total number of unit volumes in which a material will be dissolved. The diluted material must then be thoroughly mixed to achieve the true dilution. For example, a 1:5 dilution entails combining 1 unit volume of diluent (the material to be diluted) plus 4 unit volumes of the solvent medium; 1 + 4 = 5 is the dilution factor. The fact that the number of moles or equivalents of solute does not change during dilution enables us to calculate the new concentration.

14.1.2.8.1 Simple Dilutions

We demonstrate the simple dilution or dilution factor method in the following. Suppose we have a can of orange juice concentrate that is usually diluted with 4 additional cans of cold water (the dilution solvent), giving a dilution factor of 5; that is, the orange juice concentrate represents 1 unit volume to which we have added 4 more cans (same unit volumes) of water. So, the orange juice concentrate is now distributed through 5 unit volumes. This would be called a 1:5 dilution, and the orange juice is now 1/5 as concentrated at it was originally. So, in a simple dilution, add one less unit volume of solvent than the desired dilution factor value.

14.1.2.8.2 Serial Dilutions

A serial dilution is simply a series of simple dilutions amplifying the dilution factor quickly, beginning with a small initial quantity of material (e.g., bacterial culture, a chemical, orange juice). The source of dilution material for each step comes from the diluted material of the previous. In a serial dilution, the *total dilution factor* at any point is the product of the *individual dilution factor* in each step up to it:

$$\text{Final dilution factor (DF)} = (DF_1)(DF_2)(DF_3) \ldots \qquad (14.8)$$

To demonstrate the final dilution factor calculation, a typical lab experiment involves a three-step 1:100 serial dilution of a bacterial culture. The initial step combines 1 unit volume culture (10 μL) with 99 unit volumes of broth (990 μL) for a 1:100 dilution. In the next step, 1 unit volume of the 1:100 dilution is combined with 99 unit volumes of broth to yield a total dilution of $1{:}100 \times 100 = 1{:}10{,}000$ dilution. Repeated again (the third step), the total dilution would now be $1{:}100 \times 10{,}000 = 1{:}1{,}000{,}000$ total dilution. The concentration of bacteria is now 1 million times less than in the original sample.

14.1.2.8.3 $C_iV_i = C_fV_f$ Method (Fixed Volumes of Specific Concentrations from Liquid Reagents)

When we are diluting solutions, the product of the concentration and volume of the initial solution must be equal to the product of the concentration and volume of the diluted solution when the same system of units is used in both solutions. This can be expressed as the following relationship:

$$C_i \times V_i = C_f \times V_f \tag{14.9}$$

where
  C_i = Concentration of initial solution.
  V_i = Volume of initial solution.
  C_f = Concentration of final solution.
  V_f = Volume of final solution.

■ EXAMPLE 14.11

Problem: How much water must be added to 60 mL of 1.3-*M* HCl solution to produce a 0.5-*M* HCl solution?

Solution:

$$C_i \times V_i = C_f \times V_f$$

$$1.3 \times 60 = 0.5 \times x$$

$$x = \frac{1.3 \times 60}{0.5} = 156 \text{ mL}$$

The volume of the final solution is 156 mL. The volume of water to be added is then the difference between the volumes of the two solutions:

$$156 - 60 = 96 \text{ mL of water}$$

14.1.2.8.4 Molar Solutions

Sometimes it may be more efficient to use molarity when calculating concentrations. A 1.0-*M* solution is equivalent to 1 *formula weight* (FW) of chemical dissolved in 1 liter of solvent (usually water). Formula weight is always given on the label of a chemical bottle (use molecular weight if it is not given), and is expressed in grams per mole.

■ EXAMPLE 14.12

Problem: Given the following data, determine how many grams of reagent to use: chemical FW = 195 g/mole; making a 0.15-*M* solution.

Solution:

$$195 \text{ g/mol} \times 0.15 \text{ mol/L} = 29.25 \text{ g/L}$$

■ **EXAMPLE 14.13**

Problem: A chemical has a FW of 190 g/mol and we need 25 mL (0.025 L) of a 0.15-*M* solution. How many grams of the chemical must be dissolved in 25 mL water to make this solution?

Solution:

$$\text{Number grams/desired volume (L)} = \text{Desired molarity (mol/L)} \times \text{FW (g/mol)}$$

Rearranging,

$$\text{Number grams} = \text{Desired volume (L)} \times \text{desired molarity (mol/L)} \times \text{FW (g/mol)}$$

$$\text{Number grams} = 0.025 \text{ L} \times 0.15 \text{ mol/L} \times 190 \text{ g/mol} = 0.7125 \text{ g/25 mL}$$

14.1.2.8.5 Percent Solutions

Many reagents are mixed as *percent solutions*. When working with a dry chemical it is mixed as dry mass (g) per volume, where the number of grams per 100 mL is the percent concentration. A 10% solution is equal to 10 g dissolved in 100 mL of solvent. In addition, if we want to make 3% NaCl we would dissolve 3 g NaCl in 100 mL water (or the equivalent for whatever volume we need). When using liquid reagents, the percent concentration is based upon volume per volume (e.g., number mL/100 mL). For example, if we want to make 70% ethanol we would mix 70 mL of 100% ethanol with 30 mL water (or the equivalent for whatever volume we need). To convert from percent solution to molarity, multiply the percent solution value by 10 to get grams per liter, then divide by the formula weight:

$$\text{Molarity} = \frac{(\% \text{ Solution}) \times 10}{\text{Formula weight}} \tag{14.10}$$

■ **EXAMPLE 14.14**

Problem: Convert a 6.5% solution of a chemical with FW = 351 to molarity.

Solution:

$$\frac{\left(6.5 \text{ g/100 mL}\right) \times 10}{351 \text{ g/L}} = 0.1852 \text{ } M$$

To convert from molarity to percent solution, multiply the molarity by the FW and divide by 10.

$$\% \text{ Solution} = \frac{\text{Molarity} \times \text{Formula weight}}{10} \tag{14.11}$$

■ **EXAMPLE 14.15**

Problem: Convert a 0.0045-*M* solution of a chemical having FW 176.5 to percent solution.

Solution:

$$\frac{0.0045 \text{ mol/L} \times 176.5 \text{ g/mol}}{10} = 0.08\%$$

14.1.2.9 Chemical Reactions

A fundamental tool in any environmental engineer's toolbox is a basic understanding of chemical reactions, reaction rates, and physical reactions; however, before discussing chemical reactivity, it is important that we review the basics of electron distribution and chemical and physical changes and to discuss the important role heat plays in both chemical and physical reactions. Simply stated, electron distribution around the nucleus is key to understanding chemical reactivity. Only electrons are involved in chemical changes; the nuclei of atoms are not altered in any way during chemical reactions. Electrons are arranged around the nucleus in a definite pattern or series of shells. In general, only the outer shell, or *valence*, electrons (i.e., the ones farthest from the nucleus) are affected during chemical change. The valence number or valence of an element indicates the number of electrons involved in forming a compound and the number of electrons it tends to gain or lose when combining with other elements.

Note: Positive valence indicates giving up electrons. Negative valence indicates accepting electrons.

- If the valence electrons are shared with other atoms, then a *covalent bond* is formed when the compound is produced.
- If the valence electrons are donated to another atom, then an *ionic bond* is formed when the compound is produced.
- A *hydrogen bond* occurs when an atom of hydrogen is attracted by rather strong forces to two or atoms instead of only one, so that it is considered to act as a bond between them. If an atom gains or loses one or more valence electrons, it becomes an ion (charged particle).
- *Cations* are positively charged particles.
- *Anions* are negatively charged particles.

In physical and chemical changes, recall that a chemical change is the change physical substances undergo when they become new or different substances. To identify a chemical change look for observable signs such as color change, light production, smoke, bubbling or fizzing, and presence of heat. A physical change occurs when objects undergo a change that does not change their chemical nature. A physical change involves a change in physical properties. Physical properties can be observed without changing the type of matter. Examples of physical properties include texture, size, shape, color, odor, mass, volume, density, and weight.

An *endothermic reaction* is a chemical reaction that absorbs energy and where the energy content of the products is greater than that of the reactant heat taken in by the system. An *exothermic reaction* is a chemical reaction that gives out energy and where the energy content of the products is less than that of the reactants; heat is given out from the system.

14.1.2.9.1 Types of Chemical Reactions

Chemical reactions are of fundamental importance throughout chemistry and related technologies. Although experienced chemists can sometimes predict the reactions that will occur in a new chemical system, they may overlook some alternatives. Further, they are usually unable to make reliable predictions when the chemistry is unfamiliar to them.

Key Point: There are few tools available to assist in predicting chemical reactions, and none at all for predicting the novel reactions that are of greatest interest. For more information on predicting chemical reactions, see the work by Irikura and Johnson (2000).

Just as it is convenient to classify elements as gaseous and nongaseous, it is convenient to classify chemical reactions. There are so many chemical reactions that it is helpful to classify them into four general types. For example, there are over 450 named reactions listed in the Merck Index. The four general types of chemical reaction are

- *Combination reactions*—In a combination (or synthesis) reaction, two or more simple substances combine to form a more complex compound.

    Copper (an element) + Oxygen (an element) = Copper oxide (a compound)

Note: There are many pairs of reactants that combine to give a single product. The reactions happen when it is energetically favorable to do so.

- *Decomposition reactions*—Decomposition reactions occur when one compound breaks down (decomposes) into two or more substances or its elements. Basically, combination and decomposition reactions are opposites.

    Hydrogen peroxide $\Rightarrow H_2O + O_2$

- *Replacement reactions*—Replacement reactions involve the substitution of one uncombined element for another in a compound. Two reactants yield two products.

    Iron (Fe) + Sulfuric acid (H_2SO_4) = Hydrogen (H_2) + Iron sulfate $(FeSO_4)$

- *Double replacement reactions*—In a double replacement reaction parts of two compounds exchange places to form two new compounds. Two reactants yield two products:

    Sodium hydroxide + Acetic acid = Sodium acetate + Water

Key Point: Note that these four classifications of chemical reactions are not based on the type of bonding, which generally can be either covalent or ionic, depending on the reactants involved.

Note: According to the law of conservation of mass, no mass is added or removed in chemical reactions.

14.1.2.9.2 Specific Types of Chemical Reactions

- *Hydrolysis* (*hydro*, water; *lysis*, to break) is a decomposition reaction involving the splitting of water into its ions and the formation of a weak acid or base or both.
- *Neutralization* is a double replacement reaction that unites the H^- ion and an acid with the OH^- ion of a base, forming water and a salt:

    Acid + Acid $\Rightarrow$ Salt + Water

- *Oxidation/reduction (redox)* reactions are combination reactions, replacement, or double-replacement reactions that involve the gain and loss of electrons (i.e., changes in valence). Oxidation and reduction always occur simultaneously, such that one reacting species is oxidized while the other is reduced.

Note: When an atom, either free or in a molecule or ion, loses electrons, it is oxidized, and its oxidation number increases. When an atom, either free or in a molecule or ion, gains electrons, it is reduced, and its oxidation number decreases.

- *Chelation* is a combination reaction in which a ligand (such as a solvent molecule or simple ion) forms more than one bond to a central ion, giving rise to complex ions or coordination compounds.
- *Free radical reaction* is any type of reaction that involves any species with an unpaired electron. Free radical reactions frequently occur in the gas phase. They often proceed by chain reaction and are initiated by light, heat, or reagents. They contain unpaired electrons (such as oxides or peroxide decomposition products) and often are very reactive.

DID YOU KNOW?

The term *chelate* was first applied in 1920 by Sir Gilbert T. Morgan and H.D.K Drew (1920), who stated: "The adjective chelate, derived from the great claw or chela (*chely*—Greek) of the lobster or other crustaceans, is suggested for the caliperlike groups which function as two associating units and fasten to the central atom so as to produce heterocyclic rings."

DID YOU KNOW?

A common free radical reaction in aqueous solution is *electron transfer*, especially to the hydroxyl radical and to ozone.

- *Photolysis* (*photo*, light; *lysis*, to break) is generally a decomposition reaction in which the adsorption of light produces a photochemical reaction. Photolysis reactions form free radicals that can undergo other reactions. For example, the chlorine molecule can dissociate in the presence of high-energy light (for example, ultraviolet light).

$$Cl_2 + UV \text{ Energy} \Rightarrow Cl^- + Cl^-$$

Key Point: Photolysis is an important nonthermal technology used for treating dioxin and furan hazardous wastes.

- *Polymerization* is a combination reaction in which small organic molecules are linked together to form long chains, or complex two- and three-dimensional networks. Polymerization occurs only in molecules having double or triple bonds and is usually dependent on temperature, pressure, and a suitable catalyst. Free radical chain reactions are a common mode of polymerization. The term "chain reaction" is used because each reaction produces another reactive species (i.e., another free radical) to continue the process.

Note: Catalysts are agents that change the speed of a chemical reaction without affecting the yield or undergoing permanent chemical change.

Key Point: Plastics are perhaps the most common polymers, but there are also many important biopolymers such as polysaccharides.

- *Biochemical reactions* are reactions that occur in living organisms.
- *Biodegradation* is a decomposition reaction that occurs in microorganisms to create smaller, less complex inorganic and organic molecules. Usually the products of biodegradation are molecular forms that tend to occur in nature.

14.1.2.9.3   Reaction Rates (Kinetics)

The rate of a chemical reaction is a measure of how fast the reaction proceeds—that is, how fast reactants are consumed and products are formed.

$$A + B \text{ (reactants)} \Rightarrow C + D \text{ (products)}$$

The rate of a given reaction depends on many variables, including temperature, concentration of the reactants, catalysts, structure of the reactants, and the pressure of gaseous reactants or products. Without considering extreme conditions, in general reaction rates increase due to the following:

- *Increasing temperature*—Two molecules will only react if they have enough energy. By heating the mixture, the energy levels of the molecules involved in the reaction are raised. Raising the temperature means that the molecules move faster. This is kinetic theory.
- *Increasing concentration of reactants*—Increasing the concentration of the reactants increases the frequency of collisions between the two reactants (collision theory again).
- *Introducing catalysts*—Catalysts speed up chemical reactions. Catalysts speed up reactions by lowering the activation energy. Only very small quantities of the catalyst are required to produce a dramatic change in the rate of the reaction. This is the case because the reaction proceeds by a different pathway when the catalyst is present. Note that adding *more* catalyst will make absolutely no difference.
- *Increasing surface area*—The larger the surface area of a solid, the faster the reaction will be. Smaller particles have a bigger surface area than larger particles for the same mass of solid. There is a simple way to visualize this so-called "bread and butter" theory. If we take a loaf of bread and cut it into slices, we get an extra surface onto which we can spread butter. The thinner we cut the slices, the more slices we get and the more butter we can spread on them. Also, by chewing our food we increase the surface area so that digestion goes faster.
- *Increasing the pressure on a gas to increase the frequency of collisions between them*— By increasing pressure, molecules are squeezed together so that the frequency of collisions between them is increased.

Reaction rates are also affected by each reaction's activation energy—the energy the reactants must reach before they can react.

Note: A catalyst may be recovered unaltered at the end of the reaction.

Forward and backward reactions can occur, each with a different reaction rate and associated activation energy:

$$A + B \Leftrightarrow C + D$$

For example, in a dissociation reaction that occurs readily, initially the dissociation takes place at a faster rate than recombination. Eventually, as the concentration of dissociated ions build up, the rate of recombination catches up with the rate of dissociation. When the forward and backward reactions eventually occur at the same rate, a state of equilibrium is reached. The apparent effect is no change, even though both the forward and backward reactions are still occurring.

Key Point: Note that the point at which equilibrium is reached is not fixed, but it is also dependent on such variables as temperature, reactant concentration, pressure, and reactant structure.

14.1.2.9.4 Types of Physical Reactions

Knowledge of the physical behavior of wastes and hazardous wastes has been used to develop various unit processes for waste treatment that are based on physical reactions. These operations include the following:

- *Phase separation* involves separation of components of a mixture that is already in two different phases. Types of phase separation include filtration, settling, decanting, and centrifugation.
- *Phase transition* is a physical reaction in which a material changes from one physical phase to another. Types of phase transition include distillation, evaporation, precipitation, and freeze drying (lyophilization).
- *Phase transfer* consists of the transfer of a solute in a mixture from one phase to another. Two examples of phase transfer include extraction and sorption (i.e., transfer of a substance from a solution to a solid phase).

14.1.2.9.5   Chemical Equations Encountered in Water/Wastewater Operations

$Cl_2 + H_2O \leftrightarrow HCl + HOCl$

$NH_3 + HOCl \leftrightarrow NH_2Cl + H_2O$

$NH_2Cl + HOCl \leftrightarrow NHCl_2 + H_2O$

$NHCl_2 + HOCl \leftrightarrow NCl_3 + H_2O$

$Ca(OCl)_2 + Na_2CO_3 \leftrightarrow 2NaOCl + CaCO_3$

$Al_2(SO_4)_3 + 3CaCO_3 + 3H_2O \leftrightarrow Al_2(OH)_6 + 3CaSO_4 + 3CO_2$

$CO_2 + H_2O \leftrightarrow H_2CO_3$

$H_2CO_3 + CaCO_3 \leftrightarrow Ca(HCO_3)_2$

$Ca(HCO_3)_2 + Na_2CO_3 \leftrightarrow CaCO_3 + 2NaHCO_3$

$CaCO_3 + H_2SO_4 \leftrightarrow CaSO_4 + 2H_2CO_3$

$Ca(HCO_3)_2 + H_2SO_4 \leftrightarrow CaSO_4 + 2H_2CO_3$

$H_2S + Cl_2 \rightarrow 2HCl + S°\downarrow$

$H_2S + 4Cl_2 + 4H_2O \rightarrow H_2SO_4 + 8HCl$

$SO_2 + H_2O \rightarrow H_2SO_3$

$HOCl + H_2SO_3 \rightarrow H_2SO_4 + HCl$

$NH_2Cl + H_2SO_3 + H_2O \rightarrow NH_4HSO_4 + HCl$

$Na_2SO_4 + Cl_2 + H_2O \rightarrow Na_2SO_4 + 2HCl$

14.1.2.10   Chemical Dosages (Water and Wastewater Treatment)

Chemicals are used extensively in water/wastewater treatment plant operations. Water/wastewater treatment plant operators add chemicals to various unit processes for slime growth control, corrosion control, odor control, grease removal, biochemical oxygen demand (BOD) reduction, pH control, sludge bulking control, ammonia oxidation, bacterial reduction, and fluoridation, among other reasons. To apply any chemical dose correctly it is important to be able to make certain dosage calculations. One of the most frequently used calculations in water/wastewater mathematics is the conversion of milligrams per liter (mg/L) to pounds per day (lb/day) or pounds (lb) dosage or loading. The general types of mg/L to lb/day or lb calculations are for chemical dosage, BOD, chemical oxygen demand (COD), or suspended solids (SS) loading/removal; pounds of solids under aeration; and waste activated sludge (WAS) pumping rate. These calculations are usually made using either of the following equations:

$$\text{mg/L} \times \text{Flow (MGD)} \times 8.34 \text{ lb/gal} = \text{lb/day} \qquad (14.12)$$

$$\text{mg/L} \times \text{Volume (MG)} \times 8.34 \text{ lb/gal} = \text{lb} \qquad (14.13)$$

If mg/L concentration represents a concentration in a flow, then million gallons per day (MGD) flow is used as the second factor. However, if the concentration pertains to a tank or pipeline volume, then million gallons (MG) volume is used as the second factor.

14.1.2.10.1   Chlorine Dosage

Chlorine is a powerful oxidizer commonly used in water treatment for purification and in wastewater treatment for disinfection, odor control, bulking control, and other applications. When chlorine is added to a unit process, we want to ensure that a measured amount is added. The amount of chemical added or required can be specified in two ways:

- Milligrams per liter (mg/L)
- Pounds per day (lb/day)

To convert from mg/L (or ppm) concentration to lb/day, we use the following equation:

$$\text{mg/L} \times \text{MGD} \times 8.34 \text{ lb/gal} = \text{lb/day} \qquad (14.14)$$

Note: In previous years, it was normal practice to use the expression *parts per million* (ppm) as an expression of concentration, as 1 mg/L = 1 ppm; however, current practice is to use mg/L as the preferred expression of concentration.

■ EXAMPLE 14.16

Problem: Determine the chlorinator setting (lb/day) required to treat a flow of 8 MGD with a chlorine dose of 6 mg/L.

Solution:

$$mg/L \times MGD \times 8.34\ lb/gal = 6\ mg/L \times 8\ MGD \times 8.34\ lb/gal = 400\ lb/day$$

■ EXAMPLE 14.17

Problem: What should the chlorinator setting be (lb/day) to treat a flow of 3 MGD if the chlorine demand is 12 mg/L and a chlorine residual of 2 mg/L is desired?

Note: The chlorine demand is the amount of chlorine used in reacting with various components of the wastewater such as harmful organisms and other organic and inorganic substances. When the chlorine demand has been satisfied, these reactions stop.

To find the unknown value of lb/day, we must first determine chlorine dose. To do this we must use Equation 14.15:

$$\text{Chlorine dose (mg/L)} = \text{Chlorine demand (mg/L)} + \text{Chlorine residual (mg/L)} \qquad (14.15)$$

$$\text{Chlorine dose} = 12\ mg/L + 2\ mg/L = 14\ mg/L$$

Then we can make the mg/L to lb/day calculation:

$$mg/L \times MGD \times 8.34\ lb/gal = lb/day$$

$$12\ mg/L \times 3\ MGD \times 8.34\ lb/gal = 300\ lb/day$$

14.1.2.10.2 Hypochlorite Dosage

At many wastewater facilities, sodium hypochlorite or calcium hypochlorite is used instead of chlorine. The reasons for substituting hypochlorite for chlorine vary; however, due to the passage of stricter hazardous chemicals regulations by the Occupational Safety and Health Administration (OSHA) and the U.S. Environmental Protection Agency (USEPA), many facilities are deciding to substitute the hazardous chemical chlorine with nonhazardous hypochlorite. Obviously, the potential liability involved with using deadly chlorine is also a factor involved in the decision to substitute it with a less toxic chemical substance. For whatever reason, when a wastewater treatment plant decides to substitute chlorine for hypochlorite, the wastewater operator needs to be aware of the differences between the two chemicals. Chlorine is a hazardous material. Chlorine gas is used in wastewater treatment applications as 100% available chlorine. This is an important consideration to keep in mind when making or setting chlorine feed rates. For example, if the chlorine demand and residual require 100 lb/day chlorine, the chlorinator setting would be just that—100 lb/24 hr. Hypochlorite is less hazardous than chlorine; it is similar to strong bleach and comes in two forms: dry calcium hypochlorite (often referred to as HTH) and liquid sodium hypochlorite. Calcium hypochlorite contains about 65% available chlorine; sodium hypochlorite contains about 12 to 15% available chlorine (in industrial strengths).

Note: Because neither type of hypochlorite is 100% pure chlorine, more lb/day must be fed into the system to obtain the same amount of chlorine for disinfection. This is an important economical consideration for those facilities thinking about substituting hypochlorite for chlorine. Some studies indicate that such a switch can increase overall operating costs by up to three times the cost of using chlorine.

To calculate the lb/day hypochlorite required, a two-step calculation is necessary:

$$mg/L \times MGD \times 8.34 \ lb/gal = lb/day$$

$$\frac{Chlorine \ (lb/day)}{\% \ Available} \times 100 = Hypochlorite \ (lb/day)$$

■ EXAMPLE 14.18

Problem: A total chlorine dosage of 10 mg/L is required to treat a particular wastewater. If the flow is 1.4 MGD and the hypochlorite has 65% available chlorine, how many pounds per day of hypochlorite will be required?

Solution:

1. Calculate the lb/day chlorine required using the mg/L to lb/day equation:

$$mg/L \times MGD \times 8.34 \ lb/gal = lb/day$$

$$10 \ mg/L \times 1.4 \ MGD \times 8.34 \ lb/gal = 117 \ lb/day$$

2. Calculate the lb/day hypochlorite required. Because only 65% of the hypochlorite is chlorine, more than 117 lb/day will be required:

$$\frac{117 \ lb/day \ chlorine}{65\% \ available \ chlorine} \times 100 = 180 \ lb/day \ hypochlorite$$

■ EXAMPLE 14.19

Problem: A wastewater flow of 840,000 gpd requires a chlorine dose of 20 mg/L. If sodium hypochlorite (15% available chlorine) is to be used, how many pounds per day of sodium hypochlorite are required? How many gallons per day of sodium hypochlorite is this?

Solution:

1. Calculate the lb/day chlorine required:

$$mg/L \times MGD \times 8.34 \ lb/gal = lb/day$$

$$20 \ mg/L \times 0.84 \ MGD \times 8.34 \ lb/gal = 140 \ lb/day \ chlorine$$

2. Calculate the lb/day sodium hypochlorite:

$$\frac{140 \ lb/day \ chlorine}{15\% \ available \ chlorine} \times 100 = 933 \ lb/day \ hypochlorite$$

3. Calculate the gpd sodium hypochlorite:

$$\frac{933 \text{ lb/day}}{8.34 \text{ lb/gal}} = 112 \text{ gal/day hypochlorite}$$

■ EXAMPLE 14.20

Problem: How many pounds of chlorine gas are necessary to treat 5,000,000 gallons of wastewater at a dosage of 2 mg/L?

Solution:

1. Calculate the pounds of chlorine required:

Volume (10^6 gal) × Chlorine concentration (mg/L) × 8.34 lb/gal = lb chlorine

2. Substitute:

(5×10^6 gal) × 2 mg/L × 8.34 lb/gal = 83 lb chlorine

14.1.2.10.3 Additional Dosage Calculations

■ EXAMPLE 14.21

Problem: Chlorine dosage at a treatment plant averages 112.5 lb/day. Its average flow is 11.5 MGD. What is the chlorine dosage in mg/L?

Solution:

$$\text{Dosage} = \frac{112.5 \text{ lb/day}}{11.5 \times 10^6 \text{ gal} \times 8.34 \text{ lb/gal}} = \frac{1.2}{10^6} = 1.2 \text{ ppm} = 1.2 \text{ mg/L}$$

■ EXAMPLE 14.22

Problem: To treat 700,000 gal of water, 25 lb of chlorine gas is used. The chlorine demand of the water is measured to be 2.4 mg/L. What is the residual chlorine concentration in the treated water?

Solution:

Total dosage = 25 lb/0.70 MG = 36 lb/MG

$$36 \text{ lb/MG} \times \frac{1 \text{ mg/L}}{8.34 \text{ lb/MG}} = 4.3 \text{ mg/L}$$

Residual chlorine = 4.3 mg/L − 2.4 mg/L = 1.9 mg/L

■ EXAMPLE 14.23

Problem: What is the daily amount of chlorine needed to treat a 10 MGD of water to satisfy a chlorine demand of 2.9 mg/L and to provide 0.6 mg/L residual chlorine?

Solution:

Total chlorine needed = 2.9 + 0.6 = 3.5 mg/L

Daily weight = 10 × 10^6 gal/day × 8.34 lb/gal × 3.5 mg/L × 1 L/10^6 = 292 lb/day

■ **EXAMPLE 14.24**

Problem: At a 12-MGD waterworks, a pump at the rate of 0.20 gpm feeds hydrofluosilicic acid (H_2SiF_6) as a 23% by weight solution. The specific gravity of the H_2SiF_6 solution is 1.191. What is the fluoride (F) dosage?

Solution:

$$\text{Pump rate} = 0.20 \text{ gpm} \times 1440 \text{ min/day} = 288 \text{ gal/day}$$

$$\text{Fluoride applied rate} = 288 \text{ gal/day} \times 0.23 = 66.2 \text{ gal/day}$$

$$\text{Weight of fluoride} = 66.2 \text{ gal/day} \times 8.34 \text{ lb/day} \times 1.191$$

$$\text{Weight of water} = 12 \times 10^6 \text{ gal/day} \times 8.34 \text{ lb/gal} \times 1.0$$

$$\text{Dosage} = \frac{\text{Weight of fluoride}}{\text{Weight of water}} = \frac{66.2 \times 8.34 \times 1.191}{12 \times 10^6 \times 8.34 \times 1.0} = \frac{6.58}{10^6} = 6.58 \text{ mg/L}$$

■ **EXAMPLE 14.25**

Problem: 10 mg/L of liquid alum with 60% strength is continuously fed to a raw water flow that averages 8.6 MGD. How much liquid alum will be used in a month (30 days)?

Solution:

$$1 \text{ mg/L} = 1 \text{ gal/MG}$$

$$\text{Required per day} = \frac{10}{0.60} \frac{\text{gal}}{\text{MG}} \times \frac{\text{MG}}{\text{day}} = 143 \text{ gal/day}$$

14.2 FUNDAMENTAL HYDRAULICS

Water/wastewater operators make pumpage and flow rate calculations during daily operations. In this section, we describe and perform fundamental pumping and flow rate calculations to review the foundational principles of advanced hydraulics operations for environmental professionals (introduced later in the text).

14.2.1 PRINCIPLES OF WATER HYDRAULICS

Hydraulics is defined as the study of fluids at rest and in motion. While basic principles apply to all fluids, for the moment we consider only those principles that apply to water/wastewater. Although much of the basic information that follows is concerned with the hydraulics of distribution systems (e.g., piping), it is important for the operator to understand (and environmental professional to review) these basics in order to more fully appreciate the function of pumps.

14.2.1.1 Weight of Air

Our study of basic water hydraulics begins with air. A blanket of air, many miles thick, surrounds the Earth. The weight of this blanket on a given square inch of the Earth's surface will vary according to the thickness of the atmospheric blanket above that point. At sea level, the pressure exerted is 14.7 pounds per square inch (psi). On a mountaintop, air pressure decreases because the blanket is not as thick.

14.2.1.2 Weight of Water

Because water must be both stored and moved in water supplies and because wastewater must be collected, processed in unit processes, and outfalled to its receiving body we must consider some basic relationships in the weight of water. One cubic foot of water weighs 62.4 lb and contains 7.48 gal, and 1 cubic inch of water weighs 0.0362 lb. Water 1 foot deep will exert a pressure of 0.43 psi on the bottom area (12 in. × 0.062 lb/in.^{3}). A column of water 2 feet high exerts 0.86 psi, one 10 feet high exerts 4.3 psi, and one 52 feet high exerts

$$52 \text{ ft} \times 0.43 \text{ psi/ft} = 22.36 \text{ psi}$$

A column of water 2.31 feet high will exert 1.0 psi. To produce a pressure of 40 psi requires a water column that is

$$40 \text{ psi} \times 2.31 \text{ ft/psi} = 92.4 \text{ ft}$$

The term *head* is used to designate water pressure in terms of the height of a column of water in feet. For example, a 10-foot column of water exerts 4.3 psi. This can be called 4.3-psi pressure or 10 feet of head. Another example: If the static pressure in a pipe leading from an elevated water storage tank is 37 psi, what is the elevation of the water above the pressure gauge? Remembering that 1 psi = 2.31 and that the pressure at the gauge is 37 psi,

$$37 \text{ psi} \times 2.31 \text{ ft/psi} = 85.5 \text{ ft (rounded)}$$

14.2.1.3 Weight of Water Related to the Weight of Air

The theoretical atmospheric pressure at sea level (14.7 psi) will support a column of water 34 feet high:

$$14.7 \text{ psi} \times 2.31 \text{ ft/psi} = 33.957, \text{ or } 34 \text{ ft}$$

At an elevation of 1 mile above sea level, where the atmospheric pressure is 12 psi, the column of water would be only 28 feet high (12 psi × 2.31 ft/psi = 27.72, or 28 ft).

If a tube is placed in a body of water at sea level (e.g., a glass, a bucket, water storage reservoir, lake, pool), water will rise in the tube to the same height as the water outside the tube. The atmospheric pressure of 14.7 psi will push down equally on the water surface inside and outside the tube. However, if the top of the tube is tightly capped and all of the air is removed from the sealed tube above the water surface, forming a *perfect vacuum*, the pressure on the water surface inside the tube will be 0 psi. The atmospheric pressure of 14.7 psi on the outside of the tube will push the water up into the tube until the weight of the water exerts the same 14.7 psi pressure at a point in the tube even with the water surface outside the tube. The water will rise 14.7 psi × 2.31 ft/psi = 34 ft. In practice, it is impossible to create a perfect vacuum, so the water will rise somewhat less than 34 feet; the distance it rises depends on the amount of vacuum created.

■ EXAMPLE 14.26

Problem: If enough air was removed from the tube to produce an air pressure of 9.7 psi above the water in the tube, how far will the water rise in the tube?

Solution: To maintain the 14.7 psi at the outside water surface level, the water in the tube must produce a pressure of 14.7 psi – 9.7 = 5.0 psi. The height of the column of water that will produce 5.0 psi is

$$5.0 \text{ psi} \times 2.31 \text{ ft/psi} = 11.5 \text{ ft (rounded)}$$

14.2.1.4 Water at Rest

Stevin's law deals with water at rest. Specifically, it states: "The pressure at any point in a fluid at rest depends on the distance measured vertically to the free surface and the density of the fluid." Stated as a formula, this becomes

$$p = w \times h \tag{14.16}$$

where

  p = Pressure in pounds per square foot (lb/ft^2 or psf).
  w = Density in pounds per cubic foot (lb/ft^3).
  h = Vertical distance in feet.

■ EXAMPLE 14.27

Problem: What is the pressure at a point 15 ft below the surface of a reservoir?

Solution: To calculate this, we must know that the density of water (w) is 62.4 lb/ft^3. Thus,

$$p = w \times h = 62.4 \ lb/ft^3 \times 15 \ ft = 936 \ lb/ft^2, \text{ or } 936 \text{ psf}$$

Waterworks/wastewater operators generally measure pressure in pounds per square inch rather than pounds per square foot; to convert, divide by 144 $in.^2/ft^2$ (12 in. × 12 in. = 144 $in.^2$):

$$p = \frac{936 \ lb/ft^2}{144 \ in.^2/ft^2} = 6.5 \ lb/in.^2 \text{ or psi}$$

14.2.1.5 Gauge Pressure

Recall that head is the height that a column of water would rise due to the pressure at its base. We demonstrated that a perfect vacuum plus atmospheric pressure of 14.7 psi would lift the water 34 feet. If we now open the top of the sealed tube to the atmosphere and enclose the reservoir, then increase the pressure in the reservoir, the water will again rise in the tube. Because atmospheric pressure is essentially universal, we usually ignore the first 14.7 psi of actual pressure measurements and measure only the difference between the water pressure and the atmospheric pressure; we call this *gauge pressure*.

■ EXAMPLE 14.28

Problem: Water in an open reservoir is subjected to the 14.7 psi of atmospheric pressure, but subtracting this 14.7 psi leaves a gauge pressure of 0 psi. This shows that the water would rise 0 ft above the reservoir surface. If the gauge pressure in a water main is 100 psi, how far would the water rise in a tube connected to the main?

Solution:

$$100 \text{ psi} \times 2.31 \text{ ft/psi} = 231 \text{ ft}$$

14.2.1.6 Water in Motion

The study of water flow is much more complicated than that of water at rest. It is important to have an understanding of these principles because the water/wastewater in a treatment plant and/or distribution/collection system is nearly always in motion (much of this motion is the result of pumping, of course).

14.2.1.7 Discharge

Discharge is the quantity of water passing a given point in a pipe or channel during a given period. It can be calculated by the formula:

$$Q = A \times V \tag{14.17}$$

where
Q = Flow, or discharge in cubic feet per second (cfs).
A = Cross-sectional area of the pipe or channel (ft^2).
V = Water velocity in feet per second (fps).

Discharge can be converted from cubic feet per second to other units such as gallons per minute (gpm) or million gallons per day (MGD) by using appropriate conversion factors.

■ EXAMPLE 14.29

Problem: A pipe 12 in. in diameter has water flowing through it at 10 fps. What is the discharge in (a) cfs, (b) gpm, and (c) MGD?

Solution: Before we can use the basic formula, we must determine the area (A) of the pipe. The formula for the area is

$$A = \pi \times \frac{D^2}{4} \tag{14.18}$$

where
π = Constant value 3.14159.
D = Diameter of the circle in feet.

Thus, the area of the pipe is

$$A = \pi \times \frac{D^2}{4} = 3.14159 \times \frac{1 \text{ ft}^2}{4} = 0.785 \text{ ft}^2$$

Now, we can determine the discharge in cubic feet per second for part (a):

$$Q = V \times A = 10 \text{ ft/sec} \times 0.785 \text{ ft}^2 = 7.85 \text{ ft}^3/\text{sec (cfs)}$$

For part (b), we need to know that 1 cfs is equal to 449 gpm, so 7.85 cfs × 449 gpm/cfs = 3525 gpm. Finally, for part (c), 1 MGD is equal to 1.55 cfs, so

$$\frac{7.85 \text{ cfs}}{1.55 \text{ cfs/MGD}} = 5.06 \text{ MGD}$$

14.2.1.8 Law of Continuity

The law of continuity states that the discharge at each point in a pipe or channel is the same as the discharge at any other point (if water does not leave or enter the pipe or channel). That is, under the assumption of steady-state flow, the flow that enters the pipe or channel is the same flow that exits the pipe or channel. In equation form, this becomes

$$Q_1 = Q_2 \text{ or } A_1V_1 = A_2V_2 \tag{14.19}$$

■ EXAMPLE 14.30

Problem: A pipe 12 in. in diameter is connected to a 6-in.-diameter pipe. The velocity of the water in the 12-in. pipe is 3 fps. What is the velocity in the 6-in. pipe?

Solution: Using the equation $A_1V_1 = A_2V_2$, we need to determine the area of each pipe:

$$\text{12-inch pipe:} \quad A = \pi \times \frac{D^2}{4} = 3.14159 \times \frac{1 \text{ ft}^2}{4} = 0.785 \text{ ft}^2$$

$$\text{6-inch pipe:} \quad A = \pi \times \frac{D^2}{4} = 3.14159 \times \frac{0.5 \text{ ft}^2}{4} = 0.196 \text{ ft}^2$$

The continuity equation now becomes

$$0.785 \text{ ft}^2 \times 3 \text{ ft/sec} = 0.196 \text{ ft}^2 \times V_2$$

Solving for V_2,

$$V_2 = \frac{0.785 \text{ ft}^2 \times 3 \text{ ft/sec}}{0.196 \text{ ft}^2} = 12 \text{ ft/sec or fps}$$

14.2.1.9 Pipe Friction

The flow of water in pipes is caused by the pressure applied behind it either by gravity or by hydraulic machines (pumps). The flow is retarded by the friction of the water against the inside of the pipe. The resistance of flow offered by this friction depends on the diameter of the pipe, the roughness of the pipe wall, and the number and type of fittings (bends, valves, etc.) along the pipe. It also depends on the speed of the water through the pipe—the more water you try to pump through a pipe, the more pressure it will take to overcome the friction. The resistance can be expressed in terms of the additional pressure needed to push the water through the pipe, in either pounds per square inch or feet of head. Because it is a reduction in pressure, it is often referred to as *friction loss* or *head loss*.

Friction loss increases as

- Flow rate increases.
- Pipe diameter decreases.
- Pipe interior becomes rougher.
- Pipe length increases.
- Pipe is constricted.
- Bends, fittings, and valves are added.

The actual calculation of friction loss is beyond the scope of this text. Many published tables give the friction loss in different types and diameters of pipe and standard fittings. What is more important here is recognition of the loss of pressure or head due to the friction of water flowing through a pipe.

One of the factors in friction loss is the roughness of the pipe wall. A number called the *C* factor indicates pipe wall roughness; the higher the *C* factor, the smoother the pipe.

Note: *C* factor is derived from the letter *C* in the Hazen–Williams equation for calculating water flow through a pipe.

Some of the roughness in the pipe will be due to the material; cast iron pipe will be rougher than plastic, for example. Additionally, the roughness will increase with corrosion of the pipe material and deposits sediments in the pipe. New water pipes should have a *C* factor of 100 or more; older pipes can have *C* factors very much lower than this. To determine the *C* factor, published tables are usually used. In addition, when the friction losses for fittings are factored in, other published tables are available to make the proper determinations. It is standard practice to calculate the head loss from fittings by substituting the *equivalent length of pipe*, which is also available from published tables.

14.2.2　Basic Pumping Calculations*

Certain computations used for determining various pumping parameters are important to environmental practitioners responsible for water/wastewater treatment plant operation.

Key Point: The rate of flow produced by a pump is expressed as the volume of water pumped during a given period.

14.2.2.1　Pumping Rates

The mathematical problems most often encountered by water/wastewater operators with regard to determining pumping rates are often determined by using either of the following equations:

$$\text{Pumping rate (gpm)} = \text{gallons/minutes} \tag{14.20}$$

$$\text{Pumping rate (gph)} = \text{gallons/hours} \tag{14.21}$$

■ EXAMPLE 14.31

Problem: The meter on the discharge side of the pump reads in hundreds of gallons. If the meter shows a reading of 110 at 2:00 p.m. and 320 at 2:30 p.m., what is the pumping rate expressed in gallons per minute?

Solution: The problem asks for pumping rate in gallons per minute (gpm), so we use Equation 14.20:

$$\text{Pumping rate (gpm)} = \text{gallons/minutes}$$

To solve this problem, we must first find the total gallons pumped (determined from the meter readings):

$$32,000 \text{ gal} - 11,000 \text{ gal} = 21,000 \text{ gal}$$

The volume was pumped between 2:00 p.m. and 2:30 p.m., for a total of 30 minutes. From this information, calculate the gpm pumping rate:

$$\text{Pumping rate} = \frac{21,000 \text{ gal}}{30 \text{ min}} = 700 \text{ gpm}$$

■ EXAMPLE 14.32

Problem: During a 15-minute pumping test, 16,400 gal were pumped into an empty rectangular tank. What is the pumping rate in gallons per minute?

Solution: The problem asks for the pumping rate in gallons per minute, so again we use Equation 14.20:

$$\text{Pumping rate} = \frac{16,400 \text{ gal}}{15 \text{ min}} = 1093 \text{ gpm (rounded)}$$

---

* The following examples are adapted from Wahren, U., *Practical Introduction to Pumping Technology*, Gulf Publishing, Houston, TX, 1997.

■ **EXAMPLE 14.33**

Problem: A tank 50 ft in diameter is filled with water to depth of 4 ft. To conduct a pumping test, the outlet valve to the tank is closed, and the pump is allowed to discharge into the tank. After 80 minutes, the water level is 5.5 ft. What is the pumping rate in gallons per minute?

Solution: We must first determine the volume pumped in cubic feet:

Volume pumped = Area of circle × depth = 0.785×50 ft $\times 50$ ft $\times 1.5$ ft = 2944 ft^3 (rounded)

Now convert the cubic-foot volume to gallons:

$$2944 \text{ ft}^3 \times 7.48 \text{ gal/ft}^3 = 22{,}021 \text{ gal (rounded)}$$

The pumping test was conducted over a period of 80 minutes. Using Equation 14.20, calculate the pumping rate in gallons per minute:

$$\text{Pumping rate} = \frac{22{,}021 \text{ gal}}{80 \text{ min}} = 275.3 \text{ gpm (rounded)}$$

14.2.3 CALCULATING HEAD LOSS

Pump head measurements are used to determine the amount of energy a pump can or must impart to the water; they are measured in feet. One of the principle calculations used in pumping problems is determining head loss. The following formula is used to calculate head loss:

$$H_f = K(V^2/2g) \tag{14.22}$$

where
  H_f = Friction head.
  K = Friction coefficient.
  V = Velocity in pipe.
  g = Gravity (32.17 ft/sec/sec).

14.2.4 CALCULATING HEAD

For centrifugal pumps and positive displacement pumps, several other important formulae are used in determining head. In centrifugal pump calculations, the conversion of the discharge pressure to discharge head is the norm. Positive displacement pump calculations often leave given pressures in psi. In the following formulas, the specific weight of a liquid is expressed in pounds per cubic foot. For water at 68°F, the specific weight is 62.4 lb/ft^3. A water column 2.31 ft high exerts a pressure of 1 psi on 64°F water.

Use the following formulas to convert discharge pressure in psi to head in feet:

- Centrifugal pumps

$$\text{Head (ft)} = \frac{\text{Pressure (psig)} \times 2.31}{\text{Specific gravity}} \tag{14.23}$$

- Positive displacement pumps

$$\text{Head (ft)} = \frac{\text{Pressure (psig)} \times 144}{\text{Specific weight}} \tag{14.24}$$

Use the following formulas to convert head into pressure:

- Centrifugal pumps

$$\text{Pressure (psi)} = \frac{\text{Head (ft)} \times \text{specific gravity}}{2.31} \tag{14.25}$$

- Positive displacement pumps

$$\text{Pressure (psi)} = \frac{\text{Head (ft)} \times \text{specific weight}}{144} \tag{14.26}$$

14.2.5 CALCULATING HORSEPOWER AND EFFICIENCY

When considering work being done, we consider the rate at which work is being done. This is called *power* and is labeled as foot-pounds per second. At some point in the past, it was determined that the ideal work animal, the horse, could move 550 pounds 1 foot in 1 second. Because large amounts of work are also to be considered, this unit became known as *horsepower*. When pushing a certain amount of water at a given pressure, the pump performs work. One horsepower equals 33,000 ft-lb/min. The two basic terms for horsepower are

- Hydraulic horsepower (WHP)
- Brake horsepower (BHP)

14.2.5.1 Hydraulic Horsepower
One hydraulic horsepower equals

- 550 ft-lb/sec
- 33,000 ft-lb/min
- 2545 British thermal units per hour (Btu/hr)
- 0.746 kw
- 1014 metric HP

To calculate the hydraulic horsepower (WHP) using flow in gpm and head in feet, use the following formula for centrifugal pumps:

$$\text{WHP} = \frac{\text{Flow (gpm)} \times \text{head (ft)} \times \text{specific gravity}}{3960} \tag{14.27}$$

When calculating horsepower for positive displacement pumps, common practice is to use psi for pressure. Then the hydraulic horsepower becomes

$$\text{WHP} = \frac{\text{Flow (gpm)} \times \text{pressure (psi)}}{3960} \tag{14.28}$$

14.2.6 PUMP EFFICIENCY AND BRAKE HORSEPOWER (BHP)

When a motor–pump combination is used (for any purpose), neither the pump nor the motor will be 100% efficient. Simply put, not all the power supplied by the motor to the pump (brake horsepower) will be used to lift the water (hydraulic horsepower); some of the power is used to overcome friction within the pump. Similarly, not all of the power of the electric current driving the motor (motor horsepower) will be used to drive the pump; some of the current is used to overcome friction within the motor, and some current is lost in the conversion of electrical energy to mechanical power.

Note: Depending on size and type, pumps are usually 50 to 85% efficient, and motors are usually 80 to 95% efficient. The efficiency of a particular motor or pump is given in the manufacturer's technical manual accompanying the unit.

A pump's brake horsepower equals its hydraulic horsepower divided by the pump's efficiency. Thus, the brake horsepower formula becomes

$$BHP = \frac{Flow\ (gpm) \times head\ (ft) \times specific\ gravity}{3960 \times efficiency} \tag{14.29}$$

or

$$BHP = \frac{Flow\ (gpm) \times pressure\ (psig)}{1714 \times efficiency} \tag{14.30}$$

■ EXAMPLE 14.34

Problem: Calculate the BHP requirements for a pump handling salt water and having a flow of 600 gpm with 40-psi pressure differential. The specific gravity of saltwater at 68°F equals 1.03. The pump efficiency is 85%.

Solution: Convert the pressure differential to total differential head (TDH) = 40 × 2.31/1.03 = 90 ft (rounded).

$$BHP = \frac{600 \times 90 \times 1.03}{3960 \times 0.85} = 16.5\ HP\ (rounded)$$

$$BHP = \frac{600 \times 40}{1714 \times 0.85} = 16.5\ HP\ (rounded)$$

Note: Horsepower requirements vary with flow. Generally, if the flow is greater, the horsepower required to move the water would be greater.

When the motor, brake, and hydraulic horsepower are known and the efficiency is unknown, a calculation to determine motor or pump efficiency must be done. Equation 14.31 is used to determine percent efficiency:

$$Percent\ efficiency = \frac{HP\ output}{HP\ input} \times 100 \tag{14.31}$$

From Equation 14.31, the specific equations to be used for motor, pump, and overall efficiency equations are

$$Percent\ motor\ efficiency = \frac{BHP}{MHP} \times 100 \tag{14.32}$$

$$Percent\ pump\ efficiency = \frac{WHP}{BHP} \times 100 \tag{14.33}$$

$$Percent\ overall\ efficiency = \frac{WHP}{MHP} \times 100 \tag{14.34}$$

■ **EXAMPLE 14.35**

Problem: A pump has a water horsepower requirement of 8.5 WHP. If the motor supplies the pump with 12 HP, what is the efficiency of the pump?

Solution:

$$\text{Percent pump efficiency} = \frac{\text{WHP output}}{\text{BHP supplied}} \times 100 = \frac{8.5}{12} \times 100 = 0.71 \times 100 = 71\%$$

■ **EXAMPLE 14.36**

Problem: What is the efficiency if an electric power equivalent to 25 HP is supplied to the motor and 14 HP of work is accomplished by the pump?

Solution: Calculate the percent of overall efficiency:

$$\text{Percent overall efficiency} = \frac{\text{WHP}}{\text{MHP}} \times 100 = \frac{14}{25} \times 100 = 0.56 \times 100 = 56\%$$

■ **EXAMPLE 14.37**

Problem: The motor is supplied with 12 kW of power. If the brake horsepower is 14 HP, what is the efficiency of the motor?

Solution: First, convert the kilowatt power to horsepower. Based on the fact that 1 HP = 0.746 kW, the equation becomes

$$\frac{12\text{ kW}}{0.746\text{ kW/HP}} = 16.09\text{ HP}$$

Now calculate the percent efficiency of the motor:

$$\text{Percent efficiency} = \frac{\text{BHP}}{\text{MHP}} \times 100 = \frac{14}{16.09} \times 100 = 0.87 \times 100 = 87\%$$

REFERENCES AND RECOMMENDED READING

Davies, P. (2013). *Types of Colloids.* University of Bristol, http://www.chm.bris.ac.uk/webprojects2002/pda-vies/types.html.

Irikura, K.K. and Johnson III, R.D. (2000). Predicting unexpected chemical reactions by isopotential searching. *Journal of Physical Chemistry A,* 104(11), 2191–2194.

Missen, R.W., Mims, C.A., and Saville, B.A. (1999). *Introduction to Chemical Reactions Engineering and Kinetics.* John Wiley & Sons, New York.

Morgan, G.T. and Drew, H.K. (1920). Researches on residual affinity and coordination. Part II. Acetylacetones of selenium and tellurium. *J. Chem. Soc.,* 117, 1456.

Oxlade, C. (2002). *Materials Changes and Reactions (Chemicals in Action).* Heinemann, Portsmouth, NH.

Spellman, F.R. (2013). *Handbook of Water and Wastewater Treatment Plant Operations,* 3rd ed. CRC Press, Boca Raton, FL.

Wahren, U. (1997). *Practical Introduction to Pumping Technology.* Gulf Publishing, Houston, TX.

Section VIII

Environmental Health and Safety Computations

"If seven maids with seven mops
Swept it for half a year,
Do you suppose," the Walrus said,
"That they could get it clear?"
"I doubt it," said the Carpenter,
And shed a bitter tear.

**—Lewis Carroll (*Through the Looking-
Glass and What Alice Found There*)**

Environmental practitioners who are involved in professional practice of environmental health, safety, or industrial hygiene must be well versed as generalists. Moreover, they must also become expert in an area of specialization within the broader field of industrial health and safety. An important element contributing to their total skill set is mathematical computations.

15 Basic Calculations for Occupational Safety and Environmental Health Professionals

General Manager:	Well, in my opinion, safety in the workplace is relative.
My response:	Safety is relative? Relative to what? Relative to how many are injured, made ill, or killed on-the-job?
General Manager:	[total silence]
My observation:	When dealing with mental midgets, sometimes silence is all that needs to be said.

—**F.R. Spellman (1989)**

15.1 INDUSTRIAL HYGIENE: WHAT IS IT?*

According to the Occupational Safety and Health Administration (OSHA), industrial hygiene is the science of anticipating, recognizing, evaluating, and controlling workplace conditions that may cause workers injury or illness. Industrial hygienists use environmental monitoring and analytical methods to detect the extent of worker exposure and employee engineering, administrative controls, and other methods, such as personal protective equipment (PPE), to control potential health hazards (OSHA, 1998).

Do you remember 9/11? How about the post office anthrax mess? Dumb questions, right? The proper question should be "How can we ever forget?" It started with the word being passed around about airplanes crashing into the Twin Towers. News coverage was everywhere. Remember those TV shots with the planes crashing, the towers falling? Over and over again those shots were replayed, etched into our human memory chips. We watched mesmerized, like fire watchers or falling water gazers, hypnotized. Our defaulted perception and memory chips could not believe what our eyes were showing us.

Later, during the frantic hunt for survivors, TV coverage continued. We saw the brave police, fire, and emergency responders doing what they do best—rescuing survivors. We saw construction workers and unidentified others climbing over and crawling through the tangled, smoking mess, the warped-steel, jackstraw-like mess, helping where they could. We saw others, too; for example, do you recall seeing folks walking around in what looked like space suits, instruments in hand? The average TV viewer who saw these space-suited people moving cautiously and deliberately through the smoking mass of debris and death and destruction had no idea who those dedicated professionals were. They were professionals doing what they do best: monitoring and testing the area to make sure it was safe for the responders and everyone else and ensuring that it was safe for the President, arm around a hero and standing on the rubble in the smoldering mess, to speak those resolute words we all needed to hear, words the terrorists needed to hear, words they are still hearing—loud and clear.

---

* Portions of this chapter are adapted from Spellman, F.R., *Industrial Hygiene Simplified*, Government Institutes Press, Lanham, MD, 2006.

Who were those space-suited individuals who not only appeared on our TV screens during the aftermath of 9/11 but were also prominent figures in footage of post offices victimized by anthrax? They were industrial hygienists. *Terrorism* and *bioterrorism* might have been new buzzwords a decade ago, but responding to hazard sites is nothing new to those space-suited folks whose function most people can't even identify. Times have changed, but the need for fully trained professional industrial hygienists has not.

On a different note, one might ask whether the industrial hygienist is a safety professional or an environmental professional. Either. It depends. The safety profession and environmental health and industrial hygiene have commonly been thought of as separate entities (this is especially the view taken by many safety professionals and industrial hygienists). In fact, over the years, a considerable amount of debate has risen among those in the safety and industrial hygiene professions regarding safety and health issues in the workplace, including exactly who is best qualified to administer a workplace safety program. Who is really the safety professional?

Historically, the safety professional had the upper hand in this argument—that is, prior to enactment of the Occupational Safety and Health Act (OSH Act), which mandated formation of the administrative entity OSHA. Until then, industrial hygiene was not a topic that many professionals thought about, cared about, or had any understanding of. Safety was safety—and the job title "safety professional" included environmental health protection—and that was that.

After passage of the OSH Act, however, things changed, and so did perceptions; in particular, people began to look at work injuries and work-related illnesses differently. In the past, they were regarded as separate problems. Why? The primary reason for this view was obvious and perhaps not so obvious. The obvious reason was work-related injuries. Work injuries occur suddenly and the agent (e.g., electrical source, chemical spill, machine, tool, work or walking surface, careless co-worker) is usually readily obvious. Not so obvious are the agents of work-related illnesses (e.g., lead, asbestos, formaldehyde). Why? Because most occupational illnesses develop rather slowly, over time; they are often insidious, like cancer. In asbestos exposure, for example, workers who remove asbestos-containing materials without the proper training (awareness) and personal protective equipment (PPE) are subject to exposure. Typically, asbestos exposure may be either a one-time exposure event (the silver bullet syndrome) or the exposure may go on for several years. No matter the length of exposure, pathological changes due to asbestos contamination occur slowly—some time will pass before the worker notices a difference in his pulmonary function. Disease from asbestos exposure has a latency period that may be as long as 20 to 30 years before the effects are realized (or diagnosed, in some cases).

The point? Any exposure to asbestos, short term or long term, may eventually lead to a chronic disease (i.e., asbestosis) that is irreversible. Of course, many other types of workplace toxic exposures can affect workers' health. The prevention, evaluation, and control of such occurrences comprise the role of the industrial hygienist.

Because of the OSH Act and because of increasing public awareness and involvement by unions in industrial health matters, the role of the industrial hygienist has continued to grow over the years. Certain colleges and universities have incorporated industrial hygiene majors into environmental health programs. Another result of the OSH Act has been, in effect (though many practitioners in the field disagree with this view), an ongoing tendency toward uniting safety and industrial hygiene into one entity, into one profession.

This trend presents a problem with definition. When we combine safety and industrial hygiene, do we combine them into one specific title or profession? Debate on this issue continues. What is the solution to this problem—how do we end the debate? Does it matter? Do we care? Not in this text. However, whether your title is safety professional, industrial hygienist, or environmental health professional, one thing is certain: You can't possibly perform your assigned duties correctly unless you are well versed in statistical and mathematical operations.

After introducing pertinent foundational information on stressors and standards, we describe areas in occupational health where mathematical methods and operations (other than statistical methods, already presented) are standard to the profession.

15.2   WORKPLACE STRESSORS

To ensure a healthy workplace environment and associated environs, the environmental professional focuses on the recognition, evaluation, and control of chemical, physical, or biological and ergonomic stressors that can cause sickness, impaired health, or significant discomfort to humans, animals, and plants. The key word just mentioned was *stressors*, or simply, *stress*—the stress caused by the external or internal environmental demands placed upon us. Increases in external stressors beyond a person's tolerance level affect his or her overall health and on-the-job performance.

The environmental practitioner must understand not only that environmental stressors exist but that they also are sometimes cumulative. For example, workplace studies have shown that some assembly-line processes are little affected by either low illumination or vibration; however, when these two stressors are combined, assembly-line performance deteriorates. Other cases have shown just the opposite effect. A worker who has had little sleep and then is exposed to a work area where noise levels are high actually benefits (to a degree, depending on the intensity of the noise level and the worker's exhaustion level) from the increased arousal level; a lack of sleep combined with a high noise level is compensatory.

In order to recognize environmental stressors and other factors that influence an individual's health, the environmental health practitioner must be familiar with work operations, home environment, lifestyle factors, and other processes. In the workplace, for example, an essential part of the new environmental professional's orientation process should include an overview of all pertinent company work operations and processes. Obviously, the newly hired environmental profession responsible for safety and health in the workplace who has not been fully indoctrinated on company work operations and processes is not qualified to study the environmental effects of such processes and suffers from another disability—lack of credibility with supervisors and workers (if you have not been there, then you are not one of us). This point cannot be emphasized strongly enough—if your intention is to correct or remove environmental stressors you must know your organization and what it is all about.

What are the workplace and general environmental stressors the industrial hygienist, safety professional, and environmental professional should be concerned with? The stressors of concern should be those that are likely to accelerate the aging process; they can cause significant discomfort and inefficiency, as well as chronic illness, or they may be immediately dangerous to life and health (Spellman, 1998). Several stressors fall into these categories; the most important work-related health stressors include chemical, physical, biological, and ergonomic stressors.

15.2.1   CHEMICAL STRESSORS

Chemical stressors are harmful chemical compounds in the form of solids, liquids, gases, mists, dusts, fumes, and vapors that exert toxic effects by inhalation (breathing), absorption (direct contact with the skin), or ingestion (eating or drinking). Airborne chemical hazards exist as concentrations of mists, vapors, gases, fumes, or solids. Some are toxic through inhalation and some of them irritate the skin on contact; some can be toxic by absorption through the skin or through ingestion, and some are corrosive to living tissue. The degree of individual risk from exposure to any given substance depends on the nature and potency of the toxic effects and the magnitude and duration of exposure.

15.2.2 Physical Stressors

Physical stressors include excessive levels of ionizing and nonionizing electromagnetic radiation, noise, vibration, illumination, and temperature. In occupations where there is exposure to *ionizing* radiation, factors such as time, distance, and shielding are important tools for ensuring worker safety. Danger from radiation increases with the amount of time one is exposed to it; hence, the shorter the time of exposure the smaller the radiation danger. Distance also is a valuable tool in controlling exposure to both ionizing and nonionizing radiation. Radiation levels from some sources can be estimated by comparing the squares of the distances between the work and the source. For example, at a reference point of 10 feet from a source, the radiation is 1/100 of the intensity at 1 foot from the source. Shielding also is a way to protect against radiation. The greater the protective mass between a radioactive source and the worker, the lower the radiation exposure. Nonionizing radiation is also dealt with by shielding workers from the source. Sometimes limiting exposure times to nonionizing radiation or increasing the distance is not effective. Laser radiation, for example, cannot be controlled effectively by imposing time limits. An exposure faster than the blink of an eye can be hazardous. Increasing the distance from a laser source may require miles before the energy level reaches a point where the exposure would not be harmful.

Noise, another significant physical hazard, can be controlled by various measures. Noise can be reduced by installing equipment and systems that have been engineered, designed, and built to operate quietly; by enclosing or shielding noisy equipment; by making certain that equipment is in good repair and properly maintained with all worn or unbalanced parts replaced; by mounting noisy equipment on special mounts to reduce vibration; and by installing silencers, mufflers, or baffles. Substituting quiet work methods for noisy ones is another significant way to reduce noise, for example, welding parts rather than riveting them. Also, treating floors, ceilings, and walls with acoustical material can reduce reflected or reverberant noise. In addition, erecting sound barriers at adjacent work stations around noisy operations will reduce worker exposure to noise generated at adjacent work stations.

It is also possible to reduce noise exposure by increasing the distance between the source and the receiver by isolating workers in acoustical booths, limiting workers' exposure time to noise, and by providing hearing protection. OSHA requires that workers in noisy surroundings be periodically tested as a precaution against hearing loss. Another physical hazard is radiant heat exposure in factories such as steel mills. Radiant heat can be controlled by installing reflective shields and by providing protective clothing.

15.2.3 Biological Stressors

Biological stressors include bacteria, viruses, fungi, and other living organisms that can cause acute and chronic infections by entering the body either directly or through breaks in the skin. Occupations that deal with plants or animals or their products or with food and food processing may expose workers to biological hazards. Laboratory and medical personnel also can be exposed to biological hazards. Any occupations that result in contact with bodily fluids pose a risk to workers from biological hazards. In occupations where animals are involved, biological hazards are dealt with by preventing and controlling diseases in the animal population as well as proper care and handling of infected animals. Also, effective personal hygiene, particularly proper attention to minor cuts and scratches, especially those on the hands and forearms, helps keep worker risks to a minimum. In occupations where there is potential exposure to biological hazards, workers should practice proper personal hygiene, particularly hand washing. Hospitals should provide proper ventilation, proper personal protective equipment such as gloves and respirators, adequate infectious waste disposal systems, and appropriate controls including isolation in instances of particularly contagious diseases such as tuberculosis.

15.2.4 Ergonomic Stressors

The science of ergonomics studies and evaluates a full range of tasks including, but not limited to, lifting, holding, pushing, walking, and reaching. Many ergonomic problems result from technological changes such as increased assembly line speeds, adding specialized tasks, and increased repetition; some problems arise from poorly designed job tasks. Any of those conditions can cause ergonomic hazards such as excessive vibration and noise, eye strain, repetitive motion, and heavy lifting problems. Improperly designed tools or work areas also can be ergonomic hazards. Repetitive motions or repeated shocks over prolonged periods of time as in jobs involving sorting, assembling, and data entry can often cause irritation and inflammation of the tendon sheath of the hands and arms, a condition known as carpal tunnel syndrome.

Ergonomic hazards are avoided primarily by the effective design of a job or jobsite and better designed tools or equipment that meet workers' needs in terms of physical environment and job tasks. Through thorough worksite analyses, employers can set up procedures to correct or control ergonomic hazards by using the appropriate engineering controls (e.g., designing or redesigning work stations, lighting, tools, and equipment); teaching correct work practices (e.g., proper lifting methods); employing proper administrative controls (e.g., shifting workers among several different tasks, reducing production demand, increasing rest breaks); and, if necessary, providing and mandating personal protective equipment. Evaluating working conditions from an ergonomics standpoint involves looking at the total physiological and psychological demands of the job on the worker.

Overall, environmental health professionals point out that the benefits of a well-designed, ergonomic work environment can include increased efficiency, fewer accidents, lower operating costs, and more effective use of personnel. In the workplace, the environmental health professional should review the following to anticipate potential health stressors:

- Raw materials
- Support materials
- Chemical reactions
- Chemical interactions
- Products
- Byproducts
- Waste products
- Equipment
- Operating procedures

15.3 ENVIRONMENTAL HEALTH AND SAFETY TERMINOLOGY

Every branch of science, every profession, and every engineering process has its own language for communication. Environmental health and safety practice is no different.

Abatement period—The amount of time given an employer to correct a hazardous condition that has been cited.

Absorption—The taking up of one substance by another, such as a liquid by a solid or a gas by a liquid.

Accident—This term is often misunderstood and is often mistakenly used interchangeably with injury. The meanings of the two terms are different, of course. Let's look at the confusion caused by the different definitions supplied for the term accident. The dictionary defines an accident as "a happening or event that is not expected, foreseen or intended." Defined differently, "an accident is an event or condition occurring by chance or arising from an unknown or remote cause." The legal definition is "an unexpected happening causing loss or injury which is not due to any fault or misconduct on the part of the person injured, yet entitles some kind of legal relief." Are you confused? Stand by. The following definition will help clear your heads (Haddon et al., 1964, p. 28):

> With rare exception, an accident is defined, explicitly or implicitly, by the unexpected occurrence of physical or chemical change to an animate or inanimate structure. It is important to note that the term covers only damage of certain types. Thus, if a person is injured by inadvertently ingesting poison, an accident is said to have taken place; but if the same individual is injured by inadvertently ingesting poliovirus, the result is but rarely considered accidental. This illustrates a curious inconsistency in the approach to accidents as opposed to other sources of morbidity, one which continues to delay progress in the field. In addition, although accidents are defined by the unexpected occurrence of damage, it is the unexpectedness, rather than the production and prevention of that damage per se, that has been emphasized by much of accident research. The approach is not justified by present knowledge and is in sharp contrast to the approach to the causation and prevention of other forms of damage, such as those produced by infectious organisms, where little, if any, attention is paid to the unexpectedness of the insults involved, and only their physical and biological nature is emphasized—with notable success.

Now you should have a better feel for what an accident really is; however, another definition, perhaps one more applicable to our needs, is provided by safety experts, the authors of the ASSE *Dictionary of Safety Terms.* Let's see how they define accident:

> An accident is an unplanned and sometimes injurious or damaging event which interrupts the normal progress of an activity and is invariably preceded by an unsafe act or unsafe condition thereof. An accident may be seen as resulting from a failure to identify a hazard or from some inadequacy in an existing system of hazard controls. Based on applications in casualty insurance, an event that is definite in point of time and place but unexpected as to either its occurrence or its results.

In this text we use the ASSE definition of accident.

Accident analysis—A comprehensive, detailed review of the data and information compiled from an accident investigation. An accident analysis should be used to determine causal factors only, and not to point the finger of blame at any one. Once the causal factors have been determined, corrective measures should be prescribed to prevent recurrence.

Accident prevention—The act of averting a circumstance that could cause loss or injury to a person.

Accommodation—The ability of the eye to quickly and easily readjust to other focal points after viewing a video display terminal (VDT) so as to be able to focus on other objects, particularly objects at a distance.

Acoustics—In general, the experimental and theoretical science of sound and its transmission; in particular, that branch of the science that has to do with the phenomena of sound in a particular space such as a room or theater. Safety engineering is concerned with the technical control of sound and involves architecture and construction, studying control of vibration, soundproofing, and the elimination of noise to engineer out the noise hazard.

Action level—Term used by OSHA and the National Institute for Occupational Safety and Health (NIOSH), a federal agency that conducts research on safety and health concerns, and defined in the Code of Federal Regulations (CFR), Title 40, Protection of Environment. Under OSHA, an action level is the level of toxicant that requires medical surveillance, usually 50% of the permissible exposure level (PEL). Note that OSHA also uses action levels in ways other than setting the level of a toxicant. For example, in its hearing conservation standard, 29 CFR 1910.95, OSHA defines the action level as an 8-hour time-weighted average (TWA) of 85 decibels measured on the A-scale, slow response, or equivalently, a dose of 50%. Under 40 CFR 763.121, action level means an airborne concentration of asbestos of 0.1 fiber per cubic centimeter (f/cc) of air calculated as an 8-hour time-weighted average.

Acute—Health effects that show up a short length of time after exposure. An acute exposure runs a comparatively short course and its effects are easier to reverse than those of a chronic exposure.

Acute toxicity—The discernible adverse effects induced in an organism within a short period of time (days) of exposure to an agent.

Adsorption—The taking up of a gas or liquid at the surface of another substance, usually a solid (e.g., activated charcoal adsorbs gases).

Aerosols—Liquid or solid particles so small they can remain suspended in air long enough to be transported over a distance.

Air contamination—The result of introducing foreign substances into the air so as to make the air contaminated.

Air pollution—Contamination of the atmosphere (indoor or outdoor) caused by the discharge (accidental or deliberate) of a wide range of toxic airborne substances.

Air sampling—Safety engineers are interested in knowing what contaminants workers are exposed to and the contaminant concentrations. Determining the quantities and types of atmospheric contaminants is accomplished by measuring and evaluating a representative sample of air. The types of air contaminants that occur in the workplace depend upon the raw materials used and the processes employed. Air contaminants can be divided into two broad groups, depending upon physical characteristics: (1) gases and vapors, and (2) particulates.

Allergens—Because of the presence of allergens on spores, all molds studied to date have the potential to cause allergic reaction in susceptible people. Allergic reactions are believed to be the most common exposure reaction to molds (Rose, 1999).

Ambient—Descriptive of any condition of the environment surrounding a given point. For example, ambient air means that portion of the atmosphere, external to buildings, to which the general public has access. Ambient sound is the sound generated by the environment.

Asphyxiation—Suffocation from lack of oxygen. A substance (e.g., carbon monoxide) that combines with hemoglobin to reduce the blood's capacity to transport oxygen produces chemical asphyxiation. Simple asphyxiation is the result of exposure to a substance (such as methane) that displaces oxygen.

Atmosphere—In physics, a unit of pressure whereby 1 atmosphere (atm) equals 14.7 pounds per square inch (psi).

Attenuation—The reduction of intensity at a designated first location as compared with intensity at a second location farther from the source (reducing the level of noise by increasing distance from the source is a good example).

Audible range—The frequency range over which normal hearing occurs—approximately 20 Hz through 20,000 Hz. Above the range of 20,000 Hz, the term *ultrasonic* is used. Below 20 Hz, the term *subsonic* is used.

Audiogram—A record of hearing loss or hearing level measured at several different frequencies, usually 500 to 6000 Hz. The audiogram may be presented graphically or numerically. Hearing level is shown as a function of frequency.

Audiometric testing—Objective measuring of a person's hearing sensitivity. By recording the response to a measured signal, a person's level of hearing sensitivity can be expressed in decibels, as related to an audiometric zero, or no-sound base.

Authorized person—A person designated or assigned by an employer or supervisor to perform a specific type of duty or duties, to use specified equipment, or to be present in a given location at specified times (e.g., an authorized or qualified person is used in confined space entry).

Auto-ignition temperature—The lowest temperature at which a vapor-producing substance or a flammable gas will ignite even without the presence of a spark or flame.

Baghouse—Term commonly used for the housing containing bag filters for recovery of fumes from arsenic, lead, sulfa, etc.

Baseline data—Data collected prior to a project for later use in describing conditions before the project began. Also commonly used to describe the first audiogram given (within 6 months) to a worker after he or she has been exposed to the action level (85 dBA) to establish his or her baseline for comparison to subsequent audiograms.

Behavior-based safety (BBS) management models—A management theory based on the work of B.F. Skinner, it explains behavior in terms of stimulus, response, and consequences. BBS refers to a wide range of programs which focus almost entirely on changing the behavior of workers to prevent occupational injuries and illnesses.

Bel—A unit equal to 10 decibels (see decibel).

Benchmarking—A process for rigorously measuring company performance vs. "best-in-class" companies and using analysis to meet and exceed the best in class.

Biohazard—Organisms or products of organisms that present a risk to humans.

Biological aerosols—Naturally occurring biologically generated and active particles that are small enough to become suspended in air. These include mold spores, pollen, viruses, bacteria, insect parts, and animal dander.

Boiler code—ANSI/ASME Pressure Vessel Code, a set of standards prescribing requirements for the design, construction, testing, and installation of boilers and unfired pressure vessels.

Boyle's law—The product of a given pressure and volume is constant with a constant temperature.

Carcinogen—A cancer-producing agent.

Carpal tunnel syndrome—An injury to the median nerve inside the wrist, frequently caused by ergonomically incorrect repetitive motion.

Catalyst—A substance that alters the speed of, or makes possible, a chemical or biochemical reaction but remains unchanged at the end of the reaction.

Catastrophe—A loss of extraordinarily large dimensions in terms of injury, death, damage, and destruction.

Causal factor—A person, thing, or condition that contributes significantly to an accident or to a project outcome.

Charles' law—The volume of a given mass of gas at constant pressure is directly proportional to its absolute temperature (in Kelvin).

Chemical change—Change that occurs when two or more substances (reactants) interact with each other, resulting in the production of different substances (products) with different chemical compositions. A simple example of chemical change is the burning of carbon in oxygen to produce carbon dioxide.

Chemical hazards—Includes hazardous chemicals conveyed in various forms—mists, vapors, gases, dusts, and fumes.

Chemical spill—An accidental dumping, leakage, or splashing of a harmful or potentially harmful substance.

Chronic—Persistent, prolonged, repeated exposure. Chronic exposure occurs when repeated exposure to or contact with a toxic substance occurs over a period of time, the effects of which become evident only after multiple exposures.

Coefficient of friction—A numerical correlation of the resistance of one surface against another surface.

Combustible gas indicator—An instrument that samples air and indicates whether an explosive mixture is present, and the percentage of the lower explosive limit (LEL) of the air–gas mixture that has been reached.

Combustible liquid—Liquids having a flash point at or above 37.8°C (100°F).

Combustion—Burning, defined in chemical terms as the rapid combination of a substance with oxygen, accompanied by the evolution of heat and usually light.

Competent person—As defined by OSHA, one who is capable of recognizing and evaluating employee exposure to hazardous substances or to unsafe conditions and who is capable of specifying protective and precautionary measures to be taken to ensure the safety of employees as required by particular OSHA regulations under the conditions to which such regulations apply.

Confined space—A vessel, compartment, or any area having limited access and (usually) no alternative escape route, having severely limited natural ventilation or an atmosphere containing less that 19.5% oxygen, and having the capability of accumulating a toxic, flammable, or explosive atmosphere, or of being flooded (engulfing a victim).

Containment—In fire terminology, restricting the spread of fire. For chemicals, restricting chemicals to an area that is diked or walled off to protect personnel and the environment.

Contingency plan (emergency response plan)—Under 40 CFR 260.10, a document that sets forth an organized, planned, and coordinated course of action to be followed in the event of an emergency that could threaten human health or the environment.

Convection—The transfer of heat from one location to another by way of a moving medium, including air and water.

Corrosive material—Any material that dissolves metals or other materials or that burns the skin.

Cumulative injury—Any physical or psychological disability that results from the combined effects of related injuries or illnesses in the workplace.

Cumulative trauma disorder—A disorder caused by the highly repetitive motion required of one or more parts of a worker's body, which in some cases, can result in moderate to total disability.

Dalton's law of partial pressures—In a mixture of theoretically ideal gases, the pressure exerted by the mixture is the sum of the pressures exerted by each component gas of the mixture.

Decibel (dB)—A unit of measure used originally to compare sound intensities and subsequently electrical or electronic power outputs; now also used to compare voltages. In hearing conservation, it is a logarithmic unit used to express the magnitude of a change in level of sound intensity.

Decontamination—The process of reducing or eliminating the presence of harmful substances such as infectious agents, to reduce the likelihood of disease transmission from those substances.

Density—A measure of the compactness of a substance; it is equal to its mass per unit volume and is measured in kilogram per cubic meter or pounds per cubic foot (Density = Mass/Volume).

Dermatitis—Inflammation or irritation of the skin from any cause. Industrial dermatitis is an occupational skin disease.

Design load—The weight that can be safely supported by a floor, equipment, or structure, as defined by its design characteristics.

Dike—An embankment or ridge of either natural or manmade materials used to prevent the movement of liquids, sludges, solids, or other materials.

Dilute—Adding material to a chemical by the user or manufacturer to reduce the concentration of active ingredient in the mixture.

Dose—An exposure level. Exposure is expressed as weight or volume of test substance per volume of air (mg/L), or as parts per million (ppm).

Dosimeter—Measuring tool that provides a time-weighted average over a period of time such as one complete work shift.

Dusts—Various types of solid particles produced when a given type of organic or inorganic material is scraped, sawed, ground, drilled, heated, crushed, or otherwise deformed.

Electrical grounding—Precautionary measures designed into an electrical installation to eliminate dangerous voltages in and around the installation and to operate protective devices in case of current leakage from energized conductors to their enclosures.

Emergency response—The response made by firefighters, police, healthcare personnel, and other emergency service workers upon notification of a fire, chemical spill, explosion, or other incident in which human life or property may be in jeopardy.

Emergency response plan—See contingency plan.

Energize—The conductors of an electrical circuit; having voltage applied to such conductors and to surfaces that a person might touch; having voltage between such surfaces and other surfaces that might complete a circuit and allow current to flow.

Energy—The capacity for doing work. Potential energy (PE) is energy deriving from position; thus, a stretched spring has elastic PE, and an object raised to a height above the surface of the Earth or the water in an elevated reservoir has gravitational PE. A lump of coal and a tank of oil, together with the oxygen required for their combustion, have chemical energy. Other sorts of energy include electrical and nuclear energy, light, and sound. Moving bodies possess kinetic energy (KE). Energy can be converted from one form to another, but the total quantity stays the same (in accordance with the conservation of energy principle). For example, as an orange falls, it loses gravitational PE, but gains KE.

Engineering—The application of scientific principles to the design and construction of structures, machines, apparatus, manufacturing processes, and power generation and utilization, for the purpose of satisfying human needs. Safety engineering is concerned with control of environment and humankind's interface with it, especially safety interaction with machines, hazardous materials, and radiation.

Engineering controls—Methods of controlling employee exposures by modifying the source or reducing the quantity of contaminants released into the workplace environment.

Epidemiological theory—This theory holds that the models used for studying and determining epidemiological relationships can also be used to study causal relationships between environmental factors and accidents or diseases.

Ergonomics—A multidisciplinary activity dealing with interactions between man and his total working environment, plus stresses related to such environmental elements as atmosphere, heat, light, and sound, as well as all tools and equipment of the workplace.

Etiology—The study or knowledge of the causes of disease.

Exposure—Contact with a chemical, biological, or physical hazard.

Exposure ceiling—The concentration level of a given substance that should not be exceeded at any point during an exposure period.

Fall arresting system—A system consisting of a body harness, a lanyard or lifeline, and an arresting mechanism with built-in shock absorber, designed for use by workers performing tasks in locations from which falls would be injurious or fatal, or where other kinds of protection are not practical.

Fire—A chemical reaction between oxygen and a combustible fuel.

Flammable liquid—Any liquid having a flash point below 37.8°C (100°F).

Flammable solid—A non-explosive solid liable to cause fire through friction, absorption of moisture, spontaneous chemical change, or heat retained from a manufacturing process, or that can be ignited readily and when ignited, burns so vigorously and persistently as to create a serious hazard.

Flash point—The lowest temperature at which a liquid gives off enough vapor to form ignitable moisture with air, and produce a flame when a source of ignition is present. Two tests are used: open cup and closed cup.

Foot-candle—The illumination at a point on a surface 1 foot from and perpendicular to a uniform point source of 1 candle.

Fume—Airborne particulate matter formed by the evaporation of solid materials (e.g., metal fume emitted during welding); usually less that 1 micron in diameter.

Gas—A state of matter in which the material has very low density and viscosity, can expand and contract greatly in response to changes in temperature and pressure, easily diffuses into other gases, and readily and uniformly distributes itself throughout any container.

Ground-fault circuit interrupter (GFCI)—A sensitive device intended for shock protection, which functions to de-energize an electrical circuit or portion thereof within a fraction of a second, in case of leakage to ground of current sufficient to be dangerous to persons but less than that required to operate the overcurrent protective device of the circuit.

Grounded system—A system of conductors in which at least one conductor or point is intentionally grounded, either solidly or through a current-limiting (current transformer) device.

Hazard—The potential for an activity, condition, circumstance, or changing conditions or circumstances to produce harmful effects. Also an unsafe condition.

Hazard analysis—A systematic process for identifying hazards and recommending corrective action.

Hazard and operability (HAZOP) analysis—A systematic method in which process hazards and potential operating problems are identified using a series of guide words to investigate process deviations.

Hazard assessment—A qualitative evaluation of potential hazards in the interrelationships between and among the elements of a system, upon the basis of which the occurrence probability of each identified hazard is rated.

Hazard Communication Standard (HAZCOM)—An OSHA workplace standard found in 29 CFR 1910.1200 that requires all employers to become aware of the chemical hazards in their workplace and relay that information to their employees. In addition, a contractor conducting work at a client's site must provide chemical information to the client regarding the chemicals that are brought onto the work site.

Hazard control—A means of reducing the risk from exposure to a hazard.

Hazard identification—The pinpointing of material, system, process, and plant characteristics that can produce undesirable consequences through the occurrence of an accident.

Hazardous material—Any material possessing a relatively high potential for harmful effects upon persons.

Hazardous substance—Any substance that has the potential for causing injury by reason of its being explosive, flammable, toxic, corrosive, oxidizing, irritating, or otherwise harmful to personnel.

Hazardous waste—A solid, liquid, or gaseous waste that may cause or significantly contribute to serious illness or death, or that poses a substantial threat to human health or the environment when the waste is improperly managed.

Hearing conservation—The prevention of, or minimizing of noise-induced deafness through the use of hearing protection devices, the control of noise through engineering controls, annual audiometric tests, and employee training.

Heat cramps—A type of heat stress (a possible side-effect of dehydration) that occurs as a result of salt and potassium depletion.

Heat exhaustion—A condition usually caused by loss of body water from exposure to excess heat. Symptoms include headache, tiredness, nausea, and sometimes fainting.

Heat stroke—A serious disorder resulting from exposure to excess heat. It results from sweat suppression and increased storage of body heat, characterized by high fever, collapse, and sometimes convulsions or coma.

Homeland Security—Federal cabinet-level department created to protect the United States and her citizens as a result of 9/11. The new Department of Homeland Security (DHS) has three primary missions: prevent terrorist attacks within the United States, reduce America's vulnerability to terrorism, and minimize the damage from potential attacks and natural disasters.

Hot work—Work involving electric or gas welding, cutting, brazing, or similar flame or spark-producing operations.

Human factor engineering/ergonomics—For practical purposes, the terms are synonymous, and focus on human beings and their interaction with products, equipment, facilities, procedures, and environments used in work and everyday living. The emphasis is on human beings (as opposed to engineering, where the emphasis is more strictly on technical engineering considerations) and how the design of things influences people. Human factors, then, seek to change the things people use and the environments in which they use these things to better match the capabilities, limitations, and needs of people (Sanders and McCormick, 1993).

Ignition temperature—The temperature at which a given fuel bursts into flame.

Illumination—The amount of light flux a surface receives per unit area; may be expressed in lumens per square foot or in foot-candles.

Impulse noise—A noise characterized by rapid rise time, high peak value, and rapid decay.

Incident—An undesired event that, under slightly different circumstances, could have resulted in personal harm or property damage; any undesired loss of resources.

Indoor air quality (IAQ)—The effect, good or bad, of the contents of the air inside a structure on its occupants. While usually temperature (too hot and cold), humidity (too dry or too damp), and air velocity (draftiness or motionless) are considered "comfort" rather than indoor air quality issues, IAQ refers to such problems as asbestosis, sick building syndrome, biological aerosols, and ventilation issues concerning dusts and fumes, among others.

Industrial hygiene—The American Industrial Hygiene Association (AIHA) defines industrial hygiene as "that science and art devoted to the anticipation, recognition, evaluation, and control of those environmental factors or stresses—arising in the workplace—which may cause sickness, impaired health and well-being, or significant discomfort and inefficiency among workers or among citizens of the community."

Ingestion—Entry of a foreign substance into the body through the mouth.

Injury—A wound or other specific damage.

Interlock—A device that interacts with another device or mechanism to govern succeeding operations, such as the interlock on an elevator door that prevents the car from moving unless the door is properly closed.

Ionizing radiation—Radiation that becomes electrically charged (i.e., changed into ions).

Irritant—A substance that produces an irritating effect when it contacts skin, eyes, nose, or respiratory system.

Job hazard (safety) analysis—The breaking down into its component parts of any method or procedure to determine the hazards connected therewith and the requirements for performing it safely.

Kinetic energy—The energy resulting from a moving object.

Laboratory Safety Standard—A specific hazard communication program for laboratories, found in 29 CFR 1910.1450. These regulations are essentially a blend of hazard communication and emergency response for laboratories. The cornerstone of the Laboratory Safety Standard is the requirement for a written chemical hygiene plan.

Lockout/tagout procedure—An OSHA procedure found in 29 CFR 1910.147. A tag or lock is used to "tag out" or "log out" a device, so that no one can inadvertently actuate the circuit, system, or equipment that is temporarily out of service.

Log and Summary of Occupational Injuries and Illnesses (OSHA-300 Log)—A cumulative record that employers (generally of more than 10 employees) are required to maintain, showing essential facts of all reportable occupational injuries and illnesses.

Loss—The degradation of a system or component. Loss is best understood when related to dollars lost. Examples include death or injury to a worker, destruction or impairment of facilities or machines, destruction or spoiling of raw materials, and creation of delay. In the insurance business, loss connotes dollar loss, and we have seen underwriters who write it as LO$$ to make that point.

Lower explosive limit (LEL)—The minimum concentration of a flammable gas in air required for ignition in the presence of an ignition source; listed as a percent by volume in air.

Material Safety Data Sheet (MSDS)—Chemical information sheets provided by the chemical manufacturer that include such information as chemical and physical characteristics; long- and short-term health hazards; spill control procedures; personal protective equipment (PPE) to be used when handling the chemical; reactivity with other chemicals; incompatibility with other chemicals; and manufacturer's name, address, and phone number. Employee access to and understanding of MSDS are important parts of the HAZCOM program.

Medical monitoring—The initial medical exam of a worker, followed by periodic exams. The purpose of medical monitoring is to assess workers' health, determine their fitness to wear personal protective equipment, and maintain records of their health.

Metabolic heat—Produced within a body as a result of activity that burns energy.

Mists—Minute liquid droplets suspended in air.

Molds—The most typical forms of fungus found on earth, comprising approximately 25% of the earth's biomass (McNeel and Kreutzer, 1996).

Monitoring—Periodic or continuous surveillance or testing to determine the level of compliance with statutory requirements and/or pollutant levels in various media or in humans, animals, or other living things.

Mycotoxins—Some molds are able to produce *mycotoxins,* natural organic compounds that are capable of initiating a toxic response in vertebrates (McNeel and Kreutzer, 1996).

Nonionizing radiation—Radiation on the electromagnet spectrum that has a frequency of 10^{15} or less and a wavelength in meters of 3×10^{-7}.

Occupational Safety and Health Act (OSH Act)—A federal law passed in 1970 to ensure, so far as possible, that every working man and woman in the nation has safe and healthful working conditions. To achieve this goal, the Act authorizes several functions, such as encouraging safety and health programs in the workplace and encouraging labor-management cooperation in health and safety issues.

OSHA Form 300—Log and Summary of Occupational Injuries and Illnesses; formerly OSHA Form 200.

Oxidation—When a substance either gains oxygen or loses hydrogen or electrons in a chemical reaction; a chemical treatment method.

Oxidizer—Also known as an oxidizing agent, a substance that oxidizes another substance. Oxidizers are a category of hazardous materials that may assist in the production of fire by readily yielding oxygen.

Oxygen-deficient atmospheres—The legal definition of an atmosphere where the oxygen concentration is less than 19.5% by volume of air.

Particulate matter—Substances (such as diesel soot and combustion products resulting from the burning of wood) released directly into the air; any minute, separate particle of liquid or solid material.

Performance standards—OSHA regulation standards that list the ultimate goal of compliance but do not explain exactly how compliance is to be accomplished. Compliance is usually based on accomplishing the act or process in the safest manner possible, based on experience (past performance).

Permissible exposure limit (PEL)—The time-weighted average concentration of an airborne contaminant that a healthy worker may be exposed to 8-hours per day or 40-hours per week without suffering any adverse health effects. Established by legal means and enforceable by OSHA.

Personal protective equipment (PPE)—Any material or device worn to protect a worker from exposure to or contact with any harmful substance or force.

Preliminary assessment—A quick analysis to determine how serious the situation is, and to identify all potentially responsible parties. The preliminary assessment uses readily available information; for instance, forms, records, aerial photographs, and personnel interviews.

Pressure—The force exerted against an opposing fluid or thrust, distributed over a surface.

Radiant heat—The result of electromagnetic nonionizing energy that is transmitted through space without the movement of matter within that space.

Radiation—Energetic nuclear particles, including alpha rays, beta rays, gamma rays, neutrons, high-speed electrons, and high-speed protons.

Reactive—A substance that reacts violently by catching on fire, exploding, or giving off fumes when exposed to water, air, or low heat.

Reactivity hazard—The ability of a material to release energy when in contact with water. Also, the tendency of a material, when in its pure state or as a commercially produced product, to vigorously polymerize, decompose, condense, or otherwise self-react and undergo violent chemical change.

Reportable quantity (RQ)—The minimum amount of a hazardous material that, if spilled while in transport, must be reported immediately to the National Response Center. Minimum reportable quantities range from 1 pound to 5000 pounds per 24-hour day.

Resource Conservation and Recovery Act (RCRA)—A federal law enacted in 1976 to deal with both municipal and hazardous waste problems and to encourage resource recovery and recycling.

Risk—The combination of the expected frequency (event/year) and consequence (effects/event) of a single accident or a group of accidents; the result of a loss-probability occurrence and the acceptability of that loss.

Risk assessment—A process that uses scientific principles to determine the level of risk that actually exists in a contaminated area.

Risk characterization—The final step in the risk assessment process, it involves determining a numerical risk factor. This step ensures that exposed populations are not at significant risk.

Risk management—The professional assessment of all loss potentials in an organization's structure and operations, leading to the establishment and administration of a comprehensive loss control program.

Safety—A general term denoting an acceptable level of risk of, relative freedom from, and low probability of harm.

Safety factor—Based on experimental data, the amount added (e.g., 1000-fold) to ensure worker health and safety.

Safety standard—A set of criteria specifically designed to define a safe product, practice, mechanism, arrangement, process, or environment; a standard produced by a body representative of all concerned interests and based upon currently available scientific and empirical knowledge concerning the subject or scope of the standard.

Secondary containment—A method using two containment systems so that if the first is breached, the second will contain all of the fluid in the first. For USTs, secondary containment consists of either a double-walled tank or a liner system.

Security assessment—A security test intensified in scope and effort, the purpose of which is to obtain an advanced and very accurate idea of how well the organization has implemented security mechanisms, and to some degree, policy.

Sensitizers—Chemicals that in very low dose trigger an allergic response.

Short-term exposure limit (STEL)—The time-weighted average concentration to which workers can be exposed continuously for a short period of time (typically 15 minutes) without suffering irritation, chronic or irreversible tissue damage, or impairment for self-rescue.

Silica (SIO₂)—A major component of the Earth's crust and the cause of silicosis.

Specific gravity—The ratio of the density of a substance to water.

Threshold limit value (TLV)—The same concept as PEL, except that TLVs do not have the force of governmental regulations behind them but are based on recommended limits established and promoted by the American Conference of Governmental Industrial Hygienists.

Time-weighted average (TWA)—A mathematical average of exposure concentration over a specific time: Exposure (ppm) × Time (hours) × 1/4 Time (hours).

Total quality management (TQM)—A way of managing a company that entails a total and willing commitment of all personnel at all levels to quality.

Toxicity—The relative property of a chemical agent with reference to a harmful effect on some biologic mechanism and the condition under which this effect occurs. The quality of being poisonous.

Toxicology—The study of poisons, which are substances that can cause harmful effects to living things.

Unsafe condition—Any physical state that deviates from that which is acceptable, normal, or correct in terms of past production or potential future production of personal injury and/ or damage to property; any physical state that results in a reduction in the degree of safety normally present.

Upper explosive limit (UEL)—The maximum concentration of a flammable gas in air required for ignition in the presence of an ignition source.

Vulnerability assessment—A very regulated, controlled, cooperative, and documented evaluation of an organization's security posture from outside-in and inside-out, for the purpose of defining or greatly enhancing security policy.

Workers' compensation—A system of insurance required by state law and financed by employers that provides payments to employees and their families for occupational illnesses, injuries, or fatalities incurred while at work and resulting in loss of wage income, usually regardless of the employer's or employee's negligence.

Zero energy state—The state of equipment in which every power source that can produce movement of a part of the equipment, or the release of energy, has been rendered inactive.

15.4 INDOOR AIR QUALITY

Unless we are confronted by an unusual odor or are for some reason having trouble breathing, most of us rarely think about the air we inhale and exhale with such regularity. For us, "The air is the air" (*Star Trek*), and few of us realize the risks conveyed by the air we breathe and that we all face a variety of risks to our health as we go about out day-to-day lives (USEPA, 2003). Driving our cars, flying in planes, engaging in recreational activities, and being exposed to environmental pollutants all pose varying degrees of risk. Some risks are simply unavoidable. Some we choose to accept because to do otherwise would restrict our ability to lead our lives the way we want. And some are risks we might decide to avoid if we had the opportunity to make informed choices. Indoor air pollution is one risk that we can do something about. In the last several years, a growing body of scientific evidence has indicated that the air within homes and other buildings can be more seriously polluted than the outdoor air, in even the largest and most industrialized cities. Other research indicates that people spend approximately 90% of their time indoors. Thus, for many people, the risks to health may be greater from exposure to air pollution indoors than outdoors (USEPA, 2003). In addition, people who may be exposed to indoor air pollutants for the longest periods of time are often those most susceptible to the effects of indoor air pollution. Such groups include the young, the elderly, and the chronically ill, especially those suffering from respiratory or cardiovascular disease. The impact of energy conservation on inside environments may be substantial, particularly with respect to decreases in ventilation rates (Hollowell et al., 1979a) and "tight" buildings constructed to minimize infiltration of outdoor air (Hollowell et al., 1979b; Woods, 1980).

The bottom line on indoor air quality is the recognition that the outdoor air we breathe is not always of the same quality as the air we breathe indoors. This recognition that the indoor air environment is not an exact reflection of outdoor conditions is of relatively recent emergence. The impact of cigarette smoking, stove and oven operation, and emanations from certain types of particleboard, cement, and other building materials are often the most significant determinants of indoor air quality. There is a continuing need to characterize human exposures both from the standpoint of meeting pertinent ambient and occupational standards and the recognition of potential hazardous levels of pollutants which do not have applicable standards. The implications of indoor air concentrations for epidemiological studies, where exposures are based on outdoor measurements, have been recognized and, in the recent past, partially investigated (Dockery and Spengler, 1981; Spengler, et al., 1979).

So, what exactly is indoor air quality? According to Byrd (2003), indoor air quality refers to the effect, good or bad, of the contents of the air inside a structure on its occupants. Usually, temperature (too hot or too cold), humidity (too dry or too damp), and air velocity (draftiness or motionlessness) are considered "comfort" rather than indoor air quality issues, unless they are extreme, not within normal range. They may make someone uncomfortable, but they won't make a person ill. Nevertheless, most environmental professionals must take these factors into account in investigating air quality situations. What is good indoor air quality? Simply put, good IAQ is a characteristic of air that has no unwanted gases or particles in it at concentrations that will adversely affect someone. Poor IAQ occurs when gases or particles are present at an excessive concentration so as to affect the satisfaction of health of occupants (Byrd, 2003). In the workplace, poor IAQ may only be annoying to one person; however, at the extreme, it could be fatal to all of the occupants in the workplace. The concentration of the contaminant is crucial. Potentially infectious, toxic, allergenic, or irritating substances are always present in the air. Note that a threshold level below which no effect occurs is nearly always a factor.

15.4.1 Sources of Indoor Air Pollutants

Air quality is affected by the presence of various types of contaminants in the air. Some are in the form of gases, and are generally classified as toxic chemicals. Such contaminants can include combustion products (e.g., carbon monoxide, nitrogen dioxide), volatile organic compounds (formaldehyde, solvents, perfumes, fragrances), and semivolatile organic compounds (pesticides). Other pollutants are in particulate form, including various forms of animal dander; soot; particles from buildings, furnishings, and occupants (e.g., fiberglass, gypsum powder, paper dust, lint from clothing, carpet fibers); and dirt, among others. Specific sources for contaminants that result in adverse health effects in the workplace include the workers themselves (e.g., contagious diseases, allergens and other agents on clothing); building compounds (e.g., VOCs, particles, fibers); contamination of building components (e.g., allergens, microbial agents, pesticides); and outdoor air (e.g., microorganisms, allergens, chemical air pollutants) (Burge and Hoyer, 1998). When workers complain of IAQ problems, the environmental professional is called upon to determine if the problem really is an IAQ problem. If it is determined that some form of contaminant is present in the workplace, proper remedial action is required. This usually includes removing the source of the contamination.

15.5 AIR POLLUTION FUNDAMENTALS

In the last 40 years, the environmental engineering profession has expanded its responsibilities to society to include the control of air pollution in the workplace (and homes) but also from industrial sources that pollute our atmosphere. Though not exactly "seven maids with seven mops trying to get it clear," increasing numbers of environmental engineers and practitioners are being confronted with problems in this most vital area. Although the design and construction of air pollution control equipment today are accomplished with some degree of success, air pollution problems still exist. The environmental engineer of today and tomorrow must develop proficiency and improved understanding of the design and selection of air pollution control equipment in order to cope with these problems and challenges—"to get it clear."

It is in this spirit that this section is presented. In short, we simply feel that the present situation is not that grim. It is not time to start shedding tears. Why do we feel this way? Simply, we know we can do something to control environmental air pollution. Environmental professionals who are well trained and well equipped with the proper mathematical tools can make a difference when it comes time to clear the air we breathe.

The USEPA devotes (as it should) enormous amounts of time to research topics related to air pollution control. In this section, we heavily excerpt from USEPA publications on the topic. Moreover, much of the general information provided is excerpted from Spellman (2008). The excerpted materials have been rearranged and edited to make the materials more concise for environmental professionals in training and general readers to understand the basic concepts of air pollution control mathematics and processes.

Controlling environmental air pollution begins with understanding what environmental pollution is. We define *environmental air pollution* as the contamination of atmospheric air in such a manner as to cause real or potential harm to human health or well being or to damage or harm the natural surroundings without justification. Contaminants may include almost any natural or artificial composition of matter capable of being airborne (e.g., friable asbestos). Contaminants may be in the form of solids particles, liquid droplets, gases, or in combinations of these forms. Contaminants fall into two main groups: (1) those emitted directly from identifiable sources, and (2) those produced in the air by interaction between two or more primary contaminants, or by reaction with normal atmospheric constituents, with or without photoactivation.

The Clean Air Act (CAA) established two types of National Ambient Air Quality Standards (NAAQS):

- *Primary standards* are designed to establish limits to protect public health, including the health of "sensitive" populations such as asthmatics, children, and the elderly.
- *Secondary standards* set limits to protect public welfare, including protection against decreased visibility and damage to animals, crops, vegetation, and buildings.

15.5.1 Six Common Air Pollutants

National air quality standards have been set for six common pollutants (also referred to as "criteria" pollutants). These common air pollutants are discharged from various sources and include

- Ground-level ozone
- Nitrogen dioxide
- Particulate matter
- Sulfur dioxide
- Carbon monoxide
- Lead

15.5.1.1 Ground-Level Ozone

Ozone (O_3) is a highly reactive photochemically produced gas composed of three oxygen atoms. It is not usually emitted directly into the air (i.e., it is a secondary air pollutant) but at ground level is created by a chemical reaction between oxides of nitrogen (NO_x) and volatile organic compounds (VOCs) in the presence of heat and sunlight. We can characterize ozone as the Dr. Jeckel and Mr. Hyde of air pollutants. Why? Ozone has the same chemical structure whether it occurs miles above the Earth or at ground level and can be either "good" or "bad," depending on its location in the atmosphere. "Good" (Dr. Jeckel) ozone occurs naturally in the stratosphere approximately 10 to 30 miles above the Earth's surface and forms a layer that protects life on Earth from the sun's harmful rays. In the Earth's lower atmosphere, ground-level ozone is considered "bad" (Mr. Hyde).

$$VOC + NO_x + Heat + Sunlight = Ozone$$

Motor vehicle exhaust and industrial emissions, gasoline vapors, and chemical solvents are some of the major sources of NO_x and VOCs that help to form ozone. Sunlight and hot weather cause ground-level ozone to form in harmful concentrations in the air. As a result, it is known as a summertime air pollutant. Many urban areas tend to have high levels of "bad" ozone, but even rural areas are also subject to increased ozone levels because wind carries ozone and pollutants that form it hundreds of miles away from their original sources.

15.5.1.2 Nitrogen Oxides (NO_x)

Nitrogen oxides (NO_x) is the generic term for a group of highly reactive gases all of which contain nitrogen and oxygen in varying amounts. This group includes NO, NO_2, NO_3, N_2O, N_2O_3, N_2O_4, and N_2O_5—but only two are important in the study of air pollution: nitric oxide (NO) and nitrogen dioxide (NO_2). Many of the nitrogen oxides are colorless and odorless; however, nitrogen dioxide (NO_2), a common pollutant, along with particles in the air can often be seen as a reddish-brown layer over many urban areas. Nitrogen oxides form when fuel is burned at high temperatures, as in a combustion process. The primary sources of NO_x are motor vehicles, electric utilities, and other industrial, commercial, and residential sources that burn fuels.

15.5.1.3 Particulate Matter

Particulate matter (PM) is the term for particles found in the air, including dust, dirt, soot, smoke, and liquid droplets. Particles can be suspended in the air for long periods of time. Some particles are large or dark enough to be seen as soot or smoke. Others are so small that individually they can only be detected with an electron microscope. Some particles are directly emitted into the air. They come from a variety of sources such as cars, trucks, buses, factories, construction sites, tilled fields, unpaved roads, stone crushing, and burning of wood. Other particles may be formed in the air from the chemical change of gases. They are indirectly formed when gases from burning fuels react with sunlight and water vapor. These can result from fuel combustion in motor vehicles, at power plants, and in other industrial processes.

15.5.1.4 Sulfur Dioxide (SO_2)

Sulfur dioxide (SO_2) belongs to the family of sulfur oxide gases (SO_x). These gases dissolve easily in water. Sulfur enters the atmosphere in the form of corrosive sulfur dioxide gas, a colorless gas possessing the sharp, pungent odor of burning rubber. Also, sulfur is prevalent in all raw materials, including crude oil, coal, and ore that contains common metals such as aluminum, copper, zinc, lead, and iron. SO_x gases are formed when sulfur-containing fuel, such as coal and oil, is burned, and when gasoline is extracted from oil or metals are extracted from ore. SO_2 dissolves in water vapor to form acid, and interacts with other gases and particles in the air to form sulfates and other products that can be harmful to people and their environment.

Over 65% of SO_2 released to the air, or more than 13 million tons per year, come from electric utilities, especially those that burn coal. Other sources of SO_2 are industrial facilities that derive their products from raw materials such as metallic ore, coal, and crude oil, or that burn coal or oil to produce heat. Examples are petroleum refineries, cement manufacturing, and metal processing facilities. Also, large ships, locomotives, and some nonroad diesel equipment currently burn high sulfur fuel and release SO_2 emissions to the air in large quantities.

Two major environmental problems have developed in highly industrialized regions of the world where the atmospheric sulfur dioxide concentration has been relatively high: sulfurous smog and acid rain. Sulfurous smog is the haze that develops in the atmosphere when molecules of sulfuric acid serve as light screeners. The second problem, acid rain, is precipitation contaminated with dissolved acids such as sulfuric acid. Acid rain has posed a threat to the environment by causing certain lakes to become devoid of aquatic life. Sulfur dioxide produces blotches that are white to straw-colored on broad-leafed plants.

15.5.1.5 Carbon Monoxide (CO)

Carbon monoxide (CO) is a colorless, odorless, tasteless gas that is by far the most abundant of the primary pollutants; it is formed when carbon in fuel is not burned completely. It is a component of motor vehicle exhaust, which contributes about 56% of all CO emissions nationwide. Other non-road engines and vehicles (such as construction equipment and boats) contribute about 22% of all CO emissions nationwide. Higher levels of CO generally occur in areas with heavy traffic congestion. In cities, 85 to 95% of all CO emissions come from motor vehicle exhaust. Other sources of CO emissions include industrial processes (such as metals processing and chemical manufacturing), residential wood burning, and natural sources such as forest fires. Woodstoves, gas stoves, cigarette smoke, and unvented gas and kerosene space heaters are sources of CO indoors. The highest levels of CO in the outside air typically occur during the colder months of the year when inversion conditions are more frequent. The air pollution becomes trapped near the ground beneath a layer of warm air.

15.5.1.6 Lead

Lead is a metal found naturally in the environment as well as in manufactured products. The major sources of lead emissions have historically been motor vehicles (such as cars and trucks) and industrial sources. At present, because of the phase out of leaded gasoline, metals processing is the major source of lead emissions to the air. The highest levels of lead in air are generally found near lead smelters. Other stationary sources are waste incinerators, utilities, and lead-acid battery manufacturers. In high concentrations, lead can damage human health and the environment. Once lead enters the ecosystem, it remains there permanently. The good news is that, since the 1970s, stricter emission standards have caused a dramatic reduction in lead output.

15.5.2 Gases

Gases are important not only from the standpoint that a gas can be a pollutant, but also because gases convey the particulate and gaseous pollutants. For most air pollution work, expressing pollutant concentrations in volumetric terms is customary. For example, the concentration of a gaseous pollutant in parts per million (ppm), is the volume of pollutant per million parts of the air mixture. That is,

$$\text{ppm} = \frac{\text{Parts of contamination}}{\text{Million parts of air}} \tag{15.1}$$

Note that calculations for gas concentrations are based on the gas laws:

- The volume of gas under constant temperature is inversely proportional to the pressure.
- The volume of a gas under constant pressure is directly proportional to the Kelvin temperature. The Kelvin temperature scale is based on absolute zero (0°C = 273 K).
- The pressure of a gas of a constant volume is directly proportional to the Kelvin temperature.

Thus, when measuring contaminant concentrations, we must know the atmospheric temperature and pressure under which the samples were taken. At standard temperature and pressure (STP), 1 g-mol of an ideal gas occupies 22.4 L. The STP are 0°C and 760 mmHg. If the temperature is increased to 25°C (room temperature) and the pressure remains the same, 1 g-mol of gas occupies 24.45 L.

Sometimes it is necessary to convert milligrams per cubic meter (mg/m³)—a weight-per-volume ratio—into a volume-per-unit weight ratio. If it is understood that 1 g-mol of an ideal gas at 25°C occupies 24.45 L, the following relationships can be calculated:

$$\text{ppm} = \frac{24.45}{\text{Molecular weight}} \, \text{mg/m}^3 \tag{15.2}$$

$$\text{mg/m}^3 = \frac{\text{Molecular weight}}{24.45} \, \text{ppm} \tag{15.3}$$

15.5.2.1 Gas Laws

As mentioned, gases can be pollutants as well as the conveyors of pollutants. Air (which is mainly nitrogen) is usually the main gas stream. To understand the gas laws, it is imperative that we have an understanding of various terms:

- *Ideal gas*—An imaginary model of a gas that has a few very important properties. First, the gases are assumed to be infinitely small. Second, the particles move randomly in straight lines until they collide into something (another gas molecule or the side of whatever container they are in). Third, the gas particles do not interact with each other (they do not attract or repel one another like real molecules do). Finally, the energy of the particles is directly proportional to the temperature in Kelvin (in other words, the higher the temperature, the more energy the particles have). These assumptions are made because they make equations a lot simpler than they would be otherwise, and because these assumptions cause negligible deviation from the ways that actual gases behave.
- *Kelvin*—A temperature scale in which the degrees are the same size as degrees Celsius but where 0 is defined as absolute zero, the temperature at which molecules are at their lowest energy. To convert from degrees Celsius to Kelvin (not degrees Kelvins), add 273.
- *Pressure*—A measure of the amount of force that a gas exerts on whatever container we put it into. Units of pressure include atmospheres (1 atm is the average atmospheric pressure at sea level), Torr (equal to 1/760 of an atmosphere), millimeters of mercury (1 mmHg = 1 Torr = 1/760 atm), and kilopascals (101,325 kPa = 1 atm).
- *Standard temperature and pressure*—A set of conditions defined as 273 K and 1 atm.
- *Standard conditions (SC)*—More commonly used than STP, standard conditions are typical room conditions of 20°C (70°F) and 1 atm. SC units of volume are commonly given as normal cubic meters, or standard cubic feet (scf).
- *Temperature*—A measure of how much energy the particles in a gas have. It is defined as that property of a body which determines the flow of heat. Heat will flow from a warm body to a cold body.
- *Volume*—The amount of space that some object occupies. The unit of volume can be cubic centimeters (cc, cm³), milliliters (mL; 1 mL = 1 cm³), liters (L; 1 L = 1000 mL), or cubic meters (m³; 1 m³ = 1 million cm³).

There are several types of temperature scales in general use. These scales depend on the freezing and boiling points of water as boundary markers for the scale. In a conventional laboratory thermometer, the boundary points are conveniently selected to relate to the known properties of water.

TABLE 15.1

Comparison of Temperature Scales

Temperature Scale	Celsius (°C)	Kelvin (K)	Fahrenheit (°F)	Rankine (°R)
Boiling point of water	+100	+373.15	+212	+671.67
	↑	↑	↑	↑
	100 equal divisions	100 equal divisions	180 equal divisions	180 equal divisions
	↓	↓	↓	↓
Freezing point of water	0	273.15	+32	+491.67
Absolute zero	−273.15	0	−459.67	0

Note: Units of temperature that we will run into include degrees Celsius and Kelvin (equal to 273 plus degrees Celsius). The degree symbol (°) is not used for the Kelvin temperature scale.

On the Celsius scale, the freezing point of water is assigned a value of 0 and the boiling point a value of 100; the distance between these two points is divided into 100 equal increments, with each increment labeled in Celsius degree (Table 15.1). On the Kelvin scale, the freezing point of water is assigned a value of 273.15 K and the boiling point a value of 373.15; the distance between these two points is divided into 100 equal increments, and each increment is labeled as a Kelvin (Table 15.1). On the Fahrenheit scale, the freezing point of water is assigned a value of 32 and the boiling point a value of 212; the distance between these two points is divided into 180 equal increments, and each increment is labeled as a Fahrenheit degree (Table 15.1).

15.5.2.1.1   Boyle's Law

Circa 1662, Robert Boyle stated what has come to be known as Boyle's law: The volume of any definite quantity of gas at constant temperature varies inversely as the pressure on the gas:

$$P_1 \times V_1 = P_2 \times V_2 \tag{15.4}$$

In this equation, P_1 is the initial pressure of the gas, and V_1 is the initial volume of the gas. P_2 is the final pressure of the gas, and V_2 is the final volume of the gas. If we know the initial pressure and volume of a gas and know what the final pressure will be, we can predict what the volume will be after we put the pressure on it.

■ EXAMPLE 15.1

Problem: If we have 4 L of methane gas at a pressure of 1.0 atm, what will be the pressure of the gas if we compress it so is has a volume of 2.5 L?

Solution:

$$1.0 \text{ atm} \times 4 \text{ L} = x \text{ atm} \times 2.5 \text{ L}$$

$$x = 1.6 \text{ atm}$$

15.5.2.1.2   Charles' Law

Charles observed that hydrogen (H_2), carbon dioxide (CO_2), oxygen (O_2), and air expanded by an equal amount when heated from 0°C to 80°C at a constant pressure:

$$V_1/T_1 = V_2/T_2 \tag{15.5}$$

In this equation, the subscript 1 indicates the initial volume and temperature, and the subscript 2 indicates the volume and temperature after the change. Temperature, incidentally, needs to be given in Kelvin, not Celsius, because if we have a temperature below 0°C the calculation works out so the volume of the gas is negative, and we can't have a negative volume.

■ EXAMPLE 15.2

Problem: If we have 2 L of methane gas at a temperature of 40°C, what will be the volume be if we heat the gas to 80°C?

Solution: The first thing we have to do is convert the temperatures to Kelvin, because Celsius cannot be used in this equation:

$$40°C + 273 = 313 \text{ K}$$

$$80°C + 273 = 353 \text{ K}$$

We are now ready to insert these numbers into the equation:

$$V_1/T_1 = V_2/T_2$$

$$2 \text{ L}/313 \text{ K} = x \text{ L}/353 \text{ K}$$

$$x = 2.26 \text{ L}$$

15.5.2.1.3 Gay-Lussac's Law

Gay-Lussac (1802) found that all gases increase in volume for each 1°C rise in temperature, and this increase is equal to approximately 1/273.15 of the volume of the gas at 0°C.

$$P_1/T_1 = P_2/T_2 \tag{15.6}$$

If we increase the temperature of a container with fixed volume, the pressure inside the container will increase.

15.5.2.1.4 Combined Gas Law

This law combines the parameters of the preceding equations, forming

$$(P_1 \times V_1)/T_1 = (P_2 \times V_2)/T_2 \tag{15.7}$$

The advantage of this equation is that whenever we are changing the conditions of pressure, volume, and/or temperature for a gas, we just insert the numbers into this equation.

■ EXAMPLE 15.3

Problem: If we have 2 L of a gas at a temperature of 420 K and decrease the temperature to 350 K, what will the new volume of the gas be?

Solution: To solve this problem, we use the combined gas law to find the answer. Because pressure was never mentioned in this problem, we ignore it. As a result, the equation will be

$$V_1/T_1 = V_2/T_2$$

which is the same thing as Charles' law. Thus,

$$2 \text{ L}/420 \text{ K} = x \text{ L}/350 \text{ K}$$

$$x = 1.67 \text{ L}$$

15.5.2.1.5 Ideal Gas Law

The ideal gas law combines Boyle's and Charles' laws because air cannot be compressed without its temperature changing. The ideal gas law is an equation of state, which means that we use the basic properties of the gas to find out more about it without having to change it in any way. Because it's an equation of state, it allows us to not only find out what the pressure, volume, and temperature are, but also to find out how much gas is present in the first place. The ideal gas law is expressed by the following equation:

$$P \times V = n \times R \times T \tag{15.8}$$

where
  P = Pressure of the gas (atm, kPa).
  V = Volume (L).
  n = Number of moles.
  R = Ideal gas constant.
  T = Temperature (Kelvin).

The two common values for the ideal gas constant are 0.08206 L·atm/mol·K and 8.314 L·kPa/mol·K. The question is, which one do we use? The value of R used depends on the pressure given in the problem. If the pressure is given in atmospheres, use the 0.08206 value because it includes the unit atm. If the pressure is given in kilopascals, use the second value because it includes the unit kPa. The ideal gas law allows us to figure out how many grams and moles of the gas are present in a sample. After all, the number of moles is the n term in the equation, and we already know how to convert grams to moles.

■ EXAMPLE 15.4

Problem: Given 4 L of a gas at a pressure of 3.4 atm and a temperature of 300 K, how many moles of gas are present?

Solution: First, figure out what value of the ideal gas constant should be used. Because pressure is given in atmospheres, use 0.206 L·atm/mol·K. After inserting the given terms for pressure, volume, and temperature, we end up with

$$3.4 \text{ atm} \times 4 \text{ L} = n \times (0.08206 \text{ L·atm/mol·K}) \times (300 \text{ K})$$

$$n = 0.55 \text{ moles}$$

15.5.2.1.6 Composition of Air

The air mixture that surrounds us and that we breathe is a dynamic mixture of many components (see Table 15.2). The mixture is dynamic in several respects. The moisture content of water vapor, the temperature, the pressure, and the trace gas constituents all can and do vary over time and in space. The bulk of the air in the biosphere is composed of nitrogen and oxygen with various other trace gases mixed in it (see Table 15.2).

TABLE 15.2

Approximate Composition of Dry Air (by Volume)

Component	Symbol	Percent (%)	Parts per Million (ppm)
		Concentration	
Nitrogen	N_2	78.084	780,840
Oxygen	O_2	20.9476	209,476
Argon	Ar	0.934	9340
Carbon dioxide	CO_2	0.0314	314
Neon	Ne	0.001818	18.18
Helium	He	0.000524	5.24
Methane	CH_4	0.0002	2
Sulfur dioxide	SO_2	0–0.0001	0–1
Hydrogen	H_2	0.00005	0.5
Krypton	Kr	0.0002	2
Xenon	Xe	0.0002	2
Ozone	O_3	0.0002	2

Note: Figures are taken from p. 6.1 of ASHRAE's *Handbook of Fundamentals*, based on an atomic weight of carbon of 12.0000. The handbook also reports that the molecular weight of dry air is 28.9645 g/mol based on the carbon-12 scale.

15.5.3 PARTICULATE MATTER

Typically, in actual practice, the terms *particulates* (or *particles*) and *particulate matter* are used interchangeably. According to 40 CFR 51.100–190, particulate matter is defined as any airborne finely divided solid or liquid material with an aerodynamic diameter smaller than 100 micrometers (micro = 10^{-6}). Along with gases and water vapor, Earth's atmosphere is literally a boundless arena for particulate matter of many sizes and types. Atmospheric particulates vary in size from 0.0001 to 10,000 microns. Particulate size and shape have direct bearing on visibility. For example, a spherical particle in the 0.6-micron range can effectively scatter light in all directions, reducing visibility.

The types of airborne particulates in the atmosphere vary widely, with the largest sizes derived from volcanoes, tornadoes, waterspouts, burning embers from forest fires, seed parachutes, spider webs, pollen, soil particles, and living microbes. The smaller particles (the ones that scatter light) include fragments of rock, salt and spray, smoke, and particles from forested areas. The largest portion of airborne particulates is invisible. They are formed by the condensation of vapors, chemical reactions, photochemical effects produced by ultraviolet radiation, and ionizing forces that come from radioactivity, cosmic rays, and thunderstorms. Airborne particulate matter is produced either by mechanical weathering, breakage, and solution or by the vapor-to-condensation-to-crystallization process (typical of particulates from a furnace of a coal-burning power plant).

We know very well that anything that goes up must eventually come down. This is typical of airborne particulates also. Fallout of particulate matter depends, obviously, mostly on their size—less obvious on their shape, density, weight, airflow, and injection altitude. The residence time of particulate matter also is dependent on the atmosphere's cleanup mechanisms (formation of clouds and precipitation) that work to remove them from their airborne suspended state. Some large particulates may only be airborne for a matter of seconds or minutes. Intermediate sizes may be able to stay afloat for hours or days. The finer particulates may stay airborne for a much longer duration: days, weeks, months, and even years.

Particles play an important role in atmospheric phenomena; for example, particulates provide the nuclei upon which ice particles and cloud condensation are formed, and they are essential for condensation to take place. The most important role airborne particulates play is in cloud formation. Simply put, without clouds, life would be much more difficult, and the cloudbursts that eventually erupted would cause devastation so extreme that it is difficult to imagine or contemplate.

15.5.4 Pollution Emission Measurement Parameters

Because of the gaseous and particulate emissions that can be produced, combustion sources constitute a significant air quality control problem. Combustion processes can add carbon dioxide, water vapor, and heat to the atmosphere, and can produce a residue that must be disposed of in concentrated form. In the past, these environmental costs were tolerated in the interest of producing useful energy. However, it is becoming increasingly clear that the presence of these emissions in the atmosphere can result indirectly in a greenhouse effect and exacerbate the problem of acid rain. Because of the environmental impact of combustion emissions, USEPA has developed emission standards for the combustion or incineration industry. These standards usually establish the maximum allowable limit, based on volume or mass flows at specified conditions of temperature and pressure, for the discharge of specific pollutants. Emissions are measured in terms of the concentration of pollutant per volume or mass of stack (flue) gas, the pollutant mass rates, or a rate applicable to a given process. Standards fall into the following six categories:

1. *Pollutant mass rate standards* are based on the fixed rate of emissions (i.e., the mass of pollutant emitted per unit time, expressed in lb/hr or kg/hr).
2. *Process rate standards* establish the allowable emissions in terms of either the input energy or the raw material feed of process.
3. *Concentration standards* limit either the mass (weight) or volume of the pollutants in the gas leaving the stack.
4. *Ambient concentration standards* address such pollutants as toxic metals, organics, and hydrogen chloride (measured in $\mu g/m^3$).
5. *Reduction standards* are expressed as a percent reduction of the pollutants.
6. *Opacity standards* address the degree to which the stack emissions are visible and block the visibility of objects in the background. Stack emissions of 100% opacity totally block the view of background objects and indicate high pollutant levels. Zero percent opacity provides a clear view of the background and indicates no detectable particulate matter emissions.

15.5.5 Standard Corrections

Because combustion systems always produce stack gas that is at a higher temperature and pressure than those of the standards, and because actual levels of pollutants emitted can be made to appear smaller if excess air is added to the stream, corrections for these differences must be made. With regard to increases or decreases in gas temperature and pressure and the subsequent effect on gas volume, the USEPA recommends using the ideal gas law. For excess air correction calculations, various federal USEPA and state regulations give procedures for calculating percent excess air based on dry gas (Orsat type) analyses. Based on USEPA's Method 3B, Gas Analysis for Carbon Dioxide, Oxygen, Excess Air, and Dry Molecular Weight, the percent excess air can be determined by any of the following three equivalent relationships:

$$\% \text{ Excess Air} = \left(\frac{\text{Total Air} - \text{Theoretical Air}}{\text{Theoretical Air}} \right) \times 100 \tag{15.9}$$

$$\% \text{ Excess Air} = \left(\frac{\text{Excess Air}}{\text{Theoretical Air}} \right) \times 100 \qquad (15.10)$$

$$\% \text{ Excess Air} = \left(\frac{\text{Excess Air}}{\text{Total Air} - \text{Excess Air}} \right) \times 100 \qquad (15.11)$$

Theoretical air is that amount required to stoichiometrically convert all combustible species (mainly carbon, hydrogen, sulfur) to complete normal products of combustion (CO_2, H_2O, and SO_2). These relationships are stated as mole ratios of air, which equal volume ratios. In Method 3B, USEPA (2003b) provides the following explanation for determining percent excess air.

15.5.5.1 Example USEPA Calculation for Percent Excess Air

Determine the percentage of the gas that is N_2 by subtracting the sum of the percent CO_2, percent CO, and percent O_2 from 100%. Calculate the percent excess by substituting the appropriate values of percent O_2, CO, and N_2 into Equation 15.12.

$$\% \text{ Excess Air} = \left(\frac{\%O_2 - 0.5\%CO}{0.264\%N_2 - (\%O_2 - 0.5\%CO)} \right) \times 100 \qquad (15.12)$$

Equation 15.12 assumes that ambient air is used as the source of O_2 and that the fuel does not contain appreciable amounts of N_2 (as do coke oven or blast furnace gases). For those cases when appreciable amounts of N_2 are present (coal, oil, and natural gas do not contain appreciable amounts of N_2) or when oxygen enrichment is used, alternative methods are required.

15.5.6 ATMOSPHERIC STANDARDS

To perform various indoor and outdoor air pollution calculations, knowledge of the following atmospheric standards is important.

15.5.6.1 Standard Temperature and Pressure

Standard temperature and pressure (STP) is the designation given to an ambient condition in which the barometric pressure and the ambient temperature are as follows:

Barometric pressure
   1 atm, 760 mmHg, 14.70 psia (pressure per square inch absolute), 0.00 psig (pressure per square inch gauge), 1013.25 millibars, or 760 Torr

and

Ambient temperature
   0°C, 32°F, 273.16 K, or 491.67°R

15.5.6.2 Normal Temperature and Pressure

Normal temperature and pressure (NTP) is the designation given to an ambient condition in which the barometric pressure and the ambient temperature are as follows:

Barometric pressure
   1 atm, 760 mmHg, 14.70 psia, 0.00 psig, 1013.25 millibars, or 760 Torr

and

Ambient temperature
   25°C, 77°F, 298.16 K, or 576.67°R

15.5.7 AIR MONITORING AND SAMPLING

Air monitoring is widely used to measure human exposure and to characterize emission sources. It is often employed within the context of the general survey, investigating a specific complaint, or simply for regulatory compliance. It is also used for more fundamental purposes, such as in confined space entry operations. Although it is true that just about any confined space entry team member can be trained to properly calibrate and operate air monitors for safe confined space entry, it is also true that a higher level of knowledge and training is often required in the actual evaluation of confined spaces for possible oxygen deficiency and/or air contaminant problems.

In the practice of industrial hygiene, the terms *air monitoring* and *air sampling* are often used interchangeably to mean the same thing, but in reality they are different; that is, air monitoring and air sampling are separate functions. The difference is related to time: real time vs. time integration. Air monitoring is real-time monitoring and generally includes monitoring with hand-held, direct-reading units such as portable gas chromatographs (GCs), photoionization detectors (PIDs), flame ionization detectors (FIDs), dust monitors, and colorimetric tubes. Real-time air monitoring instrumentation is generally easily portable and allows the user to collect multiple samples in a relatively short sample period—ranging from a few seconds to a few minutes. Most portable real-time instruments measure low parts per million (ppm) of total volatile organics. Real-time monitoring methods have higher detection limits than time-integrated sampling methods, react with entire classes of compounds and, unless real-time monitoring is conducted continuously, provide only a snapshot of the monitored ambient air concentration. Air monitoring instruments and methods provide results that are generally used for evaluation of short-term exposure limits and can be useful in providing timely information to those engaged in various activities such as confined space entry operations. That is, in confined space operations, proper air monitoring can detect the presence or absence of life-threatening contaminants and/or insufficient oxygen levels within the confined space, alerting the entrants not to enter before making the space safe for entry (e.g., by using forced air ventilation).

On the other hand, time-integrated air sampling is intended to document actual exposure for comparison to long-term exposure limits. Air sampling data are collected at fixed locations along the perimeter of the sample area and at locations adjacent to other sensitive receptors. Because most contaminants are present in ambient air at relatively low levels, some type of sample concentrating is necessary to meet detection limits normally required in evaluating long-term health risks. Air sampling is accomplished using air-monitoring instrumentation designed to continuously sample large volumes of air over extended periods of time (typically from 8 to 24 hours). Air sampling methods involve collecting air samples on sampling media designed specifically for collection of the compounds of interest or as whole air samples. Upon completion of the sampling period the sampling media is collected, packaged, and transported for subsequent analysis. Analysis of air samples usually requires a minimum of 48 hours to complete.

Now you should have a basic understanding of air monitoring and air sampling and the difference, though in some cases subtle, between the two. Both procedures are important and both are significant tools in the industrial hygienist's toolbox.

To effectively evaluate a potentially hazardous worksite, an industrial hygienist must obtain objective and quantitative data. To do this, the environmental professional must perform some form of air sampling, dependent upon, of course, the airborne contaminant in question. Moreover, sampling operations involve the use of instruments to measure the concentration of the particulate, gas, or vapor of interest. Many instruments perform both sampling and analysis. The instrument of choice in conducting sampling and analysis typically is a direct-reading-type instrument. The environmental professional must be familiar with the uses, advantages, and limitations of such instruments. In addition, the environmental professional must use math calculations to calculate sample volumes, sample times, TLVs, and air concentrations from vapor pressures and to determine the additive effects of chemicals when multiple agents are used in the workplace. These calculations must take into account changing conditions, such as temperature and pressure change in the workplace. Finally, the environmental professional must understand how particulates, gases, and vapors are generated, how they enter the human body, how they impact worker's health, and how to evaluate particulate-, gas-, and vapor-laden workplaces.

Because air sampling is integral to just about everything the environmental professional does and is about, in the following we include basic air sampling calculations that go hand-in-hand with the conduction of air sampling protocols.

15.5.7.1 Air Sampling Calculations

- 1 mole = 22.4 L at STP.
- 1 mole = 24.45 L at NTP.
- Standard pressure is 760 Torr (760 mmHg at sea or atmosphere level).
- Standard temperature is 0°C or 273 K.

15.5.7.1.1 Boyle's Law Problems

Problem: If 500 mL of oxygen is collected at a pressure of 780 mmHg, what volume will the gas occupy if the pressure is changed to 740 mmHg?

Solution:

$$\text{New volume } (V_2) = 500 \text{ mL} \times (780 \text{ mmHg}/740 \text{ mmHg}) = 527 \text{ mL}$$

Problem: What is the volume of a gas at a pressure of 90 cmHg if 300 mL of the gas was collected at a pressure of 86 cmHg?

Solution:

$$\text{New volume } (V_2) = 300 \text{ mL} \times (86 \text{ cmHg}/90 \text{ cmHg}) = 287 \text{ mL}$$

Problem: Calculate the pressure of a gas that occupies a volume of 110 mL if it occupies a volume of 300 mL at a pressure of 80 cmHg.

Solution:

$$\text{New pressure } (P_2) = 80 \text{ cmHg} \times 300 \text{ mL}/110 \text{ mL} = 218 \text{ cmHg}$$

15.5.7.1.2 Charles' Law Problems

Problem: What volume will an amount of gas occupy at 25°C if the gas occupies a volume of 500 mL at a temperature of 0°C? Assume that the pressure remains constant.

Solution:

$$K = 273° + °C$$

$$\text{New volume } (V_2) = 500 \text{ mL} \times (298 \text{ K}/273 \text{ K}) = 546 \text{ mL}$$

Problem: What is the volume of a gas at –25°C if the gas occupied 48 mL at a temperature of 0°C?

Solution:

$$\text{New volume } (V_2) = 48 \text{ mL} \times (248 \text{ K}/273 \text{ K}) = 43.6 \text{ mL}$$

Problem: If a gas occupies a volume of 800 mL at 15°C, at what temperature will it occupy a volume of 1000 mL if the pressure remains constant?

Solution:

$$\text{New absolute pressure } (T_2) = 288 \text{ K} \times (1000 \text{ mL}/800 \text{ mL}) = 360 \text{ K}$$

15.5.7.1.3 Boyle's Law and Charles' Law Combined Problems

Problem: Calculate the volume of a gas at STP if 600 mL of the gas is collected at 25°C and 80 cmHg.

Solution:

$$\text{New volume } (V_2) = 600 \text{ mL} \times (80 \text{ cmHg}/76 \text{ cmHg}) \times (273 \text{ K}/298 \text{ K}) = 578 \text{ mL}$$

Problem: If a gas occupies a volume of 100 mL at a pressure of 76 mmHg and 25°C, what volume will the gas occupy at 900 mmHg and 40°C?

Solution:

$$\text{New volume } (V_2) = 100 \text{ mL} \times (760 \text{ mmHg}/900 \text{ mmHg}) \times (313 \text{ K}/298 \text{ K}) = 88.7 \text{ mL}$$

Problem: If 500 mL of oxygen is collected at 20°C, and the atmospheric pressure is 725.0 mmHg, what is the volume of the dry oxygen at STP?

Solution:

$$\text{New volume } (V_2) = 500 \text{ mL} \times (725.0 \text{ mmHg}/760 \text{ mmHg}) \times (273 \text{ K}/293 \text{ K}) = 443.5 \text{ mL}$$

Problem: 2.50 g of a gas occupy 240 mL at 20°C and 740 Torr. What is the gram molecular weight of the gas?

Solution:

$$240 \text{ mL} \times (273 \text{ K}/293 \text{ K}) \times (740 \text{ Torr}/760 \text{ Torr}) = 217 \text{ mL}$$

$$(2.50 \text{ g}/217 \text{ mL}) \times (1000 \text{ mL}/1 \text{ L}) \times (22.4 \text{ L}/1 \text{ mol}) = 258 \text{ g/mol}$$

15.6 VENTILATION

Simply put, ventilation is the classic method, and the most powerful tool of control used in safety engineering, to control airborne environmental hazards. Experience has shown that the proper use of ventilation as a control mechanism can ensure that the workplace air remains free of potentially hazardous levels of airborne contaminants. In accomplishing this, ventilation works in two ways: (1) by physically removing the contaminated air from the workplace, or (2) by diluting the workplace atmospheric environment to a safe level by the addition of fresh air (Spellman, 2013). A ventilation system is all very well and good (virtually essential, actually), but an improperly designed ventilation system can make the hazard worse. This essential point cannot be overemphasized. At the heart of an efficient ventilation system are proper design, proper maintenance, and proper monitoring. The environmental professional plays a critical role in ensuring that installed ventilation systems are operating at their optimum level.

Because of the importance of ventilation in the workplace, the environmental professional must be well versed in the general concepts of ventilation, principles of air movement, and monitoring practices. Environmental professionals responsible for indoor air quality must be properly prepared (through training and experience) to evaluate existing systems and design new systems to control the workplace environment. This section presents general principles of ventilation system design, evaluation, and control and basic computations. This material should provide the basic concepts and principles necessary for ensuring the proper operation of industrial ventilation systems. This material also serves to refresh the knowledge of the practitioner in the field. Probably the best source of information on ventilation is ACGIH's *Industrial Ventilation: A Manual of Recommended Practice*; this text is a must-have reference for every environmental professional responsible for ventilation systems.

15.6.1 DEFINITIONS

Because ventilation is one of the most important environmental and engineering control techniques used by environmental practitioners, they must be familiar with and understand the following ventilation terms and definitions.

acfm (actual cubic feet per minute)—Measure of gas flowing at existing temperature and pressure.

ACH, AC/H (air changes per hour)—The number of times air is replaced in an hour.

Air density—The weight of air in pounds per cubic foot. Dry standard air at a temperature of 68°F (20°C) and barometric pressure 2992 inHg (760 mmHg) has a density of 0.075 lb/ft³.

Anemometer—A device that measures the velocity of air. Common types include the swinging vane and the hot-wire anemometer.

Area (A)—The cross-sectional area through which air moves. Area may refer to the cross-sectional area of a duct, a window, a door, or any space through which air moves.

Atmospheric pressure—The pressure exerted in all directions by the atmosphere. At sea level, mean atmospheric pressure is 29.92 inHg, 14.7 psi, 407 in. wg (water gauge), or 760 mmHg.

Branch—In a junction of two ducts, the branch is the duct with the lowest volume flow rate. The branch usually enters the main at an angle of less than 90.

Canopy hood (receiving hood)—A one- or two-sided overhead hood that receives rising hot air or gas.

Capture velocity—The velocity of air induced by a hood to capture emitted contaminants external to the hood.

Coefficient of entry (C_e)—A measure of the efficiency of a hood's ability to convert static pressure to velocity pressure; the ratio of actual flow to ideal flow.

Density correction factor—A factor applied to correct or convert dry air density of any temperature to velocity pressure; the ratio of actual flow to ideal flow.

Dilution ventilation (general exhaust ventilation)—A form of exposure control that involves providing enough air in the workplace to dilute the concentration of airborne contaminants to acceptable levels.

Evase—A cone-shaped exhaust stack that recaptures static pressure from velocity pressure.

Fan—A mechanical device that moves air and creates static pressure.

Fan curve—A curve relating pressure and volume flow rate of a given fan at a fixed fan speed (rpm).

Fan laws—Relationships that describe theoretical, mutual performance changes in pressure, flow rate, rpm of the fan, horsepower, density of air, fan size, and power.

Flow rate—Volume flow rates are described by the conservation of mass formula: $Q = V \times A$, where Q is volume, V is velocity, and A is the cross-sectional area of air flow.

Friction loss—The static pressure loss in a system caused by friction between moving air and the duct wall, expressed in wg/100 ft, or fractions of velocity pressure per 100 ft of duct (mm wg/m; kPa/m).

Gauge pressure—The difference between two absolute pressures, one of which is usually atmospheric pressure.

Head—Pressure (e.g., "the head is 1 in. wg"),

Hood—A device that encloses, captures, or receives emitted contaminants.

Hood entry loss (H_e)—The static pressure lost (in inches of water) when air enters a duct through a hood. The majority of the loss usually is associated with a *vena contracta* formed in the duct.

Hood static pressure (SP_h)—The sum of the duct velocity pressure and the hood entry loss; hood static pressure is the static pressure required to accelerate air at rest outside the hood into the duct at velocity.

HVAC (heating, ventilation, and air conditioning) system—Ventilating systems designed primarily to control temperature, humidity, odors, and air quality.

in. wg (inches of water)—A unit of pressure. One inch of water is equal to 0.0735 in. of mercury, or 0.036 psi. Atmospheric pressure at standard conditions is 407 in. wg.

Indoor air quality (IAQ), sick-building syndrome, tight-building syndrome—The study, examination, and control of air quality related to temperature, humidity, and airborne contaminants.

Industrial ventilation (IV)—The equipment or operation associated with the supply or exhaust of air by natural or mechanical means to control occupational hazards in the industrial setting.

Laminar flow (also *streamline flow*)—Air flow in which air molecules travel parallel to all other molecules; laminar flow is characterized by the absence of turbulence.

Local exhaust ventilation—An industrial ventilation system that captures and removes emitted contaminants before dilution into the ambient air of the workplace.

Loss—Usually refers to the conversion of static pressure to heat in components of the ventilation system (e.g., hood entry loss).

Manometer—A device that measures pressure difference; usually a U-shaped glass tube containing water or mercury.

Minimum transport velocity (MTV)—The minimum velocity that will transport particles in a duct with little settling; MTV varies with air density, particulate loading, and other factors.

Outdoor air (OA)—Outdoor air is the "fresh" air mixed with return air (RA) to dilute contaminants in the supply air.

Pitot tube—A device used to measure total and static pressures in an airstream.

Plenum—A low-velocity chamber used to distribute static pressure throughout its interior.

Pressure—Air moves under the influence of differential pressures. A fan is commonly used to create a difference of pressure in duct systems.

Pressure drop—The loss of static pressure across a point (e.g., "the pressure drop across an orifice is 2.0 in. wg").

Replacement air (compensating air, make-up air)—Air supplied to a space to replace exhausted air.

Return air—Air that is returned from the primary space to the fan for recirculation.

scfm (standard cubic feet per minute)—A measure of air flow at standard conditions: dry air at 29.92 in. HG (760 mmHg) (gauge) and 68°F (20°C).

Slot velocity—The average velocity of air through a slot. Slot velocity is calculated by dividing the total volume flow rate by the slot area (usually, $V_s = 2000$ fpm).

Stack—A device on the end of a ventilation system that disperses exhaust contaminants for dilution by the atmosphere.

Standard air, standard conditions—Dry air at 70°F (20°C), 29.92 in. Hg (760 mmHg), 14.7 psi, 407 in. wg.

Static pressure (SP)—The pressure developed in a duct by a fan; the force in inches of water measured perpendicular to flow at the wall of the duct; the difference in pressure between atmospheric pressure and the absolute pressure inside a duct, cleaner, or other equipment. SP exerts influence in all directions.

Suction pressure—An archaic term that refers to static pressure on the upstream side of the fan. See static pressure.

Total pressure (TP)—The pressure exerted in a duct; the sum of the static pressure and the velocity pressure. Also called *impact pressure, dynamic pressure.*

Turbulent flow—Air flow characterized by transverse velocity components as well as velocity in the primary direction of flow in a duct; mixing velocities.

Velocity (V)—The time rate of movement of air; usually expressed as feet per minute.

Velocity pressure (VP)—The pressure attributed to the velocity of air.

Volume flow rate (Q)—Quantity of air flow expressed in cfm, scfm, or acfm.

15.6.2 Concept of Ventilation

The purpose of industrial ventilation is essentially to (under control) recreate what occurs in natural ventilation. Natural ventilation results from differences in pressure. Air moves from high-pressure areas to low pressure areas. This difference in pressure is the result of thermal conditions. We know that hot air rises, which (for example) allows smoke to escape from the smokestack in an industrial process, rather than disperse into areas where workers operate the process. Hot air rises because air expands as it is heated, becoming lighter. The same principle is in effect when air in the atmosphere becomes heated. The air rises and is replaced by air from a higher pressure area. Thus, convection currents cause a natural ventilation effect through the resulting winds.

What does all of this have to do with industrial ventilation? Actually, quite a lot. Simply put, industrial ventilation is installed in a workplace to circulate the air within and to provide a supply of fresh air to replace air that has undesirable characteristics. Could this be accomplished simply by natural workplace ventilation? That is, couldn't we just heat the air in the workplace so that it will rise and escape through natural ports—windows, doors, cracks in walls, or mechanical ventilators in the roof (installed wind-powered turbines, for example)? Yes, we could design a natural system like this, but in such a system, air does not circulate fast enough to remove contaminants before a hazardous level is reached, which defeats our purpose in providing a ventilation system in the first place. Thus, we use fans to provide an artificial, mechanical means of moving the air.

Along with controlling or removing toxic airborne contaminants from the air, installed ventilation systems perform several other functions within the workplace. These functions include

1. Ventilation is often used to maintain an adequate oxygen supply in an area. In most workplaces, this is not a problem because natural ventilation usually provides an adequate volume of oxygen; however, some work environments (deep mining and thermal processes which use copious amounts of oxygen for combustion) the need for oxygen is the major reason for an installed ventilation system.
2. An installed ventilation system can remove odors from a given area. This type of system (as you might guess) has applications in such places as athletic locker rooms, rest rooms, and kitchens. In performing this function, the noxious air may be replaced with fresh air, or odors may be masked with a chemical masking agent.
3. One of the primary uses of installed ventilation is one that we are familiar with providing heat, cooling, and humidity control.
4. A ventilation system can remove undesirable contaminants at their source, before they enter the workplace air (e.g., from a chemical dipping or stripping tank). Obviously, this technique is an effective way to ensure that certain contaminants never enter the breathing zone of the worker—exactly the kind of function safety engineering is intended to accomplish.

A mechanical fan is the heart of any ventilation system, but like the human heart, certain ancillaries are required to make it function as a system. Ventilation is no different. Four major components make up a ventilation system.

1. The fan forces the air to move.
2. An inlet or some type of opening allows air to enter the system.
3. An outlet must be provided for air to leave the system.
4. A conduit or pathway (ducting) not only directs the air in the right direction, but also limits the amount of flow to a predetermined level.

An important concept regarding ventilation systems is the difference between exhaust and supply ventilation. An *exhaust ventilation system* removes air and airborne contaminants from the workplace. Such a system may be designed to exhaust an entire work area, or it may be placed at the source to remove the contaminant prior to its release into the workplace air. The second type of ventilation system is the *supply ventilation system*, which (as the name implies) adds air to the work area, usually to dilute work area contaminants to lower the concentration of these contaminants. However, a supplied-air system does much more; it also provides movement to air within the space (especially when an area is equipped with both an exhaust and supply system—a usual practice, because it allows movement of air from inlet to outlet and is important in replenishing exhausted air with fresh air).

Air movement in a ventilation system is a result of differences in pressure. Note that pressures in a ventilation system are measured in relation to atmospheric pressure. In the workplace, the existing atmospheric pressure is assumed to be the zero point. In the supply system, the pressure created by the system is *in addition* to the atmospheric pressure that exists in the workplace (i.e., a positive pressure). In an exhaust system, the objective is to lower the pressure in the system below the atmospheric pressure (i.e., a negative pressure).

When we speak of increasing and decreasing pressure levels within a ventilation system, what we are really talking about is creating small differences in pressure—small when compared to the atmospheric pressure of the work area. For this reason, these differences are measured in terms of *inches of water* or *water gauge*, which results in the desired sensitivity of measurement. Air can be assumed to be incompressible, because of the small-scale differences in pressure.

Let's get back to the water gauge or inches of water. Because 1 psi of pressure is equal to 27 in. of water, 1 in. of water is equal to 0.036 lb pressure, or 0.24% of standard atmospheric pressure. Remember the potential for error introduced by considering air to be incompressible is very small at the pressure that exists with a ventilation system. The environmental professional responsible for ventilation in the workplace must be familiar with the three pressures important in ventilation: velocity pressure, static pressure, and the total pressure. To understand these three pressures and their function in ventilation systems, you must first be familiar with pressure itself. In fluid mechanics, the energy of a fluid (air) that is flowing is termed *head*. Head is measured in terms of unit weight of the fluid or in foot-pounds/pound of fluid flowing. Note that the usual convention is to describe head in terms of feet of fluid that is flowing.

So what is pressure? Pressure is the force per unit area exerted by the fluid. In the English system of measurement, this force is measured in lb/ft². Because we have stated that the fluid in a ventilation system is incompressible, the pressure of the fluid is equal to the head. Velocity pressure (VP) is created as air travels at a given velocity through a ventilation system. Velocity pressure is only exerted in the direction of airflow and is always positive (i.e., above atmospheric pressure). When you think about it, velocity pressure has to be positive, and obviously the force or pressure that causes it also must be positive. Note that the velocity of the air moving within a ventilation system is directly related to the velocity pressure of the system. This relationship can be derived into the standard equation for determining velocity (and clearly demonstrates the relationship between velocity of moving air and the velocity pressure):

$$V = 4005\sqrt{VP} \tag{15.13}$$

Static pressure (SP) is the pressure that is exerted in all directions by the air within the system, which tends to burst or collapse the duct. It is expressed in inches of water gauge (in. wg). A simple example may help you grasp the concept of static pressure. Consider the balloon that is inflated at a given pressure. The pressure within the balloon is exerted equally on all sides of the balloon. No air velocity exits within the balloon itself. The pressure in the balloon is totally the result of static pressure. Note that static pressure can be both negative and positive with respect to the local atmospheric pressure.

Total pressure (TP) is defined as the algebraic sum of the static and velocity pressures or

$$TP = SP + VP \tag{15.14}$$

The total pressure of a ventilation system can be either positive or negative (i.e., above or below atmospheric pressure). Generally, the total pressure is positive for a supply system, and negative for an exhaust system.

For the environmental professional to evaluate the performance of any installed ventilation system, he or she must make measurements of pressures in the ventilation system. Measurements are normally made using instruments such as a manometer or a Pitot tube. The manometer is often used to measure the static pressure in the ventilation system. The manometer is a simple, U-shaped tube, open at both ends, and usually constructed of clear glass or plastic so that the fluid level within can be observed. To facilitate measurement, a graduated scale is usually present on the surface of the manometer. The manometer is filled with a liquid (water, oil, or mercury). When pressure is exerted on the liquid within the manometer, the pressure causes the level of liquid to change as it relates to the atmospheric pressure external to the ventilation system. The pressure measured, therefore, is relative to atmospheric pressure as the zero point. When manometer measurements are used to obtain positive pressure readings in a ventilation system, the leg of the manometer that opens to the atmosphere will contain the higher level of fluid. When a negative pressure is being read, the leg of the tube open to the atmosphere will be lower, thus indicating the difference between the atmospheric pressure and the pressure within the system.

The Pitot tube is another device used to measure static pressure in ventilation systems. The Pitot tube is constructed of two concentric tubes. The inner tube forms the impact portion, while the outer tube is closed at the end and has static pressure holes normal to the surface of the tube. When the inner and outer tubes are connected to opposite legs of a single manometer, the velocity pressure is obtained directly. If the engineer wishes to measure static pressure separately, two manometers can be used. Positive and negative pressure measurements are indicated on the manometer as above.

15.6.2.1 Local Exhaust Ventilation

Local exhaust ventilation (the most predominant method of controlling workplace air) is used to control air contaminants by trapping and removing them near the source. In contrast to dilution ventilation (which lets the contamination spread throughout the workplace, later to be diluted by exhausting quantities of air from the workspace), local exhaust ventilation surrounds the point of emission with an enclosure, and attempts to capture and remove the emissions before they are released into the worker's breathing zone. The contaminated air is usually drawn through a system of ducting to a collector, where it is cleaned and delivered to the outside through the discharge end of the exhauster. A typical local exhaust system consists of a hood, ducting, an air-cleaning device, fan, and a stack. A local exhaust system is usually the proper method of contaminant control if

- The contaminant in the workplace atmosphere constitutes a health, fire, or explosion hazard.
- National or local codes require local exhaust ventilation at a particular process.
- Maintenance of production machinery would otherwise be difficult.
- Housekeeping or employee comfort will be improved.
- Emission sources are large, few, fixed and/or widely dispersed.
- Emission rates vary widely by time.
- Emission sources are near the worker-breathing zone.

The environmental professional must remember that determining beforehand precisely the effectiveness of a particular system is often difficult. Thus, measuring exposures and evaluating how much control has been achieved after a system is installed is essential. A good system may collect 80 to 90% or more, but a poor system may capture only 50% or less. Without total enclosure of the contaminant sources (where capture is obviously very much greater), the environmental professional must be aware of the limitations and must be familiar with handling problems like these.

Once the system is installed, and has demonstrated that it is suitable for the task at hand, the system must be well maintained. Careful maintenance is a must. In dealing with ventilation problems, the industrial hygienist soon finds out that his or her worst headache in maintaining the system is poor—or no—maintenance. A phenomenon that many environmental practitioners in the industrial hygiene field forget (or never knew in the first place) is that ventilation, when properly designed, installed, and maintained, can go a long way to ensure a healthy working environment. However, ventilation does have limitations. For example, the effects of blowing air from a supply system and removing air through an exhaust system are different. To better understand the difference and its significance, let's take an example of air supplied through a standard exhaust duct.

When air is exhausted through an opening, it is gathered equally from all directions around the opening. This includes the area behind the opening itself. Thus, the cross-sectional area of airflow approximates a spherical form, rather than the conical form that is typical when air is blown out of a supply system. To correct this problem, a flange is usually placed around the exhaust opening, which reduces the air contour, from the large spherical contour to that of a hemisphere. As a result, this increases the velocity of air at a given distance from the opening. This basic principle is used in designing exhaust hoods. Remember that the closer the exhaust hood is to the source, and the less uncontaminated air it gathers, the more efficient the hood's percentage of capture will be. Simply put, it is easier for a ventilation system to blow air than it is for one to exhaust (or suck) it. Keep this in mind whenever you are dealing with ventilation systems and/or problems. Moreover, pollutants that are not captured by the hood are considered to be fugitive emissions, which includes emissions that (1) escape capture by process equipment exhaust hoods; (2) are emitted during material transfer; (3) are emitted to the atmosphere from the source area; and (4) are emitted directly from process equipment (40 CFR Part 60, Electronic Code of Federal Regulations).

15.2.2.2 General and Dilution Ventilation

Along with local exhaust ventilation are two other major categories of ventilation systems: *general ventilation* and *dilution ventilation*. Each of these systems has a specific purpose, and finding all three types of systems present in a given workplace location is not uncommon. General ventilation systems (sometimes referred to as heat control ventilation systems) are used to control indoor atmospheric conditions associated with hot industrial environments (such as those found in foundries, laundries, bakeries, and other workplaces that generate excess heat) for the purpose of preventing acute discomfort or injury. General ventilation also functions to control the comfort level of the worker in just about any indoor working environment. Along with the removal of air that has become process-heated beyond a desired temperature level, a general ventilation system supplies air to the work area to condition (by heating or cooling) the air, or to make up for the air that has been exhausted by dilution ventilation in a local exhaust ventilation system.

A dilution ventilation system dilutes contaminated air with uncontaminated air, to reduce the concentration below a given level (usually the threshold limit value of the contaminant) to control potential airborne health hazards, fire and explosive conditions, odors, and nuisance type contaminants. This is accomplished by removing or supplying air, to cause the air in the workplace to move, and as a result, mix the contaminated with incoming uncontaminated air. This mixing operation is essential. To mix the air there must be, of course, air movement. Air movement can be accomplished by natural draft caused by prevailing winds moving through open doors and windows of the work area. Thermal draft can also move air. Whether the thermal draft is the result of natural causes or is generated from process heat, the heated air rises, carrying any contaminant present upward with it. Vents in the roof allow this air to escape into the atmosphere. Makeup air is supplied to the work area through doors and windows. A mechanical air moving device provides the most reliable source for air movement in a dilution ventilation system. Such a system is rather simple. It requires a source of exhaust for contaminated air, a source of air supply to replace the air mixture that has been removed with uncontaminated air, and a duct system to supply or remove air throughout the workplace. Dilution ventilation systems often are equipped with filtering systems to clean and temper the incoming air.

15.6.3 COMMON VENTILATION MEASUREMENTS

Duct diameters are measured to calculate duct areas. Inside duct diameter is the most important measurement, but an outside measurement is often sufficient for a sheet metal duct. To measure the duct, the tape should be thrown around the duct to obtain the duct circumference, and the number should be divided by π (3.142) to obtain the diameter of the duct. Hood and duct dimensions can be estimated from plans, drawings, and specifications. Measurements can be made with measuring tape. If a duct is constructed of 2.5- or 4-foot sections, the sections can be counted (elbows and tees should be included in the length). *Hood-face velocities* outside the hood or at the hood face can be estimated with velometers, smoke tubes, and swinging-vane anemometers, all of which are portable and reliable and require no batteries.

1. The minimum velocity that can be read by an anemometer is 50 feet per minute (fpm). The meter should always be read in the upright position, and only the tubing supplied with the equipment should be used.
2. Anemometers often cannot be used if the duct contains dust or mist because air must actually pass through the instrument for it to work. The instrument requires periodic cleaning and calibration at least once per year. Hot-wire anemometers should not be used in air-streams containing aerosols.
3. Hood-face velocity measurement involves the following steps:
   - Mark off imaginary areas.
   - Measure velocity at center of each area.
   - Average all measured velocities.
4. Smoke is useful for measuring face velocity because it is visible. Nothing convinces management and employees more quickly that the ventilation is not functioning properly than to show smoke drifting away from the hood, escaping the hood, or traveling into the worker's breathing zone, Smoke can be used to provide a rough estimate of face velocity. Squeeze off a quick blast of smoke. Time the smoke plume's travel over a 2-ft distance. Calculate the velocity in feet per minute. For example, if it takes 2 seconds for the smoke to travel 2 feet, the velocity is 60 fpm.

$$\text{Velocity} = \text{Distance/Time, or } V = D/T \qquad (15.15)$$

Hood static pressure (SP_h) should be measured about four to six duct diameters downstream in a straight section of the hood take-off duct. The measurement can be made with a Pitot tube or by a static pressure tap into the duct sheet metal.

1. Pressure gauges come in a number of varieties, the simplest being the U-tube manometer.
2. Inclined manometers offer greater accuracy and greater sensitivity at low pressures than U-tube manometers. However, manometers rarely can be used for velocities less than 800 fpm (i.e., velocity pressures less than 0.05 in. wg). Aneroid-type manometers use a calibrated bellows to measure pressures. They are easy to read and portable but require regular calibration and maintenance.

Duct velocity measurements may be made directly (with velometers and anemometers) or indirectly (with manometers and Pitot tubes) using duct velocity pressure.

1. Air flow in industrial ventilation ducts is almost always turbulent, with a small, nonmoving boundary layer at the surface of the duct.
2. Because velocity varies with distance from the edge of the duct, a single measurement may not be sufficient. However, if the measurement is taken in a straight length of round duct,

four to six diameters downstream and two to three diameters upstream from obstructions or directional changes, then the average velocity can be estimated at 90% of the centerline velocity. The average velocity pressure is about 81% of centerline velocity pressure.

3. A more accurate method is the traverse method, which involves taking six to ten measurements on each of two or three passes across the duct, 90° or 60° opposed. Measurements are made in the center of concentric circles of equal area.
4. Density corrections (e.g., temperature) for instrument use should be made in accordance with the manufacturer's instrument instruction manual and correction formulas.

Air cleaner and fan condition measurements can be made with a Pitot tube and manometer.

15.6.4 GOOD PRACTICES

Hood placement must be close to the emission source to be effective. Maximum distance from the emission source should not exceed 1.5 duct diameters.

1. Keep in mind the relationship between capture velocity (V_c) and duct velocity (V_d) for simple plain or narrow flanged hoods. For example, if an emission source is one duct diameter in front of the hood and the duct velocity (V_d) = 3000 fpm, then the expected capture velocity (V_c) is 300 fpm. At two duct diameters from the hood opening, the capture velocity decreases by a factor of 10, to 30 fpm.
2. A rule of thumb that can be used with simple capture hoods is that if the duct diameter (D) is 6 in., then the maximum distance of the emission source from the hood should not exceed 9 in. Similarly, the minimum capture velocity should not be less than 50 fpm. Simply, for simple capture hoods, maximum capture distance should not be more than 1.5 times the duct diameter.

System effect loss, which occurs at the fan, can be avoided if the necessary ductwork is in place.

1. Use of the *six-and-three rule* ensures better design by providing for a minimum loss at six diameters of straight duct at the fan inlet and a minimum loss at three diameters of straight duct at the fan outlet.
2. System effect loss is significant if any elbows are connected to the fan at inlet or outlet. For each 2.5 diameters of straight duct between the fan inlet and any elbow, CFM loss will be 20%.

Stack height should be 10 ft higher than any roof line or air intake located within 50 ft of the stack. For example, a stack placed 30 ft away from an air intake should be at least 10 ft higher than the center of the intake.

Ventilation system drawings and specifications usually follow standard forms and symbols, such as those described in the Uniform Construction Index (UCI).

1. Plan sections include electrical, plumbing, structural, or mechanical drawings. The drawings come in several views: plan (top), elevation (side and front), isometric, or section.
2. Elevations (side and front views) give the most detail. An isometric drawing is one that illustrates the system in three dimensions. A sectional drawing provides duct or component detail by showing a cross-section of the component.
3. Drawings are usually drawn to scale. (Check dimensions and lengths with a ruler or a scale to be sure that this is the case. For example, 1/8 in. on the sheet may represent 1 ft on the ground.)

15.6.5 Facts, Concepts, and Example Math Problems

Ventilation is an important form of emission/exposure control. It also provides for health, comfort, and well-being. All human occupancies require ventilation. Ventilation facts and concepts to know follow:

- Properties of air
  - Molecular weight = 29
  - Weight density = 0.075 lb/ft³ at STP
- Standard temperature and pressure for ventilation
  - Temperature = 70°F
  - Barometric pressure = 29.92 in. Hg, dry air
  - Weight density = 0.075 lb/ft³ at STP
- Density correction factor (d) is a factor derived from the ideal gas equations (where BP is absolute barometric pressure):

$$d = \left(\frac{T_{TSP} \text{ (absolute)}}{T_{actual} \text{ (absolute)}} \right) \times \frac{BP_{actual}}{BP_{STP}}$$

- Degrees Rankine (°R) = °F + 460 (absolute temperature)
- Degrees Kelvin (K) = °C + 273 (absolute temperature)

■ EXAMPLE 15.5

Problem: The temperature is 90°F and the barometric pressure is 27.50 in. Hg. What is density correction factor d?

Solution:

$$d = \left(\frac{460 + 70}{460 + 90} \right) \times \frac{27.50}{29.92} = \frac{530}{550} \times \frac{27.50}{29.92} = 0.96 \times 0.92 = 0.88$$

- Local exhaust ventilation systems are made up of five components: hood, ductwork, air cleaner, fan, and stack

15.6.5.1 Pressure

Air moves under the influence of pressure differentials. A fan is commonly used to create the pressure difference. At sea level, the standard static (barometric) pressure (SP) is 14.7 psia = 29.92 in. Hg = 407 in. wg. If a fan is capable of creating 1 in. of negative static pressure (e.g., 1 in. wg, or "one inch water gauge"), then the absolute static pressure in the duct will be reduced to 406 in. wg.

- Manometers are used to measure pressure differences (attach both legs to the manometer to measure velocity pressure).
- The Pitot tube measures both TP and SP.
- S = side = SP.
- TP = tip = top = TP.

Pressures are related as follows:

$$TP = SP + VP$$

	TP	SP	VP
Upstream	–	–	+
Downstream	+	+	+

■ **EXAMPLE 15.6**

Problem: Determine the velocity pressure (VP), when

$$TP = -0.35 \text{ in. wg}, \quad SP = -0.5 \text{ in. wg}$$

Solution:

$$TP = SP + VP$$

$$VP = TP - SP = -0.35 - (-0.50) = 0.15 \text{ in. wg}$$

15.6.5.2 Velocity Pressure

The velocity pressure is related to the velocity (V) of air in the duct. The relationship is given by

$$V = 4005 \sqrt{\frac{VP}{d}}$$

where
VP = Velocity pressure (in. wg, measured with a Pitot tube).
d = Density correction factor.

■ **EXAMPLE 15.7**

Problem: The velocity pressure of an airstream in a lab fume hood duct is 0.33 in. wg What is the velocity? (Density correction factor d is 1.)

Solution:

$$V = 4005 \times (VP/d)^{1/2} = 4005 \times (0.33/1)^{1/2} = 2300 \text{ fpm}$$

15.6.5.3 Static Pressure

Static pressure is the potential energy of the ventilation system. It is converted to kinetic energy (VP) and other (less useful) forms of energy (heat, vibration, and noise). These are the losses of the system. Volume flow rate can be described by

$$Q = V \times A$$

where
Q = Volume flow rate (cfm).
A = Cross-sectional area of duct (ft²).
V = Velocity (fpm).

■ **EXAMPLE 15.8**

Problem: The cross-sectional area of a duct is 0.7854 ft². The velocity of air flowing in the duct is 2250 fpm. What is flow rate Q?

Solution:

$$Q = V \times A = 2250 \text{ fpm} \times 0.7854 \text{ ft}^2 = 1770 \text{ scfm}$$

■ **EXAMPLE 15.9**

Problem: The static pressure is measured in a 10-in. square duct at –1.15 in. wg. The average total pressure is –0.85 in. wg. Find the velocity and volume flow rate of the air flowing in the duct at STP and $d = 1$.

Solution:

$$A = (10 \text{ in.} \times 10 \text{ in.})/144 \text{ in./ft}^2 = 0.6944 \text{ ft}^2$$

$$VP = TP - SP = -0.85 \text{ in.} - (-1.15 \text{ in.}) = 0.30 \text{ in. wg}$$

$$V = 4005 \times (VP/d)^{1/2} = 4005 \times (0.30)^{1/2} = 2194$$

$$Q = V \times A = 2194 \times 0.6944 = 1524 \text{ scfm}$$

15.6.5.4 Pressure Losses and Velocity Pressure

As the air moves through a duct, losses are created (i.e., static pressure is converted to heat, vibration, noise.) The loss is usually directly related to velocity pressure:

$$SP_{Loss} = K \times VP \times d$$

where

SP_{Loss} = Loss of static pressure (in. wg)
K = Loss factor (unitless).
VP = Average velocity pressure in duct.
d = Density correction factor.

15.6.5.5 Types of Pressure Loss

Types of pressure loss include hood entry, friction, elbow, branch entry, system effect, air cleaner, and others. The hood captures, contains, or receives contaminants generated at an emission source. The hood converts duct static pressure to velocity pressure and hood entry losses (i.e., slot and duct entry losses):

$$H_e = K \times VP \times d = |SP_h| - VP$$

where

H_e = Hood entry loss.
K = Loss factor (unitless).
VP = Velocity pressure in duct (in. wg).
d = Density correction factor
$|SP_h|$ = Absolute static pressure about 5 duct diameters down the duct from the hood (in. wg).

A hood's ability to convert static pressure to velocity pressure is given by the coefficient of entry (C_e):

$$C_e = \frac{Q_{actual}}{Q_{ideal}} = \sqrt{\frac{VP}{|SP_h|}} = \sqrt{\frac{1}{(1 + K_h)}}$$

■ EXAMPLE 15.10

Problem: What is the hood static pressure (SP_h) when the duct velocity pressure is 0.33 in. wg, and the hood entry loss is $H_e = 0.44$ in. wg? What is C_e?

Solution:

$$|SP_h| = VP + H_e = 0.33 + 0.44 = 0.77 \text{ in. wg}$$

$$C_e = \frac{Q_{actual}}{Q_{ideal}} = \left[\frac{VP}{|SP_h|} \right]^{0.5} = \left[\frac{0.33}{0.77} \right]^{0.05} = 0.65$$

Hood entry losses normally occur at the hood slots and at the entrance to the duct, due to the *vena contracta* formed. The most narrowed portion of the *vena contracta* is usually found about one-half duct diameter inside the duct or plenum. The hood static pressure is the sum total of the acceleration and all losses from the hood face to the point of measurement in the duct. Head loss is

$$H_e = K \times VP_d \times d$$

where K is the hood entry loss factor.

Hood entry loss factors have been estimated over the years and have been reported in the literature for many types of hoods, including lab fume hoods. Three types of hoods are the enclosing, capture (active, external), and receiving (passive, often canopy type hood).

■ EXAMPLE 15.11

Problem: What is the hood entry loss (H_e) for a laboratory fume hood when the average velocity pressure in the duct is 0.30 in. wg? Assume $K = 2.0$ and $d = 1$.

Solution:

$$H_e = K \times VP_d \times d = 2.0 \times 0.30 \times 1 = 0.60 \text{ in. wg}$$

The area approach is used too determine Q (flow) for capture hoods. Basically, air approaches from all directions toward the source of negative pressure. Imagine a three-dimensional sphere around the end of a small, plain duct hood. Air molecules don't know if they are in the front, to the side, or to the back of the opening. All they know is that they are experiencing a big push to get over to that spot of negative pressure. The velocity of air moving toward the opening is equal at all points on the surface of the sphere. The surface area of a sphere is given by

$$A = 4\pi x^2$$

Knowing the area and the desired capture velocity at x, we can estimate the volume flow rate from $Q = V \times A$.

■ EXAMPLE 15.12

Problem: Air enters an ideal 4-in. plain duct hood. What is the required volume flow rate for capture 6 inches in front of the hood if we need $V_c = 100$ fpm?

Solution:

$$Q = V_c \times A = 100 \times 4\pi(0.5 \text{ ft})^2 = 315 \text{ cfm}$$

where $A = 4\pi x^2$; 6 in. = 0.5 ft.

15.6.5.6 Fans

Types of fans include the centrifugal (forward curved, backward inclined, radial) and axial.

15.6.5.6.1 Fan Curves

Fan characteristic curves plot volume flow rate Q against static pressure, horsepower, noise, and efficiency.

15.6.5.6.2 Fan Specifications

Fans are specified by pressure and flow rate—the *system operating point* (SOP). The pressure is found across the fan—at the inlet and outlet of the fan in the ductwork.

15.6.5.6.3 Fan Total Pressure

The fan total pressure (FTP) represents all energy requirements for moving air through the ventilation system. FTP is calculated by adding the absolute values of the average total pressures found at the fan. If the sign convention is followed, then a formula for FTP is

$$FTP = TP_{out} - TP_{in}$$

Substituting for $TP = SP + VP$ gives

$$FTP = SP_{out} + VP_{out} - SP_{in} - VP_{in}$$

If VP_{out} equals VP_{in} (i.e., if the average inlet and outlet velocities are equal), then the VP terms in the above equation cancel, leaving

$$FTP = SP_{out} - SP_{in}$$

The FTP is often referred to as the *fan total static pressure drop.*

■ **EXAMPLE 15.13**

Problem: The outlet and inlet conditions at a fan are $SP_{out} = 0.10$ in. and $SP_{in} = -0.75$ in. What is the FTP?

Solution:

$$FTP = SP_{out} - SP_{in} = 0.10 \text{ in. wg} - (-0.75 \text{ in. wg}) = 0.85 \text{ in. wg}$$

15.7 NOISE

High noise levels in the workplace are a hazard to employees. High noise levels are a physical stressor that may produce psychological effects by annoying, startling, or disrupting the worker's concentration, which can lead to accidents. High levels can also result in damage to worker's hearing, resulting in hearing loss. In this section, we discuss the basics of noise, including those elements environmental practitioners responsible for worker safety need to know to ensure that their organization's hearing conservation program is in compliance with OSHA.

15.7.1 DEFINITIONS

There are many specialized terms used to express concepts in noise, noise control, and hearing loss prevention. Environmental engineering practitioners responsible for ensuring proper engineering design to reduce workplace noise levels and the in-house safety engineers responsible for compliance with OSHA's Hearing Conservation Program requirements must be familiar with these terms. The NIOSH (2005) definitions below were written in as non-technical a fashion as possible.

Acoustic trauma—A single incident which produces an abrupt hearing loss. Welding sparks (to the eardrum), blows to the head, and blast noise are examples of events capable of providing acoustic trauma.

Action level—The sound level which when reached or exceeded necessitates implementation of activities to reduce the risk of noise-induce hearing loss. OSHA currently uses an 8-hour time weighted average of 85 dBA as the criterion for implementing an effective hearing conservation program.

Attenuate—To reduce the amplitude of sound pressure (noise).

Attenuation—See real ear attenuation at threshold (REAT), real-world attentuation

Audible range—The frequency range over which normal ears hear: approximately 20 Hz through 20,000 Hz.

Audiogram—A chart, graph, or table resulting from an audiometric test showing an individual's hearing threshold levels as a function of frequency.

Audiologist—A professional, specializing in the study and rehabilitation of hearing, who is certified by the American Speech–Language–Hearing Association or licensed by a state board of examiners.

Background noise—Noise coming from sources other than the particular noise sources being monitored.

Baseline audiogram—A valid audiogram against which subsequent audiograms are compared to determine if hearing thresholds have changed. The baseline audiogram is preceded by a quiet period so as to obtain the best estimate of the person's hearing at that time.

Continuous noise—Noise of a constant level as measured over at least one second using the "slow" setting on a sound level meter. Note that a noise which is intermittent (e.g., on for over a second and then off for a period) would be both variable and continuous.

Controls, administrative—Efforts, usually by management, to limit workers' noise exposure by modifying workers' schedules or locations, or by modifying the operating schedule of noisy machinery.

Controls, engineering—Any use of engineering methods to reduce or control the sound level of a noise source by modifying or replacing equipment, making any physical changes at the noise source or along the transmission path (with the exception of hearing protectors).

Criterion sound level—A sound level of 90 decibels.

dB (decibel)—The unit used to express the intensity of sound. The decibel was named after Alexander Graham Bell. The decibel scale is a logarithmic scale in which 0 dB approximates the threshold of hearing in the mid frequencies for young adults and in which the threshold of discomfort is between 85 and 95 dB SPL and the threshold for pain is between 120 and 140 dB SPL.

Double hearing protection—A combination of both ear plug and ear muff type hearing protection devices is required for employees who have demonstrated temporary threshold shift during audiometric examination and for those who have been advised by a medical doctor to wear double protection in work areas that exceed 104 dBA.

Dosimeter—When applied to noise, refers to an instrument that measures sound levels over a specified interval, stores the measurements, and calculates the sound as a function of sound level and sound duration. It describes the results in terms of dose, time-weighted average, and (perhaps) other parameters such as peak level, equivalent sound level, sound exposure level, etc.

Equal-energy rule—The relationship between sound level and sound duration based upon a 3-dB exchange rate; that is, the sound energy resulting from doubling or halving a noise exposure's duration is equivalent to increasing or decreasing the sound level by 3 dB, respectively.

Exchange rate—The relationship between intensity and dose. OSHA uses a 5-dB exchange rate. Thus, if the intensity of an exposure increases by 5 dB, the dose doubles. Sometimes this is also referred to as the doubling rate. The U.S. Navy uses a 4-dB exchange rate; the U.S. Army and Air Force uses a 3-dB exchange rate. NIOSH recommends a 3-dB exchange rate. Note that the equal-energy rule is based on a 3-dB exchange rate.

Frequency—Rate in which pressure oscillations are produced. It is measured in hertz (Hz).

Hazardous noise—Any sound for which any combination of frequency, intensity, or duration is capable of causing permanent hearing loss in a specified population.

Hazardous task inventory—A concept based on using work tasks as the central organizing principle for collecting descriptive information on a given work hazard. It consists of a list(s) of specific tasks linked to a database containing the prominent characteristics relevant to the hazard(s) of interest which are associated with each task.

Hearing conservation record—Employee's audiometric record, which includes name, age, job classification, TWA exposure, date of audiogram, and name of audiometric technician. It is to be retained for duration of employment for OSHA and is kept indefinitely for workers' compensation.

Hearing damage risk criteria—A standard which defines the percentage of a given population expected to incur a specified hearing loss as a function of exposure to a given noise exposure.

Hearing handicap—A specified amount of permanent hearing loss usually averaged across several frequencies which negatively impacts employment and/or social activities. Handicap is often related to an impaired ability to communicate. The degree of handicap will also be related to whether the hearing loss is in one or both ears, and whether the better ear has normal or impartial hearing.

Hearing loss—Often characterized by the area of the auditory system responsible for the loss. For example, when injury or a medical condition affects the outer ear or middle ear (i.e., from the pinna, ear canal, and ear drum to the cavity behind the ear drum—which includes the ossicles) the resulting hearing loss is referred to as a *conductive* loss. When an injury or medical condition affects the inner ear or the auditory nerve that connects the inner ear to the brain (i.e., the cochlea and the VIII cranial nerve) the resulting hearing loss is referred to as a *sensorineural* loss. Thus, a welder's spark which damaged the ear drum would cause a conductive hearing loss. Because noise can damage the tiny hair cells located in the cochlea, it causes a sensorineural hearing loss.

Hearing loss prevention program audit—An assessment performed prior to putting a hearing loss prevention program into place or before changing an existing program. The audit should be a top-down analysis of the strengths and weaknesses of each aspect of the program.

Hearing threshold level (HTL)—The hearing level, above a reference value, at which a specified sound or tone is heard by an ear in a specified fraction of the trials. Hearing threshold levels have been established so that a 0-dB HTL reflects the best hearing of a group of persons.

Hz (Hertz)—The unit of measurement for audio frequencies. The frequency range for human hearing lies between 20 Hz and approximately 20,000 Hz. The sensitivity of the human ear drops off sharply below about 500 Hz and above 4,000 Hz.

Impulsive noise—Used to generally characterize impact or impulse noise which is typified by a sound which rapidly rises to a sharp peak and then quickly fades. The sound may or may not have a "ringing" quality (such as a striking a hammer on a metal plate or a gunshot in a reverberant room). Impulsive noise may be repetitive, or may be a single event (as with a sonic boom). If impulses occurring in very rapid succession (such as with some jack hammers), the noise would not be described as impulsive.

Loudness—The subjective attribute of a sound by which it would be characterized along a continuum from soft to loud. Although this is a subjective attribute, it depends primarily upon sound pressure level, and to a lesser extent, the frequency characteristics and duration of the sound.

Material hearing impairment—As defined by OSHA, a material hearing impairment is an average hearing threshold level of 25 dB HTL as the frequencies of 1000, 2000, and 3000 Hz.

Medical pathology—A disorder or disease. For purposes of this program, a condition or disease affecting the ear, which a physician specialist should treat.

Noise—Noise is any unwanted sound.

Noise dose—The noise exposure expressed as a percentage of the allowable daily exposure. For OSHA, a 100% dose would equal an 8-hour exposure to a continuous 90-dBA noise; a 50% dose would equal an 8-hour exposure to an 85-dBA noise or a 4-hour exposure to a 90-dBA noise. If 85 dBA is the maximum permissible level, then an 8-hour exposure to a continuous 85-dBA noise would equal a 100% dose. If a 3-dB exchange rate is used in conjunction with an 85-dBA maximum permissible level, a 50% dose would equal a 2-hour exposure to 88 dBA, or an 8-hour exposure to 82 dBA.

Noise dosimeter—An instrument that integrates a function of sound pressure over a period of time to directly indicate a noise dose.

Noise hazard area—Any area where noise levels are equal to or exceed 85 dBA. OSHA requires employers to designate work areas, post warning signs, and warn employees when work practices exceed 90 dBA as a "noise hazard area." Hearing protection must be worn whenever 90 dBA is reached or exceeded.

Noise hazard work practice—Performing or observing work where 90 dBA is equaled or exceeded. Some work practices will be specified; however, as a rule of thumb, whenever attempting to hold a normal conversation with someone who is 1 foot away and shouting must be employed to be heard, one can assume that a 90-dBA noise level or greater exists and hearing protection is required. Typical examples of work practices where hearing protection is required are jack hammering, heavy grinding, heavy equipment operations, and similar activities.

Noise-induced hearing loss—A sensorineural hearing loss that is attributed to noise and for which no other etiology can be determined.

Noise level measurement—Total sound level within an area. Includes workplace measurements indicating the combined sound levels of tool noise (from ventilation systems, cooling compressors, circulation pumps, etc.).

Noise reduction rating (NRR)—The NRR is a single-number rating method which attempts to describe a hearing protector based on how much the overall noise level is reduced by the hearing protector. When estimating A-weighted noise exposures, it is important to remember to first subtract 7 dB from the NRR and then subtract the remainder from the A-weighted noise level. The NRR theoretically provides an estimate of the protection that should be met or exceeded by 98% of the wearers of a given device. In practice, this does not prove to be the case, so a variety of methods for "de-rating" the NRR have been discussed.

Ototoxic—A term typically associated with the sensorineural hearing loss resulting from therapeutic administration of certain prescription drugs.

Ototraumatic—A broader term than ototoxic. As used in hearing loss prevention, refers to any agent (e.g., noise, drugs, or industrial chemicals) which has the potential to cause permanent hearing loss subsequent to acute or prolong exposure.

Presbycusis—The gradual increase in hearing loss that is attributable to the effects of aging, and not related to medical causes or noise exposure.

Real ear attenuation at threshold (REAT)—A standardized procedure for conducting psychoacoustic tests on human subjects designed to measure sound protection features of hearing protective devices. Typically, these measures are obtained in a calibrated sound field, and represent the difference between subjects' hearing thresholds when wearing a hearing protector vs. not wearing the protector.

Real-world attenuation—Estimated sound protection provided by hearing protective devices as worn in "real-world" environments.

Sensorineural hearing loss—A hearing loss resulting from damage to the inner ear (from any source).

Sociacusis—A hearing loss related to non-occupational noise exposure.

Sound intensity (I)—Sound intensity at a specific location is the average rate at which sound energy is transmitted through a unit area normal to the direction of sound propagation.

Sound level meter (SLM)—A device which measures sound and provides a readout of the resulting measurement. Some provide only A-weighted measurements, others provide A- and C-weighted measurements, and some can provide weighted, linear, and octave (or narrower) ban measurements. Some SLMs are also capable of providing time-integrated measurements.

Sound power—The total sound energy radiated by a source per unit time. Sound power cannot be measured directly.

Sound pressure level (SPL)—A measure of the ratio of the pressure of a sound wave relative to a reference sound pressure. Sound pressure level in decibels is typically referenced to 20 mPa. When used alone (e.g., 90 dB), a given decibel level implies an unweighted sound pressure level.

Standard threshold shift (STS)—(1) OSHA uses the term to describe a change in hearing threshold relative to the baseline audiogram of an average of 10 dB or more at 2000, 3000, and 4000 Hz in either ear; used by OSHA to trigger additional audiometric testing and related follow up. (2) NIOSH uses this term to describe a change of 15 dB or more at any frequency, 5000 through 6000 Hz, from baseline levels that are present on an immediate retest in the same ear and at the same frequency. NIOSH recommends a confirmation audiogram within 30 days with the confirmation audiogram preceded by a quiet period of at least 14 hours.

Threshold shift—Audiometric monitoring programs will encounter two types of changes in hearing sensitivity: a permanent threshold shift (PTS) and a temporary threshold shift (TTS). As the names imply, any change in hearing sensitivity which is persistent is considered a PTS. Persistence may be assumed if the change is observed on a 30-day follow-up exam. Exposure to loud noise may cause a temporary worsening in hearing sensitivity (i.e., a TTS) that may persist for 14 hours (or even longer in cases where the exposure duration exceeded 12 to 16 hours). Hearing health professionals need to recognize that not all threshold shifts represent decreased sensitivity, and not all temporary or permanent threshold shifts are due to noise exposure. When a permanent threshold shift can be attributable to noise exposure, it may be referred to as a noise-induced permanent threshold shift (NIPTS).

Velocity—Speed at which the regions of sound producing pressure changes move away from the sound source.

Wavelength—Distance required for one complete pressure cycle to be completed (1 wavelength) and is measured in feet or meters.

Weighted measurements—Two weighting curves are commonly applied to measures of sound levels to account for the way the ear perceives the "loudness" of sounds.

 A-weighting—A measurement scale that approximates the "loudness" of tones relative to a 40-db SPL 1000 Hz reference tone. A-weighting has the added advantage of being correlated with annoyance measures and is most responsive to the mid frequencies, 500 to 4000 Hz.

 C-weighting—A measurement scale that approximates the "loudness" of tones relative to a 90-dB SPL 1000 Hz reference tone. C-weighting has the added advantage of providing a relatively flat measurement scale that includes very low frequencies.

15.7.2 Occupational Noise Exposure

As mentioned earlier, noise is commonly defined as any unwanted sound. Noise literally surrounds us every day, and is with us just about everywhere we go. However, the noise we are concerned with here is that produced by industrial processes. Excessive amounts of noise in the work environment (and outside it) cause many problems for workers, including increased stress levels, interference with communication, disrupted concentration, and most importantly, varying degrees of hearing loss. Exposure to high noise levels also adversely affects job performance and increases accident rates. One of the major problems with attempting to protect workers' hearing acuity is the tendency of many workers to ignore the dangers of noise. Because hearing loss, like cancer, is insidious, it's easy to ignore. It sneaks up slowly and is not apparent (in many cases) until after the damage is done. Alarmingly, hearing loss from occupational noise exposure has been well documented since the eighteenth century, yet since the advent of the industrial revolution, the number of exposed workers has greatly increased (Mansdorf, 1993). However, today the picture of hearing loss is not as bleak as it has been in the past, as a direct result of OSHA's requirements. Now that noise exposure must be controlled in all industrial environments, that well-written and well-managed hearing conservation programs must be put in place, and that employee awareness must be raised to the dangers of exposure to excessive levels of noise, job-related hearing loss is coming under control.

15.7.3 DETERMINING WORKPLACE NOISE LEVELS

The unit of measurement for sound is the decibel. Decibels are the preferred unit for measuring sound, derived from the bel, a unit of measure in electrical communications engineering. The decibel is a dimensionless unit used to express the logarithm of the ratio of a measured quantity to a reference quantity. In regard to noise control in the workplace, the safety engineer's primary concern is first to determine if any "noise-makers" in the facility exceed the OSHA limits for worker exposure—exactly which machines or processes produce noise at unacceptable levels. Making this determination is accomplished by conducting a noise level survey of the plant or facility. Sound measuring instruments are used to make this determination. These include noise dosimeters, sound level meters, and octave-band analyzers. The uses and limitations of each kind of instrument are discussed below.

15.7.3.1 Noise Dosimeter

The noise dosimeters used by OSHA meet the American National Standards Institute (ANSI) Standard S1.25-1978, "Specifications for Personal Noise Dosimeter," which set performance and accuracy tolerances. For OSHA use, the dosimeter must have a 5-dB exchange rate, use a 90-dBA criterion level, be set at slow response, and use either an 80-dBA or 90-dBA threshold gate, or a dosimeter that has both capabilities, whichever is appropriate for evaluation.

15.7.3.2 Sound Level Meter (SLM)

When conducting the noise level survey, the industrial hygienist or survey technician should use an ANSI-approved sound level meter (SLM)—a device used most commonly to measure sound pressure. The SLM measures in decibels. One decibel is one-tenth of a bel and is the minimum difference in loudness that is usually perceptible. The SLM consists of a microphone, an amplifier, and an indicating meter that responds to noise in the audible frequency range of about 20 to 20,000 Hz. Sound level meters usually contain weighting networks designated "A", "B", or "C". Some meters have only one weighting network; others are equipped with all three. The A-network approximates the equal loudness curves at low sound pressure levels, the B-network is used for medium sound pressure levels, and the C-network is used for high levels.

In conducting a routine workplace sound level survey, using the A-weighted network (referenced dBA) in the assessment of the overall noise hazard has become common practice. The A-weighted network is the preferred choice because it is thought to provide a rating of industrial noise that indicates the injurious effects such noise has on the human ear (gives a frequency response similar to that of the human ear at relatively low sound pressure levels).

With an approved and freshly calibrated (always calibrate test equipment prior to use) sound level meter in hand, the environmental professional is ready to begin the sound level survey. In doing so, the industrial hygienist is primarily interested in answering the following questions:

1. What is the noise level in each work area?
2. What equipment or process is generating the noise?
3. Which employees are exposed to the noise?
4. How long are they exposed to the noise?

To answer these questions, environmental professionals record their findings as they move from workstation to workstation, following a logical step-by-step procedure. The first step involves using the sound level meter set for A-scale slow response mode to measure an entire work area. When making such measurements, restrict the size of the space being measured to less than 1,000 square feet. If the maximum sound level does not exceed 80 dBA, it can be assumed that all workers in this work area are working in an environment with a satisfactory noise level. However, a note of caution is advised: The key words in the preceding statement are "maximum sound level." To assure an accurate measurement, the industrial hygienist must ensure that all "noisemakers" are actually in operation when measurements are taken. Measuring an entire work area does little good when only a small percentage of the noisemakers are actually in operation.

The next step depends on the readings recorded when the entire work area was measured. For example, if the measurements indicate sound levels greater than 80 dBA, then another set of measurements need to be taken at each worker's workstation. The purpose here, of course, is to determine two things: which machine or process is making noise above acceptable levels (i.e., >80 dBA), and which workers are exposed to these levels. Remember that the worker who operates the machine or process might not be the only worker exposed to the noisemaker. You need to inquire about other workers who might, from time to time, spend time working in or around the machine or process. Our experience in conducting workstation measurements have shown us noise levels usually fluctuate. If this is the case, you must record the minimum and maximum noise levels. If you discover that the noise level is above 90 (dBA) (and it remains above this level), you have found a noisemaker that exceeds the legal limit (90 dBA). However, if your measurements indicate that the noise level is never greater than 85 (dBA) (OSHA's action level), the noise exposure can be regarded as satisfactory.

If workstation measurements indicate readings that exceed the 85-dBA level, you must perform another step. This step involves determining the length of time of exposure for workers. The easiest, most practical way to make this determination is to have the worker wear a noise dosimeter, which records the noise energy to which the worker was exposed during the workshift.

Note: This parameter assumes that the worker has good hearing acuity with no loss. If the worker has documented hearing loss, exposure to 95 dBA or higher, without proper hearing protection, may be unacceptable under any circumstances.

15.7.3.3 Octave-Band Noise Analyzers

Several Type 1 sound level meters (such as the GenRad 1982 and 1983 and the Quest 155) used by OSHA have built-in octave band analysis capability. These devices can be used to determine the feasibility of controls for individual noise sources for abatement purposes and to evaluate hearing protectors. Octave-band analyzers segment noise into its component parts. The octave-band filter sets provide filters with the following center frequencies: 31.5, 63, 125, 250, 500, 1000, 2000, 4000, 8000, and 16,000 Hz. The special signature of a given noise can be obtained by taking sound level meter readings at each of these settings (assuming that the noise is fairly constant over time). The results may indicate those octave-bands that contain the majority of the total radiated sound power. Octave-band noise analyzers can assist industrial hygienists in determining the adequacy of various types of frequency-dependent noise controls. They also can be used to select hearing protectors because they can measure the amount of attenuation offered by the protectors in the octave-bands responsible for most of the sound energy in a given situation.

15.7.4 ENGINEERING CONTROL FOR INDUSTRIAL NOISE

When environmental professionals engineers investigate the possibility of using engineering controls to control noise, the first thing they recognize is that reducing and/or eliminating all noise is virtually impossible. And this should not be the focus in the first place, eliminating or reducing the "hazard" is the goal. While the primary hazard may be the possibility of hearing loss, the distractive effect (or its interference with communication) must also be considered. The distractive effect of excessive noise can certainly be classified as hazardous whenever the distraction might affect the attention of the worker. The obvious implication of noise levels that interfere with communications is emergency response. If ambient noise is at such a high level that workers can't hear fire or other emergency alarms, this is obviously an unacceptable situation.

So what does all this mean? Environmental professionals must determine the "acceptable" level of noise. Then he or she can look into applying the appropriate noise control measures. These include making alterations in engineering design (obviously this can only be accomplished in the design phase) or making modifications after installation. Unfortunately, this latter method is the one the environmental professional is usually forced to apply—and also the most difficult, depending upon circumstances.

Let's assume that the environmental professional is trying to reduce noise levels generated by an installed air compressor to a safe level. Obviously, the first place to start is at the *source*: the air compressor. Several options are available to employ at the source. First, the safety engineer would look at the possibility of modifying the air compressor to reduce its noise output. One option might be to install resilient vibration mounting devices. Another might be to change the coupling between the motor and the compressor—install an insulator-cushioning-device between the couplings to dampen noise and vibration.

If the options described for use at the source of the noise are not feasible or are only partially effective, the next component the environmental professional would look at is the *path* along which the sound energy travels. Increasing the distance between the air compressor and the workers could be a possibility. (Sound levels decrease with distance.) Another option might be to install acoustical treatments on ceilings, floors, and walls. The best option available (in this case) probably is to enclose the air compressor, so that the dangerous noise levels are contained within the enclosure, and the sound leaving the space is attenuated to a lower, safer level. If total enclosure of the air compressor is not practicable, then erecting a barrier or baffle system between the compressor and the open work area might be an option.

The final engineering control component the environmental professional might incorporate to reduce the air compressor's noise problem is to consider the *receiver* (the worker/operator). An attempt should be made to isolate the operator by providing a noise reduction or soundproof enclosure or booth for the operator.

15.7.5 NOISE UNITS, RELATIONSHIPS, AND EQUATIONS

A number of noise units, relationships, and equations that are important to safety engineers involved with controlling noise hazards in the workplace are discussed below.

15.7.5.1 Sound Power

Sound power of a source is the total sound energy radiated by the source per unit time. It is expressed in terms of the sound power level (L_w) in decibels referenced to 10^{-12} watts (w_0). The relationship to decibels is shown below:

$$L_w = 10 \log_{10} w/w_0$$

where

L_w = Sound power level (decibels).
w = Sound power (watts).
w_0 = Reference power (10^{-12} watts).
$\log_{10}$ = Logarithm to the base 10.

15.7.5.2 Sound Pressure

Units used to describe sound pressures are

$$1 \; \mu bar = 1 \; dyne/cm^2 = 0.1 \; N/cm^2 = 0.1 \; Pa$$

15.7.5.3 Sound Pressure Level

$$SPL = 10 \log(p^2/p_0)$$

where

SPL = Sound pressure level (decibels)
p = Measured rms sound pressure (N/m², μbars).[*]
p_0 = Reference rms sound pressure (20 μPa, N/m², μbars)

15.7.5.4 Speed of Sound

$$c = f\lambda$$

15.7.5.5 Wavelength

$$\lambda = c/f$$

15.7.5.6 Frequency of Octave Bands

$$\text{Upper frequency band} = f_2 = 2f_1$$

where

f_2 = Upper frequency band.
f_1 = Lower frequency band.

$$\text{One-half octave band} = f_2 = \sqrt{2f_1}$$

where

f_2 = 1/2 octave band.
f_1 = Lower frequency band.

$$\text{One-third octave band} = f_2 = \sqrt[3]{2f_1}$$

where

f_2 = 1/3 octave band
f_1 = Lower frequency band

---

[*] The root-mean-square (rms) value of a changing quantity, such as sound pressure, is the square root of the mean of the squares of the instantaneous values of the quantity.

15.7.5.7 Adding Noise Sources When Sound Power Is Known

$$L_w = 10 \log (w_1 + w_2)/(w_0 + w_0)$$

where
  L_w = Sound power (watts).
  w_1 = Sound power of noise source 1 (watts).
  w_2 = Sound power of noise source 2 (watts).
  w_0 = Reference sound power (reference 10^{-12}) (watts).

15.7.5.8 Sound Pressure Additions When Sound Pressure Is Known

$$SPL = 10 \log p^2/p_0^2$$

where
  $p^2/p_0^2 = 10^{SPL/10}$.
  SPL = Sound pressure level (decibels).
  p = Measured root-mean-square (rms) sound pressure (N/m², μbars).
  p_0 = Reference rms sound pressure (20 μPa, N/m², μbars).

For three sources, the equation becomes

$$SPL = 10 \log \left(10^{SPL_1/10}\right) + \left(10^{SPL_2/10}\right) + \left(10^{SPL_3/10}\right)$$

When adding any number of sources, whether the sources are identical or not, the equation becomes

$$SPL = 10 \log \left(10^{SPL_1/10}\right) + \ldots + \left(10^{SPL_n/10}\right)$$

To determine the sound pressure level from multiple identical sources, use the following equation:

$$SPL_f = SPL_i + 10 \log n$$

where
  SPL_f = Total sound pressure level (dB).
  SPL_i = Individual sound pressure level (dB).
  n = Number of identical sources.

15.7.5.9 Noise Levels in a Free Field

$$SPL = L_w - 20 \log r - 0.5$$

where
  SPL = Sound pressure (reference 0.00002 N/m²).
  L_w = Sound power (reference 10^{-12} watts).
  r = Distance (feet).

15.7.5.10 Noise Levels with Directional Characteristics

$$SPL = L_w - 20 \log r - 0.5 + \log Q$$

where
  SPL = Sound pressure (reference 0.00002 N/m²).
  L_w = Sound power (reference 10^{-12} watts).
  r = Distance (feet).

Q = Directivity factor.
 = 2 for one reflecting plane.
 = 4 for two reflecting planes.
 = 8 for three reflecting planes.

15.7.5.11 Noise Level at a New Distance from the Noise Source

$$SPL = SPL_1 + 20 \log (d_1)/(d_2)$$

where
 SPL = Sound pressure level at new distance (d_2).
 SPL_1 = Sound pressure level at d_1.
 d_n = Distance from source.

15.7.5.12 Calculating Daily Noise Dose

The following formula combines the effects of different sound pressure levels and allowable exposure times:

$$\text{Daily noise dose} = \frac{C_1 + C_2 + C_3 + \ldots + C_n}{T_1 + T_2 + T_3 + \ldots + T_n}$$

where
 C_i = Number of hours exposed at given SPL_i.
 T_i = Number of hours exposed at given SPL_i.

15.7.5.13 OSHA Permissible Noise Levels

$$T_{SPL} = 8/2^{(SPL-90)/5}$$

where
 T_{SPL} = Time in hours at given SPL.
 SPL = Sound pressure level (dBA).

15.7.5.14 Converting Noise Dose Measurements to 8-Hour Equivalent TWA

$$TWA_{eq} = 90 + 16.61 \log (D)/100$$

where
 TWA_{eq} = Eight-hour equivalent TWA (dBA).
 D = Noise dosimeter reading (%).

15.7.5.15 Noise Reduction in Duct System

$$NR = 12.6 \, P\alpha^{1.4}/A$$

where:
 NR = Noise reduction (dB/ft).
 P = Perimeter of duct (in.).
 α = Absorption coefficient of the lining material at frequency of interest.
 A = Cross-sectional area of duct (in.^{2}).

15.7.5.16 Sound Intensity Level

Sound intensity level is the power passing through a unit area as the sound power radiates in free space.

$$L_1 = 10 \log I/I_o$$

where
- L_1 = Sound pressure level (dB).
- I = Sound intensity (W/m^2).
- I_o = Reference sound intensity (W/m^2).

15.7.5.17 Noise Reduction by Absorption

The amount of noise absorption from room surfaces is measured in sabins.

$$\text{Noise reduction (dB)} = 10 \log_{10}(A_2/A_1)$$

where
- A_2 = Total amount of absorption in room after treatment (sabins).
- A_1 = Total amount of absorption in room before treatment (sabins).

15.8 THERMAL STRESS

Appropriately controlling the temperature, humidity, and air distribution in work areas is an important part of providing a safe and healthy workplace. A work environment in which the temperature is not properly controlled can be uncomfortable. Extremes of either heat or cold can be more than uncomfortable—they can be dangerous. Operations involving high air temperatures, radiant heat sources, high humidity, direct physical contact with hot objects, or strenuous physical activities have a high potential for inducing heat stress in employees engaged in such operations. Such places include: iron and steel foundries, nonferrous foundries, brick-firing and ceramic plants, glass products facilities, rubber products factories, electrical utilities (particularly boiler rooms), bakeries, confectioneries, commercial kitchens, laundries, food canneries, chemical plants, mining sites, smelters, and steam tunnels (Spellman and Whiting, 2005). Outdoor activities conducted in hot weather, such as construction, refining, asbestos abatement, and hazardous waste site activities, especially those that require workers to wear semipermeable or impermeable protective clothing, are also likely to cause heat stress among exposed workers.

15.8.1 Causal Factors

The occurrence of thermal stress to workers in the workplace can be attributed to various causal factors as pointed out below.

- Age, weight, degree of physical fitness, degree of acclimatization, metabolism, use of alcohol or drugs, and a variety of medical conditions such as hypertension all affect a person's sensitivity to heat. However, even the type of clothing worn must be considered. Prior heat injury predisposes an individual to additional injury.
- It is difficult to predict just who will be affected and when, because individual susceptibility varies. In addition, environmental factors include more than the ambient air temperature. Radiant heat, air movement, conduction, and relative humidity all affect an individual's response to heat.

Heat stress and cold stress are major concerns of modern health and environmental professionals. This section provides the information they need to know in order to overcome the hazards associated with extreme temperatures.

15.8.2 Thermal Comfort

Thermal comfort in the workplace is a function of a number of different factors. Temperature, humidity, air distribution, personal preference, and acclimatization are all determinants of comfort in the workplace. Determining optimum conditions, however, is not a simple process. To fully understand the hazards posed by temperature extremes, industrial hygienists must be familiar with several basic concepts related to thermal energy and comfort. The most important of these are summarized here:

- *Conduction* is the transfer of heat between two bodies that are touching or from one location to another within a body. For example, if an employee touches a workpiece that has just been welded and is still hot, heat will be conducted from the workpiece to the hand. Of course, the result of this heat transfer is a burn.
- *Convection* is the transfer of heat from one location to another by way of a moving medium (a gas or a liquid). Convection ovens use this principle to transfer heat from an electrode by way of gases in the air to whatever is being baked.
- *Metabolic heat* is produced within a body as a result of activity that burns energy. All humans produce metabolic heat. This is why a room that is comfortable when occupied by just a few people may become uncomfortable when it is crowded. Unless the thermostat is lowered to compensate, the metabolic heat of a crowd will cause the temperature of a room to an uncomfortable level.
- *Environmental heat* is produced by external sources. Gas or electric heating systems produce environmental heat, as do sources of electricity and a number of industrial processes.
- *Radiant heat* is the result of electromagnetic nonionizing energy that is transmitted through space without the movement of matter within that space.
- Most experts state that workers should not be permitted to work when their deep body temperature exceeds 38°C (100.4°F).
- *Heat* is a measure of energy in terms of quantity.
- A *calorie* is the amount of energy in terms of quantity.
- *Evaporative cooling* takes place when sweat evaporates from the skin. High humidity reduces the rate of evaporation and thus reduces the effectiveness of the body's primary cooling mechanism.

15.8.3 Body's Response to Heat

Operations involving high air temperatures, radiant heat sources, high humidity, direct physical contact with hot objects, or strenuous physical activities have a high potential for inducing heat stress in employees engaged in such operations. Such places include: Iron and steel foundries, nonferrous foundries, brick-firing and ceramic plants, glass products facilities, rubber products factories, electrical utilities (particularly boiler rooms), bakeries, confectioneries, commercial kitchens, laundries, food canneries, chemical plants, mining sites, smelters, and steam tunnels. Outdoor operations conducted in hot weather, such as construction, refining, asbestos removal, and hazardous waste site activities, especially those that require workers to wear semipermeable or impermeable protective clothing, are also likely to cause heat stress among exposed workers (Spellman and Whiting, 2005).

The human body is equipped to maintain an appropriate balance between the metabolic heat it produces and the environmental heat to which it is exposed. Sweating and the subsequent evaporation of the sweat are the body's way of trying to maintain an acceptable temperature balance. This balance can be expressed as a function of the various factors in the following equation.

$$H = M \pm R \pm C - E \qquad (15.16)$$

where
H = Body heat.
M = Internal heat gain (metabolic).
R = Radiant heat gain.

C = Convection heat gain.
E = Evaporation (cooling).

The ideal balance when applying the equation is no new heat gain. As long as heat gained from radiation, convection, and metabolic processes do not exceed that lost through the evaporation induced by sweating, the body experiences no stress or hazard. However, when heat gain from any source is more than the body can compensate for by sweating, the result is *heat stress*.

15.8.4 WORK-LOAD ASSESSMENT

Under conditions of high temperature and heavy workload, the industrial hygienist should determine the work-load category of each job (refer to applicable ACGIH tables). Work-load category is determined by averaging metabolic rates for the tasks and then ranking them:

- Light work—up to 200 kcal/hour
- Medium work—200 to 350 kcal/hour
- Heavy work—350 to 500 kcal/hour

15.8.4.1 Cool Rest Area

Where heat conditions in the rest area are different from those in the work area, the metabolic rate (M) should be calculated using a time-weighted average, as shown in Equation 15.17:

$$\text{Average metabolic rate} = \frac{(M_1)(t_1)+(M_2)(t_2)+\ldots+(M_n)(t_n)}{t_1+t_2+\ldots+t_n} \tag{15.17}$$

where
M = Metabolic rate.
t = Time in minutes.

15.8.5 SAMPLING METHODS

Currently used thermal exposure sampling methods are discussed below.

- *Body temperature measurements*—Although instruments are available to estimate deep body temperature by measuring the temperature in the ear canal or on the skin, these instruments are not sufficiently reliable to use in compliance evaluations.
- *Environmental measurements*—Environmental heat measurements should be made at, or as close as possible to, the specific work area where the worker is exposed. When a worker is not continuously exposed in a single hot area but moves between two or more areas having different levels of environmental heat, or when the environmental heat varies substantially at a single hot area, environmental heat exposures should be measured for each area and for each level of environmental heat to which employees are exposed.
- *Wet bulb globe temperature index*—Wet bulb globe temperature (WBGT) should be calculated using the appropriate formula (see Equation 15.17). The WBGT for continuous all-day or several hour exposures should be averaged over a 60-minute period. Intermittent exposures should be averaged over a 120-minute period. These averages should be calculated using Equation 15.18.

$$\text{Average } WBGT = \frac{(WBGT_1)(t_1)+(WBGT_2)(t_2)+\ldots+(WBGT_n)(t_n)}{t_1+t_2+\ldots+t_n} \tag{15.18}$$

For indoor and outdoor conditions with no solar load, WBGT is calculated using Equation 15.19:

$$WBGT = 0.7NWB + 0.3GT \tag{15.19}$$

For outdoors with a solar load, WBGT is calculated using Equation 15.20:

$$WBGT = 0.7NWB + 0.2GT + 0.1DB \tag{15.20}$$

- *Measurement*—Portable heat stress meters or monitors are used to measure heat conditions. These instruments can calculate both the indoor and outdoor WBGT index according to established ACGIH threshold limit value equations. With this information, in addition to information on the type of work being performed, heat stress meters can determine how long a person can safely work or remain in a particular hot environment. It should be noted that measurement is often required for those environmental factors that most nearly correlate with deep body temperature and other physiological responses to heat. The WBGT is the technique used most often for measuring these environmental factors.

15.8.5.1 Determining WBGT

The determination of WBGT requires the use of a black globe thermometer, a natural (static) wet bulb thermometer, and a dry bulb thermometer. The measure of environmental factors is performed as follows:

- The range of the dry and the natural wet bulb thermometers should be within –5°C to +50°C, with an accuracy of ± 0.5°C. The dry bulb thermometer must be shielded from the sun and the other radiant surfaces of the environment without restricting the airflow around the bulb. The wick of the natural wet bulb thermometer should be kept wet with distilled water for at least one-half hour before the temperature reading is made. It is not enough to immerse the other end of the wick into a reservoir of distilled water and wait until the whole wick becomes wet by capillarity. The wick must be wetted by direct application of water from a syringe one-half hour before each reading. The wick must cover the bulb of the thermometer and an equal length of additional wick must cover the stem above the bulb. The wick should always be clean, and new wicks should be washed before using.
- A globe thermometer, consisting of a 15 cm (6-in.) in diameter hollow copper sphere painted on the outside with a matte black finish, or equivalent, must be used. The bulb or sensor of a thermometer (range –5°C to + 100°C with an accuracy of ±0.5°C) must be fixed in the center of the sphere. The globe thermometer should be exposed at least 25 minutes before it is read.
- A stand should be used to suspend the three thermometers so air flow around the bulbs is not restricted, and the wet bulb and globe thermometers are not shaded.
- It is permissible to use any other type of temperature sensor that gives a reading similar to that of a mercury thermometer under the same conditions. The thermometers must be placed so that the readings are representative of the employee's work or rest areas, as appropriate.

Once the WBGT has been estimated, environmental professionals can estimate workers' metabolic heat load using the ACGIH method to determine the appropriate work/rest regimen, clothing, and equipment to use to control the heat exposures of workers in their facilities.

15.8.5.2 Additional Thermal Stress Indices

1. The *effective temperature* (ET) index combines the temperature, the humidity of the air, and air velocity. This index has been used extensively in the field of comfort ventilation and air-conditioning. ET remains a useful measurement technique in mines and other places where humidity is high and radiant heat is low.
2. The *heat stress index* (HSI) was developed by Belding and Hatch in 1965. Although the HSI considers all environmental factors and work rate, it is not completely satisfactory for determining an individual worker's heat stress and it is also difficult to use.

15.8.6 HEAT STRESS SAMPLE MEASUREMENT AND CALCULATION

The wet bulb globe temperature index (usually abbreviated WBGT) is the most widely used algebraic approximation of an "effective temperature" currently in use today. It is an index that can be determined quickly, requiring a minimum of effort and operator skill. As an approximation to an "effective temperature," the WBGT takes into account virtually all the commonly accepted mechanisms of heat transfer (e.g., radiant, evaporative). It does not account for the cooling effect of wind speed. Because of its simplicity, WBGT has been adopted by the American Conference of Government Hygienists as its principal index for use in specifying a heat stress related threshold limit value (TLV).

■ EXAMPLE 15.10

Problem: What would be the wet globe temperature index (in °C) for a quarry worker in Connecticut, who must work on a sunny morning when the outdoor dry bulb temperature is 88°F, the wet bulb temperature is 72°F, and the globe temperature is 102°F?

Solution:

$$WBGT = 0.7(NWB) + 0.2(GT) + 0.1(DB) = 0.7(72) + 0.2(102) + 8.8 = 50.4 + 20.4 + 8.8 = 78.8°F \ (26°C)$$

■ EXAMPLE 15.14

Problem: Later is the same afternoon, at the same quarry identified above, rain clouds have gathered, and rain has commenced to fall. The quarry manager has covered the work area in the quarry pit with a large tarpaulin to protect his employees. If the wet bulb temperature under the tarp has increased to 78°F, while the globe temperature has remained unchanged, what will be the new WBGT for this slightly different situation? (*Hint:* This is an indoor environment.)

Solution:

$$WBGT = 0.7(NWB) + 0.3(GT) = 0.7(78°) + 0.3(102) = 54.6 + 30.6 = 85.2°F$$

15.8.7 COLD HAZARDS

Temperature hazards are generally thought of as relating to extremes of heat. This is natural because most workplace temperature hazards do relate to heat. However, temperature extremes at the other end of the spectrum—cold—can also be hazardous. Employees who work outdoors in colder climates and employees who work indoors in such jobs as meatpacking are subjected to cold hazards. There are four factors that contribute to cold stress: cold temperature, high or cold wind, dampness and cold water. These factors, alone or in combination, draw heat away from the body. Other cold stress factors include age, disease, and overall physical condition. OSHA expresses cold stress though its cold stress equation:

$$\text{Low Temperature} + \text{Wind Speed} + \text{Wetness} = \text{Injuries and Illness} \qquad (15.21)$$

The ACGIH recommends protective clothing for temperatures less than 41°F (5°C). To estimate the amount of clothing insulation required (in clo units), the following formula is used:

$$I_{clo} = 11.5(33 - T_{db})/M \qquad (15.22)$$

where
I_{clo} = Insulation (clo units, where 1 clo = 0.155 K·m²/W).
T_{db} = Dry bulb temperature (°C).
M = Metabolic rate (watts).

The major injuries associated with extremes of cold can be classified as being either generalized or localized. A generalized injury from extremes of cold is hypothermia. Localized injuries include frostbite, frostnip, and trenchfoot.

- *Hypothermia* results when the body is unable to produce enough heat to replace the heat loss to the environment. It may occur at air temperatures up to 65°F. The body uses its defense mechanisms to help maintain its core temperature.
- *Frostbite* is an irreversible condition in which the skin freezes, causing ice crystals to form between cells. The toes, fingers, nose, ears, and cheeks, are the most common sites of freezing cold injury.
- *Frostnip* is less severe than frostbite. It causes the skin to turn white and typically occurs on the face and other exposed parts of the body. There is not tissue damage; however, if the exposed area is not either covered or removed from exposure to the cold, frostnip can become frostbite.
- *Trenchfoot* is caused by continuous exposure to cold water. It may occur in wet, cold environments or through actual immersion in water

15.9   RADIATION

15.9.1   DEFINITIONS

Absorbed dose is the energy imparted by ionizing radiation per unit mass of irradiated material. The units of absorbed dose are the rad and the gray (Gy).
Activity is the rate of disintegration (transformation) or decay of radioactive material. The units of activity are the curie (Ci) and the becquerel (Bq).
Alpha particle is a strongly ionizing particle emitted from the nucleus of an atom during radioactive decay, containing two protons and neutrons and having a double positive charge.
Alternate authorized user serves in the absence of the authorized user and can assume any duties as assigned.
Authorized user is an employee who is approved by the radiation safety officer and radiation safety committee and is ultimately responsible for the safety of those who use radioisotopes under his/her supervision.
Beta particle is an ionizing charge particle emitted from the nucleus of an atom during radioactive decay, equal in mass and charge to an electron.
Bioassay means the determination of kinds, quantities or concentrations, and, in some cases, the locations of radioactive material in the human body, whether by direct measurement (in vivo counting) or by analysis and evaluation of materials excreted or removed from the human body.
Biological half-life is the length of time required for one-half of a radioactive substance to be biologically eliminated from the body.

Bremsstrahlung is electromagnetic (x-ray) radiation associated with the deceleration of charged particles passing through matter.

Contamination is the deposition of radioactive material in any place where it is not wanted.

Controlled area means an area, outside of a restricted area but inside the site boundary, access to which can be limited by the licensee for any reason.

Counts per minute (cpm) is the number of nuclear transformations from radioactive decay able to be detected by a counting instrument in a one minute time interval.

Curie (Ci) is a unit of activity equal to 37 billion disintegrations per second.

Declared pregnant woman means a woman who has voluntary informed her employer, in writing, of her pregnancy and the estimated date of conception.

Disintegrations per minute (dpm) is the number of nuclear transformation from radioactive decay in a one minute time interval.

Dose equivalent is a quantity of radiation dose expressing all radiation on a common scale for calculating the effective absorbed dose. The units of dose equivalent are the rem and sievert (SV).

Dosimeter is a device used to determine the external radiation dose a person has received.

Effective half-life is the length of time required for a radioactive substance in the body to lose one-half of its activity present through a combination of biological elimination and radioactive decay.

Exposure means the amount of ionization in air from x-rays and gamma rays.

Extremity means hand, elbow, and arm below the elbow, foot, knee, or leg below the knee.

Gamma rays are very penetrating electromagnetic radiations emitted from a nucleus and an atom during radioactive decay.

Half-life is the length of time required for a radioactive substance to lose one-half of its activity by radioactive decay.

Limits (dose limits) means the permissible upper bounds of radiation doses.

Permitted worker is a laboratory worker who does not work with radioactive materials but works in a radiation laboratory.

Photon means a type of radiation in the form of an electromagnetic wave.

Rad is a unit of radiation absorbed dose. One rad is equal to 100 ergs per gram.

Radioactive decay is the spontaneous process of unstable nuclei in an atom disintegrating into stable nuclei, releasing radiation in the process.

Radiation (ionizing radiation) means alpha particles, beta particles, gamma rays, x-rays, neutrons, high-speed electrons, high-speed protons, and other particles capable of producing ions.

Radiation workers are those personnel listed on the Authorized User Form of the supervisor to conduct work with radioactive materials.

Radioisotope is a radioactive nuclide of a particular element.

Rem is a unit of dose equivalent. One rem is approximately equal to one rad of beta, gamma, or x-ray radiation, or 1/20 of alpha radiation.

Restricted area means an area, access to which is limited by the licensee for the purpose of protecting individuals against undue risks from exposure to radiation and radioactive materials.

Roentgen is a unit of radiation exposure. One roentgen is equal to 0.00025 Coulombs of electrical charge per kilogram of air.

Thermoluminescent dosimeter (TLD) is a dosimeter worn by radiation workers to measure their radiation dose. The TLD contains crystalline material which stores a fraction of the absorbed ionizing radiation and releases this energy in the form of light photons when heated.

Total effective dose equivalent (TEDE) means the sum of the deep-dose equivalent (for external exposures) and the committed effective dose equivalent (for internal exposures).

Unrestricted area means an area, access to which is neither limited nor controlled by the licensee.

X-rays is a penetrating type of photon radiation emitted from outside the nucleus of a target atom during bombardment of a metal with fast electrons.

15.9.2 IONIZING RADIATION

Ionization is the process by which atoms are made into ions by the removal or addition of one or more electrons; they produce this effect by the high kinetic energies of the quanta (discrete pulses) they emit. Simply, ionizing radiation is any radiation capable of producing ions by interaction with matter. Direct ionizing particles are charged particles (e.g., electrons, protons, alpha particles) having sufficient kinetic energy to produce ionization by collision. Indirect ionizing particles are uncharged particles (e.g., photons, neutrons) that can liberate direct ionizing particles. Sources of ionizing radiation can be found in a wide range of occupational settings, including healthcare facilities, research institutions, nuclear reactors and their support facilities, nuclear weapon production facilities, and other various manufacturing settings, just to name a few. These sources of ionizing radiation can pose a considerable health risk to affected workers if not properly controlled. Ionization of cellular components can lead to functional changes in the tissues of the body. Alpha, beta, and neutron particles; x-rays; gamma rays; and cosmic rays are all ionizing radiations.

Three mechanisms for external radiation protection include time, distance, and shielding. A shorter time in a radiation field means less dose. From a point source, dose rate is reduced by the square of the distance and expressed by the inverse square law:

$$I_1(d_1)^2 = I_2(d_2)^2 \tag{15.23}$$

where

I_1 = Dose rate or radiation intensity at distance d_1.
I_2 = Dose rate or radiation intensity at distance d_2.

Radiation is reduced exponentially by thickness of shielding material.

15.9.3 EFFECTIVE HALF-LIFE

The half-life is the length of time required for one-half of a radioactive substance to disintegrate. The formula depicted below is used when the industrial hygienist is interested in determining how much radiation is left in a worker's stomach after a period of time. Effective half-life is a combination of radiological and biological half-lives and is expressed as

$$T_{eff} = \frac{T_b \times T_r}{T_b + T_r} \tag{15.24}$$

where:

T_b = Biological half-life.
T_r = Radiological half-life.

It is important to point out that T_{eff} will always be shorter than either T_b or T_r. T_b may be modified by diet and physical activity.

15.9.4   ALPHA RADIATION

Alpha radiation is used for air ionization—elimination of static electricity (polonium-210), clean room applications, and smoke detectors (americium-241). It is also used in air density measurement, moisture meters, non-destructive testing, and oil well logging. Naturally occurring alpha particles are also used for physical and chemical properties, including uranium (coloring of ceramic glaze, shielding) and thorium (high temperature materials). The characteristics of alpha radiation are listed below.

- Alpha (α) radiation is a particle composed of two protons and neutrons.
- Alpha radiation is not able to penetrate skin.
- Alpha-emitting materials can be harmful to humans if the materials are inhaled, swallowed, or absorbed through open wounds.
- A variety of instruments have been designed to measure alpha radiation. Special training in use of these instruments is essential for making accurate measurements.
- A civil defense instrument (CDV-700) cannot detect the presence of radioactive materials that produce alpha radiation unless the radioactive materials also produce beta and/or gamma radiation.
- Instruments cannot detect alpha radiation through even a thin layer of water, blood, dust, paper, or other material, because alpha radiation is not penetrating.
- Alpha radiation travels a very short distance through air.
- Alpha radiation is not able to penetrate turnout gear, clothing, or a cover on a probe. Turnout gear and dry clothing can keep alpha emitters off of the skin.

15.9.4.1   Alpha Radiation Detectors

The types of high-sensitivity portable equipment used to evaluate alpha radiation in the workplace include

- Geiger–Mueller counter
- Scintillators
- Solid-state analysis
- Gas proportional devices

15.9.5   BETA RADIATION

Beta radiation is used for thickness measurements for coating operations, radioluminous signs, tracers for research, and for air ionization (gas chromatograph, nebulizers). The characteristics of Beta radiation are listed below.

- Beta (β) is a high energy electron particle.
- Beta radiation may travel meters in air and is moderately penetrating.
- Beta radiation can penetrate human skin to the "germinal layer," where new skin cells are produced. If beta-emitting contaminants are allowed to remain on the skin for a prolonged period of time, they may cause skin injury.
- Beta-emitting contaminants may be harmful if deposited internally.
- Most beta emitters can be detected with a survey instrument (such as a CDV-700, provided the metal probe cover is open). Some beta emitters, however, produce very low energy, poorly penetrating radiation that may be difficult or impossible to detect. Examples of these are carbon-14, tritium, and sulfur-35
- Beta radiation cannot be detected with an ionization chamber such as CDV-715.

- Clothing and turnout gear provide some protection against most beta radiation. Turnout gear and dry clothing can keep beta emitters off of the skin.
- Exposure to beta radiation can be external or internal. External beta radiation hazards are primarily skin burns. Internal beta radiation hazards are similar to alpha emitters.

15.9.5.1　Beta Detection Instrumentation

The types of equipment used to evaluate beta radiation in the workplace include

- Geiger–Mueller counter
- Gas proportional devices
- Scintillators
- Ion chambers
- Dosimeters

15.9.5.2　Shielding for Beta Radiation

Shielding for beta radiation is best accomplished by using materials with a low atomic number (low z materials) to reduce Bremsstrahlung radiation (i.e., secondary x-radiation produced when a beta particle is slowed down or stopped by a high-density surface). The thickness is critical to stop maximum energy range, and varies with the type of material used. Typical shielding materials include lead, water, wood, plastics, cement, Plexiglas, and wax.

15.9.6　Gamma Radiation and X-Rays

Gamma radiation and x-rays are used for sterilization of food and medical products; radiography of welds, castings, and assemblies; gauging of liquid levels and material density; oil well logging; and material analysis. The characteristics of gamma radiation and x-rays are listed below:

- Gamma (γ) radiation and x-rays are highly penetrating elecromagnetic radiation.
- X-rays are composed of photons (generated by electrons leaving an orbit).
- Gamma radiation and x-rays are electromagnetic radiation like visible light, radiowaves, and ultraviolet light. These various types of electromagnetic radiation differ only in the amount of energy they have. Gamma rays and x-rays are the most energetic of these.
- Gamma radiation is able to travel many meters in air and many centimeters in human tissue. It readily penetrates most materials and is sometimes called *penetrating radiation*.
- X-rays are also penetrating radiation.
- Radioactive materials that emit gamma radiation and x-rays constitute both an external and internal hazard to humans.
- Dense materials are needed for shielding from gamma radiation. Clothing and turnout gear provide little shielding from penetrating radiation but will prevent contamination of the skin by radioactive materials.
- Gamma radiation is detected with survey instruments, including civil defense instruments. Low levels can be measured with a standard Geiger counter, such as the CDV-700. High levels can be measure with an ionization chamber, such as the CDV-715.
- Gamma radiation or x-rays frequently accompany the emission of alpha and beta radiation.
- Instruments designed solely for alpha detection (such as an alpha scintillation counter) will not detect gamma radiation.
- Pocket chamber (pencil) dosimeters, film badges, thermoluminescent, and other types of dosimeters can be used to measure accumulated exposure to gamma radiation.
- The principal health concern associated with gamma radiation is external exposure by penetrating radiation and physically strong source housing. Sensitive organs include the lens of the eye, the gonads, and damage to the bone marrow.

15.9.6.1 Gamma Detection Instrumentation

The types of equipment used to evaluate gamma radiation in the workplace include:

- Ion chambers
- Gas proportional devices
- Geiger–Mueller counter

15.9.6.2 Shielding for Gamma and X-Rays

Shielding gamma and x-radiation depends on energy level. Protection follows an exponential function of shield thickness. At low energies, absorption can be achieved with millimeters of lead. At high energies, shielding can attenuate gamma radiation.

15.9.7 RADIOACTIVE DECAY EQUATIONS

Radioactive materials emit alpha particles, beta particles, and photon energy, and lose a proportion of their radioactivity with a characteristic half-life. This is known as radioactive decay. To calculate the amount of radioactivity remaining after a given period of time, use the following basic formulae for decay calculations:

$$\text{Later activity} = (\text{Earlier activity}) \times e^{-\lambda} \times (\text{Elapsed time}) \qquad (15.25)$$

$$A = A_i \times e^{-\lambda} \times t$$

where

λ = Decay constant (probability of an atom decaying in a unit time) = $\ln 2/T$.
A = New or later radioactivity level
A_i = Initial radioactivity level
t = Time.
$\ln 2 = 0.693$.
T = Radioactive half-life (time period in which half of a radioactive isotope decays)

Use the following to determine the time required for a radioactive material to decay (A_o to A):

$$t = (-\ln A/A_i)(T/\ln 2) \qquad (15.26)$$

where

t = Time.
A = New or later radioactivity level.
A_i = Initial radioactivity level.
T = Radioactive half-life (time period in which half of a radioactive isotope decays).
$\ln 2 = 0.693$.

Basic rule of thumb: In 7 half-lives, radioactivity is reduced to <1%; in 10 half-lives, to <0.1%.

To determine the rate of radioactive decay, keep in mind that radioactive disintegration is directly proportional to the number of nuclei present. Thus, the radioactive decay rate is expressed in nuclei disintegrated per unit time.

$$A_i = (0.693/T)(N_i) \qquad (15.27)$$

where

A_i = Initial rate of decay.
N_i = Initial number of radionuclei.
T = Half life.

Half-life is defined as the time it takes for a material to lose 50% of its radioactivity. The following equation can be used to determine half-life.

$$A = A_i(0.5)^{t/T} \tag{15.28}$$

where

A = Activity at time t.
A_i = Initial activity.
t = Time.
T = Half-life.

15.9.8 RADIATION DOSE

In the United States, radiation *absorbed dose, dose equivalent*, and *exposure* are often measured and stated in the traditional units called *rad, rem*, or *roentgen*. For practical purposes with gamma and x-rays, these units of measurement for exposure or dose are considered equal. This exposure can be from an external source irradiating the whole body, an extremity, or other organ or tissue resulting in an *external radiation dose*. Alternatively, internally deposited radioactive material may cause an *internal radiation dose* to the whole body or to an organ or tissue.

A prefix is often used for smaller measured fractional quantities; for example, milli- (m) means 1/1000 and 1 rad = 1000 mrad. Micro- (μ) means 1/1,000,000, so 1,000,000 μrad = 1 rad, or 10 μR = 0.000010 R.

The SI system for radiation measurement is the official system of measurement and uses the gray (Gy) and sievert (Sv) for absorbed dose and equivalent dose, respectively. Conversions are as follows:

1 Gy = 100 rad
1 mGy = 100 mrad
1 Sv = 100 rem
1 mSv = 100 mrem

Radioactive transformation events (radiation counting systems) can be measured in units of disintegrations per minute (dpm) and, because instruments are not 100% efficient, counts per minute (cpm). Background radiation levels are typically less than 10 μR per hour, but due to differences in detector size and efficiency, the cpm reading on fixed monitors and various handheld survey meters will vary considerably.

REFERENCES AND RECOMMENDED READING

ACGIH. (2010). *Industrial Ventilation: A Manual of Recommended Practice*, 27th ed. American Conference of Governmental Industrial Hygienists, Cincinnati, OH.
ASSE. (1998). *Dictionary of Terms Used in the Safety Profession*, 3rd ed. American Society of Safety Engineers, Des Plaines, IL.
Bird, F. E. and Germain, G. L. (1966). *Damage Control*. American Management Association, New York.
Boyce, A. (1997). *Introduction to Environmental Technology*. Van Nostrand Reinhold, New York.
Burge, H.A. (1997). The fungi: how they grow and their effects on human health. *HPAC*, July 69–75.
Burge, H.A. and Hoyer, M.E. (1998). Indoor air quality, in DiNardi, S.R., Ed., *The Occupational Environment— Its Evaluation and Control*. American Industrial Hygiene Association, Fairfax, VA.
Byrd, R.R. (2003). *IAQ FAG Part 1*. Machado Environmental Corporation, Glendale, CA.
CDC. (1999). *Reports of Members of the CDC External Expert Panel on Acute Idiopathic Pulmonary Hemorrhage in Infants: A Synthesis*. U.S. Centers for Disease Control and Prevention, Atlanta, GA.
CCPS. (2008). *Guidelines for Hazard Evaluation Procedures*, 2nd ed. American Institute of Chemical Engineers, New York.

Davis, P.J. (2001). *Molds, Toxic Molds, and Indoor Air Quality*. California Research Bureau, California State Library, Sacramento.

Dockery, D.W. and Spengler, J.D. (1981). Indoor–outdoor relationships of respirable sulfates and particles. *Atmospheric Environment*, 15, 335–343.

Fletcher, J. A. (1972). *The Industrial Environment—Total Loss Control*. National Profile Limited, Ontario.

Goetsch, D.L. (1996). *Occupational Safety and Health in the Age of High Technology for Technologists, Engineers, and Managers*, 2nd ed. Prentice-Hall, Englewood Cliffs, NJ.

Haddon, Jr., W., Suchm, E. A., and Klein, D. (1964). *Accident Research*. Harper & Row, New York, p. 28.

Hitchcock, R.T. and Patterson, R.M. (1995). *Radiofrequency and ELF Electromagnetic Energies*. Van Nostrand Reinhold, New York.

Hollowell, C.D. et al. (1979a). Impact of infiltration and ventilation on indoor air quality. *ASHRAE Journal*, July, 49–53.

Hollowell, C.D. et al. (1979b). Impact of energy conservation in buildings on health, in Razzolare, R.A. and Smith, C.B., Eds., *Changing Energy Use Futures: Conference Proceedings*, Pergamon Press, New York.

Mansdorf, S.Z. (1993). *Complete Manual of Industrial Safety*. Prentice Hall, New York.

McNeel, S. and Kreutzer, R. (1996). Fungi and indoor air quality. *Health and Environment Digest*, 10(2), 9–12.

NIOSH. (2005). *Common Hearing Loss Prevention Terms*. National Institute for Occupational Safety and Health, Washington, DC.

Olishifski, J.B. (1998). Overview of industrial hygiene, in Olishifski, J.B., *Fundamentals of Industrial Hygiene*, 3rd ed. National Safety Council, Chicago, IL.

OSHA. (2005). *Informational Booklet on Industrial Hygiene*, OSHA 3143. U.S. Department of Labor, Washington, DC.

Passon, T., Brown, J.W., and Mante, S. (1996) "Sick-Building Syndrome and Building-Related Illnesses." *Medical Laboratory Observer*, 28(7), 84–95.

Plog, B.A., Ed., (2001). *Fundamentals of Industrial Hygiene*, 5th ed. National Safety Council, Chicago, IL.

Rose, C.F. (1999). Antigens, in Macher, J., Ed., *Bioaerosols Assessment and Control*. American Conference of Governmental Industrial Hygienists, Cincinnati, OH.

Sanders, M.S. and McCormick, E.J. (1993). *Human Factors in Engineering and Design*, 7th ed. McGraw-Hill, New York.

Shlein, B., Slaback, L., and Birky, B. (1998). *Handbook of Health Physics and Radiological Health*. Lippincott Williams & Wilkins, Baltimore, MD.

Spellman, F.R. (1998). *Surviving an OSHA Audit*. CRC Press, Boca Raton, FL.

Spellman, F.R. (2013). *Safe Work Practices for Green Energy Jobs*. DEStech Publications, Lancaster, PA.

Spellman, F.R. and Whiting, N. (2005). *Safety Engineering: Principles and Practices*, 2nd ed. Government Institutes Press, Lanham, MD.

Spengler, J.D. et al. (1979), Sulfur dioxide and nitrogen dioxide levels inside and outside homes and the implications on health effects research. *Environmental Science & Technology*, 13, 1276–1280.

USEPA. (2001). *Indoor Air Facts No. 4 (Revised): Sick Building Syndrome*. U.S. Environmental Protection Agency, Washington, DC, http://www.epa.gov/iaq/pdfs/sick_building_factsheet.pdf.

USEPA. (2003). *A Guide to Indoor Air Quality*. U.S. Environmental Protection Agency, Washington, DC.

Woods, J.E. (1980). *Environmental Implications of Conservation and Solar Space Heating*. Engineering Research Institute, Iowa State University, Ames.

WSDOH. (2003). *Formaldehyde*. Washington State Department of Health, Office of Environmental Health and Safety, http://www.doh.wa.gov/YouandYourFamily/HealthyHome/Contaminants/Formaldehyde.aspx.

Section IX

Math Concepts:
Air Pollution Control

Birds are indicators of the environment. If they are in trouble, we know we'll soon be in trouble.

—Roger Tory Peterson

16 Gas Emission Control

Be it known to all within the sound of my voice, whoever shall be found guilty of burning coal shall suffer the loss of his head.

—King Edward II

16.1 INTRODUCTION

Limiting gaseous emissions into the air is both technically difficult and expensive. Although rain is Nature's vacuum cleaner—the only air-cleansing mechanism available—it is not very efficient. Good air quality depends on pollution prevention (i.e., limiting what is emitted) and sound engineering policies, procedures, and practices. The control of gaseous air emissions may be realized in a number of ways. In this chapter, we discuss many of these technologies and the sources of gaseous pollutants emitted from various sources and their control points (see Figure 16.1). The applicability of a given technique depends on the properties of the pollutant and the discharge system. In making the difficult and often complex decision of which gaseous air pollution control to employ, it is helpful to follow the guidelines based on experience and set forth by Buonicore and Davis (1992) in their prestigious engineering text, *Air Pollution Engineering Manual*. Table 16.1 summarizes the main techniques and technologies used to control gaseous emissions. After defining key air emission and pollution terms, we discuss the air control technologies given in Table 16.1. Much of the information contained in this chapter is adapted from Spellman (1999) and USEPA (1981). The excerpted materials have been rearranged and edited to make the materials more concise for the reader's use.

16.2 DEFINITIONS

Absolute pressure—The total pressure in a system, including both the pressure of a substance and the pressure of the atmosphere (about 14.7 psi, at sea level).

Acid—Any substance that releases hydrogen ions (H^+) when it is mixed into water.

Acid precipitation—Rain, snow, or fog that contains higher than normal levels of sulfuric or nitric acid, which may damage forests, aquatic ecosystems, and cultural landmarks.

Acid surge—A period of short, intense acid deposition in lakes and streams as a result of the release (by rainfall or spring snowmelt) of acids stored in soil or snow.

Acidic solution—A solution that contains significant numbers of (H^+) ions.

Airborne toxins—Hazardous chemical pollutants that have been released into the atmosphere and are carried by air currents.

Albedo—Reflectivity, or the fraction of incident light that is reflected by a surface.

Arithmetic mean—A measurement of average value, calculated by summing all terms and dividing by the number of terms.

Arithmetic scale—A scale is a series of intervals (marks or lines), usually made along the side or bottom of a graph, that represents the range of values of the data. When the marks or lines are equally spaced, it is called an arithmetic scale.

Atmosphere—A 500-km-thick layer of colorless, odorless gases known as air that surrounds the Earth and is composed of nitrogen, oxygen, argon, carbon dioxide, and other gases in trace amounts.

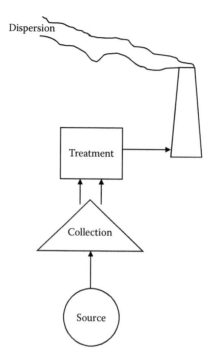

FIGURE 16.1   Air pollution control points.

Atom—The smallest particle of an element that still retains the characteristics of that element.

Atomic number—The number of protons in the nucleus of an atom.

Atomic weight—The sum of the number of protons and the number of neutrons in the nucleus of an atom.

Base—Any substance that releases hydroxyl ions (OH^-) when it dissociates in water.

Chemical bond—The force that holds atoms together within molecules. A chemical bond is formed when a chemical reaction takes place. Two types of chemical bonds are ionic bonds and covalent bonds.

Chemical reaction—A process that occurs when atoms of certain elements are brought together and combine to form molecules, or when molecules are broken down into individual atoms.

Climate—The long-term weather pattern of a particular region.

Covalent bond—A type of chemical bond in which electrons are shared.

TABLE 16.1
Comparison of Air Control Technologies

Treatment Technology	Concentration	Efficiency	Comments
Absorption	<200 ppmv	90–95%	Can blowdown stream be accomplished at site?
	>200 ppmv	95+%	
Carbon adsorption	>200 ppmv	90+%	Recovered organics may need additional treatment; can
	>1000 ppmv	95+%	increase cost.
Incineration	<100 ppmv	90–95%	Incomplete combustion may require additional controls
	>100 ppmv	95–99%	
Condensation	>2000 ppmv	80+%	Must have low temperature or high pressure for efficiency

Source: Spellman, F.R., *The Science of Air*, 2nd ed., CRC Press, Boca Raton, FL, 2008, p. 221.

Density—The weight of a substance per unit of its volume (e.g., pounds per cubic foot).

Dewpoint—The temperature at which a sample of air becomes saturated; that is, it has a relative humidity of 100%.

Element—Any of more than 100 fundamental substances that consist of atoms of only one kind and that constitute all matter.

Emission standards—The maximum amount of a specific pollutant permitted to be legally discharged from a particular source in a given environment.

Emissivity—The relative power of a surface to reradiate solar radiation back into space in the form of heat, or long-wave infrared radiation.

Energy—The ability to do work, to move matter from place to place, or to change matter from one form to another.

First law of thermodynamics—Natural law that dictates that during physical or chemical change energy is neither created not destroyed, but it may be changed in form and moved from place to place.

Global warming—The increase in global temperature predicted to arise from increased levels of carbon dioxide, methane, and other greenhouse gases in the atmosphere.

Greenhouse effect—The prevention of the reradiation of heat waves to space by carbon dioxide, methane, and other gases in the atmosphere. The greenhouse effect makes possible the conditions that enable life to exist on Earth.

Insolation—Solar radiation received by the Earth and its atmosphere (incoming solar radiation).

Ion—An atom or radical in solution carrying an integral electrical charge either positive (cation) or negative (anion).

Lapse rate—The rate of temperature change with altitude. In the troposphere, the normal lapse rate is –3.5°F per 1000 ft.

Matter—Anything that exists in time, occupies space, and has mass.

Mesosphere—A region of the atmosphere based on temperature that is between approximately 35 to 60 miles in altitude.

Meteorology—The study of atmospheric phenomena.

Mixture—Two or more elements, compounds, or both, mixed together with no chemical reaction occurring.

Ozone—The compound O_3. It is found naturally in the atmosphere in the ozonosphere and is also a constituent of photochemical smog.

pH—A means of expressing hydrogen ion concentration in terms of the powers of 10; measurement of how acidic or basic a substance is. The pH scale runs from 0 (most acidic) to 14 (most basic). The center of the range (7) indicates the substance is neutral.

Photochemical smog—An atmospheric haze that occurs above industrial sites and urban areas resulting from reactions, which take place in the presence of sunlight, between pollutants produced in high temperature and pressurized combustion processes (such as the combustion of fuel in a motor vehicle). The primary component of smog is ozone.

Photosynthesis—The process of using the sun's light energy by chlorophyll-containing plants to convert carbon dioxide (CO_2) and water (H_2O) into complex chemical bonds forming simple carbohydrates such as glucose and fructose.

Pollutant—A contaminant at a concentration high enough to endanger the environment.

Pressure—The force pushing on a unit area. Normally, in air applications, pressure is measured in atmospheres (atm), Pascals (Pa), or pounds per square inch (psi).

Primary pollutants—Pollutants that are emitted directly into the atmosphere where they exert an adverse influence on human health or the environment. The six primary pollutants are carbon dioxide, carbon monoxide, sulfur oxides, nitrogen oxides, hydrocarbons, and particulates. All but carbon dioxide are regulated in the United States.

Radon—A naturally occurring radioactive gas arising from the decay of uranium 238, which may be harmful to human health in high concentrations.

Rain shadow effect—The phenomenon that occurs as a result of the movement of air masses over a mountain range. As an air mass rises to clear a mountain, the air cools and precipitation forms. Often, both the precipitation and the pollutant load carried by the air mass will be dropped on the windward side of the mountain. The air mass is then devoid of most of its moisture; consequently, the lee side of the mountain receives little or no precipitation and is said to lie in the rain shadow of the mountain range.

Raleigh scattering—The preferential scattering of light by air molecules and particles that accounts for the blueness of the sky. The scattering is proportional to $1/\lambda^4$.

Relative humidity—The concentration of water vapor in the air. It is expressed as the percentage that its moisture content represents of the maximum amount that the air could contain at the same temperature and pressure. The higher the temperature the more water vapor the air can hold.

Second law of thermodynamics—Natural law that dictates that with each change in form some energy is degraded to a less useful form and given off to the surroundings, usually as low-quality heat.

Secondary pollutants—Pollutants formed from the interaction of primary pollutants with other primary pollutants or with atmospheric compounds such as water vapor.

Solute—The substance dissolved in a solution.

Solution—A liquid containing a dissolved substance.

Specific gravity—The ratio of the density of a substance to a standard density. For gases, the density is compared with the density of air (= 1).

Stratosphere—Atmospheric layer extending from 6 to 30 miles above the Earth's surface.

Stratospheric ozone depletion—The thinning of the ozone layer in the stratosphere; occurs when certain chemicals (such as chlorofluorocarbons) capable of destroying ozone accumulate in the upper atmosphere.

Thermosphere—An atmospheric layer that extends from 56 miles to outer space.

Troposphere—The atmospheric layer that extends from the Earth's surface to 6 to 7 miles above the surface.

Weather—The day-to-day pattern of precipitation, temperature, wind, barometric pressure, and humidity.

Wind—Horizontal air motion.

16.3 ABSORPTION

Absorption (or scrubbing) is a major chemical engineering unit operation that involves bringing contaminated effluent gas into contact with a liquid absorbent so that one or more constituents of the effluent gas are selectively dissolved into a relatively nonvolatile liquid. Key terms used when discussing the absorption process include the following:

Absorbent—The liquid, usually water mixed with neutralizing agents, into which the contaminant is absorbed

Solute—The gaseous contaminant being absorbed (e.g., SO_2, H_2S)

Carrier gas—The inert portion of the gas stream, usually flue gas, from which the contaminant is to be removed

Interface—The area where the gas phase and the absorbent contact each other

Solubility—The capability of a gas to be dissolved in a liquid

Absorption units are designed to transfer the pollutant from a gas phase to a liquid phase. The absorption unit accomplishes this by providing intimate contact between the gas and the liquid, providing optimum diffusion of the gas into the solution. The actual mechanism of removal of a pollutant from the gas stream takes place in three steps: (1) diffusion of the pollutant gas to the surface

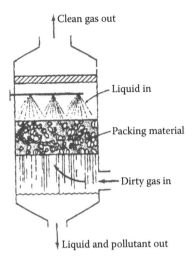

Clean gas out

Liquid in

Packing material

Dirty gas in

Liquid and pollutant out

FIGURE 16.2 Typical countercurrent-flow packed tower (From USEPA, *Control Techniques of Gases and Particulates*, U.S. Environmental Protection Agency, Washington, DC, 1971.)

of the liquid, (2) transfer across the gas–liquid interface, and (3) diffusion of the dissolved gas away from the interface into the liquid (Davis and Cornwell, 1991). Several types of scrubbing towers are available, including spray chambers (towers or columns), plate or tray towers, packed towers, and Venturi scrubbers. Pollutant gases commonly controlled by absorption include sulfur dioxide, hydrogen sulfide, hydrogen chloride, ammonia, and oxides of nitrogen.

The two most common absorbent units in use today are *plate towers* and *packed towers*. Plate towers contain perforated horizontal plates or trays designed to provide large liquid–gas interfacial areas. The polluted airstream is usually introduced at one side of the bottom of the tower or column and rises up through the perforations in each plate; the rising gas prevents the liquid from draining through the openings rather than through a downpipe. During continuous operation, contact is maintained between air and liquid, allowing gaseous contaminants to be removed, with clean air emerging from the top of the tower.

The packed tower scrubbing system (Figure 16.2) is predominately used to control gaseous pollutants in industrial applications, where it typically demonstrates a removal efficiency of 90% to 95%. Usually configured in a vertical fashion, the packed tower is literally "packed" with devices (see Figure 16.3) with a large surface-to-volume ratio and a large void ratio offering minimum resistance to gas flow. In addition, packing should provide even distribution of both fluid phases, be sturdy enough to support itself in the tower, and be low cost, available, and easily handled (Hesketh, 1991). The flow through a packed tower is typically countercurrent, with gas entering at the bottom of the tower and liquid entering at the top. Liquid flows over the surface of the packing in a thin film, affording continuous contact with the gases. Though highly efficient for removal of gaseous contaminants, packed towers may create liquid disposal problems, become easily clogged when gases with high particulate loads are introduced, and have relatively high maintenance costs.

16.3.1 Solubility

Solubility is a function of both the temperature and to a lesser extent the pressure of the system. As temperature increases, the amount of gas that can be absorbed by a liquid decreases (gases are more soluble in cold liquids than in hot liquids). Gas phase pressure can also influence solubility. Pressure affects the solubility of a gas in the opposite manner. By increasing the pressure of a system the amount of gas absorbed generally increases. However, this is not a major variable in absorbers used for air pollution control because they operate at close to atmospheric pressure (USEPA, 1981).

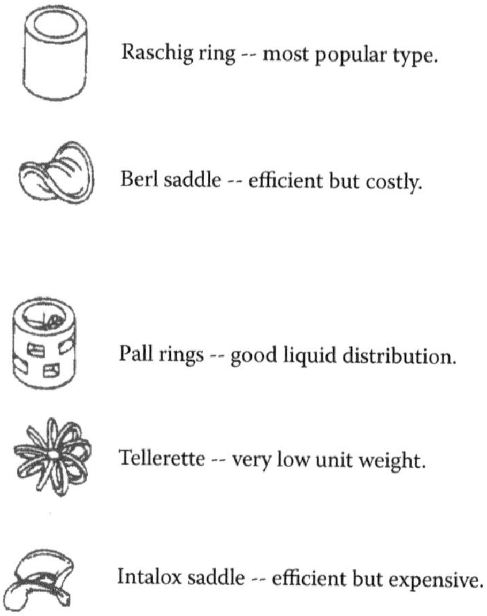

Raschig ring -- most popular type.

Berl saddle -- efficient but costly.

Pall rings -- good liquid distribution.

Tellerette -- very low unit weight.

Intalox saddle -- efficient but expensive.

FIGURE 16.3 Various packing used in packed tower scrubbers. (Adapted from AIHA, *Air Pollution Manual: Control Equipment, Part II*, American Industrial Hygiene Association, Detroit, MI, 1968.)

16.3.2 EQUILIBRIUM SOLUBILITY AND HENRY'S LAW

Under certain conditions, Henry's law can express the relationship between the gas phase concentration and the liquid phase concentration of the contaminant at equilibrium. Henry's law states that for dilute solutions, where the components do not interact, the resulting partial pressure (p) of a component A in equilibrium with other components in a solution can be expressed as

$$p = Hx_A \tag{16.1}$$

where

p = Partial pressure of contaminant in gas phase at equilibrium.
H = Henry's law constant.
x_A = Mole fraction of contaminant or concentration of A in liquid phase.

Equation 16.1 is the equation of a straight line where the slope (m) is equal to H. Henry's law can be used to predict solubility only if the equilibrium line is straight—that is, when the solute concentrations are very dilute. In air pollution control applications this is usually the case. For example, an exhaust stream that contains a 1000-ppm SO_2 concentration corresponds to a mole fraction of SO_2 in the gas phase of only 0.001. Another restriction on using Henry's law is that it does not hold true for gases that react or dissociate upon dissolution. If this happens, the gas no longer exists as a simple molecule. For example, scrubbing HF or HCl gases with water causes both compounds to dissociate in solution. In these cases, the equilibrium lines are curved rather than straight. Data on systems that exhibit curved equilibrium lines must be obtained from experiments.

The units of Henry's law constants are atm/mole fraction. The smaller the Henry's law constant, the more soluble the gaseous compound is in the liquid. The following example from USEPA (1981) illustrates how to develop an equilibrium diagram from solubility data.

■ EXAMPLE 16.1

Problem: Given the following data for the solubility of SO_2 in pure water at 303°K (30°C) and 101.3 kPa (760 mmHg), plot the equilibrium diagram and determine if Henry's law applies.

Solubility of SO$_2$ in Pure Water: Equilibrium Data

Concentration SO$_2$ (g of SO$_2$ per 100 g/H$_2$O)	p (Partial Pressure of SO$_2$)
0.5	6 kPa (42 mmHg)
1.0	11.6 kPa (85 mmHg)
1.5	18.3 kPa (129 mmHg)
2.0	24.3 kPa (176 mmHg)
2.5	30.0 kPa (224 mmHg)
3.0	36.4 kPa (273 mmHg)

Solution: The data must first be converted to mole fraction units. The mole fraction in the gas phase (*y*) is obtained by dividing the partial pressure of SO$_2$ by the total pressure of the system. From the first entry in the data table:

$$y = p/P = (6 \text{ kPa}/101.3 \text{ kPa}) = 0.06 \text{ kPa}$$

The mole fraction in the liquid phase (*x*) is obtained by dividing the moles of SO$_2$ by the total moles of liquid:

$$x = \frac{\text{Moles of SO}_2 \text{ in solution}}{\text{Moles of SO}_2 \text{ in solution} + \text{Moles H}_2\text{O}}$$

For the first entry (*x*) of the data table:

$$x = 0.0078/(0.0078 + 5.55) = 0.0014$$

The table on the next page has been completed. The data from this table are plotted in Figure 16.4. Henry's law applies in the given concentration range with Henry's law constant equal to 42.7 mole fraction SO$_2$ in air/mole fraction SO$_2$ in water.

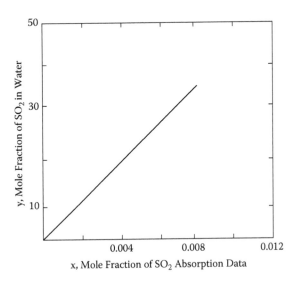

FIGURE 16.4 Sulfur dioxide absorption data for Example 16.1. (Adapted from USEPA, *APTI Course 415: Control of Gaseous Emissions*, EPA 450/2-81-005, U.S. Environmental Protection Agency Air Pollution Training Institute, Washington, DC, 1981, p. 4-8.)

Solubility Data for SO$_2$

C = (g of SO$_2$)/(100 g H$_2$O)	p (kPa)	y = p/101.3	x = (C/64)/(C/64 + 5.55)
0.5	6	0.06	0.0014
1.0	11.6	0.115	0.0028
1.5	18.3	0.18	0.0042
2.0	24.3	0.239	0.0056
2.5	30	0.298	0.007
3.0	36.4	0.359	0.0084

16.3.3 Material (Mass) Balance

The simplest way to express the fundamental engineering concept/principle of material or mass balance is to say, "Everything has to go somewhere." More precisely, the law of conservation of mass says that when chemical reactions take place, matter is neither created nor destroyed. What this important concept allows us to do is track materials, that is, pollutants, microorganisms, chemicals, and other materials from one place to another. The concept of material balance plays an important role in environmental treatment technologies where we assume a balance exists between the material entering and leaving the treatment process: "What comes in must equal what goes out." The concept is very helpful in evaluating process operations. In air pollution control of gas emissions using a typical countercurrent flow absorber, the solute (contaminant compound) is the material balance. Figure 16.5 illustrates a typical countercurrent flow absorber in which a material balance is drawn. The following equation can be derived for material balance:

$$Y_1 - Y_2 = (L_m/G_m)(X_1 - X_2) \tag{16.2}$$

where
  Y_1 = Inlet source concentration.
  Y_2 = Outlet solute concentration.
  L_m = Liquid flow rate (g-mol/hr).
  G_m = Gas flow rate (g-mol/hr).
  X_1 = Outlet composition of scrubbing liquid.
  X_2 = Inlet composition of scrubbing liquid.

Equation 16.2 is the equation of a straight line. When this line is plotted on an equilibrium diagram, it is referred to as an *operating line* (see Figure 16.5). This line defines operating conditions within the absorber; that is, what is going in and what is coming out. The slope of the operating line is the liquid mass flow rate divided by the gas mass flow rate, which is the liquid-to-gas ratio or (L_m/G_m). When describing or comparing absorption systems, the liquid-to-gas ratio is used extensively. The following example (using Henry's law) illustrates how to compute the minimum liquid rate required to achieve desired removal efficiency.

■ EXAMPLE 16.2

Problem: Using the data and results from Example 16.1, compute the minimum liquid rate of pure water required to remove 90% of the SO$_2$ from a gas stream of 84.9 m^3/min (3000 acfm) containing 3% SO$_2$ by volume. The temperature is 293 K and the pressure is 101.3 kPa (USEPA, 1981, p. 4-20).

Given:
  Inlet gas solute concentration (Y_1) = 0.03
  Minimum acceptable standards (outlet solute concentration) (Y_2) = 0.003

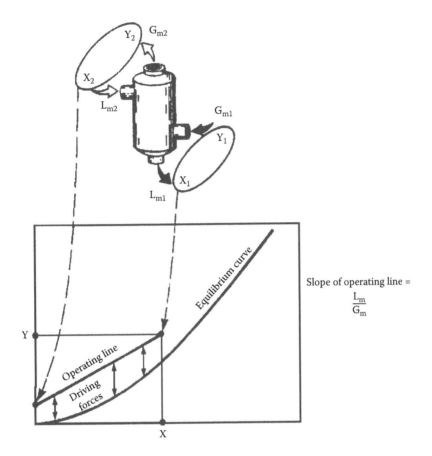

FIGURE 16.5 Operating line for a countercurrent flow absorber. (From USEPA, *APTI Course 415: Control of Gaseous Emissions*, EPA 450/2-81-005, U.S. Environmental Protection Agency Air Pollution Training Institute, Washington, DC, 1981, p. 4-17.)

Composition of the liquid into the absorber $(X_2) = 0$
Gas flow rate $(Q) = 84.9$ m³/min
Outlet liquid concentration $(X_1) = ?$
Liquid flow rate $(L) = ?$
H = Henry's constant

Solution: Sketch and label a drawing of the system (see Figure 16.6). $Y_1 = 3\%$ by volume $= 0.03$, and $Y_2 = 90\%$ reduction from Y_1, or only 10% of Y_1; therefore, $Y_2 = (0.10)(0.03) = 0.003$. At the minimum liquid rate, Y_1 and X_1 will be in equilibrium. The liquid will be saturated with SO_2.

$$Y_1 = H \times X_1$$

From Figure 16.4, $H = 42.7$ (mole fraction SO_2 in air ÷ mole fraction SO_2 in water); thus,

$$0.03 = 42.7 \times X_1$$

$$X_1 = 0.000703 \text{ mole fraction}$$

The minimum liquid-to-gas ratio is

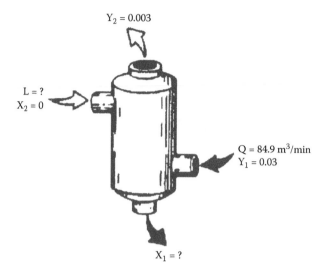

FIGURE 16.6 Material balance for absorber. (From USEPA, *APTI Course 415: Control of Gaseous Emissions*, EPA 450/2-81-005, U.S. Environmental Protection Agency Air Pollution Training Institute, Washington, DC, 1981, p. 4-20.)

$$Y_1 - Y_2 = \left(L_m/G_m \right)\left(X_1 - X_2 \right)$$

$$\left(L_m/G_m \right) = \frac{\left(Y_1 - Y_2 \right)}{\left(X_1 - X_2 \right)}$$

$$= \frac{0.03 - 0.003}{0.000703 - 0}$$

$$= 38.4 \text{ (g-mol water/g-mol air)}$$

Compute the minimum required liquid flow rate. First, convert m³ of air to g-mol:

At 0°C (273 K) and 101.3 kPa, there are 0.0224 m³/g-mol of an ideal gas.
At 20°C (293 K),

$$0.0224 \left(\frac{m^3}{\text{g-mol}} \right)\left(\frac{293 \text{ K}}{273 \text{ K}} \right) = 0.024 \text{ m}^3/\text{g-mol}$$

$$G_m = 84.9 \frac{m^3}{\text{min}} \left(\frac{\text{g-mol air}}{0.024 \text{ m}^3} \right) = 3538 \text{ g-mol air/min}$$

$$L_m/G_m = 38.4 \text{ (g-mol water/g-mol air) at minimum conditions}$$

$$L_m = 38.4 \times 3538 = 136.0 \text{ kg-mol water minimum}$$

In mass units,

$$L = 136 \text{ kg-mol/min} \times 18 \text{ kg/kg-mol} = 2448 \text{ kg/min}$$

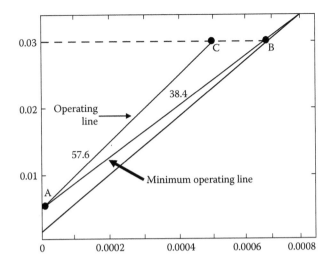

FIGURE 16.7 Solution to Example 16.2. (From USEPA, *APTI Course 415: Control of Gaseous Emissions*, EPA 450/2-81-005, U.S. Environmental Protection Agency Air Pollution Training Institute, Washington, DC, 1981, p. 4-21.)

Figure 16.7 illustrates the graphical solution to this problem. The slope of the minimum operating line multiplied by 1.5 is the slope of the actual operating line (line *AC*):

$$38.4 \times 1.5 = 57.6$$

16.3.4 PACKED TOWERS AND PLATE TOWERS

16.3.4.1 Sizing the Packed Tower Absorber Diameter

The main parameter that affects the size of a packed tower is the gas velocity at which liquid droplets become entrained in the existing gas stream. Consider a packed tower operating at set gas and liquid flow rates. By decreasing the diameter of the column, the gas flow rate (m/s or ft/s) through the column will increase. If the gas flow rate through the tower is gradually increased by using smaller and smaller diameter towers a point will be reached where the liquid flowing down over the packing begins to be held in the void spaces between the packing. This gas-to-liquid flow ratio is termed the *loading point*. The pressure drop over the column begins to increase, and the degree of mixing between the phases decreases. A further increase in gas velocity causes the liquid to completely fill the void spaces in the packing. The liquid forms a layer over the top of the packing, and no more liquid can flow down through the tower. The pressure drop increases substantially, and mixing between the phases is minimal. This condition is referred to as *flooding* and the gas velocity at which it occurs is the *flooding velocity*. Using an extremely large diameter tower would eliminate this problem. However, as the diameter increases, the cost of the tower increases (USEPA, 1981, p. 4-22).

Normal practice is to size a packed column diameter to operate at a certain percent of the flooding velocity. A typical operating range for the gas velocity through the towers is 50 to 75% of the flooding velocity. It is assumed that by operating in this range, the gas velocity will also be below the loading point. A common and relatively simple procedure to estimate the flooding velocity (thus minimum column diameter) is to use a generalized flooding and pressure drop correlation. One version of the flooding and pressure drop relationship in a packed tower is shown in Figure 16.8. This correlation is based on the physical properties of the gas and liquid streams and tower packing characteristics. The procedure to determine the tower diameter is summarized in the following set of calculations:

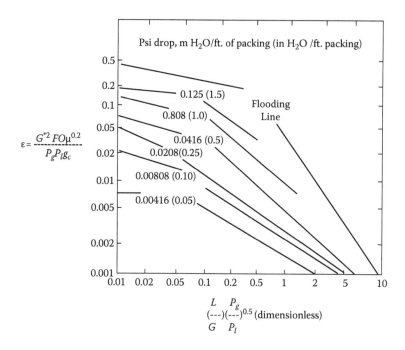

$$\varepsilon = \frac{G^{*2}\,FO\mu^{0.2}}{P_g P_l g_c}$$

FIGURE 16.8 Generalized flooding and psi drop correlation. (From USEPA, *APTI Course 415: Control of Gaseous Emissions*, EPA 450/2-81-005, U.S. Environmental Protection Agency Air Pollution Training Institute, Washington, DC, 1981, p. 4-23.)

1. Calculate the value of the horizontal axis (abscissa) of Figure 16.8 using Equation 16.3:

$$\text{Abscissa} = \left(\frac{L}{G}\right) \times \left(\frac{P_g}{P_l}\right)^{0.5} \tag{16.3}$$

where
- L = Mass flow rate of liquid stream.
- G = Mass flow rate of gas stream.
- P_g = Gas density.
- P_l = Liquid density.

2. From the point on the abscissa, calculated in Equation 16.3, proceed up the graph to the flooding line and read the ordinate ε.
3. Using Equation 16.4, calculate the gas flow rate at flooding and solve for G':

$$G' = \left(\frac{\varepsilon \times P_g \times P_l \times g_c}{F \times \phi \times (\mu_l)^{0.2}}\right)^{0.5} \tag{16.4}$$

where
- G' = Mass flow rate of gas per unit cross-sectional area at flooding (lb/s-ft²).
- ε = Ordinate value in Figure 16.8.
- P_g = Density of gas stream (lb/ft³).
- P_l = Density of absorbing liquid (lb/ft³).
- g_c = Gravitational constant = 9.82 m/s² (32.2 ft/s²).

F = Packing factor (dimensionless).
ϕ = Ratio of specific gravity of the scrubbing liquid to that of water (dimensionless).
μ_l = Viscosity of liquid (water = 0.8 cP = 0.0008 Pa-s; use Pa-s in this equation).

4. G' at operating is a fraction of G' at flooding:

$$G'_{operating} = fG'_{flooding} \tag{16.5}$$

5. Calculate the cross-sectional area (A):

$$Area = \frac{Total\ gas\ flow\ rate}{Gas\ flow\ rate\ per\ unit\ area} \tag{16.6}$$

or

$$A = G/G'_{operating} \tag{16.7}$$

6. The diameter of the tower is obtained from Equation 16.8:

$$d = (4A/\pi)^{0.5} \tag{16.8}$$

The problem in Example 16.3 illustrates the calculation procedure for estimating the packed bed tower chamber.

■ EXAMPLE 16.3. TOWER DIAMETER

Problem: For the scrubber in Example 16.2, determine the tower diameter if the operating liquid rate is 1.5 times the minimum. The gas velocity should be no greater than 75% of the flooding velocity and the packing material is 2-in. ceramic Intalox™ saddles.

Solution: Calculate the value of the abscissa in Figure 16.8. From Example 16.2:

G_m = 3538 g-mol/min
L_m = 2448 kg/min

Convert gas molar flow to a mass flow, assuming the molecular weight of the gas is 29 kg/mole:

$$G = 3538\ kg\text{-}mol/min \times 29\ kg/mol = 102.6\ kg/min$$

Adjust the liquid flow to 1.5 times the minimum:

$$L = 1.5 \times 2448 = 3672\ kg/min$$

The densities of water and air at 20°C are

P_l = 1000 kg/m³
P_g = 1.17 kg/m³

Calculate the abscissa using Equation 16.3:

$$Abscissa = \left(\frac{L}{G}\right) \times \left(\frac{P_g}{P_l}\right)^{0.5} = \left(\frac{3672}{102.6}\right) \times \left(\frac{1.17}{1000}\right)^{0.5} = 1.22$$

From Figure 16.8, determine the flooding line from 1.22. The ordinate ε is 0.019. From Equation 16.4, calculate G':

$$G' = \left(\frac{\varepsilon \times P_g \times P_l \times g_c}{F \times \phi \times (\mu_l)^{0.2}} \right)^{0.5}$$

For water, $\phi = 1.0$, and the liquid viscosity is equal to 0.0008 Pa-s. For 2-inch Intalox™ saddles, $F = 40$ ft^2/ft^3 or 131 m^2/m^3 and $g_c = 9.82$ kg/m^3·sec.

$$G' = \left(\frac{0.019 \times 1.17 \times 1000 \times 9.82}{131 \times 1.0 \times (0.0008)^{0.2}} \right)^{0.5} = 2.63 \text{ kg/m}^2 \cdot \text{sec at flooding}$$

Now calculate the actual gas flow rate per unit area:

$$G'_{operating} = fG'_{flooding} = 0.75 \times 2.63 = 1.97 \text{ kg/(m}^2\text{-s)}$$

Finally, calculate the tower diameter:

$$\text{Tower area} = \frac{\text{Gas flow rate}}{G_{operating}} = \frac{102.6 \text{ kg/min} \times 1 \text{ min/60 sec}}{1.97 \text{ kg/m}^2 \cdot \text{sec}}$$

$$\text{Tower diameter} = 1.13 \, A^{0.5} = 1.05 \text{ m, or at least 1 m (3.5 ft)}$$

16.3.4.2 Sizing the Packed Tower Absorber Height

The height of a packed tower refers to the depth of packing material needed to accomplish the required removal efficiency. The more difficult the separations, the larger the packing height required. For example, a much larger packing height would be required to remove SO_2 than to remove Cl_2 from an exhaust stream using water as the absorbent. This is because Cl_2 is more soluble in water than SO_2. Determining the proper height of packing is important since it affects both the rate and efficiency of absorption (USEPA, 1981, p. 4-26). The required packing height of the tower can be expressed as

$$Z = HTU \times NTU \tag{16.9}$$

where
  Z = Height of packing.
  HTU = Height of a transfer unit.
  NTU = Number of transfer units.

The concept of a transfer unit comes from the operation of tray (tray/plate) tower absorbers. Discrete stages (trays or plates) of separation occur in tray/plate tower units. These stages can be visualized as a transfer unit with the number and height of each giving the total tower height. Although packed columns operate as one continuous separation process, in design terminology it is treated as if it were broken into discrete sections (height of a transfer unit). The number and the height of a transfer unit are based on either the gas or liquid phase. Equation 16.9 can be modified to yield Equation 16.10:

$$Z = N_{OG}H_{OG} = N_{OL}H_{OL} \tag{16.10}$$

where

Z = Height of packing (m).
N_{OG} = Number of transfer units based on overall gas film coefficient.
H_{OG} = Height of a transfer unit based on overall gas film coefficient (m).
N_{OL} = Number of transfer units based on overall liquid film coefficient.
H_{OL} = Height of a transfer unit based on overall liquid film coefficient (m).

Values for the height of a transfer unit used in designing absorption systems are usually obtained from experimental data. To ensure the greatest accuracy, vendors of absorption equipment normally perform pilot plant studies to determine the height of a transfer unit. When no experimental data are available, or if only a preliminary estimate of absorber efficiency is needed, there are generalized correlations available to predict the height of a transfer unit. The correlations for predicting the H_{OG} or the H_{OL} are empirical in nature and are a function of

- Type of packing
- Liquid and gas flow rates
- Concentration and solubility of the contaminant
- Liquid properties
- System temperature

These correlations can be found in engineering texts. For most applications, the height of a transfer unit ranges between 0.305 and 1.22 m (1 and 4 ft). As a rough estimate, 0.6 m (2.0 ft) can be used.

The number of transfer units (NTUs) can be obtained experimentally or calculated from a variety of methods. When the solute concentration is very low and the equilibrium line is straight, Equation 16.11 can be used to determine the number of transfer units (N_{OG}) based on the gas phase resistance:

$$N_{OG} = \frac{\ln\left[\left(\dfrac{Y_1 - mX_2}{Y_2 - mX_2}\right)\left(1 - \dfrac{mG_m}{L_m}\right) + \left(\dfrac{mG_m}{L_m}\right)\right]}{\left(1 - \dfrac{mG_m}{L_m}\right)} \tag{16.11}$$

where

N_{OG} = Transfer units.
Y_1 = Mole fraction of solute in entering gas.
m = Slope of the equilibrium line.
X_2 = Mole fraction of solute entering the absorber in the liquid.
Y_2 = Mole fraction of solute in exiting gas.
G_m = Molar flow rate of gas (kg-mol/hr).
L_m = Molar flow rate of liquid (kg-mol/hr).

Equation 16.9 may be solved directly or graphically by using the Colburn diagram presented in Figure 16.9. The Colburn diagram is a plot of N_{OG} vs. $\ln[Y_1 - mX_2)/(Y_2 - mX_2)]$, reading up the graph to the line corresponding to (mG_m/L_m), and then reading across to obtain the value for N_{OG}.

Equation 16.11 can be further simplified for situations where a chemical reaction occurs or if the solute is extremely soluble. In these cases, the solute exhibits almost no partial pressure, and, therefore, the slope of the equilibrium line approaches zero ($m = 0$). For either of these cases, Equation 16.11 reduces to Equation 16.12:

$$N_{OG} = \ln(Y_1/Y_2) \tag{16.12}$$

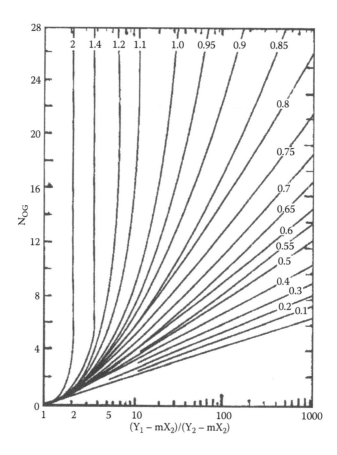

FIGURE 16.9   Coburn diagram. (From USEPA, *APTI Course 415: Control of Gaseous Emissions*, EPA 450/2-81-005, U.S. Environmental Protection Agency Air Pollution Training Institute, Washington, DC, 1981, p. 4-30.)

The number of transfer units depends only on the inlet and outlet concentration of the solute (contaminant or pollutant). For example, if the conditions in Equation 16.12 are met, 2.3 transfer units are required to achieve 90% removal of any pollutant. Equation 16.12 only applies when the equilibrium line is straight (low concentrations) and the slope approaches zero (very soluble or reactive gases). Example 16.4 illustrates a procedure to calculate the packed tower height.

■ EXAMPLE 16.4. TOWER HEIGHT

Problem: From pilot plant studies of the absorption system in Example 16.2, it was determined that the H_{OG} for the SO_2–water system is 0.829 (2.72 ft). Calculate the total height of packing required to achieve 90% removal. The following data were taken from the previous examples.

Given:
  $H_{OG} = 0.829$ m
  $m = 42.7$ kg-mol $H2O$/kg mole air
  $G_m = 3.5$ kg-mol/min
  $L_m = (3672$ kg/min)(kg-mol/18 kg) $= 204$ kg-mol/min
  $X_2 = 0$ (no recirculated liquid)
  $Y_1 = 0.03$
  $Y_2 = 0.003$

Solution: Compute N_{OG} from Equation 16.11:

$$N_{OG} = \frac{\ln\left[\left(\dfrac{Y_1 - mX_2}{Y_2 - mX_2}\right)\left(1 - \dfrac{mG_m}{L_m}\right) + \left(\dfrac{mG_m}{L_m}\right)\right]}{\left(1 - \dfrac{mG_m}{L_m}\right)}$$

$$= \frac{\ln\left[\left(\dfrac{0.03}{0.003}\right)\left(1 - \dfrac{42.7 \times 3.5}{204}\right) + \dfrac{42.7 \times 3.5}{204}\right]}{\left(1 - \dfrac{42.7 \times 3.5}{204}\right)}$$

$$= 4.58$$

Calculate the total packing height:

$$Z = H_{OG} \times N_{OG} = 0.829 \times 4.58 = 3.79 \text{ m}$$

16.3.4.3 Sizing the Plate (Tray) Tower Absorber

In a plate tower absorber, the scrubbing liquid enters at the top of the tower, passes over the top plate, and then down over each lower plate until it reaches the bottom. Absorption occurs as the gas, which enters at the bottom, passes up through the plate and contacts the liquid. In a plate tower, absorption occurs in a stepwise or stage process (USEPA, 1981, p. 4-32). The minimum diameter of a single-pass plate tower is determined by using the gas velocity through the tower. If the gas velocity is too great, liquid droplets are entrained, causing a condition known as *priming*. Priming occurs when the gas velocity through the tower is so great that it causes liquid on one plate to foam and rise to the plate above. Priming reduces absorber efficiency by inhibiting gas and liquid contact. For the purpose of determining tower diameter, priming in a plate tower is analogous to the flooding point in a packed tower. It determines the minimum acceptable diameter. The actual diameter should be larger. The smallest allowable diameter for a plate tower is expressed by Equation 16.13:

$$d = \Psi[Q(\rho_g)^{0.5}]^{0.5} \tag{16.13}$$

where
   d = Plate tower diameter.
   Ψ = Empirical correlation $(m^{0.25} \text{ hr})^{0.25}/kg^{0.25})$.
   Q = Volumetric gas glow (m^3/hr).
   ρ_g = Gas density (kg/m^3).

The term Ψ is an empirical correlation and is a function of both the tray spacing and the densities of the gas and liquid streams. Values of Ψ shown in Table 16.2 are for a tray spacing of 61 cm (24 in.) and a liquid specific gravity of 1.05. If the specific gravity of a liquid varies significantly from 1.05, the values for Ψ in the table cannot be used.

Depending on operating conditions, trays are spaced at a minimum distance between plates to allow the gas and liquid phases to separate before reaching the plate above. Trays should be spaced to allow for easy maintenance and cleaning. Trays are normally spaced 45 to 70 cm (18 to 28 in.) apart. In using the information in Table 16.2 for tray spacing different than 61-cm, a correction factor must be used. Figure 16.10 is used to determine the correction factor that is multiplied by the estimated diameter. Example 16.5 illustrates how the minimum diameter of a tray tower absorber is estimated.

TABLE 16.2
Empirical Parameters for Equation 16.13

Tray	Metric Ψ	Engineering Ψ
Bubble cap	0.0162	0.1386
Sieve	0.0140	0.1196
Valve	0.0125	0.1069

Note: Metric Ψ is expressed in $m^{0.25} \, hr^{0.5} \, kg^{0.25}$ for use with Q expressed in m^3/hr and ρ_g expressed in kg/m^3. Engineering Ψ is expressed in $ft^{0.25} \, min^{0.5} \, lb^{0.25}$ for use with Q expressed in cfm and ρ_g expressed in lb/ft^3.

Source: Adapted from USEPA, *Wet Scrubber System Study*, NTIS Report PB-213016, U.S. Environmental Protection Agency, Research Triangle Park, NC, 1972.

■ **EXAMPLE 16.5. PLATE TOWER DIAMETER**

Problem: For the conditions described in Example 16.2, determine the minimum acceptable diameter if the scrubber is a bubble cap tray tower absorber. The trays are spaced 0.53 m (21 in.) apart (USEPA, 1981, p. 4-34).

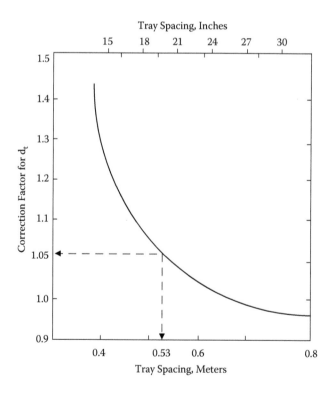

FIGURE 16.10 Tray spacing correction factor. (Adapted from USEPA, *APTI Course 415: Control of Gaseous Emissions*, EPA 450/2-81-005, U.S. Environmental Protection Agency Air Pollution Training Institute, Washington, DC, 1981, p. 4-33.)

Solution: From Examples 16.2 and 16.3:

Gas flow rate = Q = 84.9 m³/min
Density = ρ_g = 1.17 kg/m³

From Table 16.2 for a bubble cap tray:

$$\Psi = 0.0162 \text{ m}^{0.25} \text{ hr}^{0.50}/\text{kg}^{0.25}$$

Before Equation 16.13 can be used, Q must be converted to m³/hr:

$$Q = 84.9 \text{ m}^3/\text{min} \times 60 \text{ min/hr} = 5094 \text{ m}^3/\text{hr}$$

Substituting these values into Equation 16.13 for a minimum diameter d,

$$d = \Psi \left[Q \left(\rho_g \right)^{0.5} \right]^{0.5} = 0.0162 \left[5094(1.17)^{0.5} \right]^{0.5} = 1.2 \text{ m}$$

Correct this diameter for a tray spacing of 0.53 m. From Figure 16.10, read a correction factor of 1.05. Therefore, the minimum diameter is

$$d = 1.2 \times 1.05 = 1.26 \text{ m} \ (4.13 \text{ ft})$$

Note: This estimated diameter is a minimum acceptable diameter based on actual conditions. In practice, a larger diameter (based on maintenance and economic considerations) is usually chosen.

16.3.4.4 Theoretical Number of Absorber Plates or Trays

The several methods used to determine the number of ideal plates or trays required for a given removal efficiency can become quite complicated. One method used is a graphical technique (USEPA, 1981, p. 4-34). The number of ideal plates is obtained by drawing "steps" on an operating diagram. This procedure is illustrated in Figure 16.11. This method can be rather time consuming, and inaccuracies can result at both ends of the graph. Equation 16.14 is a simplified method of estimating the number of plates. It can only be used if both the equilibrium and operating lines for the system are straight. This is a valid assumption for most air pollution control systems.

$$N_p = \frac{\ln\left[\left(\dfrac{Y_1 - mX_2}{Y_2 - mX_2} \right) \left(1 - \dfrac{mG_m}{L_m} \right) + \left(\dfrac{mG_m}{L_m} \right) \right]}{\ln\left(\dfrac{L_m}{mG_m} \right)} \qquad (16.14)$$

where
N_p = Number of theoretical plates.
Y_1 = Mole fraction of solute in entering gas.
X_2 = Mole fraction of solute entering the tower.
Y_2 = Mole fraction of solute in exiting gas.
m = Slope of equilibrium line.
G_m = Molar flow rate of gas (kg-mol/hr).
L_m = Molar flow rate of liquid (kg-mol/hr).

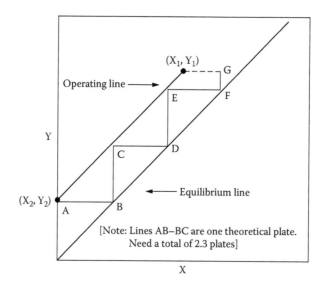

FIGURE 16.11 Graphical determination of the number of theoretical plates. (From USEPA, *APTI Course 415: Control of Gaseous Emissions*, EPA 450/2-81-005, U.S. Environmental Protection Agency Air Pollution Training Institute, Washington, DC, 1981, p. 4-35.)

Equation 16.14 is used to predict the number of theoretical plates (N_p) required to achieve a given removal efficiency. The operating conditions for a theoretical plate assume that the gas and liquid stream leaving the plate are in equilibrium with each other. This ideal condition is never achieved in practice. A larger number of actual trays are required to compensate for this decreased tray efficiency.

Three types of efficiencies are used to describe absorption efficiency for a plate tower: (1) an overall efficiency, which is concerned with the entire column; (2) Murphree efficiency, which is applicable with a single plate; and (3) local efficiency, which pertains to a specific location on a plate. The simplest of the tray efficiency concepts, the overall efficiency, is the ratio of the number of theoretical plates to the number of actual plates. Because overall tray efficiency is an oversimplification of the process, reliable values are difficult to obtain. For a rough estimate, overall tray efficiencies for absorbers operating with low-viscosity liquid normally fall in a range from 65 to 80%. Example 16.6 shows a procedure to calculate the number of theoretical plates.

■ EXAMPLE 16.6

Problem: Calculate the number of theoretical plates required for the scrubber in Example 16.5, using the same conditions as in Example 16.4. Estimate the total height of the tower if the trays are spaced at 0.53-m intervals and assume an overall tray efficiency of 70%.

Solution: From Example 16.5 and the previous examples the following data are obtained:

$m = 42.7$
$G_m = 3.5$ kg-mol/min
$L_m = 204$ kg-mol/min
$X_2 = 0$ (no recycle liquid)
$Y_1 = 0.03$
$Y_2 = 0.003$

The number of theoretical plates from Equation 16.14 is

$$N_p = \cfrac{\ln\left[\left(\cfrac{Y_1 - mX_2}{Y_2 - mX_2}\right)\left(1 - \cfrac{mG_m}{L_m}\right) + \left(\cfrac{mG_m}{L_m}\right)\right]}{\ln\left(\cfrac{L_m}{mG_m}\right)}$$

$$N_p = \cfrac{\ln\left[\left(\cfrac{0.03 - 0}{0.003 - 0}\right)\left(1 - \cfrac{42.7 \times 3.5}{204}\right) + \left(\cfrac{42.7 \times 3.5}{204}\right)\right]}{\ln\left(\cfrac{204}{42.7 \times 3.5}\right)}$$

$$= 3.94 \text{ theoretical plates}$$

Assuming that the overall plate efficiency is 70%, the actual number of plates is

$$\text{Actual number of plates} = 3.94/0.7 = 5.6 \text{ (or 6) plates}$$

The height of the tower is given by

$$Z = N_p + \text{Top height}$$

where N_p is the tray spacing. The top height is the distance (freeboard) over the top plate that allows the gas-vapor mixture to separate. This distance is usually the same as the tray spacing.

$$Z = (6 \text{ plates} \times 0.53 \text{ m}) + 0.53 \text{ m} = 3.71 \text{ m}$$

Note that this height is approximately the same as that predicted for the packed tower in Example 16.4. This is logical since both the packed and plate towers are efficient gas absorption devices. However, due to many assumptions, no generalization should be made.

16.4   ADSORPTION

Adsorption is a mass transfer process that involves passing a stream of effluent gas through the surface of prepared porous solids (*adsorbents*). The surfaces of the porous solid substance attract and hold the gas (*adsorbate*) by either physical or chemical adsorption. Adsorption occurs on the internal surfaces of the particles (USEPA, 1981, p. 5-1). In physical adsorption (a readily reversible process), a gas molecule adheres to the surface of the solid because of an imbalance of electron distribution. In chemical adsorption (not readily reversible), once the gas molecule adheres to the surface, it reacts chemically with it.

Several materials possess adsorptive properties. These materials include activated carbon, alumina, bone char, magnesia, silica gel, molecular sieves, strontium sulfate, and others. The most important adsorbent for air pollution control is activated charcoal. The surface area of activated charcoal will preferentially adsorb hydrocarbon vapors and odorous organic compounds from an airstream.

In an adsorption system (in contrast to the absorption system where the collected contaminant is continuously removed by flowing liquid), the collected contaminant remains in the adsorption bed. The most common adsorption system is the fixed-bed adsorber, which can be contained in either a vertical or horizontal cylindrical shell. The adsorbent (usually activated carbon) is arranged in beds or trays in layers about 0.5 inch thick. Multiple-beds may be used. In multiple-bed systems, one or more beds are adsorbing vapors, while the other bed is being regenerated.

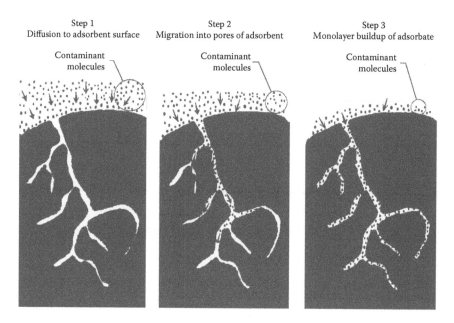

FIGURE 16.12   Gas collection by adsorption. (Adapted from USEPA, *APTI Course 415: Control of Gaseous Emissions*, EPA 450/2-81-005, U.S. Environmental Protection Agency Air Pollution Training Institute, Washington, DC, 1981, p. 5-2.)

The efficiency of most adsorbers is near 100% at the beginning of the operation and remains high until a breakpoint or breakthrough occurs. When the adsorbent becomes saturated with adsorbate, contaminant begins to leak out of the bed, signaling that the adsorber should be renewed or regenerated. Although adsorption systems are high-efficiency devices that may allow recovery of product, have excellent control and response to process changes, and have the capability of being operated unattended, they also have some disadvantages, including the need for expensive extraction schemes if product recovery is required, relatively high capital cost, and gas stream prefiltering needs (to remove any particulate capable of plugging the adsorbent bed)(Spellman, 2008).

16.4.1   ADSORPTION STEPS

Adsorption occurs in a series of three steps (USEPA, 1981, p. 5-2). In the first step, the contaminant diffuses from the major body of the air stream to the external surface of the adsorbent particle. In the second step, the contaminant molecule migrates from the relatively small area of the external surface (a few square meters per gram) to the pores within each adsorbent particle. The bulk of adsorption occurs in these pores because the majority of available surface area is there (hundreds of square meters per gram). In the third step, the contaminant molecule adheres to the surface in the pore. Figure 16.12 illustrates this overall diffusion and adsorption process.

16.4.2   ADSORPTION FORCES—PHYSICAL AND CHEMICAL

The adsorption process is classified as either *physical* or *chemical*. The basic difference between physical and chemical adsorption is the manner in which the gas molecule is bonded to the adsorbent. In physical adsorption, the gas molecule is bonded to the solid surface by weak forces of intermolecular cohesion. The chemical nature of the adsorbed gas remains unchanged. Therefore, physical adsorption is a readily reversible process. In chemical adsorption, a much stronger bond is formed between the gas molecule and adsorbent. A sharing or exchange of electrons takes place. Chemical adsorption is not easily reversible.

- *Physical adsorption*—The forces active in physical adsorption are electrostatic in nature. These forces are present in all states of matter: gas, liquid, and solid. They are the same forces of attraction that cause gases to condense and cause gases to deviate from ideal behavior under extreme conditions. Physical adsorption is also sometimes referred to as *van der Waals' adsorption*. The electrostatic effect, which produces the Van der Waal's forces, depends on the polarity of both the gas and solid molecules. Molecules in any state are either polar or nonpolar depending on their chemical structure. *Polar substances* exhibit a separation of positive and negative charges within the compound. This separation of positive and negative charges is referred to as a *permanent dipole*. Water is a prime example of a polar substance. *Nonpolar substances* have both their positive and negative charges in one center so they have no permanent dipole. Because of their symmetry, most organic compounds are nonpolar.
- *Chemical adsorption*—Chemical adsorption or chemisorption results from the chemical interaction between the gas and the solid. The gas is held to the surface of the adsorbent by the formation of a chemical bond. Adsorbents used in chemisorption can be either pure substances or chemicals deposited on an inert carrier material. One example is using pure iron oxide chips to adsorb H_2S gases. Another example is the use of activated carbon that has been impregnated with sulfur to remove mercury vapors.

All adsorption processes are exothermic whether adsorption occurs from chemical or physical forces. In adsorption, molecules are transferred from the gas to the surface of a solid. The fast-moving gas molecules lose their kinetic energy of motion to the adsorbent in the form of heat.

16.4.3 ADSORPTION EQUILIBRIUM RELATIONSHIPS

Most available data on adsorption systems are determined at equilibrium conditions. Adsorption equilibrium is the set of conditions at which the number of molecules arriving on the surface of the adsorbent equals the number of molecules leaving. The adsorbent bed is said to be "saturated with vapors" and can remove no more vapors from the exhaust stream. The equilibrium capacity determines the maximum amount of vapor that can be adsorbed at a given set of operating conditions. Although a number of variables affect adsorption, gas temperature and pressure are the two most important in determining adsorption capacity for a given system. Three types of equilibrium graphs are used to describe adsorption capacity:

- Isotherm at constant temperature
- Isostere at constant amount of vapors adsorbed
- Isobar at constant pressure

16.4.3.1 Isotherm

The most common and useful adsorption equilibrium data is the adsorption isotherm. The isotherm is a plot of the adsorbent capacity versus the partial pressure of the adsorbate at a constant temperature. Adsorbent capacity is usually given in weight percent expressed as grams of adsorbate per 100 grams of adsorbent. Figure 16.13 is a typical example of an adsorption isotherm for carbon tetrachloride on activated carbon. Graphs of this type are used to estimate the size of adsorption systems as illustrated in Example 16.7.

■ EXAMPLE 16.7

Problem: A dry-cleaning process exhausts a 15,000-cfm air stream containing 680-ppm carbon tetrachloride. Given Figure 16.13 and assuming that the exhaust stream is at approximately 140°F and 14.7 psia, determine the saturation capacity of the carbon (USEPA, 1981, p. 5-8).

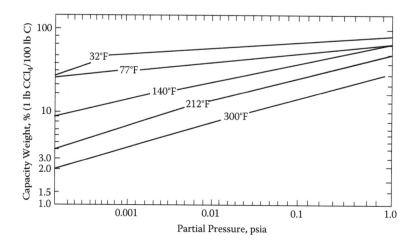

FIGURE 16.13 Adsorption isotherm for carbon tetrachloride on activated carbon (Adapted from USEPA, *APTI Course 415: Control of Gaseous Emissions*, EPA 450/2-81-005, U.S. Environmental Protection Agency Air Pollution Training Institute, Washington, DC, 1981, p. 5-8.)

Solution: In the gas phase, mole fraction Y is equal to the percent by volume:

$$Y = \% \text{ Volume} = 680 \text{ ppm} = 680/(10)^6 = 0.00068$$

Obtain the partial pressure:

$$p = YP = 0.00068 \times 14.7 \text{ psia} = 0.01 \text{ psia}$$

From Figure 16.13, at a partial pressure of 0.01 psia and a temperature of 140°F, the carbon capacity is read as 30%. This means that, at saturation, 30 pounds of vapor are removed per 100 pounds of carbon in the adsorber (30 kg/100 kg).

16.4.3.2 Isostere

The isostere is a plot of the ln p vs. $1/T$ at a constant amount of vapor adsorbed. Adsorption isostere lines are usually straight for most adsorbate–absorbent systems. Figure 16.14 is an adsorption isostere graph for the adsorption of H_2S gas onto molecular sieves. The isostere is important in that the slope of the isostere corresponds to the heat of adsorption.

16.4.3.3 Isobar

The *isobar* is a plot of the amount of vapor adsorbed vs. temperature at a constant pressure. Figure 16.15 shows an isobar line for the adsorption of benzene vapors on activated carbon. Note that—as is always the case for physical adsorption—the amount adsorbed decreases with increasing temperature. Because these three relationships (i.e., isotherm, isostere, and isobar) were developed at equilibrium conditions, they depend on each other. By determining one, such as the isotherm, the other two relationships can be determined for a given system. In the design of an air pollution control system, the adsorption isotherm is by far the most commonly used equilibrium relationship.

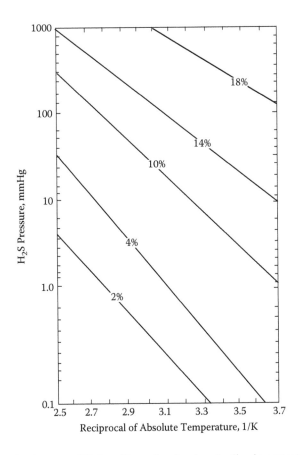

FIGURE 16.14 Adsorption isoteres of H_2S on 13× molecular sieve loading in percent weight. (Adapted from USEPA, *APTI Course 415: Control of Gaseous Emissions*, EPA 450/2-81-005, U.S. Environmental Protection Agency Air Pollution Training Institute, Washington, DC, 1981, p. 5-11.)

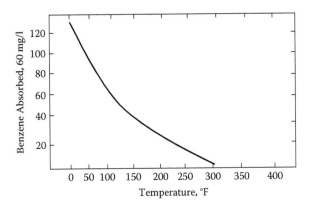

FIGURE 16.15 Adsorption isobar for benzene on carbon (benzene at 10.0 mm Hg). (Adapted from USEPA, *APTI Course 415: Control of Gaseous Emissions*, EPA 450/2-81-005, U.S. Environmental Protection Agency Air Pollution Training Institute, Washington, DC, 1981, p. 5-8.)

16.4.4 FACTORS AFFECTING ADSORPTION

A number of factors or system variables influence the performance of an adsorption system, including the following (USEPA, 1981, p. 5-18)

- Temperature
- Pressure
- Gas velocity
- Bed depth
- Humidity
- Contamination

We discuss these variables and their effects on the adsorption process in this subsection.

16.4.4.1 Temperature

For physical adsorption processes, the capacity of an adsorbent decreases as the temperature of the system increases. As the temperature increases, the vapor pressure of the adsorbate increases, raising the energy level of the adsorbed molecules. Adsorbed molecules now have sufficient energy to overcome the van der Waal's attraction and migrate back to the gas phase. Molecules already in the gas phase tend to stay there due to their high vapor gas phase. As a general rule, adsorber temperatures are kept below 130°F (54°C) to ensure adequate bed capacities. Temperatures above this limit can be divided by cooling the exhaust stream to be treated.

16.4.4.2 Pressure

Adsorption capacity increases with an increase in the partial pressure of the vapor. The partial pressure of a vapor is proportional to the total pressure of the system. Any increase in pressure will increase the adsorption capacity of a system. The increase in capacity occurs because of a decrease in the mean free path of vapor at higher pressures. Simply, the molecules are packed more tightly together. More molecules have a chance to hit the available adsorption sites, increasing the number of molecules adsorbed.

16.4.4.3 Gas Velocity

The gas velocity through the adsorber determines the contact or residence time between the contaminant stream and adsorbent. The residence time directly affects capture efficiency. The slower the contaminant stream flows through the adsorbent bed, the greater the probability of a contaminant molecule hitting an available site. Once a molecule has been captured, it will stay on the surface until the physical conditions of the system are changed. In order to achieve 90% + capture efficiency, most carbon adsorption stems are designed for a maximum gas flow velocity of 30 m/min (100 ft/min) through the adsorber. A lower limit of at least 6 m/min (20 ft/min) is maintained to avoid flow distribution problems, such as channeling.

16.4.4.4 Bed Depth

Providing a sufficient depth of adsorbent is very important in achieving efficient gas removal. If the adsorber bed depth is shorter than the required mass transfer zone, breakthrough will occur immediately, rendering the system ineffective. Computing the length of the mass transfer zone (MTZ) is very difficult since it depends upon six factors: adsorbent particle size, gas velocity, adsorbate concentration, fluid properties of the gas stream, temperature, and pressure. The relationship between breakthrough capacity and MTZ is

$$C_B = \frac{0.5C_s(MTZ) + C_s(D - MTZ)}{D} \tag{16.15}$$

where

C_B = Breakthrough capacity.
C_s = Saturation capacity.
MTZ = Mass transfer zone length.
D = Adsorption bed depth.

Equation 16.15 is used mainly as a check to ensure that the proposed bed depth is longer than the MTZ. Actual bed depths are usually many times longer than the length of the MTZ.

The total amount of adsorbent required is usually determined from the adsorption isotherm, as illustrated in Example 16.8. Once this has been set, the bed depth can then be estimated by knowing the tower diameter and density of the adsorbent. Example 16.8 illustrates how this is done. Generally, the adsorbent bed is sized to the maximum length allowed by the pressure drop across the bed.

■ EXAMPLE 16.8

Problem: Assume the same conditions as stated earlier in Example 16.7. Estimate the amount of carbon that would be required if the adsorber were to operate on a 4-hour cycle. The molecular weight of CCl_4 is 154 lb/lb-mole (USEPA, 1981, p. 5-18).

Solution: From Example 16.7, we know that the carbon used will remove 30 lb of vapor for every 100 lb of carbon at saturation conditions. First compute the flow rate (Q) of CCl_4:

$$Q = 15{,}000 \text{ scfm} \times 0.00068 = 10.2 \text{ scfm } CCl_4$$

Convert to pounds per hour:

$$10.2 \left(\frac{ft^3}{min} \right) \times \left(\frac{\text{lb-mol}}{359 \text{ ft}^3} \right) \times \left(\frac{154 \text{ lb}}{\text{lb-mol}} \right) \times \left(\frac{60 \text{ min}}{hr} \right) = 262.5 \text{ lb/hr}$$

The amount of carbon (at saturation) required (assuming that the working charge is twice the saturation capacity). (Note that this gives only a rough estimate of the amount of carbon needed.)

$$2 \times 3500 = 7000 \text{ lb} \times 3182 \text{ kg carbon per 4-hour cycle per adsorber}$$

■ EXAMPLE 16.9

The following example is based on a number of adsorber design maximum and minimum rules of thumb. It is intended as a guide to illustrate how to "red flag" any parameters that may be greatly exceeded.

Problem: A solvent degreaser is designed to recover toluene from a 3.78 m³/s (8000 acfm) air stream at 25°C (77°F) and atmosphere pressure. The company is planning to use a two-bed carbon adsorption system with a cycle time of 4 hours. The maximum concentration of toluene is kept below 50% of the lower explosive limit (LEL) for safety purposes. Using Figure 16.16, the adsorption isotherm for toluene, and the additional operational data, estimate (USEPA, 1981, p. 5-36):

1. The amount of carbon required for a 4-hour cycle
2. Square feet of surface area required based on a 0.58 m/s (100 fpm) maximum velocity
3. Depth of the carbon bed

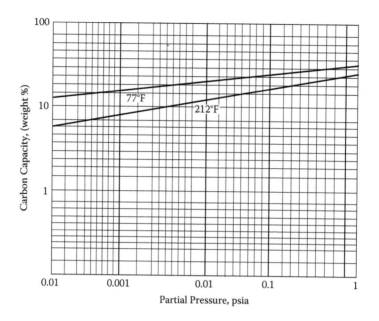

FIGURE 16.16 Adsorption isotherm for toluene. (Adapted from USEPA, *APTI Course 415: Control of Gaseous Emissions*, EPA 450/2-81-005, U.S. Environmental Protection Agency Air Pollution Training Institute, Washington, DC, 1981.)

Given:
  LEL for toluene = 1.2%
  Molecular weight of toluene = 92.1 kg/kg-mol
  Carbon density = 480 kg/m³ (30 lb/ft³)

Solution: First calculate the toluene flow rate:

$$3.78 \text{ m}^3\text{/s} \times 50\% \times 1.2\% = 0.023 \text{ m}^3\text{/s toluene}$$

To determine the saturation capacity of the carbon, calculate the partial pressure of toluene at the adsorption conditions:

$$p = YP = \left(\frac{0.023 \text{ m}^3\text{/s}}{3.78 \text{ m}^3\text{/s}} \right)(14.7 \text{ psia}) = 0.089 \text{ psia}$$

From Figure 16.16, the saturation capacity of the carbon is 40% or 40-kg toluene per 100kg of carbon. The flow rate of toluene is

$$\left(0.023 \text{ m}^3\text{/s} \right)\left(\frac{\text{kg-mol}}{22.4 \text{ m}^3} \right)\left(\frac{273 \text{ K}}{350 \text{ K}} \right)\left(\frac{92.1 \text{ kg}}{\text{kg-mol}} \right) = 0.074 \text{ kg/s}$$

The amount of carbon at saturation for a 4-hour cycle is

$$\left(0.074 \text{ kg/s toluene} \right)\left(\frac{100 \text{ kg carbon}}{40 \text{ kg toluene}} \right)\left(\frac{3600 \text{ s}}{\text{hr}} \right)(4 \text{ hr}) = 2664 \text{ kg of carbon}$$

The working charge of carbon can be estimated by doubling the saturation capacity. Therefore, the working charge is

$$\text{Working charge} = 2 \times 2664 \text{ kg} = 5328 \text{ kg}$$

The square feet of superficial surface area is the surface area set by the maximum velocity of 0.508 m/s (100 ft/min) through the adsorber. The required surface area is

$$A = \frac{Q}{\text{Maximum velocity}} = \frac{3.78 \text{ m}^3/\text{s}}{0.508 \text{ m/s}} = 7.44 \text{ m}^2$$

For a horizontal flow adsorber this would correspond to a vessel approximately 2 m (6.6 ft) in width and 4 m (13.1 ft) in length to give 8.0 m² (87 ft²) surface area. This would supply more than the required area. The flow rate is too high to be handled by a single vertical flow adsorber. An alternative would be to use three vessels, two adsorbing while one is being regenerated. Each vessel then must be sized to handle 1.89 m³/s (4000 acfm). The area required for a limiting velocity of 0.508 m/s is

$$\text{Area to handle half flow} = \frac{1.89 \text{ m}^3/\text{s}}{0.508 \text{ m/s}} = 3.72 \text{ m}^2$$

This cross-sectional are corresponds to a vessel diameter of

$$d = \left[\left(\frac{4}{\pi}\right)A\right]^{0.5} = \left[\left(\frac{4}{\pi}\right)3.72\right]^{0.5} = 2.18 \text{ m (7 ft)}$$

Determine the volume that the carbon would occupy in horizontal bed:

$$\text{Volume of carbon} = \text{Weight} \times \frac{1}{\text{Density}} = 5328 \times \frac{\text{m}^3}{480 \text{ kg}} = 11.1 \text{ m}^3$$

Note: For the three-bed vertical system, the volume of each bed is half of this, or 5.55 m³.

Now estimate bed depth:

$$\text{Depth of carbon} = \frac{\text{Volume of carbon}}{\text{Cross-sectional area of adsorber}} = \frac{11.1 \text{ m}^3}{7.44 \text{ m}^2} = 1.49 \text{ m}$$

Note: The depth for the three-bed vertical system is the same, because both the volume and area are halved.

16.4.4.5 Humidity

Activated carbon will preferentially adsorb nonpolar hydrocarbons over polar water vapor. The water vapor molecules in the exhaust stream exhibit strong attractions for each other rather than the adsorbent. At high relative humidity, over 50%, the number of water molecules increases such that they begin to compete with the hydrocarbon molecule for active adsorption sites. This reduces the capacity and the efficiency of the adsorption system. Exhaust streams with humidity greater than 50% may require installation of additional equipment to remove some of the moisture. Coolers to remove the water are one solution. Dilution air with significantly less moisture in it than the process stream has also been used. The contaminant stream may also be heated to reduce the humidity as long as the increase in temperature does not greatly affect adsorption efficiency.

16.4.4.6 Contaminants

Particulate matter, entrained liquid droplets, and organic compounds that have high boiling points can also reduce adsorber efficiency if present in the air stream. Any micron-sized particle of dust or lint, which is not filtered, can cover the surface of the adsorbent. This greatly reduces the surface area of the adsorbent available to the gas molecule for adsorption. Covering active adsorption sites by an inert material is referred to as "blinding" or "deactivation." To avoid this situation, almost all industrial adsorption systems are equipped with some type of particulate matter removal device. Entrained liquid droplets can also cause operational problems. Liquid droplets that are nonadsorbing can act the same as particulate matter: The liquid covers the surface, blinding the bed. If the liquid is the same as the adsorbate, high heats of adsorption occur. This is especially a problem in activated carbon systems where liquid organic droplets carried over from the process can cause bed fires from the released heat. When liquid droplets are present, some type of entrainment separator may be required.

16.5   INCINERATION

Incineration (or combustion) is a major source of air pollution; however, if properly operated, it can be a beneficial air pollution control system in which the object is to convert certain air contaminants (usually organic compounds classified as volatile organic compounds (VOCs) and/or air toxic compounds (Spellman, 1999; USEPA, 1973). Incineration is a chemical process defined as rapid, high-temperature gas-phase oxidation. The incineration equipment used to control air pollution emissions is designed to push these oxidation reactions as close to complete incineration as possible, leaving a minimum of unburned residue. Depending upon the contaminant being oxidized, equipment used to control waste gases by combustion can be divided into three categories: direct-flame incineration (or flaring), thermal incineration (afterburners), or catalytic incineration.

16.5.1   Factors Affecting Incineration for Emission Control

The operation of any incineration system used for emission control is governed by seven variables: temperature, residence time, turbulence, oxygen, combustion limit, flame combustion, and heat. For complete incineration to occur, oxygen must be available and put into contact with sufficient temperature (turbulence) and held at this temperature for a sufficient time. These seven variables are not independent—changing one affects the entire process.

16.5.1.1   Temperature

The rate at which a combustible compound is oxidized is greatly affected by temperature. The higher the temperature, the faster the oxidation reaction will proceed. The chemical reactions involved in the combination of a fuel and oxygen can occur even at room temperature, but very slowly. For this reason, a pile of oily rags can be a fire hazard. Small amounts of heat are liberated by the slow oxidation of the oils. This in turn raises the temperature of the rags and increases the oxidation rate, liberating more heat. Eventually a fully engulfed fire can break out (USEPA, 1981, p. 3-2). Most incinerators operate at higher temperature than the ignition temperature, which is a minimum temperature. Thermal destruction of most organic compounds occurs between 590 and 650°F (1100 and 1200°F). However, most incinerators are operated at 700 to 820°C (1300 to 1500°F) to convert CO to CO_2, which occurs only at these higher temperatures.

16.5.1.2   Residence Time

Much in the same manner that higher temperature and pressure affect the volume of a gas, time and temperature affect combustion. When one variable is increased, the other may be decreased with the same end result. With a higher temperature, a shorter residence time can achieve the same degree of oxidation. The reverse is also true—a higher residence time allows the use of a lower temperature. Simply, the residence time needed to complete the oxidation reactions in the incinerator depends partly on the rate of the reactions at the prevailing temperature and partly on the mixing of the waste stream and the hot combustion gases from the burner or burners. The residence time of gases in the incinerator may be calculated from a simple ratio of the volume of the refractory-lined combustion chamber and the volumetric flow rate of combustion products through the chamber.

$$t = V/Q \qquad\qquad (16.16)$$

where
  t = Residence time (sec).
  V = Chamber volume (m^3).
  Q = Gas volumetric flow rate at combustion conditions (m^3/s).

Q is the total flow of hot gases in the combustion chamber. Adjustments to the flow rate must include any outside air added for combustion. Example 16.10 below shows the determination of residence time from the volumetric flow rate of gases.

16.5.1.3 Turbulence

Proper mixing is important in combustion processes for two reasons. First, mixing of the burner fuel with air is needed to ensure complete combustion of the fuel. If not, unreacted fuel will be exhausted from the stack. Second, the organic compound-containing waste gases must be thoroughly mixed with the burner combustion gases to ensure that the entire waste gas stream reaches the necessary combustion temperatures. Otherwise, incomplete combustion will occur. A number of methods are available to improve mixing of the air and combustion streams. Some of these include the use of refractory baffles, swirl fired burners, or baffle plates. It is not easy to obtain complete mixing. Unless properly designed, many of these mixing devices may create dead spots and reduce operating temperatures. Inserting obstructions to increase turbulence may not be sufficient. In afterburner systems, the process of mixing the flame and the fume streams to obtain a uniform temperature for decomposition of pollutants is the most difficult part in the design of the afterburner (USEPA, 1973).

16.5.1.4 Oxygen Requirement

Not only is oxygen necessary for combustion to occur, oxygen requirements for the supplemental heat burner used in thermal incinerators must be taken into account when sizing the burner system and the combustion chamber. To achieve complete combustion of a compound or the fuel (propane, No. 2 fuel oil, natural gas, for example), a sufficient supply of oxygen must be present in the burner flame to convert all of the carbon to CO_2. This quantity of oxygen is referred to as the stoichiometric or theoretical amount. The stoichiometric amount of oxygen is determined from a balanced chemical equation summarizing the oxidation reactions; for example, 1 mole of methane requires 2 moles of oxygen for complete combustion (USEPA, 1981, p. 3-4):

$$CH_4 + 2O_2 \rightarrow CO_2 + 2H_2O \qquad\qquad (16.17)$$

If an insufficient amount of oxygen is supplied, the mixture is referred to as rich. Incomplete combustion occurs under these conditions. This reduces the peak flame temperature and creates black smoke emissions. If more than the stoichiometric amount of oxygen is supplied, the mixture is referred to as lean. The added oxygen plays no part in the oxidation reaction and passes through the incinerator. To ensure combustion, more than the stoichiometric amount of air is used. This extra volume is referred to as excess air.

16.5.1.5   Combustion Limits

Not all mixtures of fuel and air are able to support combustion. Incinerators usually operate at organic vapor concentrations below 25% of the *lower explosive limit* (LEL) or *lower flammability limit* (LFL). Combustion does not occur readily in this range. The explosive or flammable limits for a mixture are the maximum and minimum concentrations of fuel in air that will support combustion. The *upper explosive limit* (UEL) is defined as the concentration of fuel that produces a non-burning mixture due to a lack of oxygen. The *lower explosive limit* (LEL) is defined as the concentration of fuel below which combustion will not be self-sustaining.

16.5.1.6   Flame Combustion

When mixing fuel and air, two different mechanisms of combustion can occur (USEPA, 1981, p. 3-7). When air and fuel flowing through separate ports are ignited at the burner nozzle, a luminous yellow flame results. The yellow flame results from thermal cracking of the fuel. Cracking occurs when hydrocarbons are intensely heated before they have a chance to combine with oxygen. The cracking releases both hydrogen and carbon, which diffuse into the flame to form carbon dioxide and water. The carbon particles give the flame the yellow appearance. If incomplete combustion occurs from flame temperature cooling or if there is insufficient oxygen, soot and black smoke will form. When the fuel and air are premixed in front of the burner nozzle, blue flame combustion occurs. The reason for the different flame is that the fuel-air mixture is gradually heated. The hydrocarbon molecules are slowly oxidized, going from aldehydes and ketones to carbon dioxide and water. No cracking occurs and no carbon particles are formed. Incomplete combustion results in the release of the intermediate partially oxidized compounds. Blue haze and odors are emitted from the stack.

16.5.1.7   Heat

The fuel requirement (for burners) is one of the main parameters of concern in incineration systems. Moreover, the amount of fuel required to raise the temperature of the waste stream to the temperature required for complete oxidation is another area of concern. The burner fuel requirement can be estimated based on a simple heat balance of the unit and information concerning the waste gas stream. The first step in computing the heat required is to perform a heat balance around the oxidation system. From the first law of conservation of energy:

$$\text{Heat in} = \text{Heat out} + \text{Heat loss} \qquad (16.18)$$

Heat is a relative term that is compared at a reference temperature. In order to calculate the heat that exits the incinerator with the waste gas stream, the enthalpies of the inlet and outlet waste gas streams must be determined. *Enthalpy* is a thermodynamic term that includes the sensible heat and latent heat of a material. The heat content of a substance is arbitrarily taken as zero at a specified reference temperature. In the natural gas industry, the reference temperature is normally 16°C (60°F).

The enthalpy of the waste gas stream can be computed from Equation 16.19:

$$H = C_p(T - T_0) \qquad (16.19)$$

where
H = Enthalpy (J/kg, Btu/lb).
C_p = Specific heat at temperature T (J/kg, °C; Btu/lb, °F).
T = Temperature of the substance (°C or °F).
T_0 = Reference temperature (°C or °F).

Subtracting the enthalpy of the waste gas stream exiting the incinerator from the waste gas stream entering the incinerator gives the heat that must be supplied by the fuel. This is referred to as a *change in enthalpy* or *heat content*. Using Equation 16.19, the enthalpy entering (T_1) is subtracted from the enthalpy leaving (T_2), giving

$$q = m\Delta H = mC_p(T_2 - T_1) \qquad (16.20)$$

where

q = Heat rate (Btu/hr).
m = Mass flow rate (lb/hr).
ΔH = Enthalpy change.
C_p = Specific heat (J/kg, °C; Btu/lb, °F).
T_2 = Exit temperature of the substance (°C or °F).
T_1 = Initial temperature (°C or °F).

16.5.2 INCINERATION EXAMPLE CALCULATIONS

■ EXAMPLE 16.10. RESIDENCE TIME (METRIC UNITS)

Problem: A thermal incinerator controls emissions from a paint-baking oven. The cylindrical unit diameter is 1.5 m (5 ft) and it is 3.5 m (11.5 ft) long. The exhaust from the oven is 3.8 m³s (8050 scfm). The incinerator uses 300 scfm of natural gas and operates at a temperature of 1,400°F. If all the oxygen necessary for combustion is supplied from the process stream, no outside air added, what is the residence time in the combustion chamber (USEPA, 1981, p. 3-7)?

Solution: To solve this problem, we use the approximation that 11.5 m³ of combustion products are formed for every 1.0 m³ of natural gas burned at standard conditions (16°C and 101.3 kPa). Moreover, 10.33 m³ of theoretical air is required to combust 1 m³ of natural gas at standard conditions. First determine the volume of combustion products from burning the natural gas:

$$\left(0.14 \text{ m}^3/\text{s}\right) \times \left(\frac{11.5 \text{ m}^3 \text{ of product}}{1.0 \text{ m}^3 \text{ of gas}}\right) = 1.61 \text{ m}^3/\text{s}$$

Determine the air required for combustion:

$$\left(0.14 \text{ m}^3/\text{s}\right) \times \left(\frac{10.3 \text{ m}^3 \text{ of air}}{1.0 \text{ m}^3 \text{ of gas}}\right) = 1.45 \text{ m}^3/\text{s}$$

Add the volumes and subtract the air required for combustion:

Flow from paint bake oven = 3.8
Products from combustion = 1.61
Air required for combustion = 1.45

$$\text{Total volume} = 3.8 + 1.61 - 1.45 = 3.96 \text{ m}^3/\text{s}$$

Convert the m³/s calculated under the standard conditions to actual conditions:

$$\left(3.96 \text{ m}^3/\text{s}\right) \times \left(\frac{273°C + 760°C}{273°C}\right) = 14.98 \text{ m}^3/\text{s}$$

Determine the chamber volume:

$$V = \rho \times r \times {}^2L = 3.14 \times 0.75 \text{ m}^2 \times 3.5 \text{ m} = 6.18 \text{ m}^3$$

Calculate the residence time:

$$t = V/Q = (6.18 \text{ m}^3)/(14.98 \text{ m}^3/\text{s}) = 0.41 \text{ s}$$

■ EXAMPLE 16.11. INCINERATOR FUEL REQUIREMENT

Problem: The exhaust from a meat smokehouse contains obnoxious odors and fumes. The company plans to incinerate the 5000-acfm-exhaust stream. What quantity of natural gas is required to raise the waste gas stream from a temperature 90°F to the required temperature of 1200°F? The gross heating value of natural gas is 1059 Btu/scf. Assume no heat losses (USEPA, 1981, p. 3-14).

Given:
  Standard condition (state 1) $T_1 = 60°F$
  Exhaust gas flow rate $V_2 = 5000$ acfm at temperature $T = 90°F$
  Exhaust gas initial temperature $T_2 = 90°F$
  Natural gas gross heating value = 1059 Btu/scf
  Combustion temperature $T_3 = 1200°F$

Solution: Correct the actual waste gas volume (V_a) to standard condition volume (in standard cubic feet per hour). The correction equation is

$$\frac{V_1}{T_1} = \frac{V_2}{T_2}$$

$$V_1 = V_2 \left(\frac{T_1}{T_2} \right) = 5000 \left(\frac{460 + 60}{460 + 90} \right) = 4727 \text{ scfm} = 283{,}620 \text{ scfh}$$

Convert the volumetric flow rate to a mass flow rate by multiplying by the density:

$$\text{Mass Flow Rate} = \text{Volume Rate} \times \text{Density}$$

The standard volume of an ideal gas at 60°F = 379.64 ft³/lb-mol. Assume that the waste gas molecular weight is the same as air, 29 lb/lb-mol:

$$\text{Density} = (\text{Molecular Weight})/\text{Volume} = 29/379.64 = 0.076388 \text{ lb/ft}^3$$

$$\text{Mass Flow Rate} = 4727 \times 0.076388 = 361 \text{ lb/min}$$

Calculate the heat rate by using the ideal gas equation:

$$Q = m \times C_p \times (T_3 - T_2) = 361 \times 0.26 \times (1200 - 90) = 104{,}185 \text{ Btu/min}$$

Determine the heating value of natural gas. For natural gas, 1 scf contains 1059 Btu. Finally, determine the natural gas quantity:

$$W = 104{,}185/1059 = 98 \text{ scfm}$$

16.6 CONDENSATION

Condensation is a process by which volatile gases are removed from the contaminant stream and changed into a liquid. That is, it is a process that reduces a gas or vapor to a liquid. Condensers condense vapors to a liquid phase by either increasing system pressure without a change in temperature or by decreasing the system temperature to its saturation temperature without a pressure change. The common approach is to reduce the temperature of the gas stream, because increasing the pressure of a gas is very expensive (USEPA, 1981, p. 6-1). Condensation is affected by the composition of the contaminant gas stream. When different gases are present in the streams that condense under different conditions, condensation is hindered.

In air pollution control, a condenser can be used in two ways: either for pretreatment to reduce the load problems with other air pollution control equipment or for effectively controlling contaminants in the form of gases and vapors. There are two basic types of condensation equipment—contact and surface condensers. In a contact condenser (which resembles a simple spray scrubber), spraying liquid directly on the vapor stream (see Figure 16.17) cools the vapor. The cooled vapor condenses, and the water and condensate mixture are removed, treated, and disposed of.

A surface condenser is normally a shell-and-tube heat exchanger (see Figure 16.18). It uses a cooling medium of air or water where the vapor to be condensed is separated from the cooling medium by a metal wall. Coolant flows through the tubes, while the vapor is passed over the tubes, condenses on the outside of the tubes, and drains off to storage (USEPA, 1971).

In general, condensers are simple, relatively inexpensive devices that normally use water or air to cool and condense a vapor stream. Condensers are used in a wide range of industrial applications, including petroleum refining, petrochemical manufacturing, basic chemical manufacturing, dry cleaning, and degreasing.

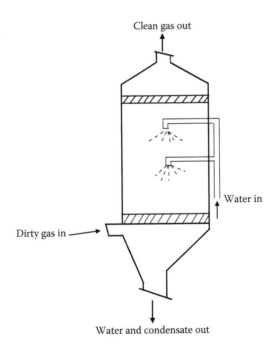

FIGURE 16.17 Contact condenser. (From USEPA, *Control Techniques for Gases and Particulates*, U.S. Environmental Protection Agency, Washington, DC, 1971.)

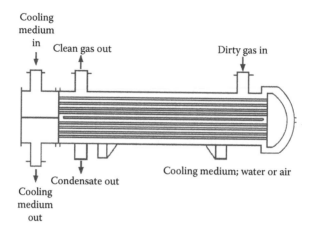

FIGURE 16.18 Surface condenser. (From USEPA, *Control Techniques for Gases and Particulates*, U.S. Environmental Protection Agency, Washington, DC, 1971.)

16.6.1 CONTACT CONDENSER CALCULATIONS

In a contact condenser (see Figure 16.17) the coolant and vapor stream are physically mixed. They leave the condenser as a single exhaust stream. Simplified heat balance calculations are used to estimate the important parameters. The first step in analyzing any heat transfer process is to set up a heat balance relationship. For a condensation system, the heat balance can be expressed as

$$\text{Heat in} = \text{Heat out}$$

$$\left(\begin{array}{c}\text{Heat required to reduce}\\\text{vapors to the dewpoint}\end{array}\right) + \left(\begin{array}{c}\text{Heat required to}\\\text{condense vapors}\end{array}\right) = \left(\begin{array}{c}\text{Heat to be removed}\\\text{by the coolant}\end{array}\right)$$

This heat balance is written in equation form as

$$q = mC_p(T_{G1} - T_{dewpoint}) + mH_v = LC_p(T_{L2} - T_{L1}) \qquad (16.21)$$

where
    q = Heat transfer rate (Btu/hr).
    m = Mass flow rate of vapor (lb/hr).
    C_p = Average specific heat of a gas or liquid (Btu/lb-°F).
    T = Temperature of the streams (subscripts G for gas and L for liquid coolant).
    H_v = Heat of condensation or vaporization (Btu/lb).
    L = Mass flow rate of liquid coolant (lb/hr).

In Equation 16.21, mass flow rate m and inlet temperature T_{G1} of the vapor stream are set by the process exhaust stream. The temperature of the coolant entering the condenser (T_{L1}) is also set. The average specific heats (C_p) of both streams, the heat of condensation (H_v), and the dewpoint temperature can be obtained from chemistry handbooks. Therefore, only the amount of coolant L and its outlet temperature are left to be determined. If either one of these terms is set by process restrictions (e.g., only x pounds an hour of coolant are available or the outlet temperature) then the other term can be solved for directly. Equation 16.21 is applicable for direct contact condensers and should be used only to obtain rough estimates. The equation has a number of limitations:

1. The specific heat (C_p) of a substance is dependent on temperature, and the temperature throughout the condenser is constantly changing.
2. The dewpoint of a substance is dependent on its concentration in the gas phase, and because the mass flow rate is constantly changing (vapors being condensed) the dewpoint temperature is constantly changing.
3. No provision is made for cooling the vapors below the dewpoint. An additional term would have to be added to the left side of the equation to account for this amount of cooling.

16.6.2 SURFACE CONDENSER CALCULATIONS

In a surface condenser or heat exchanger (see Figure 16.18), heat is transferred from the vapor stream to the coolant through a heat exchange surface (USEPA, 1981, p. 6-3). The rate of heat transfer depends on three factors: total cooling surface available, resistance to heat transfer, and mean temperature difference between condensing vapor and coolant. This can be expressed mathematically by

$$q = UA\Delta T_m \qquad (16.22)$$

where

q = Heat transfer rate (Btu/hr).
U = Overall heat transfer coefficient (Btu/°F·ft²·hr).
A = Heat transfer surface area (ft²).
ΔT_m = Mean temperature difference (°F).

The overall heat transfer coefficient U is a measure of the total resistance that heat experiences while being transferred from a hot body to a cold body. In a shell-and-tube condenser, cold water flows through the tubes causing vapor to condense on the outside surface of the tube wall. Heat is transferred from the vapor to the coolant. The ideal situation for heat transfer is when heat is transferred from the vapor to the coolant without any heat loss (heat resistance).

Every time heat moves through a different medium, it encounters a different and additional heat resistance. These heat resistances occur throughout the condensate, through any scale or dirt on the outside of the tube (fouling), through the tube itself, and through the film on the inside of the tube (fouling). Each of these resistances are individual heat transfer coefficients and must be added together to obtain an overall heat transfer coefficient. An estimate of an overall heat transfer coefficient can be used for preliminary calculations. The overall heat transfer coefficients shown in Table 16.3 should be used only for preliminary estimating purposes.

TABLE 16.3
Typical Heat Transfer Coefficients in Tubular Heat Exchangers

Condensing Vapor (Shell Side)	Cooling Liquid	U (Btu/°F·ft²·hr)
Organic solvent vapor with high percent of noncondensable gases	Water	20–60
High-boiling hydrocarbon vapor (vacuum)	Water	20–50
Low-boiling hydrocarbon vapor	Water	80–200
Hydrocarbon vapor and steam	Water	80–100
Steam	Feedwater	400–1000
Naphtha	Water	50–75
Water	Water	200–250

Source: Adapted from USEPA, *APTI Course 415: Control of Gaseous Emissions*, EPA 450/2-81-005, U.S. Environmental Protection Agency Air Pollution Training Institute, Washington, DC, 1981, p. 6-12.

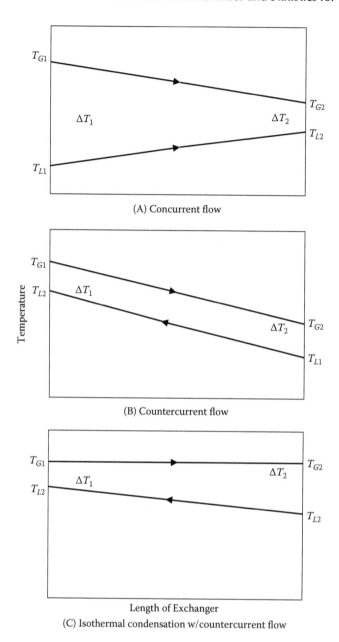

FIGURE 16.19 Temperature profiles in a heat exchanger. (From USEPA, *APTI Course 415: Control of Gaseous Emissions*, EPA 450/2-81-005, U.S. Environmental Protection Agency Air Pollution Training Institute, Washington, DC, 1981, p. 6-13.)

In a surface heat exchanger, the temperature difference between the hot vapor and the coolant usually varies throughout the length of the exchanger. Therefore, a mean temperature difference (ΔT_m) must be used. The log mean temperature difference can be used for the special cases when the flow of both streams is completely co-current, the flow of both streams is completely countercurrent, or the temperature of the fluids remains constant (as is the case in condensing a pure liquid). The temperature profiles for these three conditions are illustrated in Figure 16.19. The log mean temperature for countercurrent flow can be expressed as shown in Equation 16.23.

$$\Delta T_m = \Delta T_{lm} = (\Delta T_1 - \Delta T_2)/\ln(\Delta T_1 \Delta T_2) \tag{16.23}$$

where ΔT_{lm} is the log mean temperature.

The value calculated from Equation 16.23 is used for single-pass heat exchangers or condensers. For multiple-pass exchangers, a correction factor for the log mean temperature must be included. However, no correction factor is needed for the special case of isothermal condensation (no change in temperature) when there is a single component vapor and the gas temperature is equal to the dewpoint temperature. In order to size a condenser, Equation 16.22 must be rearranged to solve for the surface area.

$$A = q/U\Delta T_{lm} \tag{16.24}$$

where

A = Surface area of a shell-and-tube condenser (ft^2).
q = Heat transfer rate (Btu/lb).
U = Overall heat transfer coefficient (Btu/°F·ft^2·hr).
ΔT_{lm} = Log mean temperature difference (°F).

Equation 16.24 is valid only for isothermal condensation of a single component. This implies that the pollutant is a pure vapor stream comprised of only one specific hydrocarbon, such as benzene, and not a mixture of hydrocarbons. Nearly all air pollution control applications involve mulitcomponent mixtures, which complicates the design procedure for a condenser. For preliminary rough estimates of condenser size, the procedure in Example 16.12 for single-component condensation can be used for condensing multicomponents. In choosing a heat transfer coefficient, the smallest value should be used to allow as much over design as possible.

■ EXAMPLE 16.12

Problem: In a rendering plant, tallow is obtained by removing the moisture from animal matter in a cooker. Exhaust gases from the cookers contain essentially steam; however, the entrained vapors are highly odorous and must be controlled. Condensers are normally used to remove most of the moisture prior to incineration, scrubbing, or carbon adsorption (USEPA, 1981, p. 6-15). The exhaust rate from the continuous cooker is 20,000 acfm at 250°F. The exhaust gases are 95% moisture with the remaining portion consisting of air and obnoxious organic vapors. The exhaust steam is sent first to a shell-and-tube condenser to remove the moisture and then to a carbon absorption unit. If the coolant water enters at 60°F and leaves at 120°F, estimate the required surface area of the condenser. The condenser is a horizontal, countercurrent flow system with the bottom few tubes flooded to provide subcooling.

Solution: Compute the pounds of steam condensed per minute:

$$20,000 \text{ acfm} \times 0.95 = 19,000 \text{ acfm steam}$$

From the ideal gas law,

$$P \times V = n \times R \times T$$

$$n = (P \times V)/(R \times T)$$

$$n = (1 \text{ atm})(19,000 \text{ acfm})/[0.73 \text{ atm-ft}^3/\text{lb-mol-°R})(250 + 460)°R] = 36.66 \text{ lb-mol/min}$$

$$m = (36.66 \text{ lb-mole/min})(18 \text{ lb/lb-mole}) = 660 \text{ lb/min of steam to be condensed}$$

Solve the heat balance to determine q for cooling the superheated steam and condensing only:

q = (Heat needed to cool steam to condensation temperature) + (Heat of condensation)

$= mC_p \Delta T + mH_v$

The average specific heat (C_p) of steam at 250°F is roughly 0.45 Btu/lb-°F. The heat of vaporization of steam at 212°F = 970.3 Btu/lb. Substituting into the equation,

q = (660 lb/min)(0.45 Btu/lb-°F)(250 − 212°F) + (660 lb/min)(970.3 Btu/lb)

q = 11,286 Btu/min + 640,398 Btu/min = 651,700 Btu/min

Now use Equation 16.24 to estimate the surface area for this part of the condenser:

$$A = q/U\Delta T_{lm}$$

For a countercurrent condenser, the log mean temperature is given by

$$T_{lm} = \frac{(T_{G1} - T_{L2}) - (T_{G2} - T_{L1})}{\ln\left(\dfrac{T_{G1} - T_{L2}}{T_{G1} - T_{L1}}\right)}$$

Remember that the desuperheater–condenser section is designed using the saturation temperature to calculate the log mean temperature difference.

Gas entering temperature (T_{G1}) = 212°F
Coolant leaving temperature (T_{L2}) = 120°F
Gas leaving temperature (T_{G2}) = 212°F
Coolant entering temperature (T_{L1}) = 60°F

$$T_{lm} = \frac{(212 - 120) - (212 - 60)}{\ln\left(\dfrac{212 - 120}{212 - 60}\right)} = 119.5°F$$

The overall heat transfer coefficient (U) is assumed to be 100 Btu/°F·ft²·hr. Substituting the appropriate values into Equation 16.24:

$$A = \frac{(651,700 \text{ Btu/min})(60 \text{ min/hr})}{(100 \text{ Btu/°F} \cdot \text{ft}^2 \cdot \text{hr})(119.5°F)} = 3272 \text{ ft}^2$$

Estimate the total size of the condenser. Allow for subcooling of the water (212°F − 160°F). 160°F is an assumed safe margin. Refiguring the heat balance for cooling the water:

$$q = UA\Delta T_m$$

where m = 660 lb/min (assuming all the steam is condensed)

q = (660 lb/min)(1 Btu/lb-°F)(212 − 160)°F = 34,320 Btu/min

$$T_{lm} = \frac{(212-120)-(160-60)}{\ln\left(\dfrac{212-120}{160-60}\right)} = 96°F$$

For cooling water with a water coolant, U is assumed to be 200 Btu/°F·ft²·hr.

$$A = \frac{q}{U\ T_{lm}} = \frac{34,320\ \text{Btu/min}}{(200\ \text{Btu/°F} \cdot \text{ft}^2 \cdot \text{hr})(96°F)} = 1.79\ \text{ft}^2, \text{ or } 2\ \text{ft}^2$$

The total area needed is

$$A = 3272 + 2 = 3274\ \text{ft}^2$$

As illustrated by this example, the area for subcooling is very small compared to the area required for condensing.

REFERENCES AND RECOMMENDED READING

AIHA. (1968). *Air Pollution Manual: Control Equipment, Part II*. American Industrial Hygiene Association, Detroit, MI.

Buonicore, A.J. and Davis, W.T., Eds. (1992) *Air Pollution Engineering Manual*. Van Nostrand Reinhold, New York.

Davis, M.L. and Cornwell, D.A. (1991). *Introduction to Environmental Engineering*. McGraw-Hill, New York.

Hesketh, H.E. (1991). *Air Pollution Control: Traditional and Hazardous Pollutants*. Technomic, Lancaster, PA.

Spellman, F.R. (1999). The Science of Air. CRC Press, Boca Raton, FL.

Spellman, F.R. (2008). *The Science of Air: Concepts and Applications*, 2nd ed. CRC Press, Boca Raton, FL.

USEPA. (1971). *Control Techniques for Gases and Particulates*. U.S. Environmental Protection Agency, Washington, DC.

USEPA. (1972). *Wet Scrubber System Study*, NTIS Report PB-213016. U.S. Environmental Protection Agency, Research Triangle Park, NC.

USEPA. (1973). *Air Pollution Engineering Manual*, 2nd ed. U.S. Environmental Protection Agency, Research Triangle Park, NC.

USEPA. (1981). *APTI Course 415: Control of Gaseous Emissions*, EPA 450/2-81-005. U.S. Environmental Protection Agency Air Pollution Training Institute, Research Triangle Park, NC.

USEPA. (1984). *Wet Scrubber Plan Review: Self-Instructional Guidebook*, EPA 450/2-82-020. U.S. Environmental Protection Agency Air Pollution Training Institute, Research Triangle Park, NC.

17 Particulate Emission Control

Fresh air is good if you do not take too much of it; most of the achievements and pleasures of life are in bad air.

—Oliver Wendell Holmes (1809–1894)

17.1 PARTICULATE EMISSION CONTROL BASICS

Particle or particulate matter is defined as tiny particles or liquid droplets suspended in the air that can contain a variety of chemical components. Larger particles are visible as smoke or dust and settle out relatively rapidly. The tiniest particles can be suspended in the air for long periods of time and are the most harmful to human health because they can penetrate deep into the lungs. Some particles are directly emitted into the air. Constituting a major class of air pollutants, particulates have a variety of shapes and sizes, and as either liquid droplets or dry dust, they have a wide range of physical and chemical characteristics. Dry particulates are emitted from a variety of different sources in industry, mining, construction activities, incinerators, and internal combustion engines—from cars, trucks, buses, factories, construction sites, tilled fields, unpaved roads, stone crushing, and wood burning. Dry particulates also come from natural sources—volcanoes, forest fires, pollen, and windstorms. Other particles are formed in the atmosphere by chemical reactions.

When a flowing fluid (engineering and science applications consider both liquid and gaseous states as a fluid) approaches a stationary object such as a metal plate, a fabric thread, or a large water droplet, the fluid flow will diverge around that object. Particles in the fluid (because of inertia) will not follow stream flow exactly but will tend to continue in their original directions. If the particles have enough inertia and are located close enough to the stationary object they will collide with the object, and can be collected by it. This is an important phenomenon and is depicted in Figure 17.1.

17.1.1 Interaction of Particles with Gas

To understand the interaction of particles with the surrounding gas, knowledge of certain aspects of the kinetic theory of gases is necessary. This kinetic theory explains temperature, pressure, mean free path, viscosity, and diffusion in regard to the motion of gas molecules (Hinds, 1986). The theory assumes gases—along with molecules being rigid spheres that travel in straight lines—contain a large number of molecules that are small enough so that the relevant distances between them are discontinuous. Air molecules travel at an average of 1519 ft/s (463 m/s) at standard conditions. Speed decreases with increased molecule weight. As the square root of absolute temperature increases, molecular velocity increases. Thus, temperature is an indication of the kinetic energy of gas molecules. When molecular impact on a surface occurs, pressure develops and is directly related to concentration. Gas viscosity represents the transfer of momentum by randomly moving molecules from a faster moving layer of gas to an adjacent slower moving layer of gas. Viscosity of a gas is independent of pressure but will increase as temperature increases. Finally, diffusion is the transfer of molecular mass without any fluid flow (Hinds, 1986). Diffusion transfer of gas molecules is from a higher to a lower concentration. Movement of gas molecules by diffusion is directly proportional to the concentration gradient, inversely proportional to concentration, and proportional to the square root of absolute temperature.

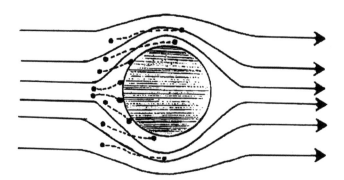

FIGURE 17.1   Particle collection of a stationary object. (Adapted from USEPA, *APTI Course SI:412C: Wet Scrubber Plan Review: Self-Instructional Guidebook*, EPA 450/2-82-020, U.S. Environmental Protection Agency Air Pollution Training Institute, Research Triangle Park, NC, 1984.)

The *mean free path*, kinetic theory's most critical quantity, is the average distance a molecule travels in a gas between collisions with other molecules. The mean free path increases with increasing temperature and decreases with increasing pressure (Hinds, 1986).

The Reynolds number characterizes gas flow, which is a dimensionless index that describes the flow regime. The Reynolds number for a gas is determined by the following equation:

$$\text{Re} = \frac{pU_g D}{\eta} \tag{17.1}$$

where
   Re = Reynolds number.
   p = Gas density (lb/ft^3, kg/m^3).
   U_G = Gas velocity (ft/s, m/s).
   D = Characteristic length (ft, m).
   η = Gas viscosity (lb$_m$/ft·s, kg/m·s).

The Reynolds number helps to determine the flow regime, the application of certain equations, and geometric similarity (Baron and Willeke, 1993). Flow is laminar at low Reynolds numbers, and viscous forces predominate. Inertial forces dominate the flow at high Reynolds numbers when mixing causes the streamlines to disappear.

17.1.2   PARTICULATE COLLECTION

Particles are collected by gravity, centrifugal force, electrostatic force, and by impaction, interception, and diffusion. Impaction occurs when the center of mass of a particle that is diverging from the fluid strikes a stationary object. Interception occurs when the particle's center of mass closely misses the object, but because of its size, the particle strikes the object. Diffusion occurs when small particulates happen to "diffuse" toward the object while passing near it. Particles that strike the object by any of these means are collected if short-range forces (chemical, electrostatic, and so forth) are strong enough to hold them to the surface (Copper and Alley, 1990). Different classes of particulate control equipment include gravity settlers, cyclones, electrostatic precipitators, wet (Venturi) scrubbers, and baghouses (fabric filters). In the following sections we discuss many of the calculations used in particulate emission control operations. Many of the calculations presented are excerpted from USEPA (1984a).

17.2 PARTICULATE SIZE CHARACTERISTICS AND GENERAL CHARACTERISTICS

As we have said, particulate air pollution consists of solid and/or liquid matter in air or gas. Airborne particles come in a range of sizes. From near molecular size, the size of particulate matter ranges upward and is expressed in micrometers, μm—one millionth of a meter. For control purposes, the lower practical limit is about 0.01 μm. Because of the increased difficulty in controlling their emission, particles 3 μm or smaller are defined as *fine particles*. Unless otherwise specified, concentrations of particulate matter are by mass. Liquid particulate matter and particulates formed from liquids (very small particles) are likely to be spherical in shape. In order to express the size of nonspherical (irregular) particles as a diameter, several important relationships are discussed. These include

- Aerodynamic diameter
- Equivalent diameter
- Sedimentation diameter
- Cut diameter
- Dynamic shape factor

17.2.1 AERODYNAMIC DIAMETER

Aerodynamic diameter (d_a) is the diameter of a unit density sphere (i.e., density = 1.00 g/cm³) that would have the same settling velocity as the particle or aerosol in question. Note that because the USEPA is interested in how deeply a particle penetrates into the lung, it is more interested in nominal aerodynamic diameter than in the other methods of assessing size of nonspherical particles. Nevertheless, a particle's nominal aerodynamic diameter is generally similar to its conventional, nominal physical diameter.

17.2.2 EQUIVALENT DIAMETER

Equivalent diameter (d_e) is the diameter of a sphere that has the same value of a physical property as that of the nonspherical particle and is given by

$$d_e = \left(\frac{6V}{\pi} \right)^{1/3} \tag{17.2}$$

where V is volume of the particle.

17.2.3 SEDIMENTATION DIAMETER

Sedimentation diameter or Stokes' diameter (d_s) is the diameter of a sphere having the same terminal settling velocity and density as the particle. In regards to density particles, it is called the reduced sedimentation diameter, making it the same as aerodynamic diameter. The dynamic shape factor accounts for a nonspherical particle settling more slowly than a sphere of the same volume.

17.2.4 CUT DIAMETER

Cut diameter (d_c) is the diameter of particles collected with 50% efficiency (i.e., individual efficiency $\varepsilon_i = 0.5$) and that half penetrate through the collector (i.e., penetration $P_t = 0.5$):

$$Pt_i = 1 - \varepsilon_i \tag{17.3}$$

17.2.5 DYNAMIC SHAPE FACTOR

Dynamic shape factor, χ, is a dimensionless proportionality constant relating the equivalent and sedimentation diameters:

$$\chi = \left(\frac{d_e}{d_s}\right)^2 \tag{17.4}$$

The d_e equals d_s for spherical particles, so χ for spheres is 1.0.

17.3 FLOW REGIME OF PARTICLE MOTION

Air pollution control devices collect solid or liquid particles via the movement of a particle in the gas (fluid) stream. For a particle to be captured, the particle must be subjected to external forces large enough to separate it from the gas stream. Forces acting on a particle include three major forces and other forces (USEPA, 1984a):

- Gravitational force
- Buoyant force
- Drag force
- Other forces (magnetic, inertial, electrostatic, and thermal forces)

The consequence of acting forces on a particle results in the settling velocity for a particle to settle. The settling velocity (also known as the terminal settling velocity) is a constant value of velocity reached when all forces (gravity, drag, buoyancy, etc.) acting on a body are balanced. That is, the sum of all the forces is equal to zero (no acceleration). In order to solve for an unknown particle settling velocity, we must determine the flow regime of particle motion. Once the flow regime has been determined, the settling velocity of a particle can be calculated.

The flow regime can be calculated using the following equation (USEPA, 1984a, p. 3-10):

$$K = d_p \left(\frac{g p_p p_a}{\mu^2}\right)^{0.33} \tag{17.5}$$

where
K = Dimensionless constant that determines the range of the fluid-particle dynamic laws.
d_p = Particle diameter (cm, ft).
g = Gravity force (cm/s^2, ft/s^2).
p_p = Particle density (g/cm^3, lb/ft^3).
p_a = Fluid (gas) density (g/cm^3, lb/ft^3).
μ = Fluid (gas) viscosity (g/cm-s, lb/ft-s).

The K values corresponding to different flow regimes are as follows (USEPA, 1984a, p. 3-10):

- Laminar regime (also known as Stokes' law range): $K < 3.3$
- Transition regime (also known as intermediate law range): $3.3 < K < 43.6$
- Turbulent regime (also known as Newton's law range): $K > 43.6$

The K value determines the appropriate range of the fluid-particle dynamic laws that apply:

- For a laminar regime (Stokes' law range), the terminal settling velocity is

$$v = \frac{g p_p (d_p)^2}{18\mu} \tag{17.6}$$

- For a transition regime (intermediate law range), the terminal settling velocity is

$$v = \frac{0.153 g^{0.71} (d_p)^{1.14} (p_p)^{0.71}}{(\mu)^{0.43} (p_a)^{0.29}} \tag{17.7}$$

- For a turbulent regime (Newton's law range), the terminal settling velocity is

$$v = 1.74 \left(\frac{g d_p p_p}{p_a} \right)^{0.5} \tag{17.8}$$

When particles approach sizes comparable with the mean free path of the fluid molecules (also known as the Knudsen number, Kn), the medium can no longer be regarded as continuous, because particles can fall between the molecules at a faster rate than predicted by aerodynamic theory. Cunningham's correction factor, which includes thermal and momentum accommodation factors based on the Millikan oil-drop studies and is empirically adjusted to fit a wide range of Kn values, is introduced into Stoke's law to allow for this slip (Hesketh, 1991; USEPA, 1984b, p. 58):

$$v = \frac{g p_p (d_p)^2 C_f}{18\mu} \tag{17.9}$$

where
C_f = Cunningham correction factor = $1 + (2Al/d_p)$.
$A = 1.257 + 0.40e^{-1.10 d_p/2l}$.
l = Free path of the fluid molecules (6.53×10^{-6} cm for ambient air).

■ EXAMPLE 17.1

Problem: Calculate the settling velocity of a particle moving in a gas stream. Assume the following information (USEPA, 1984a, p. 3-11):

Given:
d_p = Particle diameter = 45 μm
g = Gravity forces = 980 cm/s^2
p_p = Particle density = 0.899 g/cm^3
p_a = Fluid (gas) density = 0.012 g/cm^3
μ = Fluid (gas) viscosity = 1.82×10^{-4} g/cm-s
C_f = 1.0 (if applicable)

Solution: Use Equation 17.5 calculate the K parameter to determine the proper flow regime:

$$K = \left[\frac{2.5}{25,400 \times 12} \right] \left[\left(\frac{32.2 \times (122.3 - 0.0764) \times 0.0764}{(1.22 \times 10^{-5})^2} \right) \right]^{0.33} = 0.104$$

The result demonstrates that the flow regime is laminar. Now use Equation 17.9 to determine the settling velocity:

$$v = \frac{g p_p (d_p)^2 C_f}{18\mu} = \frac{980 \times 0.899 \times (45 \times 10^{-4})^2 \times 1}{18 \times (1.82 \times 10^{-4})} = 5.38 \text{ cm/sec}$$

■ EXAMPLE 17.2

Problem: Three differently sized fly ash particles settle through air. Calculate the particle terminal settling velocity (assume the particles are spherical) and determine how far each will fall in 30 seconds.

Given:
   Fly ash particle diameters = 0.4, 40, and 400 μm
   Air temperature and pressure = 238°F, 1 atm
   Specific gravity of fly ash = 2.31

Because the Cunningham correction factor is usually applied to particles equal to or smaller than 1 micron, check how it affects the terminal settling velocity for the 0.4-μm particle.

Solution: Determine the value for K for each fly ash particle size settling in air by first calculating the particle density using the specific gravity given:

$$p_p = \text{Specific gravity of fly ash} \times \text{Density of water} = 2.31 \times 62.4 = 144.14 \text{ lb/ft}^3$$

Calculate the air density and air viscosity:

$$p = \frac{PM}{RT} = \frac{1 \times 29}{0.7302 \times (238 + 460)} = 0.0569 \text{ lb/ft}^3$$

$$\mu = 0.021 \text{ cp} = 1.41 \times 10^{-5} \text{ lb/ft} \cdot \text{s}$$

Determine the flow regime (K):

$$K = d_p \left(\frac{g p_p p_a}{\mu^2} \right)^{0.33}$$

For $d_p = 0.4$ μm:

$$K = \left(\frac{0.4}{25,000 \times 12} \right) \times \left[\left(\frac{32.2 \times 144.14 \times 0.0569}{(1.41 \times 10^{-5})^2} \right) \right]^{0.33} = 0.0144$$

where

$$1 \text{ ft} = 25,400 \times 12 \text{ μm} = 304,800 \text{ μm}$$

For $d_p = 40$ μm:

$$K = \left(\frac{40}{25,000 \times 12} \right) \times \left[\left(\frac{32.2 \times 144.14 \times 0.0569}{(1.41 \times 10^{-5})^2} \right) \right]^{0.33} = 1.44$$

For $d_p = 400$ μm:

$$K = \left(\frac{400}{25,000 \times 12}\right) \left[\times \left(\frac{32.2 \times 144.14 \times 0.0569}{\left(1.41 \times 10^{-5}\right)^2}\right)\right]^{0.33} = 14.4$$

Now select appropriate law. The numerical value of K determines the appropriate law:

- $K < 3.3$, Stokes' law range
- $3.3 < K < 43.6$, intermediate law range
- $43.6 < K < 2360$, Newton's law range

- For $d_p = 0.4$ μm, the flow regime is laminar.
- For $d_p = 40$ μm, the flow regime is also laminar.
- For $d_p = 400$ μm, the flow regime is the transition regime.

For $d_p = 0.4$ μm:

$$v = \frac{g p_p (d_p)^2}{18\mu} = \frac{32.2 \times 144.14 \times [(0.4/25,400) \times 12]^2}{18 \times (1.41 \times 10^{-5})} = 3.15 \times 10^{-5} \text{ ft/s}$$

For $d_p = 40$ μm:

$$v = \frac{g p_p (d_p)^2}{18\mu} = \frac{32.2 \times 144.14 \times [(40/25,400) \times 12]^2}{18 \times (1.41 \times 10^{-5})} = 3.15 \times 10^{-5} \text{ ft/s}$$

For $d_p = 400$ μm (use transition regime equation):

$$v = \frac{0.153 g^{0.71} (d_p)^{1.14} (p_p)^{0.71}}{(\mu)^{0.43} (p_a)^{0.29}} = \frac{0.153(32.2)^{0.71} \times [(400/25,400) \times 12]^{1.14} \times (144.14)^{0.71}}{(1.41 \times 10^{-5})^{0.43} \times (0.0569)^{0.29}} = 8.90 \text{ ft/s}$$

For $d_p = 40$ μm:

$$\text{Distance} = \text{Time} \times \text{Velocity} = 30 \times 0.315 = 9.45 \text{ ft}$$

For $d_p = 400$ μm:

$$\text{Distance} = \text{Time} \times \text{Velocity} = 30 \times 8.90 = 267 \text{ ft}$$

For $d_p = 0.4$ μm without Cunningham correction factor:

$$\text{Distance} = \text{Time} \times \text{Velocity} = 30 \times (3.15 \times 10^{-5}) = 94.5 \times 10^{-5} = 94.5 \times 10^{-5} \text{ ft}$$

For $d_p = 0.4$ μm with Cunningham correction factor, the velocity term must be corrected. For our purposes, assume particle diameter = 0.5 micron and temperature = 212°F to find the C_f value. Thus, C_f is approximately equal to 1.446, and

$$\text{Corrected velocity} = v C_f = 3.15 \times 10^{-5}(1.446) = 4.55 \times 10^{-5} \text{ ft/s}$$

$$\text{Distance} = 30(4.55 \times 10^{-5}) = 1.365 \times 10^{-3} \text{ ft}$$

■ **EXAMPLE 17.3**

Problem: Determine the minimum distance downstream from a cement dust-emitting source that will be free of cement deposit. The source is equipped with a cyclone (USEPA, 1984b, p. 59).

Given:
Particle size range of cement dust = 2.5 to 50.0 μm
Specific gravity of the cement dust = 1.96
Wind speed = 3.0 mph .

The cyclone is located 150 ft above ground level. Assume ambient conditions are at 60°F and 1 atm and disregard meteorological aspects:

$$\mu = \text{Air viscosity at } 60°F = 1.22 \times 10^{-5} \text{ lb/ft-s}$$

$$\mu m \ (1 \text{ micron} = 10^{-6} \text{ m}) = 3.048 \times 10^{5} \text{ ft}$$

Solution: A particle diameter of 2.5 μm is used to calculate the minimum distance downstream free of dust since the smallest particle will travel the greatest horizontal distance. Determine the value of K for the appropriate size of the dust, and calculate the particle density (p_p) using the specific gravity given.

$$p_p = \text{Specific gravity of fly ash} \times \text{Density of water} = 1.96 \times 62.4 = 122.3 \text{ lb/ft}^3$$

Calculate the air density (P) by using the modified ideal gas equation:

$$PV = nR_uT = (m/M)R_uT$$

$$P = \text{Mass} \times \text{Volume} = PM/R_uT = (1 \times 29)/[0.73 \times (60 + 460)] = 0.0764 \text{ lb/ft}^3$$

Determine flow regime K:

$$K = d_p \left(\frac{g p_p p_a}{\mu^2} \right)^{0.33}$$

For $d_p = 2.5$ μm:

$$K = d_p \left(\frac{g(p_p - p)p}{\mu^2} \right)^{0.33} = \left[\frac{2.5}{25,400 \times 12} \right] \left[\left(\frac{32.2 \times (122.3 - 0.0764) \times 0.0764}{\left(1.22 \times 10^{-5}\right)^2} \right) \right]^{0.33} = 0.104$$

where 1 ft = 25,400 × 12 μm = 304,800 μm. Now determine which fluid–particle dynamic law applies for the above value of K. Compare the K value of 0.104 with the following range:

• $K < 3.3$, Stokes' law range
• $3.3 < K < 43.6$, intermediate law range
• $43.6 < K < 2360.0$, Newton's law range

The flow is in the Stokes' law range; thus, it is laminar. Now calculate the terminal settling velocity in ft/s. For Stokes' law range, the velocity is

$$v = \frac{gp_p(d_p)^2}{18\mu} = \frac{32.2 \times 122.3 \times [(2.5/25,400) \times 12]^2}{18 \times (1.22 \times 10^{-5})} = 1.21 \times 10^{-3} \text{ ft/s}$$

Calculate the time for settling:

$$t = \frac{\text{Outlet height}}{\text{Terminal velocity}} = \frac{150}{1.21 \times 10^{-3}} = 1.24 \times 10^5 \text{ s} = 34.4 \text{ hr}$$

Calculate the horizontal distance traveled:

$$\text{Distance} = \text{Time for descent} \times \text{Wind speed} = (1.24 \times 10^5)(3.0/3600) = 103.3 \text{ miles}$$

17.4   PARTICULATE EMISSION CONTROL EQUIPMENT CALCULATIONS

Different classes of particulate control equipment include gravity settlers, cyclones, electrostatic precipitators, wet (Venturi) scrubbers (discussed in Chapter 18), and baghouses (fabric filters). In the following section, calculations used for each of the major types of particulate control equipment is discussed.

17.4.1   GRAVITY SETTLERS

Gravity settlers have long been used by industry for removing solid and liquid waste materials from gaseous streams. Simply constructed (see Figures 17.2 and 17.3), a gravity settler is actually nothing more than an enlarged chamber in which the horizontal gas velocity is slowed, allowing particles to settle out by gravity. Gravity settlers have the advantage of having low initial cost and are relatively inexpensive to operate—there's not a lot to go wrong. Although simple in design, gravity settlers require a large space for installation and have relatively low efficiency, especially for removal of small particles (<50 μm).

17.4.1.1   Gravity Settling Chamber Theoretical Collection Efficiency

The theoretical collection efficiency of the gravity-settling chamber is given by (USEPA, 1984a, p. 5-4)

$$\eta = (v_y L)(v_x H) \tag{17.10}$$

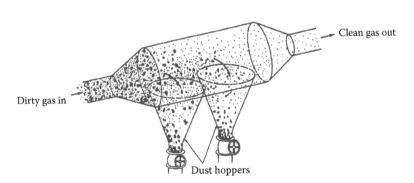

FIGURE 17.2   Gravitational settling chamber. (From USEPA, *Control Techniques for Gases and Particulates*, U.S. Environmental Protection Agency, Washington, DC, 1971.)

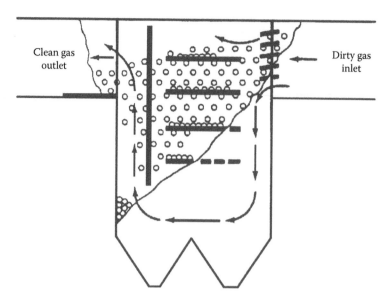

FIGURE 17.3 Baffled gravitational settling chamber. (From USEPA, *Control Techniques for Gases and Particulates*, U.S. Environmental Protection Agency, Washington, DC, 1971.)

where

    η = Fractional efficiency of particle size d_p (one size).
    v_y = Vertical settling velocity.
    v_x = Horizontal gas velocity.
    L = Chamber length.
    H = Chamber height (greatest distance a particle must fall to be collected).

The settling velocity can be calculated from Stokes' law. As a rule of thumb, Stokes' law applies when particle size d_p is less than 100 μm in size. The settling velocity is

$$v_t = \frac{g(d_p)^2(p_p - p_a)}{18\mu}$$

(17.11)

where

    v_t = Settling velocity in Stokes' law range (m/s, ft/s).
    g = Acceleration due to gravity (9.8 m/s², 32.1 ft/s²).
    d_p = Diameter of the particle (μm).
    p_p = Particle density (kg/m³, lb/ft³).
    p_a = Gas density (kg/m³, lb/ft³).
    μ = Gas viscosity (Pa-s, lb/ft-s).
    p_a = N/m², where N = kg-m/s².

Equation 17.11 can be rearranged to determine the minimum particle size that can be collected in the unit with 100% efficiency. The minimum particle size $(d_p)^*$ in μm is given as

$$(d_p)^* = \left[\frac{v_t(18\mu)}{g(p_p - p_a)}\right]^{0.5}$$

(17.12)

Because the density of particle p_p is usually much greater than the density of gas p_a, the quantity $p_p - p_a$ reduces to p_p. The velocity can be written as

$$V = Q/BL \tag{17.13}$$

where
Q = Volumetric flow.
B = Chamber width.
L = Chamber length.

Equation 17.12 is reduced to

$$\left(d_p\right)^* = \left[\frac{18\mu Q}{g p_p BL}\right]^{0.5} \tag{17.14}$$

The efficiency equation can also be expressed as

$$\eta = \left(\frac{g p_p BL N_c}{18\mu Q}\right)\left(d_p\right)^2 \tag{17.15}$$

where N_c is the number of parallel chambers: 1 for simple settling chamber and (N trays + 1) for a Howard settling chamber. Equation 17.15 can be written as

$$\eta = 0.5\left(\frac{g p_p BL N_c}{18\mu Q}\right)\left(d_p\right)^2 \tag{17.16}$$

and the overall efficiency can be calculated using

$$\eta_{tot} = \sum \eta_i w_i \tag{17.17}$$

where
η_{tot} = Overall collection efficiency.
η_i = Fractional efficiency of specific size particle.
w_i = Weight fraction of specific size particle.

When flow is turbulent, Equation 17.18 is used:

$$\eta = \exp\left[-\left(\frac{L v_y}{H v_x}\right)\right] \tag{17.18}$$

When using Equations 17.10 to 17.16, note that Stokes' law does not work for particles greater than 100 μm.

17.4.1.2 Minimum Particle Size
Most gravity settlers are precleaners that remove the relatively large particles (>60 μm) before the gas stream enters a more efficient particulate control device such as a cyclone, baghouse, electrostatic precipitator, or scrubber.

■ **EXAMPLE 17.4**

Problem: A hydrochloric acid mist in air at 25°C is to be collected in a gravity settler. Calculate the smallest mist droplet (spherical in shape) that will be collected by the settler. Stokes' law applies; assume the acid concentration to be uniform through the inlet cross-section of the unit (USEPA, 1984b, p. 61).

Given:
    Dimensions of gravity settler = 30 ft wide, 20 ft high, 50 ft long
    Actual volumetric flow rate of acid gas in air = 50 ft³/s
    Specific gravity of acid = 1.6
    Viscosity of air = 0.0185 cp = 1.243×10^{-5} lb/ft-s
    Density of air = 0.076 lb/ft³

Solution: Calculate the density of the acid mist using the specific gravity given:

Particle density (p_p) = Specific gravity of fly ash × Density of water = $1.6 \times 62.4 = 99.84$ lb/ft³

Calculate the minimum particle diameter both in feet and microns assuming Stokes' law applies. For Stokes' law range,

$$\text{Minimum } d_p = \left(\frac{18\mu Q}{g p_p BL} \right)^{0.5}$$

$$= \left(\frac{18 \times \left(1.243 \times 10^{-5}\right) \times 50}{32.2 \times 99.84 \times 30 \times 50} \right)^2 = 4.82 \times 10^{-5} \text{ ft}$$

$$4.82 \times 10^{-5} \text{ ft} \times \left(3.048 \times 10^5 \text{ } \mu m/ft\right) = 14.7^5 \text{ } \mu m/ft$$

■ **EXAMPLE 17.5**

Problem: A settling chamber is installed in a small heat plant that uses a traveling grate stoker. Determine the overall collection efficiency of the settling chamber given the operating conditions, chamber dimensions, and particle size distribution data (USEPA, 1984b, p. 62).

Given:
    Chamber width = 10.8 ft
    Chamber height = 2.46 ft
    Chamber length = 15.0 ft
    Volumetric flow rate of contaminated air stream = 70.6 scfs
    Flue gas temperature = 446°F
    Flue gas pressure = 1 atm
    Particle concentration = 0.23 grains/scf
    Particle specific gravity = 2.65
    Standard conditions = 32°F, 1 atm

Particle size distribution data of the inlet dust from the traveling grate stoker are shown in Table 17.1. Assume that the actual terminal settling velocity is one-half of the Stokes' law velocity.

Solution: Plot the size efficiency curve for the settling chamber. The size efficiency curve is needed to calculate the outlet concentration for each particle size (range). These outlet concentrations are then used to calculate the overall collection efficiency of the settling chamber. The collection

TABLE 17.1
Particle Size Distribution Data

Particle Size Range (μm)	Average Particle Diameter (μm)	Inlet Grains/scf	wt%
0–20	10	0.0062	2.7
20–30	25	0.0159	6.9
30–40	35	0.0216	9.4
40–50	45	0.0242	10.5
50–60	55	0.0242	10.5
60–70	65	0.0218	9.5
70–80	75	0.0161	7.0
80–94	85	0.0218	9.5
94	94	0.0782	34.0
Total	—	0.2300	100.0

efficiency for a settling chamber can be expressed in terms of the terminal settling velocity, volumetric flow rate of contaminated stream, and chamber dimensions:

$$\eta = v\left(\frac{BL}{Q}\right) = \left(\frac{gp_p(d_p)^2}{18\mu}\right)\left(\frac{BL}{Q}\right)$$

where
η = Fractional collection efficiency.
v = Terminal settling velocity.
B = Chamber width.
L = Chamber length.
Q = Volumetric flow rate of the stream.

Express the collection efficiency in terms of particle diameter d_p. Replace the terminal settling velocity in the above equation with Stokes' law. Because the actual terminal settling velocity is assumed to be one-half of the Stokes' law velocity (according to the given problem statement), the velocity equation becomes

$$v = \frac{gp_p(d_p)^2}{36\mu}$$

$$\eta = \left(\frac{gp_p(d_p)^2}{36\mu}\right)\left(\frac{BL}{Q}\right)$$

Determine the viscosity of the air in lb/ft-s:

Viscosity of air at 446°F = 1.75×10^{-5} lb/ft-s

Determine the particle density in lb/ft³:

$$p_p = 2.65(62.4) = 165.4 \text{ lb/ft}^3$$

TABLE 17.2

**Collection Efficiency
for Each Particle Size**

d_p (µm)	η (%)
94	100
90	92
80	73
60	41
40	18.2
20	4.6
10	1.11

Determine the actual flow rate in acfs. In order to calculate the collection efficiency of the system at the operating conditions, the standard volumetric flow rate of contaminated air of 70.6 scfs is converted to actual volumetric flow of 130 acfs:

$$Q_a = Q_s(T_a/T_s) = 70.6(446 + 460)/(32 + 460) = 130 \text{ acfs}$$

Express the collection efficiency in terms of d_p with d_p in feet. Also express the collection efficiency in terms of d_p with d_p in microns. Use the following equation, substituting values for p_p, g, B, L, μ, and Q in consistent units. Use the conversion factor for feet to microns. To convert d_p from ft^2 to µ^2, d_p is divided by $(304,800)^2$:

$$\eta = \left(\frac{gp_p(d_p)^2}{36\mu}\right)\left(\frac{BL}{Q}\right) = \frac{32.2 \times 165.4 \times 10.8 \times 15 \times (d_p)^2}{36 \times (1.75 \times 10^{-5}) \times 130 \times (304,800)^2} = 1.134 \times 10^{-4}(d_p)^2$$

where d_p is in µm. Calculate the collection efficiency for each particle size. For a particle diameter of 10 µm:

$$\eta = (1.134 \times 10\text{-}4)(d_p)^2 = (1.134 \times 10^{-4})(10)^2 = 1.1 \times 10^{-2} = 1.1\%$$

Table 17.2 provides the collection efficiency for each particle size. The size efficiency curve for the settling chamber is shown in Figure 17.4. Read off the collection efficiency of each particle size from Figure 17.4, and calculate the overall collection efficiency (Table 17.3):

$$\eta = \sum w_i \eta_i$$

$$= (0.027 \times 1.1) + (0.069 \times 7.1) + (0.094 \times 14.0) + (0.105 \times 23.0) + (0.105 \times 34.0)$$

$$+ (0.095 \times 48.0) + (0.070 \times 64.0) + (0.095 \times 83.0) + (0.340 \times 100.0)$$

$$= 59.0\%$$

17.4.2 CYCLONES

Cyclones—the most common dust removal devices used within industry (Strauss, 1975)—remove particles by causing the entire gas stream to flow in a spiral pattern inside a tube. They are the collector of choice for removing particles greater than 10 µm in diameter. By centrifugal force, the larger particles move outward and collide with the wall of the tube. The particles slide down the

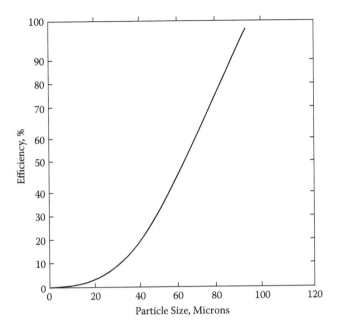

FIGURE 17.4   Size efficiency curve for settling chamber. (Adapted from USEPA, *Control of Gaseous and Particulate Emissions: Self-Instructional Problem Workbook*, EPA 450/2-84-007, U.S. Environmental Protection Agency Air Pollution Training Institute, Research Triangle Park, NC, 1984.)

TABLE 17.3
Data for Calculation of Overall Collection Efficiency

d_p (μm)	Weight Fraction (w_i)	η_i
10	0.027	1.1
25	0.069	7.1
35	0.094	14
45	0.105	23
55	0.105	34
65	0.095	48
75	0.07	64
85	0.095	83
94	0.34	100.11
Total	1	

wall and fall to the bottom of the cone, where they are removed. The cleaned gas flows out the top of the cyclone (see Figure 17.5). Cyclones have low construction costs and have relatively small space requirements for installation. Cyclones are much more efficient in particulate removal than that of settling chambers. However, note that the cyclone's overall particulate collection efficiency is low, especially on particles below 10 μm in size, and they do not handle sticky materials well. The most serious problems encountered with cyclones are with airflow equalization and their tendency to plug (Spellman, 1999). They are often installed as precleaners before more efficient devices such as electrostatic precipitators and baghouses (USEPA, 1984a, p. 6-1). Cyclones have been used successfully at feed and grain mills, cement plants, fertilizer plants, petroleum refineries, and other applications involving large quantities of gas containing relatively large particles.

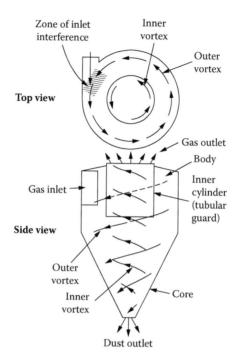

FIGURE 17.5 Convection reverse-flow cyclone. (From USEPA, *Control of Gaseous and Particulate Emissions: Self-Instructional Problem Workbook*, EPA 450/2-84-007, U.S. Environmental Protection Agency Air Pollution Training Institute, Research Triangle Park, NC, 1984.)

17.4.2.1 Factors Affecting Cyclone Performance

The factors that affect cyclone performance include centrifugal force, cut diameter, pressure drop, collection efficiency, and a variety of performance characteristics. Of these parameters, it is the cut diameter that is the most convenient way of defining efficiency for a control device because it gives an idea of the effectiveness for a particle size range. The cut diameter is defined as the size (diameter) of particles collected with 50% efficiency. Note that the cut diameter, $[d_p]_{cut}$, is a characteristic of the control device and should not be confused with the geometric mean particle diameter of the size distribution. A frequently used expression for cut diameter, where collection efficiency is a function of the ratio of particle diameter to cut diameter, is known as the Lapple cut diameter equation (method) and is given below (Copper and Alley, 1986):

$$\left[d_p\right]_{cut} = \left[\frac{9\mu B}{2\pi n_t v_i\left(p_p - p_g\right)}\right] \tag{17.19}$$

where
$[d_p]_{cut}$ = Cut diameter (ft, μm).
μ = Viscosity (lb/s-ft, Pa-s, kg/s-m).
B = Inlet width (ft, m).
n_t = Effective number of turns (5 to 10 for common cyclones).
v_i = Inlet gas velocity (ft/s, m/s).
p_p = Particle density (lb/ft³, kg/m³).
p_g = Gas density (lb/ft³, kg/m³).

TABLE 17.4
Particle Size Distribution Data

Average Particle Size in Range d_p (μm)	wt%
1	3
5	20
10	15
20	20
30	16
40	10
50	6
60	3
>60	7

■ **EXAMPLE 17.6**

Problem: Determine the cut size diameter and overall collection efficiency of a cyclone given the particle size distribution of a dust from a cement kiln (USEPA, 1984b, p. 66).

Given:
   Gas viscosity (μ) = 0.02 cp = 0.02(6.72 × 10⁻⁴) lb/ft-s
   Specific gravity of the particle = 2.9
   Inlet gas velocity to cyclone = 50 ft/s
   Effective number of turns within cyclone = 5
   Cyclone diameter = 10 ft
   Cyclone inlet width = 2.5 ft

Particle size distribution data is shown in Table 17.4.

Solution: Calculate the cut diameter, $[d_p]_{cut}$. The cut diameter is the particle collected at 50% efficiency. For cyclones,

$$[d_p]_{cut} = \left[\frac{9\mu B_c}{2\pi n_t v_i (p_p - p_g)}\right]^{0.5}$$

where
   μ = Gas viscosity (lb/ft-s)
   B_c = Cyclone inlet width (ft)
   n_t = Number of turns
   v_i = Inlet gas velocity (ft/s)
   p_p = Particle density (lb/ft³)
   p_g = Gas density (lb/ft³).

Determine the value of $p_p - p_g$. Because the particle density is much greater than the gas density, $p_p - p_g$ can be assumed to be p_p:

$$p_p - p_g = p_p = 2.9 \times 62.4 = 180.96 \text{ lb/ft}^3$$

Calculate the cut diameter:

TABLE 17.5

Size Efficiency Table

d_p (μm)	w_i	$d_p/[d_p]_{cut}$	η_i (%)	$w_i\eta_i$ (%)
1	0.03	0.1	0	0
5	0.20	0.5	20	4
10	0.15	1	50	7.5
20	0.20	2	80	16
30	0.16	3	90	14.4
40	0.10	4	93	9.3
50	0.06	5	95	5.7
60	0.03	6	98	2.94
>60	0.07	—	100	7

$$[d_p]_{cut} = \left[\frac{9\mu B_c}{2\pi n_t v_i (p_p - p_g)}\right]^{0.5} = \left[\frac{9\times 0.02\times(6.72\times10^{-4})\times 2.5}{2\pi\times5\times50\times180.96}\right] = 3.26\times10^{-5} \text{ ft} = 9.94 \text{ μm}$$

Complete size efficiency table (Table 17.5) using Lapple's method (Lapple, 1951), which provides the collection efficiency as a function of the ratio of particle diameter to cut diameter. Use the equation

$$\eta = \frac{1-(1.0)}{1.0+\left(\dfrac{d_p}{[d_p]_{cut}}\right)^2}$$

to determine overall collection efficiency.

$$\Sigma w_i n_i\ (\%) = 0 + 4 + 7.5 + 16 + 14.4 + 9.3 + 5.7 + 2.94 + 7 = 66.84\%$$

■ **EXAMPLE 17.7**

Problem: An air pollution control officer has been asked to evaluate a permit application to operate a cyclone as the only device on the ABC Stoneworks plant's gravel drier (USEPA, 1984b, p. 68).

Given (design and operating data from permit application):
   Average particle diameter = 7.5 μm
   Total inlet loading to cyclone = 0.5 grains/ft³
   Cyclone diameter = 2.0 ft
   Inlet velocity = 50 ft/s
   Specific gravity of the particle = 2.75
   Number of turns = 4.5 turns
   Operating temperature = 70°F
   Viscosity of air at operating temperature = 1.21 × 10⁻⁵ lb/ft-s
   Conventional cyclone

Air Pollution Control Agency criteria:
   Maximum total outlet loading = 0.1 grains/ft³
   Cyclone efficiency as a function of particle size ratio is provided in Figure 17.6 (Lapple's curve)

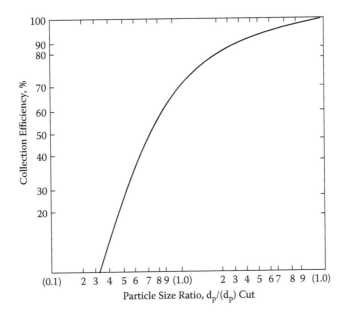

FIGURE 17.6 Lapple's curve: cyclone efficiency vs. particle size ratio. (Adapted from USEPA, *Control of Gaseous and Particulate Emissions: Self-Instructional Problem Workbook*, EPA 450/2-84-007, U.S. Environmental Protection Agency Air Pollution Training Institute, Research Triangle Park, NC, 1984.)

Solution: Determine the collection efficiency of the cyclone. Use Lapple's method, which provides collection efficiency, values from a graph relating efficiency to the ratio of average particle diameter to the cut diameter. Again, the cut diameter is the particle diameter collected at 50% efficiency (see Figure 17.6). Calculate the cut diameter using the Lapple method (Equation 17.19):

$$\left[d_p\right]_{cut} = \left[\frac{9\mu B_c}{2\pi n_t v_i \left(p_p - p_g\right)}\right]^{0.5}$$

Determine the inlet width of the cyclone (B_c). The permit application has established this cyclone as conventional. The inlet width of a conventional cyclone is 1/4 the cyclone diameter:

$$B_c = \text{cyclone diameter}/4 = 2.0/4 = 0.5 \text{ ft}$$

Determine the value of $p_p - p_g$. Because the particle density is much greater than the gas density, $p_p - p_g$ can be assumed to be p_p:

$$p_p - p_g = p_p = 2.75 \times 62.4) = 171.6 \text{ lb/ft}^3$$

Calculate the cut diameter:

$$\left[d_p\right]_{cut} = \left[\frac{9\mu B_c}{2\pi n_t v_i \left(p_p - p_g\right)}\right]^{0.5} = \left[\frac{9 \times \left(1.21 \times 10^{-4}\right) \times 0.5}{2\pi \times 4.5 \times 50 \times 171.6}\right] = 4.57 \text{ μm}$$

Calculate the ratio of average particle diameter to the cut diameter:

$$d_p/[d_p]_{cut} = 7.5/4.57 = 1.64$$

Determine the collection efficiency utilizing Lapple's curve (see Figure 17.6):

$$\eta = 72\%$$

Calculate the required collection efficiency for the approval of the permit:

$$\eta = \left(\frac{\text{Inlet loading} - \text{Outlet loading}}{\text{Inlet loading}} \right) \times 100 = \left(\frac{0.5 - 0.1}{0.5} \right) \times 100 = 80\%$$

Should the permit be approved? Because the collection efficiency of the cyclone is lower than the collection efficiency required by the agency, the permit should not be approved.

17.4.3 ELECTROSTATIC PRECIPITATORS

The electrostatic precipitator (ESP) has been used as an effective particulate-control device for many years. It is usually used to remove small particles from moving gas streams at high collection efficiencies. ESPs are used extensively where dust emissions are less than 10 to 20 µm in size with a predominant portion in the submicron range (USEPA, 1984a, p. 7-1). Widely used in power plants for removing fly ash from the gases prior to discharge, an electrostatic precipitator applies electrical force to separate particles from the gas stream. A high voltage drop is established between electrodes, and particles passing through the resulting electrical field acquire a charge. The charged particles are attracted to and collected on an oppositely charged plate, and the cleaned gas flows through the device. Periodically, the plates are cleaned by rapping to shake off the layer of dust that accumulates, and the dust is collected in hoppers at the bottom of the device (see Figure 17.7). Although electrostatic precipitators have the advantages of low operating costs, capability for operation in high-temperature applications (to 1300°F), low pressure drop, and extremely high particulate (coarse and fine) collection efficiencies; they have the disadvantages of high capital costs and space requirements.

17.4.3.1 Collection Efficiency

The ESP collection efficiency can be expressed by the following two equations (USEPA, 1984a, p. 7-9):

- Migration velocity equation
- Deutsch–Anderson equation

The migration velocity (w) (sometimes referred to as the *drift velocity*) represents the parameter at which a group of dust particles in a specific process can be collected in a precipitator and is

$$w = \frac{d_p E_o E_p}{4\pi\mu} \tag{17.20}$$

where
  w = Migration velocity.
  d_p = Diameter of the particle (µm).
  E_o = Strength of field in which particles are charged, volts per meter (represented by peak voltage).
  E_p = Strength of field in which particles are collected, volts per meter (normally the field close to the collecting plates).
  μ = Viscosity of gas (Pa-s).

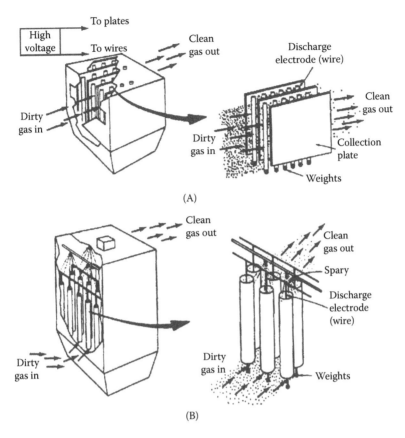

FIGURE 17.7 Typical simple fabric filter baghouse design. (From USEPA, *Control Techniques for Gases and Particulates*, U.S. Environmental Protection Agency, Washington, DC, 1971.)

Migration velocity is quite sensitive to the voltage, because the electric field appears twice in Equation 17.20. Therefore, the precipitator must be designed using the maximum electric field for maximum collection efficiency. The migration velocity is also dependent on particle size; larger particles are collected more easily than smaller ones. Particle migration velocity can also be determined by the following equation:

$$w = \frac{qE_p}{4\pi\mu r} \tag{17.21}$$

where
   w = Migration velocity.
   q = Particle charge (charges).
   E_p = Strength of field in which particles are collected, volts per meter (normally the field close to the collecting plates).
   μ = Viscosity of gas (Pa-s).
   r = Radius of the particle (μm).

The Deutsch–Anderson equation or derivatives thereof are used extensively throughout the precipitator industry. Specifically, it has been used to determine the collection efficiency of the precipitator under ideal conditions. Though scientifically valid, there is a number of operating parameters that can cause the results to be in error by a factor of two of more. Therefore, this equation should be used only for making preliminary estimates of precipitation collection efficiency. The simplest form of the equation is

$$\eta = 1 - \exp(-wA/Q) \qquad (17.22)$$

where

    η = Fractional collection efficiency.
    w = Drift velocity.
    A = Collection surface area of the plates.
    Q = Gas volumetric flow rate.

17.4.3.2 Precipitator Example Calculations

■ **EXAMPLE 17.8**

Problem: A horizontal parallel-plate electrostatic precipitator consists of a single duct 24 ft high and 20 ft deep with an 11-inch plate-to-plate spacing. Given collection efficiency at a gas flow rate of 4200 acfm, determine the bulk velocity of the gas, outlet loading, and drift velocity of this electrostatic precipitator. Also calculate revised collection efficiency if the flow rate and the plate spacing are changed (USEPA, 1984, p. 71).

    Given:
        Inlet loading = 2.82 g/ft^3
        Collection efficiency at 4200 acfm = 88.2%
        Increased (new) flow rate = 5400 acfm
        New plate spacing = 9 in.

Solution: Calculate the bulk flow (throughput) velocity V. The equation for calculating throughput velocity is

$$V = Q/S$$

where

    Q = Gas volumetric flow rate.
    S = Cross-sectional area through which the gas passes.

Thus,

$$V = Q/S = (4200)/[(11/12) \times 24)] = 191 \text{ ft/min} = 3.2 \text{ ft/s}$$

Calculate outlet loading. Remember that

$$\eta \text{ (fractional)} = \frac{\text{Inlet loading} - \text{Outlet loading}}{\text{Inlet loading}}$$

Therefore,

$$\text{Outlet loading} = \text{Inlet loading} \times (1 - \eta) = 2.82 \times (1 - 0.882) = 0.333 \text{ grains/ft}^3$$

Calculate the drift velocity. The drift velocity is the velocity at which the particle migrates toward the collection electrode with the electrostatic precipitator. Recall that Equation (17.22), the Deutsch-Anderson equation, describes the collection efficiency of an electrostatic precipitator:

$$\eta = 1 - \exp\left(\frac{-wA}{Q}\right)$$

where
   η = Fractional collection efficiency.
   A = Collection surface area of the plates.
   Q = Gas volumetric flow rate.
   w = Drift velocity.

Calculate the collection surface area (A). Remember that the particles will be collected on both sides of the plate.

$$A = 2 \times 24 \times 20 = 960 \text{ ft}^2$$

Calculate the drift velocity w. Because the collection efficiency, gas flow rate, and collection surface area are now known, the drift velocity can easily be found from the Deutsch-Anderson equation:

$$\eta = 1 - \exp\left(\frac{-wA}{Q}\right)$$

$$0.882 = 1 - \exp\left(\frac{-960 \times w}{4200}\right)$$

Solving for w,

$$w = 9.36 \text{ ft/min}$$

Calculate the revised collection efficiency when the gas volumetric flow rate is increased to 5400 cfm. Assume the drift velocity remains the same.

$$\eta = 1 - \exp\left(\frac{-wA}{Q}\right) = 1 - \exp\left(\frac{-960 \times 9.36}{5400}\right) = 0.812 = 81.2\%$$

Does the collection efficiency change with changed plate spacing? No. Note that the Deutsch–Anderson equation does not contain a plate-spacing term.

■ **EXAMPLE 17.9**

Problem: Calculate the collection efficiency of an electrostatic precipitator containing three ducts with plates of a given size, assuming a uniform distribution of particles. Also, determine the collection efficiency if one duct is fed 50% of the gas and the other passages 25% each (USEPA, 1984, p. 73).

   Given:
      Volumetric flow rate of contaminated gas = 4000 acfm
      Operating temperature and pressure = 200°C and 1 atm
      Drift velocity = 0.40 ft/s
      Size of the plate = 12 ft long and 12 ft high
      Plate-to-plate spacing = 8 in.

Solution: What is the collection efficiency of the electrostatic precipitator with a uniform volumetric flow rate to each duct? Use the Deutsch–Anderson equation to determine the collection efficiency of the electrostatic precipitator:

$$\eta = 1 - \exp\left(\frac{-wA}{Q}\right)$$

Calculate the collection efficiency of the electrostatic precipitator using the Deutsch–Anderson equation. The volumetric flow rate (Q) through a passage is one-third of the total volumetric flow rate:

$$Q = 4000/(3 \times 60) = 22.22 \text{ acfs}$$

$$\eta = 1 - \exp\left(\frac{-wA}{Q}\right) = 1 - \exp\left(\frac{-288 \times 0.4}{22.22}\right) = 0.9944 = 99.44\%$$

What is the collection efficiency of the electrostatic precipitator if one duct is fed 50% of the gas and the others 25% each? The collection surface area per duct remains the same. First calculate the volumetric flow rate of gas through the duct in acfs:

$$Q = 4000/(2 \times 60) = 33.33 \text{ acfs}$$

and then calculate the collection efficiency of the duct with 50% of gas:

$$\eta_1 = 1 - \exp\left(\frac{(-288 \times 0.04)}{33.33}\right) = 0.9684 = 96.84\%$$

To calculate the collection efficiency of the duct with 25% of gas flow, first calculate the volumetric flow rate of gas through the duct in actual cubic feet per second (acfs):

$$Q = 4000/(4 \times 60) = 16.67 \text{ acfs}$$

and then calculate the collection efficiency (η_2) of the duct with 25% of gas:

$$\eta_1 = 1 - \exp\left(\frac{(-288 \times 0.04)}{16.67}\right) = 0.9990 = 99.90\%$$

Now calculate the new overall collection efficiency. The equation becomes

$$\eta_t = (0.5 \times \eta_1) + (2 \times 0.25 \times \eta_2) = (0.5 \times 96.84) + (2 \times 0.25 \times 99.90) = 98.37\%$$

■ **EXAMPLE 17.10**

Problem: A vendor has compiled fractional efficiency curves describing the performance of a specific model of an electrostatic precipitator. Although these curves are not available, the cut diameter is known. The vendor claims that this particular model will perform with a given efficiency under particular operating conditions. Verify this claim and make certain the effluent loading does not exceed the standard set by USEPA (1984b, p. 75).

Given:
  Plate-to-plate spacing = 10 in.
  Cut diameter = 0.9 μm
  Collection efficiency claimed by the vendor = 98%
  Inlet loading = 14 grains/ft³
  USEPA standard for the outlet loading = 0.2 grains/ft³ (maximum)
  Particle size distribution as given in Table 17.6

TABLE 17.6
Particle Size Distribution

Weight Range	Average Particle Size d_p (µm)
0–20%	3.5
20–40%	8
40–60%	13
60–80%	19
80–100%	45.5

A Deutsch–Anderson type of equation describing the collection efficiency of an electrostatic precipitator is

$$\eta = 1 - \exp(-Kd_p)$$

where
η = Fractional collection efficiency.
K = Empirical constant.
d_p = Particle diameter.

Solution: Is the overall efficiency of the electrostatic precipitator equal to or greater than 98%? Since the weight fractions are given, collection efficiencies of each particle size are needed to calculate the overall collection efficiency. Determine the value of K by using the given cut diameter. Because the cut diameter is known, we can solve the Deutsch–Anderson type equation directly for K:

$$\eta = 1 - \exp(-Kd_p)$$

$$0.5 = 1 - \exp[-K(0.9)]$$

Solving for K, $K = 0.77$. Now calculate the collection efficiency using the Deutsch–Anderson equation where $d_p = 3.5$:

$$\eta = 1 - \exp[(-0.77)(3.5)] = 0.9325$$

Table 17.7 shows the collection efficiency for each particle size. Calculate overall collection efficiency:

$$\eta = \sum w_i \eta_i$$

$$= (0.2 \times 0.9325) + (0.2 \times 0.9979) + (0.2 \times 0.9999) + (0.2 \times 0.9999) + (0.2 \times 0.9999)$$

$$= 0.9861 = 98.61\%$$

TABLE 17.7
Collection Efficiency for Each Particle Size

Weight Fraction w_i	Average Particle Size d_p (µm)	η_i
0.2	3.5	0.9325
0.2	8	0.9979
0.2	13	0.9999
0.2	19	0.9999
0.2	45	0.9999

where

η = Overall collection efficiency.

w_i = Weight fraction of the ith particle size.

η_i = Collection efficiency of the ith particle size.

Is the overall collection efficiency greater than 98%? Yes.

Now determine whether the outlet loading meets the USEPA standard. First calculate the outlet loading in grains/ft³:

$$\text{Outlet loading} = (1.0 - \eta) \times (\text{inlet loading})$$

where η is the fractional efficiency for the above equation.

$$\text{Outlet loading} = (1.0 - 0.9861) \times 14 = 0.195 \text{ grains/ft}^3$$

Is the outlet loading less than 0.2 grains/ft³? Yes. Is the vendor's claim verified? Yes.

17.4.4 BAGHOUSES (FABRIC FILTERS)

"The term *baghouse* encompasses an entire family of collectors with several types of filter bag shapes, cleaning mechanisms, and body shapes" (Heumann, 1997). Baghouse filters (or fabric filters or more properly called tube filters) are the most commonly used air pollution control filtration system. In much the same manner as the common vacuum cleaner, fabric filter material, capable of removing most particles as small as 0.5 μm and substantial quantities of particles as small as 0.1 μm, is formed into cylindrical or envelope bags and suspended in the baghouse (see Figure 17.8). The particulate-laden gas stream is forced through the fabric filter, and as the air passes through the fabric, particulates accumulate on the cloth, providing a cleaned airstream. As particulates build up on the inside surfaces of the bags, the pressure drop increases. Before the pressure drop becomes too severe, the bags must be relieved of some of the particulate layer. The particulates are periodically removed from the cloth by shaking or by reversing the airflow.

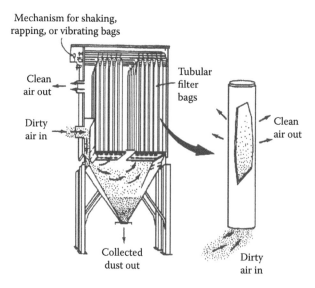

FIGURE 17.8 Typical simple fabric baghouse design. (From USEPA, *Control Techniques for Gases and Particulates*, U.S. Environmental Protection Agency, Washington, DC, 1971.)

Fabric filters are relatively simple to operate, provide high overall collection efficiencies up to 99+%, and are very effective in controlling submicrometer particles, but they do have limitations. These include relatively high capital costs, high maintenance requirements (bag replacement, etc.), high space requirements, and flammability hazards for some dusts.

17.4.4.1 Air-to-Filter (Media) Ratio

The air-to-filter (cloth) ratio is a measurement of the velocity (filtration velocity) of the air passing through the filter media. The ratio definition is the volume of air expressed in cubic feet per minute or cubic meters per hour divided by the filter media area expressed in square feet or square meters. Generally, the smaller a particle diameter, the more difficult it is to filter and thereby requires a lower A/C value. The formula used to express air-to-filter (cloth) ratio (A/C ratio), filtration velocity, or face velocity (terms can be used interchangeably) is

$$v_f = Q/A_c \qquad (17.23)$$

where
v_f = Filtration velocity (ft/min, cm/s).
Q = Volumetric air flow rate (ft³/min, cm³/s).
A_c = Area of cloth filter (ft², cm²).

17.4.4.2 Baghouse Example Calculations

■ EXAMPLE 17.11

Problem: The manufacturer does baghouse sizing. A simple check or estimate of the amount of baghouse cloth needed for a given process flow rate can be computed by using the A/C ratio, Equation 17.23 (USEPA, 1984a, p. 8-34):

$$v_f = Q/A_c \text{ or } A_c = Q/v_f$$

For example, if the process gas exhaust rate is given as 4.72×10^6 cm³/s (10,000 ft³/min) and the filtration velocity is 4 cm/s (A/C is 4:1 (cm³/s)/cm²), the cloth area would be

$$A_c = (4.72 \times 10^6)/4 = 118 \text{ m}^2 \text{ (cloth required)}$$

To determine the number of bags required in the baghouse, use the formula:

$$A_b = \pi \times D \times h$$

where
A_b = Area of bag (m, ft).
D = Bag diameter (m, ft).
h = Bag height (m, ft).

If the bag diameter is 0.203 m (8 in.) and the bag height is 3.66 m (12 ft), the area of each bag is

$$A_b = 3.14 \times 0.203 \times 3.66 = 2.33 \text{ m}^2$$

The calculated number of bags in the baghouse is

$$\text{Number of bags} = 118/2.33 = 51 \text{ bags}$$

■ **EXAMPLE 17.12**

Problem: A proposal to install a pulse jet fabric filter system to clean an air stream containing particulate matter must be evaluated. Select the most appropriate filter bag considering the performance and cost (USEPA, 1984b, p. 84).

Given:

Volumetric flow rate of polluted air stream = 10,000 scfm (60°F, 1 atm)
Operating temperature = 250°F
Concentration of pollutants = 4 grains/ft^3
Average air-to-cloth ratio (A/C ratio) = 2.5 cfm/ft^2 cloth
Collection efficiency requirement = 99%

Table 17.8 lists information given by filter bag manufacturers. Assume no bag has an advantage from the standpoint of durability under the operating conditions for which the bag is to be designed.

Solution: Eliminate from consideration bags that, on the basis of given characteristics, are unsatisfactory. Considering operating temperature and bag tensile strength required for a pulse jet system,

• Bag D is eliminated because its recommended maximum temperature (220°F) is below the operating temperature of 250°F.
• Bag C is also eliminated because a pulse jet fabric filter system requires the tensile strength of the bag to be at least above average.

Determine comparative costs of the remaining bags. Total cost for each bag type is the number of bags × cost per bag. No one type of bag is more durable than the other. Establish the cost per bag. From the information given in Table 17.8, the cost per bag is $26 for bag A and $38 for bag B. Now determine the number of bags (N) for each type. The number of bags required (N) is the total filtering area required divided by the filtering area per bag. Calculate the total filter area (A_t) and the given flow rate to acfm (Q_a):

$$Q_a = (10,000)(250 + 460)/(60 + 460) = 13,654 \text{ acfm}$$

The *A/C* ratio, expressed in cfm/ft^2, is the same as the filtering velocity, which is given above as 2.5 cfm/ft^2 cloth. From the information given in Table 17.8, the filtering velocity is

$$v_f = 2.5 \text{ cfm/ft}^2 = 2.5 \text{ ft/min}$$

TABLE 17.8
Filter Bag Properties

Property	Filter Bag			
	A	**B**	**C**	**D**
Tensile strength	Excellent	Above average	Fair	Excellent
Recommended maximum temperature	260°F	275°F	260°F	220°F
Resistance factor	0.9	1	0.5	0.9
Cost per bag	$26	$38	$10	$20
Standard size	8 in. × 16 ft	10 in. × 16 ft	1 ft × 16 ft	1 ft × 20 ft

Calculate the total filtering cloth area (A_c) from the acfm and filtering velocity determined above:

$$A_c = Q_a/v_f = 13,654/2.5 = 5461.6 \text{ ft}^2$$

Calculate the filtering area per bag. Bags are assumed to be cylindrical; the bag area is $A = \pi \times D \times h$, where $D =$ bag diameter and $h =$ bag length.

For bag A, $A = \pi \times D \times h = \pi \times (8/12) \times 16 = 33.5 \text{ ft}^2$

For bag B, $A = \pi \times D \times h = \pi \times (10/12) \times 16 = 41.9 \text{ ft}^2$

Determine the number of bags required (N):

$$N = (\text{Filtering cloth area of each bag, } A_c)/(\text{Bag area, } A)$$

For bag A, $N = 5461.6/33.5 = 163$

For bag B, $N = 5461.6/41.9 = 130$

Determine the total cost for each bag:

$$\text{Total cost} = N \times \text{Cost per bag}$$

For bag A, total cost = $163 \times 26.00 = \$4238$

For bag B, total cost = $130 \times 38.00 = \$4940$

Select the most appropriate filter bag considering the performance and cost. Because the total cost for bag A is less than bag B, select bag A.

■ EXAMPLE 17.13

Problem: Determine the number of filtering bags required and cleaning frequency for a plant equipped with a fabric filter system. Operating and design data are given below (USEPA, 1984b, p. 86).

Given:
  Volumetric flow rate of the gas stream = 50,000 acfm
  Dust concentration = 5.0 grains/ft³
  Efficiency of the fabric filter system = 98.0%
  Filtration velocity = 10 ft/min
  Diameter of filtering bag = 1.0 ft
  Length of filtering bag = 15 ft

The system is designed to begin cleaning when the pressure drop reaches 8.9 inches of water. The pressure drop is given by

$$\Delta p = 0.2v_f + 5c(v_f)^2 t$$

where
  $\Delta p =$ Pressure drop (in. H_2O).
  $v_f =$ Filtration velocity (ft/min).
  $c =$ Dust concentration (lb/ft³).
  $t =$ Time since the bags were cleaned (min).

Solution: To calculate the number of bags needed (N), we need the total required surface area of the bags and the surface area of each bag. Calculate the total required surface area of the bags A_c in ft²:

$$A_v = Q/v_f$$

where
A_v = Total surface area of the bags.
Q = Volumetric flow rate.
v_f = Filtering velocity.

Thus,

$$A_v = Q/v_t = 50{,}000/10 = 5000 \text{ ft}^2$$

Calculate the surface area of each bag (A) in ft²:

$$A = \pi \times D \times h$$

where
A = Surface area of a bag.
D = Diameter of the bag.
h = Length of the bag.

Thus,

$$A = \pi \times D \times h = \pi \times 1.0 \times 15 = 47.12 \text{ ft}^2$$

Finally, calculate the number of bags N required:

$$N = A_c/A = 5000/47.12 = 106$$

Now calculate the required frequency of cleaning:

$$\Delta p = 0.2v_f + 5c(v_f)^2 t$$

Because Δp is given as 8.0 in. H_2O, the time since the bags were cleaned is calculated by solving the above equation:

$$5.0 \text{ grains/ft}^3 = 0.0007143 \text{ lb/ft}^3 \text{ and } \Delta p = 0.2v_f + 5c(v_f)^2 t$$

$$8.0 = (0.2 \times 10) + (5 \times 0.0007143 \times 10^2)t$$

Thus, $t = 16.8$ min.

■ EXAMPLE 17.14

Problem: An installed baghouse is presently treating a contaminated gas stream. Suddenly some of the bags are broken. We are now requested to estimate the new outlet loading of this baghouse system (USEPA, 1984b, p. 88).

Given:
Operation conditions of the system = 60°F, 1 atm
Inlet loading = 4.0 grains/acf
Outlet loading before bag failure = 0.02 grains/acf

Volumetric flow rate of contaminated gas = 50,000 acfm
Number of compartments = 6
Number of bags per compartment = 100
Bag diameter = 6 inches
Pressure drop across the system = 6 in. H_2O
Number of broken bags = 2 bags

Assume that all the contaminated gas emitted through the broken bags is the same as that passing through the tube sheet thimble.

Solution: Calculate the collection efficiency and penetration before the bag failures. Collection efficiency is a measure of degree of performance of a control device; it specifically refers to degree of removal of pollutants. *Loading* refers to the concentration of pollutants, usually in grains of pollutants per cubic feet of contaminated gas streams. Mathematically, the collection efficiency is defined as

$$\eta = \left(\frac{\text{Inlet loading} - \text{Outlet loading}}{\text{Inlet loading}} \right) \times 100$$

From the preceding equation, the collected amount of pollutants by a control unit is the product of collection efficiency η and inlet loading. The inlet loading minus the amount collected gives the amount discharged to the atmosphere.

Another term used to describe the performance or collection efficiency of control devices is penetration (P_t):

$$P_t = 1 - \eta/100 \qquad \text{(fractional basis)}$$

$$P_t = 100 - \eta \qquad \text{(percent basis)}$$

The effect of bag failure on baghouse efficiency is described by the following:

$$P_{t1} = P_{t2} + P_{tc}$$

$$P_{tc} = 0.582(\Delta p)^{0.5}/\phi$$

$$\phi = Q/(LD^2[T + 460]^{0.5})$$

where
P_{t1} = Penetration after bag failure.
P_{t2} = Penetration before bag failure.
P_{tc} = Penetration correction term, contribution of broken bags to P_{t1}.
Δp = Pressure drop (in. H_2O).
ϕ = Dimensionless parameter.
Q = Volumetric flow rate of contaminated gas (acfm).
L = Number of broken bags.
D = Bag diameter (in.).
T = Temperature (°F).

Calculate collection efficiency η:

η = (Inlet loading − Outlet loading)/(Inlet loading) = (4.0 − 0.02)/(4.0) = 0.005 = 99.5%

Penetration is

$$P_t = 1.0 - \eta = 0.005$$

Now calculate the bag failure parameter (ϕ) (a dimensionless number):

$$\phi = Q/(LD^2(T + 460)^{0.5}) = 50,000/(2)(6)^2(60 + 460)^{0.5} = 30.45$$

Calculate penetration correction P_{tc}, which determines penetration from bag failure:

$$P_{tc} = 0.582(\Delta p)^{0.5}/\phi = (0.582)(6)^{0.5}/30.45 = 0.0468$$

Calculate the penetration and efficiency after the two bags failed. Use the earlier results to calculate P_{t1}:

$$P_{t1} = P_{t2} + P_{tc} = 0.005 + 0.0468 = 0.0518$$

$$\eta^* = 1 - 0.0518 = 0.948$$

Calculate the new outlet loading after the bag failures. Relate inlet loading and new outlet loading to the revised efficiency or penetration:

$$\text{New outlet loading} = (\text{Inlet loading})P_{t1} = (4.0)(0.0518) = 0.207 \text{ grains/acf}$$

■ EXAMPLE 17.15

Problem: A plant emits 50,000 acfm of gas containing a dust loading of 2.0 grains/ft³. A particulate control device is employed for particle capture and the dust captured from the unit is worth $0.01/lb of dust. We are required to determine at what collection efficiency is the cost of power equal to the value of the recovered material. Also determine the pressure drop in inches of H_2O at this condition (USEPA, 1984b, p. 122).

Given:
Overall fan efficiency = 55%
Electric power cost = $0.06/kWh

For this control device, assume that the collection efficiency is related to the system pressure drop Δp through the following equation:

$$\eta = \Delta p/(\Delta p + 5.0)$$

where
η = Fractional collection efficiency.
Δp = Pressure drop (lb/ft²).

Solution: Express the value of the dust collected in terms of collection efficiency η:

$$\text{Amount of dust collected} = (Q)(\text{Inlet loading})(\eta)$$

Note that there are 7000 grains per pound.

$$\text{Value of dust collected} = (50,000 \text{ ft}^3/\text{min}) \times (2 \text{ grains/ft}^3) \times (1/7000 \text{ lb/grain}) \times (0.01 \text{ \$/lb})\eta$$
$$= 0.143\eta \text{ \$/min}$$

Express the value of the dust collected in terms of pressure drop Δp. Recall that

$$\eta = \Delta p/(\Delta p + 5.0)$$

Thus,

$$\text{Value of dust collected} = 0.143[\Delta p/(\Delta p + 5.0)](\$/\text{min})$$

Express the cost of power in terms of pressure drop Δp:

$$\text{Brake horsepower } (Bhp) = Q\Delta p/\eta'$$

where
  Q = Volumetric flow rate.
  Δp = Pressure drop (lb/ft^2).
  η' = Fan efficiency.

$$\text{Cost of power} = \left(p \text{ lb/ft}^2 \right) \times \left(50{,}000 \text{ ft}^3/\text{min}\right) \times (1/44{,}200 \text{ kW-min/ft-lb})$$
$$\times(1/0.55) \times \left(0.06 \text{ \$/kW-h}\right) \times (1/60 \text{ hr/min})$$
$$= 0.002 \ p \ \$/\text{min}$$

Set the cost of power equal to the value of dust collected and solve for Δp in lb/ft^2. This represents breakeven operation. Then, convert this pressure drop to in. H$_2$O. To convert from lb/ft^2 to in. H$_2$O, divide by 5.2.

$$(0.143)\Delta p/(\Delta p + 5) = 0.002\Delta p$$

Solving for Δp,

$$\Delta p = 66.5 \text{ lb/ft}^2 = 12.8 \text{ in. H}_2\text{O}$$

Calculate the collection efficiency using the value of Δp calculated above:

$$\eta = 66.5/(66.5 + 5) = 0.93 = 93.0\%$$

■ EXAMPLE 17.16

Problem: Determine capital, operating, and maintenance costs on an annualized basis for a textile dye and finishing plant (with two coal-fired stoker boilers) where a baghouse is employed for particulate control. Operating, design, and economics factors age given (USEPA, 1984b, p. 123).

Given:
  Exhaust volumetric flow from two boilers = 70,000 acfm
  Overall fan efficiency = 60%
  Operating time = 6240 hr/yr
  Surface area of each bag = 12.0 ft^2
  Bag type = Teflon® felt
  Air-to-cloth ratio = 5.81 acfm/ft^2
  Total pressure drop across the system = 17.16 lb$_f$/ft^2
  Cost of each bag = $75.00
  Installed capital costs = $2.536/acfm
  Cost of electrical energy = $0.03/kWh
  Yearly maintenance cost = $5000 plus yearly cost to replace 25% of the bags

Salvage value = 0
Interest rate (i) = 8%
Lifetime of baghouse (m) = 15 yr

$$\text{Annual installed capital cost (AICC)} = (\text{Installed capital cost})\left(\frac{i(1+i)^m}{(1+i)^m - 1}\right)$$

Solution: Determine the annual maintenance cost by first calculating the number of bags (N):

$$N = Q/(\text{Air-to-cloth ratio} \times A)$$

where
Q = Total exhaust volumetric flow rate.
A = Surface area of a bag.

$$N = Q/(\text{Air-to-cloth ratio} \times A) = (70,000)/(5.81 \times 12) = 1004 \text{ bags}$$

Calculate the annual maintenance cost in dollars per year:

$$\text{Cost of replacing 25\% of bags each year} = 0.25 \times 1004 \times 75.00 = \$18,825$$

$$\text{Annual maintenance cost} = \$5000 + \$18,825 = \$23,825/\text{yr}$$

Determine the annualized installed capital cost (AICC) by first calculating the installed capital cost in dollars:

$$\text{Installed capital cost} = (Q)(\$2.536/\text{acfm}) = 70,000 \times 2.536 = \$177,520$$

Calculate the AICC using the equation given above:

$$\text{AICC} = (\text{Installed capital cost})\left(\frac{i(1+i)^m}{(1+i)^m - 1}\right)$$

$$= (177,520)\left(\frac{0.08(1+0.08)^{15}}{(1+0.08)^{15} - 1}\right)$$

$$= \$20,740/\text{yr}$$

Calculate the operating cost in dollars per year:

$$\text{Operating cost} = Q\Delta p(\text{Operating time})(0.03/\text{kWh/E})$$

Because 1 ft-lb/s = 0.0013558 kW,

$$\text{Operating cost} = [(70,000/60)(17.16)(6240)(0.03)(0.0012558)]/0.6 = \$8470/\text{yr}$$

Calculate the total annualized cost in dollars per year:

$$\text{Total annualized cost} = (\text{Maintenance cost}) + \text{AICC} + (\text{Operating cost})$$

$$= 23,825 + 20,740 + 8470 = \$53,035/\text{yr}$$

REFERENCES AND RECOMMENDED READING

Baron, P.A. and Willeke, K. (1993). Gas and particle motion, in Willeke, K. and Baron, P.A., Eds., *Aerosol Measurement: Principles, Techniques and Applications*. Van Nostrand Reinhold, New York.

Cooper, C.D. and Alley, F.C. (1986). *Air Pollution Control*. Waveland Press, Philadelphia, PA.

Hesketh, H.E. (1991). *Air Pollution Control: Traditional and Hazardous Pollutants*. Technomic, Lancaster, PA.

Heumann, W.L. (1997). *Industrial Air Pollution Control Systems*. McGraw-Hill, New York.

Hinds, W.C. (1986). *Aerosol Technology: Properties, Behavior, and Measurement of Airborne Particulates*. John Wiley & Sons, New York.

Lapple, C.E. (1951). *Fluid and Particle Mechanics*. University of Delaware, Newark.

Spellman, F.R. (1999). *The Science of Air: Concepts and Applications*. Technomic, Lancaster, PA.

Spellman, F.R. (2008). *The Science of Air: Concepts and Applications*, 2nd ed. CRC Press, Boca Raton, FL.

Strauss, W. (1975). *Industrial Gas Cleaning*, 2nd ed. Pergamon Press, Oxford, U.K.

USEPA. (1969). *Control Techniques for Particulate Air Pollutants*. U.S. Environmental Protection Agency, Washington, DC.

USEPA. (1971). *Control Techniques for Gases and Particulates*. U.S. Environmental Protection Agency, Washington, DC.

USEPA. (1984a). *APTI Course 413: Control of Particulate Emissions*, EPA 450/2-80-066. U.S. Environmental Protection Agency Air Pollution Training Institute, Research Triangle Park, NC.

USEPA. (1984b). *Control of Gaseous and Particulate Emissions: Self-Instructional Problem Workbook*, EPA 450/2-84-007. U.S. Environmental Protection Agency Air Pollution Training Institute, Research Triangle Park, NC.

USEPA. (1984c). *APTI Course SI:412C: Wet Scrubber Plan Review: Self-Instructional Guidebook*, EPA 450/2-82-020. U.S. Environmental Protection Agency Air Pollution Training Institute, Research Triangle Park, NC.

18 Wet Scrubbers for Emission Control

I durst not laugh for fear of opening my lips and receiving the bad air.

— William Shakespeare (*Julius Caesar*)

18.1 INTRODUCTION

How do scrubbers work? To answer this question we need only look to a critical part of Earth's natural pollution control system—that is, the cleaning of the lower atmosphere by rain. Obviously, this is most evident by the freshness of the air following a rainstorm. The simplicity of spraying water into a gas stream to remove a relatively high percentage of contaminants has contributed to scrubbers having been used within industry extensively since the early 1900s. Heumann and Subramania (1997) pointed out that "most pollution control problems are solved by the selection of equipment based upon two simple questions … (1) Will the equipment meet the pollution control requirements? And (2) which selection will cost the least?"

How are scrubbing systems capabilities evaluated? They are evaluated based on empirical relationships, theoretical models, and pilot-scale test data. Two important parameters in the design and operation of wet scrubbing systems that are a function of the process being controlled are dust properties and exhaust gas characteristics. Particle size distribution is the most critical parameter in choosing the most effective scrubber design and determining the overall collection efficiency.

In operation, scrubbers are considered universal control devices because they can control either or both particulate and gaseous contaminants. There are numerous types of scrubbers, including wet scrubbers, wet–dry scrubbers, and dry–dry scrubbers. Scrubbers use chemicals to accomplish contaminant removal; the gaseous contaminants are absorbed or converted to particles and then wasted or removed from the stream. In this chapter, our focus is on the calculations used for wet scrubber systems. Much of the information is excerpted from Spellman (1999, 2008) and USEPA (1984a,b,c).

18.1.1 WET SCRUBBERS

Wet scrubbers (or collectors) have found widespread use in cleaning contaminated gas streams (e.g., acid mists, foundry dust emissions, furnace fumes) because of their ability to effectively remove particulate and gaseous contaminants. Wet scrubbers vary in complexity from simple spray chambers that remove coarse particles to high-efficiency systems (Venturi types) that remove fine particles. Whichever system is used, its operation employs the same basic principles of inertial impingement or impaction and interception of dust particles by droplets of water. The larger, heavier water droplets are easily separated from the gas by gravity. The solid particles can then be independently separated from the water, or the water can be otherwise treated before reuse or discharge. Increasing either the gas velocity or the liquid droplet velocity in a scrubber increases the efficiency because of the greater number of collisions per unit time. For the ultimate in wet scrubbing, where high collection efficiency is desired, Venturi scrubbers are used. They are primarily used to control particulate matter (PM), including PM less than or equal to

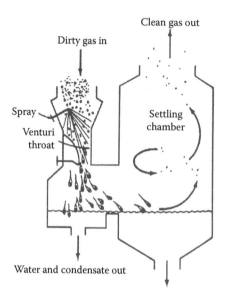

FIGURE 18.1   Venturi wet scrubber. (From USEPA, *Control Techniques for Gases and Particulates*, U.S. Environmental Protection Agency, Washington, DC, 1971.)

10 micrometers (μm) in aerodynamic diameter (PM_{10}), and PM less than or equal to 2.5 μm in aerodynamic diameter ($PM_{2.5}$). Though capable of some incidental control of volatile organic compounds (VOCs), generally Venturi scrubbers are limited to controlling PM and high solubility gases (USEPA, 1992, 1996). Venturi scrubbers operate at extremely high gas and liquid velocities with a very high pressure drop across the Venturi throat. Venturi scrubbers such as the one shown in Figure 18.1 are most efficient for removing particulate matter in the size range of 0.5 to 5 μm, which makes them especially effective for the removal of submicron particulates associated with smoke and fumes.

The Venturi scrubber shown in Figure 18.1 accelerates the waste gas stream to atomize the scrubbing liquid and to improve gas–liquid contact. In a Venturi scrubber, a "throat" section is built into the duct that forces the gas stream to accelerate as the duct narrows and then expands. As the gas enters the Venturi throat, both gas velocity and turbulence increase. Depending upon the scrubber design, the scrubbing liquid is sprayed into the gas stream before the gas encounters the Venturi throat, or in the throat, or upwards against the gas flow in the throat. The scrubbing liquid is then atomized into small droplets by the turbulence in the throat and droplet-particle interaction is increased. Some designs use supplemental hydraulically or pneumatically atomized sprays to augment droplet creation. The disadvantage of these designs is that clean liquid feed is required to avoid clogging (Corbitt, 1990; USEPA, 1998). After the throat section, the mixture decelerates, and further impacts occur causing the droplets to agglomerate. Once the particles have been captured by the liquid, the wetted PM and excess liquid droplets are separated from the gas stream by an entrainment section which usually consists of a cyclonic separator and/or a mist separator (Corbitt, 1990; USEPA, 1998).

Current designs for Venturi scrubbers generally use the vertical downflow of gas through the Venturi throat and incorporate three features: (1) a wet approach, or flooded-wall entry section, to avoid a dust buildup at a wet–dry junction; (2) an adjustable Venturi throat to provide for adjustment of the gas velocity and the pressure drop; and (3) a flooded elbow that is located below the Venturi and ahead of the entrainment separator to reduce wear by abrasive particles. The Venturi throat is sometimes fitted with a refractory lignin to resist abrasion by dust particles (Perry and Green, 1984).

Wet scrubbers have relatively small space requirements, have low capital costs, and can handle high-temperature, high-humidity gas streams; however, their power and maintenance costs are relatively high, they may create water disposal problems, their corrosion problems are more severe than dry systems, and the final product they produce is collected wet (Cooper and Alley, 1994; Perry and Green, 1984; Spellman, 2008).

18.2 WET SCRUBBER COLLECTION MECHANISMS AND EFFICIENCY (PARTICULATES)

As mentioned, wet scrubbers capture relatively small dust particles with large liquid droplets. This is accomplished by generating easily collected large particles by combining liquid droplets with relatively small dust particles. In this process the dust particles are grown into larger particles by several methods. These methods include impaction, diffusion, direct interception, electrostatic attraction, condensation, centrifugal force, and gravity.

18.2.1 COLLECTION EFFICIENCY

Collection efficiency is commonly expressed in terms of penetration, which is defined as the fraction of particles in the exhaust system that passes through the scrubber uncollected. Simply, penetration is the opposite of the fraction of particles collected. It is expressed as (USEPA, 1984c, p. 9-3)

$$P_t = 1 - \eta \tag{18.1}$$

where
  P_t = Penetration.
  η = Collection efficiency (expressed as a fraction).

Wet scrubbers usually have an efficiency curve that fits the relationship of:

$$\eta = 1 - e^{-f(system)} \tag{18.2}$$

where
  η = Collection efficiency.
  e = Exponential function.
  $f(system)$ = Some function of the scrubbing system variables.

Substituting for efficiency, penetration can be expressed as

$$P_t = 1 - \eta = 1 - (1 - e^{-f(system)}) = e^{-f(system)}$$

18.2.2 IMPACTION

In a wet scrubbing system, particulate matter tends to follow the streamlines of the exhaust system. When liquid droplets are introduced into the exhaust stream, however, particulate matter cannot always follow these streamlines as they diverge around the droplet. Instead, because of the particle's mass, it breaks away from the streamlines and impacts on the droplet. Where gas stream velocity exceeds 0.3 m/s (1 ft/s), impaction is the predominant collection mechanism. Most scrubbers do operate with gas stream velocities well above 0.3 m/s, allowing particles having diameters greater than 1.0 μm to be collected by this mechanism (USEPA, 1984c, p. 1-4). Impaction increases as the velocity of the particles in the exhaust stream increases relative to the liquid droplet's velocity. In addition, as the size of the liquid droplet decreases impaction also increases.

For impaction, after the design, the key parameter is known as the impaction parameter Ψ and is expressed by (USEPA, 1984a, p. 9-5)

$$\Psi = \frac{C_f p_p v(d_p)^2}{18 d_d \mu} \tag{18.3}$$

where

p_p = Particle density.
v = Gas velocity at Venturi throat (ft/s).
d_p = Particle diameter (ft).
d_d = Droplet diameter (ft).
μ = Gas viscosity (lb/ft-s).
C_f = Cunningham correction factor.

The collection efficiency associated with this impaction effect is expressed as (USEPA, 1984a, p. 9-7)

$$\eta_{impaction} = f(\Psi)$$

18.2.3 INTERCEPTION

If a small particle is moving around an obstruction (e.g., water droplet) in the flow stream, it may come into contact with that object due to the particle's physical size. This interception of particles on the collector usually occurs on the sides before reaching the top or bottom. That is, because the center of a particle follows the streamlines around the droplet, collision occurs if the distance between the particle and droplet is less than the radius of the particle. Collection of particles by interception results in an increase in overall collection efficiency. This effect is characterized by the separation number, which is the ratio of the particle diameter to the droplet diameter and is expressed as (USEPA, 1984a, p. 9-8)

$$d_p/d_d$$

where

d_p = Particle diameter (ft).
d_d = Droplet diameter (ft).

The collection efficiency associated with this interception effect as a function is d_p and d_d, or

$$\eta_{impaction} = f(d_p/d_d)$$

18.2.4 DIFFUSION

Very small particles with aerodynamic particle diameters of less than 0.1 μm are primarily collected by Brownian diffusion as they have little inertial impaction (bumping) due to their small mass, and interception is limited by their reduced physical size. This bumping causes them to first move one way and then another in a random manner (diffused) through the gas. The Brownian diffusion process leading to particle capture is most often described by a dimensionless parameter called the Peclet number (Pe) (USEPA, 1984a, p. 9-8):

$$Pe = \frac{3\pi\mu v d_p d_d}{C_f k_B T} \tag{18.4}$$

where
  Pe = Peclet number.
  μ = Gas viscosity.
  v = Gas stream velocity.
  d_p = Particle diameter.
  d_d = Liquid droplet diameter.
  C_f = Cunningham correction factor.
  k_B = Boltzmann's constant.
  T = Temperature of gas stream.

Equation 18.4 shows that, as temperature increases, Pe decreases; that is, as temperature increases, gas molecules move around faster than at lower temperatures. This action leads to increased bombing of the small particles, increased random motion, and increased collection efficiency by this mechanism. Collection efficiency by the diffusion process is generally expressed as

$$\eta_{diffusion} = f(1/Pe) \tag{18.5}$$

Equation 18.5 shows that as the Peclet number decreases collection efficiency by diffusion increases.

18.2.5 CALCULATION OF VENTURI SCRUBBER EFFICIENCY

Several models are available for the calculation of Venturi particle collection efficiency (USEPA, 1984a, p. 9-1; USEPA, 1984c, p. 9-1):

- Johnstone equation
- Infinite throat model
- Cut power method
- Contact power theory
- Pressure drop

18.2.5.1 Johnstone Equation

The collection efficiency for liquid Venturi scrubbers, considering only the predominant mechanism of inertial impaction, is often determined with an equation from Johnstone. The Johnstone equation is given as

$$\eta = 1 - \exp\left[-k \left(\frac{Q_L}{Q_G} \right) \sqrt{\Psi} \right]$$

where
  η = Fractional collection efficiency.
  k = Correlation coefficient, the value of which depends on the system geometry and operating conditions (typically 0.1 to 0.2 acf/gal).
  Q_L/Q_G = Liquid-to-gas ratio (gal/1000 acf).

Ψ is an internal impaction parameter given by

$$\frac{C_f p_p v (d_p)^2}{18 d_d \mu}$$

where

C_f = Cunningham correction factor.
p_p = Particle density.
v = Gas velocity at Venturi throat (ft/s).
d_p = Particle diameter (ft).
d_d = Droplet diameter (ft).
μ = Gas viscosity (lb/ft-s).

18.2.5.2 Infinite Throat Model

Another method for predicting particle collection efficiency in a Venturi scrubber is the infinite throat model (Yung et al., 1977). This model is a refined version of the Calvert correlation given in Calvert et al. (1972). The equations presented in the infinite throat model assume that the water in the throat section of the Venturi captures all particles. Two studies found that this method correlated very well with actual Venturi scrubber operating data (Calvert et al., 1972; Yung et al., 1977). The equations listed in the model can be used to predict the penetration (P_t) for one particle size or for the overall penetration (P_t^*), which is obtained by integrating over the entire particle-size distribution. The equations are provided below (USEPA, 1984c, p. 9-4):

$$\ln P_t(d_p) = -B\left(\frac{4K_{po} + 4.2 - 5.02(K_{po})^{0.5}\left(1 + \frac{0.7}{K_{po}}\right)\tan^{-1}\left(\frac{K_{po}}{0.7}\right)^{0.5}}{K_{po} + 0.7}\right) \tag{18.6}$$

where

$P_t(d_p)$ = Penetration for one particle size.
B = Parameter characterizing the liquid-to-gas ratio (dimensionless).
K_{po} = Inertial parameter at throat entrance (dimensionless).

Note: Equation 18.6 was developed assuming that the Venturi scrubber has an infinite-sized throat length (l). This is valid only when l in the following equation is greater than 2.0:

$$l = \frac{3l_t C_D p_g}{2d_d p_l}$$

where

l = Throat length parameter (dimensionless).
l_t = Venturi throat length (cm).
C_D = Drag coefficient for the liquid at the throat entrance (dimensionless).
p_g = Gas density (g/cm³).
d_d = Droplet diameter (cm).
p_l = Liquid density (g/cm³).

The following equation is known as the Nukiyama–Tanasawa equation:

$$d_d = \frac{50}{v_{gt}} + 91.8(L/G)^{1.5} \tag{18.7}$$

where

d_d = Droplet diameter (cm).
v_{gt} = Gas velocity in the throat (cm/s).
L/G = Liquid-to-gas ratio (dimensionless).

The parameter characterizing the liquid-to-gas ratio (B) can be calculated using Equation 18.8:

$$B = (L/G)\frac{p_l}{p_g C_d} \tag{18.8}$$

where
  B = Parameter characterizing liquid-to-gas (dimensionless).
  L/G = Liquid-to-gas ratio (dimensionless).
  p_l = Liquid density (g/cm^3).
  p_g = Gas density (g/cm^3).
  C_D = Drag coefficient for the liquid at the throat entrance (dimensionless).

The inertial parameter at the throat entrance is calculated in Equation 18.9:

$$K_{po} = \frac{(d_p)^2 v_{gt}}{9\mu_g d_d} \tag{18.9}$$

where
  K_{po} = Inertial parameter at throat entrance (dimensionless).
  d_p = Particle aerodynamic resistance diameter (cm).
  v_{gt} = Gas velocity in the throat (cm/s).
  μ_g = Gas viscosity (g/s-cm).
  d_d = Droplet diameter (cm).

$$K_{pg} = \frac{(d_{pg})^2 v_{gt}}{9\mu_g d_d} \tag{18.10}$$

where
  K_{pg} = Inertial parameter for mass-median diameter (dimensionless).
  d_{pg} = Particle aerodynamic geometric mean diameter (cm).
  v_{gt} = Gas velocity in the throat (cm/s).
  μ_g = Gas viscosity (g/s-cm).
  d_d = Droplet diameter (cm).

The value for C_D is calculated using Equation 18.11:

$$C_D = 0.22 + \left(\frac{24}{N_{\text{Re}o}}\right)\left(1 + 0.15(N_{\text{Re}o})^{0.6}\right) \tag{18.11}$$

where
  C_D = Drag coefficient for the liquid at the throat entrance (dimensionless).
  $N_{\text{Re}o}$ = Reynolds number for the liquid droplet at the throat inlet (dimensionless).

The Reynolds number is determined in Equation 18.12:

$$N_{\text{Re}o} = \frac{v_{gt} d_d}{v_g} \tag{18.12}$$

where

N_{Reo} = Reynolds number for the liquid droplet at the throat inlet (dimensionless).
v_{gt} = Gas velocity in the throat (cm/s).
d_d = Droplet diameter (cm).
v_g = Gas kinematic viscosity (cm²/s).

$$d_{pg} = d_{ps}(C_f \times p_p)^{0.5} \tag{18.13}$$

where

d_{pg} = Particle aerodynamic geometric mean diameter (μmA, where A has units of (g/cm³)$^{0.5}$).
d_{ps} = Particle physical or Stokes' diameter (μm).
C_f = Cunningham slip correction factor (dimensionless).
p_p = Particle density (g/cm³).

The Cunningham slip correction factor (C_f) can be found by solving Equation 18.14:

$$C_f = 1 + \frac{\left(6.21 \times 10^{-4}\right) T}{d_{pg}} \tag{18.14}$$

where

C_f = Cunningham slip correction factor (dimensionless).
T = Absolute temperature (K).
d_{ps} = Particle physical or Stokes' diameter (μm).

Example 18.1 illustrates how to use the infinite throat model to predict the performance of a Venturi scrubber. When using the equations given in the model, make sure that the units for each equation are consistent.

■ **EXAMPLE 18.1**

Problem: Cheeps Disposal, Inc., is planning to install a hazardous-waste incinerator that will burn both liquid and solid waste materials. The exhaust gas from the incinerator will pass through a quench spray and then into a Venturi scrubber and finally though a packed bed scrubber. Caustic will be added to the scrubbing liquor to remove any HCl from the flue gas and to control the pH of the scrubbing liquor. The uncontrolled particulate emissions leaving the incinerator are estimated to be 1100 kg/hr (maximum average). The local air pollution regulation states that particulate emissions must not exceed 10 kg/hr. Using the following data, estimate the particulate collection efficiency of the Venturi scrubber (USEPA, 1984c, p. 9-8).

Given:
d_{ps} = Mass-median particle size (physical) = 9.0 μm
σ_{gm} = Geometric standard deviation = 2.5
p_p = Particle density = 1.9 g/cm³
μ_g = Gas viscosity = 2.0 × 10^{-4} g/cm-s
v_g = Gas kinematic viscosity = 0.2 cm²/s
p_g = Gas density = 1.0 kg/m³
Q_G = Gas flow rate = 15 m³/s
v_{gt} = Gas velocity in Venturi throat = 9000 cm/s
T_g = Gas temperature (in Venturi) = 80°C
T_l = Water temperature = 30°C

p_l = Liquid density = 1000 kg/m³
Q_L = Liquid flow rate = 0.014 m³/s
L/G = Liquid-to-gas ratio = 0.0009 L/m³

Solution: Calculate the Cunningham slip correction factor. Mass-median particle size (physical) d_{ps} is 9.0 μm. Since the particle aerodynamic geometric mean diameter (d_{pg}) is not known, we must use Equation 18.13 to calculate d_{pg} and Equation 18.14 to calculate the Cunningham slip correction factor (C_f). From Equation (18.14),

$$C_f = 1 + \frac{\left(6.21 \times 10^{-4}\right)T}{d_{pg}} = 1 + \frac{\left(6.21 \times 10^{-4}\right)(273 + 80)}{9} = 1.024$$

From Equation (18.13),

$$d_{pg} = d_{ps}(C_f \times p_p)0.5 = 9 \ \mu m(1.024 \times 1.9 \ g/cm^3)^{0.5} = 12.6 \ \mu mA = 12.6 \times 10^{-4} \ cmA$$

Note: This step would not have been required if the particle diameter had been given as the aerodynamic geometric mean diameter d_{pg} and expressed in units of μmA.

Now calculate droplet diameter d_d from Equation 18.7 (Nukiyama and Tanasawa equation):

$$d_d = \frac{50}{v_{gr}} + 91(L/G)^{1.5}$$

where
d_d = Droplet diameter (cm).
v_{gr} = Gas velocity in the throat (cm/s).
L/G = Liquid-to-gas ratio (dimensionless).

$$d_d = \frac{50}{9000 \ cm/s} + 91(0.0009)^{1.5} = 0.00080 \ cm$$

Calculate the inertial parameter for mass-median diameter K_{pg} using Equation 18.10:

$$K_{pg} = \frac{\left(d_{pg}\right)^2 v_{gr}}{9\mu_g d_d}$$

where
K_{pg} = Inertial parameter for mass-median diameter (dimensionless).
d_{pg} = Particle aerodynamic geometric mean diameter (cmA).
v_{gr} = Gas velocity in the throat (cm/s).
μ_g = Gas viscosity (g/s-cm).
d_d = Droplet diameter (cm).

$$K_{pg} = \frac{\left(12.6 \times 10^{-4} \ cm\right)^2 (9000 \ cm/s)}{9\left(2.0 \times 10^{-4} \ g/cm\text{-}s\right)^2 (0.008 \ cm)} = 992$$

Calculate the Reynolds number, N_{Reo}, using Equation 18.12:

$$N_{Reo} = \frac{v_{gr}d_d}{v_g}$$

where

N_{Reo} = Reynolds number for the liquid droplet at the throat inlet (dimensionless).
v_{gr} = Gas velocity in the throat (cm/s).
d_d = Droplet diameter (cm).
v_g = Gas kinematic viscosity, (cm²/s).

$$N_{Reo} = \frac{v_{gr}d_d}{v_g} = \frac{(9000 \text{ cm/s})(0.008 \text{ cm})}{0.2 \text{ cm}^2/s} = 360$$

Calculate the drag coefficient for the liquid at the throat entrance, C_D, using Equation 18.11:

$$C_D = 0.22 + (24/N_{Reo})[1 + 0.15(N_{Reo})^{0.6}]$$

where

C_D = Drag coefficient for the liquid at the throat entrance (dimensionless).
N_{Reo} = Reynolds number for the liquid droplet at the throat inlet (dimensionless).

$$C_D = 0.22 + (24/N_{Reo})[1 + 0.15(N_{Reo})^{0.6}] = 0.22 + (24/360)[(1 + 0.15(360)^{0.6}] = 0.628$$

Now calculate the parameter characterizing the liquid-to-gas ratio B, using Equation 18.8:

$$B = (L/G)\left(\frac{p_l}{p_g C_D}\right)$$

where

B = Parameter characterizing liquid-to-gas ratio (dimensionless).
L/G = Liquid-to-gas ratio (dimensionless).
p_l = Liquid density (g/cm³).
p_g = Gas density (g/cm³).
C_D = Drag coefficient for the liquid at the throat entrance (dimensionless).

$$B = (L/G)\left(\frac{p_l}{p_g C_D}\right) = (0.0009)\left(\frac{1000 \text{ kg/m}^3}{(1.0 \text{ kg/m}^3)(0.628)}\right) = 1.43$$

The geometric standard deviation (σ_{gm}) is 2.5, and overall penetration (P_t^*) is 0.008. The collection efficiency can be calculated as follows:

$$\eta = 1 - P_t^* = 1 - 0.008 = 0.992 = 99.2\%$$

Finally, determine whether the local regulations for particulate emissions are being met. The local regulations state that the particulate emissions cannot exceed 10 kg/hr. The required collection efficiency can be calculated by using the following equation:

$$\eta_{required} = \frac{dust_{in} - dust_{out}}{dust_{in}}$$

where
    $dust_{in}$ = Dust concentration leading into the Venturi.
    $dust_{out}$ = Dust concentration leaving the Venturi.

Thus,

$$\eta_{required} = \frac{dust_{in} - dust_{out}}{dust_{in}} = \frac{1100 \text{ kg/hr} - 10 \text{ kg/hr}}{1100 \text{ kg/hr}} = 0.991 = 99.1\%$$

The estimated efficiency of the Venturi scrubber is slightly higher than the required efficiency.

18.2.5.3 Cut Power Method

The cut power method is an empirical correlation used to predict the collection efficiency of a scrubber. In this method, penetration is a function of the cut diameter of the particles to be collected by the scrubber. Recall that the cut diameter is the diameter of the particles that are collected by the scrubber with at least 50% efficiency. Because scrubbers have limits to the size of particles they can collect, knowledge of the cut diameter is useful in evaluating the scrubbing system (USEPA, 1984c, p. 9-11). In the cut power method, penetration is a function of the particle diameter and is given as

$$P_t = \exp[-A_{cut}d_p(B_{cut})] \tag{18.15}$$

where
    P_t = Penetration.
    A_{cut} = Parameter characterizing the particle size distribution.
    d_p = Aerodynamic diameter of the particle.
    B_{cut} = Empirically determined constant, depending on the scrubber design.

Penetration, calculated by Equation 18.15, is given for only one particle size (d_p). To obtain the overall penetration, the equation can be integrated over the log-normal particle size distribution. By mathematically integrating P_t over a log-normal distribution of particles and by varying the geometric standard deviation (σ_{gm}) and the geometric mean particle diameter (d_{pg}), the overall penetration (P_t^*) can be obtained.

■ EXAMPLE 18.2

Problem: Given similar conditions used in the example in the infinite throat section, estimate the cut diameter for a Venturi scrubber. The data below are approximate (USEPA, 1984c, p. 9-12).

    Given:
        σ_{gm} = Geometric standard deviation = 2.5
        d_{pg} = Particle aerodynamic geometric mean diameter = 12.6 μmA
        η = Required efficiency = 99.1% or 0.991

Solution: For an efficiency of 99.1%, the overall penetration can be calculated from

$$P_t^* = 1 - \eta = 1 - 0.991 = 0.009$$

The overall penetration is 0.009, and the geometric standard deviation is 2.5. The figures in USEPA (1984c, p. 9-12) provide us with the following information:

$$P_t^* = 0.009$$
$$\sigma_{gm} = 2.5$$
$$(d_p)_{cut}/d_{pg} = 0.09$$

The cut diameter $(d_p)_{cut}$ is calculated from

$$(d_p)_{cut} = 0.09(12.6 \ \mu mA) = 1.134 \ \mu mA$$

■ **EXAMPLE 18.3**

Problem: A particle size analysis indicated the following (USEPA, 1984a, 9-14):

d_{gm} = Geometric mean particle diameter = 12 μm
σ_{gm} = Standard deviation of the distribution = 3.0
η = Wet collector efficiency = 99%

If a collection efficiency of 99% were required to meet emission standards, what would the cut diameter of the scrubber have to be?

Solution: Write the penetration (P_t) equation:

$$P_t^* = 1 - \eta = 1 - 0.99 = 0.01$$

From the figures in USEPA (1984c, p. 9-12), we read $(d_p)_{cut}/d_{gm}$, for $P_t^* = 0.01$ and $\sigma_{gm} = 3.0$; $[d_p]_{cut}/d_{gm}$ equals 0.063. Because $d_{gm} = 12$ μm, the scrubber must be able to collect particles of size 0.063 × 12 = 0.76 μm with at least 50% efficiency to achieve an overall scrubber efficiency of 99%.

18.2.5.4 Contact Power Theory

A more general theory for estimating collection efficiency is the contact power theory. This theory is based on a series of experimental observations made by Lapple and Kamack (1955). The fundamental assumption of the theory can be expressed as: "When compared at the same power consumption, all scrubbers give substantially the same degree of collection of a given dispersed dust, regardless of the mechanism involved and regardless of whether the pressure drop is obtained by high gas flow rates or high water flow rates" (Lapple and Kamack, 1955). In other words, collection efficiency is a function of how much power the scrubber uses, and not of how the scrubber is designed. This has a number of implications in the evaluation and selection of wet collectors. Once we know the amount of power needed to attain a certain collection efficiency, the claims about specially located nozzles, baffles, etc. can be evaluated more objectively. The choice between two different scrubbers with the same power requirements may depend primarily on ease of maintenance (USEPA, 1984a, p. 9-16; USEPA, 1984c, p. 9-13).

Semrau (1960, 1963) developed the contact power theory from the work of Lapple and Kamack (1955). The theory, as developed by Semrau, is empirical in approach and relates the total pressure loss (P_T) of the system to the collection efficiency. The total pressure loss is expressed in terms of the power expended to inject the liquid into the scrubber plus the power needed to move the process gas through the system (USEPA, 1984c, p. 9-13):

$$P_T = P_G + P_L, \quad P_G = 0.157\Delta p, \quad P_L = 0.583p_t(Q_L/Q_G) \tag{18.16}$$

where

   P_T = Total contacting power (total pressure loss) (kWh/100 m³, hp/1000 acfm).
   P_G = Power input from gas stream (kWh/100 m³, hp/1000 acfm).
   P_L = Contacting power from liquid injection (kWh/100 m³, hp/1000 acfm).

Note: The total pressure loss (P_T) should not be confused with penetration (P_t).

The power expended in moving the gas through the system (P_G) is expressed in terms of the scrubber pressure drop:

$$P_G = (2.724 \times 10^{-4})\Delta p \text{ (k/Wh/1000 m}^3) \tag{18.17a}$$

or

$$P_G = 0.1575\Delta p \text{ (hp/1000 acfm)} \tag{18.17b}$$

where Δp is the pressure drop (kPa, in. H$_2$O).
   The power expended in the liquid stream (P_L) is expressed as

$$P_L = 0.28 p_L(Q_L/Q_G) \text{ (kWh/1000 m}^3) \tag{18.18a}$$

or

$$P_L = 0.583 p_L(Q_L/Q_G) \text{ (hp/1000 acfm)} \tag{18.18b}$$

where

   p_L = Liquid inlet pressure (100 kPa, lb/in²).
   Q_L = Liquid feed rate (m³/h, gal/min).
   Q_G = Gas flow rate (m³/h, ft³/min).

The constants given in the expressions for P_G and P_L incorporate conversion factors to put the terms on a consistent basis. The total power can therefore be expressed as

$$P_T = P_G + P_L = (2.724 \times 10^{-4})\Delta p + 0.28 p_L(Q_L/Q_G) \text{ (kWh/1000 m}^3) \tag{18.19a}$$

or

$$P_T = 0.1575\Delta p + 0.583 p_L(Q_L/Q_G) \text{ (hp/1000 acfm)} \tag{18.19b}$$

Correlate this with scrubber efficiency using the following equations:

$$\eta = 1 - \exp[-f(system)] \tag{18.20}$$

where $f(system)$ is defined as

$$f(system) = N_t = \alpha(P_T)^\beta \tag{18.21}$$

where

   N_t = Number of transfer units.
   P_T = Total contacting power.
   α and β = Empirical constants determined from experimentation that depend on characteristics of the particles.

The efficiency then becomes

$$\eta = 1 - \exp[-\alpha(P_T)^\beta] \tag{18.22}$$

Note: The values of α and β can be in either metric or British units and can be found in USEPA (1984c, p. 9-15).

Scrubber efficiency is also expressed as the number of transfer units (USEPA, 1984a, p. 9-17):

$$N_t = \alpha(P_T)^\beta = \ln[1/(1-\eta)] \tag{18.23}$$

where
  N_t = Number of transfer units.
  η = Fractional collection efficiency.
  α and β = Characteristic parameters for the type of particulates being collected.

Unlike the cut power and Johnstone theories, the contact power theory cannot predict efficiency from a given particle size distribution. The contact power theory gives a relationship, which is independent of the size of the scrubber. With this observation, a small pilot scrubber could first be used to determine the pressure drop needed for the required collection efficiency. The full-scale scrubber design could then be scaled up from the pilot information. Consider the following example.

■ EXAMPLE 18.4

Problem: A wet scrubber is to be used to control particulate emissions from a foundry cupola. Stack test results reveal that the particulate emissions must be reduced by 85% to meet emission standards. If a 100-acfm pilot unit is operated with a water flow rate of 0.5 gal/min at a water pressure of 80 psi, what pressure drop (Δp) would be needed across a 10,000-acfm scrubber unit (USEPA, 1984c, p. 9-15; USEPA, 1984a, p. 9-18)?

Solution: From the table in USEPA (1984c, p. 9-15), read the α and β parameters for foundry cupola dust.

  $\alpha = 1.35$
  $\beta = 0.621$

Calculate the number of transfer units N_t using Equation 18.20:

$$\eta = 1 - \exp(-N_t)$$

$$N_t = \ln[1/(1-\eta)] = \ln[1/(1-0.85)] = 1.896$$

Now calculate the total contacting power (P_T) using Equation 18.21:

  $N_t = \alpha(P_T)^\beta$
  $1.896 = 1.35(P_T)^{0.621}$
  $1.404 = (P_T)^{0.621}$
  $\ln 1.404 = 0.621(\ln P_T)$
  $0.3393 = 0.621(\ln P_T)$
  $0.5464 = \ln P_T$
  $P_T = 1.73$ hp/1000 acfm

Calculate the pressure drop (Δp) using Equation 18.19:

$$P_T = 0.1575\Delta p + 0.583 p_L(Q_L/Q_G)$$

$$1.73 = 0.1575\Delta p + 0.583(80)(0.5/100)$$

$$\Delta p = 9.5 \text{ in. } H_2O$$

18.2.5.5 Pressure Drop

A number of factors affect particle capture in a scrubber. One of the most important for many scrubber types is pressure drop. Pressure drop is the difference in pressure between the inlet and the outlet of the scrubbing process. Static pressure drop of a system is dependent on the mechanical design of the system and collection efficiency required. It is the sum of the energy required to accelerate and move the gas stream and the frictional losses as the gases move through the scrubbing system (UESPA, 1984c, p. 9-17). The following factors affect the pressure drop in a scrubber:

- Scrubber design and geometry
- Gas velocity
- Liquid-to-gas ratio

As with calculating collection efficiency, no one equation can predict the pressure drop for all scrubbing systems.

Many theoretical and empirical relationships are available for estimating the pressure drop across a scrubber. Generally, the most accurate are those developed by scrubber manufacturers for *their* particular scrubbing systems. Due to the lack of validated models, it is recommended that users consult the vendor's literature to estimate pressure drop for the particular scrubbing device of concern. One widely accepted expression was developed for Venturi scrubbers. The correlation proposed by Calvert (Yung et al. 1977) is

$$\Delta p = (8.24 \times 10^{-4})(v_{gt})^2 (L/G) \text{ (metric units)} \tag{18.24a}$$

or

$$\Delta p = (4.0 \times 10^{-5})(v_{gt})^2 (L/G) \text{ (English units)} \tag{18.24b}$$

where
  Δp = Pressure drop (cmH$_2$O (in. H$_2$O).
  v_{gt} = Velocity of gas in the Venturi throat (cm/s, ft/s).
  L/G = Liquid-to-gas ratio (dimensionless, but actually in L/m^3 or gal/1000 ft^3).

Using Equation 18.24a for the conditions given in the example in the infinite throat model section, we obtain

$$v_{gt} = 9000 \text{ cm/s}$$

$$L/G = 0.0009 \text{ L/m}^3$$

$$\Delta p = 8.24 \times 10^{-4}(9000)^2(0.0009) = 60 \text{ cmH}_2\text{O}$$

18.3 WET SCRUBBER COLLECTION MECHANISMS AND EFFICIENCY (GASEOUS EMISSIONS)

Although Venturi scrubbers are used predominantly for control of particulate air pollutants, these devices can simultaneously function as absorbers. Consequently, absorption devices used to remove gaseous contaminants are referred to as *absorbers* or *wet scrubbers* (USEPA, 1984c, p. 1-7). To remove a gaseous pollutant by absorption, the exhaust stream must be passed through (brought into contact with) a liquid. The process involves three steps:

- The gaseous pollutant diffuses from the bulk area of the gas phase to the gas–liquid interface.
- The gas moves (transfers) across the interface to the liquid phase. This step occurs extremely rapidly once the gas molecules (pollutants) arrive at the interface area.
- The gas diffuses into the bulk area of the liquid, thus making room for additional gas molecules to be absorbed. The rate of absorption (mass transfer of the pollutant from the gas phase to the liquid phase) depends on the diffusion rates of the pollutant in the gas phase (first step) and in the liquid phase (third step).

To enhance gas diffusion and, therefore, absorption, the steps include the following:

- Provide a large interfacial contact area between the gas and liquid phases.
- Provide good mixing of the gas and liquid phases (turbulence).
- Allow sufficient residence, or contact, time between the phases for adsorption to occur.

18.4 ASSORTED VENTURI SCRUBBER EXAMPLE CALCULATIONS

In this section, we provide several example calculations environmental engineers would be expected to perform with regard to problems dealing with Venturi scrubber design, scrubber overall collection efficiency, scrubber plan review, spray tower, packed tower review, and tower height and diameter.

18.4.1 CALCULATIONS FOR SCRUBBER DESIGN OF A VENTURI SCRUBBER

■ EXAMPLE 18.5

Problem: Calculate the throat area of a Venturi scrubber to operate at specified collection efficiency (USEPA, 1984b, p. 77).

Given:
     Volumetric flow rate of process gas stream = 11,040 acfm (at 68°F)
     Density of dust = 187 lb/ft^3
     Liquid-to-gas ratio = 2 gal/1000 ft^3
     Average particle size = 3.2 μm (1.05×10^{-5} ft)
     Water droplet size = 48 μm (1.575×10^{-4} ft)
     Scrubber coefficient $k = 0.14$
     Required collection efficiency = 98%
     Viscosity of gas = 1.23×10^{-5} lb/ft-s
     Cunningham correction factor = 1.0

Solution: Calculate the inertial impaction parameter (Ψ) from Johnstone's equation, which describes the collection efficiency of a Venturi scrubber (USEPA, 1984a, p. 9-11):

$$\eta = 1 - \exp\left(-k\left(Q_L/Q_G\right)\sqrt{\Psi}\right)$$

where
     η = Fractional collection efficiency.
     k = Correlation coefficient, the value of which depends on the system geometry and operating conditions (typically 0.1 to 0.2 acf/gal).
     Q_L/Q_G = Liquid-to-gas ratio (gal/1000 acf).
     Ψ = Inertial impaction parameter:

$$\frac{C_f p_p v (d_p)^2}{18 d_d \mu}$$

where
  C_f = Cunningham correction factor.
  p_p = Particle density.
  v = Gas velocity at Venturi throat (ft/s).
  d_p = Particle diameter (ft).
  d_d = Droplet diameter (ft).
  μ = Gas viscosity (lb/ft-s).

From the calculated value of Ψ (inertial impaction parameter) above, back calculate the gas velocity at the Venturi throat. First calculate Ψ:

$$\eta = 1 - \exp\left(-k \left(Q_L/Q_G\right)\sqrt{\Psi}\right)$$

$$0.98 = 1 - \exp\left(-0.14(2)\sqrt{\Psi}\right)$$

Thus,

$$\Psi = 195.2$$

Now calculate v:

$$\Psi = \frac{C_f p_p v (d_p)^2}{18 d_d \mu}$$

$$v = \frac{18 \Psi d_d \mu}{C_f p_p (d_p)^2}$$

$$= (18)(195.2)\left(1.575 \times 10^{-4}\right)\left(1.23 \times 10^{-5}\right)(1)(187)\left(1.05 \times 10^{-5}\right)^2$$

$$= 330.2 \text{ ft/s}$$

Finally, calculate throat area S, using gas velocity at Venturi throat v:

$$S = (\text{Volumetric flowrate})/(\text{Velocity}) = 11{,}040/(60 \times 330.2) = 0.557 \text{ ft}^2$$

■ EXAMPLE 18.6

Problem: Calculate the overall collection efficiency of a Venturi scrubber that cleans a fly ash-laden gas stream, given the liquid-to-gas ratio, throat velocity, and particle size distribution (USEPA, 1984b, p. 79).

  Given:
    Liquid-to-gas ratio = 8.5 gal/1000 ft^3
    Throat velocity = 227 ft/s
    Particle density of fly ash = 43.7 lb/ft^3
    Gas viscosity = 1.5 × 10^{-5} lb/ft-s

The particle size distribution data are given in Table 18.1. Use Johnstone's equation, with a k value of 0.2, to calculate the collection efficiency. Neglect the Cunningham correction factor effect.

TABLE 18.1

Particle Size Distribution Data

d_p (μm)	Weight (%)
< 0.1	0.01
0.1–0.5	0.21
0.5–1.0	0.78
1.0–5.0	13.00
5.0–10.0	16.00
10.0–15.0	12.00
15.0–20.0	8.00
> 20.0	50.00

Solution: What are the parameters used in Johnstone's equation?

$$\eta = 1 - \exp\left(-k\left(Q_L/Q_G\right)\sqrt{\Psi}\right)$$

where

η = Fractional collection efficiency.

k = Correlation coefficient, the value of which depends on the system geometry and operating conditions (typically 0.1 to 0.2 acf/gal).

Q_L/Q_G = Liquid-to-gas ratio (gal/1000 acf).

Ψ = Inertial impaction parameter:

$$\Psi = \frac{C_f p_p v (d_p)^2}{18 d_d \mu}$$

where

C_f = Cunningham correction factor.

p_p = Particle density.

v = Gas velocity at Venturi throat (ft/s).

d_p = Particle diameter (ft).

d_d = Droplet diameter (ft).

μ = Gas viscosity (lb/ft-s).

Calculate the average droplet diameter in feet. The average droplet diameter may be calculated using as follows:

$$d_d = (16,400/V) + 1.45(Q_L/Q_G)^{1.5}$$

where d_d = droplet diameter (μm):

$$d_d = (16,400/272) + 1.45(8.5)^{1.5} = 96.23 \text{ μm} = 3.156 \times 10^{-4} \text{ ft}$$

Express the inertial impaction parameter in terms of d_p (ft):

$$\Psi = \frac{C_f p_p v (d_p)^2}{18 d_d \mu} = \frac{(1)(43.7)(272)(d_p)^2}{(18)\left(3.156 \times 10^{-4}\right)\left(1.5 \times 10^{-5}\right)} = \left(1.3945 \times 10^{11}\right)(d_p)^2$$

Express the fractional collection efficiency (η_i) in terms of d_{pi} (d_p in ft):

$$\eta = 1 - \exp\left(-k\left(Q_L/Q_G\right)\sqrt{\Psi}\right)$$

$$\eta_i = 1 - \exp\left(-(0.2)(8.5)\left(1.3945\times10^{11}\right)\left(Q_L/Q_G\right)\left(d_p\right)^2\right)$$

$$= 1 - \exp\left[-\left(6.348\times10^5\right)d_{pi}\right]$$

Now calculate the collection efficiency for each particle size appearing in Table 18.1. For $d_p = 0.05$ μm (1.64×10^{-7} ft), for example:

$$\eta = 1 - \exp\left[-\left(6.348\times10^5\right)d_{pi}\right] = 1 - \exp\left[-\left(6.348\times10^5\right)\left(1.64\times10^{-7}\right)\right] = 0.0989$$

Table 18.2 shows the results (rounded) of the above calculations for each particle size. Now calculate the overall collection efficiency:

$$\eta = \Sigma w_i\eta_i = (9.89\times10^{-4}) + 0.0975 + 0.6325 + 12.980 + 16.00 + 12.00 + 8.00 + 50.00 = 99.71\%$$

■ EXAMPLE 18.7

Problem: A vendor proposes to use a spray tower on a lime kiln operation to reduce the discharge of solids to the atmosphere. The inlet loading is to be reduced in order to meet state regulations. The vendor's design calls for a certain water pressure drop and gas pressure drop across the tower. Determine whether this spray tower will meet state regulations. If the spray tower does not meet state regulations, propose a set of operating conditions that will meet the regulations (USEPA, 1984b, p. 81).

Given:
  Gas flow rate = 10,000 acfm
  Water rate = 50 gal/min
  Inlet loading = 5.0 grains/ft³
  Maximum gas pressure drop across the unit = 15 in. H_2O
  Maximum water pressure drop across the unit = 100 psi
  Water pressure drop = 80 psi
  Gas pressure drop across the tower = 5.0 in. H_2O

State regulations require a maximum outlet loading of 0.05 grains/ft³. Assume that the contact power theory applies (USEPA, 1984a, p. 9-15).

TABLE 18.2
Particle Size Data

d_p (ft)	w_i (%)	η_i	$w_i\eta_i$ (%)
1.64×10^{-7}	0.01	0.0989	9.89×10^{-4}
9.84×10^{-7}	0.21	0.4645	0.0975
2.62×10^{-6}	0.78	0.8109	0.6325
9.84×10^{-6}	13	0.9981	12.98
2.62×10^{-5}	16	1	16
4.27×10^{-5}	12	1	12
5.91×10^{-5}	8	1	8
6.56×10^{-5}	50	1	50

Solution: Calculate the collection efficiency based on the design data given by the vendor. The contact power theory is an empirical approach relating particulate collection efficiency and pressure drop in wet scrubber systems. It assumes that particulate collection efficiency is a sole function of the total pressure loss for the unit.

$$P_T = P_G + P_L$$

$$P_G = 0.157\Delta p$$

$$P_L = 0.583 p_L (Q_L/Q_G)$$

where

P_T = Total pressure loss (hp/1000 acfm).
P_G = Contacting power based on gas stream energy input (hp/1000 acfm).
P_L = Contacting power based on liquid stream energy input (hp/1000 acfm).
Δp = Pressure drop across the scrubber (in. H_2O).
p_L = Liquid inlet pressure (psi).
Q_L = Liquid feed rate (gal/min).
Q_G = Gas flow rate (ft^3/min).

The scrubber collection efficiency is also expressed as the number of transfer units:

$$N_t = \alpha(P_T)^\beta = \ln[1/(1 - \eta)]$$

where

N_t = Number of transfer units.
α and β = Characteristic parameters for the type of particulates being collected.
P_T = Total pressure loss (hp/1000 acfm).
η = Fractional collection efficiency.

Calculate the total pressure loss (P_T). To calculate the total pressure loss, we need the contacting power for both the gas stream energy input and liquid stream energy input. Calculate the contacting power based on the gas stream energy input (P_G) in hp/1000 acfm. Because the vendor gives the pressure drop across the scrubber, we can calculate P_G:

$$P_G = 0.157\Delta p = (0.157 \times 5.0) = 0.785 \text{ hp/1000 acfm}$$

Calculate the contacting power based on the liquid stream energy input (P_L) in hp/1000 acfm. Because the liquid inlet pressure and liquid-to-gas ratio are given, we can calculate P_L:

$$P_L = 0.583 p_L (Q_L/Q_G) = 0.583 \times 80 \times (50/10,000) = 0.233 \text{ hp/1000 acfm}$$

Calculate the total pressure loss (P_T) in hp/1000 acfm:

$$P_T = P_G + P_L = 0.785 + 0.233 = 1.018 \text{ hp/1000 acfm}$$

Calculate the number of transfer units N_t:

$$N_t = \alpha(P_T)^\beta$$

The values of α and β for a lime kiln operation are 1.47 and 1.05, respectively. These coefficients have been previously obtained from field test data. Therefore,

$$N_t = \alpha(P_T)\beta = 1.47 \times (1.018)^{1.05} = 1.50$$

Calculate the collection efficiency based on the design data given by the vendor:

$$N_t = \ln[1/(1 - \eta)]$$

$$1.50 = \ln[1/(1 - \eta)]$$

Thus,

$$\eta = 0.777 = 77.7\%$$

Now calculate collection efficiency required by state regulations. Because the inlet loading is known and the outlet loading is set by the regulations, the collection efficiency can be readily calculated.

$$\text{Collection efficiency} = \left(\frac{\text{Inlet loading} - \text{Outlet loading}}{\text{Inlet loading}}\right) \times 100 = \left(\frac{5.0 - 0.05}{5.0}\right) \times 100 = 99.0\%$$

Does the spray tower meet the regulations? No. The collection efficiency based on the design data given by the vendor should be higher than the collection efficiency required by the state regulations. Assuming that the spray tower does not meet the regulations, propose a set of operating conditions that will meet the regulations. Note that the calculation procedure is now reversed.

Calculate the total pressure loss (P_T) in hp/1000 acfm using the collection efficiency required by the regulations. Calculate the number of transfer units for the efficiency required by the regulations:

$$N_t = \ln[1/(1 - \eta)] = \ln[1(1 - 0.99)] = 4.605$$

Calculate the total pressure loss, P_T, in hp/1000 acfm:

$$N_t = \alpha(P_T)^\beta$$

$$4605 = 1.47(P_T)^{1.05}$$

Thus,

$$P_T = 2.96 \text{ hp/1000 acfm}$$

Calculate the contacting power based on the gas stream energy input (P_G) using a Δp of 15 in. H_2O. A pressure drop Δp of 15 in. H_2O is the maximum value allowed by the design:

$$P_G = 0.157\Delta p = 0.157 \times 15 = 2.355 \text{ hp/1000 acfm}$$

Calculate the contacting power based on the liquid stream energy input P_L:

$$P_L = P_T - P_G = 2.96 - 2.355 = 0.605 \text{ hp/1000 acfm}$$

Calculate Q_L/Q_G in gal/acf using a p_L of 100 psi:

$$P_L = 0.583 p_L(Q_L/Q_G)$$

$$Q_L/Q_G = P_L/0.583 p_L = 0.605/(0.583 \times 100) = 0.0104$$

Determine the new water flow rate (Q_L') in gal/min:

$$(Q_L') = (Q_L/Q_G)(10,000 \text{ acfm}) = 0.0104 \times 10,000 \text{ acfm} = 104 \text{ gal/min}$$

What are the new set of operating conditions that will meet the regulations?

$$Q_L' = 104 \text{ gal/min}$$

$$P_T = 2.96 \text{ hp/1000 acfm}$$

18.4.2   SPRAY TOWER CALCULATIONS

■ EXAMPLE 18.8

Problem: A steel pickling operation emits 300 ppm HCl (hydrochloric acid) with peak values of 500 ppm 15% of the time. The airflow is a constant 25,000 acfm at 75°F and 1 atm. Only sketchy information was submitted with the scrubber permit application for a spray tower. We are requested to determine if the spray unit is satisfactory (USEPA, 1984b, p. 100).

Given:
Emission limit = 25 ppm HCl
Maximum gas velocity allowed through the water = 3 ft/s
Number of sprays = 6
Diameter of the tower = 14 ft

The plans show a countercurrent water spray tower. For a very soluble gas (Henry's law constant approximately zero), the number of transfer units (N_{OG}) can be determined by the following equation:

$$N_{OG} = \ln(y_1/y_2)$$

where
y_1 = Concentration of inlet gas.
y_2 = Concentration of outlet gas.

In a spray tower, the number of transfer units (N_{OG}) for the first (or top) spray will be about 0.7. Each lower spray will have only about 60% of the N_{OG} of the spray above it. The final spray, if placed in the inlet duct, has an N_{OG} of 0.5. The spray sections of a tower are normally spaced at 3-ft intervals. The inlet duct spray adds no height to the column.

Solution: Calculate the gas velocity through the tower:

$$V = Q/S = Q/(\pi D^2/4)$$

where
V = Velocity.
Q = Actual volumetric gas flow.
S = Cross-sectional area.
D = Diameter of the tower.

$$V = Q/(\pi D^2/4) = 25,000/[\pi(14)^2/4] = 162.4 \text{ ft/min} = 2.7 \text{ ft/s}$$

Does the gas velocity meet the requirement? Yes, because the gas velocity is less than 3 ft/s. Now calculate the number of overall gas transfer units (N_{OG}) required to meet the regulation. Recall that

$$N_{OG} = \ln(y_1/y_2)$$

where

y_1 = Concentration of inlet gas.
y_2 = Concentration of outlet gas.

Use the peak value for inlet gas concentration:

$$N_{OG} = \ln(y_1/y_2) = \ln(500/25) = 3.0$$

Determine the total number of transfer units provided by a tower with six spray sections. Remember that each lower spray has only 60% of the efficiency of the section above it (due to a back mixing of liquids and gases from adjacent sections). Spray section N_{OG} values are derived accordingly:

Top spray N_{OG} = 0.7 (given)
2nd spray N_{OG} = 0.7 × 0.6 = 0.42
3rd spray N_{OG} = 0.42 × 0.6 = 0.252
4th spray N_{OG} = 0.252 × 0.6 = 0.1512
5th spray N_{OG} = 0.1512 × 0.6 = 0.0907
Inlet N_{OG} = 0.5 (given)

$$\text{Total } N_{OG} = 0.7 + 0.42 + 0.252 + 0.1512 + 0.0907 + 0.5 = 2.114$$

This value is below the required value of 3.0. Now calculate the outlet concentration of gas:

$$N_{OG} = \ln(y_1/y_2)$$

$$y_1/y_2 = \exp(N_{OG}) = \exp(2.114) = 8.28$$

$$y_2 = 500/8.28 = 60.4 \text{ ppm}$$

Does the spray tower meet the HCl regulation? Because y_2 is greater than the required emission limit of 25 ppm, the spray unit is not satisfactory.

18.4.3 PACKED TOWER CALCULATIONS

■ EXAMPLE 18.9

Problem: Pollution Unlimited, Inc., has submitted plans for a packed ammonia scrubber on an air stream containing NH_3. The operating and design data are given by Pollution Unlimited, Inc. We remember approving plans for a nearly identical scrubber for Pollution Unlimited, Inc., in 1978. After consulting our old files we find all the conditions were identical except for the gas flow rate. What is our recommendation (USEPA, 1984b, p. 102)?

Given:
Tower diameter = 3.57 ft
Packed height of column = 8 ft
Gas and liquid temperature = 75°F
Operating pressure = 1.0 atm
Ammonia-free liquid flow rate (inlet) = 1000 lb/ft²-hr
Gas flow rate = 1575 acfm
Gas flow rate in the 1978 plan = 1121 acfm
Inlet NH_3 gas composition = 2.0 mol%

Outlet NH_3 gas composition = 0.1 mol%
Air density = 0.0743 lb/ft^3
Molecular weight of air = 29
Henry's law constant m = 0.972
Molecular weight of water = 18
Emission regulation = 0.1% NH_3

Solution: What is the number of overall gas transfer units, N_{OG}? The number of overall gas transfer units N_{OG} is used in calculating the packing height requirements. This number is a function of the extent of the desired separation and the magnitude of the driving force through the column (the displacement of the operating line from the equilibrium line). Calculate the gas molar flow rate (G_m) and liquid molar flow rate (L_m) in lb-mol/ft^2-hr. Calculate the cross-sectional area of the tower S in ft^2:

$$S = \pi D^2/4$$

where
   S = Cross-sectional area of the tower.
   D = Diameter of the tower.

$$S = \pi D^2/4 = [\pi \times (3.57)^2]/4 = 10.0 \text{ ft}^2$$

Calculate gas molar flow rate (G_m) in lb-mol/ft^2-hr:

$$G_m = Qp/SM$$

where
   G_m = Gas molar flow rate (lb-mol/ft^2-hr).
   Q = Volumetric flow rate of gas stream.
   p = Density of air.
   S = Cross-sectional area of the tower.
   M = Molecular weight of air.

$$G_m = Qp/SM = (1575 \times 0.0743)/(10.0 \times 29) = 0.404 \text{ lb-mol/ft}^2\text{-min} = 24.2 \text{ lb-mol/ft}^2\text{-hr}$$

Calculate the liquid molar flow rate (L_m) in lb-mol/ft^2-hr:

$$L_m = L/M_L$$

where
   L_m = Liquid molar flow rate (lb-mol/ft^2-hr).
   L = Liquid mass velocity (lb/ft^2-hr).
   M_L = Liquid molecular weight.

$$L_m = L/M_L = (1000)/(18) = 55.6 \text{ lb mol/ft}^2\text{-hr}$$

Calculate the value of mG_m/L_m, where

   m = Henry's law constant.
   G_m = Gas molar flow rate (lb-mol/ft^2-hr).
   L_m = Liquid molar flow rate (lb-mol/ft^2-hr).

$$mG_m/L_m = (0.972)(24.2/55.6) = 0.423$$

Calculate the value of $(y_1 - mx_1)/(y_2 - mx_2)$, where

y_1 = Inlet gas mole fraction.
m = Henry's law constant.
x_1 = Outlet liquid mole fraction.
y_2 = Outlet gas mole fraction.
x_2 = Inlet liquid mole fraction.

$$(y_1 - mx_1)/(y_2 - mx_2) = [0.02 - (0.972)(0)]/[0.001 - (0.972)(0)] = 20.0$$

Determine the value of N_{OG} and use the values of $(y_1 - mx_2)/(y_2 - mx_2)$ and mG_m/L_m to find the value of N_{OG}:

$$N_{OG} = 4.3$$

What is the height of an overall gas transfer unit H_{OG}? The height of an overall gas transfer unit H_{OG} is also used in calculating packing height requirements. H_{OG} values in air pollution are almost always based on experience. H_{OG} is a strong function of the solvent viscosity and difficulty of separation, increasing with increasing values of both. Calculate the gas mass velocity (G) in lb/ft²-hr:

$$G = pQ/S$$

where
G = Gas mass velocity (lb/ft²-hr).
p = Density of air.
S = Cross-sectional area of the tower.

$$G = pQ/S = (1575 \times 0.0743)/10.0 = 11.7 \text{ lb/ft}^2\text{-min} = 702 \text{ lb/ft}^2\text{-hr}$$

The H_{OG} value is 2.2 ft.

What is the required packed column height (Z) in ft?

$$Z = (N_{OG})(H_{OG})$$

where
Z = Height of packing.
H_{OG} = Height of an overall gas transfer unit.
N_{OG} = Number of transfer units.

$$Z = N_{OG} \times H_{OG} = 4.3 \times 2.2 = 9.46 \text{ ft}$$

Compare the packed column height of 8 ft specified by Pollution Unlimited, Inc., to the height calculated above. What is our recommendation? Disapproval, because the calculated height (9.46 ft) is higher than that (8 ft) proposed by the company.

18.4.4 PACKED COLUMN HEIGHT AND DIAMETER CALCULATIONS

■ EXAMPLE 18.10

Problem: A packed column is designed to absorb ammonia from a gas stream. Given the operating conditions and type of packing below (see Figure 18.2), calculate the height of packing and column diameter (USEPA, 1984b, p. 106).

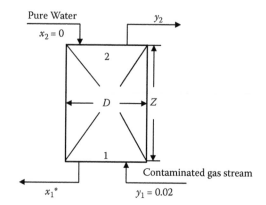

FIGURE 18.2 Graphical representation of the packed column. (Adapted from USEPA, *Control of Gaseous and Particulate Emissions: Self-Instructional Problem Workbook*, EPA 450/2-84-007, U.S. Environmental Protection Agency Air Pollution Training Institute, Research Triangle Park, NC, 1984, p. 107.)

Given:

Gas mass flow rate = 5000 lb/hr
NH_3 concentration in inlet gas stream = 2.0 mol%
Scrubbing liquid = pure water
Packing type = 1-inch Raschig rings
Packing factor $F = 160$
H_{OG} of the column = 2.5 ft
Henry's law constant $m = 1.20$
Density of air = 0.075 lb/ft^3
Density of water = 62.4 lb/ft^3
Viscosity of water = 1.8 cp
Generalized flooding and pressure drop correction graph (USEPA, 1984b, p. 107)

The unit operates at 60% of the flooding gas mass velocity, the actual liquid flow rate is 25% more than the minimum, and 90% of ammonia is to be collected based on state regulations.

Solution: What is the number of overall gas transfer units N_{OG}? Remember that the height of packing Z is given by

$$Z = (H_{OG})(N_{OG})$$

where

Z = Height of packing.
H_{OG} = Height of an overall gas transfer unit.
N_{OG} = Number of transfer units.

Because H_{OG} is given, we only need the value of N_{OG} to calculate Z. N_{OG} is a function of both the liquid and gas flow rates; however, the value is usually available for most air pollution applications. What is the equilibrium outlet liquid composition (x_1) and the outlet gas composition (y_2) for 90% removal? Recall that we need the inlet and outlet concentrations (mole fractions) of both streams. Calculate the equilibrium outlet concentration x_1^* at $y_1 = 0.02$. According to Henry's law, x_1^* at y_1/m. The equilibrium outlet liquid composition is needed to calculate the minimum L_m/G_m, where

x_1^* = Outlet concentration.
y_1 = Inlet gas mole fraction.
m = Henry's law constant.
L_m = Liquid molar flow rate (lb-mol/ft^2-hr).
G_m = Gas molar flow rate (lb-mol/ft^2-hr).

$$x_1^* = y_1/m = (0.02)/(1.20) = 0.0167$$

Calculate y_2 for 90% removal. Because it is required to remove 90% of NH_3, there will be 10% of NH_3 remaining in the outlet gas stream; therefore, by material balance,

$$y_2 = (0.1)(y_1)/[(1 - y_1) + (0.1)(y_1)]$$

where
  y_1 = Inlet gas mole fraction.
  y_2 = Outlet gas mole fraction.

$$y_2 = (0.1)(y_1)/[(1 - y_1) + (0.1)(y_1)] = (0.1)(0.02)/[(1 - 0.02) + (0.1)(0.02)] = 0.00204$$

Determine the minimum ratio of molar liquid flow rate to molar gas flow rate $(L_m/G_m)_{min}$ by a material balance. Material balance around the packed column is calculated as

$$G_m(y_1 - y_2) = L_m(x_1^* - x_2)$$

$$(L_m/G_m)_{min} = (y_1 - y_2)/(x_1^* - x_2)$$

where
  L_m = Liquid molar flow rate (lb mol/ft^2-hr).
  G_m = Gas molar flow rate (lb mol/ft^2-hr).
  y_1 = Inlet gas mole fraction.
  y_2 = Outlet gas mole fraction.
  x_1^* = Outlet concentration.
  x_2 = Inlet liquid mole fraction.

$$(L_m/G_m)_{min} = (y_1 - y_2)/(x_1^* - x_2) = (0.02 - 0.00204)/(0.0167 - 0) = 1.08$$

Calculate the actual ratio of molar liquid flow rate to molar gas flow rate (L_m/G_m). Remember that the actual liquid flow rate is 25% more than the minimum based on the given operating conditions:

$$(L_m/G_m) = 1.25(L_m/G_m)_{min} = 1.25 \times 1.08 = 1.35$$

Calculate the value of $(y_1 - mx_1)/(y_2 - mx_2)$, where

  y_1 = Inlet gas mole fraction.
  m = Henry's law constant.
  x_1 = Outlet liquid mole fraction.
  y_2 = Outlet gas mole fraction.
  x_2 = Inlet liquid mole fraction.

$$(y_1 - mx_1)/(y_2 - mx_2) = [(0.02) - (1.2)(0)]/[(0.00204) - (1.2)(0)] = 9.80$$

Calculate the value of mG_m/L_m, where

> m = Henry's law constant.
> G_m = Gas molar flow rate (lb-mol/ft²-hr).
> L_m = Liquid molar flow rate (lb-mol/ft²-hr).

Even though the individual values of G_m and L_m are not known, the ratio of the two has been previously calculated:

$$mG_m/L_m = 1.2/1.35 = 0.889$$

Determine the number of overall gas transfer units (N_{OG}) using the values that we calculated previously (i.e., 9.80 and 0.899):

$$N_{OG} = 6.2$$

Now calculate the height of packing Z:

$$Z = (N_{OG})(H_{OG})$$

where

> Z = Height of packing.
> H_{OG} = Height of an overall gas transfer unit.
> N_{OG} = Number of transfer units.

$$Z = (N_{OG})(H_{OG}) = 6.2 \times 2.5 = 15.5 \text{ ft}$$

What is the diameter of the packed column? The actual gas mass velocity must be determined. To calculate the diameter of the column, we need the flooding gas mass velocity. USEPA's generalized flooding and pressure drop correction graph (USEPA, 1984b, p. 107) is used to determine the flooding gas mass velocity. The mass velocity is obtained by dividing the mass flow rate by the cross-sectional area. Calculate flooding gas mass velocity (G_f), and calculate the abscissa of USEPA's generalized flooding and pressure drop correction graph $(L/G)(p/p_L)^{0.5}$:

$$(L/G)(p/p_L)^{0.5} = (L_m/G_m)(18/29)(p/p_L)^{0.5}$$

where

> L = Liquid mass velocity (lb/s-ft²).
> G = Gas mass velocity, (lb/s-ft²).
> p = Gas density.
> p_L = Liquid density.
> L_m = Liquid molar flow rate (lb-mol/ft²-hr).
> G_m = Gas molar flow rate (lb-mol/ft²-hr).
> 18/29 = Ratio of molecular weight of water to air.

Note that the L and G terms (in USEPA's generalized flooding and pressure drop correction graph) are based on mass and not moles.

$$(L/G)(p/p_L)^{0.5} = (1.35)(18/29)(0.075/62.4)^{0.5} = 0.0291$$

Determine the value of the ordinate at the flooding line using the calculated value of the abscissa:

$$\text{Ordinate} = \frac{G^2 F \Psi \left(\mu_L\right)^{0.2}}{p_L p g_c}$$

where
   G = Gas mass velocity, (lb/ft^2-s).
   F = Packing factor = 160 for 1-inch Raschig rings.
   Ψ = Ratio, density of water/density of liquid.
   μ_L = Viscosity of liquid (cp).
   p_L = Liquid density.
   p = Gas density.
   g_c = 32.2 lb-ft/lb-s^2.

From USEPA's generalized flooding and pressure drop correction graph,

$$\frac{G^2 F \Psi (\mu_L)^{0.2}}{p_L p g_c} = 0.19$$

Solve the abscissa for the flooding gas mass velocity (G_f) in lb/ft^2-s. The G value becomes G_f for this case. Thus,

$$G_f = \left(\frac{(0.19)(p_L p g_c)}{F \Psi (\mu_L)^{0.2}} \right)^{0.5} = \left(\frac{(0.19)(62.4)(0.075)(32.2)}{(160)(1)(1.8)^{0.2}} \right)^{0.5} = 0.400 \text{ lb/ft}^2\text{-s}$$

Calculate the actual gas mass velocity (G_{act}) in lb/ft^2-s:

$$G_{act} = 0.6 G_f = 0.6 \times 0.400 = 0.240 \text{ lb/ft}^2\text{-s} = 864 \text{ lb/ft}^2\text{hr}$$

Calculate the diameter of the column in feet:

$$S = \text{(mass flowrate of gas stream)}/G_{act} = 5000/G_{act}$$

$$S = \pi D^2/4$$

$$\pi D^2/4 = 5000/G_{act}$$

$$D = [(4(5000))/(\pi G_{act})]^{0.5} = 2.71 \text{ ft}$$

REFERENCES AND RECOMMENDED READING

Calvert, S.J., Goldschmid, D., Leith, D., and Metha, D. (1972). *Wet Scrubber System Study*. Vol 1. *Scrubber Handbook*, EPA R2-72-118a. U.S. Environmental Protection Agency, Washington, DC.

Cooper, D. and Alley, F. (1994). *Air Pollution Control: A Design Approach*, 2nd ed. Waveland Press, Prospect Heights, IL.

Corbitt, R.A. (1990). *Standard Handbook of Environmental Engineering*. McGraw-Hill, New York.

Heumann, W.L. and Subramania, V. (1997). Particle scrubbing, in Heumann, W.L., Ed., *Industrial Air Pollution Control Systems*. McGraw-Hill, New York.

Lapple, C.E. and Kamack, H.J. (1955). Performance of wet dust scrubbers. *Chemical Engineering Progress*, 51, 110–121.

Nukiyama, S. and Tanasawa, Y. (1983). An experiment on atomization of liquid by means of air stream (in Japanese). *Transactions of the Japan Society of Mechanical Engineers*, 4, 86.

Perry, R. and Green, D., Eds. (1984). *Perry's Chemical Engineers' Handbook*, 6th ed.. McGraw-Hill, New York.

Semrau, K.T. (1960). Correlation of dust scrubber efficiency. *Journal of the Air Pollution Control Association*, 10, 200–207.

Semrau, K.T. (1963). Dust scrubber design—a critique on the state of the art. *Journal of the Air Control Association*, 13, 587–593.

Spellman, F.R. (1999). *The Science of Air: Concepts and Applications*. Technomic, Lancaster, PA.

Spellman, F.R. (2008). *The Science of Air: Concepts and Applications*, 2nd ed. CRC Press, Boca Raton, FL.

USEPA. (1971). *Control Techniques for Gases and Particulates*. U.S. Environmental Protection Agency, Washington, DC.

USEPA. (1984a). *APTI Course 413: Control of Particulate Emissions*, EPA 450/2-80-066. U.S. Environmental Protection Agency Air Pollution Training Institute, Research Triangle Park, NC.

USEPA. (1984b). *Control of Gaseous and Particulate Emissions: Self-Instructional Problem Workbook*, EPA 450/2-84-007. U.S. Environmental Protection Agency Air Pollution Training Institute, Research Triangle Park, NC.

USEPA. (1984c). *APTI Course SI:412C: Wet Scrubber Plan Review: Self-Instructional Guidebook*, EPA 450/2-82-020. U.S. Environmental Protection Agency Air Pollution Training Institute, Research Triangle Park, NC.

USEPA. (1992). *Control Technologies for Volatile Organic Compound Emissions from Stationary Sources*, EPA 453/R-92-018. U.S. Environmental Protection Agency, Washington, DC.

USEPA. (1996). *OAQPS Control Cost Manual*, 5th ed., EPA 45/B-96-001. U.S. Environmental Protection Agency, Washington, DC.

USEPA. (1998). *Stationary Source Control Techniques Document for Fine Particulate Matter*, EPA-452/R-97-001. U.S. Environmental Protection Agency, Washington, DC.

Yung, S., Calvert, S., and Barbarika, J.F. (1977). *Venturi Scrubber Performance Model*, EPA 600/2-77-172. U.S. Environmental Protection Agency, Cincinnati, OH.

Section X

Math Concepts: Water Quality

Nothing in the world is more flexible and yielding than water. Yet when it attacks the firm and the strong, none can withstand it, because they have no way to change it. So the flexible overcome the adamant, the yielding overcome the forceful. Everyone knows this, but no one can do it.

—Lao Tzu, Chinese philosopher (600–532 BC)

19 Running Waters

In terms of practical usefulness the waste assimilation capacity of streams as a water resource has its basis in the complex phenomenon termed *stream self-purification*. This is a dynamic phenomenon reflecting hydrologic and biologic variations, and the interrelations are not yet fully understood in precise terms. However, this does not preclude applying what is known. Sufficient knowledge is available to permit quantitative definition of resultant stream conditions under expected ranges of variation to serve as practical guides in decisions dealing with water resource use, development, and management.

— **Velz (1970)**

Running water is hypnotic, like fire watching, wave watching, or dust-mote watching.

19.1 BALANCING THE AQUARIUM

An outdoor excursion to the local stream can be a relaxing and enjoyable undertaking. However, when the wayfarer arrives at the local stream, spreads a blanket on its bank, and then looks out upon its flowing mass and discovers a parade of waste and discarded rubble bobbing along cluttering the adjacent shoreline and downstream areas, he quickly loses any feeling of relaxation or enjoyment. The sickening sensation only increases as our observer closely scrutinizes the putrid flow. He recognizes the rainbow-colored shimmer of an oil slick, interrupted here and there by dead fish and floating refuse, and the slimy fungal growth that prevails. At the same time, the observer's sense of smell is also alerted to the noxious conditions. Along with the fouled water and the stench of rot-filled air, the observer notices the ultimate insult and tragedy: a sign warning, "DANGER—NO SWIMMING or FISHING." The observer soon realizes that the stream before him is not a stream at all; it is little more than an unsightly drainage ditch. He has discovered what ecologists have known and warned about for years—that is, contrary to popular belief, rivers and streams do not have an infinite capacity for taking care of pollution (Spellman, 1996).

Before the early 1970s, occurrences such as the one just described were common along the rivers and streams near main metropolitan areas throughout most of the United States. Many aquatic habitats were fouled during the past because of industrialization, but our streams and rivers were not always in such deplorable condition. Before the Industrial Revolution of the 1800s, metropolitan areas were small and sparsely populated. Thus, river and stream systems within or next to early communities received insignificant quantities of discarded waste. Early on, these river and stream systems were able to compensate for the small amount of wastes they received; when wounded by pollution, nature has a way of fighting back. In the case of rivers and streams, nature provides their flowing waters with the ability to restore themselves through their own self-purification process. It was only when humans gathered in great numbers to form great cities that the stream systems were not always able to recover from having received great quantities of refuse and other wastes.

What exactly is it that human populations do to rivers and streams? Halsam (1990) pointed out that man's actions are determined by his expediency. Add to this the fact that most people do not realize that we have the same amount of water as we did millions of years ago. Through the water cycle, we continually reuse that same water; in fact, we are using water that was used by the ancient Romans and Greeks. Increased demand has put enormous stress on our water supply. Humans are the cause of this stress, as we upset the delicate balance between pollution and the purification process. Anyone who has kept fish knows what happens when the aquarium or pond water becomes too fouled. In a sense, we tend to unbalance the aquarium for our own water supplies.

With the advent of industrialization, local rivers and streams became deplorable cesspools that worsened with time. During the Industrial Revolution, the removal of horse manure and garbage from city streets became a pressing concern. Moran and colleagues (1986) pointed out that, "None too frequently, garbage collectors cleaned the streets and dumped the refuse into the nearest river." As late as 1887, river keepers were employed full time to remove a constant flow of dead animals from a river in London. Moreover, the prevailing attitude of that day was "I don't want it anymore, throw it into the river" (Halsam, 1990).

Once we came to understand the dangers of unclean waters, any threat to the quality of water destined for use for drinking and recreation has quickly angered those affected. Fortunately, since the 1970s we have moved to correct the stream pollution problem. Through scientific study and incorporation of wastewater treatment technology, we have started to restore streams to their natural condition.

Fortunately, through the phenomenon of self-purification, the stream aids us in this effort to restore a steam's natural water quality. A balance of biological organisms is normal for all streams. Clean, healthy streams have certain characteristics in common; for example, as mentioned, one property of streams is their ability to dispose of small amounts of pollution. However, if streams receive unusually large amounts of waste, the stream life will change and attempt to stabilize such pollutants; that is, the biota will attempt to balance the aquarium. If the stream biota are not capable of self-purifying, then the stream may become a lifeless body. This self-purification process relates to the purification of organic matter only. In this chapter, we briefly discuss only organic stream pollution and self-purification.

19.1.1 Sources of Stream Pollution

Sources of stream pollution are normally classified as point or nonpoint sources. A *point source* (PS) is a source that discharges effluent, such as wastewater from sewage treatment and industrial plants. Simply put, a point source is usually easily identified as "end of the pipe" pollution; that is, it emanates from a concentrated source or sources. In addition to organic pollution received from the effluents of sewage treatment plants, other sources of organic pollution include runoffs and dissolution of minerals throughout an area and are not from one or more concentrated sources. Non-concentrated sources are known as *nonpoint sources*. Nonpoint source (NPS) pollution, unlike pollution from industrial and sewage treatment plants, comes from many diffuse sources. Rainfall or snowmelt moving over and through the ground causes NPS pollution. As the runoff moves, it picks up and carries away natural and manmade pollutants, finally depositing them into streams, lakes, wetlands, rivers, coastal waters, and even our underground sources of drinking water. Some of these pollutants are listed below:

- Excess fertilizers, herbicides, and insecticides from agricultural lands and residential areas
- Oil, grease, and toxic chemicals from urban runoff and energy production
- Sediment from improperly managed construction sites, crop and forest lands, and eroding streambanks
- Salt from irrigation practices and acid drainage from abandoned mines
- Bacteria and nutrients from livestock, pet wastes, and faulty septic systems

Atmospheric deposition and hydromodification are also sources of nonpoint source pollution (USEPA, 1994).

Of particular interest to environmental practitioners are agricultural effluents. As a case in point, take, for example, farm silage effluent, which has been estimated to be more than 200 times as potent (in terms of biochemical oxygen demand, or BOD) as treated sewage (Mason, 1990).

Nutrients are organic and inorganic substances that provide food for microorganisms such as bacteria, fungi, and algae. Nutrients are supplemented by the discharge of sewage. The bacteria, fungi, and algae are consumed by the higher trophic levels in the community. Each stream, due to a limited amount of dissolved oxygen (DO), has a limited capacity for aerobic decomposition of organic matter without becoming anaerobic. If the organic load received is above that capacity, the stream becomes unfit for normal aquatic life, and it is not able to support organisms sensitive to oxygen depletion (Smith, 1974).

Effluent from a sewage treatment plant is most commonly disposed of in a nearby waterway. At the point of entry of the discharge, there is a sharp decline in the concentration of DO in the stream. This phenomenon is known as *oxygen sag*. Unfortunately (for the organisms that normally occupy a clean, healthy stream), when the DO decreases, there is a concurrent massive increase in BOD because microorganisms utilize the DO as they break down the organic matter. When the organic matter is depleted, the microbial population and BOD decline, while the DO concentration increases, assisted by stream flow (in the form of turbulence) and by the photosynthesis of aquatic plants. This self-purification process is very efficient, and the stream will suffer no permanent damage as long as the quantity of waste is not too high. Obviously, an understanding of this self-purification process is important in preventing stream ecosystem overload.

As urban and industrial centers continue to grow, waste disposal problems also grow. Because wastes have increased in volume and are much more concentrated than earlier, natural waterways must have help in the purification process. Wastewater treatment plants provide this help. Wastewater treatment plants function to reduce the organic loading that raw sewage would impose when discharged into streams. Wastewater treatment plants utilize three stages of treatment: primary, secondary, and tertiary treatment. In breaking down the wastes, a secondary wastewater treatment plant uses the same type of self-purification process found in any stream ecosystem. Small bacteria and protozoans (one-celled organisms) begin breaking down the organic material. Aquatic insects and rotifers are then able to continue the purification process. Eventually, the stream will recover and show little or no effects of the sewage discharge. This phenomenon is known as natural stream purification (Spellman and Whiting, 1999).

19.2   IS DILUTION THE SOLUTION?

In the early 1900s, wastewater disposal practices were based on the premise that "the solution is dilution." The most economical means of dealing with wastewater was to dispose of it into running waters (primarily rivers), and doing so was considered good engineering practice (Clark et al., 19977; Velz, 1970). Early practices in the field evolved around mixing-zone concepts based on the lateral, vertical, and longitudinal dispersion characteristics of the receiving waters (Peavy et al., 1975). Various formulas predicting space and dispersion characteristics for diluting certain pollutants to preselected concentrations were developed. Highly polluted discharged water was viewed as acceptable because the theory prevailed that the stream or river would eventually purify itself. Actually, a more accurate perception was that discharged wastewater out of sight was out of mind— leave it to the running water body to dilute the wastestream.

Although dilution is a powerful factor in the self-cleansing mechanisms of surface waters, it has its limitations. Dilution is a viable tool that takes advantage of the ability of running water to self-purify, but only if the discharges are limited to relatively small quantities of waste into relatively large bodies of water. One factor that impedes the ability of running water to self-purify is growth in the number of dumpers. "Growth in population and industrial activity, with attendant increases in water demand and wastewater quantities, precludes the use of many streams for dilution of raw or poorly treated wastewaters" (Peavy et al, 1975).

19.2.1 DILUTION CAPACITY OF RUNNING WATERS

Within limits, dilution is an effective means of dealing with a discharged wastestream. Immediately beyond the point of discharge, the process of mixing and dilution begins. However, complete mixing does not take place at the outfall. Instead, a waste plume is formed that gradually widens. The length and width (dispersion) of the plume depend upon running water geometry, flow velocity, and flow depth (Gupta, 1997). Beyond the mixing zone, the dilution capacity of running waters can be calculated using the principles of mass balance relation, using worst-case scenario conditions, and a 7-day, 10-year low flow for stream flow condition. A simplistic dilution equation can be written as

$$C_d = \frac{Q_s C_s + Q_w C_w}{Q_s + Q_w} \tag{19.1}$$

where
C_d = Completely mixed constituent concentration downstream of the effluent (mg/L).
Q_s = Stream flow upstream of the effluent, cubic feet per second (cfs).
C_s = Constituent concentration of upstream flow (mg/L).
Q_w = Flow of the effluent (cfs).
C_w = Constituent concentration of the effluent (mg/L).

■ EXAMPLE 19.1

Problem: A power plant pumps 25 cfs from a stream with a flow of 180 cfs. The discharge of the plant's ash pond is 22 cfs. The boron concentrations for upstream water and effluent are 0.053 and 8.7 mg/L, respectively. Compute the boron concentration in the stream after complete mixing.

Solution:

$$C_d = \frac{Q_s C_s + Q_w C_w}{Q_s + Q_w} = \frac{(180 - 25)(0.053) + (22)(8.7)}{(180 - 25) + 22} = 1.13 \text{ mg/L}$$

19.3 DISCHARGE MEASUREMENT

The total discharge for a running water body can be estimated by the float method with wind and other surface effects, by die study, or by actual subsection flow measurement, depending on cost, time, personnel, local conditions, etc. The discharge in a stream cross-section is typically measured from a subsection using the following equation:

$$Q = \text{Sum (mean depth} \times \text{width} \times \text{mean velocity)}$$

$$Q = \sum_{n=1}^{n} \frac{1}{2}(h_n + h_{n-1})(w_n - w_{n-1}) \times \frac{1}{2}(v_n + v_{n-1}) \tag{19.2}$$

where
Q = Discharge (cfs).
w_n = nth distance from initial point 0 (ft).
h_n = nth water depth (ft).
v_n = nth velocity (ft/s).

A velocity meter measures velocity (v).

19.4 TIME OF TRAVEL

Dye study or computation is used to determine the time of travel of running water. Running water time of travel and river or stream geometry characteristics can be computed using a volume displacement model. The time of travel is determined at any specific reach as the channel volume of the reach divided by the flow as follows:

$$t = V/Q \times 1/86,400 \qquad (19.3)$$

where
  t = Time of travel at a stream reach (days).
  V = Stream reach volume (ft^3 or m^3).
  Q = Average stream flow in each (ft^3/sec, cfs; m^3/m).

■ EXAMPLE 19.2

Problem: The cross-section areas at river miles 63.5, 64.0, 64.5, 65.0, and 65.7 are, respectively, 270, 264, 263, 258, 257, and 260 ft^2 at a surface water elevation. The average flow is 32.3 cfs. Find the time of travel for a reach between river miles 63.5 and 65.7.

Solution: Find the area in the reach:

$$\text{Average area} = 1/6(270 + 264 + 263 + 258 + 257 + 260) = 262 \text{ ft}^2$$

Now find the volume:

$$\text{Distance of the reach} = 65.7 \text{ miles} - 63.5 \text{ miles} = 2.2 \text{ miles} \times 5280 \text{ ft/mile} = 11,616 \text{ ft}$$

$$\text{Volume} = 262 \text{ ft}^2 \times 11,616 \text{ ft} = 3,043,392 \text{ ft}^3$$

Now find time of travel (t):

$$t = \frac{V}{Q} \times \frac{1}{86,400} = \frac{3,043,392}{32.3 \times 86,400} = 1.1 \text{ days}$$

19.5 DISSOLVED OXYGEN

A running water system both produces and consumes oxygen. It gains oxygen from the atmosphere and from plants as a result of photosynthesis. Running water, because of its churning, dissolves more oxygen than does still water, such as that in a lake, reservoir or pond. Respiration by aquatic animals, decomposition, and various chemical reactions consume oxygen. Dissolved oxygen (DO) in running waters is as critical to the good health of stream organisms as is gaseous oxygen to humans. Simply, DO is essential to the respiration of aquatic organisms, and its concentration in streams is a major determinant of the species composition of biota in the water and underlying sediments. Moreover, DO in streams has a profound effect on the biochemical reactions that occur in water and sediments, which in turn affect numerous aspects of water quality, including the solubility of many lotic elements and aesthetic qualities of odor and taste. For these reasons, DO historically has been one of the most frequently measured indicators of water quality (Hem, 1985).

In the absence of substances that cause its depletion, the DO concentration in running waters approximates the saturation level for oxygen in water in contact with the atmosphere and decreases with increasing water temperature from about 14 mg/L (milligrams per liter) at freezing to about 7 mg/L at 86°F (30°C). For this reason, in ecologically healthy streams, the DO concentration depends primarily on temperature, which varies with season and climate.

Criteria for defining desirable DO concentration often are differentiated as applicable to cold-water biota, such as trout and their insect prey, and the more low-oxygen-tolerant species of warm-water ecosystems. Moreover, because of the critical respiratory function of DO in aquatic animals, criteria often are expressed in terms of the short-term duration and frequency of occurrence of minimum concentration rather than long-term average concentrations. Studies cited by the U.S. Environmental Protection Agency (USEPA) of the dependence of freshwater biota on DO suggest that streams in which the concentration is less than 6.5 mg/L for more than about 20% of the time generally are not capable of supporting trout or other coldwater fish, and such concentrations could impair population growth among some warmwater game fish, such as largemouth bass (USEPA, 1986). Streams in which the DO deficit concentration is greater than 4 mg/L for more than 20% of the time generally cannot support either cold- or warmwater game fish. DO deficit refers to the difference between the saturation and measured concentrations of DO in a water sample and is a direct measure of the effects of oxygen-demanding substances on DO in streams.

Major sources of substances that cause depletion of DO in streams are discharges from municipal and industrial wastewater treatment plants; leaks and overflows from sewage lines and septic tanks; stormwater runoff from agricultural and urban land; and decaying vegetation, including aquatic plants from the stream itself and detrital terrestrial vegetation. DO is added to stream water by the process of aeration (waterfalls, riffles) and the photosynthesis of plants.

Dissolved oxygen saturation (DO_{sat}) values for various water temperatures can be computed using the American Society of Civil Engineer's equation (Elmore and Hayes, 1960). This equation represents saturation values for distilled water (β, 1.0) at sea-level pressure. Even though water impurities can increase or decrease the saturation level, for most cases β is assumed to be unity, or 1.0.

$$DO_{sat} = 14.652 - 0.41022T + 0.0079910T^2 - 0.0000777747T^3 \qquad (19.4)$$

where

DO_{sat} = Dissolved oxygen saturation concentration (mg/L).
T = Water temperature (°C).

■ EXAMPLE 19.3

Problem: Calculate the DO saturation concentration for water temperatures (T) of 0, 10, 20, and 30°C, assuming $\beta = 1.0$.

Solution:

At $T = 0$°C,

$$DO_{sat} = 14.652 - 0 + 0 - 0 = 14.652 \text{ mg/L}$$

At $T = 10$°C,

$$DO_{sat} = 14.652 - (0.41022 \times 10) + (0.0079910 \times 10^2) - (0.000077774 \times 10^3) = 11.27 \text{ mg/L}$$

At $T = 20$°C,

$$DO_{sat} = 14.652 - (0.41022 \times 20) + (0.0079910 \times 20^2) - (0.000077774 \times 20^3) = 9.02 \text{ mg/L}$$

At $T = 30$°C,

$$DO_{sat} = 14.652 - (0.41022 \times 30) + (0.0079910 \times 30^2) - (0.000077774 \times 30^3) = 7.44 \text{ (mg/L)}$$

19.5.1 Dissolved Oxygen Correction Factor

Because of differences in air pressure caused by air temperature changes and for elevation above the mean sea level (MSL), the DO saturation concentrations generated by the formula must be corrected. The correction factor (f) can be calculated as follows:

$$f = \frac{2116.8 - (0.08 - 0.000115A)E}{2116.8} \qquad (19.5)$$

where
  f = Correction factor for above MSL.
  A = Air temperature (°C).
  E = Elevation of the site (feet above MSL).

■ **EXAMPLE 19.4**

Problem: Find the correction factor for the DO_{sat} value for water at 640 ft above the MSL and air temperature of 25°C. What is the DO_{sat} at a water temperature of 20°C?

Solution:

$$f = \frac{2116.8 - (0.08 - 0.000115A)E}{2116.8}$$

$$= \frac{2116.8 - (0.08 - 0.000115 \times 25)640}{2116.8} = \frac{2116.8 - 49.4}{2116.8} = 0.977$$

DO_{sat} at $T = 20°C = 9.02$ mg/L. With an elevation correction factor of 0.977,

$$DO_{sat} = 9.02 \text{ mg/L} \times 0.977 = 8.81 \text{ mg/L}$$

19.6 BIOCHEMICAL OXYGEN DEMAND

Biochemical oxygen demand (BOD) measures the amount of oxygen consumed by microorganisms in decomposing organic matter in stream water. BOD also measures the chemical oxidation of inorganic matter (i.e., the extraction of oxygen from water via chemical reaction). A test is used to measure the amount of oxygen consumed by these organisms during a specified period of time (usually 5 days at 20°C). The rate of oxygen consumption in a stream is affected by a number of variables: temperature, pH, the presence of certain kinds of microorganisms, and the type of organic and inorganic material in the water. BOD directly affects the amount of dissolved oxygen in running waters. The greater the BOD, the more rapidly oxygen is depleted in the system. This means less oxygen is available to higher forms of aquatic life. The consequences of high BOD are the same as those for low dissolved oxygen: Aquatic organisms become stressed, suffocate, and die. Sources of BOD include leaves and wood debris; dead plants and animals; animal manure; effluents from pulp and paper mills, wastewater treatment plants, feedlots, and food-processing plants; failing septic systems; and urban stormwater runoff.

19.6.1 BOD Test Procedure

Standard BOD test procedures can be found in *Standard Methods for the Examination of Water and Wastewater* (Clesceri et al., 1999). When the dilution waste is seeded, oxygen uptake (consumed) is assumed to be the same as the uptake in the seeded blank. The difference between the sample BOD and the blank BOD, corrected for the amount of seed used in the sample, is the true BOD. Formulas for calculation of BOD are as follows (Clesceri et al., 1995):

- When dilution water is not seeded:

$$\text{BOD (mg/L)} = \frac{D_1 - D_2}{P} \tag{19.6}$$

- When dilution water is seeded:

$$\text{BOD (mg/L)} = \frac{(D_i - D_e) - (B_i - B_e)f}{P} \tag{19.7}$$

where
D_1, D_i = DO of diluted sample immediately after preparation (mg/L).
D_2, D_e = DO of diluted sample after incubation at 20°C (mg/L).
P = Decimal volumetric fraction of sample used (mL of sample/300 mL).
B_i = DO of seed control before incubation (mg/L).
B_e = DO of seed control after incubation (mg/L).
f = Ratio of seed in diluted sample to seed in seed control.
P = Percent seed in diluted sample/percent seed in seed control.

If seed material is added directly to the sample and to control bottles, f = volume of seed in diluted sample/volume of seed in seed control.

19.6.2 PRACTICAL BOD CALCULATION PROCEDURE

The following BOD calculation procedures for unseeded and seeded samples are commonly used in practice.

19.6.2.1 Unseeded BOD Procedure

1. Select the dilutions that meet the test criteria.
2. Calculate the BOD for each selected dilution using the following formula:

$$\text{BOD (mg/L)} = \frac{(DO_{start} - DO_{final}) \times 300 \text{ mL}}{\text{Sample volume (mL)}} \tag{19.8}$$

where DO_{start} and DO_{final} are in mg/L.

■ EXAMPLE 19.5

Problem: Determine BOD in mg/L given the following data:

Initial DO = 8.2 mg/L
Final DO = 4.4 mg/L
Sample size = 5 mL

Solution:

$$\text{BOD (mg/L)} = \frac{(8.2 - 4.4) \times 300 \text{ mL}}{5} = 228 \text{ mg/L}$$

19.6.2.2 Seeded BOD Procedure

1. Select those dilutions that meet the test criteria.
2. Calculate the BOD for each selected dilution using the following formula:

$$BOD\ (mg/L) = \frac{(DO_{start} - DO_{final}) - Seed\ correction \times 300\ mL}{Sample\ volume\ (mL)} \qquad (19.9)$$

where DO_{start} and DO_{final} are in mg/L.
Seed correction is calculated using

$$Seed\ correction\ (mg/L) = \frac{BOD_{start}}{300\ mL} \times Seed\ in\ sample\ dilution\ (mL) \qquad (19.10)$$

■ EXAMPLE 19.6

Problem: A series of seed dilutions were prepared in 300-mL BOD bottles using seed material (settled raw wastewater) and unseeded dilution water. The average BOD for the seed material was 204 mg/L. One milliliter of the seed material was also added to each bottle of a series of sample dilutions. Given the data for two samples in the table below, calculate the seed correction factor (SC) and BOD of the sample.

Bottle No.	Sample (mL)	Seed/Bottle (mL)	DO (mg/L) Initial	Final	Depletion (mg/L)
12	50	1	8.0	4.6	3.4
13	75	1	7.7	3.9	2.8

Solution: First calculate the BOD of each milliliter of seed material:

$$BOD\ per\ mL\ seed = \frac{204\ mg/L}{300\ mL} = 0.68\ mg/L\ BOD\ per\ mL\ seed$$

Calculate the seed correction (SC) factor:

$$Seed\ correction\ factor = (0.68\ mg/L\ BOD\ per\ mL\ seed) \times (1\ mL\ seed\ per\ bottle) = 0.68\ mg/L$$

Calculate the BOD of each sample dilution:

$$Bottle\ 12\ BOD\ (mg/L) = \frac{3.4 - 0.68}{50\ mL} \times 300 = 16.3\ mg/L$$

$$Bottle\ 13\ BOD\ (mg/L) = \frac{3.8 - 0.68}{75\ mL} \times 300 = 12.5\ mg/L$$

Finally, calculate reported BOD:

$$Reported\ BOD = \frac{16.3 + 12.5}{2} = 14.4\ mg/L$$

19.7 OXYGEN SAG (DEOXYGENATION)

Biochemical oxygen demand is the amount of oxygen required to decay or breakdown a certain amount of organic matter. Measuring the BOD of a stream is one way to determine how polluted it is. When too much organic waste, such as raw sewage, is added to the stream, all of the available oxygen will be used up. The high BOD reduces DO levels because they are interrelated. A typical DO-vs.-time or -distance curve is somewhat spoon-shaped due to the reaeration process. This spoon-shaped curve, commonly called the *oxygen sag curve*, is obtained using the Streeter–Phelps equation (to be discussed later).

Simply stated, an oxygen sag curve is a graph of the measured concentration of DO in water samples collected (1) upstream from a significant point source of readily degradable organic material (i.e., pollution), (2) from the area of the discharge, and (3) from some distance downstream from the discharge, plotted by sample location. The amount of DO is typically high upstream, diminishes at and just downstream from the discharge location (causing a sag in the line graph), and returns to the upstream levels at some distance downstream from the source of pollution or discharge. The oxygen sag curve is illustrated in Figure 19.1. From the figure, we can see that the percentage of DO vs. time or distance shows a characteristic sag, which occurs because the organisms breaking down the wastes use up the DO in the decomposition process. When the wastes are decomposed, recovery takes place and the DO rate rises again.

Several factors determine the extent of recovery. First of all, the minimum level of dissolved oxygen found below a sewage outfall depends on the BOD strength and quantity of the waste, as well as other factors. These other factors include velocity of the stream, stream length, biotic content, and the initial DO content (Porteous, 1992). The rates of reaeration and deoxygenation determine the amount of DO in the stream. If there is no reaeration, the DO will reach zero in a short period of time after the initial discharge of sewage into the stream. But, due to reaeration, the rate of which is influenced directly by the rate of deoxygenation, there is enough compensation for aerobic decomposition of organic matter. When the velocity of the stream is too low and the stream is too deep, the DO level may reach zero.

Depletion of oxygen causes a deficit in oxygen, which in turn causes absorption of atmospheric oxygen at the air–liquid interface. Thorough mixing due to turbulence brings about effective reaeration. A shallow, rapid stream will have a higher rate of reaeration (will be constantly saturated with oxygen) and will purify itself faster than a deep sluggish one (Smith, 1974).

Key Point: Reoxygenation of a stream is effected through aeration, absorption, and photosynthesis. Riffles and other natural turbulence in streams enhance aeration and oxygen absorption. Aquatic plants add oxygen to the water through transpiration. Oxygen production from photosynthesis of

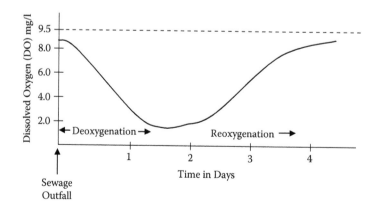

FIGURE 19.1 Oxygen-sag curve. (From Spellman, F.R., *Stream Ecology and Self-Purification*, Technomic, Lancaster, PA, 1996.)

aquatic plants, primarily blue–green algae, slows down or ceases at night, creating a diurnal or daily fluctuation in DO levels in streams. The amount of DO a stream can retain increases as water temperatures cool and concentration of dissolved solids diminishes.

19.8 STREAM PURIFICATION: A QUANTITATIVE ANALYSIS*

Before sewage is dumped into a stream it is important to determine the maximum BOD loading for the stream to avoid rendering it septic. The most common method of ultimate wastewater disposal is discharge into a selected body of water. The receiving water, stream, lake, or river is given the final job of purification. The degree of purification that takes place depends on the flow or volume, oxygen content, and reoxygenation ability of the receiving water. Moreover, self-purification is a dynamic variable, changing from day to day and closely following the hydrological variations characteristic to each stream. Additional variables include stream runoff, water temperature, reaeration, and the time of passage down the stream.

The purification process is carried out by several different aquatic organisms. As mentioned, during the purification of the waste, a sag in the oxygen content of the stream occurs. Mathematical expressions help in determining the oxygen response of the receiving stream. One must keep in mind, however, that since the biota and conditions in various parts of the stream change (that is, decomposition of organic matter in a stream is a function of degradation by microorganisms and oxygenation by reaeration, which are competing processes that are working simultaneously), it is difficult to quantify variables and results.

Streeter and Phelps first described the most common and well-known mathematical equation for oxygen sag for streams and rivers in 1925. The Streeter–Phelps equation is presented as follows:

$$D = \frac{k_1 \times L_a}{k_2 - k_1}\left(e^{-k_1 t} - e^{-k_2 t}\right) + D_a e^{-k_2 t} \qquad (19.11)$$

where
  D = Dissolved oxygen deficit (ppm).
  k_1 = BOD rate coefficient (per day).
  L_a = Ultimate BOD of the stream after the waste enters.
  k_2 = Reaeration constant (per day).
  t = Time of flow (days).
  D_a = Initial oxygen deficit (before discharge) (ppm).

Note: The deoxygenation constant, k_1, is the rate at which microbes consume oxygen for aerobic decomposition of organic matter. The following equation is used to calculate k_1:

$$y = L\left(1 - 10^{-kt}\right) \quad or \quad k_1 = \frac{-\log(1 - y/L)}{5t} \qquad (19.12)$$

where
  y = BOD$_5$ (5 days BOD).
  L = Ultimate or BOD$_{21}$.
  k_1 = Deoxygenation constant.
  t = Time in days (5 days).

---

* The important concepts presented in this section are excerpted from Spellman, F.R., *Stream Ecology and Self-Purification*, Technomic, Lancaster, PA, 1996.

TABLE 19.1

Typical Reaeration Constants (k_2) for Water Bodies

Water Body	Ranges of k_2 @ 20°C
Backwaters	0.10–0.23
Sluggish streams	0.23–0.35
Large streams (low velocity)	0.35–0.46
Large streams (normal velocity)	0.46–0.69
Swift streams	0.69–1.15
Rapids	>1.15

Source: Spellman, F.R., *Stream Ecology and Self-Purification*, Technomic, Lancaster, PA, 1996.

The reaeration constant, k_2, is the reaction characteristic of the stream and varies from stretch to stretch, depending on the velocity of the water, the depth, the surface area exposed to the atmosphere, and the amount of biodegradable organic matter in the stream. Reaeration constants are given in Table 19.1. The reaeration constant for a fast-moving, shallow stream is higher than for a sluggish stream or a lake. The reaeration constant for shallow streams, where vertical gradient and shear stress exist, is commonly found using the following formulation:

$$k_2 \ (20°C) = \frac{48.6 S^{1/4}}{H^{5/4}} \tag{19.13}$$

where

k_2 = Reaeration constant.
S = Slope of stream bed (ft/ft).
H = Stream depth (ft).

The reaeration constant for turbulence that is typical in deep streams can be found by using the following equation:

$$k_2 \ (20°C) = \frac{1.30 V^{1/2}}{H^{3/2}} \tag{19.14}$$

where:

k_2 = Reaeration constant.
V = Velocity of stream (ft/sec).
H = Stream depth (ft).

The Streeter–Phelps equation should be used with caution. It should be remembered that the Streeter–Phelps equation assumes that conditions (flow, BOD removal and oxygen demand rate, depth, and temperature throughout the stream) are constant. In other words, their equation assumes that all conditions are the same or constant for every stream. However, this is seldom true; a stream, from reach to reach, changes. Additionally, because rivers and streams are usually longer than they are wide, organic pollution mixes rapidly in these surface waters. Further, some rivers and streams are wider than others are. Thus, the mixing of organic pollutants with river or stream water does not occur at the same rate in different rivers and streams.

■ EXAMPLE 19.7

Problem: Calculate the oxygen deficit in a stream after pollution. Use the following equation and parameters for a stream to calculate the oxygen deficit (*D*) in the stream after pollution.

Parameters:

Pollution enters stream at point *X*.

$t = 2.13$.

$L_a = 22$ mg/L (pollution in stream at point *X*).

$D_a = 2$ mg/L.

$k_1 = 0.280$/day (base e).

$k_2 = 0.550$/day (base e).

Note: To convert log base e to base 10, divide by 2.31.

Solution:

$$D = \frac{k_1 \times L_a}{k_2 - k_1}\left(e^{-k_1 t} - e^{-k_2 t}\right) + D_a e^{-k_2 t}$$

$$D = \frac{0.280 \times 22}{0.550 - 0.280}\left(e^{-0.280 \times 2.13} - e^{-0.550 \times 2.13}\right) + 2e^{-0.550 \times 2.13}$$

$$= \frac{6.16}{0.270} \times \left(10^{-0.258} - 10^{-0.510}\right) + \left(2 \times 10^{-0.510}\right)$$

$$= 22.8 \times (0.5520 - 0.3090) + (2 \times 0.3090)$$

$$= 22.81 \times 0.243 + 0.6180$$

$$= 6.16 \text{ mg/L}$$

■ EXAMPLE 19.8

Problem: Calculate deoxygenation constant k_1 for a domestic sewage with BOD_5 of 135 mg/L and BOD_{21} of 400 mg/L.

Solution:

$$k_1 = \frac{-\log\left(1 - \frac{BOD_1}{400}\right)}{t} = \frac{-\log\left(1 - \frac{135}{400}\right)}{5} = \frac{-\log 0.66}{5} = \frac{0.1804}{5} = 0.361\text{/day}$$

Biochemical oxygen demand occurs in two different phases (see Figure 19.2). The first phase is *carbonaceous BOD* (CBOD), when mainly the organic or carbonaceous material is broken down. The second phase is the *nitrogenous BOD* phase. Here, nitrogen compounds are decomposed, which requires oxygen. This is of particular concern when conducting tests for discharge permit compliance, especially if nitrification is known to occur.

The Streeter–Phelps equation provides a rough estimate of the ecological conditions in a stream. Small variations in the stream may cause the DO to be higher or lower than the equation indicates; however, this equation may be used in several different ways. As mentioned, the quantity of the

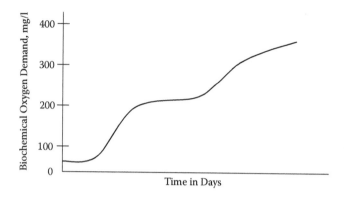

FIGURE 19.2 Carbonaceous and nitrogenous BOD in a waste sample.

waste is important in determining the extent of environmental damage. The Streeter–Phelps equation may be used to determine whether a certain stream has the capability to handle the estimated flow of wastes. In addition, if the stream receives wastes from other sites, then the equation may be helpful in determining whether the stream has recovered fully before it receives the next batch of wastes and, therefore, if it will be able to recover from the succeeding waste effluents.

REFERENCES AND RECOMMENDED READING

Clark, J.W., Viessman, Jr., W., and Hammer, M.J. (1977). *Water Supply and Pollution Control*, 3rd ed. New York: Harper & Row.

Clesceri, L.S., Greenberg, A.E., and Eaton, A.D. (1999). *Standards Methods for the Examination of Water and Wastewater*, 20th ed. American Public Health Association, Washington, DC.

Elmore, H.L. and Hayes, T.W. (1960). Solubility of atmospheric oxygen in water. *Proceedings of the American Society of Civil Engineers*, 86(SA4), 41–53.

Gupta, R.S. (1997). *Environmental Engineering and Science: An Introduction*. Government Institutes, Rockville, MD.

Halsam, S.M. (1990). *River Pollution: An Ecological Perspective*. Bellhaven Press, New York.

Hem, J.D. (1985). *Study and Interpretation of the Chemical Characteristics of Natural Water*, 3rd ed. U.S. Geological Survey, Washington, DC.

Mason, C.F. (1990). Biological aspects of freshwater pollution, in Harrison, R.M., Ed., *Pollution: Causes, Effects, and Control*. Royal Society of Chemistry, Cambridge, U.K.

Moran, J.M., Morgan, M.D., and Wiersma, J.H. (1986). *Introduction to Environmental Science*. W.H. Freeman, New York.

Peavy, H.S., Rowe, D.R., and Tchobanoglous, G. (1975). *Environmental Engineering*. McGraw-Hill, New York.

Porteous, A. (1992). *Dictionary of Environmental Science and Technology*, revised ed. John Wiley & Sons, New York.

Smith, R.I. (1974). *Ecology and Field Biology*. Harper & Row, New York.

Spellman, F.R. (1996). *Stream Ecology and Self-Purification*. Technomic, Lancaster, PA.

Spellman, F.R. and Whiting, N.E. (1999). *Water Pollution Control Technology: Concepts and Applications*. Government Institutes, Rockville, MD.

Streeter, J.P. and Phelps, E.B. (1925). *A study of the Pollution and Natural Purification of the Ohio River*, Bulletin No. 146. U.S. Public Health Service, Cincinnati, OH.

USEPA. (1986). *Quality Criteria for Water 1986*. U.S. Environmental Protection Agency, Washington, DC.

USEPA. (1994). *What Is Nonpoint Source Pollution?*, EPA-F-94-005. U.S. Environmental Protection Agency, Washington, DC.

Velz, C.J. (1970). *Applied Stream Sanitation*. Wiley-Interscience, New York.

20 Still Waters

Consider a river pool, isolated by fluvial processes and time from the main stream flow. We are immediately struck by one overwhelming impression: It appears so still ... so very still ... still enough to soothe us. The river pool provides a kind of poetic solemnity, if only at the pool's surface. No words of peace, no description of silence or motionless can convey the perfection of this place, in this moment stolen out of time.

We consider that the water is still, but does the term *still* correctly describe what we are viewing? Is there any other term we can use besides *still*—is there any other kind of still? Yes, of course, we know many ways to characterize still. *Still* can mean inaudible, noiseless, quiet, or silent. *Still* can also mean immobile, inert, motionless, or stationary—which is how the pool appears to the casual visitor on the surface. The visitor sees no more than water and rocks.

The rest of the pool? We know very well that a river pool is more than just a surface. How does the rest of the pool (the subsurface, for example) fit the descriptors we tried to use to characterize its surface? Maybe they fit, maybe they don't. In time, we will go beneath the surface, through the liquid mass, to the very bottom of the pool to find out. For now, remember that images retained from first glances are almost always incorrectly perceived, incorrectly discerned, and never fully understood.

On second look, we see that the fundamental characterization of this particular pool's surface is correct enough. Wedged in a lonely riparian corridor—formed by a river bank on one side and sand bar on the other—between a youthful, vigorous river system on its lower end and a glacier- and artesian-fed lake on its headwater end, almost entirely overhung by mossy old Sitka spruce, the surface of the large pool, at least at this particular location, is indeed still. In the proverbial sense, the pool's surface is as still and as flat as a flawless sheet of glass.

The glass image is a good one, because like perfect glass, the pool's surface is clear, crystalline, unclouded, definitely transparent, and yet perceptively deceptive as well. The water's clarity, accentuated by its bone-chilling coldness, is apparent at close range. Further back, we see only the world reflected in the water—the depths are hidden and unknown. Quiet and reflective, the polished surface of the water perfectly reflects in mirror-image reversal the spring greens of the forest at the pond's edge, without the slightest ripple. Up close, looking straight into the depths of the pool we are struck by the water's transparency. In the motionless depths, we do not see a deep, slow-moving reach with the muddy bottom typical of a river or stream pool; instead, we clearly see the warm variegated tapestry of blues, greens, blacks stitched together with threads of fine, warm-colored sand that carpets the bottom, at least 12 feet below. Still waters can run deep.

No sounds emanate from the pool. The motionless, silent water does not, as we might expect, lap against its bank or bubble or gurgle over the gravel at its edge. Here, the river pool, held in temporary bondage, is patient, quiet, waiting, withholding all signs of life from its surface visitor.

Then the reality check: This stillness, like all feelings of calm and serenity, could be fleeting, momentary, temporary, you think. And you would be correct, of course, because there is nothing still about a healthy river pool. At this exact moment, true clarity is present, it just needs to be perceived ... and it will be.

We toss a small stone into the river pool and watch the concentric circles ripple outward as the stone drops through the clear depths to the pool bottom. For a brief instant, we are struck by the obvious: The stone sinks to the bottom, following the laws of gravity, just as the river flows according to those same inexorable laws—downhill in its search for the sea. As we watch, the ripples die away, leaving as little mark as the usual human lifespan creates in the waters of the world, then disappear as if they had never been. Now the river water is as before, still. At the pool's edge, we peer down through the depth to the very bottom—the substrate.

We determine that the pool bottom is not flat or smooth but instead is pitted and mounded occasionally with discontinuities. Gravel mounds alongside small corresponding indentations—small, shallow pits—make it apparent to us that gravel was removed from the indentations and piled into slightly higher mounds. From our topside position, as we look down through the cool, quiescent liquid, the exact height of the mounds and the depth of the indentations are difficult for us to judge; our vision is distorted through several feet of water.

However, we can detect near the low gravel mounds (where female salmon buried their eggs and where their young grow until they are old enough to fend for themselves), and actually through the gravel mounds, movement—water flow—an upwelling of groundwater. This water movement explains our ability to see the variegated color of pebbles. The mud and silt that would normally cover these pebbles have been washed away by the water's subtle, inescapable movement. Obviously, in the depths, our still water is not as still as it first appeared.

The slow, steady, inexorable flow of water in and out of the pool, along with the upflowing of groundwater through the pool's substrate and through the salmon redds (nests) is only a small part of the activities occurring within the pool, including the air above it, the vegetation surrounding it, and the damp bank and sandbar forming its sides.

Let's get back to the pool itself. If we could look at a cross-sectional slice of the pool, at the water column, the surface of the pool may carry those animals that can literally walk on water. The body of the pool may carry rotifers and protozoa and bacteria—tiny microscopic animals—as well as many fish. Fish will also inhabit hidden areas beneath large rocks and ledges, to escape predators. Going down further in the water column, we come to the pool bed. This is the benthic zone, and certainly the greatest number of creatures live here, including larvae and nymphs of all sorts, worms, leeches, flatworms, clams, crayfish, dace, brook lampreys, sculpins, suckers, and water mites.

We need to go down even farther, down into the pool bed, to see the whole story. How far we have to go and what lives here, beneath the water, depends on whether it is a gravelly bed or a silty or muddy one. Gravel will allow water, with its oxygen and food, to reach organisms that live underneath the pool. Many of the organisms that are found in the benthic zone may also be found underneath, in the hyporheic zone. But to see the rest of the story we need to look at the pool's outlet, where its flow enters the main river. This is a riffle area—a shallow place where water runs fast and is disturbed by rocks. Only organisms that cling very well, such as net-winged midges, caddisflies, stoneflies, some mayflies, dace, and sculpins can spend much time here, and the plant life is restricted to diatoms and small algae. Riffles are a good place for mayflies, stoneflies, and caddisflies to live because they offer plenty of gravel to hide in.

Earlier, we struggled to find the right words to describe the river pool. Eventually, we settled on *still waters*. We did this because of our initial impression, and because of our lack of understanding and lack of knowledge. Even knowing what we know now, we might still describe the river pool as still waters. However, in reality, we must call the pool what it really is—a dynamic habitat. Each river pool has its own biological community, the members interwoven with each other in complex fashion, all depending on each other. Thus, our river pool habitat is part of a complex, dynamic ecosystem. On reflection, we realize, moreover, that anything dynamic certainly cannot be accurately characterized as still—including our river pool (Spellman and Drinan, 2001).

20.1 STILL WATER SYSTEMS

Freshwater systems may be conveniently considered in two classes: running water and still or standing water systems. There is no sharp distinction between the two classes. Lakes are defined as basins filled with water with no immediate means of flowing to the sea, containing relatively still waters. Ponds are small lakes in which rooted plants on the top layer reach to the bottom. Reservoirs are usually manmade impoundments of potable water. Lakes, ponds, and reservoirs are sensitive to pollution inputs because they flush out their contents relatively slowly. Lakes undergo eutrophication, an aging process caused by the inputs of organic matters and siltation. Simply, we can state that lakes, ponds, and reservoirs—that is, all still waters—are temporary holding basins.

20.2 STILL WATER SYSTEM CALCULATIONS

Environmental professionals involved with still water system management are generally concerned with determining and measuring lake, pond, or reservoir morphometric data, which are commonly recorded on pre-impoundment topographic maps. Determining and maintaining water quality in still water systems is also a major area of concern for environmental engineers. Water quality involves the physical, chemical, and biological integrity of water resources. The USEPA and other regulatory agencies promulgate water quality goals for protection of water resources in watershed management. Again, most still water data are directly related to the morphological features of the water basin.

Mapping the water basin should be the centerpiece of any comprehensive study of a still water body. Calculations made from the map allow the investigator to accumulate and relate a lot of data concerning the still water body system. When determining and measuring the water quality of a still water body, several different models are used. The purpose of modeling is to help the environmental engineer organize an extended project. Modeling is a direct measurement method, intended for a smaller body of water (e.g., lake, pond, reservoir); for example, water budget models and energy budget (lake evaporation) models can be used.

20.2.1 STILL WATER BODY MORPHOMETRY CALCULATIONS

20.2.1.1 Volume

The volume (V) of a still water body can be calculated when the area circumscribed by each isobath (i.e., each subsurface contour line) is known. The formula for water body volume is as follows (Wetzel, 1975):

$$V = \sum_{i=0}^{n} \frac{h}{3}\left(A_i + A_{i+1} + \sqrt{A_i \times A_{i+1}}\right) \tag{20.1}$$

where
  V = Volume (ft^3, acre-ft, m^3).
  h = Depth of the stratum (ft, m).
  i = Number of depth stratum.
  A_i = Area at depth i (ft^2, acre, m^2).

The formula for the volume of water between the shoreline contour (z_0) and the first subsurface contour (z_1) is as follows (Cole, 1994):

$$V_{z_1 - z_0} = \frac{1}{3}\left(A_{z_0} + A_{z_1} + \sqrt{\left(A_{z_0} + A_{z_1}\right)\left(z_1 - z_0\right)}\right) \tag{20.2}$$

where
  z_0 = Shoreline contour.
  z_1 = First subsurface contour.
  A_{z_0} = Total area of the water body.
  A_{z_1} = Area limited by the z_1 line.

20.2.1.2 Shoreline Development Index

The shoreline development index (D_L) is a comparative figure relating the shoreline length to the circumference of a circle that has the same area as the still water body. The smallest possible index would be 1.0. For the following formula, both L and A must be in consistent units for this comparison—meters and square meters:

$$D_L = \frac{L}{s\sqrt{\pi A}}$$ (20.3)

where
    D_L = Shoreline development index.
    L = Length of shoreline (miles or m).
    A = Surface area of lake (acre, ft^2, m^2).

20.2.1.3 Mean Depth

The still water body volume divided by its surface area will yield the mean depth. Remember to keep the units the same. If volume is in cubic meters, then area must be in square meters. The equation is as follows:

$$\bar{D} = \frac{V}{A}$$ (20.4)

where
    $\bar{D}$ = Mean depth (ft, m).
    V = Volume of lake (ft^3, acre-ft, m^3).
    A = Surface area (ft^2, acre, m^2).

■ EXAMPLE 20.1

Problem: A pond has a shoreline length of 8.6 miles. Its surface area is 510 acres. Its maximum depth is 8.0 feet. The areas for each foot depth are 460, 420, 332, 274, 201, 140, 110, 75, 30, and 1. Calculate the volume of the lake, shoreline development index, and mean depth of the pond.

Solution: Compute the volume of the pond:

$$V = \sum_{i=0}^{n} \frac{h}{3}\left(A_i + A_{i+1} + \sqrt{A_i \times A_{i+1}}\right)$$

$$= \frac{1}{3}\begin{bmatrix} \left(510+460+\sqrt{510\times460}\right)+\left(460+420+\sqrt{460\times420}\right)+\left(420+332+\sqrt{420\times332}\right) \\ +\left(332+274+\sqrt{332\times274}\right)+\left(274+201+\sqrt{274\times201}\right)+\left(201+140+\sqrt{201\times140}\right) \\ +\left(140+110+\sqrt{140\times110}\right)+\left(110+75+\sqrt{110\times75}\right)+\left(75+30+\sqrt{75\times30}\right) \\ +\left(30+1+\sqrt{30\times0}\right) \end{bmatrix}$$

$$= 1/3 \times 6823 = 2274 \text{ acre-ft}$$

Compute the shoreline development index:

$$A = 510 \text{ acres} = 510 \text{ acres} \times \frac{1 \text{ m}^2}{640 \text{ acres}} = 0.7969 \text{ m}^2$$

$$D_L = \frac{L}{2\sqrt{\pi A}} = \frac{8.60}{2\sqrt{3.14\times0.7969}} = \frac{8.60}{3.16} = 2.72$$

Compute the mean depth:

$$\bar{D} = \frac{V}{A} = \frac{2274 \text{ acre-ft}}{510 \text{ acres}} = 4.46 \text{ ft}$$

20.2.1.4   Bottom Slope

$$S = \frac{\bar{D}}{D_m} \tag{20.5}$$

where
    S = Bottom slope.
    $\bar{D}$ = Mean depth (ft, m).
    D_m = Maximum depth (ft, m).

20.2.1.5   Volume Development

Another morphometric parameter is volume development (D_v) (Cole, 1994). This metric compares the shape of the still water basin to an inverted cone with a height equal to D_m and a base equal to the still water body's surface area:

$$D_v = 3 \times \frac{\bar{D}}{D_m} \tag{20.6}$$

20.2.1.6   Water Retention Time

$$RT = \frac{\text{Storage capacity } \left(\text{acre-ft, m}^3\right)}{\text{Annual runoff } \left(\text{acre-ft/yr, m}^3\text{/yr}\right)} \tag{20.7}$$

where RT is retention time (years).

20.2.1.7   Ratio of Drainage Area to Still Water Body Capacity

$$R = \frac{\text{Drainage area } \left(\text{acre, m}^2\right)}{\text{Storage capacity } \left(\text{acre-ft, m}^3\right)} \tag{20.8}$$

■ EXAMPLE 20.2

Problem: Assume annual rainfall is 38.8 inches and watershed drainage is 10,220 acres. Using the data provided in Example 20.1, calculate the bottom slope, volume development, water retention time, and ratio of drainage area to lake capacity.

Solution: First determine the bottom slope:

$$S = \frac{\bar{D}}{D_m} = \frac{4.46}{8.0} = 0.56 \text{ ft}$$

Calculate volume development:

$$D_v = 3 \times \frac{\bar{D}}{D_m} = 3 \times 0.56 = 1.68 \text{ ft}$$

Calculate water retention time:

Storage capacity = 2274 acre-ft

Annual runoff = 38.8 in./yr × 10,220 acres

= 38.8 in./yr × 1 ft/12 in. × 10,220 acres = 33,045 acre-ft/year

$$\text{Retention time} = \frac{\text{Storage capacity}}{\text{Annual runoff}} = \frac{2274}{33,045} = 0.069 \text{ years}$$

Now calculate the ratio of drainage area to lake capacity:

$$R = \frac{\text{Drainage area}}{\text{Storage capacity}} = \frac{10,220}{2274} = \frac{4.49}{1}$$

20.3 STILL WATER SURFACE EVAPORATION

In lake, reservoir, and pond management, knowledge of evaporative processes is important to the environmental professional in understanding how water losses through evaporation are determined. Evaporation increases the storage requirement and decreases the yield of lakes and reservoirs. Several models and empirical methods are used for calculating lake and reservoir evaporative processes. The following sections discuss applications of the water budget and energy budget models, along with four empirical methods: Priestly–Taylor, Penman, DeBruin–Keijman, and Papadakis.

20.3.1 WATER BUDGET MODEL

The water budget model for lake evaporation is used to estimate lake evaporation in some areas. It depends on accurate measurement of the inflow and outflow of the lake and is expressed as

$$\Delta S = P + R + G_I - G_O - E - T - O \tag{20.9}$$

where
    ΔS = Change in lake storage (mm).
    P = Precipitation (mm).
    R = Surface runoff or inflow (mm).
    G_I = Groundwater inflow (mm).
    G_O = Groundwater outflow (mm).
    E = Evaporation (mm).
    T = Transpiration (mm).
    O = Surface water release (mm).

If a lake has little vegetation and negligible groundwater inflow and outflow, lake evaporation (E) can be estimated by

$$E = P + R - O \pm \Delta S \tag{20.10}$$

20.3.2   Energy Budget Model[*]

The energy budget (Lee and Swancar, 1996) is recognized as the most accurate method for determining lake evaporation. It is also the most costly and time-consuming method (Mosner and Aulenbach, 2003). The evaporation rate is given by

$$E_{EB} = \frac{Q_s - Q_r + Q_a + Q_{ar} - Q_{bs} + Q_v - Q_x}{L(1 + BR) + T_0}$$ (20.11)

where
  E_{EB} = Evaporation (cm/day).
  Q_s = Incident shortwave radiation (cal/cm²/day).
  Q_r = Reflected shortwave radiation (cal/cm²/day).
  Q_a = Incident longwave radiation from atmosphere (cal/cm²/day).
  Q_{ar} = Reflected longwave radiation (cal/cm²/day).
  Q_{bs} = Longwave radiation emitted by lake (cal/cm²/day).
  Q_v = Net energy advected by streamflow, ground water, and precipitation (cal/cm²/day).
  Q_x = Change in heat stored in water body (cal/cm²/day).
  L = Latent heat of vaporization (cal/g).
  BR = Bowen ratio (dimensionless).
  T_0 = Water surface temperature (°C).

20.3.3   Priestly–Taylor Equation

The Priestly–Taylor equation (Winter et al., 1995) is used to calculate potential evapotranspiration, which is a measure of the maximum possible water loss from an area under a specified set of weather conditions or evaporation as a function of latent heat of vaporization and heat flux in a water body. It is defined as

$$PET = \alpha \times \left(\frac{s}{s+\gamma}\right)\left[\frac{(Q_n - Q_x)}{L}\right]$$ (20.12)

where
  PET = Potential evapotranspiration (cm/day).
  α = 1.26, a Priestly–Taylor empirically derived constant (dimensionless).
  s = Slope of the saturated vapor pressure gradient (dimensionless).
  γ = Psychrometric constant (dimensionless).
  Q_n = Net radiation (cal/cm²/day).
  Q_x = Change in heat stored in water body (cal/cm²/day).
  L = Latent heat of vaporization (cal/g).

Note that s and γ are parameters derived from the slope of the saturated vapor pressure–temperature curve at the mean air temperature.

---

[*] Much of the following information is adapted from Mosner, M.S. and Aulenbach, B.T., *Comparison of Methods Used to Estimate Lake Evaporation for a Water Budget of Lake Seminole, Southwestern Georgia and Northwestern Florida*, U.S. Geological Survey, Atlanta, GA, 2003.

20.3.4 PENMAN EQUATION

The Penman equation (Winter et al., 1995) estimates potential evapotranspiration:

$$E_0 = \frac{\left(\Delta/\gamma\right)H_e + E_a}{\left(\Delta/\gamma\right)+1} \tag{20.13}$$

where

E_0 = Evapotranspiration.

Δ = Slope of the saturation absolute humidity curve at the air temperature.

γ = Psychrometric constant.

H_e = Evaporation equivalent of the net radiation.

E_a = Aerodynamic expression for evaporation.

20.3.5 DeBRUIN–KEIJMAN EQUATION

The DeBruin–Keijman equation (Winter et al., 1995) determines evaporation rates as a function of the moisture content of the air above the water body, the heat stored in the still water body, and the psychrometric constant, which is a function of atmospheric pressure and latent heat of vaporization:

$$PET = \left(\frac{SVP}{0.95SVP + 0.63\gamma}\right) \times \left(Q_n - Q_x\right) \tag{20.14}$$

where SVP is the saturated vapor pressure at mean air temperature (millibars/K). All other terms have been defined previously.

20.3.6 PAPADAKIS EQUATION

The Papadakis equation (Winger et al., 1995) does not account for the heat flux that occurs in the still water body to determine evaporation. Instead, the equation depends on the difference in the saturated vapor pressure above the water body at maximum and minimum air temperatures, and evaporation is defined by the equation

$$PET = 0.5625\left[e_0 \max - \left(e_0 \min - 2\right)\right] \tag{20.15}$$

where all terms have been defined previously.

REFERENCES AND RECOMMENDED READING

Cole, G.A. (1994). *Textbook of Limnology*, 4th ed. Waveland Press, Prospect Heights, IL.

Lee, T.M. and Swancar, A. (1996). *Influence of Evaporation, Ground Water, and Uncertainty in the Hydrologic Budget of Lake Lucerne, a Seepage Lake in Polk County, Florida*. U.S. Geological Survey, Atlanta, GA.

Mosner, M.S. and Aulenbach, B.T. (2003). *Comparison of Methods Used to Estimate Lake Evaporation for a Water Budget of Lake Seminole, Southwestern Georgia and Northwestern Florida*. U.S. Geological Survey, Atlanta, GA.

Rosenberry, D.O., Sturrock, A.M., and Winter, T.C. (1993). Evaluation of the energy budget method of determining evaporation at Williams Lake, Minnesota, using alternative instrumentation and study approaches. *Water Resources Research*, 29(8), 2473–2483.

Spellman, F.R. and Drinan, J. (2001). *Stream Ecology and Self-Purification*, 2nd ed. Technomic, Lancaster, PA.

Wetzel, R.G. (1975). *Limnology*. W.B. Saunders, Philadelphia, PA.

Winter, T.C., Rosenberry, D.O., and Sturrock, A.M. (1995). Evaluation of eleven equations for determining evaporation for a small lake in the north central United States. *Water Resources Research*, 31(4), 983–993.

21 Groundwater

... Anyone who has learned the language of running water will see its character in the dark.

—John Muir (1872)

Once polluted, groundwater is difficult, if not impossible, to clean up, since it contains few decomposing microbes and is not exposed to sunlight, strong water flow, or any of the other natural purification processes that cleanse surface water.

—Odum (1993)

21.1 GROUNDWATER AND AQUIFERS

Part of the precipitation that falls on land may infiltrate the surface, percolate downward through the soil under the force of gravity, and become what is known as *groundwater*. Groundwater, like surface water, is an extremely important part of the hydrologic cycle. Almost half of the people in the United States obtain their public water supply from groundwater. Overall, there is more groundwater than surface water in the United States, including the water in the Great Lakes, but sometimes it is not economical to pump it to the surface for use, and in recent years the pollution of groundwater supplies from improper disposal has become a significant problem (Spellman, 1996).

Groundwater is found in saturated layers under the Earth's surface called *aquifers* that lie under the Earth's surface. Aquifers are made up of a combination of solid material, such as rock and gravel, and open spaces called *pores*. Regardless of the type of aquifer, the groundwater in the aquifer is always in motion. The aquifer that lies just under the Earth's surface is the zone of saturation, an *unconfined aquifer* (see Figure 21.1). The top of the zone of saturation is the water table. An unconfined aquifer is not contained, except on the bottom, and is dependent on local precipitation for recharge. This type of aquifer is often referred to as a *water table aquifer.*

The actual amount of water in an aquifer is dependent upon the amount of space available between the various grains of material that make up the aquifer. The amount of space available is called *porosity*. The ease of movement through an aquifer is dependent upon how well the pores are connected. The ability of an aquifer to pass water is its *permeability*. Types of aquifers include the following:

1. *Unconfined aquifers* are a primary source of shallow well water (see Figure 21.1). However, because these wells are shallow they are subject to local contamination from hazardous and toxic material such as fuel and oil, agricultural runoff containing nitrates and microorganisms, and septic tanks that provide increased levels of nitrates and microorganisms. This type of well may be classified as groundwater under the direct influence of surface water (GUDISW) and may, therefore, require treatment for control of microorganisms (disinfection).
2. *Confined aquifers* are sandwiched between two impermeable layers that block the flow of water. The water in a confined aquifer is under hydrostatic pressure. It does not have a free water table (see Figure 21.2).

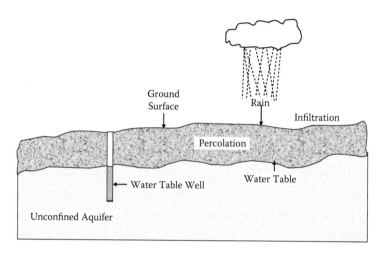

FIGURE 21.1   Unconfined aquifer. (From Spellman, F.R., *Stream Ecology and Self-Purification*, Technomic, Lancaster, PA, 1996.)

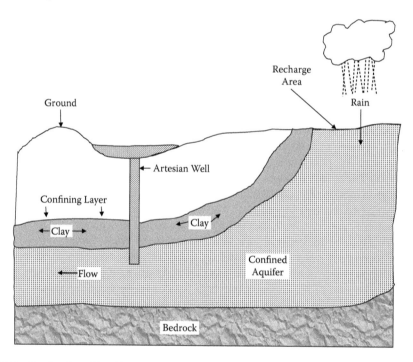

FIGURE 21.2   Confined aquifer. (From Spellman, F.R., *Stream Ecology and Self-Purification*, Technomic, Lancaster, PA, 1996.)

Confined aquifers are artesian aquifers. A well drilled in an artesian aquifer is called an *artesian well* and commonly yields large quantities of high-quality water. A well in a confined aquifer is normally referred to as a *deep well* and is not generally affected by local hydrological events. A confined aquifer is recharged by rain or snow in the mountains where it is close to the surface of the Earth. Because the recharge area is some distance from the area of possible contamination of the confined aquifer, the possibility of contamination is usually very low. However, once contaminated, it may take centuries before it recovers. When groundwater exits the Earth's crust it is called a *spring*. The water in a spring can originate from a water table aquifer or from a confined aquifer. Only water from a confined aquifer spring is considered desirable for a public water system.

21.1.1 GROUNDWATER QUALITY

Generally, groundwater possesses high chemical, bacteriological, and physical quality. When pumped from an aquifer composed of a mixture of sand and gravel, groundwater is often used without filtration (i.e., if not directly influenced by surface water). It can also be used without disinfection if it has a low coliform count. However (as pointed out earlier), groundwater can become contaminated when, for example, septic systems fail, saltwater intrudes, wastes are disposed of improperly, chemicals are improperly stockpiled, underground storage tanks leak, hazardous materials are spilled, fertilizers and pesticides are misapplied, and when mines are improperly abandoned. When groundwater is removed from its underground water-bearing stratum via a well, water flows toward the center of the well. In a water table aquifer, this movement will cause the water table to sag toward the well. This sag is called the *cone of depression*. The shape and size of the cone are dependent on the relationship between the pumping rate and the rate at which water can move toward the well. If the rate is high, the cone will be shallow and its growth will stabilize. The area that is included in the cone of depression is called the *zone of influence*; any contamination in this zone will be drawn into the well.

21.1.2 GROUNDWATER UNDER THE DIRECT INFLUENCE OF SURFACE WATER

Groundwater under the direct influence of surface water (GUDISW) is not classified as a groundwater supply. When a supply is designated as GUDISW, the state's surface water rules apply to the source rather than the groundwater rules. The Surface Water Treatment Rule of the Safe Drinking Water Act requires each site to determine which groundwater supplies are influenced by surface water (i.e., when surface water can infiltrate a groundwater supply and could contaminate it with *Giardia*, viruses, turbidity, or organic material from the surface water source). To determine whether a groundwater supply is under the direct influence of surface water, the U.S. Environmental Protection Agency (USEPA) has developed procedures that focus on significant and relatively rapid shifts in water quality characteristics such as turbidity, temperature, and pH. When these shifts can be closely correlated with rainfall or other surface water conditions or when certain indicator organisms associated with surface water are found, the source is said to be under the direct influence of surface water.

21.2 AQUIFER PARAMETERS

Certain aquifer parameters are relevant to determining the available volume of water and the ease of its withdrawal. We identify and define these relevant parameters in this section.

21.2.1 AQUIFER POROSITY

Aquifer porosity is defined as the ratio of the volume of voids (open spaces) in the soil to the total volume. Simply stated, porosity is the volume of open space and is often determined using the following equation:

$$\Phi = \frac{V_{\text{void}}}{V_{\text{total}}} \tag{21.1}$$

Two basic types of porosity are common: *primary*, formed at the time the rock was deposited, and *secondary*, formed later (e.g., by dissolution of carbonate in caverns). Well-sorted materials tend to have higher porosities than poorly sorted materials. Fine-grained sediments tend to have higher porosities than coarse-grained sediments, although they are often poorly connected. Some typical values of porosity are clay, 55%; fine sand, 45%; sand and gravel, 20%; sandstone, 15%; and limestone, 15%. The interconnected or effective porosity (Φ_e) is the most important in hydrology and is always $\leq \Phi$.

21.2.2 SPECIFIC YIELD (STORAGE COEFFICIENT)

Specific yield is the percentage of water that is free to drain from the aquifer under the influence of gravity. It is not equal to porosity because the molecular and surface tension forces in the pore spaces keep some of the water in the voids. Specific yield reflects the amount of water available for development (Davis and Cornwell, 1991). Specific yield and the storage coefficient may be used interchangeably for unconfined aquifers.

21.2.3 PERMEABILITY

Permeability (K) is the property of an aquifer that is a measure of its ability to transmit water under a sloping piezometric surface. It is defined as the discharge that occurs through a unit cross-section of aquifer.

21.2.4 TRANSMISSIVITY

Transmissivity (T) describes the capacity of an aquifer to transmit water. It is the product of hydraulic conductivity (permeability) and the aquifer's saturated thickness:

$$T = K \times b \tag{21.2}$$

where
    T = Transmissivity of an aquifer (gpd/ft, m²/d).
    K = Permeability (gpd/ft², m³/(d·m)).
    b = Thickness of aquifer (ft, m).

A rough estimation of T can be obtained by multiplying specific capacity by 2000 (USEPA, 1994).

■ EXAMPLE 21.1

Problem: If an aquifer's thickness is 60 ft, estimate the permeability of the aquifer with transmissivity of 30,000 gpm/ft.

Solution: Rearranging Equation 21.2,

$$K = T/b = (30{,}000 \text{ gpm/ft})/60 \text{ ft} = 500 \text{ gpm/ft}^2$$

21.2.5 HYDRAULIC GRADIENT AND HEAD

The height of the potentiometric surface at any point in an aquifer is the hydraulic gradient. Stated differently, the hydraulic gradient is the slope of the piezometric surface. The difference in elevation from one point to another along the hydraulic gradient is a measure of pressure. This elevation difference is called pressure head. We usually express the amount of mechanical energy that groundwater contains in terms of the hydraulic head (h), the total mechanical energy per unit weight. The hydraulic head is conveniently measured as the elevation to which water will rise in an open pipe, relative to some reference level. The hydraulic head therefore has units of length and is the elevation to which water will rise.

Two main components contribute to the mechanical energy or the hydraulic head of groundwater: potential energy due to gravity and pressure that is exerted on the water. Kinetic energy, a third energy, is due to the movement of water and is very small compared to the other two energies because groundwater flows very slowly and can therefore be neglected. In terms of hydraulic head, the potential energy is expressed as the elevation head (z) or simply the elevation of the point of interest relative to some reference level. The energy of fluid pressure is expressed as the pressure head (h_p). The pressure head is equivalent to the height of the water column overlying the point of interest. The total hydraulic head is then given by

$$h = z + h_p \tag{21.3}$$

Groundwater will move from areas of high mechanical energy to areas with low mechanical energy (Baron, 2003). The hydraulic gradient of a flow system of interest is defined as the difference in hydraulic head between two points of interest (dh) and the flow distance between these two points (dl), or in mathematical terms

$$\text{Gradient } h = dh/dl \tag{21.4}$$

21.2.6 FLOW LINES AND FLOW NETS

A flow line is an imaginary line that follows the path that a parcel of groundwater would follow as it flows through an aquifer. Flow lines are useful tools for visualizing the flow of groundwater. Flow lines can be constructed from equipotential lines or lines of equal hydraulic head. The combination of equipotential lines and flow lines results in a flow net—basically, a graphical solution of the two-dimensional Laplace's equation (Fetter, 1994).

21.3 GROUNDWATER FLOW

Groundwater flow for a steady-state condition in which the water table or piezometric head does not change within a specified time is expressed by the following equations (Gupta, 1997):

$$\text{Pore velocity or advection, } v = \frac{K(h_1 - h_2)}{nL} \tag{21.5}$$

where pore area of flow $A_v = \eta A$.

Because $Q = A_v$, by combining we can obtain Darcy's law:

$$\text{Rate of groundwater flow, } Q = \frac{K(h_1 - h_2)A}{L} \tag{21.6}$$

where

Q = Rate of groundwater flow.
K = Hydraulic conductivity.
h_1 = Water head at upstream end.
h_2 = Water head at downstream end.
A = Aquifer cross-section area through which flow takes place.
L = Distance between h_1 and h_2.
η = Porosity.

Note: The term $(h_1 - h_2)/L$ is the hydraulic gradient.

■ EXAMPLE 21.2

Problem: An irrigation ditch runs parallel to a pond. They are 2200 ft apart. A pervious formation of 40-ft average thickness connects them. Hydraulic conductivity and porosity of the pervious formation are 12 ft/day and 0.55, respectively. The water level in the ditch is at an elevation of 120 ft and at 110 ft in the pond. Determine the rate of seepage from the ditch to the pond.

Solution:

$$\text{Hydraulic gradient, } I = \frac{h_1 - h_2}{L} = \frac{120 - 110}{2200} = 0.0045$$

For each 1 ft width,

$$A = 1 \times 40 = 40 \text{ ft}^2$$

From Equation 21.6:

$$Q = 12 \text{ ft/day} \times 0.0045 \times 40 \text{ ft}^2) = 2.16 \text{ ft}^3/\text{day/ft width}$$

From Equation 21.5:

$$\text{Seepage velocity, } v = \frac{K(h_1 - h_2)}{nL} = \frac{12 \times 0.0045}{0.55} = 0.098 \text{ ft/day}$$

21.4   GENERAL EQUATIONS OF GROUNDWATER FLOW

The combination of Darcy's law and a statement of mass conservation results in general equations describing the flow of groundwater through a porous medium. These general equations are partial differential equations in which the spatial coordinates in all three dimensions, x, y, and z, and the time are all independent variables. To derive the general equations, Darcy's law and the law of mass conservation are applied to a small volume of aquifer, the *control volume*. The law of mass conservation is basically an account of all the water that goes into and out of the control volume. That is, all the water that goes into the control volume has to come out, or there has to be a change in water storage in the control volume. Applying these two laws to a confined aquifer results in Laplace's equation, a famous partial differential equation that can also be used to describe many other physical phenomena—for example, the flow of heat through a solid (Baron, 2003):

$$\frac{\partial^2 h}{\partial x^2} + \frac{\partial^2 h}{\partial y^2} + \frac{\partial^2 h}{\partial z^2} = 0$$

Applying Darcy's law and the law of mass conservation to two-dimensional flow in an unconfined aquifer results in the Boussinesq equation:

$$\partial/\partial x \left(h \, \partial h/\partial s \right) + \partial/\partial x \left(h \, \partial h/\partial y \right) = \frac{S_y}{K} \frac{\partial h}{\partial T}$$

where S_y is the specific yield of the aquifer. If the drawdown in the aquifer is very small compared with the saturated thickness, the variable thickness, h, can be replaced with an average thickness that is assumed to be constant over the aquifer. The Boussinesq equation can then be simplified to

$$\frac{\partial^2 h}{\partial x^2} + \frac{\partial^2 h}{\partial y^2} = \frac{S_y}{Kb} \frac{\partial h}{\partial t}$$

Describing groundwater flow in confined and unconfined aquifers, using the general partial differential equations, is difficult to solve directly. However, these differential equations can be simplified to algebraic equations for the solution of simple cases (e.g., one-dimensional flow in a homogeneous porous medium). Another approach is to use a flow net (described earlier) to graphically solve Laplace's equation for relatively simple cases. More complex cases, however, have to be solved mathematically, most commonly with computerized groundwater modeling programs. The most popular of these programs is MODFLOW-2000, published by the USGS.

21.4.1 STEADY FLOW IN A CONFINED AQUIFER

If steady movement of groundwater in a confined aquifer and the hydraulic heads do not change over time, we can use another derivation of Darcy's law directly to determine how much water is flowing through a unit width of aquifer, using the following equation:

$$q' = -Kb\left(\frac{dh}{dl}\right) \tag{21.7}$$

where
q' = Flow per unit width (L^2/T).
K = Hydraulic conductivity (L/T).
b = Aquifer thickness (L).
dh/dl = Hydraulic gradient (dimensionless).

21.4.2 STEADY FLOW IN AN UNCONFINED AQUIFER

Steady flow of water through an unconfined aquifer can be described by

$$q' = -\frac{1}{2}K\left(\frac{h_1^2 - h_2^2}{L}\right) \tag{21.8}$$

where h_1 and h_2 are the water level at two points of interest, and L is the distance between these two points. This equation states that the hydraulic gradient is equal to the slope of the water table; the streamlines are horizontal and the equipotential lines are vertical. This equation is useful, particularly in field evaluations of the hydraulic characteristics of aquifer materials.

REFERENCES AND RECOMMENDED READING

Baron, D. (2003). *Water: California's Precious Resource*. California State University, Bakersfield.

Davis, M.L. and Cornwell, D.A. (1985). *Introduction to Environmental Engineering*, 2nd ed. McGraw-Hill, New York.

Fetter, C.W. (1994). *Applied Hydrology*, 3rd ed. Prentice-Hall, New York.

Gupta, R.S. (1997). *Environmental Engineering and Science: An Introduction*. Government Institutes, Rockville, MD.

Odum, E.P. (1993). *Ecology and Our Endangered Life Support Systems*. Sinauer Associates, Sunderland, MA.

USEPA. (1994). *Handbook: Ground Water and Wellhead Protection*, EPA/625/R-94/001. U.S. Environmental Protection Agency, Washington, DC.

22 Water Hydraulics

Evidence conforms to conceptions just as often as conceptions conform to evidence.

— Ludwik Fleck (1896–1961)

22.1 INTRODUCTION

The word "hydraulics" is derived from the Greek words *hydro* ("water") and *aulis* ("pipe"). Originally, the term referred only to the study of water at rest and in motion (flowing through pipes or channels). Today, it is taken to mean the flow of *any* liquid in a system. *Hydraulics*—the study of fluids at rest and in motion—is essential for an understanding of how water/wastewater systems work, especially water distribution and wastewater collection systems.

22.2 BASIC CONCEPTS

$$\text{Air pressure (at sea level)} = 14.7 \text{ pounds per square inch (psi)}$$

This relationship is important because our study of hydraulics begins with air. A blanket of air many miles thick surrounds the Earth. The weight of this blanket on a given square inch of the Earth's surface will vary according to the thickness of the atmospheric blanket above that point. As shown above, at sea level the pressure exerted is 14.7 pounds per square inch (psi). On a mountain top, air pressure decreases because the blanket is not as thick.

$$1 \text{ ft}^3 \text{ H}_2\text{O} = 62.4 \text{ lb}$$

This relationship is also important; note that both cubic feet and pounds are used to describe a volume of water. A defined relationship exists between these two methods of measurement. The specific weight of water is defined relative to a cubic foot. One cubic foot of water weighs 62.4 lb (see Figure 22.1). This relationship is true only at a temperature of 4°C and at a pressure of 1 atmosphere, conditions which are referred to as *standard temperature and pressure* (STP). Note that 1 atmosphere = 14.7 lb/in.^{2} at sea level and 1 ft^3 of water contains 7.48 gal.

The weight varies so little that, for practical purposes, this weight is used for temperatures ranging from 0 to 100°C. One cubic inch of water weighs 0.0362 lb. Water 1 ft deep will exert a pressure of 0.43 lb/in.^{2} on the bottom area (12 in. × 0.0362 lb/in.^{3}). A column of water 2 ft high exerts 0.86 psi (2 ft × 0.43 psi/ft); one 10 ft high exerts 4.3 psi (10 ft × 0.43 psi/ft); and one 55 ft high exerts 23.65 psi (55 ft × 0.43 psi/ft). A column of water 2.31 feet high will exert 1.0 psi (2.31 ft × 0.43 psi/ft). To produce a pressure of 50 psi requires a 115.5-ft water column (50 psi × 2.31 ft/psi).

Another relationship is also important:

$$1 \text{ gal } H_2O = 8.34 \text{ lb}$$

At standard temperature and pressure, 1 ft³ of water contains 7.48 gal. With these two relationships, we can determine the weight of 1 gal of water. This is accomplished by

$$\text{Weight of 1 gal of water} = 62.4 \text{ lb} \div 7.48 \text{ gal} = 8.34 \text{ lb/gal}$$

Thus,

$$1 \text{ gal } H_2O = 8.34 \text{ lb}$$

■ EXAMPLE 22.1

Problem: Find the number of gallons in a reservoir that has a volume of 855.5 ft³.

Solution:

$$855.5 \text{ ft}^3 \times 7.48 \text{ gal/ft}^3 = 6399 \text{ gal (rounded)}$$

The term *head* is used to designate water pressure in terms of the height of a column of water in feet; for example, a 10-ft column of water exerts 4.3 psi. This can be referred to as 4.3 psi pressure or 10 ft of head. Another example: If the static pressure in a pipe leading from an elevated water storage tank is 45 pounds per square inch (psi), what is the elevation of the water above the pressure gauge?

Remembering that 1 psi = 2.31 feet and that the pressure at the gauge is 45 psi,

$$45 \text{ psi} \times 2.31 \text{ ft/psi} = 104 \text{ ft}$$

When demonstrating the relationship of the weight of water to the weight of air we can say, theoretically, that the atmospheric pressure at sea level (14.7 psi) will support a column of water 34 feet high:

$$14.7 \text{ psi} \times 2.31 \text{ ft/psi} = 34 \text{ ft}$$

At an elevation of one mile above sea level, where the atmospheric pressure is 12 psi, the column of water would be only 28 feet high: 12 psi × 2.31 ft/psi = 28 ft (rounded).

If a glass or clear plastic tube is placed in a body of water at sea level, the water will rise in the tube to the same height as the water outside the tube. The atmospheric pressure of 14.7 psi will push down equally on the water surface inside and outside the tube. However, if the top of the tube is tightly capped and all of the air is removed from the sealed tube above the water surface, forming a *perfect vacuum*, the pressure on the water surface inside the tube will be zero psi. The atmospheric pressure of 14.7 psi on the outside of the tube will push the water up into the tube until the weight of the water exerts the same 14.7 psi pressure at a point in the tube even with the water surface outside the tube. The water will rise 14.7 psi × 2.31 ft/psi = 34 feet.

In practice, it is impossible to create a perfect vacuum, thus the water will rise somewhat less than 34 feet; the distance it rises depends on the amount of vacuum created. If, for example, enough air was removed from the tube to produce an air pressure of 9.7 psi above the water in the tube, how far will the water rise in the tube? To maintain the 14.7 psi at the outside water surface level, the water in the tube must produce a pressure of 14.7 psi – 9.7 psi = 5.0 psi. The height of the column of water that will produce 5.0 psi is

$$5.0 \text{ psi} \times 2.31 \text{ ft/psi} = 11.5 \text{ ft}$$

22.2.1 Stevin's Law

Stevin's law deals with water at rest. Specifically, it states: "The pressure at any point in a fluid at rest depends on the distance measured vertically to the free surface and the density of the fluid." Stated as a formula, this becomes

$$p = w \times h \tag{22.1}$$

where
p = Pressure in pounds per square foot (lb/ft² or psf).
w = Density in pounds per cubic foot (lb/ft³).
h = Vertical distance in feet.

■ EXAMPLE 22.2

Problem: What is the pressure at a point 18 ft below the surface of a reservoir?

Solution: To calculate this, we must know that the density of the water (w) is 62.4 lb/ft³:

$$p = w \times h = 62.4 \text{ lb/ft}^3 \times 18 \text{ ft} = 1123 \text{ lb/ft}^2 \text{ (psf)}$$

Environmental professionals involved with water and wastewater treatment generally measure pressure in pounds per square inch rather than pounds per square foot; to convert, divide by 144 in.²/ft² (12 in. × 12 in. = 144 in.²):

$$P = \frac{1123 \text{ psf}}{144 \text{ in.}^2/\text{ft}^2} = 7.8 \text{ lb/in.}^2 \text{ or psi (rounded)}$$

22.2.2 Density and Specific Gravity

When we say that iron is heavier than aluminum, we say that iron has greater density than aluminum. In practice, what we are really saying is that a given volume of iron is heavier than the same volume of aluminum. *Density* is the *mass per unit volume* of a substance. Consider a tub of lard and a large box of cold cereal, each having a mass of 600 grams. The density of the cereal would be much less than the density of the lard because the cereal occupies a much larger volume than the lard occupies. The density of an object can be calculated by using the formula:

$$\text{Density} = \frac{\text{Mass}}{\text{Volume}} \tag{22.2}$$

In general use, perhaps the most common measures of density are pounds per cubic foot (lb/ft³) and pounds per gallon (lb/gal):

- 1 ft³ of water weighs 62.4 lb; density = 62.4 lb/ft³.
- 1 gal of water weighs 8.34 lb; density = 8.34 lb/gal.

The density of a dry material, such as cereal, lime, soda, or sand, is usually expressed in pounds per cubic foot. The densities of plain and reinforced concrete are 144 and 150 lb/ft³, respectively. The density of a liquid, such as liquid alum, liquid chlorine, or water, can be expressed either as pounds per cubic foot or as pounds per gallon. The density of a gas, such as chlorine gas, methane, carbon dioxide, or air, is usually expressed in pounds per cubic foot.

The density of a substance such as water changes slightly as the temperature of the substance changes. This occurs because substances usually increase in volume as they become warmer. Because of this expansion with warming, the same weight is spread over a larger volume, so the density is lower when a substance is warm than when it is cold.

Specific gravity is defined as the weight (or density) of a substance compared to the weight (or density) of an equal volume of water. (The specific gravity of water is 1.) This relationship is easily seen when a cubic foot of water, which weighs 62.4 lb, is compared to a cubic foot of aluminum, which weights 178 lb. Aluminum is 2.7 times as heavy as water.

Finding the specific gravity of a piece of metal is not difficult. All we have to do is to weigh the metal in air, then weigh it under water. Its loss of weight is the weight of an equal volume of water. To find the specific gravity, divide the weight of the metal by its loss of weight in water:

$$\text{Specific Gravity} = \frac{\text{Weight of substance}}{\text{Weight of equal volume of water}} \tag{22.3}$$

■ EXAMPLE 22.3

Problem: Suppose a piece of metal weighs 150 lb in air and 85 lb under water. What is the specific gravity?

Solution:

$$150 \text{ lb} - 85 \text{ lb} = 65 \text{ lb loss of weight in water}$$

$$\text{Specific Gravity} = \frac{150}{65} = 2.3$$

Note: In a calculation of specific gravity, it is *essential* that the densities be expressed in the same units.

The specific gravity of water is 1, which is the standard, the reference against which all other liquid or solid substances are compared. Specifically, any object that has a specific gravity greater than 1 (e.g., rocks, steel, iron, grit, floc, sludge) will sink in water. Substances with a specific gravity of less than 1 (e.g., wood, scum, gasoline) will float. Considering the total weight and volume of a ship, its specific gravity is less than 1; therefore, it can float.

The most common use of specific gravity in water/wastewater treatment operations is in gallon-to-pound conversions. In many cases, the liquids being handled have a specific gravity of 1 or very nearly 1 (between 0.98 and 1.02), so 1 may be used in the calculations without introducing significant error. However, in calculations involving a liquid with a specific gravity of less than 0.98 or greater than 1.02, the conversions from gallons to pounds must consider the exact specific gravity. This technique is illustrated in the following example.

■ EXAMPLE 22.4

Problem: A basin contains 1455 gal of a liquid. If the specific gravity of the liquid is 0.94, how many pounds of liquid are in the basin?

Solution: Normally, for a conversion from gallons to pounds, we would use the factor 8.34 lb/gal (the density of water) if the specific gravity of the substance is between 0.98 and 1.02. In this instance, however, the substance has a specific gravity outside this range, so the 8.34 factor must be adjusted by multiplying 8.34 lb/gal by the specific gravity to obtain the adjusted factor:

$$8.34 \text{ lb/gal} \times 0.94 = 7.84 \text{ lb/gal (rounded)}$$

Then convert 1455 gal to pounds using the corrected factor:

$$1455 \text{ gal} \times 7.84 \text{ lb/gal} = 11,407 \text{ lb (rounded)}$$

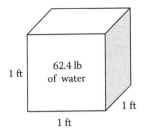

FIGURE 22.1 One cubic foot of water weights 62.4 lb. (From Spellman, F.R. and Drinan, J., *Water Hydraulics*, Technomic, Lancaster, PA, 2001.)

22.2.3 FORCE AND PRESSURE

Water exerts force and pressure against the walls of its container, whether it is stored in a tank or flowing in a pipeline. Force and pressure are different, although they are closely related. *Force* is the push or pull influence that causes motion. In the English system, force and weight are often used in the same way. The weight of 1 ft³ of water is 62.4 lb. The force exerted on the bottom of a 1-ft cube is 62.4 lb (Figure 22.1). If we stack two 1-ft cubes on top of one another, the force on the bottom will be 124.8 lb. *Pressure* is the force per unit of area. In equation form, this can be expressed as

$$P = \frac{F}{A} \tag{22.4}$$

where
  P = Pressure.
  F = Force.
  A = Area over which the force is distributed.

Pounds per square inch (lb/in.² or psi) or pounds per square foot (lb/ft²) are common expressions of pressure. The pressure on the bottom of the cube is 62.4 lb/ft² (see Figure 22.1). It is normal to express pressure in pounds per square inch. This is easily accomplished by determining the weight of 1 in.² of a 1-ft cube. If we have a cube that is 12 in. on each side, the number of square inches on the bottom surface of the cube is 12 × 12 = 144 in². Dividing the weight by the number of square inches determines the weight on each square inch:

$$\text{psi} = \frac{62.4 \ \text{lb/ft}}{144 \ \text{in.}^2} = 0.433 \ \text{psi/ft}$$

This is the weight of a column of water 1 in. square and 1 ft tall. If the column of water was 2 ft tall, the pressure would be 2 ft × 0.433 psi/ft = 0.866. With the above information, feet of head can be converted to psi by multiplying the feet of head by 0.433 psi/ft.

■ EXAMPLE 22.5

Problem: A tank is mounted at a height of 90 ft. Find the pressure at the bottom of the tank.

Solution:

$$90 \ \text{ft} \times 0.433 \ \text{psi/ft} = 39 \ \text{psi (rounded)}$$

If we wanted to make the conversion of psi to feet, we would divide the psi by 0.433 psi/ft.

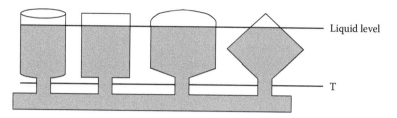

FIGURE 22.2 Hydrostatic pressure. (From Spellman, F.R. and Drinan, J., *Water Hydraulics*, Technomic, Lancaster, PA, 2001.)

■ **EXAMPLE 22.6**

Problem: Find the height of water in a tank if the pressure at the bottom of the tank is 22 psi.

Solution:

$$\text{Height} = \frac{22 \text{ psi}}{0.433 \text{ psi/ft}} = 51 \text{ ft (rounded)}$$

22.2.4 HYDROSTATIC PRESSURE

Figure 22.2 shows a number of differently shaped, connected, open containers of water. Note that the water level is the same in each container, regardless of the shape or size of the container. This occurs because pressure is developed within a liquid by the weight of the liquid above. If the water level in any one container is momentarily higher than that in any of the other containers, the higher pressure at the bottom of this container would cause some water to flow into the container having the lower liquid level. In addition, the pressure of the water at any level (such as line T) is the same in each of the containers. Pressure increases because of the weight of the water. The farther down from the surface, the more pressure is created. This illustrates that the weight, not the volume, of water contained in a vessel determines the pressure at the bottom of the vessel.

1. The pressure depends only on the depth of water above the point in question (not on the water surface area).
2. The pressure increases in direct proportion to the depth.
3. The pressure in a continuous volume of water is the same at all points that are at the same depth.
4. The pressure at any point in the water acts in all directions at the same depth.

22.2.5 HEAD

Head is defined as the vertical distance the water/wastewater must be lifted from the supply tank to the discharge or as the height a column of water would rise due to the pressure at its base. A perfect vacuum plus atmospheric pressure of 14.7 psi would lift the water 34 ft. When the top of the sealed tube is open to the atmosphere and the reservoir is enclosed, the pressure in the reservoir is increased; the water will rise in the tube. Because atmospheric pressure is essentially universal, we usually ignore the first 14.7 psi of actual pressure measurements and measure only the difference between the water pressure and the atmospheric pressure; we call this *gauge pressure*. Consider water in an open reservoir subjected to 14.7 psi of atmospheric pressure; subtracting this 14.7 psi leaves a gauge pressure of 0 psi, indicating that the water would rise 0 feet above the reservoir surface. If the gauge pressure in a water main were 120 psi, the water would rise in a tube connected to the main:

$$120 \text{ psi} \times 2.31 \text{ ft/psi} = 277 \text{ ft (rounded)}$$

The *total head* includes the vertical distance the liquid must be lifted (*static head*), the loss to friction (*friction head*), and the energy required to maintain the desired velocity (*velocity head*):

$$\text{Total Head} = \text{Static Head} + \text{Friction Head} + \text{Velocity Head} \qquad (22.5)$$

22.2.5.1 Static Head

Static head is the actual vertical distance the liquid must be lifted:

$$\text{Static Head} = \text{Discharge Elevation} - \text{Supply Elevation} \qquad (22.6)$$

■ EXAMPLE 22.7

Problem: A supply tank is located at elevation 118 ft. The discharge point is at elevation 215 ft. What is the static head in feet?

Solution:
$$\text{Static head} = 215 \text{ ft} - 118 \text{ ft} = 97 \text{ ft}$$

22.2.5.2 Friction Head

Friction head is the equivalent distance of the energy that must be supplied to overcome friction. Engineering references include tables showing the equivalent vertical distance for various sizes and types of pipes, fittings, and valves. The total friction head is the sum of the equivalent vertical distances for each component:

$$\text{Friction Head (ft)} = \text{Energy losses due to friction} \qquad (22.7)$$

22.2.5.3 Velocity Head

Velocity head is the equivalent distance of the energy consumed to achieve and maintain the desired velocity in the system:

$$\text{Velocity Head (ft)} = \text{Energy losses to maintain velocity} \qquad (22.8)$$

or

$$V_h = \frac{V^2}{2g}$$

22.2.5.4 Total Dynamic Head (Total System Head)

$$\text{Total Head} = \text{Static Head} + \text{Friction Head} + \text{Velocity Head} \qquad (22.9)$$

22.2.5.5 Pressure and Head

The pressure exerted by water/wastewater is directly proportional to its depth or head in the pipe, tank, or channel. If the pressure is known, the equivalent head can be calculated:

$$\text{Head (ft)} = \text{Pressure (psi)} \times 2.31 \text{ (ft/psi)} \qquad (22.10)$$

or

$$PE = pAV \text{ (for water flow in pipe)}$$

where
   p = Pressure at a cross-section.
   A = Pipe cross-sectional area (cm², in.²).
   V = Mean velocity.

■ **EXAMPLE 22.8**

Problem: The pressure gauge on the discharge line from the influent pump reads 72.3 psi. What is the equivalent head in feet?

Solution:

$$\text{Head} = 72.3 \times 2.31 \text{ ft/psi} = 167 \text{ ft}$$

22.2.5.6   Head and Pressure

If the head is known, the equivalent pressure can be calculated by

$$\text{Pressure (psi)} = \frac{\text{Head (ft)}}{2.31 \text{ ft/psi}} \tag{22.11}$$

■ **EXAMPLE 22.9**

Problem: A tank is 22 ft deep. What is the pressure in psi at the bottom of the tank when it is filled with water?

Solution:

$$\text{Pressure} = \frac{22 \text{ ft}}{2.31 \text{ ft/psi}} = 9.52 \text{ psi}$$

22.3   FLOW AND DISCHARGE RATE: WATER IN MOTION

The study of fluid flow is much more complicated than that of fluids at rest, but it is important to have an understanding of these principles because the water in a waterworks and distribution system and in a wastewater treatment plant and collection system is nearly always in motion. *Discharge* (or flow) is the quantity of water passing a given point in a pipe or channel during a given period. Stated another way for open channels, the flow rate through an open channel is directly related to the velocity of the liquid and the cross-sectional area of the liquid in the channel:

$$Q = A \times V \tag{22.12}$$

where
   Q = Flow, or discharge in cubic feet per second (cfs).
   A = Cross-sectional area of the pipe or channel (ft²).
   V = Water velocity in feet per second (fps, ft/sec).

■ **EXAMPLE 22.10**

Problem: A channel is 6 ft wide and the water depth is 3 ft. The velocity in the channel is 4 fps. What is the discharge or flow rate in cubic feet per second?

Solution:

$$\text{Flow} = 6 \text{ ft} \times 3 \text{ ft} \times 4 \text{ ft/sec} = 72 \text{ cfs}$$

Discharge or flow can be recorded as gal/day (gpd), gal/min (gpm), or cubic feet per second (cfs). Flows treated by many waterworks or wastewater treatment plants are large and are often referred to in million gallons per day (MGD). The discharge or flow rate can be converted from cfs to other units such as gpm or MGD by using appropriate conversion factors.

■ EXAMPLE 22.11

Problem: A 12-in.-diameter pipe has water flowing through it at 10 fps. What is the discharge in (a) cfs, (b) gpm, and (c) MGD?

Solution: Before we can use the basic formula, we must determine the area (*A*) of the pipe. The formula for the area of a circle is

$$A = \pi \times \frac{D^2}{4} \tag{22.13}$$

where
  π = Constant value 3.14159.
  D = Diameter of the circle in feet.

Thus, the area of the pipe is

$$A = \pi \times \frac{D^2}{4} = 3.14 \times \frac{(1 \text{ ft})^2}{4} = 0.785 \text{ ft}^2$$

Now, we can determine the discharge in cubic feet per second for part (a):

$$Q = V \times A = 10 \text{ ft/sec} \times 0.785 \text{ ft}^2 = 7.85 \text{ ft}^3/\text{sec (cfs)}$$

For part (b), we need to know that 1 cfs is equal to 449 gpm, so 7.85 cfs × 449 gpm/cfs = 3525 gpm. Finally, for part (c), 1 MGD is equal to 1.55 cfs, so

$$\frac{7.85 \text{ cfs}}{1.55 \text{ cfs/MGD}} = 5.06 \text{ MGD}$$

22.3.1 AREA AND VELOCITY

The *law of continuity* states that the discharge at each point in a pipe or channel is the same as the discharge at any other point (if water does not leave or enter the pipe or channel). That is, under the assumption of steady-state flow, the flow that enters the pipe or channel is the same flow that exits the pipe or channel. In equation form, this becomes

$$Q_1 = Q_2 \text{ or } A_1V_1 = A_2V_2 \tag{22.14}$$

■ EXAMPLE 22.12

Problem: A pipe 12 in. in diameter is connected to a 6-in.-diameter pipe. The velocity of the water in the 12-in. pipe is 3 fps. What is the velocity in the 6-in. pipe?

Solution: Using the equation $A_1V_1 = A_2V_2$, we need to determine the area of each pipe:

$$\text{12-inch pipe:} \quad A = \pi \times \frac{D^2}{4} = 3.14159 \times \frac{1 \text{ ft}^2}{4} = 0.785 \text{ ft}^2$$

$$\text{6-inch pipe:} \quad A = \pi \times \frac{D^2}{4} = 3.14159 \times \frac{0.5 \text{ ft}^2}{4} = 0.196 \text{ ft}^2$$

The continuity equation now becomes

$$0.785 \text{ ft}^2 \times 3 \text{ ft/sec} = 0.196 \text{ ft}^2 \times V_2$$

Solving for V_2,

$$V_2 = \frac{0.785 \text{ ft}^2 \times 3 \text{ ft/sec}}{0.196 \text{ ft}^2} = 12 \text{ ft/sec or fps}$$

22.3.2 Pressure and Velocity

In a closed pipe flowing full (under pressure), the pressure is indirectly related to the velocity of the liquid. This principle, when combined with the principle discussed in the previous section, forms the basis for several flow measurement devices (Venturi meters and rotameters), as well as the injector used for dissolving chlorine into water and for dissolving chlorine, sulfur dioxide, or other chemicals into wastewater:

$$\text{Velocity}_1 \times \text{Pressure}_1 = \text{Velocity}_2 \times \text{Pressure}_2 \tag{22.15}$$

or

$$V_1 P_1 = V_2 P_2$$

22.4 BERNOULLI'S THEOREM

Swiss physicist and mathematician Samuel Bernoulli developed the calculation for the total energy relationship from point to point in a steady-state fluid system in the 1700s (Nathanson, 1997). Before discussing Bernoulli's energy equation, it is important to understand the basic principle behind Bernoulli's equation. Water (and any other hydraulic fluid) in a hydraulic system possesses two types of energy—kinetic and potential. *Kinetic energy* is present when the water is in motion. The faster the water moves, the more kinetic energy is used. *Potential energy* is a result of the water pressure. The *total energy* of the water is the sum of the kinetic and potential energy. Bernoulli's principle states that the total energy of the water (fluid) always remains constant; therefore, when the water flow in a system increases, the pressure must decrease. When water starts to flow in a hydraulic system, the pressure drops. When the flow stops, the pressure rises again. The pressure gauges shown in Figure 22.3 illustrate this balance more clearly.

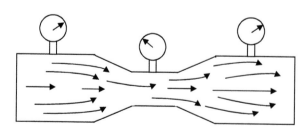

FIGURE 22.3 Demonstration of Bernoulli's principle. (From Spellman, F.R. and Drinan, J., *Water Hydraulics*, Technomic, Lancaster, PA, 2001.)

22.4.1   Bernoulli's Equation

In a hydraulic system, total energy head is equal to the sum of three individual energy heads. This can be expressed as

Total Head = Elevation Head + Pressure Head + Velocity Head

This can be expressed mathematically as

$$E = z + \frac{p}{w} + \frac{V^2}{2g} \qquad (22.16)$$

where
  E = Total energy head.
  z = Height of the water above a reference plane (ft).
  p = Pressure (psi).
  w = Unit weight of water (62.4 lb/ft^3).
  v = Flow velocity (ft/sec).
  g = Acceleration due to gravity (32.2 ft/sec^2).

Consider the constriction in the section of pipe shown in Figure 22.4. We know, based on the law of energy conservation, that the total energy head at section A (E_1) must equal the total energy head at section B (E_2), and using Equation 22.16 we get Bernoulli's equation:

$$z_A + \frac{P_A}{w} + \frac{V_A^2}{2g} = z_B + \frac{P_B}{w} + \frac{V_B^2}{2g} \qquad (22.17)$$

The pipeline system shown in Figure 22.4 is horizontal; therefore, we can simplify Bernoulli's equation because $z_A = z_B$. Because they are equal, the elevation heads cancel out from both sides, leaving

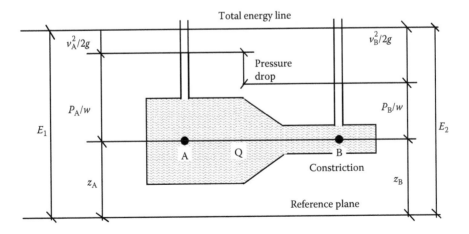

FIGURE 22.4   The law of conservation: Because the velocity and kinetic energy of the water flowing in the constricted section must increase, the potential energy will decrease. This is observed as a pressure drop in the constriction. (Adapted from Nathanson, J.A., *Basic Environmental Technology: Water Supply, Waste Management, and Pollution Control*, 2nd ed., Prentice-Hall, Upper Saddle River, NJ, 1997, p. 29.)

$$\frac{P_A}{w} + \frac{V_A^2}{2g} \cdot \frac{P_B}{w} + \frac{V_B^2}{2g} \qquad (22.18)$$

In Figure 22.4, as water passes through the constricted section of the pipe (section B), we know from continuity of flow that the velocity at section B must be greater than the velocity at section A, because of the smaller flow area at section B. This means that the velocity head in the system increases as the water flows into the constricted section. However, the total energy must remain constant. For this to occur, the pressure head, and therefore the pressure, must drop. In effect, pressure energy is converted into kinetic energy in the constriction.

The fact that the pressure in the narrower pipe section (constriction) is less than the pressure in the bigger section seems to defy common sense; however, it does follow logically from continuity of flow and conservation of energy. The fact that there is a pressure difference allows measurement of flow rate in the closed pipe.

■ EXAMPLE 22.13

Problem: In Figure 22.4, the diameter at section A is 8 in., and at section B it is 4 in. The flow rate through the pipe is 3.0 cfs and the pressure at section A is 100 psi. What is the pressure in the constriction at section B?

Solution: Compute the flow area at each section, as follows:

$$A_A = \frac{\pi(0.666 \text{ ft})^2}{4} = 0.349 \text{ ft}^2 \text{ (rounded)}$$

$$A_B = \frac{\pi(0.333 \text{ ft})^2}{4} = 0.087 \text{ ft}^2 \text{ (rounded)}$$

From $Q = A \times V$ or $V = Q/A$, we get

$$V_A = \frac{3.0 \text{ ft}^3/\text{s}}{0.349 \text{ ft}^2} = 8.6 \text{ ft/s (rounded)}$$

and

$$V_B = \frac{3.0 \text{ ft}^3/\text{s}}{0.087 \text{ ft}^2} = 34.5 \text{ ft/s (rounded)}$$

22.5   CALCULATING MAJOR HEAD LOSS

Darcy, Weisbach, and others developed the first practical equation used to determine pipe friction in about 1850. The equation or formula now known as the *Darcy–Weisbach* equation for circular pipes is

$$h_f = f\frac{LV^2}{D^2 g} \qquad (22.19)$$

$$h_f = \frac{8fLQ^2}{\pi^2 g D^5} \qquad (22.20)$$

where

h_f = Head loss (ft).
f = Coefficient of friction.
L = Length of pipe (ft).
V = Mean velocity (ft/sec).
D = Diameter of pipe (ft).
g = Acceleration due to gravity (32.2 ft/sec^2).
Q = Flow rate, (ft^3/sec).

The Darcy–Weisbach formula was meant to apply to the flow of any fluid, and into this friction factor was incorporated the degree of roughness and an element known as the *Reynolds number*, which is based on the viscosity of the fluid and the degree of turbulence of flow. The Darcy–Weisbach formula is used primarily for head loss calculations in pipes. For open channels, the *Manning* equation was developed during the later part of the 19th century. Later, this equation was used for both open channels and closed conduits.

In the early 1900s, a more practical equation, the *Hazen–Williams* equation, was developed for use in making calculations related to water pipes and wastewater force mains:

$$Q = 0.435 \times CD^{2.63} \times S^{0.54} \tag{22.21}$$

where

Q = Flow rate (ft^3/sec).
C = Coefficient of roughness (C decreases with roughness).
D = Hydraulic radius R (ft).
S = Slope of energy grade line (ft/ft).

22.5.1 C FACTOR

As mentioned in Chapter 14, the C factor, as used in the Hazen–Williams formula, designates the coefficient of roughness. C does not vary appreciably with velocity, and by comparing pipe types and ages it includes only the concept of roughness, ignoring fluid viscosity and Reynolds number. Based on experience (experimentation), accepted tables of C factors have been established for pipe and are given in engineering tables. Generally, the C factor decreases by one with each year of pipe age. Flow for a newly designed system is often calculated with a C factor of 100, based on averaging it over the life of the pipe system.

Note: A high C factor means a smooth pipe; a low C factor means a rough pipe.

22.6 CHARACTERISTICS OF OPEN-CHANNEL FLOW

Basic hydraulic principles apply in open-channel flow (with water depth constant) although there is no pressure to act as the driving force (McGhee, 1991). Velocity head is the only natural energy this water possesses, and at normal water velocities it is a small value ($V^2/2g$). Several parameters can be (and often are) used to describe open-channel flow; however, we begin our discussion by addressing several characteristics of open-channel flow, including whether it is laminar or turbulent, uniform or varied, or subcritical, critical, or supercritical.

22.6.1 LAMINAR AND TURBULENT FLOW

Laminar and turbulent flows in open channels are analogous to those in closed pressurized conduits (e.g., pipes). It is important to point out, however, that flow in open channels is usually turbulent. In addition, laminar flow essentially never occurs in open channels in either water or wastewater unit processes or structures.

22.6.2 UNIFORM AND VARIED FLOW

Flow can be a function of time and location. If the flow quantity is invariant, it is said to be steady. Uniform flow is flow in which the depth, width, and velocity remain constant along a channel; that is, if the flow cross-section does not depend on the location along the channel, the flow is said to be uniform. Varied or nonuniform flow involves a change in these variables, with a change in one producing a change in the others. Most circumstances of open-channel flow in water and wastewater systems involve varied flow. The concept of uniform flow is valuable, however, in that it defines a limit that the varied flow may be considered to be approaching in many cases.

22.6.3 CRITICAL FLOW

Critical flow (i.e., flow at the critical depth and velocity) defines a state of flow between two flow regimes. Critical flow coincides with minimum specific energy for a given discharge and maximum discharge for a given specific energy. Critical flow occurs in flow measurement devices at or near free discharges and establishes controls in open-channel flow. Critical flow occurs frequently in water/wastewater systems and is very important in their operation and design.

22.6.4 PARAMETERS USED IN OPEN CHANNEL FLOW

The three primary parameters used in open channel flow are *hydraulic radius*, *hydraulic depth*, and *slope (S)*.

22.6.4.1 Hydraulic Radius

The hydraulic radius is the ratio of area in flow to wetted perimeter:

$$r_H = \frac{A}{P} \tag{22.22}$$

where
   r_H = Hydraulic radius.
   A = Cross-sectional area of the water.
   P = Wetted perimeter.

Consider, for example, that in open channels it is of primary importance to maintain the proper velocity. This is the case, of course, because if velocity is not maintained then flow stops (theoretically). To maintain velocity at a constant level, the channel slope must be adequate to overcome friction losses. As with other flows, calculation of head loss at a given flow is necessary, and the Hazen–Williams equation is useful ($Q = 0.435 \times C \times d^{2.63} \times S^{.54}$). Keep in mind that the concept of slope has not changed. The difference? We are now measuring, or calculating for, the physical slope of a channel (ft/ft), equivalent to head loss.

The preceding seems logical and makes sense, but there is a problem. The problem is with the diameter. In conduits that are not circular (e.g., grit chambers, contact basins, streams, rivers) or in pipes only partially full (e.g., drains, wastewater gravity mains, sewers), where the cross-sectional area of the water is not circular, there is no diameter. Without a diameter, what do we do? Another good question. Because we do not have a diameter in situations where the cross-sectional area of the water is not circular, we must use another parameter to designate the size of the cross-section and the amount of it that contacts the sides of the conduit. This is where the hydraulic radius (r_H) comes in. The hydraulic radius is a measure of the efficiency with which the conduit can transmit water. Its value depends on pipe size, and amount of fullness. We use the hydraulic radius to measure how much of the water is in contact with the sides of the channel or how much of the water is not in contact with the sides (see Figure 22.5).

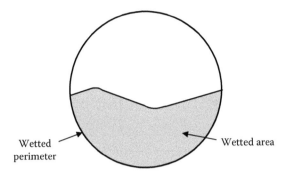

Wetted
perimeter

Wetted area

FIGURE 22.5 Hydraulic radius. (From Spellman, F.R. and Drinan, J., *Water Hydraulics*, Technomic, Lancaster, PA, 2001.)

22.6.4.2 Hydraulic Depth

The hydraulic depth is the ratio of area in flow to the width of the channel at the fluid surface (note that other names for hydraulic depth are *hydraulic mean depth* and *hydraulic radius*):

$$d_H = \frac{A}{w} \tag{22.23}$$

where
  d_h = Hydraulic depth.
  a = Area in flow.
  w = Width of the channel at the fluid surface.

22.6.4.3 Slope

The slope (S) in open-channel equations is the slope of the energy line. If the flow is uniform, the slope of the energy line will parallel the water surface and channel bottom. In general, the slope can be calculated from the Bernoulli equation as the energy loss per unit length of channel:

$$S = \frac{D_h}{D_l} \tag{22.24}$$

22.7 OPEN-CHANNEL FLOW CALCULATIONS

The calculation for head loss at a given flow is typically accomplished by using the Hazen–Williams equation. In addition, in open-channel flow problems, although the concept of slope has not changed, a problem again rises with the diameter. In pipes only partially full where the cross-sectional area of the water is not circular, we have no diameter to work with, and the hydraulic radius is used for these noncircular areas. In the original version of the Hazen–Williams equation, the hydraulic radius was incorporated. Moreover, similar versions developed by Chezy (pronounced "Shay-zee"), Manning, and others incorporated the hydraulic radius. For use in open channels, Manning's formula has become the most commonly used:

$$Q = \frac{1.5}{n} A \times R^{0.66} \times s^{0.5} \tag{22.25}$$

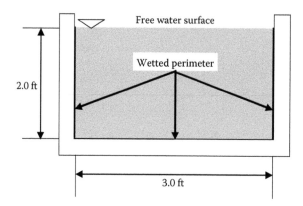

FIGURE 22.6   For Example 22.14.

where

   Q = Channel discharge capacity (ft³/sec).
   1.5 = Constant.
   n = Channel roughness coefficient.
   A = Cross-sectional flow area (ft²).
   R = Hydraulic radius of the channel (ft).
   S = Slope of the channel bottom (dimensionless).

Recall that we defined the hydraulic radius of a channel as the ratio of the flow area to the wetted perimeter P. In formula form, $R = A/P$. The new component here is n (the roughness coefficient), and it depends on the material and age of a pipe or lined channel and on topographic features for a natural streambed. It approximates roughness in open channels and can range from a value of 0.01 for a smooth clay pipe to 0.1 for a small natural stream. The value of n commonly assumed for concrete pipes or lined channels is 0.013. The n values decrease as the channels become smoother. The following example illustrates the application of Manning's formula for a channel with a rectangular cross-section.

■ EXAMPLE 22.14

Problem: A rectangular drainage channel is 3 ft wide and is lined with concrete, as illustrated in Figure 22.6. The bottom of the channel drops in elevation at a rate of 0.5 per 100 ft. What is the discharge in the channel when the depth of water is 2 ft?

Solution: Assume that $n = 0.013$. Referring to Figure 22.6, we see that the cross-sectional flow area (A) = 3 ft × 2 ft = 6 ft², and the wetted perimeter (P) = 2 ft + 3 ft + 2 ft = 7 ft. The hydraulic radius $(R) = A/P$ = 6 ft²/7 ft = 0.86 ft. The slope (S) = 0.5/100 = 0.005. Applying Manning's formula, we get

$$Q = \frac{2.0}{0.013} \times 6 \times 0.86^{0.66} \times 0.005^{0.5} = 59 \text{ cfs}$$

REFERENCES AND RECOMMENDED READING

McGhee, T.J. (1991). *Water Supply and Sewerage*, 2nd ed. McGraw-Hill, New York.
Nathanson, J.A. (1997). *Basic Environmental Technology: Water Supply Waste Management, and Pollution Control*, 2nd ed. Prentice-Hall, Upper Saddle River, NJ.

23 Water Treatment Processes

Pure water is the world's best and foremost medicine.

—**Slovakian Proverb**

23.1 INTRODUCTION

Because of huge volume and flow conditions, the quality of natural water cannot be modified significantly within the body of water (Gupta, 1997). Consequently, the quality control approach is directed to the water withdrawn from a source for a specific use. The drawn water is treated prior to its use. Typically, the overall treatment of water (for potable use) consists of physical and chemical methods of treatment—unlike wastewater treatment where physical, chemical, and/or biological unit processes are used, depending on the desired quality of the effluent and operational limitations. The physical unit operations used in water treatment include

- *Screening*—This process removes large floating and suspended debris.
- *Mixing*—Chemicals known as coagulants (e.g., alum) are mixed to make tiny particles stick together.
- *Flocculation*—Water mixed with coagulants is given a low motion to allow particles to meet and floc together.
- *Sedimentation (settling)*—Water is detained for a sufficient time so flocculated particles settle by gravity.
- *Filtration*—Fine particles still remaining in water after settling and some microorganisms present are filtered through a bed of sand and coal.

Chemical unit processes used in treating raw water, depending on regulatory requirements and the need for additional chemical treatment, include disinfection, precipitation, adsorption, ion exchange, and gas transfer. A flow diagram of a conventional water treatment system is shown in Figure 23.1.

23.2 WATER SOURCE AND STORAGE CALCULATIONS

Approximately 40 million cubic miles of water cover or reside within the Earth. The oceans contain about 97% of all water on Earth. The other 3% is freshwater: (1) snow and ice on the surface of the Earth contain about 2.25% of the water, (2) usable ground water is approximately 0.3%, and (3) surface freshwater is less than 0.5%. In the United States, for example, average rainfall is approximately 2.6 ft (a volume of 5900 km^3). Of this amount, approximately 71% evaporates (about 4200 km^3), and 29% goes to stream flow (about 1700 km^3). Beneficial freshwater uses include manufacturing, food production, domestic and public needs, recreation, hydroelectric power production, and flood control. Stream flow withdrawn annually is about 7.5% (440 km^3). Irrigation and industry use almost half of this amount (3.4%, or 200 km^3/yr). Municipalities use only about 0.6% (35 km^3/yr) of this amount. Historically, in the United States, water usage has been increasing (as might be expected); for example, in 1900, 40 billion gallons of freshwater were used. In 1975, usage increased to 455 billion gallons. Estimated use in 2000 was about 720 billion gallons.

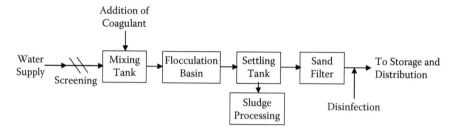

FIGURE 23.1 Conventional water treatment model.

The primary sources of freshwater include the following:

- Captured and stored rainfall in cisterns and water jars
- Groundwater from springs, artesian wells, and drilled or dug wells
- Surface water from lakes, rivers, and streams
- Desalinized seawater or brackish groundwater
- Reclaimed wastewater

23.2.1 WATER SOURCE CALCULATIONS

Water source calculations covered in this section apply to wells and pond or lake storage capacity. Specific well calculations discussed include well drawdown, well yield, specific yield, well casing disinfection, and deep-well turbine pump capacity.

23.2.1.1 Well Drawdown

Drawdown is the drop in the level of water in a well when water is being pumped (Figure 23.2). Drawdown is usually measured in feet or meters. One of the most important reasons for measuring drawdown is to make sure that the source water is adequate and not being depleted. The data collected to calculate drawdown can indicate if the water supply is slowly declining. Early detection can give the system time to explore alternative sources, establish conservation measures, or obtain any special funding that may be needed to get a new water source. Well drawdown is the difference between the pumping water level and the static water level:

$$\text{Drawdown (ft)} = \text{Pumping water level (ft)} - \text{Static water level (ft)} \qquad (23.1)$$

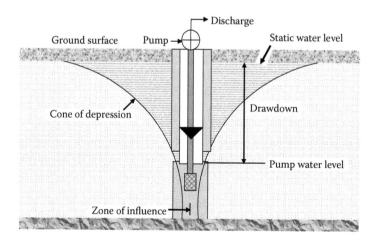

FIGURE 23.2 Hydraulic characteristics of a well.

■ **EXAMPLE 23.1**

Problem: The static water level for a well is 70 ft. If the pumping water level is 90 ft, what is the drawdown?

Solution:

Drawdown = Pumping water level (ft) – Static water level (ft) = 90 ft – 70 ft = 20 ft

■ **EXAMPLE 23.2**

Problem: The static water level of a well is 122 ft. The pumping water level is determined using the sounding line. The air pressure applied to the sounding line is 4.0 psi, and the length of the sounding line is 180 ft. What is the drawdown?

Solution: First calculate the water depth in the sounding line and the pumping water level:

1. Water depth in sounding line = 4.0 psi × 2.31 ft/psi = 9.2 ft
2. Pumping water level = 180 ft – 9.2 ft = 170.8 ft

Then calculate drawdown as usual:

Drawdown = Pumping water level (ft) – Static water level (ft = 170.8 ft – 122 ft = 48.8 ft

23.2.1.2 Well Yield

Well yield is the volume of water per unit of time that is produced from the well pumping. Usually, well yield is measured in terms of gallons per minute (gpm) or gallons per hour (gph). Sometimes, large flows are measured in cubic feet per second (cfs). Well yield is determined as follows:

$$\text{Well yield (gpm)} = \text{Gallons produced/Duration of test (min)} \tag{23.2}$$

■ **EXAMPLE 23.3**

Problem: When the drawdown level of a well was stabilized, it was determined that the well produced 400 gal during a 5-min test. What was the well yield?

Solution:

Well yield = Gallons produced/Duration of test (min) = 400 gal/5 min = 80 gpm

■ **EXAMPLE 23.4**

Problem: During a 5-min test for well yield, a total of 780 gal was removed from the well. What was the well yield in gpm? In gph?

Solution:

Well yield = Gallons produced/Duration of test (min) = 780 gal/5 min = 156 gpm

Then convert gpm flow to gph flow:

156 gpm × 60 min/hr = 9360 gph

23.2.1.3 Specific Yield

Specific yield is the discharge capacity of the well per foot of drawdown. The specific yield may range from 1 gpm/ft drawdown to more than 100 gpm/ft drawdown for a properly developed well. Specific yield is calculated as follows:

$$\text{Specific yield (gpm/ft)} = \text{Well yield (gpm)/Drawdown (ft)} \tag{23.3}$$

■ **EXAMPLE 23.5**

Problem: A well produces 260 gpm. If the drawdown for the well is 22 ft, what is the specific yield in gpm/ft?

Solution:

Specific yield = Well yield (gpm)/Drawdown (ft) = 260 gpm/22 ft = 11.8 gpm/ft

■ **EXAMPLE 23.6**

Problem: The yield for a particular well is 310 gpm. If the drawdown for this well is 30 ft, what is the specific yield in gpm/ft?

Solution:

Specific yield = Well yield (gpm)/Drawdown (ft) = 310 gpm/30 ft = 10.3 gpm/ft

23.2.1.4 Well Casing Disinfection

A new, cleaned, or a repaired well normally contains contamination that may remain for weeks unless the well is thoroughly disinfected. This may be accomplished by using ordinary bleach at a concentration of 100 parts per million (ppm) of chlorine. The amount of disinfectant required is determined by the amount of water in the well. The following equation is used to calculate the pounds of chlorine required for disinfection:

$$\text{Chlorine (lb)} = \text{Chlorine (mg/L)} \times \text{Casing volume (MG)} \times 8.34 \text{ lb/gal} \qquad (23.4)$$

■ **EXAMPLE 23.7**

Problem: A new well is to be disinfected with chlorine at a dosage of 50 mg/L. If the well casing diameter is 8 in. and the length of the water-filled casing is 110 ft, how many pounds of chlorine will be required?

Solution: First calculate the volume of the water-filled casing:

$$0.785 \times 0.67 \times 67 \times 110 \text{ ft} \times 7.48 \text{ gal/ft}^3 = 290 \text{ gal}$$

Then determine the pounds of chlorine required using the mg/L to lb equation:

$$\text{Chlorine} = \text{Chlorine (mg/L)} \times \text{Volume (MG)} \times 8.34 \text{ lb/gal}$$

$$= 50 \text{ mg/L} \times 0.000290 \text{ MG} \times 8.34 \text{ lb/gal} = 0.12 \text{ lb}$$

23.2.1.5 Deep Well Turbine Pump Calculations

The deep well turbine pump is used for high-capacity deep wells. The pump, usually consisting of more than one stage of centrifugal pump, is fastened to a pipe called the *pump column*; the pump is located in the water. The pump is driven from the surface through a shaft running inside the pump column. The water is discharged from the pump up through the pump column to the surface. The pump may be driven by a vertical shaft, electric motor at the top of the well, or some other power source, usually through a right-angle gear drive located at the top of the well. A modern version of the deep-well turbine pump is the submersible type of pump, where the pump (as well as a close-coupled electric motor built as a single unit) is located below water level in the well. The motor is built to operate submerged in water.

23.2.2   VERTICAL TURBINE PUMP CALCULATIONS

The calculations pertaining to well pumps include head, horsepower, and efficiency calculations. *Discharge head* is measured by the pressure gauge located close to the pump discharge flange. The pressure (psi) can be converted to feet of head using the equation:

$$\text{Discharge head (ft)} = \text{Pressure (psi)} \times 2.31 \text{ ft/psi} \qquad (23.5)$$

Total pumping head (*field head*) is a measure of the lift below the discharge head pumping water level (*discharge head*). Total pumping head is calculated as follows:

$$\text{Pumping head (ft)} = \text{Pumping water level (ft)} + \text{Discharge head (ft)} \qquad (23.6)$$

■ EXAMPLE 23.8

Problem: The pressure gauge reading at a pump discharge head is 4.1 psi. What is this discharge head expressed in feet?

Solution:

$$4.1 \text{ psi} \times 2.31 \text{ ft/psi} = 9.5 \text{ ft}$$

■ EXAMPLE 23.9

Problem: The static water level of a pump is 100 ft. The well drawdown is 26 ft. If the gauge reading at the pump discharge head is 3.7 psi, what is the total pumping head?

Solution:

$$\text{Total pumping head (ft)} = \text{Pumping water level (ft)} + \text{Discharge head (ft)} \qquad (23.7)$$

$$\text{Total pumping head} = (100 \text{ ft} + 26 \text{ ft}) + (3.7 \text{ psi} \times 2.31 \text{ ft/psi}) = 126 \text{ ft} + 8.5 \text{ ft} = 134.5 \text{ ft}$$

Five types of *horsepower* calculations are used for vertical turbine pumps; it is important to have a general understanding of these five horsepower types:

- *Motor horsepower* refers to the horsepower supplied to the motor. The following equation is used to calculate motor horsepower:

$$\text{Motor horsepower (input horsepower)} = \frac{\text{Field brake horsepower}}{\text{Motor efficiency}/100} \qquad (23.8)$$

- *Total brake horsepower* refers to the horsepower output of the motor. The following equation is used to calculate total brake horsepower:

$$\text{Total bhp} = \text{Field bhp} + \text{Thrust bearing loss (hp)} \qquad (23.9)$$

- *Field horsepower* refers to the horsepower required at the top of the pump shaft. The following equation is used to calculate field horsepower:

$$\text{Field bhp} = \text{Bowl bhp} + \text{Shaft loss (hp)} \qquad (23.10)$$

- *Bowl or laboratory horsepower* refers the horsepower at the entry to the pump bowls. The following equation is used to calculate bowl horsepower:

$$\text{Bowl horsepower (lab horsepower)} = \frac{\text{Bowl head (ft)} \times \text{Capacity (gpm)}}{3960 \times \left(\text{Bowl efficiency}/100\right)} \qquad (23.11)$$

- *Water horsepower* refers to the horsepower at the pump discharge. The following equation is used to calculate water hp:

$$\text{Water horsepower} = \frac{\text{Field head (ft)} \times \text{Capacity (gpm)}}{3960} \qquad (23.12)$$

or the equivalent equation

$$\text{Water horsepower} = \frac{\text{Field head (ft)} \times \text{Capacity (gpm)}}{33{,}000 \text{ ft-lb/min}}$$

■ EXAMPLE 23.10

Problem: The pumping water level for a well pump is 150 ft and the discharge pressure measured at the pump discharge centerline is 3.5 psi. If the flow rate from the pump is 700 gpm, what is the water horsepower?

Solution: First calculate the field head. The discharge head must be converted from psi to ft:

$$3.5 \text{ psi} \times 2.31 \text{ ft/psi} = 8.1 \text{ ft}$$

The field head is, therefore,

$$150 \text{ ft} + 8.1 \text{ ft} = 158.1 \text{ ft}$$

The water horsepower can now be determined:

$$\text{Water horsepower} = \frac{158.1 \text{ ft} \times 700 \text{ gpm} \times 8.34 \text{ lb/gal}}{33{,}000 \text{ ft-lb/min}} = 28$$

■ EXAMPLE 23.11

Problem: The pumping water level for a pump is 170 ft. The discharge pressure measured at the pump discharge head is 4.2 psi. If the pump flow rate is 800 gpm, what is the water horsepower?

Solution: The field head must first be determined. In order to determine field head, the discharge head must be converted from psi to ft:

$$4.2 \text{ psi} \times 2.31 \text{ ft/psi} = 9.7 \text{ ft}$$

The field head can now be calculated:

$$170 \text{ ft} + 9.7 \text{ ft} = 179.7 \text{ ft}$$

And then the water horsepower can be calculated:

$$\text{Water horsepower} = \frac{179.7 \text{ ft} \times 800 \text{ gpm} \times 8.34 \text{ lb/gal}}{33,000 \text{ ft-lb/min}} = 36$$

■ EXAMPLE 23.12

Problem: A deep-well vertical turbine pump delivers 600 gpm. If the lab head is 185 feet and the bowl efficiency is 84%, what is the bowl horsepower?

Solution:

$$\text{Bowl horsepower} = \frac{\text{Bowl head (ft)} \times \text{Capacity (gpm)}}{3960 \times \left(\text{Bowl efficiency}/100\right)}$$

$$= \frac{185 \text{ ft} \times 600 \text{ gpm}}{3960 \times \left(84.0/100\right)} = \frac{185 \text{ ft} \times 600 \text{ gpm}}{3960 \times 0.84} = 33.4$$

■ EXAMPLE 23.13

Problem: The bowl bhp is 51.8 bhp. If the 1-inch diameter shaft is 170 ft long and is rotating at 960 rpm with a shaft fiction loss of 0.29 hp loss per 100 ft, what is the field bhp?

Solution: Before field bhp can be calculated, the shaft loss must be factored in as follows:

$$\frac{(0.29 \text{ hp loss}) \times (170 \text{ ft})}{100} = 0.5$$

Now determine the field bhp:

$$\text{Field bhp} = \text{Bowl bhp} + \text{Shaft Loss (hp)} = 51.8 \text{ bhp} + 0.5 \text{ hp} = 52.3 \text{ bhp}$$

■ EXAMPLE 23.14

Problem: The field horsepower for a deep-well turbine pump is 62 bhp. If the thrust bearing loss is 0.5 hp and the motor efficiency is 88%, what is the motor input horsepower?

Solution:

$$\text{Motor input horsepower} = \frac{\text{Field (total) bhp}}{\text{Motor efficiency}/100} = \frac{62 \text{ bhp} + 0.5 \text{ hp}}{0.88} = 71 \text{ mhp}$$

When we speak of the *efficiency* of any machine, we are speaking primarily of a comparison of what is put out by the machine (e.g., energy output) compared to its input (e.g., energy input). Horsepower efficiency, for example, is a comparison of horsepower output of the unit or system with horsepower input to that unit or system—the unit's efficiency. With regard to vertical turbine pumps, there are four types of efficiencies considered with vertical turbine pumps:

- Bowl efficiency
- Field efficiency
- Motor efficiency
- Overall efficiency

The general equation used in calculating percent efficiency is shown below:

$$\% = \frac{\text{Part}}{\text{Whole}} \times 100 \qquad (23.13)$$

Vertical turbine pump bowl efficiency is easily determined using a pump performance curve chart provided by the pump manufacturer. Field efficiency is determined using Equation 23.14:

$$\text{Field efficiency (\%)} = \frac{\text{Field head (ft)} \times \text{Capacity (gpm)}}{3960 \times \text{Total bhp}} \times 100 \qquad (23.14)$$

■ EXAMPLE 23.15

Problem: Given the data below, calculate the field efficiency of the deep well turbine pump:

Field head—180 ft
Capacity—850 gpm
Total bhp—61.3 bhp

Solution:

$$\text{Field efficiency (\%)} = \frac{\text{Field head (ft)} \times \text{Capacity (gpm)}}{3960 \times \text{Total bhp}} \times 100$$

$$= \frac{180 \text{ ft} \times 850 \text{ gpm}}{3960 \times 61.3} \times 100 = 63\%$$

Overall efficiency is a comparison of the horsepower output of the system with that entering the system. Equation 23.15 is used to calculate overall efficiency:

$$\text{Overall efficiency (\%)} = \frac{\text{Field efficiency (\%)} \times \text{Motor efficiency (\%)}}{100} \qquad (23.15)$$

■ EXAMPLE 23.16

Problem: The efficiency of a motor is 90%. If the field efficiency is 83%, what is the overall efficiency of the unit?

Solution:

$$\text{Overall efficiency (\%)} = \frac{\text{Field efficiency (\%)} \times \text{Motor efficiency (\%)}}{100}$$

$$= \frac{83\% \times 90\%}{100} = 74.7\%$$

23.3 WATER STORAGE

Water storage facilities for water distribution systems are required primarily to provide for fluctuating demands of water usage (to provide a sufficient amount of water to average or equalize daily demands on the water supply system). In addition, other functions of water storage facilities include increasing operating convenience, leveling pumping requirements (to keep pumps from running 24 hours a day), decreasing power costs, providing water during power source or pump failure, providing large quantities of water to meet fire demands, providing surge relief (to reduce the surge associated with stopping and starting pumps), increasing detention time (to provide chlorine contact time and satisfy the desired contact time value requirements), and blending water sources.

23.3.1 WATER STORAGE VOLUME CALCULATIONS

The storage capacity, in gallons, of a reservoir, pond, or small lake can be estimated using Equation 23.16:

$$\text{Capacity (gal)} = \text{Average length (ft)} \times \text{Average width (ft)} \times \text{Average depth (ft)} \times 7.48 \text{ gal/ft}^3 \quad (23.16)$$

■ EXAMPLE 23.17

Problem: A pond has an average length of 250 ft, an average width of 110 ft, and an estimated average depth of 15 ft. What is the estimated volume of the pond in gallons?

Solution:

$$\text{Volume} = \text{Average length (ft)} \times \text{Average width (ft)} \times \text{Average depth (ft)} \times 7.48 \text{ gal/ft}^3$$

$$= 250 \text{ ft} \times 110 \text{ ft} \times 15 \text{ ft} \times 7.48 \text{ gal/ft}^3 = 3,085,500 \text{ gal}$$

■ EXAMPLE 23.18

Problem: A small lake has an average length of 300 ft and an average width of 95 ft. If the maximum depth of the lake is 22 ft, what is the estimated volume (in gallons) of the lake?

Note: For small ponds and lakes, the average depth is generally about 0.4 times the greatest depth; therefore, to estimate the average depth, measure the greatest depth and multiply that number by 0.4.

Solution: First, the average depth of the lake must be estimated:

$$= 22 \text{ ft} \times 0.4 \text{ ft} = 8.8 \text{ ft}$$

Then, the lake volume can be determined:

$$\text{Volume} = \text{Average length (ft)} \times \text{Average width (ft)} \times \text{Average depth (ft)} \times 7.48 \text{ gal/ft}^3$$

$$= 300 \text{ ft} \times 95 \text{ ft} \times 8.8 \text{ ft} \times 7.48 \text{ gal/ft}^3 = 1,875,984 \text{ gal}$$

23.3.2 COPPER SULFATE DOSING

Algal control by applying copper sulfate is perhaps the most common *in situ* treatment of lakes, ponds, and reservoirs; the copper ions in the water kill the algae. Copper sulfate application methods and dosages will vary depending on the specific surface water body being treated. The desired copper sulfate dosage may be expressed in mg/L copper, lb copper sulfate per ac-ft, or lb copper sulfate per acre. For a dose expressed as mg/L copper, the following equation is used to calculate lb copper sulfate required:

$$\text{Copper sulfate (lb)} = \frac{\text{Copper (mg/L)} \times \text{Volume (MG)} \times 8.34 \text{ lb/gal}}{\% \text{ Available copper}/100} \quad (23.17)$$

■ EXAMPLE 23.19

Problem: For algae control in a small pond, a dosage of 0.5 mg/L copper is desired. The pond has a volume of 15 MG. How many pounds of copper sulfate will be required? (Copper sulfate contains 25% available copper.)

Solution:

$$\text{Copper sulfate} = \frac{\text{Copper (mg/L)} \times \text{Volume (MG)} \times 8.34 \text{ lb/gal}}{\% \text{ Available copper}/100}$$

$$= \frac{0.5 \text{ mg/L} \times 15 \text{ MG} \times 8.34 \text{ lb/gal}}{25/100} = 250 \text{ lb}$$

For calculating lb copper sulfate per ac-ft, use the following equation (assuming a desired copper sulfate dose of 0.9 lb/ac-ft):

$$\text{Copper sulfate (lb)} = \frac{0.9 \text{ lb Copper sulfate} \times \text{ac-ft}}{1 \text{ ac-ft}} \quad (23.18)$$

■ EXAMPLE 23.20

Problem: A pond has a volume of 35 ac-ft. If the desired copper sulfate dose is 0.9 lb/ac-ft, how many lb of copper sulfate will be required?

Solution:

$$\text{Copper sulfate (lb)} = \frac{0.9 \text{ lb Copper sulfate} \times \text{ac-ft}}{1 \text{ ac-ft}}$$

$$\frac{0.9 \text{ lb Copper sulfate}}{1 \text{ ac-ft}} = \frac{x \text{ lb Copper sulfate}}{35 \text{ ac-ft}}$$

Then solve for x:

$$0.9 \times 35 = x = 31.5 \text{ lb copper sulfate}$$

The desired copper sulfate dosage may also be expressed in terms of lb copper sulfate per acre. The following equation is used to determine lb copper sulfate (assuming a desired copper sulfate dose of 5.2 lb/ac):

$$\text{Copper sulfate (lb)} = (5.2 \text{ lb copper sulfate} \times \text{acres})/1 \text{ ac} \quad (23.19)$$

■ EXAMPLE 23.21

Problem: A small lake has a surface area of 6.0 ac. If the desired copper sulfate dose is 5.2 lb/ac, how many pounds of copper sulfate are required?

Solution:

$$\text{Copper sulfate (lb)} = (5.2 \text{ lb copper sulfate} \times 6.0 \text{ ac})/1 \text{ ac} = 31.2 \text{ lb copper sulfate}$$

23.4   COAGULATION, MIXING, AND FLOCCULATION

23.4.1   COAGULATION

Following screening and the other pretreatment processes, the next unit process in a conventional water treatment system is a mixer where the first chemicals are added in what is known as *coagulation*. The exception to this situation occurs in small systems using groundwater, when chlorine or other taste and odor control measures are introduced at the intake and are the extent of treatment. The coagulation process involves a series of chemical and mechanical operations by which coagulants are applied and made effective. These operations are comprised of two distinct phases: (1) rapid mixing to disperse coagulant chemicals by violent agitation into the water being treated, and (2) flocculation to agglomerate small particles into well-defined floc by gentle agitation for a much longer time. The coagulant must be added to the raw water and perfectly distributed into the liquid; such uniformity of chemical treatment is reached through rapid agitation or mixing.

Coagulation is a reaction caused by adding polymers or metallic salts such as iron or aluminum compounds to the water. Common coagulants include

- Alum (aluminum sulfate)
- Sodium aluminate
- Ferric sulfate
- Ferrous sulfate
- Ferric chloride
- Polymers

23.4.2   MIXING

To ensure maximum contact between the reagent and suspended particles, coagulants and coagulant aids must be rapidly dispersed (mixed) throughout the water; otherwise, the coagulant will react with water and dissipate some of its coagulating power. To ensure complete mixing and optimum plug flow reactor operation, proper detention time in the basin is required. Detention time can be calculated using the following procedures.

For complete mixing:

$$t = \frac{V}{Q} = \left(\frac{1}{K}\right)\left(\frac{C_i - C_e}{C_e}\right) \tag{23.20}$$

For plug flow:

$$t = \frac{V}{Q} = \frac{L}{v} = \left(\frac{1}{K}\right)\left(\ln\frac{C_i}{C_e}\right) \tag{23.21}$$

where
t = Detention time of the basin (min).
V = Volume of basin (m^3 or ft^3).
Q = Flow rate (m^3/s or cfs).
K = Rate constant.
C_i = Influent reactant concentration (mg/L).
C_e = Effluent reactant concentration (mg/L).
L = Length of rectangular basin (m or ft).
v = Horizontal velocity of flow (m/s or ft/s).

■ **EXAMPLE 23.22**

Problem: Alum dosage is 40 mg/L and $K = 90$ per day based on lab tests. Compute the detention times for complete mixing and plug flow reactor for 90% reduction.

Solution: First find C_e:

$$C_e = (1 - 0.9) \times C_i = 0.1 \times C_i = 0.1 \times 40 \text{ mg/L} = 4 \text{ mg/L}$$

Now calculate t for complete mixing (Equation 23.20):

$$t = \frac{V}{Q} = \left(\frac{1}{K}\right)\left(\frac{C_i - C_e}{C_e}\right) = \left(\frac{1}{90/\text{day}}\right)\left(\frac{40 \text{ mg/L} - 4 \text{ mg/L}}{4 \text{ mg/L}}\right) = \left(\frac{1d}{90}\right) \times \left(\frac{1440 \text{ min}}{1 \text{ day}}\right) = 144 \text{ min}$$

Finally, calculate t for plug flow using the following formula:

$$t = \left(\frac{1}{K}\right)\left(\ln \frac{C_i}{C_e}\right) = \left(\frac{1440}{90}\right)\left(\ln \frac{40}{4}\right) = 36.8 \text{ min}$$

23.4.3 FLOCCULATION

Flocculation follows coagulation in the conventional water treatment process. Flocculation is the physical process of slowly mixing the coagulated water to increase the probability of particle collision. Through experience, we see that effective mixing reduces the required amount of chemicals and greatly improves the sedimentation process, which results in longer filter runs and higher quality finished water. The goal of flocculation is to form a uniform, feather-like material similar to snowflakes—a dense, tenacious floc that traps the fine, suspended, and colloidal particles and carries them down rapidly into the settling basin. To increase the speed of floc formation and the strength and weight of the floc, polymers are often added.

23.4.4 COAGULATION AND FLOCCULATION GENERAL CALCULATIONS

Proper operation of the coagulation and flocculation unit processes requires calculations to determine chamber or basin volume, chemical feed calibration, chemical feeder settings, and detention time.

23.4.4.1 Chamber and Basin Volume Calculations

To determine the volume of a square or rectangular chamber or basin, we use Equation 23.22 or Equation 23.23:

$$\text{Volume (ft}^3) = \text{Length (ft)} \times \text{Width (ft)} \times \text{Depth (ft)} \qquad (23.22)$$

$$\text{Volume (gal)} = \text{Length (ft)} \times \text{Width (ft)} \times \text{Depth (ft)} \times 7.48 \text{ gal/ft}^3 \qquad (23.23)$$

■ **EXAMPLE 23.23**

Problem: A flash mix chamber is 4 ft square with water to a depth of 3 ft. What is the volume of water (in gallons) in the chamber?

Solution:

$$\text{Volume} = \text{Length (ft)} \times \text{Width (ft)} \times \text{Depth (ft)} \times 7.48 \text{ gal/ft}^3$$

$$= 4 \text{ ft} \times 4 \text{ ft} \times 3 \text{ ft} \times 7.48 \text{ gal/ft}^3 = 359 \text{ gal}$$

■ **EXAMPLE 23.24**

Problem: A flocculation basin is 40 ft long by 12 ft wide with water to a depth of 9 ft. What is the volume of water (in gallons) in the basin?

Solution:

$$\text{Volume} = \text{Length (ft)} \times \text{Width (ft)} \times \text{Depth (ft)} \times 7.48 \text{ gal/ft}^3$$

$$= 40 \text{ ft} \times 12 \text{ ft} \times 9 \text{ ft} \times 7.48 \text{ gal/ft}^3 = 32{,}314 \text{ gal}$$

■ **EXAMPLE 23.25**

Problem: A flocculation basin is 50 ft long by 22 ft wide and contains water to a depth of 11 ft, 6 in. How many gallons of water are in the tank?

Solution: First convert the 6-in. portion of the depth measurement to feet:

$$(6 \text{ in.})/(12 \text{ in./ft}) = 0.5 \text{ ft}$$

Then calculate basin volume:

$$\text{Volume} = \text{Length (ft)} \times \text{Width (ft)} \times \text{Depth (ft)} \times 7.48 \text{ gal/ft}^3$$

$$= 50 \text{ ft} \times 22 \text{ ft} \times 11.5 \text{ ft} \times 7.48 \text{ gal/ft}^3 = 94{,}622 \text{ gal}$$

23.4.4.2 Detention Time

Because coagulation reactions are rapid, detention time for flash mixers is measured in seconds, whereas the detention time for flocculation basins is generally between 5 and 30 min. The equation used to calculate detention time is shown below:

$$\text{Detention time (min)} = \text{Volume of tank (gal)/Flow rate (gpm)} \qquad (23.24)$$

■ **EXAMPLE 23.26**

Problem: The flow to a flocculation basin that is 50 ft long by 12 ft wide by 10 ft deep is 2100 gpm. What is the detention time in the tank (in minutes)?

Solution:

$$\text{Tank volume (gal)} = 50 \text{ ft} \times 12 \text{ ft} \times 10 \text{ ft} \times 7.48 \text{ gal/ft}^3 = 44{,}880 \text{ gal}$$

$$\text{Detention time} = \text{Volume of tank (gal)/Flow rate (gpm)} = 44{,}880 \text{ gal}/2100 \text{ gpm} = 21.4 \text{ min}$$

■ **EXAMPLE 23.27**

Problem: A flash mix chamber is 6 ft long by 4 ft with water to a depth of 3 ft. If the flow to the flash mix chamber is 6 MGD, what is the chamber detention time in seconds (assuming that the flow is steady and continuous)?

Solution: First, convert the flow rate from gpd to gps so the time units will match:

$$6{,}000{,}000/(1440 \text{ min/day} \times 60 \text{ sec/min}) = 69 \text{ gps}$$

Then calculate detention time:

$$\text{Detention time} = \text{Volume of tank (gal)/Flow rate (gpm)}$$

$$= (6 \text{ ft} \times 4 \text{ ft} \times 3 \text{ ft} \times 7.48 \text{ gal/ft}^3)/69 \text{ gps} = 7.8 \text{ sec}$$

23.4.4.3 Determining Dry Chemical Feeder Setting (lb/day)

When adding (dosing) chemicals to the water flow, a measured amount of chemical is called for. The amount of chemical required depends on such factors as the type of chemical used, the reason for dosing, and the flow rate being treated. To convert from mg/L to lb/day, the following equation is used:

$$\text{Chemical added (lb/day)} = \text{Chemical (mg/L)} \times \text{Flow (MGD)} \times 8.34 \text{ lb/gal} \qquad (23.25)$$

■ **EXAMPLE 23.28**

Problem: Jar tests indicate that the best alum dose for water is 8 mg/L. If the flow to be treated is 2,100,000 gpd, what should the lb/day setting be on the dry alum feeder?

Solution:

$$\text{Chemical added} = \text{Chemical (mg/L)} \times \text{Flow (MGD)} \times 8.34 \text{ lb/gal}$$

$$= 8 \text{ mg/L} \times 2.10 \text{ MGD} \times 8.34 \text{ lb/gal} = 140 \text{ lb/day}$$

■ **EXAMPLE 23.29**

Problem: Determine the desired lb/day setting on a dry chemical feeder if jar tests indicate an optimum polymer dose of 12 mg/L and the flow to be treated is 4.15 MGD.

Solution:

$$\text{Polymer (lb/day)} = 12 \text{ mg/L} \times 4.15 \text{ MGD} \times 8.34 \text{ lb/gal} = 415 \text{ lb/day}$$

23.4.4.4 Determining Chemical Solution Feeder Setting (gpd)

When solution concentration is expressed as pound chemical per gallon solution, the required feed rate can be determined using the following equations:

$$\text{Chemical (lb/day)} = \text{Chemical (mg/L)} \times \text{Flow (MGD)} \times 8.34 \text{ lb/gal} \qquad (23.26)$$

Then convert the lb/day dry chemical to gpd solution:

$$\text{Solution (gpd)} = \text{Chemical (lb/day)/lb Chemical per gal solution} \qquad (23.27)$$

■ **EXAMPLE 23.30**

Problem: Jar tests indicate that the best alum dose for water is 7 mg/L. The flow to be treated is 1.52 MGD. Determine the gpd setting for the alum solution feeder if the liquid alum contains 5.36 lb of alum per gallon of solution.

Solution: First calculate the lb/day of dry alum required, using the mg/L to lb/day equation:

$$\text{Dry alum} = \text{Chemical (mg/L)} \times \text{Flow (MGD)} \times 8.34 \text{ lb/gal}$$

$$= 7 \text{ mg/L} \times 1.52 \text{ MGD} \times 8.34 \text{ lb/gal} = 89 \text{ lb/day}$$

Then calculate gpd solution required:

$$\text{Alum solution} = 89 \text{ lb/day/5.36 lb alum per gal solution} = 16.6 \text{ gpd}$$

23.4.4.5 Determining Chemical Solution Feeder Setting (Milliliters/Minute)

Some solution chemical feeders dispense chemical as milliliters per minute (mL/min). To calculate the mL/min solution required, use the following procedure:

$$\text{Solution (mL/min)} = (\text{gpd} \times 3785 \text{ mL/gal})/(1440 \text{ min/day}) \qquad (23.28)$$

■ **EXAMPLE 23.31**

Problem: The desired solution feed rate was calculated to be 9 gpd. What is this feed rate expressed as mL/min?

Solution:

$$\text{Solution} = (\text{gpd} \times 3785 \text{ mL/gal})/(1440 \text{ min/day})$$

$$= (9 \text{ gpd} \times 3785 \text{ mL/gal})/(1440 \text{ min/day}) = 24 \text{ mL/min}$$

■ **EXAMPLE 23.32**

Problem: The desired solution feed rate has been calculated to be 25 gpd. What is this feed rate expressed as mL/min?

Solution:

$$\text{Solution} = (\text{gpd} \times 3785 \text{ mL/gal})/(1440 \text{ min/day})$$

$$= (25 \text{ gpd} \times 3785 \text{ mL/gal})/(1440 \text{ min/day}) = 65.7 \text{ mL/min}$$

Sometimes we will need to know the mL/min solution feed rate but we do not know the gpd solution feed rate. In such cases, calculate the gpd solution feed rate first, using the following the equation:

$$\text{gpd} = \frac{\text{Chemical (mg/L)} \times \text{Flow (MGD)} \times 8.34 \text{ lb/gal}}{\text{Chemical (lb)/Solution (gal)}} \qquad (23.29)$$

23.4.5 Determining Percent Strength of Solutions

The strength of a solution is a measure of the amount of chemical solute dissolved in the solution. We use the following equation to determine the percent strength of a solution:

$$\% \text{ Strength} = \frac{\text{Chemical (lb)}}{\text{Water (lb)} + \text{Chemical (lb)}} \times 100 \qquad (23.30)$$

■ **EXAMPLE 23.33**

Problem: If a total of 10 oz. of dry polymer is added to 15 gal of water, what is the percent strength (by weight) of the polymer solution?

Solution: Before calculating percent strength, the ounces of chemical must be converted to pounds of chemical:

$$(10 \text{ oz.})/(16 \text{ oz./lb}) = 0.625 \text{ lb chemical}$$

Now calculate percent strength:

$$\% \text{ Strength} = \frac{\text{Chemical (lb)}}{\text{Water (lb)} + \text{Chemical (lb)}} \times 100$$

$$= \frac{0.625 \text{ lb chemical}}{(15 \text{ gal} \times 8.34 \text{ lb/gal}) + 0.625 \text{ lb}} \times 100 = \frac{0.625 \text{ lb chemical}}{125.7 \text{ lb solution}} \times 100 = 0.5\%$$

■ EXAMPLE 23.34

Problem: If 90 g (1 g = 0.0022 lb) of dry polymer is dissolved in 6 gal of water, what percent strength is the solution?

Solution: First, convert grams of chemical to pounds of chemical:

$$90 \text{ g polymer} \times 0.0022 \text{ lb/g} = 0.198 \text{ lb polymer}$$

Now calculate percent strength of the solution:

$$\% \text{ Strength} = \frac{\text{lb Polymer}}{\text{lb Water} + \text{lb Polymer}} \times 100$$

$$= \frac{0.198 \text{ lb Polymer}}{(6 \text{ gal} \times 8.34 \text{ lb/gal}) + 0.198 \text{ lb Polymer}} \times 100 = 4\%$$

23.4.5.1 Determining Percent Strength of Liquid Solutions

When using liquid chemicals to make up solutions (e.g., liquid polymer), a different calculation is required, as shown below:

$$\frac{\text{Liquid polymer (lb)} \times \text{Liquid polymer (\% strength)}}{100}$$

$$= \frac{\text{Polymer solution (lb)} \times \text{Polymer solution (\% strength)}}{100} \tag{23.31}$$

■ EXAMPLE 23.35

Problem: A 12% liquid polymer is to be used in making up a polymer solution. How many pounds of liquid polymer should be mixed with water to produce 120 lb of a 0.5% polymer solution?

Solution:

$$\frac{\text{Liq. polymer (lb)} \times \text{Liq. polymer (\% strength)}}{100} = \frac{\text{Polymer sol. (lb)} \times \text{Polymer sol. (\% strength)}}{100}$$

$$\frac{x \text{ lb} \times 12}{100} = \frac{120 \text{ lb} \times 0.5}{100}$$

$$x = \frac{120 \times 0.005}{0.12}$$

$$x = 5 \text{ lb}$$

23.4.5.2 Determining Percent Strength of Mixed Solutions

The percent strength of solution mixture is determined using the following equation:

$$\% \text{ Strength} = \frac{\dfrac{\text{Sol. 1 (lb)} \times \text{Sol. 1 (\% strength)}}{100} + \dfrac{\text{Sol. 2 (lb)} \times \text{Sol. 2 (\% strength)}}{100}}{\text{Sol. 1 (lb)} + \text{Sol. 2 (lb)}} \times 100 \qquad (23.32)$$

■ **EXAMPLE 23.36**

Problem: If 12 lb of a 10% strength solution is mixed with 40 lb of a 1% strength solution, what is the percent strength of the solution mixture?

Solution:

$$\% \text{ Strength of mix} = \frac{\dfrac{\text{Sol. 1 (lb)} \times \text{Sol. 1 (\% strength)}}{100} + \dfrac{\text{Sol. 2 (lb)} \times \text{Sol. 2 (\% strength)}}{100}}{\text{Sol. 1 (lb)} + \text{Sol. 2 (lb)}} \times 100$$

$$= \frac{(12 \text{ lb} \times 0.1) + (40 \text{ lb} \times 0.1)}{12 \text{ lb} + 40 \text{ lb}} \times 100 = \frac{1.2 \text{ lb} + 0.40 \text{ lb}}{52 \text{ lb}} \times 100 = 3.1\%$$

23.4.6 DRY CHEMICAL FEEDER CALIBRATION

Occasionally, we need to perform a calibration calculation to compare the actual chemical feed rate with the feed rate indicated by the instrumentation. To calculate the actual feed rate for a dry chemical feeder, place a container under the feeder, weigh the container when empty, then weigh the container again after a specified length of time (e.g., 30 min). The actual chemical feed rate can be calculated using the following equation:

$$\text{Chemical feed rate (lb/min)} = \frac{\text{Chemical applied (lb)}}{\text{Length of application (min)}} \qquad (23.33)$$

If desired, the chemical feed rate can be converted to lb/day:

$$\text{Chemical feed rate (lb/day)} = \text{Feed rate (lb/min)} \times 1440 \text{ min/day} \qquad (23.34)$$

■ **EXAMPLE 23.37**

Problem: Calculate the actual chemical feed rate (lb/day) if a container is placed under a chemical feeder and a total of 2 lb is collected during a 30-min period.

Solution: First calculate the lb/min feed rate:

$$\text{Chemical feed rate} = \frac{\text{Chemical applied (lb)}}{\text{Length of application (min)}} = \frac{2 \text{ lb}}{30 \text{ min}} = 0.06 \text{ lb/min}$$

Then calculate the lb/day feed rate:

$$\text{Chemical feed rate} = 0.06 \text{ lb/min} \times 1440 \text{ min/day} = 86.4 \text{ lb/day}$$

■ **EXAMPLE 23.38**

Problem: Calculate the actual chemical feed rate (lb/day) if a container is placed under a chemical feeder and a total of 1.6 lb is collected during a 20-min period.

Solution: First calculate the lb/min feed rate:

$$\text{Chemical feed rate} = \frac{\text{Chemical applied (lb)}}{\text{Length of application (min)}} = \frac{1.6 \text{ lb}}{20 \text{ min}} = 0.08 \text{ lb/min}$$

Then calculate the lb/day feed rate:

$$\text{Chemical feed rate} = 0.08 \text{ lb/min} \times 1440 \text{ min/day} = 115 \text{ lb/day}$$

23.4.6.1 Solution Chemical Feeder Calibration

As with other calibration calculations, the actual solution chemical feed rate is determined and then compared with the feed rate indicated by the instrumentation. To calculate the actual solution chemical feed rate, first express the solution feed rate in MGD. When the MGD solution flow rate has been calculated, use the mg/L equation to determine chemical dosage in lb/day. If solution feed is expressed as mL/min, first convert mL/min flow rate to gpd flow rate:

$$\text{gpd} = \frac{(\text{mL/min}) \times 1440 \text{ min/day}}{3785 \text{ mL/gal}} \qquad (23.35)$$

Then calculate chemical dosage:

$$\text{Chemical dosage (lb/day)} = \text{Chemical (mg/L)} \times \text{Flow (MGD)} \times 8.34 \text{ lb/day} \qquad (23.36)$$

■ **EXAMPLE 23.39**

Problem: A calibration test is conducted for a solution chemical feeder. During a 5-min test, the pump delivered 940 mg/L of the 1.20% polymer solution. (Assume that the polymer solution weighs 8.34 lb/gal.) What is the polymer dosage rate in lb/day?

Solution: The flow rate must be expressed as MGD; therefore, the mL/min solution flow rate must first be converted to gpd and then MGD. The mL/min flow rate is calculated as

$$(940 \text{ mL})/(5 \text{ min}) = 188 \text{ mL/min}$$

Next convert the mL/min flow rate to gpd flow rate:

$$\frac{188 \text{ mL/min} \times 1440 \text{ min/day}}{3785 \text{ mL/gal}} = 72 \text{ gpd}$$

Then calculate the polymer feed rate:

$$\text{Polymer feed rate} = 12{,}000 \text{ mg/L} \times 0.000072 \text{ MGD} \times 8.34 \text{ lb/day} = 7.2 \text{ lb/day}$$

■ **EXAMPLE 23.40**

Problem: A calibration test is conducted for a solution chemical feeder. During a 24-hr period, the solution feeder delivers a total of 100 gal of solution. The polymer solution is a 1.2% solution. What is the lb/day feed rate? (Assume that the polymer solution weighs 8.34 lb/gal.)

Solution: The solution feed rate is 100 gal per day, or 100 gpd. Expressed as MGD, this is 0.000100 MGD. Use the mg/L to lb/day equation to calculate the actual feed rate:

$$\text{Chemical feed rate} = \text{Chemical (mg/L)} \times \text{Flow (MGD)} \times 8.34 \text{ lb/day}$$

$$\text{Chemical feed rate} = 12,000 \text{ mg/L} \times 0.000100 \text{ MGD} \times 8.34 \text{ lb/day} = 10 \text{ lb/day}$$

The actual pumping rates can be determined by calculating the volume pumped during a specified time frame; for example, if 60 gal are pumped during a 10-min test, the average pumping rate during the test is 6 gpm. Actual volume pumped is indicated by a drop in tank level. By using the following equation, we can determine the flow rate in gpm:

$$\text{Flow rate (gpm)} = \frac{0.785 \times D^2 \times \text{Drop in level (ft)} \times 7.48 \text{ gal/ft}^3}{\text{Duration of test (min)}} \quad (23.37)$$

■ **EXAMPLE 23.41**

Problem: A pumping rate calibration test is conducted for a 15-min period. The liquid level in the 4-ft-diameter solution tank is measured before and after the test. If the level drops 0.5 ft during the 15-min test, what is the pumping rate in gpm?

Solution:

$$\text{Flow rate (gpm)} = \frac{0.785 \times D^2 \times \text{Drop in level (ft)} \times 7.48 \text{ gal/ft}^3}{\text{Duration of test (min)}}$$

$$= \frac{0.785 \times (4 \text{ ft} \times 4 \text{ ft}) \times 0.5 \text{ ft} \times 7.48 \text{ gal/ft}^3}{15 \text{ min}} = 3.1 \text{ gpm}$$

23.4.7 DETERMINING CHEMICAL USAGE

One of the primary functions performed by water operators is the recording of data. The lb/day or gpd chemical use is part of the data from which the average daily use of chemicals and solutions can be determined. This information is important in forecasting expected chemical use, comparing it with chemicals in inventory, and determining when additional chemicals will be required. To determine average chemical use, we use Equation 23.38 (lb/day) or Equation 23.39 (gpd):

$$\text{Average use (lb/day)} = \frac{\text{Total chemical used (lb)}}{\text{Number of days}} \quad (23.38)$$

$$\text{Average use (gpd)} = \frac{\text{Total chemical used (gal)}}{\text{Number of days}} \quad (23.39)$$

Then we can calculate the days supply in inventory:

$$\text{Days supply in inventory} = \frac{\text{Total chemical in inventory (lb)}}{\text{Average use (lb/day)}} \quad (23.40)$$

$$\text{Days supply in inventory} = \frac{\text{Total chemical in inventory (gal)}}{\text{Average use (gpd)}} \quad (23.41)$$

■ **EXAMPLE 23.42**

Problem: The chemical used for each day during a week is given below. Based on these data, what was the average lb/day chemical use during the week?

Monday	88 lb/day
Tuesday	93 lb/day
Wednesday	91 lb/day
Thursday	88 lb/day
Friday	96 lb/day
Saturday	92 lb/day
Sunday	86 lb/day

Solution:

$$\text{Average use} = \frac{\text{Total chemical used (lb)}}{\text{Number of days}} = \frac{634 \text{ lb}}{7 \text{ days}} = 90.6 \text{ lb/day}$$

■ **EXAMPLE 23.43**

Problem: The average chemical use at a plant is 77 lb/day. If the chemical inventory is 2800 lb, how many days supply is this?

Solution:

$$\text{Days supply in inventory} = \frac{\text{Total chemical in inventory (lb)}}{\text{Average use (lb/day)}} = \frac{2800 \text{ lb}}{77 \text{ lb/day}} = 36.4 \text{ days}$$

23.4.7.1 Paddle Flocculator Calculations

The gentle mixing required for flocculation is accomplished by a variety of devices. Probably the most common device in use is the basin equipped with mechanically driven paddles. Paddle flocculators have individual compartments for each set of paddles. The useful power input imparted by a paddle to the water depends on the drag force and the relative velocity of the water with respect to the paddle (Droste, 1997). For paddle flocculator design and operation, environmental engineers are mainly interested in determining the velocity of a paddle at a set distance, the drag force of the paddle on the water, and the power input imparted to the water by the paddle. Because of slip, factor k, the velocity of the water will be less than the velocity of the paddle. If baffles are placed along the walls in a direction perpendicular to the water movement, the value of k decreases because the baffles obstruct the movement of the water (Droste, 1997). The frictional dissipation of energy depends on the relative velocity, v. The relative velocity can be determined using Equation 23.42:

$$v = v_p - v_t = v_p - kv_p = v_p(1-k) \tag{23.42}$$

where
v_t = Water velocity.
v_p = Paddle velocity.

To determine the velocity of the paddle at a distance r from the shaft, we use Equation 23.43:

$$v_p = \frac{2\pi N}{60}(r) \tag{23.43}$$

where N is the rate of revolution of the shaft (rpm).

To determine the drag force of the paddle on the water we use Equation 23.44:

$$F_D = (1/2)pC_DAv^2 \qquad (23.44)$$

where
  F_D = Drag force.
  C_D = Drag coefficient.
  A = Area of the paddle.

To determine the power input imparted to the water by an elemental area of the paddle the usual equation used is Equation 23.45:

$$dP = dF_Dv = 1/2pC_Dv^3dA \qquad (23.45)$$

23.5   SEDIMENTATION CALCULATIONS

Sedimentation, the solid–liquid separation by gravity, is one of the most basic processes of water and wastewater treatment. In water treatment, plain sedimentation, such as the use of a presedimentation basin for grit removal and a sedimentation basin following coagulation–flocculation, is the most commonly used approach.

23.5.1   TANK VOLUME CALCULATIONS

The two common shapes of sedimentation tanks are rectangular and cylindrical. The equations for calculating the volume for each type tank are shown below.

23.5.1.1   Calculating Tank Volume

For rectangular sedimentation basins, we use Equation 23.46:

$$\text{Volume (gal)} = \text{Length (ft)} \times \text{Width (ft)} \times \text{Depth (ft)} \times 7.48 \text{ gal/ft}^3 \qquad (23.46)$$

For cylindrical clarifiers, we use Equation 23.47:

$$\text{Volume (gal)} = 0.785 \times (\text{Diameter})^2 \times \text{Depth (ft)} \times 7.48 \text{ gal/ft}^3 \qquad (23.47)$$

■ **EXAMPLE 23.44**

Problem: A sedimentation basin is 25 ft wide by 80 ft long and contains water to a depth of 14 ft. What is the volume of water in the basin, in gallons?

Solution:

$$\text{Volume} = \text{Length (ft)} \times \text{Width (ft)} \times \text{Depth (ft)} \times 7.48 \text{ gal/ft}^3$$
$$= 80 \text{ ft} \times 25 \text{ ft} \times 14 \text{ ft} \times 7.48 \text{ gal/ft}^3 = 209,440 \text{ gal}$$

■ **EXAMPLE 23.45**

Problem: A sedimentation basin is 24 ft wide by 75 ft long. When the basin contains 140,000 gal, what would the water depth be?

Solution:

$$\text{Volume} = \text{Length (ft)} \times \text{Width (ft)} \times \text{Depth (ft)} \times 7.48 \text{ gal/ft}^3$$

$$140,000 \text{ gal} = 75 \text{ ft} \times 24 \text{ ft} \times x \text{ ft} \times 7.48 \text{ gal/ft}^3$$

$$x \text{ ft} = \frac{140,000}{75 \times 24 \times 7.48}$$

$$x \text{ ft} = 10.4 \text{ ft}$$

23.5.2 DETENTION TIME

Detention time for clarifiers varies from 1 to 3 hr. The equations used to calculate detention time are shown below.

Basic detention time equation:

$$\text{Detention time (hr)} = \frac{\text{Volume of tank (gal)}}{\text{Flow rate (gph)}} \tag{23.48}$$

Rectangular sedimentation basin equation:

$$\text{Detention time (hr)} = \frac{\text{Length (ft)} \times \text{Width (ft)} \times \text{Depth (ft)} \times 7.48 \text{ gal/ft}^3}{\text{Flow rate (gph)}} \tag{23.49}$$

Circular basin equation:

$$\text{Detention time (hr)} = \frac{0.785 \times D^2 \times \text{Depth (ft)} \times 7.48 \text{ gal/ft}^3}{\text{Flow rate (gph)}} \tag{23.50}$$

■ **EXAMPLE 23.46**

Problem: A sedimentation tank has a volume of 137,000 gal. If the flow to the tank is 121,000 gph, what is the detention time in the tank (in hours)?

Solution:

$$\text{Detention time} = \frac{\text{Volume of tank (gal)}}{\text{Flow rate (gph)}} = \frac{137,000 \text{ gal}}{121,000 \text{ gph}} = 1.1 \text{ hr}$$

■ **EXAMPLE 23.47**

Problem: A sedimentation basin is 60 ft long by 22 ft wide and has water to a depth of 10 ft. If the flow to the basin is 1,500,000 gpd, what is the sedimentation basin detention time?

Solution: First, convert the flow rate from gpd to gph so the times units will match:

$$(1,500,000 \text{ gpd})/(24 \text{ hr/day}) = 62,500 \text{ gph}$$

Then calculate detention time:

$$\text{Detention time} = \frac{\text{Volume of tank (gal)}}{\text{Flow rate (gph)}} = \frac{60 \text{ ft} \times 22 \text{ ft} \times 10 \text{ ft} \times 7.48 \text{ gal/ft}^3}{62,500 \text{ gph}} = 1.6 \text{ hr}$$

23.5.3 Surface Overflow Rate

The surface overflow rate—similar to the hydraulic loading rate (flow per unit area)—is used to determine loading on sedimentation basins and circular clarifiers. Hydraulic loading rate, however, measures the total water entering the process, whereas surface overflow rate measures only the water overflowing the process (plant flow only).

Note: Surface overflow rate calculations do not include recirculated flows. Other terms used synonymously with surface overflow rate are *surface loading rate* and *surface settling rate*.

Surface overflow rate is determined using the following equation:

$$\text{Surface overflow rate} = \frac{\text{Flow (gpm)}}{\text{Area (ft}^2)} \tag{23.51}$$

■ **EXAMPLE 23.48**

Problem: A circular clarifier has a diameter of 80 ft. If the flow to the clarifier is 1800 gpm, what is the surface overflow rate in gpm/ft²?

Solution:

$$\text{Surface overflow rate} = \frac{\text{Flow (gpm)}}{\text{Area (ft}^2)} = \frac{1800 \text{ gpm}}{0.785 \times 80 \text{ ft} \times 80 \text{ ft}} = 0.36 \text{ gpm/ft}^2$$

■ **EXAMPLE 23.49**

Problem: A sedimentation basin 70 ft by 25 ft receives a flow of 1000 gpm. What is the surface overflow rate in gpm/ft²?

Solution:

$$\text{Surface overflow rate} = \frac{\text{Flow (gpm)}}{\text{Area (ft}^2)} = \frac{1000 \text{ gpm}}{70 \text{ ft} \times 25 \text{ ft}} = 0.6 \text{ gpm/ft}^2$$

23.5.4 Mean Flow Velocity

The measure of average velocity of the water as it travels through a rectangular sedimentation basin is known as mean flow velocity. Mean flow velocity is calculated using Equation 23.52:

$$\text{Flow } (Q) \text{ (ft}^3\text{/min)} = \text{Cross-sectional area } (A) \text{ (ft}^2) \times \text{Volume } (V) \text{ (ft/min)} \tag{23.52}$$

$$Q = A \times V$$

■ **EXAMPLE 23.50**

Problem: A sedimentation basin is 60 ft long by 18 ft wide and has water to a depth of 12 ft. When the flow through the basin is 900,000 gpd, what is the mean flow velocity in the basin in ft/min?

Solution: Because velocity is desired in ft/min, the flow rate in the $Q = AV$ equation must be expressed in ft³/min (cfm):

$$\frac{900,000 \text{ gpd}}{1440 \text{ min/day} \times 7.48 \text{ gal/ft}^3} = 84 \text{ cfm}$$

Then, use $Q = A \times V$ to calculate velocity:

$$Q = A \times V$$

$$84 \text{ cfm} = (18 \text{ ft} \times 12 \text{ ft}) \times x \text{ fpm}$$

$$x = 84 \text{ cfm}/(18 \text{ ft} \times 12 \text{ ft}) = 0.4 \text{ fpm}$$

■ EXAMPLE 23.51

Problem: A rectangular sedimentation basin 50 ft long by 20 ft wide has a water depth of 9 ft. If the flow to the basin is 1,880,000 gpd, what is the mean flow velocity in ft/min?

Solution: Because velocity is desired in ft/min, the flow rate in the $Q = AV$ equation must be expressed in ft³/min (cfm):

$$\frac{1,880,000 \text{ gpd}}{1440 \text{ min/day} \times 7.48 \text{ gal/ft}^3} = 175 \text{ cfm}$$

Then, use the $Q = A \times V$ equation to calculate velocity:

$$Q = A \times V$$

$$175 \text{ cfm} = (20 \text{ ft} \times 9 \text{ ft}) \times x \text{ fpm}$$

$$x = 175 \text{ cfm}/(20 \text{ ft} \times 9 \text{ ft}) = 0.97 \text{ fpm}$$

23.5.5 WEIR OVERFLOW RATE

Weir overflow rate (weir loading rate) is the amount of water leaving the settling tank per linear foot of weir. The result of this calculation can be compared with design. Normally, weir overflow rates of 10,000 to 20,000 gal/day/ft are used in the design of a settling tank. Typically, the weir overflow rate is a measure of the flow in gallons per minute (gpm) over each foot of weir. The weir overflow rate is determined using the following equation:

$$\text{Weir overflow rate (gpm/ft)} = \frac{\text{Flow (gpm)}}{\text{Weir length (ft)}} \qquad (23.53)$$

■ EXAMPLE 23.52

Problem: A rectangular sedimentation basin has a total of 115 ft of weir. What is the weir overflow rate in gpm/ft when the flow is 1,110,000 gpd?

Solution:

$$\text{Flow} = \frac{1,110,000 \text{ gpd}}{1440 \text{ min/day}} = 771 \text{ gpm}$$

$$\text{Weir overflow rate} = \frac{\text{Flow (gpm)}}{\text{Weir length (ft)}} = \frac{771 \text{ gpm}}{115 \text{ ft}} = 6.7 \text{ gpm/ft}$$

■ **EXAMPLE 23.53**

Problem: A circular clarifier receives a flow of 3.55 MGD. If the diameter of the weir is 90 ft, what is the weir overflow rate in gpm/ft?

Solution:

$$\text{Flow} = \frac{3{,}550{,}000 \text{ gpd}}{1440 \text{ min/day}} = 2465 \text{ gpm}$$

$$\text{Weir length} = 3.14 \times 90 \text{ ft} = 283 \text{ ft}$$

$$\text{Weir overflow rate} = \frac{\text{Flow (gpm)}}{\text{Weir length (ft)}} = \frac{2465 \text{ gpm}}{283 \text{ ft}} = 8.7 \text{ gpm/ft}$$

23.5.6 PERCENT SETTLED BIOSOLIDS

The percent settled biosolids test (*volume over volume test*, or V/V test) is conducted by collecting a 100-mL slurry sample from the solids contact unit and allowing it to settle for 10 min. After 10 min, the volume of settled biosolids at the bottom of the 100-mL graduated cylinder is measured and recorded. The equation used to calculate percent settled biosolids is shown below:

$$\% \text{ Settled biosolids} = \frac{\text{Settled biosolids volume (mL)}}{\text{Total sample volume (mL)}} \times 100 \qquad (23.54)$$

■ **EXAMPLE 23.54**

Problem: A 100-mL sample of slurry from a solids contact unit is placed in a graduated cylinder and allowed to set for 10 min. The settled biosolids at the bottom of the graduated cylinder after 10 min is 22 mL. What is the percent of settled biosolids of the sample?

Solution:

$$\% \text{ Settled biosolids} = \frac{\text{Settled biosolids volume (mL)}}{\text{Total sample volume (mL)}} \times 100$$

$$= \frac{22 \text{ mL}}{100 \text{ mL}} \times 100 = 19\% \text{ settled biosolids}$$

■ **EXAMPLE 23.55**

Problem: A 100-mL sample of slurry from a solids contact unit is placed in a graduated cylinder. After 10 min, a total of 21 mL of biosolids settled to the bottom of the cylinder. What is the percent settled biosolids of the sample?

Solution:

$$\% \text{ Settled biosolids} = \frac{\text{Settled biosolids volume (mL)}}{\text{Total sample volume (mL)}} \times 100$$

$$= \frac{21 \text{ mL}}{100 \text{ mL}} \times 100 = 21\% \text{ settled biosolids}$$

23.5.7 DETERMINING LIME DOSAGE (MG/L)

During the alum dosage process, lime is sometimes added to provide adequate alkalinity (HCO_3^-) in the solids contact clarification process for the coagulation and precipitation of the solids. To determine the lime dose required, in mg/L, three steps are required. In Step 1, the total alkalinity required to react with the alum to be added and provide proper precipitation is determined using the following equation:

$$\text{Total alkalinity required} = \text{Alkalinity reacting with alum} + \text{Alkalinity in the water} \qquad (23.55)$$

$$\uparrow$$

$$(1 \text{ mg/L alum reacts with } 0.45 \text{ mg/L alkalinity})$$

■ EXAMPLE 23.56

Problem: Raw water requires an alum dose of 45 mg/L, as determined by jar testing. If a residual 30-mg/L alkalinity must be present in the water to ensure complete precipitation of alum added, what is the total alkalinity required (in mg/L)?

Solution: First calculate the alkalinity that will react with 45 mg/L alum:

$$\frac{0.45 \text{ mg/L alkalinity}}{1 \text{ mg/L alum}} = \frac{x \text{ mg/L alkalinity}}{45 \text{ mg/L alum}}$$

$$0.45 \times 45 = x$$

$$x = 20.25 \text{ mg/L alkalinity}$$

Then calculate the total alkalinity required:

$$\text{Total alkalinity required} = \text{Alkalinity reacting with alum} + \text{Alkalinity in the water}$$

$$= 20.25 \text{ mg/L} + 30 \text{ mg/L} = 50.25 \text{ mg/L}$$

■ EXAMPLE 23.57

Problem: Jar tests indicate that 36 mg/L alum is optimum for a particular raw water. If a residual 30-mg/L alkalinity must be present to promote complete precipitation of the alum added, what is the total alkalinity required (in mg/L)?

Solution: First calculate the alkalinity that will react with 36-mg/L alum:

$$\frac{0.45 \text{ mg/L alkalinity}}{1 \text{ mg/L alum}} = \frac{x \text{ mg/L alkalinity}}{36 \text{ mg/L alum}}$$

$$0.45 \times 36 = x$$

$$x = 16.2 \text{ mg/L alkalinity}$$

Then calculate the total alkalinity required:

$$\text{Total alkalinity required} = 16.2 \text{ mg/L} + 30 \text{ mg/L} = 46.2 \text{ mg/L}$$

In Step 2, we make a comparison between required alkalinity and alkalinity already in the raw water to determine how many mg/L alkalinity should be added to the water. The equation used to make this calculation is shown below:

$$\text{Alkalinity to be added} = \text{Total alkalinity required} - \text{Alkalinity present in water} \qquad (23.56)$$

■ EXAMPLE 23.58

Problem: A total of 44-mg/L alkalinity is required to react with alum and ensure proper precipitation. If the raw water has an alkalinity of 30 mg/L as bicarbonate, how much mg/L alkalinity should be added to the water?

Solution:

$$\text{Alkalinity to be added} = \text{Total alkalinity required} - \text{Alkalinity present in water}$$

$$= 44 \text{ mg/L} - 30 \text{ mg/L} = 14 \text{ mg/L}$$

In Step 3, after determining the amount of alkalinity to be added to the water, we determine how much lime (the source of alkalinity) must be added. We accomplish this by using the ratio shown in Example 23.59.

■ EXAMPLE 23.59

Problem: It has been calculated that 16 mg/L alkalinity must be added to a raw water. How much mg/L lime will be required to provide this amount of alkalinity? (1 mg/L alum reacts with 0.45 mg/L alkalinity and 1 mg/L alum reacts with 0.35 mg/L lime.)

Solution: First determine the mg/L lime required by using a proportion that relates bicarbonate alkalinity to lime:

$$\frac{0.45 \text{ mg/L alkalinity}}{0.35 \text{ mg/L lime}} = \frac{16 \text{ mg/L alkalinity}}{x \text{ mg/L lime}}$$

Then cross-multiply:

$$0.45x = 16 \times 0.35$$

$$x = \frac{16 \times 0.35}{0.45}$$

$$x = 12.4 \text{ mg/L lime}$$

In Example 23.60, we use all three steps to determine the lime dosage (mg/L) required.

■ EXAMPLE 23.60

Problem: Given the following data, calculate the lime dose required, in mg/L:

Alum dose required (determined by jar tests) = 52 mg/L
Residual alkalinity required for precipitation = 30 mg/L
1 mg/L alum reacts with 0.35 mg/L lime
1 mg/L alum reacts with 0.45 mg/L alkalinity
Raw water alkalinity = 36 mg/L

Solution: To calculate the total alkalinity required, we must first calculate the alkalinity that will react with 52 mg/L alum:

$$\frac{0.45 \text{ mg/L alkalinity}}{1 \text{ mg/L alum}} = \frac{x \text{ mg/L alkalinity}}{52 \text{ mg/L alum}}$$

$$0.45 \times 52 = x$$

$$23.4 \text{ mg/L alkalinity} = x$$

The total alkalinity requirement can now be determined:

Total alkalinity required = Alkalinity reacting with alum + Residual alkalinity

$$= 23.4 \text{ mg/L} + 30 \text{ mg/L} = 53.4 \text{ mg/L}$$

Next calculate how much alkalinity must be added to the water:

Alkalinity to be added = Total alkalinity required – Alkalinity present in water

$$= 53.4 \text{ mg/L} - 36 \text{ mg/L} = 17.4 \text{ mg/L}$$

Finally, calculate the lime required to provide this additional alkalinity:

$$\frac{0.45 \text{ mg/L alkalinity}}{0.35 \text{ mg/L lime}} = \frac{17.4 \text{ mg/L alkalinity}}{x \text{ mg/L alum}}$$

$$0.45x = 17.4 \times 0.35$$

$$x = \frac{17.4 \times 0.35}{0.45}$$

$$x = 13.5 \text{ mg/L lime}$$

23.5.8 DETERMINING LIME DOSAGE (LB/DAY)

After the lime dose has been determined in terms of mg/L, it is a fairly simple matter to calculate the lime dose in lb/day, which is one of the most common calculations in water and wastewater treatment. To convert from mg/L to lb/day lime dose, we use the following equation:

$$\text{Lime (lb/day)} = \text{Lime (mg/L)} \times \text{Flow (MGD)} \times 8.34 \text{ lb/gal} \tag{23.57}$$

■ EXAMPLE 23.61

Problem: The lime dose for a raw water has been calculated to be 15.2 mg/L. If the flow to be treated is 2.4 MGD, how many lb/day lime will be required?

Solution:

$$\text{Lime (lb/day)} = \text{Lime (mg/L)} \times \text{Flow (MGD)} \times 8.34 \text{ lb/gal}$$

$$= 15.2 \text{ mg/L} \times 2.4 \text{ MGD} \times 8.34 \text{ lb/gal} = 304 \text{ lb/day lime}$$

■ EXAMPLE 23.62

Problem: The flow to a solids contact clarifier is 2,650,000 gpd. If the lime dose required is determined to be 12.6 mg/L, how many lb/day lime will be required?

Solution:

$$\text{Lime (lb/day)} = \text{Lime (mg/L)} \times \text{flow (MGD)} \times 8.34 \text{ lb/gal}$$

$$= 12.6 \text{ mg/L} \times 2.65 \text{ MGD} \times 8.34 \text{ lb/gal} = 278 \text{ lb/day lime}$$

23.5.9 DETERMINING LIME DOSAGE (GRAMS PER MINUTE)

To convert from mg/L lime to g/min lime, use Equation 23.58:

Key Point: 1 lb = 453.6 g.

$$\text{Lime (g/min)} = \frac{\text{Lime (lb/day)} \times 453.6 \text{ g/lb}}{1440 \text{ min/day}} \tag{23.58}$$

■ EXAMPLE 23.63

Problem: A total of 275 lb/day lime will be required to raise the alkalinity of the water passing through a solids-contact clarification process. How many g/min lime does this represent?

Solution:

$$\text{Lime} = \frac{\text{Lime (lb/day)} \times 453.6 \text{ g/lb}}{1440 \text{ min/day}} = \frac{275 \text{ lb/day} \times 453.6 \text{ g/lb}}{1440 \text{ min/day}} = 86.6 \text{ g/min}$$

■ EXAMPLE 23.64

Problem: A lime dose of 150 lb/day is required for a solids-contact clarification process. How many g/min lime does this represent?

Solution:

$$\text{Lime} = \frac{\text{Lime (lb/day)} \times 453.6 \text{ g/lb}}{1440 \text{ min/day}} = \frac{150 \text{ lb/day} \times 453.6 \text{ g/lb}}{1440 \text{ min/day}} = 47.3 \text{ g/min}$$

23.5.10 PARTICLE SETTLING (SEDIMENTATION)*

Particle settling (sedimentation) may be described for a singular particle by Newton's equation (Equation 23.64) for terminal settling velocity of a spherical particle. For the engineer, knowledge of this velocity is basic in the design and performance of a sedimentation basin. The rate at which discrete particles will settle in a fluid of constant temperature is given by the following equation:

$$u = \left(\frac{4g(p_p - p)d}{3C_D p} \right)^{1/2} \tag{23.59}$$

---

* Much of the information presented in this section is based on USEPA, *Guidance Manual for Compliance with the Interim Enhanced Surface Water Treatment Rule: Turbidity Provisions*, EPA 815-R-99-012, U.S. Environmental Protection Agency, Washington, DC 1999.

where

u = Settling velocity of particles (m/s, ft/s).
g = Gravitational acceleration (m/s², ft/s²).
p_p = Density of particles (kg/m³, lb/ft³).
p = Density of water (kg/m³, lb/ft³).
d = Diameter of particles (m, ft).
C_D = Coefficient of drag.

The terminal settling velocity is derived by equating the drag, buoyant, and gravitational forces acting on the particle. At low settling velocities, the equation is not dependent on the shape of the particle and most sedimentation processes are designed so as to remove small particles, ranging from 1.0 to 0.5 µm, which settle slowly. Larger particles settle at higher velocity and will be removed whether or not they follow Newton's law or Stokes' law—the governing equation when the drag coefficient is sufficiently small (0.5 or less) as is the case for colloidal products (McGhee, 1991).

Typically, a large range of particle sizes will exist in the raw water supply. There are four types of sedimentation (Gregory and Zabel, 1990):

Type 1—Discrete particle settling (particles of various sizes, in a dilute suspension, which settle without flocculating).
Type 2—Flocculant settling (heavier particles coalesced with smaller and lighter particles).
Type 3—Hindered settling (high densities of particles in suspension resulting in an interaction of particles).
Type 4—Compression settling.

The values of the drag coefficient depend on the density of water (p), relative velocity (u), particle diameter (d), and viscosity of water (μ), which gives the Reynolds number, Re, as

$$\mathrm{Re} = \frac{p \times u \times d}{\mu} \tag{23.60}$$

As the Reynolds number increases, the value of C_D increases. For Re less than 2, C_D is related to Re by the linear expression as follows:

$$C_D = \frac{24}{\mathrm{Re}} \tag{23.61}$$

At low levels of Re, the Stokes equation for laminar flow conditions is used (Equations 23.60 and 23.61 substituted into Equation 23.59):

$$u = \frac{G(p_p - p)d^2}{18\mu}$$

In the region of higher Reynolds numbers ($2 < \mathrm{Re} < 500\text{-}1000$), C_D becomes (Fair et al. 1968)

$$C_D = \frac{24}{\mathrm{Re}} + \frac{3}{\sqrt{\mathrm{Re}}} + 0.34 \tag{23.62}$$

Note: In the region of turbulent flow ($500\text{–}1000 < \mathrm{Re} < 200{,}000$), C_D remains approximately constant at 0.44.

The velocity of settling particles results in Newton's equation (AWWA and ASCE, 1990):

$$u = 1.74 \left(\frac{(p_p - p)gd}{p} \right)^{1/2} \tag{23.63}$$

Key Point: When the Reynolds number is greater than 200,000, the drag force decreases substantially and C_D becomes 0.10. No settling occurs at this condition.

■ EXAMPLE 23.65

Problem: Estimate the terminal settling velocity in water at a temperature of 21°C of spherical particles with specific gravity 2.40 and average diameter of (a) 0.006 mm and (b) 1.0 mm.

Solution for part (a): Use Equation 23.62.

Given:
Temperature $(T) = 21°C$
$p = 998 \text{ kg/m}^3$
$\mu = 0.00098 \text{ N s/m}^2$
$d = 0.06 \text{ mm} = 6 \times 10^{-5} \text{ m}$
$g = 9.81 \text{ m/s}^2$

$$u = \frac{g(p_p - p)d^2}{18\mu} = \frac{9.81 \text{ m/s}^2 \times (2400 \text{ kg/m}^3 - 998 \text{ kg/m}^3) \times (6 \times 10^{-5} \text{ m})^2}{18\mu} = 0.00281 \text{ m/s}$$

Use Equation 23.60 to check the Reynolds number:

$$Re = \frac{p \times u \times d}{\mu} = \frac{998 \times 0.00281 \times (6 \times 10^{-5})}{0.00098} = 0.172$$

Stokes' law applies, because Re < 2.

Solution for part (b): Use Equation 23.62:

$$u = \frac{9.81 \times (2400 - 998) \times (0.001)^2}{18 \times 0.00098} = \frac{0.137536}{0.01764} = 0.779 \text{ m/s}$$

Use Equation 23.60 to check the Reynolds number (assume that the irregularities of the particles $\Phi = 0.80$):

$$Re = \frac{\Phi \times p \times u \times d}{\mu} = \frac{0.80 \times 998 \times 0.779 \times 0.001}{0.00098} = 635$$

Because Re > 2, Stokes' law does not apply. Use Equation 23.59 to calculate u:

$$C_D = \frac{24}{635} + \frac{3}{\sqrt{635}} + 0.34 = 0.50$$

$$u^2 = \frac{4g(p_p - p)d}{3C_D p} = \frac{4 \times 9.81 \times (2400 - 998) \times 0.001}{3 \times 0.50 \times 998}$$

$$u = 0.192 \text{ m/s}$$

Re-check Re:

$$\text{Re} = \frac{\Phi \times p \times u \times d}{\mu} = \frac{0.80 \times 998 \times 0.192 \times 0.001}{0.00098} = 156$$

Calculate u using the new Re value:

$$C_D = \frac{24}{156} + \frac{3}{\sqrt{156}} + 0.34 = 0.73$$

$$u^2 = \frac{4 \times 9.81 \times 1402 \times 0.001}{3 \times 0.73 \times 998}$$

$$u = 159 \text{ m/s}$$

Re-check Re:

$$\text{Re} = \frac{0.80 \times 998 \times 0.159 \times 0.001}{0.00098} = 130$$

Calculate u using the new Re value:

$$C_D = \frac{24}{130} + \frac{3}{\sqrt{130}} + 0.34 = 0.79$$

$$u^2 = \frac{4 \times 9.81 \times 1402 \times 0.001}{3 \times 0.79 \times 998}$$

$$u = 0.152 \text{ m/s}$$

Thus, the estimated velocity is approximately 0.15 m/s.

23.5.11 OVERFLOW RATE (SEDIMENTATION)

Overflow rate, detention time, horizontal velocity, and weir overflow rate are the parameters typically used for sizing sedimentation basin. The *theoretical detention time* (plug flow theory) is computed from the volume of the basin divided by average daily flow:

$$t = \frac{24V}{Q} \tag{23.64}$$

where
t = Detention time (hr).
24 = 24 hr per day.
V = Volume of basin (m^3, gal).
Q = Average daily flow (m^3/d, MGD).

The overflow rate is a standard design parameter that can be determined from discrete particle settling analysis. The overflow rate or surface loading rate is calculated by dividing the average daily flow by the total area of the sedimentation basin:

$$u = \frac{Q}{A} = \frac{Q}{lw} \qquad (23.65)$$

where
   u = Overflow rate (m^3/m^2·d, gpd/ft^2).
   Q = Average daily flow (m^3/d, gpd).
   A = Total surface area of basin (m^2, ft^2).
   l = Length of basin (m, ft).
   w = Width of basin (m, ft).

Note: All particles having a settling velocity greater than the overflow rate will settle and be removed.

Rapid particle density changes due to temperature, solids concentration, or salinity can induce density currents which can cause severe short-circuiting in horizontal tanks (Hudson, 1989).

■ EXAMPLE 23.66

Problem: A water treatment plant has two clarifiers treating 2.0 MGD of water. Each clarifier is 14 ft wide, 80 ft long, and 17 ft deep. Determine: (a) detention time, (b) overflow rate, (c) horizontal velocity, and (d) weir overflow rate, assuming the weir length is 2.5 times the basin width.

Solution:

(a) Compute detention time (t) for each clarifier:

$$Q = \frac{2\ \text{MGD}}{2} = \frac{1,000,000\ \text{gal}}{\text{day}} \times \frac{1\ \text{ft}}{7.48\ \text{gal}} \times \frac{1\ \text{day}}{24\ \text{hr}} = 5570\ \text{ft}^3/\text{hr} = 92.8\ \text{ft}^3/\text{min}$$

(b) Compute overflow rate u:

$$u = \frac{Q}{\text{Length} \times \text{Width}} = \frac{1,000,000\ \text{gpd}}{14\ \text{ft} \times 80\ \text{ft}} = 893\ \text{gpd/ft}$$

(c) Compute horizontal velocity V:

$$V = \frac{Q}{\text{Width} \times \text{Depth}} = \frac{92.8\ \text{ft}^3/\text{min}}{14\ \text{ft} \times 17\ \text{ft}} = 0.39\ \text{ft/min}$$

(d) Compute weir overflow rate u_w:

$$u_w = \frac{Q}{2.5 \times \text{Width}} = \frac{1,000,000\ \text{gpd}}{2.5 \times 14\ \text{ft}} = 28,571\ \text{gpd/ft}$$

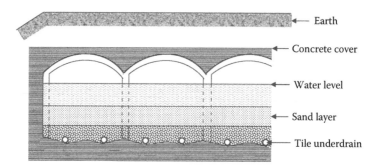

FIGURE 23.3   Slow sand flow.

23.6   WATER FILTRATION CALCULATIONS

Water filtration is a physical process of separating suspended and colloidal particles from waste by passing the water through a granular material (see Figure 23.3). The process of filtration involves straining, settling, and adsorption. As floc passes into the filter, the spaces between the filter grains become clogged, reducing this opening and increasing removal. Some material is removed merely because it settles on a media grain. One of the most important processes is adsorption of the floc onto the surface of individual filter grains. In addition to removing silt and sediment, floc, algae, insect larvae, and any other large elements, filtration also contributes to the removal of bacteria and protozoans such as *Giardia lamblia* and *Cryptosporidium*. Some filtration processes are also used for iron and manganese removal. The *Surface Water Treatment Rule* (SWTR) specifies four filtration technologies, although SWTR also allows the use of alternative filtration technologies (e.g., cartridge filters). These include slow sand filtration/rapid sand filtration, pressure filtration, diatomaceous earth filtration, and direct filtration. Of these, all but rapid sand filtration are commonly employed in small water systems that use filtration. Each type of filtration system has advantages and disadvantages. Regardless of the type of filter, however, filtration involves the processes of *straining* (where particles are captured in the small spaces between filter media grains), *sedimentation* (where the particles land on top of the grains and stay there), and *adsorption* (where a chemical attraction occurs between the particles and the surface of the media grains).

23.6.1   Flow Rate through a Filter (gpm)

Flow rate in gpm through a filter can be determined by simply converting the gpd flow rate, as indicated on the flow meter. The gpm flow rate can be calculated by taking the meter flow rate (gpd) and dividing by 1440 min/day as shown below:

$$\text{Flow rate (gpm)} = \text{Flow rate (gpd)}/1440 \text{ min/day} \qquad (23.66)$$

■ EXAMPLE 23.67

Problem: The flow rate through a filter is 4.25 MGD. What is this flow rate expressed as gpm?

Solution:

$$\text{Flow rate} = 4.25 \text{ MGD}/1440 \text{ min/day}$$

$$= 4{,}250{,}000 \text{ gpd}/1440 \text{ min/day} = 2951 \text{ gpm}$$

■ **EXAMPLE 23.68**

Problem: During a 70-hr filter run, a total of 22.4 million gal of water are filtered. What is the average flow rate through the filter in gpm during this filter run?

Solution:

$$\text{Flow rate} = \text{Total gallons produced/Filter run (min)}$$

$$= 22,400,000 \text{ gal}/(70 \text{ hr} \times 60 \text{ min/hr}) = 5333 \text{ gpm}$$

■ **EXAMPLE 23.69**

Problem: At an average flow rate of 4000 gpm, how long of a filter run (in hours) would be required to produce 25 MG of filtered water?

Solution: Write the equation as usual, filling in known data:

$$\text{Flow rate} = \text{Total gallons produced/filter run (min)}$$

$$4000 \text{ gpm} = 25,000,000 \text{ gal}/(x \text{ hr} \times 60 \text{ min/hr})$$

Then solve for *x:*

$$x = 25,000,000 \text{ gal}/(4000 \text{ gpm} \times 60 \text{ min/hr}) = 104 \text{ hr}$$

■ **EXAMPLE 23.70**

Problem: A filter box is 20 ft by 30 ft (including the sand area). If the influent valve is shut, the water drops 3 in./min. What is the rate of filtration in MGD?

Solution:

Given:
    Filter box = 20 ft × 30 ft
    Water drops = 3 in./min

Step 1. Find the volume of water passing through the filter:

$$\text{Volume} = \text{Area} \times \text{Height}$$

$$\text{Area} = \text{Width} \times \text{Length}$$

Note: The best approach to performing calculations of this type is a step-by-step one, breaking down the problem into what is given and what is to be found.

Area = 20 ft × 30 ft = 600 ft²
Convert 3.0 in. into feet: 3/12 = 0.25 ft
Volume = 600 ft² × 0.25 ft = 150 ft³ of water passing through the filter in 1 min

Step 2. Convert cubic feet to gallons:

$$150 \text{ ft}^3 \times 7.48 \text{ gal/ft}^3 = 1122 \text{ gpm}$$

Step 3. The problem asks for the rate of filtration in MGD. To find MGD, multiply the number of gallons per minute by the number of minutes per day:

$$1122 \text{ gpm} \times 1440 \text{ min/day} = 1.62 \text{ MGD}$$

■ **EXAMPLE 23.71**

Problem: The influent valve to a filter is closed for 5 min. During this time, the water level in the filter drops 0.8 ft (10 in.). If the filter is 45 ft long and 15 ft wide, what is the gpm flow rate through the filter? Water drop equals 0.16 ft/min.

Solution: First calculate cfm flow rate using the $Q = A \times V$ equation:

$$Q = \text{Length (ft)} \times \text{Width (ft)} \times \text{Drop velocity (ft/min)}$$

$$= 45 \text{ ft} \times 15 \text{ ft} \times 0.16 \text{ ft/min} = 108 \text{ cfm}$$

Then convert cfm flow rate to gpm flow rate:

$$108 \text{ cfm} \times 7.48 \text{ gal/ft}^3 = 808 \text{ gpm}$$

23.6.2 FILTRATION RATE

One measure of filter production is filtration rate (generally ranging from 2 to 10 gpm/ft²). Along with filter run time, it provides valuable information for operation of filters. It is the gallons of water filtered per minute through each square foot of filter area. Filtration rate is determined as follows:

$$\text{Filtration rate (gpm/ft}^2) = \frac{\text{Flow rate (gpm)}}{\text{Filter surface area (ft}^2)} \qquad (23.67)$$

■ **EXAMPLE 23.72**

Problem: A filter 18 ft by 22 ft receives a flow of 1750 gpm. What is the filtration rate in gpm/ft²?

Solution:

$$\text{Filtration rate} = \frac{\text{Flow rate (gpm)}}{\text{Filter surface area (ft}^2)} = \frac{1750 \text{ gpm}}{18 \text{ ft} \times 22 \text{ ft}} = 4.4 \text{ gpm/ft}^2$$

■ **EXAMPLE 23.73**

Problem: A filter 28 ft long by 18 ft wide treats a flow of 3.5 MGD. What is the filtration rate in gpm/ft²?

Solution:

$$\text{Flow rate} = 3,500,000 \text{ gpd/(1440 min/day)} = 2431 \text{ gpm}$$

$$\text{Filtration rate} = \frac{\text{Flow rate (gpm)}}{\text{Filter surface area (ft}^2)} = \frac{2431 \text{ gpm}}{28 \text{ ft} \times 18 \text{ ft}} = 4.8 \text{ gpm/ft}^2$$

■ **EXAMPLE 23.74**

Problem: A filter 45 ft long by 20 ft wide produces a total of 18 MG during a 76-hr filter run. What is the average filtration rate in gpm/ft² for this filter run?

Solution: First calculate the gpm flow rate through the filter:

$$\text{Flow rate} = 18,000,000 \text{ gal/(76 hr} \times 60 \text{ min/hr)} = 3947 \text{ gpm}$$

Then calculate the filtration rate:

$$\text{Filtration rate} = \frac{\text{Flow rate (gpm)}}{\text{Filter surface area (ft}^2)} = \frac{3947 \text{ gpm}}{45 \text{ ft} \times 20 \text{ ft}} = 4.4 \text{ gpm/ft}^2$$

■ EXAMPLE 23.75

Problem: A filter is 40 ft long by 20 ft wide. During a test of flow rate, the influent valve to the filter is closed for 6 min. The water level drop during this period is 16 in. What is the filtration rate for the filter in gpm/ft²?

Solution: First calculate the gpm flow rate using the $Q = A \times V$ equation:

$$Q \text{ (gpm)} = \text{Length (ft)} \times \text{Width (ft)} \times \text{Drop velocity (ft/min)} \times 7.48 \text{ gal/ft}^3$$

$$= (40 \text{ ft} \times 20 \text{ ft} \times 1.33 \text{ ft} \times 7.48 \text{ gal/ft}^3)/6 \text{ min} = 1326 \text{ gpm}$$

Then calculate the filtration rate:

$$\text{Filtration rate} = \frac{\text{Flow rate (gpm)}}{\text{Filter surface area (ft}^2)} = \frac{1326 \text{ gpm}}{40 \text{ ft} \times 20 \text{ ft}} = 1.6 \text{ gpm/ft}^2$$

23.6.3   UNIT FILTER RUN VOLUME (UFRV)

The unit filter run volume (UFRV) calculation indicates the total gallons passing through each square foot of filter surface area during an entire filter run. This calculation is used to compare and evaluate filter runs. UFRVs are usually at least 5000 gal/ft² and generally in the range of 10,000 gpd/ft². The UFRV value will begin to decline as the performance of the filter begins to deteriorate. The equation to be used in these calculations is shown below:

$$\text{Unit filter run volume} = \text{Total gallons filtered/Filter surface area (ft}^2) \qquad (23.68)$$

■ EXAMPLE 23.76

Problem: The total water filtered during a filter run (between backwashes) is 2,220,000 gal. If the filter is 18 ft by 18 ft, what is the unit filter run volume in gal/ft²?

Solution:

$$\text{Unit filter run volume} = \text{Total gallons filtered/filter surface area (ft}^2)$$

$$= 2,220,000 \text{ gal}/(18 \text{ ft} \times 18 \text{ ft}) = 6852 \text{ gal/ft}^2$$

■ EXAMPLE 23.77

Problem: The total water filtered during a filter run is 4,850,000 gal. If the filter is 28 ft by 18 ft, what is the unit filter run volume in gal/ft²?

Solution:

$$\text{Unit filter run volume} = \text{Total gallons filtered/filter surface area (ft}^2)$$

$$= 4,850,000 \text{ gal}/(28 \text{ ft} \times 18 \text{ ft}) = 9623 \text{ gal/ft}^2$$

Equation 23.68 can be modified as shown in Equation 23.69 to calculate the unit filter run volume given filtration rate and filter run data:

$$\text{Unit filter run volume} = \text{Filtration rate (gpm/ft}^2) \times \text{Filter run time (min)} \qquad (23.69)$$

■ EXAMPLE 23.78

Problem: The average filtration rate for a filter was determined to be 2 gpm/ft². If the filter run time was 4250 minutes, what was the unit filter run volume in gal/ft²?

Solution:

$$\text{Unit filter run volume} = \text{Filtration rate (gpm/ft}^2) \times \text{Filter run time (min)}$$

$$= 2 \text{ gpm/ft}^2 \times 4250 = 8500 \text{ gal/ft}^2$$

The problem indicates that, at an average filtration rate of 2 gal entering each square foot of filter each minute, the total gallons entering during the total filter run is 4250 times that amount.

■ EXAMPLE 23.79

Problem: The average filtration rate during a particular filter run was determined to be 3.2 gpm/ft². If the filter run time was 61.0 hr, what was the UFRV in gal/ft² for the filter run?

Solution:

$$\text{Unit filter run volume} = \text{Filtration rate (gpm/ft}^2) \times \text{Filter run (hr)} \times 60 \text{ min/hr}$$

$$= 3.2 \text{ gpm/ft}^2 \times 61.0 \text{ hr} \times 60 \text{ min/hr} = 11{,}712 \text{ gal/ft}^2$$

23.6.4 BACKWASH RATE

In filter backwashing, one of the most important operational parameters to be determined is the amount of water in gallons required for each backwash. This amount depends on the design of the filter and the quality of the water being filtered. The actual washing typically lasts 5 to 10 min and usually amounts to 1 to 5% of the flow produced.

■ EXAMPLE 23.80

Problem: A filter has the following dimensions:

Length = 30 ft
Width = 20 ft
Depth of filter media = 24 in.

Assuming that a backwash rate of 15 gal/ft²/min is recommended and 10 min of backwash is required, calculate the amount of water in gallons required for each backwash.

Solution: Given the above data, find the amount of water in gallons required:

1. Area of filter = 30 ft × 20 ft = 600 ft²
2. Gallons of water used per square foot of filter = 15 gal/ft²/min × 10 min = 150 gal/ft²
3. Gallons required for backwash = 150 gal/ft² × 600 ft² = 90,000 gal

Typically, backwash rates will range from 10 to 25 gpm/ft². The backwash rate is determined by using Equation 23.70:

$$\text{Backwash rate} = \frac{\text{Flow rate (gpm)}}{\text{Filter area (ft}^2)} \tag{23.70}$$

■ **EXAMPLE 23.81**

Problem: A filter 30 ft by 10 ft has a backwash rate of 3120 gpm. What is the backwash rate in gpm/ft²?

Solution:

$$\text{Backwash rate} = \frac{\text{Flow rate (gpm)}}{\text{Filter area (ft}^2)} = \frac{3120 \text{ gpm}}{(30 \text{ ft} \times 10 \text{ ft})} = 10.4 \text{ gpm/ft}^2$$

■ **EXAMPLE 23.82**

Problem: A filter 20 ft by 20 ft has a backwash rate of 4.85 MGD. What is the filter backwash rate in gpm/ft²?

Solution:

$$\text{Flow rate} = 4{,}850{,}000 \text{ gpd}/(1440 \text{ min/day}) = 3368 \text{ gpm}$$

$$\text{Backwash rate} = \frac{\text{Flow rate (gpm)}}{\text{Filter area (ft}^2)} = \frac{3368 \text{ gpm}}{(20 \text{ ft} \times 20 \text{ ft})} = 8.42 \text{ gpm/ft}^2$$

23.6.5 BACKWASH RISE RATE

Backwash rate is occasionally measured as the upward velocity of the water during backwashing, expressed as in./min rise. To convert from gpm/ft² backwash rate to an in./min rise rate, use either Equation 23.71 or Equation 23.72:

$$\text{Backwash rate (in./min)} = \frac{\text{Backwash rate (gpm/ft}^2) \times (12 \text{ in./ft})}{7.48 \text{ gal/ft}^3} \tag{23.71}$$

$$\text{Backwash rate (in./min)} = \text{Backwash rate (gpm/ft}^2) \times 1.6 \tag{23.72}$$

■ **EXAMPLE 23.83**

Problem: A filter has a backwash rate of 16 gpm/ft². What is this backwash rate expressed as an in./min rise rate?

Solution:

$$\text{Backwash rate} = \frac{\text{Backwash rate (gpm/ft}^2) \times (12 \text{ in./ft})}{7.48 \text{ gal/ft}^3} = \frac{16 \text{ gpm/ft}^2 \times (12 \text{ in./ft})}{7.48 \text{ gal/ft}^3} = 25.7 \text{ in./min}$$

■ **EXAMPLE 23.84**

Problem: A filter 22 ft long by 12 ft wide has a backwash rate of 3260 gpm. What is this backwash rate expressed as an in./min rise?

Solution: First calculate the backwash rate as gpm/ft²:

$$\text{Backwash rate} = \frac{\text{Flow rate (gpm)}}{\text{Filter area (ft}^2)} = \frac{3260 \text{ gpm}}{(22 \text{ ft} \times 12 \text{ ft})} = 12.3 \text{ gpm/ft}^2$$

Then convert gpm/ft^2 to the in./min rise rate:

$$\text{Rise rate} = \frac{12.3 \text{ gpm/ft}^2 \times 12 \text{ in./ft}}{7.48 \text{ gal/ft}^3} = 19.7 \text{ in./min}$$

23.6.6   VOLUME OF BACKWASH WATER REQUIRED (GAL)

To determine the volume of water required for backwashing, we must know both the desired backwash flow rate (gpm) and the duration of backwash (min):

Backwash water volume (gal) = Backwash (gpm) × Duration of backwash (min)   (23.73)

■ EXAMPLE 23.85

Problem: For a backwash flow rate of 9000 gpm and a total backwash time of 8 min, how many gallons of water will be required for backwashing?

Solution:

Backwash water volume = Backwash (gpm) × Duration of backwash (min)

= 9000 gpm × 8 min = 72,000 gal

■ EXAMPLE 23.86

Problem: How many gallons of water would be required to provide a backwash flow rate of 4850 gpm for a total of 5 min?

Solution:

Backwash water volume = Backwash (gpm) × Duration of backwash (min)

= 4850 gpm × 5 min = 24,250 gal

23.6.7   REQUIRED DEPTH OF BACKWASH WATER TANK (FT)

The required depth of water in the backwash water tank is determined from the volume of water required for backwashing. To make this calculation, simply use Equation 23.74:

Volume (gal) = 0.785 × (Diameter)^2 × Depth (ft) × 7.48 gal/ft^3 (23.74)

■ EXAMPLE 23.87

Problem: The volume of water required for backwashing has been calculated to be 85,000 gal. What is the required depth of water in the backwash water tank to provide this amount of water if the diameter of the tank is 60 ft?

Solution: Use the volume equation for a cylindrical tank, filling in known data, then solve for x:

Volume = 0.785 × (Diameter)^2 × Depth (ft) × 7.48 gal/ft^3

85,000 gal = 0.785 × (60 ft)^2 × x ft × 7.48 gal/ft^3

x = 85,000/(0.785 × 60 × 60 × 7.48) = 4 ft

■ **EXAMPLE 23.88**

Problem: A total of 66,000 gal of water will be required for backwashing a filter at a rate of 8000 gpm for a 9-min period. What depth of water is required if the backwash tank has a diameter of 50 ft?

Solution: Use the volume equation for cylindrical tanks:

$$\text{Volume (gal)} = 0.785 \times (\text{Diameter})^2 \times \text{Depth (ft)} \times 7.48 \text{ gal/ft}^3$$

$$66{,}000 \text{ gal} = 0.785 \times (50 \text{ ft})^2 \times x \text{ ft} \times 7.48 \text{ gal/ft}^3$$

$$x = 66{,}000/(0.785 \times 50 \times 50 \times 7.48) = 4.5 \text{ ft}$$

23.6.8 BACKWASH PUMPING RATE (GPM)

The desired backwash pumping rate (gpm) for a filter depends on the desired backwash rate in gpm/ft^2 and the ft^2 area of the filter. The backwash pumping rate (gpm) can be determined by using Equation 23.75:

$$\text{Backwash pumping rate (gpm)} = \text{Desired backwash rate (gpm/ft}^2) \times \text{Filter area (ft}^2) \quad (23.75)$$

■ **EXAMPLE 23.89**

Problem: A filter is 25 ft long by 20 ft wide. If the desired backwash rate is 22 gpm/ft^2, what backwash pumping rate (gpm) will be required?

Solution: The desired backwash flow through each square foot of filter area is 20 gpm. The total gpm flow through the filter is therefore 20 gpm times the entire square foot area of the filter:

$$\text{Backwash pumping rate} = \text{Desired backwash rate (gpm/ft}^2) \times \text{Filter area (ft}^2)$$

$$= 20 \text{ gpm/ft}^2 \times (25 \text{ ft} \times 20 \text{ ft}) = 10{,}000 \text{ gpm}$$

■ **EXAMPLE 23.90**

Problem: The desired backwash pumping rate for a filter is 12 gpm/ft^2. If the filter is 20 ft long by 20 ft wide, what backwash pumping rate (gpm) will be required?

Solution:

$$\text{Backwash pumping rate} = \text{Desired backwash rate (gpm/ft}^2) \times \text{Filter area (ft}^2)$$

$$= 12 \text{ gpm/ft}^2 \times (20 \text{ ft} \times 20 \text{ ft}) = 4800 \text{ gpm}$$

23.6.9 PERCENT PRODUCT WATER USED FOR BACKWASHING

Along with measuring filtration rate and filter run time, another aspect of filter operation that is monitored for filter performance is the percent of product water used for backwashing. The equation for percent of product water used for backwashing calculations used is shown below:

$$\text{Backwash water (\%)} = \frac{\text{Backwash water (gal)}}{\text{Water filtered (gal)}} \times 100 \quad (23.76)$$

■ **EXAMPLE 23.91**

Problem: A total of 18,100,000 gal of water was filtered during a filter run. If backwashing used 74,000 gal of this product water, what percent of the product water was used for backwashing?

Solution:

$$\text{Backwash water} = \frac{\text{Backwash water (gal)}}{\text{Water filtered (gal)}} \times 100 = \frac{74,000 \text{ gal}}{18,100,000 \text{ gal}} \times 100 = 0.4\%$$

■ **EXAMPLE 23.92**

Problem: A total of 11,400,000 gal of water was filtered during a filter run. If backwashing used 48,500 gal of this product water, what percent of the product water was used for backwashing?

Solution:

$$\text{Backwash water} = \frac{\text{Backwash water (gal)}}{\text{Water filtered (gal)}} \times 100 = \frac{48,500 \text{ gal}}{11,400,000 \text{ gal}} \times 100 = 0.43\%$$

23.6.10 PERCENT MUDBALL VOLUME

Mudballs are heavier deposits of solids near the top surface of the medium that break into pieces during backwash, resulting in spherical accretions of floc and sand (usually less than 12 in. in diameter). The presence of mudballs in the filter media is checked periodically. The principal objection to mudballs is that they diminish the effective filter area. To calculate the percent mudball volume we use Equation 23.77:

$$\% \text{ Mudball volume} = \frac{\text{Mudball volume (mL)}}{\text{Total sample volume (mL)}} \times 100 \tag{23.77}$$

■ **EXAMPLE 23.93**

Problem: A 3350-mL sample of filter media was taken for mudball evaluation. The volume of water in the graduated cylinder rose from 500 mL to 525 mL when mudballs were placed in the cylinder. What is the percent mudball volume of the sample?

Solution: First determine the volume of mudballs in the sample:

$$525 \text{ mL} - 500 \text{ mL} = 25 \text{ mL}$$

Then calculate the percent mudball volume:

$$\% \text{ Mudball volume} = \frac{\text{Mudball volume (mL)}}{\text{Total sample volume (mL)}} \times 100 = \frac{25 \text{ mL}}{3350 \text{ mL}} \times 100 = 0.75\%$$

■ **EXAMPLE 23.94**

Problem: A filter is tested for the presence of mudballs. The mudball sample has a total sample volume of 680 mL. Five samples were taken from the filter. When the mudballs were placed in 500 mL of water, the water level rose to 565 mL. What is the percent mudball volume of the sample?

Solution: The mudball volume is the volume that the water rose:

$$565 \text{ mL} - 500 \text{ mL} = 65 \text{ mL}$$

Because 5 samples of media were taken, the total sample volume is 5 times the sample volume:

$$5 \times 680 \text{ mL} = 3400 \text{ mL}$$

$$\% \text{ Mudball volume} = \frac{\text{Mudball volume (mL)}}{\text{Total sample volume (mL)}} \times 100 = \frac{65 \text{ mL}}{3400 \text{ mL}} \times 100 = 1.9\%$$

23.6.11 FILTER BED EXPANSION

In addition to backwash rate, it is also important to expand the filter media during the wash to maximize the removal of particles held in the filter or by the media; that is, the efficiency of the filter wash operation depends on the expansion of the sand bed. Bed expansion is determined by measuring the distance from the top of the unexpanded media to a reference point (e.g., top of the filter wall) and from the top of the expanded media to the same reference. A proper backwash rate should expand the filter 20 to 25%. Percent bed expansion is given by dividing the bed expansion by the total depth of expandable media (i.e., media depth less support gravels) and multiplied by 100, as follows:

$$\text{Expanded measurement} = \text{Depth to top of media during backwash (in.)}$$

$$\text{Unexpanded measurement} = \text{Depth to top of media before backwash (in.)}$$

$$\text{Bed expansion} = \text{Unexpanded measurement (in.)} - \text{expanded measurement (in.)}$$

$$\text{Bed expansion (\%)} = \frac{\text{Bed expansion measurement (in.)}}{\text{Total depth of expandable media (in.)}} \times 100 \tag{23.78}$$

■ EXAMPLE 23.95

Problem: The backwashing practices for a filter with 30 in. of anthracite and sand are being evaluated. While at rest, the distance from the top of the media to the concrete floor surrounding the top of filter is measured to be 41 in. After the backwash has begun and the maximum backwash rate is achieved, a probe containing a white disk is slowly lowered into the filter bed until anthracite is observed on the disk. The distance from the expanded media to the concrete floor is measured to be 34.5 in. What is the percent bed expansion?

Solution:

Given:
  Unexpanded measurement = 41 in.
  Expanded measurement = 34.5 in.

$$\text{Bed expansion} = 41 \text{ in.} - 34.5 \text{ in.} = 6.5$$

$$\text{Bed expansion} = \frac{\text{Bed expansion measurement (in.)}}{\text{Total depth of expandable media (in.)}} \times 100 = \frac{6.5 \text{ in.}}{30 \text{ in.}} \times 100 = 22\%$$

23.6.12 Filter Loading Rate

Filter loading rate is the flow rate of water applied to the unit area of the filter. It is the same value as the flow velocity approaching the filter surface and can be determined by using Equation 23.79:

$$u = Q/A \qquad\qquad (23.79)$$

where
u = Loading rate (m³/(m²·day, gpm/ft²).
Q = Flow rate (m³/day, ft³/day, gpm).
A = Surface area of filter (m², ft²).

Filters are classified as slow sand filters, rapid sand filters, or high-rate sand filters on the basis of loading rate. Typically, the loading rate for rapid sand filters is 120 m³/m²·day (83 L/m²·min or 2 gal/min/ft²). The loading rate may be up to five times this rate for high-rate filters.

■ **EXAMPLE 23.96**

Problem: A sanitation district is to install rapid sand filters downstream of the clarifiers. The design-loading rate is selected to be 150 m³/m². The design capacity of the waterworks is 0.30 m³/sec (6.8 MGD). The maximum surface per filter is limited to 45 m². Design the number and size of filters and calculate the normal filtration rate.

Solution: Determine the total surface area required:

$$A = \frac{Q}{u} = \frac{0.30 \text{ m}^3/\text{sec } (85,400 \text{ sec/day})}{150 \text{ m}^3/\text{m}^2 \cdot \text{day}} = \frac{25,920}{150} = 173 \text{ m}^2$$

Determine the number of filters:

$$\text{No. of filters} = \frac{173 \text{ m}^2}{45 \text{ m}} = 3.8$$

Select four filters. The surface area (A) for each filter is

$$A = 173 \text{ m}^2 \div 4 = 43.25 \text{ m}^2$$

We can use 6 m × 7 m or 6.4 m × 7 m or 6.42 m × 7 m. If a 6-m × 7-m filter is installed, the normal filtration rate is

$$u = \frac{Q}{A} = \frac{0.30 \text{ m}^3/\text{sec} \times 86,400 \text{ sec/day}}{4 \text{ m} \times 6 \text{ m} \times 7 \text{ m}} = 154.3 \text{ m}^3/\text{m}^2 \cdot \text{day}$$

23.6.13 Filter Medium Size

Filter medium grain size has an important effect on the filtration efficiency and on backwashing requirements for the medium. The actual medium selected is typically determined by performing a grain size distribution analysis—sieve size and percentage passing by weight relationships are plotted on logarithmic-probability paper. The most common parameters used in the United States to characterize a filter medium are effective size (ES) and uniformity coefficient (UC) of medium size distribution. The ES is that grain size for which 10% of the grains are smaller by weight; it is often abbreviated by d_{10}. The UC is the ratio of the 60-percentile (d_{60}) to the 10-percentile. The

90-percentile, d_{90}, is the size for which 90% of the grains are smaller by weight. The d_{90} size is used for computing the required filter backwash rate for a filter medium. Values of d_{10}, d_{60}, and d_{90} can be read from an actual sieve analysis curve. If such a curve is not available and if a linear log–probability plot is assumed, the values can be interrelated by Equation 23.80 (Cleasby, 1990):

$$d_{90} = d_{10}(10^{1.67\log UC}) \tag{23.80}$$

■ EXAMPLE 23.97

Problem: A sieve analysis curve of a typical filter sand gives $d_{10} = 0.52$mm and $d_{60} = 0.70$ mm. What are its uniformity coefficient and d_{90}?

Solution:

$$UC = d_{60}/d_{10} = 0.70 \text{ mm}/0.52 \text{ mm} = 1.35$$

$$d_{90} = d_{10}(10^{1.67\log UC}) = 0.52 \text{ mm} \times (10^{1.67\log 1.35}) = 0.52 \text{ mm} \times (10^{0.218}) = 0.86 \text{ mm}$$

23.6.14 Mixed Media

An innovation in filtering systems has offered a significant improvement and economic advantage to rapid rate filtration: the mixed media filter bed. Mixed media filter beds offer specific advantages in specific circumstances, but will give excellent operating results at a filtering rate of 5 gal/ft^2/min. Moreover, the mixed media filtering unit is more tolerant of higher turbidities in the settled water. For improved process performance, activated carbon or anthracite is added on the top of the sand bed. The approximate specific gravities of ilmenite (<60% TiO_2), silica sand, anthracite, and water are 4.2, 2.6, 1.5, and 1.0, respectively. The economic advantage of the mixed bed media filter is based upon filter area; it will safely produce 2-1/2 times as much filtered water as a rapid sand filter. When settling velocities are equal, the particle sizes for media of different specific gravities can be computed by using Equation 23.81:

$$\frac{d_1}{d_2} = \left(\frac{s_2 - s}{s_1 - s} \right)^{2/3} \tag{23.81}$$

where

d_1, d_2 = Diameter of particles 1 and 2, and water, respectively.
s_1, s, s_2 = Specific gravity of particle 1, water, and particle 2, respectively.

■ EXAMPLE 23.98

Problem: Estimate the particle size of ilmenite sand (specific gravity = 4.2) that has the same settling velocity as silica sand that is 0.60 mm in diameter (specific gravity = 2.6).

Solution: Find the diameter of ilmenite sand using Equation 23.81:

$$d = 0.6 \text{ mm} \times \left(\frac{2.6 - 1}{4.2 - 1} \right)^{2/3} = 0.38 \text{ mm}$$

23.6.15 Head Loss for Fixed Bed Flow

When water is pumped upward through a bed of fine particles at a very low flow rate the water percolates through the pores (void spaces) without disturbing the bed. This is a fixed bed process. The head loss (pressure drop) through a clean granular-media filter is generally less than 0.9 m (3 ft).

With the accumulation of impurities, head loss gradually increases until the filter is backwashed. The Kozeny equation, shown below, is typically used for calculating head loss through a clean fixed-bed flow filter:

$$\frac{h}{L} = \left(\frac{k\mu(1-\varepsilon)^2}{gp\varepsilon^3} \right) \left(\frac{A}{V} \right)^2 u \tag{23.82}$$

where

h = Head loss in filter depth L (m, ft).
L = Depth of filter (ft, m).
k = Dimensionless Kozeny constant (5 for sieve openings, 6 for size of separation).
μ = Absolute viscosity of water (N·s/m², lb·s/ft²).
ε = Porosity (dimensionless).
g = Acceleration of gravity (9.81 m/s or 32.2 ft/s).
p = Density of water (kg/m³, lb/ft³).
A/V = Grain surface area per unit volume of grain.
      = Specific surface S (shape factor = 6.0–7.7).
      = $6/d$ for spheres.
      = $6/\Psi d_{eq}$ for irregular grains.
Ψ = Grain sphericity or shape factor.
d_{eq} = Grain diameter of spheres of equal volume.
u = Filtration (superficial) velocity (m/s, fps).

■ EXAMPLE 23.99

Problem: A dual-medium filter is composed of 0.3 m of anthracite (mean size, 2.0 mm) that is placed over a 0.6-m layer of sand (mean size, 0.7 mm) with a filtration rate of 9.78 m/hr. Assume that the grain sphericity is $\Psi = 0.75$ and the porosity for both is 0.42. Although such values are normally taken from the appropriate table at 15°C, we provide the head loss data of the filter at 1.131×10^{-6} m²·s.

Solution: Determine head loss through the anthracite layer using Equation 23.82:

$$\frac{h}{L} = \left(\frac{k\mu(1-\varepsilon)^2}{gp\varepsilon^3} \right) \left(\frac{A}{V} \right)^2 u$$

where

$k = 6$.
$g = 9.81$ m/s².
$\mu p = v = 1.131 \times 10^{-6}$ m²·s (from the appropriate table).
$\varepsilon = 0.40$.
$A/V = 6/0.75d = 8/d = 8/0.002$.
$u = 9.78$ m/h = 0.00272 m/s.
$L = 0.3$ m.

Then,

$$h = 6 \times \frac{1.131 \times 10^{-6}}{9.81} \times \frac{(1-0.42)^2}{0.42^3} \times \left(\frac{8}{0.002} \right)^2 \times 0.00272 \times 0.3 = 0.0410 \text{ m}$$

Compute the head loss passing through the sand. Use the same data but insert:

$k = 5$
$d = 0.0007$ m
$L = 0.6$ m

$$h = 5 \times \frac{1.131 \times 10^{-6}}{9.81} \times \frac{(0.58)^2}{0.42^3} \times \left(\frac{8}{0.0007}\right)^2 \times 0.00272 \times 0.6 = 0.5579 \text{ m}$$

Compute total head loss (h):

$$h = 0.0410 \text{ m} + 0.5579 \text{ m} = 0.599 \text{ m}$$

23.6.16 Head Loss through a Fluidized Bed

If the upward water flow rate through a filter bed is very large the bed mobilizes pneumatically and may be swept out of the process vessel. At an intermediate flow rate the bed expands and is in what we call an *expanded* state. In the fixed bed the particles are in direct contact with each other, supporting each other's weight. In the expanded bed, the particles have a mean free distance between particles and the drag force of the water supports the particles. The expanded bed has some of the properties of the water (i.e., of a fluid) and is called a *fluidized bed* (Chase, 2002). Simply, *fluidization* is defined as upward flow through a granular filter bed at sufficient velocity to suspend the grains in the water. Minimum fluidizing velocity (U_{mf}) is the superficial fluid velocity required to start fluidization; it is important in determining the required minimum backwashing flow rate. Wen and Yu (1966) proposed that the U_{mf} equation include the constants (over a wide range of particles) 33.7 and 0.0408 but exclude porosity of fluidization and shape factor (Wen and Yu, 1966):

$$U_{mf} = \frac{\mu}{p d_{eq}} \times (1135.69 + 0.0408 G_n)^{0.5} - \frac{33.7\mu}{p d_{eq}} \qquad (23.83)$$

where
  μ = Absolute viscosity of water (N·s/m², lb·s/ft²).
  p = Density of water (kg/m³, lb/ft³).
  $d_{eq} = d_{90}$ sieve size is used instead of d_{eq}.
  G_n = Galileo number, which is equal to

$$d_{eq}^3 p (p_s - p) g / \mu^2 \qquad (23.84)$$

Note: Based on the studies of Cleasby and Fan (1981), we use a safety factor of 1.3 to ensure adequate movement of the grains.

■ EXAMPLE 23.100

Problem: Estimate the minimum fluidized velocity and backwash rate for a sand filter. The d_{90} size of the sand is 0.90 mm. The density of the sand is 2.68 g/cm³.

Solution: Compute the Galileo number. From the data given and the applicable table at 15°C:

$p = 0.999$ g/cm³
$\mu = 0.0113$ N s/m² = 0.00113 kg/ms = 0.0113 g/cm s
$\mu p = 0.0113$ cm²/s
$g = 981$ cm/s²
$d = 0.90$ cm
$p_s = 2.68$ g/cm³

Using Equation 23.84,

$$G_n = d_{eq}^3 p(p_s - p)g / \mu^2 = (0.090)^3 \times 0.999 \times (2.68 - 0.999) \times 981 / (0.0113)^2 = 9405$$

Compute U_{mf} using Equation 23.83:

$$U_{mf} = \frac{\mu}{pd_{eq}} \times (1135.69 + 0.0408G_n)^{0.5} - \frac{33.7\mu}{pd_{eq}}$$

$$= \frac{0.0113}{0.999 \times 0.090} \times (1135.69 + 0.0408 \times 9405)^{0.5} - \frac{33.7 \times 0.0113}{0.999 \times 0.090}$$

$$= 0.660 \text{ cm/s}$$

Compute backwash rate. Apply a safety factor of 1.3 to U_{mf} as backwash rate:

$$\text{Backwash rate} = 1.3 \times 0.660 \text{ cm/s} = 0.858 \text{ cm/s}$$

$$0.858 \times \frac{\text{cm}^3}{\text{cm}^2 \cdot \text{s}} \times \frac{\text{L}}{1000 \text{ cm}^3} \times \frac{1}{3.785} \times \frac{\text{gal}}{\text{L}} \times 929 \times \frac{\text{cm}^2}{\text{ft}^2} \times \frac{60 \text{ s}}{\text{min}} = 12.6 \text{ pgm/ft}^2$$

23.6.17 HORIZONTAL WASHWATER TROUGHS

Wastewater troughs are used to collect backwash water as well as to distribute influent water during the initial stages of filtration. Washwater troughs are normally placed above the filter media in the United States. Proper placement of these troughs is very important to ensure that the filter medium is not carried into the troughs during the backwash and removed from the filter. These backwash troughs are constructed from concrete, plastic, fiberglass, or other corrosion-resistant materials. The total rate of discharge in a rectangular trough with free flow can be calculated by using Equation 23.85:

$$Q = C \times w \times h^{1.5} \tag{23.85}$$

where
Q = Flow rate (cfs).
C = Constant (2.49).
w = Trough width (ft).
h = Maximum water depth in trough (ft).

■ EXAMPLE 23.101

Problem: Troughs are 18 ft long, 18 in. wide, and 8 ft to the center with a horizontal flat bottom. The backwash rate is 24 in./min. Estimate (a) the water depth of the troughs with free flow into the gullet, and (b) the distance between the top of the troughs and the 30-in. sand bed. Assume a 40% expansion and 6 in. of freeboard in the troughs and 6 in. of thickness.

Solution: Estimate the maximum water depth (h) in the trough:

$$v = 24 \text{ in/min} = 2 \text{ ft/60 s} = 1/30 \text{ fps}$$

$$A = 18 \text{ ft} \times 8 \text{ ft} = 144 \text{ ft}^2$$

$$Q = V/A = 144/30 \text{ cfs} = 4.8 \text{ cfs}$$

Using Equation 23.85,

$$w = 1.5 \text{ ft}$$

$$Q = Cwh^{1.5} = 2.49wh^{1.5}$$

$$h = \left(Q/2.49w\right)^{2/3} = \left[4.8(2.49 \times 1.5)\right]^{2/3} = 1.18 \text{ ft (approx. 14 in.} = 1.17 \text{ ft)}$$

Determine the distance (y) between the sand bed surface and the top troughs:

$$\text{Freeboard} = 6 \text{ in.} = 0.5 \text{ ft}$$

$$\text{Thickness} = 8 \text{ in.} = 0.67 \text{ ft (bottom of trough)}$$

$$y = 2.5 \text{ ft} \times 0.4 + 1.17 \text{ ft} + 0.5 \text{ ft} + 0.5 \text{ ft} = 3.2 \text{ ft}$$

23.6.18 FILTER EFFICIENCY

Water treatment filter efficiency is defined as the effective filter rate divided by the operation filtration rate as shown in Equation 23.86 (AWWA and ASCE, 1998):

$$E = \frac{R_e}{R_o} = \frac{UFRV - UBWV}{UFRV} \tag{23.86}$$

where
  E = Filter efficiency (%).
  R_e = Effective filtration rate (gpm/ft^2).
  R_o = Operating filtration rate (gpm/ft^2).
  $UFRV$ = Unit filter run volume (gal/ft^2).
  $UBWV$ = Unit backwash volume (gal/ft^2).

■ EXAMPLE 23.102

Problem: A rapid sand filter operates at 3.9 gpm/ft^2 for 48 hours. Upon completion of the filter run, 300 gal/ft^2 of backwash water is used. Find the filter efficiency.

Solution: Calculate the operating filtration rate (R_o):

$$R_o = 3.9 \text{ gpm/ft}^2 \times 60 \text{ min/hr} \times 48 \text{ hr} = 11,232 \text{ gal/ft}^2$$

Calculate the effective filtration rate (R_e):

$$R_e = 11,232 \text{ gal/ft}^2 - 300 \text{ gal/ft}^2 = 10,932 \text{ gal/ft}^2$$

Calculate the filter efficiency (E) using Equation 23.86:

$$E = 10,932/11,232 = 97.3\%$$

23.7 WATER CHLORINATION CALCULATIONS

Chlorine is the most commonly used substance for disinfection of water in the United States. The addition of chlorine or chlorine compounds to water is called *chlorination*. Chlorination is considered to be the single most important process for preventing the spread of waterborne disease.

23.7.1 CHLORINE DISINFECTION

Chlorine can destroy most biological contaminants by various mechanisms, including

- Damaging the cell wall
- Altering the permeability of the cell (the ability to pass water in and out through the cell wall)
- Altering the cell protoplasm
- Inhibiting the enzyme activity of the cell so it is unable to use its food to produce energy
- Inhibiting cell reproduction

Chlorine is available in a number of different forms: (1) as pure elemental gaseous chlorine (a greenish-yellow gas possessing a pungent and irritating odor that is heavier than air, nonflammable, and nonexplosive), which, when released to the atmosphere, is toxic and corrosive; (2) as solid calcium hypochlorite (in tablets or granules); or (3) as a liquid sodium hypochlorite solution (in various strengths). The choice of one form of chlorine over another for a given water system depends on the amount of water to be treated, configuration of the water system, the local availability of the chemicals, and the skill of the operator. One of the major advantages of using chlorine is the effective residual that it produces. A residual indicates that disinfection is completed and the system has an acceptable bacteriological quality. Maintaining a residual in the distribution system helps to prevent regrowth of those microorganisms that were injured but not killed during the initial disinfection stage.

23.7.2 DETERMINING CHLORINE FEED RATE

The expressions milligrams per liter (mg/L) and pounds per day (lb/day) are most often used to describe the amount of chlorine added or required. Equation 23.87 can be used to calculate either mg/L or lb/day chlorine dosage:

$$\text{Chlorine feed rate (lb/day)} = \text{Chlorine (mg/L)} \times \text{Flow (MGD)} \times 8.34 \text{ lb/gal} \qquad (23.87)$$

■ EXAMPLE 23.103

Problem: Determine the chlorinator setting (lb/day) required to treat a flow of 4 MGD with a chlorine dose of 5 mg/L.

Solution:

$$\text{Chlorine feed rate} = \text{Chlorine (mg/L)} \times \text{Flow (MGD)} \times 8.34 \text{ lb/gal}$$

$$= 5 \text{ mg/L} \times 4 \text{ MGD} \times 8.34 \text{ lb/gal} = 167 \text{ lb/day}$$

■ EXAMPLE 23.104

Problem: A pipeline that is 12 in. in diameter and 1400 ft long is to be treated with a chlorine dose of 48 mg/L. How many pounds of chlorine will this require?

Solution: First determine the gallon volume of the pipeline:

$$\text{Volume} = 0.785 \times (\text{Diameter})^2 \times \text{Length (ft)} \times 7.48 \text{ gal/ft}^3$$

$$= 0.785 \times (1 \text{ ft})^2 \times 1400 \text{ ft} \times 7.48 \text{ gal/ft}^3 = 8221 \text{ gal}$$

Now calculate the pounds chlorine required:

$$\text{Chlorine} = \text{Chlorine (mg/L)} \times \text{Volume (MG)} \times 8.34 \text{ lb/gal}$$

$$= 48 \text{ mg/L} \times 0.008221 \text{ MG} \times 8.34 \text{ lb/gal} = 3.3 \text{ lb}$$

■ **EXAMPLE 23.105**

Problem: A chlorinator setting is 30 lb per 24 hr. If the flow being chlorinated is 1.25 MGD, what is the chlorine dosage expressed as mg/L?

Solution:

$$\text{Chlorine (lb/day)} = \text{Chlorine (mg/L)} \times \text{Flow (MGD)} \times 8.34 \text{ lb/gal}$$

$$30 \text{ lb/day} = x \text{ mg/L} \times 1.25 \text{ MGD} \times 8.34 \text{ lb/gal}$$

$$x = 30/(1.25 \times 8.34) = 2.9 \text{ mg/L}$$

■ **EXAMPLE 23.106**

Problem: A flow of 1600 gpm is to be chlorinated. At a chlorinator setting of 48 lb per 24 hr, what would be the chlorine dosage in mg/L?

Solution: Convert the gpm flow rate to MGD flow rate:

$$1600 \text{ gpm} \times 1440 \text{ min/day} = 2,304,000 \text{ gpd} = 2.304 \text{ MGD}$$

Now calculate the chlorine dosage in mg/L:

$$\text{Chlorine (lb/day)} = \text{Chlorine (mg/L)} \times \text{Flow (MGD)}$$

$$48 \text{ lb/day} = x \text{ mg/L} \times 2.304 \text{ MGD} \times 8.34 \text{ lb/gal}$$

$$x = 48/(2.304 \times 8.34) = 2.5 \text{ mg/L}$$

23.7.3 CALCULATING CHLORINE DOSE, DEMAND, AND RESIDUAL

Common terms used in chlorination include the following:

- *Chlorine dose*—The amount of chlorine added to the system. It can be determined by adding the desired residual for the finished water to the chlorine demand of the untreated water. Dosage can be either milligrams per liter (mg/L) or pounds per day (lb/day). The most common is mg/L:

$$\text{Chlorine dose (mg/L)} = \text{Chlorine demand (mg/L)} + \text{Chlorine residual (mg/L)} \qquad (23.88)$$

- *Chlorine demand*—The amount of chlorine used by iron, manganese, turbidity, algae, and microorganisms in the water. Because the reaction between chlorine and microorganisms is not instantaneous, demand is relative to time; for example, the demand 5 min after applying chlorine will be less than the demand after 20 min. Demand, like dosage, is expressed in mg/L:

$$\text{Chlorine demand (mg/L)} = \text{Chlorine dose (mg/L)} - \text{Chlorine residual (mg/L)}$$

The following examples illustrate the calculation of chlorine dose, demand, and residual using Equation 23.88.

■ **EXAMPLE 23.107**

Problem: A water sample is tested and found to have a chlorine demand of 1.7 mg/L. If the desired chlorine residual is 0.9 mg/L, what is the desired chlorine dose in mg/L?

Solution:

$$\text{Chlorine dose} = \text{Chlorine demand (mg/L)} + \text{Chlorine residual (mg/L)}$$

$$= 1.7 \text{ mg/L} + 0.9 \text{ mg/L} = 2.6 \text{ mg/L}$$

■ EXAMPLE 23.108

Problem: The chlorine dosage for water is 2.7 mg/L. If the chlorine residual after 30 min of contact time is found to be 0.7 mg/L, what is the chlorine demand expressed in mg/L?

Solution:

$$\text{Chlorine dose (mg/L)} = \text{Chlorine demand (mg/L)} + \text{Chlorine residual (mg/L)}$$

$$2.7 \text{ mg/L} = x \text{ mg/L} + 0.6 \text{ mg/L}$$

$$x \text{ mg/L} = 2.7 \text{ mg/L} - 0.7 \text{ mg/L} = 2.0 \text{ mg/L}$$

■ EXAMPLE 23.109

Problem: What should the chlorinator setting (lb/day) be to treat a flow of 2.35 MGD if the chlorine demand is 3.2 mg/L and a chlorine residual of 0.9 mg/L is desired?

Solution: Determine the chlorine dosage in mg/L:

$$\text{Chlorine dose} = \text{Chlorine demand (mg/L)} + \text{Chlorine residual (mg/L)}$$

$$= 3.2 \text{ mg/L} + 0.9 \text{ mg/L} = 4.1 \text{ mg/L}$$

Calculate the chlorine dosage (feed rate) in lb/day:

$$\text{Chlorine (lb/day)} = \text{Chlorine (mg/L)} \times \text{Flow (MGD)} \times 8.34 \text{ lb/gal}$$

$$= 4.1 \text{ mg/L} \times 2.35 \text{ MGD} \times 8.34 \text{ lb/gal} = 80.4 \text{ lb/day}$$

23.7.4 BREAKPOINT CHLORINATION CALCULATIONS

To produce a free chlorine residual, enough chlorine must be added to the water to produce what is referred to as *breakpoint chlorination*, the point at which near complete oxidation of nitrogen compounds is reached. Any residual beyond breakpoint is mostly free chlorine (see Figure 23.4). When chlorine is added to natural waters, the chlorine begins combining with and oxidizing the chemicals in the water before it begins disinfecting. Although residual chlorine will be detectable in the water, the chlorine will be in the combined form with a weak disinfecting power. As we see in Figure 23.4, adding more chlorine to the water at this point actually decreases the chlorine residual as the additional chlorine destroys the combined chlorine compounds. At this stage, water may have a strong swimming pool or medicinal taste and odor. To avoid this taste and odor, add still more chlorine to produce a free residual chlorine. Free chlorine has the highest disinfecting power. The point at which most of the combined chlorine compounds have been destroyed and the free chlorine starts to form is the *breakpoint*.

Note: The actual chlorine breakpoint of water can only be determined by experimentation.

To calculate the actual increase in chlorine residual that would result from an increase in chlorine dose, we use the mg/L to lb/day equation as shown below:

$$\text{Increase in chlorine (lb/day)} = \text{Expected increase (mg/L)} \times \text{Flow (MGD)} \times 8.34 \text{ lb/gal} \qquad (23.89)$$

Note: The actual increase in residual is simply a comparison of new and old residual data.

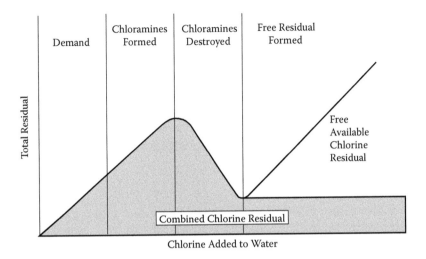

FIGURE 23.4 Breakpoint chlorination curve.

■ **EXAMPLE 23.110**

Problem: A chlorinator setting is increased by 2 lb/day. The chlorine residual before the increased dosage was 0.2 mg/L. After the increased chlorine dose, the chlorine residual was 0.5 mg/L. The average flow rate being chlorinated is 1.25 MGD. Is the water being chlorinated beyond the breakpoint?

Solution: Calculate the expected increase in chlorine residual using the mg/L to lb/day equation:

$$\text{Increase in chlorine (lb/day)} = \text{Expected increase (mg/L)} \times \text{Flow (MGD)} \times 8.34 \text{ lb/gal}$$

$$2 \text{ lb/day} = x \text{ mg/L} \times 1.25 \text{ MGD} \times 8.34 \text{ lb/gal}$$

$$x = (2 \text{ lb/day})/(1.25 \text{ MGD} \times 8.34 \text{ lb/gal}) = 0.19 \text{ mg/L}$$

The actual increase in residual chlorine is

$$0.5 \text{ mg/L} - 0.19 \text{ mg/L} = 0.31 \text{ mg/L}$$

■ **EXAMPLE 23.111**

Problem: A chlorinator setting of 18 lb chlorine per 24 hr results in a chlorine residual of 0.3 mg/L. The chlorinator setting is increased to 22 lb per 24 hr. The chlorine residual increased to 0.4 mg/L at this new dosage rate. The average flow being treated is 1.4 MGD. On the basis of these data, is the water being chlorinated past the breakpoint?

Solution: Calculate the expected increase in chlorine residual:

$$\text{Increase in chlorine (lb/day)} = \text{Expected increase (mg/L)} \times \text{Flow (MGD)} \times 8.34 \text{ lb/gal}$$

$$4 \text{ lb/day} = x \text{ mg/L} \times 1.4 \text{ MGD} \times 8.34 \text{ lb/gal}$$

$$x = (4 \text{ lb/day})/(1.4 \text{ MGD} \times 8.34 \text{ lb/gal}) = 0.34 \text{ mg/L}$$

The actual increase in residual chlorine is

$$0.4 \text{ mg/L} - 0.3 \text{ mg/L} = 0.1 \text{ mg/L}$$

23.7.5 Calculating Dry Hypochlorite Feed Rate

The most commonly used dry hypochlorite, calcium hypochlorite contains about 65 to 70% available chlorine, depending on the brand. Because hypochlorites are not 100% pure chorine, more lb/day must be fed into the system to obtain the same amount of chlorine for disinfection. The equation used to calculate the lb/day hypochlorite needed is Equation 23.90:

$$\text{Hypochlorite (lb/day)} = \frac{\text{Chlorine (lb/day)}}{\% \text{ Available chlorine}/100} \tag{23.90}$$

■ EXAMPLE 23.112

Problem: A chlorine dosage of 110 lb/day is required to disinfect a flow of 1,550,000 gpd. If the calcium hypochlorite to be used contains 65% available chlorine, how many lb/day hypochlorite will be required for disinfection?

Solution: Because only 65% of the hypochlorite is chlorine, more than 110 lb of hypochlorite will be required:

$$\text{Hypochlorite} = \frac{\text{Chlorine (lb/day)}}{\% \text{ Available chlorine}/100} = \frac{110}{65/100} = \frac{110}{0.65} = 169 \text{ lb/day}$$

■ EXAMPLE 23.113

Problem: A water flow of 900,000 gpd requires a chlorine dose of 3.1 mg/L. If calcium hypochlorite (65% available chlorine) is to be used, how many lb/day of hypochlorite are required?

Solution: Calculate the lb/day chlorine required:

$$\text{Chlorine} = \text{Chlorine (mg/L)} \times \text{Flow (MGD)} \times 8.34 \text{ lb/gal}$$
$$= 3.1 \text{ mg/L} \times 0.90 \text{ MGD} \times 8.34 \text{ lb/gal} = 23 \text{ lb/day}$$

Calculate the lb/day hypochlorite:

$$\text{Hypochlorite} = \frac{\text{Chlorine (lb/day)}}{\% \text{ Available chlorine}/100} = \frac{23}{65/100} = \frac{23}{0.65} = 35 \text{ lb/day}$$

■ EXAMPLE 23.114

Problem: A tank contains 550,000 gal of water and is to receive a chlorine dose of 2.0 mg/L. How many pounds of calcium hypochlorite (65% available chlorine) will be required?

Solution:

$$\text{Hypochlorite} = \frac{\text{Chlorine (mg/L)} \times \text{Volume (MG)} \times 8.34 \text{ lb/gal}}{\% \text{ Available chlorine}/100}$$

$$= \frac{2.0 \text{ mg/L} \times 0.550 \text{ MG} \times 8.34}{65/100} = \frac{9.2}{0.65} = 14.2 \text{ lb}$$

■ EXAMPLE 23.115

Problem: A total of 40 lb of calcium hypochlorite (65% available chlorine) is used in a day. If the flow rate treated is 1,100,000 gpd, what is the chlorine dosage in mg/L?

Solution: Calculate the lb/day chlorine dosage:

$$\text{Hypochlorite} = \frac{\text{Chlorine (lb/day)}}{\%\ \text{Available chlorine}/100}$$

$$40\ \text{lb/day} = \frac{x\ \text{lb/day}}{0.65}$$

$$0.65 \times 40 = x$$

$$26\ \text{lb/day} = x$$

Then calculate mg/L chlorine using the mg/L to lb/day equation and filling in the known information:

$$26\ \text{lb/day chlorine} = x\ \text{mg/L chlorine} \times 1.10\ \text{MGD} \times 8.34\ \text{lb/gal}$$

$$x = 26\ \text{lb/day}/(1.10\ \text{MGD} \times 8.34\ \text{lb/gal}) = 2.8\ \text{mg/L}$$

■ EXAMPLE 23.116

Problem: A flow of 2,550,000 gpd is disinfected with calcium hypochlorite (65% available chlorine). If 50 lb of hypochlorite are used in a 24-hr period, what is the mg/L chlorine dosage?

Solution: Calculate the lb/day chlorine dosage:

$$50\ \text{lb/day hypochlorite} = (x\ \text{lb/day chlorine})/0.65$$

$$x = 32.5\ \text{lb/day chlorine}$$

Calculate mg/L chlorine:

$$x\ \text{mg/L chlorine} \times 2.55\ \text{MGD} \times 8.34\ \text{lb/gal} = 32.5\ \text{lb/day}$$

$$x = 1.5\ \text{mg/L chlorine}$$

23.7.6 CALCULATING HYPOCHLORITE SOLUTION FEED RATE

Liquid hypochlorite (i.e., sodium hypochlorite) is supplied as a clear, greenish-yellow liquid in strengths varying from 5.25 to 16% available chlorine. Often referred to as *bleach*, it is, in fact, used for bleaching. Common household bleach is a solution of sodium hypochlorite containing 5.25% available chlorine. When calculating gallons per day (gpd) liquid hypochlorite, the lb/day hypochlorite required must be converted to gpd hypochlorite. This conversion is accomplished using Equation 23.91:

$$\text{Hypochlorite (gpd)} = \frac{\text{Hypochlorite (lb/day)}}{8.34\ \text{lb/gal}} \qquad (23.91)$$

■ **EXAMPLE 23.117**

Problem: A total of 50 lb/day sodium hypochlorite is required for disinfection of a 1.5-MGD flow. How many gallons per day hypochlorite is this?

Solution: Because lb/day hypochlorite has already has been calculated, we simply convert lb/day to gpd hypochlorite required:

$$\text{Hypochlorite (gpd)} = \frac{\text{Hypochlorite (lb/day)}}{8.34\ \text{lb/gal}} = \frac{50\ \text{lb/day}}{8.34\ \text{lb/gal}} = 6.0\ \text{gpd}$$

■ **EXAMPLE 23.118**

Problem: A hypochlorinator is used to disinfect the water pumped from a well. The hypochlorite solution contains 3% available chlorine. A chlorine dose of 1.3 mg/L is required for adequate disinfection throughout the system. If the flow being treated is 0.5 MGD, how many gpd of the hypochlorite solution will be required?

Solution: Calculate the lb/day chlorine required:

$$\text{Chlorine (lb/day)} = 1.3\ \text{mg/L} \times 0.5\ \text{MGD} \times 8.34\ \text{lb/gal} = 5.4\ \text{lb/day}$$

Calculate the lb/day hypochlorite solution required:

$$\text{Hypochlorite (lb/day)} = 5.4\ \text{lb/day chlorine}/0.03 = 180\ \text{lb/day}$$

Calculate the gpd hypochlorite solution required:

$$\text{Hypochlorite (gpd)} = 180\ \text{lb/day}/8.34\ \text{lb/gal} = 21.6\ \text{gpd}$$

23.7.7 PERCENT STRENGTH OF SOLUTIONS

If a teaspoon of salt is dropped into a glass of water it gradually disappears. The salt dissolves in the water, but a microscopic examination of the water would not show the salt. Only examination at the molecular level, which is not easily done, would show the salt and water molecules intimately mixed. If we taste the liquid, we would know that the salt is there. We can recover the salt by evaporating the water. In a solution, the molecules of the salt, the *solute*, are homogeneously dispersed among the molecules of water, the *solvent*. This mixture of salt and water is homogeneous on a molecular level. Such a homogeneous mixture is called a *solution*. The composition of a solution can be varied within certain limits. The three common states of matter are gas, liquid, and solid. In this discussion, we are only concerned, at the moment, with the solid (calcium hypochlorite) and liquid (sodium hypochlorite) states.

23.7.8 CALCULATING PERCENT STRENGTH USING DRY HYPOCHLORITE

To calculate the percent strength of a chlorine solution, we use Equation 23.92:

$$\% \text{ Chlorine strength} = \frac{\text{Hypochlorite (lb)}\left(\dfrac{\%\ \text{Available chlorine}}{100}\right)}{\text{Water (lb)} + \text{Hypochlorite (lb)}\left(\dfrac{\%\ \text{Available chlorine}}{100}\right)} \times 100 \qquad (23.92)$$

■ **EXAMPLE 23.119**

Problem: If a total of 72 oz. of calcium hypochlorite (65% available chlorine) is added to 15 gal of water, what is the percent chlorine strength (by weight) of the solution?

Solution: Convert the ounces of hypochlorite to pounds of hypochlorite:

$$(72 \text{ oz.})/(16 \text{ oz./lb}) = 4.5 \text{ lb chemical}$$

$$\% \text{ Chlorine strength} = \frac{\text{Hypochlorite (lb)}\left(\dfrac{\% \text{ Available chlorine}}{100}\right)}{\text{Water (lb)} + \text{Hypochlorite (lb)}\left(\dfrac{\% \text{ Available chlorine}}{100}\right)} \times 100$$

$$= \frac{4.5 \text{ lb} \times 0.65}{(15 \text{ gal} \times 8.34 \text{ lb/gal}) + (4 \text{ lb} \times 0.65)} \times 100$$

$$= \frac{2.9 \text{ lb}}{125.1 \text{ lb} + 2.9 \text{ lb}} \times 100 = \frac{2.9 \text{ lb}}{126} \times 100 = 2.3\%$$

23.7.9 CALCULATING PERCENT STRENGTH USING LIQUID HYPOCHLORITE

To calculate the percent strength of a chlorine solution, we use Equation 23.93:

$$\text{Liquid hypochlorite (gal)} \times 8.34 \text{ lb/gal} \times \left(\frac{\% \text{ Strength of hypochlorite}}{100}\right)$$

$$= \text{Hypochlorite sol. (gal)} \times 8.34 \text{ lb/gal} \times \left(\frac{\% \text{ Strength of hypochlorite}}{100}\right) \qquad (23.93)$$

■ **EXAMPLE 23.120**

Problem: A 12% liquid hypochlorite solution is to be used in making up a hypochlorite solution. If 3.3 gal of liquid hypochlorite are mixed with water to produce 25 gal of hypochlorite solution, what is the percent strength of the solution?

Solution:

$$\text{Liquid hypochlorite (gal)} \times 8.34 \text{ lb/gal} \times \left(\frac{\% \text{ Strength of hypochlorite}}{100}\right)$$

$$= \text{Hypochlorite sol. (gal)} \times 8.34 \text{ lb/gal} \times \left(\frac{\% \text{ Strength of hypochlorite}}{100}\right)$$

$$3.3 \text{ gal} \times 8.34 \text{ lb/gal} \times \left(\frac{12}{100}\right) = 25 \text{ gal} \times 8.34 \text{ lb/gal} \times \left(\frac{x}{100}\right)$$

$$x = \frac{\cancel{100} \times 3.3 \times \cancel{8.34} \times 12}{25 \times \cancel{8.34} \times \cancel{100}}$$

$$x = \frac{3.3 \times 12}{25} = 1.6\%$$

23.8   CHEMICAL USE CALCULATIONS

In a typical plant operation, chemical use is recorded each day. Such data provide a record of daily use from which the average daily use of the chemical or solution can be calculated. To calculate average use in pounds per day (lb/day), we use Equation 23.94. To calculate average use in gallons per day (gpd), we use Equation 23.95:

$$\text{Average use (lb/day)} = \text{Total chemical used (lb)/Number of days} \qquad (23.94)$$

$$\text{Average use (gpd)} = \text{Total chemical used (gal)/Number of days} \qquad (23.95)$$

To calculate the days supply in inventory, we use Equation 23.96 or Equation 23.97:

$$\text{Days supply in inventory} = \text{Total chemical in inventory (lb)/Average use (lb/day)} \quad (23.96)$$

$$\text{Days supply in inventory} = \text{Total chemical in inventory (gal)/Average use (gpd)} \quad (23.97)$$

■ EXAMPLE 23.121

Problem: Calcium hypochlorite usage for each day during a week is given below. Based on these data, what was the average lb/day hypochlorite chemical use during the week?

Monday	50 lb/day
Tuesday	55 lb/day
Wednesday	51 lb/day
Thursday	46 lb/day
Friday	56 lb/day
Saturday	51 lb/day
Sunday	48 lb/day

Solution:

$$\text{Average use (lb/day)} = \text{Total chemical used (lb)/Number of days}$$

$$= 357 \text{ lb/7 days} = 51 \text{ lb/day}$$

■ EXAMPLE 23.122

Problem: The average calcium hypochlorite use at a plant is 40 lb/day. If the chemical inventory in stock is 1100 lb, how many days' supply is this?

$$\text{Days supply in inventory} = \text{Total chemical in inventory (lb)/Average use (lb/day)}$$

$$= (1100 \text{ lb in inventory})/(40 \text{ lb/day average use}) = 27.5 \text{ days}$$

23.9   CHLORINATION CHEMISTRY

Chlorine is used in the form of free elemental chlorine or as hypochlorites. Temperature, pH, and organic content in the water influence its chemical form in water. When chlorine gas is dissolved in water, it rapidly hydrolyzes to hydrochloric acid (HCl) and hypochlorous acid (HOCl):

$$Cl_2 + H_2O \leftrightarrow H^+ + Cl^- + HOCl \qquad (23.98)$$

The equilibrium constant is (White, 1972)

$$K_H = \frac{[H^+][Cl^-][HOCl]}{[Cl_{2(aq)}]} = 4.48 \times 10^4 \text{ at } 25°C \qquad (23.99)$$

Henry's law is used to explain the dissolution of gaseous chlorine, $Cl_{2(aq)}$. Henry's law describes the effect of the pressure on the solubility of the gases: There is a linear relationship between the partial pressure of gas above a liquid and the mole fraction of the gas dissolved in the liquid (Fetter, 1999). The Henry's law constant, K_H (as shown in Equation 23.99), is a measure of the compound transfer between the gaseous and aqueous phases. K_H is presented as a ratio of the compound's concentration in the gaseous phase to that in the aqueous phase at equilibrium:

$$K_H = \frac{P}{C_{water}} \qquad (23.100)$$

where
K_H = Henry's Law constant.
P = Compound's partial pressure in the gaseous phase.
C_{water} = Compound's concentration in the aqueous solution.

Note: The unit of the Henry's law constant is dependent on the choice of measure; however, it can also be dimensionless. For our purpose, Henry's law can be expressed as (Downs and Adams, 1973)

$$Cl_{2(g)} = \frac{Cl_{2(aq)}}{H \text{ (mol/L} \cdot \text{atm)}} = \frac{[Cl_{2(aq)}]}{P_{Cl_2}} \qquad (23.101)$$

where
$[Cl_{2(aq)}]$ = Molar concentration of Cl_2.
H = Henry's law constant = $4.805 \times 10^{-6} \exp(2818.48/T)$, where T is temperature (K).
P_{Cl_2} = Partial pressure of chlorine in the atmosphere.

The disinfection capabilities of hypochlorous acid (HOCl) are generally higher than those of hypochlorite ions (OCl⁻) (Water, 1978). Hypochlorous acid is a weak acid and subject to further dissociation to hypochlorite ions (OCl⁻) and hydrogen ions:

$$HOCl \leftrightarrow OCl^- + H^+ \qquad (23.102)$$

Its acid dissociation constant K_a is

$$K_H = \frac{[OCl^-][H^+]}{[HOCl]} \qquad (23.103)$$

$$= 3.7 \times 10^{-8} \text{ at } 25°C$$

$$= 2.61 \times 10^{-8} \text{ at } 20°C$$

The value of K_a for hypochlorous acid is a function of temperature (K) as follows (Morris, 1966):

$$\ln K_a = 23.184 - 0.058T - 6908/T \qquad (23.104)$$

REFERENCES AND RECOMMENDED READING

AWWA and ASCE. (1990). *Water Treatment Plant Design*, 2nd ed. McGraw-Hill, New York.

AWWA and ASCE. (1998). *Water Treatment Plant Design*, 3rd ed. McGraw-Hill, New York.

Chase, G.L. (2002). *Solids Notes: Fluidization*. University of Akron, Akron, OH.

Cleasby, J.L. (1990). Filtration, in Pontius, F.W., Ed., *Water Quality and Treatment: A Handbook of Community Water Supplies*, 4th ed. McGraw-Hill, New York.

Cleasby, J.L. and Fan, K.S. (1981). Predicting fluidization and expansion of filter media. *Journal of the Environmental Engineering Division*, 107(EE3), 355–471.

Downs, A.J. and Adams, C.J. (1973). *The Chemistry of Chlorine, Bromine, Iodine, and Astatine*. Pergamon, Oxford, U.K.

Droste, R.L. (1997). *Theory and Practice of Water and Wastewater Treatment*. John Wiley & Sons, New York.

Fair, G.M., Geyer, J.C., and Okun, D.A. (1968). *Water and Wastewater Engineering*. Vol. 2. *Water Purification and Wastewater Treatment and Disposal*. John Wiley & Sons, New York.

Fetter, C.W. (1998). *Handbook of Chlorination*. Litton Educational, New York.

Gregory, R. and Zabel, T.R. (1990). Sedimentation and flotation, in Pontius, F.W., Ed., *Water Quality and Treatment: A Handbook of Community Water Supplies*, 4th ed. McGraw-Hill, New York.

Gupta, R.S. (1997). *Environmental Engineering and Science: An Introduction*. Government Institutes, Rockville, MD.

Hudson, Jr., H.E. (1989). Density considerations in sedimentation. *Journal of the American Water Works Association*, 64(6), 382–386.

McGhee, T.J. (1991). *Water Resources and Environmental Engineering*, 6th ed. McGraw-Hill, New York.

Morris, J.C. (1966). The acid ionization constant of HOCl from 5°C to 35°C. *Journal of Physical Chemistry*, 70(12), 3789.

Water, G.C. (1978). *Disinfection of Wastewater and Water for Reuse*. Van Nostrand Reinhold, New York.

Wen, C.Y. and Yu, Y.H. (1966). Minimum fluidization velocity. *AIChE Journal*, 12(3), 610–612.

White, G.C. (1972). *Handbook of Chlorination*. Litton Education, New York.

Section XI

Math Concepts: Wastewater

In the glory days of Empire Textiles, the solvents and dyes used on the fabrics were dumped directly into the river, staining the banks below the falls red and green and yellow, according to the day of the week and the size of the batch. The sloping banks contained rings, like those in a tree trunk, except these were in rainbow colors, they recorded not the years but the rise and fall of the river. Even now, fifty years later, only the hardiest weeds and scrub trees grew south of the pavement on Front Street, and when the brush was periodically cleared, surprising patches of fading chartreuse and magenta were revealed.

—Richard Russo (*Empire Falls*, 2001)

Whenever in Peru, an important rule of thumb to keep in mind is do not flush your toilet paper.

—Frank R. Spellman

24 Wastewater Calculations

Water is the most critical resource issue of our lifetime and our children's lifetime. The health of our waters is the principal measure of how we live on the land.

—**Luna Leopold**

24.1 INTRODUCTION

Standard wastewater treatment consists of a series of steps or unit processes (see Figure 24.1) tied together with the ultimate purpose of taking the raw sewage influent and turning it into an effluent that is often several times cleaner than the water in the outfalled water body. Like the water calculations presented in Chapter 23, this chapter presents math calculations related to wastewater at the operations level as well as the engineering level. Again, our purpose in using this format is consistent with our intention to provide a single, self-contained, ready reference source.

24.2 PRELIMINARY TREATMENT

The initial stage of treatment in the wastewater treatment process (following collection and influent pumping) is preliminary treatment. Process selection normally is based on the expected characteristics of the influent flow. Raw influent entering the treatment plant may contain many kinds of materials (trash), and preliminary treatment protects downstream plant equipment by removing these materials, which could cause clogs, jams, or excessive wear in plant machinery. In addition, the removal of various materials at the beginning of the treatment train saves valuable space within the treatment plant.

Two of the processes used in preliminary treatment include screening and grit removal; however, preliminary treatment may also include other processes, each designed to remove a specific type of material that presents a potential problem for downstream unit treatment processes. These processes include shredding, flow measurement, preaeration, chemical addition, and flow equalization. Except in extreme cases, plant design will not include all of these items. Here, we focus on and describe typical calculations used in two of these processes: screening and grit removal.

24.2.1 SCREENING REMOVAL CALCULATIONS

Screening removes large solids, such as rags, cans, rocks, branches, leaves, roots, etc. from the flow before the flow moves on to downstream processes. Wastewater operators responsible for screenings disposal are typically required to keep a record of the amount of screenings removed from the flow. To keep and maintain accurate screening records, the volume of screenings withdrawn must be determined. Two methods are commonly used to calculate the volume of screenings withdrawn:

$$\text{Screenings removed (ft}^3/\text{day}) = \frac{\text{Screenings (ft}^3)}{\text{day}} \qquad (24.1)$$

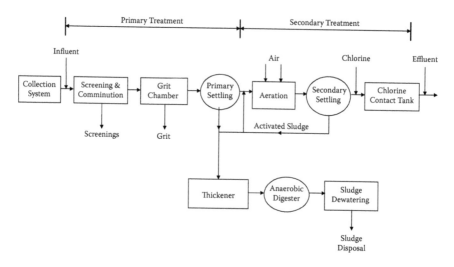

FIGURE 24.1 Illustration of a wastewater treatment process providing primary and secondary treatment using activated sludge process.

$$\text{Screenings removed (ft}^3\text{/MG)} = \frac{\text{Screenings (ft}^3)}{\text{Flow (MG)}} \quad\quad (24.2)$$

■ **EXAMPLE 24.1**

Problem: A total of 65 gal of screenings is removed from the wastewater flow during a 24-hr period. What is the screenings removal reported as ft³/day?

Solution: First convert gallon screenings to cubic feet:

$$(65 \text{ gal})/(7.48 \text{ gal/ft}^3) = 8.7 \text{ ft}^3 \text{ screenings}$$

Then calculate screenings removed as ft³/day:

$$\text{Screenings removed (ft}^3\text{/day)} = \frac{8.7 \text{ ft}^3}{1 \text{ day}} = 8.7 \text{ ft}^3\text{/day}$$

■ **EXAMPLE 24.2**

Problem: During 1 week, a total of 310 gal of screenings was removed from the wastewater screens. What is the average screenings removal in ft³/day?

Solution: First convert gallon screenings to cubic feet:

$$(310 \text{ gal}) \div (7.48 \text{ gal/ft}^3) = 41.4 \text{ ft}^3 \text{ screenings}$$

Then calculate screenings removed as ft³/day:

$$\text{Screenings removed (ft}^3\text{/day)} = \frac{41.4 \text{ ft}^3}{7 \text{ days}} = 5.9 \text{ ft}^3\text{/day}$$

24.2.2 SCREENING PIT CAPACITY CALCULATIONS

Recall that detention time may be considered the time required for flow to pass through a basin or tank or the time required to fill a basin or tank at a given flow rate. In screening pit capacity problems, the time required to fill a screening pit is being calculated. The equation used in screening pit capacity problems is given below:

$$\text{Fill time (days)} = \frac{\text{Volume of pit (ft}^3)}{\text{Screenings removed (ft}^3/\text{day})} \qquad (24.3)$$

■ EXAMPLE 24.3

Problem: A screening pit has a capacity of 500 ft³. (The pit is actually larger than 500 ft³ to accommodate soil for covering.) If an average of 3.4 ft³ of screenings is removed daily from the wastewater flow, in how many days will the pit be full?

Solution:

$$\text{Fill time (days)} = \frac{\text{Volume of pit (ft}^3)}{\text{Screenings removed (ft}^3/\text{day})}$$

$$= \frac{500 \text{ ft}^3}{3.4 \text{ ft}^3/\text{day}} = 147.1 \text{ days}$$

■ EXAMPLE 24.4

Problem: A plant has been averaging a screenings removal of 2 ft³/MG. If the average daily flow is 1.8 MGD, how many days will it take to fill the pit with an available capacity of 125 ft³?

Solution: The filling rate must first be expressed as ft³/day:

$$\frac{2 \text{ ft}^3 \times 1.8 \text{ MGD}}{\text{MG}} = 3.6 \text{ ft}^3/\text{day}$$

Then,

$$\text{Fill time (days)} = \frac{\text{Volume of pit (ft}^3)}{\text{Screenings removed (ft}^3/\text{day})}$$

$$= \frac{125 \text{ ft}^3}{3.6 \text{ ft}^3/\text{day}} = 34.7 \text{ days}$$

■ EXAMPLE 24.5

Problem: A screening pit has a capacity of 12 yd³ available for screenings. If the plant removes an average of 2.4 ft³ of screenings per day, in how many days will the pit be filled?

Solution: Because the filling rate is expressed as ft³/day, the volume must be expressed as ft³:

$$12 \text{ yd}^3 \times 27 \text{ ft}^3/\text{yd}^3 = 324 \text{ ft}^3$$

Now calculate fill time:

$$\text{Fill time (days)} = \frac{\text{Volume of pit (ft}^3)}{\text{Screenings removed (ft}^3/\text{day})} = \frac{324 \text{ ft}^3}{2.4 \text{ ft}^3/\text{day}} = 135 \text{ days}$$

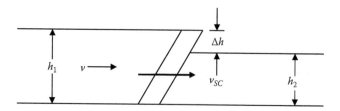

FIGURE 24.2   Water profile through a screen.

24.2.3 Head Loss through Bar Screen Calculations

Head loss through a bar screen is determined by using Bernoulli's equation (see Figure 24.2):

$$h_1 + \frac{v^2}{2g} = h_2 + \frac{v_{sc}^2}{2g} + \text{Losses} \qquad (24.4)$$

where
   h_1 = Upstream depth of flow.
   v = Upstream velocity.
   g = Acceleration of gravity.
   h_2 = Downstream depth of flow.
   v_{sc} = Velocity of flow through the screen.

The losses can be incorporated into a coefficient:

$$h = h_1 - h_2 = \frac{1}{2gC_d^2}\left(v_{sc}^2 - v^2\right) \qquad (24.5)$$

where C_d is the discharge coefficient (typical value = 0.84), a value usually supplied by the manufacturer or determined through experimentation

24.2.4 Grit Removal Calculations

The purpose of grit removal is to remove inorganic solids (sand, gravel, clay, egg shells, coffee grounds, metal filings, seeds, and other similar materials) that could cause excessive mechanical wear. Several processes or devices are used for grit removal, all based on the fact that grit is heavier than the organic solids, which should be kept in suspension for treatment in subsequent unit processes. Grit removal may be accomplished in grit chambers or by the centrifugal separation of biosolids. Processes use gravity/velocity, aeration, or centrifugal force to separate the solids from the wastewater.

Wastewater systems typically average 1 to 15 ft³ of grit per million gallons of flow (sanitary systems, 1 to 4 ft³/MG; combined wastewater systems, 4 to 15 ft³/MG of flow), with higher ranges during storm events. Generally, grit is disposed of in sanitary landfills. Because of this process, for planning purposes operators must keep accurate records of grit removal. Most often, the data are reported as cubic feet of grit removed per million gallons for flow:

$$\text{Grit removed (ft}^3\text{/MG)} = \frac{\text{Grit volume (ft}^3)}{\text{Flow (MG)}} \qquad (24.6)$$

Over a given period, the average grit removal rate at a plant (at least a seasonal average) can be determined and used for planning purposes. Typically, grit removal is calculated as cubic yards because excavation is normally expressed in terms of cubic yards:

$$\text{Grit (yd}^3) = \frac{\text{Total grit (ft}^3)}{27 \text{ ft}^3/\text{yd}} \tag{24.7}$$

■ **EXAMPLE 24.6**

Problem: A treatment plant removes 10 ft³ of grit in one day. How many ft³ of grit are removed per million gallons if the plant flow is 9 MGD?

Solution:

$$\text{Grit removed} = \frac{\text{Grit volume (ft}^3)}{\text{Flow (MG)}} = \frac{10 \text{ ft}^3}{9 \text{ MG}} = 1.1 \text{ ft}^3/\text{MG}$$

■ **EXAMPLE 24.7**

Problem: The total daily grit removed for a plant is 250 gal. If the plant flow is 12.2 MGD, how many cubic feet of grit are removed per MG flow?

Solution: First, convert gallon grit removed to ft³:

$$(250 \text{ gal}) \div (7.48 \text{ gal/ft}^3) = 33 \text{ ft}^3$$

Next, complete the calculation of ft³/MG:

$$\text{Grit removed} = \frac{\text{Grit volume (ft}^3)}{\text{Flow (MG)}} = \frac{33 \text{ ft}^3}{12.2 \text{ MG}} = 2.7 \text{ ft}^3/\text{MG}$$

■ **EXAMPLE 24.8**

Problem: The daily average grit removal is 2.5 ft³/MG. If the monthly average flow is 2,500,000 gpd, how many cubic yards must be available for grit disposal if the disposal pit is to have a 90-day capacity?

Solution: First, calculate the grit generated each day:

$$(2.5 \text{ ft}^3/\text{MG}) \times (2.5 \text{ MGD}) = 6.25 \text{ ft}^3 \text{ each day}$$

The ft³ grit generated for 90 days would be

$$(6.25 \text{ ft}^3/\text{day}) \times (90 \text{ days}) = 562.5 \text{ ft}^3$$

Convert ft³ to yd³ grit:

$$(562.5 \text{ ft}^3) \div (27 \text{ ft}^3/\text{yd}^3) = 21 \text{ yd}^3$$

24.2.5 GRIT CHANNEL VELOCITY CALCULATIONS

The optimum velocity in sewers is approximately 2 ft/sec (fps) at peak flow, because this velocity normally prevents solids from settling out in the lines; however, when the flow reaches the grit channel, the velocity should decrease to about 1 fps to permit the heavy inorganic solids to settle. In the example calculations that follow, we describe how the velocity of the flow in a channel can be determined by the float and stopwatch method and by channel dimensions.

■ EXAMPLE 24.9: VELOCITY BY FLOAT AND STOPWATCH

$$\text{Velocity (fps)} = \frac{\text{Distance traveled (ft)}}{\text{Time required (sec)}} \qquad (24.8)$$

Problem: A float requires 30 sec to travel 37 ft in a grit channel. What is the velocity of the flow in the channel?

Solution:

$$\text{Velocity (fps)} = 37 \text{ ft}/30 \text{ sec} = 1.2 \text{ fps}$$

■ EXAMPLE 24.10: VELOCITY BY FLOW AND CHANNEL DIMENSIONS

This calculation can be used for a single channel or tank or for multiple channels or tanks with the same dimensions and equal flow. If the flow through each unit of the unit dimensions is unequal, the velocity for each channel or tank must be computed individually:

$$\text{Velocity (fps)} = \frac{\text{Flow (MGD)} \times 1.55 \text{ cfs/MGD}}{\text{Channels in service} \times \text{channel width (ft)} \times \text{water depth (ft)}} \qquad (24.9)$$

Problem: The plant is currently using two grit channels. Each channel is 3 ft wide and has a water depth of 1.3 ft. What is the velocity when the influent flow rate is 4.0 MGD?

Solution:

$$\text{Velocity (fps)} = \frac{40 \text{ MGD} \times 1.55 \text{ cfs/MGD}}{2 \text{ channels} \times 3 \text{ ft} \times 1.3 \text{ ft}} = \frac{6.2 \text{ cfs}}{7.8 \text{ ft}^2} = 0.79 \text{ fps}$$

Because 0.79 is within the 0.7 to 1.4 range, the operator of this unit would not make any adjustments.

Note: The channel dimensions must always be in feet. Convert inches to feet by dividing by 12 inches per foot.

24.2.6 REQUIRED SETTLING TIME CALCULATIONS

This calculation can be used to determine the time required for a particle to travel from the surface of the liquid to the bottom at a given settling velocity. To compute the settling time, settling velocity in fps must be provided or determined by experiment in a laboratory.

$$\text{Settling time (sec)} = \frac{\text{Liquid depth (ft)}}{\text{Settling velocity (fps)}} \qquad (24.10)$$

■ EXAMPLE 24.11

Problem: The plant's grit channel is designed to remove sand, which has a settling velocity of 0.080 fps. The channel is currently operating at a depth of 2.3 ft. How many seconds will it take for a sand particle to reach the channel bottom?

Solution:

$$\text{Settling time (sec)} = \frac{2.3 \text{ ft}}{0.080 \text{ fps}} = 28.7 \text{ sec}$$

24.2.7 REQUIRED GRIT CHANNEL LENGTH CALCULATIONS

This calculation can be used to determine the length of channel required to remove an object with a specified settling velocity:

$$\text{Channel length} = \frac{\text{Channel depth (ft)} \times \text{flow velocity (fps)}}{0.080 \text{ fps}} \qquad (24.11)$$

■ EXAMPLE 24.12

Problem: The plant's grit channel is designed to remove sand, which has a settling velocity of 0.080 fps. The channel is currently operating at a depth of 3 ft. The calculated velocity of flow through the channel is 0.85 fps. The channel is 36 ft long. Is the channel long enough to remove the desired sand particle size?

Solution:

$$\text{Channel length} = \frac{3 \text{ ft} \times 0.85 \text{ fps}}{0.080 \text{ fps}} = 31.6 \text{ ft}$$

Yes, the channel is long enough to ensure that all of the sand will be removed.

24.2.8 VELOCITY OF SCOUR CALCULATIONS

The Camp–Shields equation (Camp, 1942) is used to estimate the velocity of scour necessary to resuspend settled organics:

$$v_s = \sqrt{\frac{8kgd}{f}\left(\frac{\rho_p - \rho}{\rho}\right)} \qquad (24.12)$$

where

v_s = Velocity of scour.
k = Empirically determined constant.
d = Nominal diameter of the particle.
f = Darcy–Weisbach friction factor.
ρ_p = Particle density.
ρ = Fluid density.

If the channel is rectangular and discharges over a rectangular weir, the discharge relation based on Bernoulli's equation is

$$Q = C_d A \sqrt{2gH} = C_w H^{3/2} \qquad (24.13)$$

where

Q = Width of the channel.
C_d = Discharge coefficient.
A = Cross-sectional area of the channel.
H = Depth of flow in the channel.
C_w = Equal to $C_d \sqrt{2g}$.

The horizontal velocity (v_h) is related to the discharge rate and channel velocity by

$$v_h = \frac{Q}{A} = \frac{Q}{wH} = CH^{1/2} = C\left(\frac{Q}{C_w}\right)^{1/3} \qquad (24.14)$$

24.3 PRIMARY TREATMENT

Primary treatment (primary sedimentation or clarification) should remove both settleable organic and floatable solids. Poor solids removal during this step of treatment may cause organic overloading of the biological treatment processes following primary treatment. Normally, each primary clarification unit can be expected to remove 90 to 95% of settleable solids, 40 to 60% of total suspended solids, and 25 to 35% of BOD.

24.3.1 PROCESS CONTROL CALCULATIONS

As with many other wastewater treatment plant unit processes, several process control calculations may be helpful in evaluating the performance of the primary treatment process. Process control calculations are used in the sedimentation process to determine

- Surface loading rate (surface settling rate)
- Weir overflow rate (weir loading rate)
- BOD and suspended solids removed (lb/day)
- Percent removal
- Hydraulic detention time
- Biosolids pumping
- Percent total solids (%TS)

In the following sections, we take a closer look at a few of these process control calculations and provide example problems.

Note: The calculations presented in the following sections allow us to determine values for each function performed. Again, keep in mind that an optimally operated primary clarifier should have values within an expected range. Recall that the expected ranges of percent removal for a primary clarifier are

- Settleable solids, 90–95%
- Suspended solids, 40–60%
- Biochemical oxygen demand (BOD), 25–35%

The expected range of hydraulic detention time for a primary clarifier is 1 to 3 hr. The expected range of surface loading/settling rate for a primary clarifier is 600 to 1200 gpd/ft^2 (ballpark estimate). The expected range of weir overflow rate for a primary clarifier is 10,000 to 20,000 gpd/ft.

24.3.2 SURFACE LOADING RATE (SURFACE SETTLING RATE/ SURFACE OVERFLOW RATE) CALCULATIONS

Surface loading rate is the number of gallons of wastewater passing over 1 ft^2 of tank per day. This can be used to compare actual conditions with design. Plant designs generally use a surface loading rate of 300 to 1200 gpd/ft^2:

$$\text{Surface loading rate (gpd/ft}^2) = \frac{\text{Flow (gpd)}}{\text{Tank surface area (ft}^2)} \qquad (24.15)$$

■ EXAMPLE 24.13

Problem: The circular settling tank has a diameter of 120 ft. If the flow to the unit is 4.5 MGD, what is the surface loading rate in gpd/ft^2?

Solution:

$$\text{Surface loading rate} = \frac{4.5 \text{ MGD} \times 1{,}000{,}000 \text{ gal/MGD}}{0.785 \times 120 \text{ ft} \times 120 \text{ ft}} = 398 \text{ gpd/ft}^2$$

■ EXAMPLE 24.14

Problem: A circular clarifier has a diameter of 50 ft. If the primary effluent flow is 2,150,000 gpd, what is the surface overflow rate in gpd/ft^2?

Solution:

Key Point: Remember that area = 0.785 × 50 ft × 50 ft.

$$\text{Surface overflow rate} = \frac{\text{Flow (gpd)}}{\text{Tank surface area (ft}^2)} = \frac{2{,}150{,}000 \text{ gpd}}{0.785 \times 50 \text{ ft} \times 50 \text{ ft}} = 1096 \text{ gpd/ft}$$

■ EXAMPLE 24.15

Problem: A sedimentation basin 90 ft by 20 ft receives a flow of 1.5 MGD. What is the surface overflow rate in gpd/ft^2?

$$\text{Surface overflow rate} = \frac{\text{Flow (gpd)}}{\text{Tank surface area (ft}^2)} = \frac{1{,}500{,}000 \text{ gpd}}{90 \text{ ft} \times 20 \text{ ft}} = 833 \text{ gpd/ft}^2$$

24.3.3 WEIR OVERFLOW RATE (WEIR LOADING RATE) CALCULATIONS

A weir is a device used to measure wastewater flow. *Weir overflow rate* (weir loading rate) is the amount of water leaving the settling tank per linear foot of water. The result of this calculation can be compared with design. Normally, weir overflow rates of 10,000 to 20,000 gpd/ft are used in the design of a settling tank:

$$\text{Weir overflow rate (gpd/ft)} = \frac{\text{Flow (gpd)}}{\text{Weir length (ft)}} \qquad (24.16)$$

Key Point: To calculate weir circumference, use total feet of weir = 3.14 × weir diameter (ft).

■ **EXAMPLE 24.16**

Problem: The circular settling tank is 80 ft in diameter and has a weir along its circumference. The effluent flow rate is 2.75 MGD. What is the weir overflow rate in gallons per day per foot?

Solution:

$$\text{Weir overflow rate (gpd/ft)} = \frac{2.75 \text{ MGD} \times 1{,}000{,}000 \text{ gal}}{3.14 \times 80 \text{ ft}} = 10{,}947 \text{ gpd/ft}$$

Note: 10,947 gpd/ft is above the recommended minimum of 10,000.

■ **EXAMPLE 24.17**

Problem: A rectangular clarifier has a total of 70 ft of weir. What is the weir overflow rate in gpd/ft when the flow is 1,055,000 gpd?

Solution:

$$\text{Weir overflow rate (gpd/ft)} = \frac{\text{Flow (gpd)}}{\text{Weir length (ft)}} = \frac{1{,}055{,}000 \text{ gpd}}{70 \text{ ft}} = 15{,}071 \text{ gpd}$$

24.3.4 Primary Sedimentation Basin Calculations

■ **EXAMPLE 24.18**

Problem: Two rectangular settling tanks are each 8 m wide, 26 m long, and 2.5 m deep. Each is used alternatively to treat 1800 m³ in a 12-hr period. Compute the surface overflow rate, detention time, horizontal velocity, and outlet weir overflow rate using an H-shaped weir with three times width.

Solution: First determine design flow Q:

$$Q = \frac{1800 \text{ m}^3}{12 \text{ hr}} \times \frac{24 \text{ hr}}{1 \text{ day}} = 3600 \text{ m}^3/\text{day}$$

Compute the surface overflow rate, v_o:

$$v_o = \frac{Q}{A} = \frac{3600 \text{ m}^3/\text{day}}{8 \text{ m} \times 26 \text{ m}} = 17.3 \text{ m}^3(\text{m}^2 \cdot \text{day})$$

Compute the detention time, t:

$$\text{Tank volume } (V) = 8 \text{ m} \times 26 \text{ m} \times 2.5 \text{ m} \times 2 = 1040 \text{ m}^3$$

$$t = \frac{V}{Q} = \frac{1040 \text{ m}^3}{3600 \text{ m}^3/\text{day}} = 0.289 \text{ days} = 6.9 \text{ hr}$$

Compute the horizontal velocity, v_h:

$$v_h = \frac{3600 \text{ m}^3/\text{day}}{8 \text{ m} \times 2.5 \text{ m}} = 180 \text{ m/day} = 0.125 \text{ m/min} = 0.410 \text{ ft/min}$$

Compute the outlet weir overflow rate, w_l:

$$w_l = \frac{3600 \text{ m}^3/\text{day}}{8 \text{ m} \times 3 \text{ m}} = 150 \text{ m}^3/(\text{day} \cdot \text{m}) = 12,100 \text{ gal/day}$$

24.4 BIOSOLIDS PUMPING CALCULATIONS

Determination of biosolids pumping (the quantity of solids and volatile solids removed from the sedimentation tank) provides accurate information necessary for process control of the sedimentation process.

$$\text{Solids Pumped} = \text{Pump Rate (gpm)} \times \text{Pump Time (min/day)} \times 8.34 \text{ lb/gal} \times \% \text{ Solids} \qquad (24.17)$$

$$\text{Volatile Solids (lb/day)} = \text{Pump Rate} \times \text{Pump Time} \times 8.34 \times \% \text{ Solids} \times \% \text{ Volatile Matter} \qquad (24.18)$$

■ EXAMPLE 24.19

Problem: A biosolids pump operates 30 min/hr. The pump delivers 25 gal/min of biosolids. Laboratory tests indicate that the biosolids are 5.3% solids and 68% volatile matter. How many pounds of volatile matter are transferred from the settling tank to the digester? Assume a 24-hour period.

Solution:

Pump time = 30 min/hr
Pump rate = 25 gpm
%Solids = 5.3%
%Volatile matter = 68%

Volatile Matter = 25 gpm × (30 min/hr × 24 hr/day) × 8.34 lb/gal × 0.053 × 0.68 = 5410 lb/day

24.4.1 PERCENT TOTAL SOLIDS CALCULATIONS

Problem: A settling tank biosolids sample is tested for solids. The sample and dish weigh 73.79 g. The dish alone weighs 21.4 g. After drying, the dish with dry solids weighs 22.4 g. What is the percent total solids (%TS) of the sample?

Solution:

Sample + dish	73.79 g	Dish + dry solids	22.4 g
Dish alone	–21.40 g	Dish alone	–21.4 g
	52.39 g		1.0 g

$$(1.0 \text{ g} \div 52.39 \text{ g}) \times 100\% = 1.9\%$$

24.4.2 BOD AND SS REMOVAL CALCULATIONS

To calculate the pounds of BOD or suspended solids removed each day, we need to know the mg/L BOD or SS removed and the plant flow, then we can use the mg/L to lb/day equation:

$$\text{SS removed} = \text{mg/L} \times \text{MGD} \times 8.34 \text{ lb/gal} \qquad (24.19)$$

■ **EXAMPLE 24.20**

Problem: If 120 mg/L suspended solids are removed by a primary clarifier, how many lb/day suspended solids are removed when the flow is 6,250,000 gpd?

Solution:

$$\text{SS removed} = 120 \text{ mg/L} \times 6.25 \text{ MGD} \times 8.34 \text{ lb/gal} = 6255 \text{ lb/day}$$

■ **EXAMPLE 24.21**

Problem: The flow to a secondary clarifier is 1.6 MGD. If the influent BOD concentration is 200 mg/L and the effluent BOD concentration is 70 mg/L, how many pounds of BOD are removed daily?

Solution:

$$\text{BOD removed} = 200 \text{ mg/L} - 70 \text{ mg/L} = 130 \text{ mg/L}$$

After calculating mg/L BOD removed, calculate lb/day BOD removed:

$$\text{BOD removed} = 130 \text{ mg/L} \times 1.6 \text{ MGD} \times 8.34 \text{ lb/gal} = 1735 \text{ lb/day}$$

24.5 TRICKLING FILTER

The trickling filter process (Figure 24.3) is one of the oldest forms of dependable biological treatment for wastewater. By its very nature, the trickling filter has its advantages over other unit processes; for example, it is a very economical and dependable process for treatment of wastewater prior to discharge. Capable of withstanding periodic shock loading, process energy demands are low because aeration is a natural process.

As shown in Figure 24.4, trickling filter operation involves spraying wastewater over a solid media such as rock, plastic, or redwood slats (or laths). As the wastewater trickles over the surface of the media, a growth of microorganisms (bacteria, protozoans, fungi, algae, helminths or worms, and larvae) develops. This growth is visible as a shiny slime very similar to the slime found on rocks in a stream. As wastewater passes over this slime, the slime adsorbs the organic (food) matter. This organic matter is used for food by the microorganisms. At the same time, air moving through the open spaces in the filter transfers oxygen to the wastewater. This oxygen is then transferred to the slime to keep the outer layer aerobic. As the microorganisms use the food and oxygen, they produce more organisms, carbon dioxide, sulfates, nitrates, and other stable byproducts; these materials are then discarded from the slime back into the wastewater flow and are carried out of the filter.

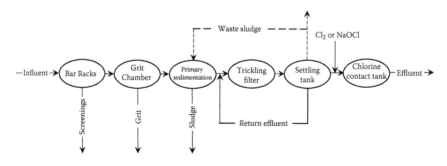

FIGURE 24.3   Simplified flow diagram of trickling filter used for wastewater treatment.

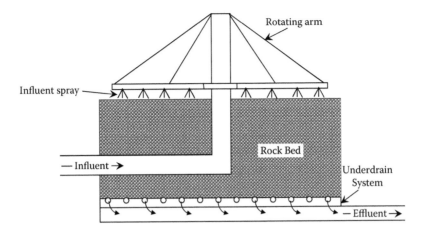

FIGURE 24.4   Schematic of cross-section of a trickling filter.

24.5.1   TRICKLING FILTER PROCESS CALCULATIONS

Several calculations are useful in the operation of trickling filters; these include hydraulic loading, organic loading, and BOD and SS removal. Each type of trickling filter is designed to operate with specific loading levels. These levels vary greatly depending on the filter classification. To operate the filter properly, filter loading must be within the specified levels. The main three loading parameters for the trickling filter are hydraulic loading, organic loading, and recirculation ratio.

24.5.1.1   Hydraulic Loading Rate

Calculating the hydraulic loading rate is important in accounting for both the primary effluent as well as the recirculated trickling filter effluent. These are combined before being applied to the filter surface. The hydraulic loading rate is calculated based on filter surface area. The normal hydraulic loading rate ranges for standard rate and high rate trickling filters are

Standard rate—25 to 100 gpd/ft^2 or 1 to 40 MGD/ac
High rate—100 to 1000 gpd/ft^2 or 4 to 40 MGD/ac

Key Point: If the hydraulic loading rate for a particular trickling filter is too low, septic conditions will begin to develop.

■ EXAMPLE 24.22

Problem: A trickling filter 80 ft in diameter is operated with a primary effluent of 0.588 MGD and a recirculated effluent flow rate of 0.660 MGD. Calculate the hydraulic loading rate on the filter in units gpd/ft^2.

Solution: The primary effluent and recirculated trickling filter effluent are applied together across the surface of the filter; therefore, 0.588 MGD + 0.660 MGD = 1.248 MGD = 1,248,000 gpd:

$$\text{Circular surface area} = 0.785 \times (\text{diameter})^2 = 0.785 \times (80 \text{ ft})^2 = 5024 \text{ ft}^2$$

$$\text{Hydraulic loading rate} = \frac{1,248,000 \text{ gpd}}{5024 \text{ ft}^2} = 248.4 \text{ gpd/ft}^2$$

■ **EXAMPLE 24.23**

Problem: A trickling filter 80 ft in diameter treats a primary effluent flow of 750,000 gpd. If the recirculated flow to the clarifier is 0.2 MGD, what is the hydraulic loading on the trickling filter?

Solution:

$$\text{Hydraulic loading rate} = \frac{750,000 \text{ gpd}}{0.785 \times 80 \text{ ft} \times 80 \text{ ft}} = 149 \text{ gpd/ft}^2$$

■ **EXAMPLE 24.24**

Problem: A high-rate trickling filter receives a daily flow of 1.8 MGD. What is the dynamic loading rate in MGD/ac if the filter is 90 ft in diameter and 5 ft deep?

Solution:

$$0.785 \times 90 \text{ ft} \times 90 \text{ ft} = 6359 \text{ ft}^2$$

$$\frac{6359 \text{ ft}^2}{43,560 \text{ ft}^2/\text{ac}} = 0.146 \text{ ac}$$

$$\text{Hydraulic loading rate} = \frac{1.8 \text{ MGD}}{0.146 \text{ ac}} = 12.3 \text{ MGD/ac}$$

Key Point: When hydraulic loading rate is expressed as MGD per acre, this is still an expression of gallon flow over surface area of the trickling filter.

24.5.1.2 Organic Loading Rate

Trickling filters are sometimes classified by the *organic loading rate* applied. The organic loading rate is expressed as a certain amount of BOD applied to a certain volume of media. In other words, the organic loading is defined as the pounds of BOD_5 or chemical oxygen demand (COD) applied per day per 1000 ft^3 of media—a measure of the amount of food being applied to the filter slime. To calculate the organic loading on the trickling filter, two things must be known: the pounds of BOD or COD being applied to the filter media per day and the volume of the filter media in 1000-ft^3 units. The BOD and COD contribution of the recirculated flow is not included in the organic loading.

■ **EXAMPLE 24.25**

Problem: A trickling filter, 60 ft in diameter, receives a primary effluent flow rate of 0.440 MGD. Calculate the organic loading rate in units of pounds of BOD applied per day per 1000 ft^3 of media volume. The primary effluent BOD concentration is 80 mg/L. The media depth is 9 ft.

Solution:

$$0.440 \text{ MGD} \times 80 \text{ mg/L} \times 8.34 \text{ lb/gal} = 293.6 \text{ lb BOD}$$

$$\text{Surface area} = 0.785 \times (60)^2 = 2826 \text{ ft}^2$$

$$\text{Area} \times \text{depth} = \text{Volume} = 2826 \text{ ft}^2 \times 9 \text{ ft} = 25,434 \text{ ft}^3$$

Note: To determine the pounds of BOD per 1000 ft^3 in a volume of thousands of cubic feet, we must set up the equation as follows:

$$\frac{293.6 \text{ lb BOD/day}}{25,434 \text{ ft}^3} \times \frac{1000}{1000}$$

Regrouping the numbers and the units together,

$$\frac{293.6 \text{ lb BOD/day} \times 1000}{25,434 \text{ ft}^3} \times \frac{\text{lb BOD/day}}{1000 \text{ ft}^3} = 11.5 \times \frac{\text{lb BOD/day}}{1000 \text{ ft}^3}$$

24.5.1.3 BOD and SS Removed

To calculate the pounds of BOD or suspended solids removed each day, we need to know the mg/L BOD or SS removed and the plant flow.

■ EXAMPLE 24.26

Problem: If 120 mg/L suspended solids are removed by a trickling filter, how many pounds per day suspended solids are removed when the flow is 4.0 MGD?

Solution:

SS (mg/L) × Flow (MGD) × 8.34 lb/gal = 120 mg/L × 4.0 MGD × 8.34 lb/gal = 4003 lb/day

■ EXAMPLE 24.27

Problem: The 3,500,000-gpd influent flow to a trickling filter has a BOD content of 185 mg/L. If the trickling filter effluent has a BOD content of 66 mg/L, how many pounds of BOD are removed daily?

Solution:

BOD removed = 185 mg/L – 66 mg/L = 119 mg/L

BOD removed (lb/day) = BOD removed (mg/L) × Flow (MGD) × 8.34 lb/gal

BOD removed = 119 mg/L × 3.5 MGD × 8.34 lb/gal = 3474 lb/day

24.5.1.4 Recirculation Flow

Recirculation in trickling filters involves the return of filter effluent back to the head of the trickling filter. It can level flow variations and assist in solving operational problems such as ponding, filter flies, and odors. The operator must check the rate of recirculation to ensure that it is within design specifications. Rates above design specifications indicate hydraulic overloading; rates under design specifications indicate hydraulic underloading. The trickling filter recirculation ratio is the ratio of the recirculated trickling filter flow to the primary effluent flow. The trickling filter recirculation ratio may range from 0.5:1 (.5) to 5:1 (5); however, the ratio is often found to be 1:1 or 2:1:

$$\text{Recirculation ratio} = \frac{\text{Recirculated flow (MGD)}}{\text{Primary effluent flow (MGD)}} \qquad (24.20)$$

■ EXAMPLE 24.28

Problem: A treatment plant receives a flow of 3.2 MGD. If the trickling filter effluent is recirculated at the rate of 4.50 MGD, what is the recirculation ratio?

Solution:

$$\text{Recirculation ratio} = \frac{\text{Recirculated flow (MGD)}}{\text{Primary effluent flow (MGD)}} = \frac{4.5 \text{ MGD}}{3.2 \text{ MGD}} = 1.4$$

■ **EXAMPLE 24.29**

Problem: A trickling filter receives a primary effluent flow of 5 MGD. If the recirculated flow is 4.6 MGD, what is the recirculation ratio?

Solution:

$$\text{Recirculation ratio} = \frac{\text{Recirculated flow (MGD)}}{\text{Primary effluent flow (MGD)}} = \frac{4.6 \text{ MGD}}{5 \text{ MGD}} = 0.92$$

24.5.1.5 Trickling Filter Design

In trickling filter design, the parameters used are hydraulic loading and BOD:

$$\text{Hydraulic loading} = \frac{Q_o + R}{A} \tag{24.21}$$

where
  Q_o = Average wastewater flow rate (MGD).
  R = Recirculated flow = Q_o × circulation ratio.
  A = Filter area (acres).

$$\text{BOD loading} = \frac{8340\,(\text{BOD}_s)(Q_o)}{V} \tag{24.22}$$

where
  BOD_s = Settled BOD_5 from primary treatment (mg/L).
  Q_o = Average wastewater flow rate (MGD).
  V = Filter volume (ft³).
  8340 = Conversion of units.

24.6 ROTATING BIOLOGICAL CONTACTORS

The rotating biological contactor (RBC) is a variation of the attached growth idea provided by the trickling filter. Still relying on microorganisms that grow on the surface of a medium, the RBC is instead a *fixed-film* biological treatment device (see Figures 24.5 and 24.6). The basic biological process, however, is similar to that occurring in trickling filters. An RBC consists of a series of circular plastic disks mounted side by side and closely spaced; they are typically about 11.5 ft in diameter. Attached to a rotating horizontal shaft, approximately 40% of each disk is submersed in a tank that contains the wastewater to be treated. As the RBC rotates, the attached biomass film (zoogleal slime) that grows on the surface of the disks moves into and out of the wastewater. While submerged in the wastewater, the microorganisms absorb organics; when they are rotated out of the wastewater, they are supplied with needed oxygen for aerobic decomposition. As the zoogleal slime reenters the wastewater, excess solids and waste products are stripped off the media as *sloughings*. These sloughings are transported with the wastewater flow to a settling tank for removal.

24.6.1 RBC Process Control Calculations

Several process control calculations may be useful in the operation of an RBC. These include soluble BOD, total media area, organic loading rate, and hydraulic loading. Settling tank calculations and biosolids pumping calculations may be helpful for evaluation and control of the settling tank following the RBC.

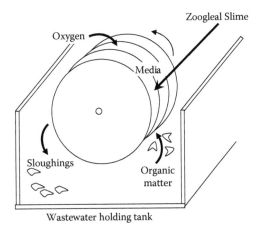

FIGURE 24.5   Rotating biological contactor (RBC) cross-section and treatment system.

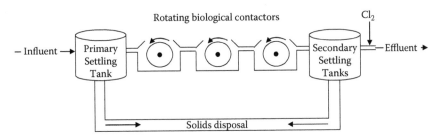

FIGURE 24.6   Rotating biological contactor (RBC) treatment system.

24.6.1.1   Hydraulic Loading Rate

The manufacturer normally specifies the RBC media surface area, and the hydraulic loading rate is based on the media surface area, usually in square feet (ft^2). Hydraulic loading is expressed in terms of gallons of flow per day per square foot of media. This calculation can be helpful in evaluating the current operating status of the RBC. Comparison with design specifications can determine if the unit is hydraulically over- or underloaded. Hydraulic loading on an RBC can range from 1 to 3 gpd/ft^2.

■ **EXAMPLE 24.30**

Problem: An RBC treats a primary effluent flow rate of 0.244 MGD. What is the hydraulic loading rate in gpd/ft^2 if the media surface area is 92,600 ft^2?

Solution:

$$\text{Hydraulic loading rate} = \frac{\text{Flow (gpd)}}{\text{Media area (ft}^2)} = \frac{244,000 \text{ gpd}}{92,000 \text{ ft}^2} = 2.65 \text{ ft}^2$$

■ **EXAMPLE 24.31**

Problem: An RBC treats a flow of 3.5 MGD. The manufacturer's data indicate a media surface area of 750,000 ft^2. What is the hydraulic loading rate on the RBC?

Solution:

$$\text{Hydraulic loading rate} = \frac{\text{Flow (gpd)}}{\text{Media area (ft}^2)} = \frac{3,500,000 \text{ gpd}}{750,000 \text{ ft}^2} = 4.7 \text{ ft}^2$$

■ **EXAMPLE 24.32**

Problem: A rotating biological contactor treats a primary effluent flow of 1,350,000 gpd. The manufacturer's data indicate that the media surface area is 600,000 ft². What is the hydraulic loading rate on the filter?

Solution:

$$\text{Hydraulic loading rate} = \frac{\text{Flow (gpd)}}{\text{Media area (ft}^2)} = \frac{1,350,000 \text{ gpd}}{600,000 \text{ ft}^2} = 2.3 \text{ ft}^2$$

24.6.1.2 Soluble BOD

The soluble BOD concentration of the RBC influent can be determined experimentally in the laboratory, or it can be estimated using the suspended solids concentration and the K factor. The K factor is used to approximate the BOD (particulate BOD) contributed by the suspended matter. The K factor must be provided or determined experimentally in the laboratory. The K factor for domestic wastes is normally in the range of 0.5 to 0.7:

$$\text{Soluble BOD}_5 = \text{Total BOD}_5 - (K \text{ factor} \times \text{Total suspended solids}) \qquad (24.23)$$

■ **EXAMPLE 24.33**

Problem: The suspended solids concentration of a wastewater is 250 mg/L. If the K value at the plant is 0.6, what is the estimated particulate biochemical oxygen demand (BOD) concentration of the wastewater?

Solution:

$$250 \text{ mg/L} \times 0.6 = 150 \text{ mg/L particulate BOD}$$

The K value of 0.6 indicates that about 60% of the suspended solids are organic suspended solids (particulate BOD).

■ **EXAMPLE 24.34**

Problem: A rotating biological contactor receives a flow of 2.2 MGD with a BOD content of 170 mg/L and suspended solids concentration of 140 mg/L. If the K value is 0.7, how many pounds of soluble BOD enter the RBC daily?

Solution:

$$\text{Total BOD} = \text{Particulate BOD} + \text{soluble BOD}$$

$$170 \text{ mg/L} = (140 \text{ mg/L} \times 0.7) + x \text{ mg/L}$$

$$170 \text{ mg/L} = 98 \text{ mg/L} + x \text{ mg/L}$$

$$170 \text{ mg/L} - 98 \text{ mg/L} = x$$

$$x = 72 \text{ mg/L soluble BOD}$$

Now the lb/day soluble BOD may be determined:

$$\text{Soluble BOD (mg/L)} \times \text{flow (MGD)} \times 8.34 \text{ lb/gal} = \text{lb/day}$$

$$72 \text{ mg/L} \times 2.2 \text{ MGD} \times 8.34 \text{ lb/gal} = 1321 \text{ lb/day}$$

■ **EXAMPLE 24.35**

Problem: The wastewater entering a rotating biological contactor has a BOD content of 210 mg/L. The suspended solids content is 240 mg/L. If the K value is 0.5, what is the estimated soluble BOD (mg/L) of the wastewater?

Solution:

$$\text{Total BOD (mg/L)} = (\text{Particulate BOD} \times K) + \text{Soluble BOD}$$
$$210 = (240 \times 0.5) + x$$
$$210 = 120 + x$$
$$\text{Soluble BOD} = 90 \text{ mg/L} = x$$

24.6.1.3 Organic Loading Rate

The organic loading rate can be expressed as total BOD loading in pounds per day per 1000 ft² of media. The actual values can then be compared with plant design specifications to determine the current operating condition of the system:

$$\text{Organic loading rate} = \frac{\text{Soluble BOD} \times \text{flow (MGD)} \times 8.34 \text{ lb/gal}}{\text{Media area (1000 ft}^2)} \qquad (24.24)$$

■ **EXAMPLE 24.36**

Problem: A rotating biological contactor has a media surface area of 500,000 ft² and receives a flow of 1,000,000 gpd. If the soluble BOD concentration of the primary effluent is 160 mg/L, what is the organic loading on the RBC in lb/day/1000 ft²?

Solution:

$$\text{Organic loading rate} = \frac{\text{Soluble BOD} \times \text{Flow (MGD)} \times 8.34 \text{ lb/gal}}{\text{Media area (1000 ft}^2)}$$
$$= \frac{160 \text{ mg/L} \times 1.0 \text{ MGD} \times 8.34 \text{ lb/gal}}{500 \times 1000 \text{ ft}^2}$$
$$= 2.7 \text{ lb/day/1000 ft}^2$$

■ **EXAMPLE 24.37**

Problem: The wastewater flow to an RBC is 3,000,000 gpd. The wastewater has a soluble BOD concentration of 120 mg/L. The RBC consists of six shafts (each 110,000 ft²), with two shafts comprising the first stage of the system. What is the organic loading rate in lb/day/1000 ft² on the first stage of the system?

Solution:

$$\text{Organic loading rate} = \frac{\text{Soluble BOD (lb/day)}}{\text{Media area (1000 ft}^2)}$$
$$= \frac{120 \text{ mg/L} \times 3.0 \text{ MGD} \times 8.34 \text{ lb/gal}}{220 \times 1000 \text{ ft}^2} = 13.6 \text{ lb/day/1000 ft}^2$$

24.6.1.4 Total Media Area

Several process control calculations for the RBC use the total surface area of all the stages within the train. As was the case with the soluble BOD calculation, plant design information or information supplied by the unit manufacturer must provide the individual stage areas (or the total train area), because physical determination of this would be extremely difficult:

$$\text{Total area} = \text{1st stage area} + \text{2nd stage area} + \dots + n\text{th stage area} \qquad (24.25)$$

24.6.1.5 Modeling RBC Performance

Although a number of semi-empirical formulations have been used, the Schultz–Germain formula for trickling filters is recommended for modeling RBC performance (Spengel and Dzombok, 1992):

$$S_e = S_i e^{[k(V/Q)^{0.5}]} \qquad (24.26)$$

where
  S_e = Total BOD of settled effluent (mg/L).
  S_i = Total BOD of wastewater applied to filter (mg/L).
  V = Filter volume (m³).
  Q = Wastewater flow (m³/s).

24.6.1.6 RBC Performance Parameter

The control parameter for RBC performance is soluble BOD (SBOD):

$$\text{SBOD} = \text{TBOD} - \text{Suspended BOD} \qquad (24.27)$$

$$\text{Suspended BOD} = c(\text{TSS}) \qquad (24.28)$$

$$\text{SBOD} = \text{TBOD} - c(\text{TSS}) \qquad (24.29)$$

where
  TBOD = Total BOD.
  TSS = Total suspended solids.
  c = Coefficient
    = 0.5 to 0.7 for domestic wastewater.
    = 0.5 for raw domestic wastewater (TSS > TBOD).
    = 0.6 for raw wastewater (TSS $\cong$ TBOD).
    = 0.6 for primary effluents.
    = 0.5 for secondary effluents.

■ EXAMPLE 24.38

Problem: Average TBOD is 152 mg/L and TSS is 132 mg/L. What is the influent SBOD concentration that can be used for the design of an RBC system? RBC is used as the secondary treatment unit.

Solution: For the primary effluent (RBC influent) $c = 0.6$. Estimate the SBOD concentration of the RBC influent using Equation 24.29:

$$\text{SBOD} = \text{TBOD} - c(\text{TSS}) = 152 \text{ mg/L} - 0.6(132 \text{ mg/L}) = 73 \text{ mg/L}$$

24.7 ACTIVATED BIOSOLIDS

The *activated biosolids process* is an artificial process that mimics the natural self-purification process that takes place in streams. In essence, we can state that the activated biosolids treatment process is a "stream in a container." In wastewater treatment, activated biosolids processes are used for both secondary treatment and complete aerobic treatment without primary sedimentation. *Activated biosolids* refers to biological treatment systems that use a suspended growth of organisms to remove BOD and suspended solids.

The basic components of an activated biosolids sewage treatment system include an aeration tank and a secondary basin, settling basin, or clarifier. Primary effluent is mixed with settled solids recycled from the secondary clarifier and is then introduced into the aeration tank. Compressed air is injected continuously into the mixture through porous diffusers located at the bottom of the tank, usually along one side. Wastewater is fed continuously into an aerated tank, where the microorganisms metabolize and biologically flocculate the organics. Microorganisms (activated biosolids) are settled from the aerated mixed liquor under quiescent conditions in the final clarifier and are returned to the aeration tank. Left uncontrolled, the number of organisms would eventually become too great; therefore, some must periodically be removed (wasted). A portion of the concentrated solids from the bottom of the settling tank must be removed from the process (waste activated sludge, or WAS). Clear supernatant from the final settling tank is the plant effluent.

24.7.1 ACTIVATED BIOSOLIDS PROCESS CONTROL CALCULATIONS

As with other wastewater treatment unit processes, process control calculations are important tools used by the operator to control and optimize process operations. In this section, we review many of the most frequently used activated biosolids process calculations.

24.7.1.1 Moving Averages

When performing process control calculations, the use of a 7-day *moving average* is recommended. The moving average is a mathematical method for leveling the impact of any one test result. The moving average is determined by adding the test results collected during the past 7 days and dividing by the number of tests:

$$\text{Moving average} = \frac{\text{Test 1} + \text{Test 2} + \text{Test 3} + \ldots + \text{Test 7}}{\text{No. of tests performed in 7 days}} \quad (24.30)$$

■ **EXAMPLE 24.39**

Problem: Calculate the 7-day moving average for days 7, 8, and 9.

Day	MLSS	Day	MLSS
1	3340	6	2780
2	2480	7	2476
3	2398	8	2756
4	2480	9	2655
5	2558	10	2396

Solution:

1. Moving average, day 7 = (3340 + 2480 + 2398 + 2480 + 2558 + 2780 + 2476)/7 = 2645
2. Moving average, day 8 = (2480 + 2398 + 2480 + 2558 + 2780 + 2476 + 2756)/7 = 2561
3. Moving average, day 9 = (2398 + 2480 + 2558 + 2780 + 2476 + 2756 + 2655)/7 = 2586

24.7.1.2 BOD, COD, or SS Loading

When calculating BOD, COD, or SS loading on an aeration process (or any other treatment process), loading on the process is usually calculated as lb/day. The following equation is used:

BOD, COD, or SS loading (lb/day) = BOD, COD, or SS (mg/L) × Flow (MGD) × 8.34 lb/gal   (24.31)

■ EXAMPLE 24.40

Problem: The BOD concentration of the wastewater entering an aerator is 210 mg/L. If the flow to the aerator is 1,550,000 gpd, what is the lb/day BOD loading?

Solution:

$$BOD \text{ (lb/day)} = BOD \text{ (mg/L)} \times Flow \text{ (MGD)} \times 8.34 \text{ lb/gal}$$

$$BOD = 210 \text{ mg/L} \times 1.55 \text{ MGD} \times 8.34 \text{ lb/gal} = 2715 \text{ lb/day}$$

■ EXAMPLE 24.41

Problem: The flow to an aeration tank is 2750 gpm. If the BOD concentration of the wastewater is 140 mg/L, how many pounds of BOD are applied to the aeration tank daily?

Solution: First convert the gpm flow to gpd flow:

$$2750 \text{ gpm} \times 1440 \text{ min/day} = 3,960,000 \text{ gpd}$$

Then calculate lb/day BOD:

$$BOD \text{ (lb/day)} = BOD \text{ (mg/L)} \times flow \text{ (MGD)} \times 8.34 \text{ lb/gal}$$

$$= 140 \text{ mg/L} \times 3.96 \text{ MGD} \times 8.34 \text{ lb/gal}$$

$$= 4624 \text{ lb/day}$$

24.7.1.3 Solids Inventory

In the activated biosolids process, it is important to control the amount of solids under aeration. The suspended solids in an aeration tank are called *mixed liquor suspended solids* (MLSS). To calculate the pounds of solids in the aeration tank, we need to know the mg/L MLSS concentration and the aeration tank volume. Then lb MLSS can be calculated as follows:

$$MLSS \text{ (lb)} = MLSS \text{ (mg/L)} \times Volume \text{ (MG)} \times 8.34 \text{ lb/gal} \qquad (24.32)$$

■ EXAMPLE 24.42

Problem: If the mixed liquor suspended solids concentration is 1200 mg/L, and the aeration tank has a volume of 550,000 gal, how many pounds of suspended solids are in the aeration tank?

Solution:

$$MLSS \text{ (lb)} = MLSS \text{ (mg/L)} \times volume \text{ (MG)} \times 8.34 \text{ lb/gal}$$

$$= 1200 \text{ mg/L} \times 0.550 \text{ MG} \times 8.34 \text{ lb/gal}$$

$$= 5504 \text{ lb}$$

24.7.1.4 Food-to-Microorganism Ratio

The food-to-microorganism ratio (F/M ratio) is a process control approach based on maintaining a specified balance between available food materials (BOD or COD) in the aeration tank influent and the aeration tank mixed liquor volatile suspended solids (MLVSS) concentration. The chemical oxygen demand (COD) test is sometimes used, because the results are available in a relatively short period of time. To calculate the F/M ratio, the following information is required:

- Aeration tank influent flow rate (MGD)
- Aeration tank influent BOD or COD (mg/L)
- Aeration tank MLVSS (mg/L)
- Aeration tank volume (MG)

$$\text{F/M ratio} = \frac{\text{Primary effluent COD/BOD (mg/L)} \times \text{flow (MGD)} \times 8.34 \text{ lb/mg/L/MG}}{\text{MLVSS (mg/L)} \times \text{aerator volume (MG)} \times 8.34 \text{ lb/mg/L/MG}} \qquad (24.33)$$

Typical F/M ratios for activated biosolids process are shown in the following:

	BOD (lb)/MLVSS (lb)	COD (lb)/MLVSS (lb)
Conventional	0.2–0.4	0.5–1.0
Contact stabilization	0.2–0.6	0.5–1.0
Extended aeration	0.05–0.15	0.2–0.5
Pure oxygen	0.25–1.0	0.5–2.0

■ EXAMPLE 24.43

Problem: The aeration tank influent BOD is 145 mg/L, and the aeration tank influent flow rate is 1.6 MGD. What is the F/M ratio if the MLVSS is 2300 mg/L and the aeration tank volume is 1.8 MG?

Solution:

$$\text{F/M ratio} = \frac{145 \text{ mg/L} \times 1.6 \text{ MGD} \times 8.34 \text{ lb/mg/L/MG}}{2300 \text{ mg/L} \times 1.8 \text{ MG} \times 8.34 \text{ lb/mg/L/MG}} = 0.0.6 \text{ lb BOD per lb MLVSS}$$

Key Point: If the MLVSS concentration is not available, it can be calculated if the percent volatile matter (% VM) of the mixed liquor suspended solids is known.

$$\text{MLVSS} = \text{MLSS} \times \% \text{ Volatile Matter (\%VM)} \qquad (24.34)$$

Note: The F value in the F/M ratio for computing loading to an activated biosolids process can be either BOD or COD. Remember, the reason for biosolids production in the activated biosolids process is to convert BOD to bacteria. One advantage of using COD over BOD for analysis of organic load is that COD is more accurate.

■ EXAMPLE 24.44

Problem: The aeration tank contains 2885 mg/L of MLSS. Lab tests indicate that the MLSS is 66% volatile matter. What is the MLVSS concentration in the aeration tank?

Solution:

$$\text{MLVSS (mg/L)} = 2885 \text{ mg/L} \times 0.66 = 1904 \text{ mg/L}$$

24.7.1.4.1   Required MLVSS Quantity

The pounds of MLVSS required in the aeration tank to achieve the optimum F/M ratio can be deter-mined from the average influent food (BOD or COD) and the desired F/M ratio:

$$\text{MLVSS (lb)} = \frac{\text{Primary effluent BOD/COD} \times \text{flow (MGD)} \times 8.34 \text{ lb/gal}}{\text{Desired F/M ratio}} \qquad (24.35)$$

The required pounds of MLVSS determined by this calculation can then be converted to a concen-tration value by

$$\text{MLVSS (mg/L)} = \frac{\text{Desired MLVSS (lb)}}{\text{Aeration volume (MG)} \times 8.34 \text{ lb/gal}} \qquad (24.36)$$

■ **EXAMPLE 24.45**

Problem: The aeration tank influent flow is 4.0 MGD and the influent COD is 145 mg/L. The aeration tank volume is 0.65 MG. The desired F/M ratio is 0.3 lb COD per lb MLVSS. How many pounds of MLVSS must be maintained in the aeration tank to achieve the desired F/M ratio?

Solution: Determine the required concentration of MLVSS in the aeration tank:

$$\text{MLVSS (lb)} = \frac{145 \text{ mg/L} \times 4.0 \text{ MGD} \times 8.34 \text{ lb/gal}}{0.3 \text{ lb COD per lb MLVSS}} = 16,124 \text{ lb}$$

$$\text{MLVSS (mg/L)} = \frac{16,124 \text{ lb}}{0.65 \text{ MG} \times 8.34 \text{ lb/gal}} = 2974 \text{ mg/L}$$

24.7.1.4.2   Calculating Waste Rates Using F/M Ratio

Maintaining the desired F/M ratio is accomplished by controlling the MLVSS level in the aeration tank. This may be achieved by adjustment of return rates; however, the most practical method is by proper control of the waste rate:

$$\text{Waste volume solids (lb/day)} = \text{Actual MLVSS (lb/day)} - \text{Desired MLVSS (lb/day)} \qquad (24.37)$$

If the desired MLVSS is greater than the actual MLVSS, wasting is stopped until the desired level is achieved. Practical considerations require that the waste quantity be converted to a required volume of waste per day. This is accomplished by converting the waste pounds to flow rate in million gal-lons per day or gallons per minute:

$$\text{Waste (MGD)} = \frac{\text{Waste volatile (lb/day)}}{\text{Waste volatile concentration (mg/L)} \times 8.34 \text{ lb/gal}} \qquad (24.38)$$

$$\text{Waste (gpm)} = \frac{\text{Waste (MGD)} \times 1,000,000 \text{ gpd/MGD}}{1440 \text{ min/day}} \qquad (24.39)$$

Key Point: When the F/M ratio is used for process control, the volatile content of the waste activated sludge should be determined.

■ EXAMPLE 24.46

Problem: Given the following information, determine the required waste rate in gallons per minute to maintain an F/M ratio of 0.17 lb COD per lb MLVSS:

Primary effluent COD	140 mg/L
Primary effluent flow	2.2 MGD
MLVSS (mg/L)	3549 mg/L
Aeration tank volume	0.75 MG
Waste volatile concentrations	4440 mg/L

$$\text{Actual MLVSS} = 3549 \text{ mg/L} \times 0.75 \text{ MG} \times 8.34 \text{ lb/gal} = 22{,}199 \text{ lb/day}$$

$$\text{Required MLVSS} = \frac{140 \text{ mg/L} \times 2.2 \text{ MGD} \times 8.34 \text{ lb/gal}}{0.17 \text{ lb COD per lb MLVSS}} = 15{,}110 \text{ lb/day}$$

$$\text{Waste (lb)} = 22{,}199 \text{ lb/day} - 15{,}110 \text{ lb/day} = 7089 \text{ lb/day}$$

$$\text{Waste (MGD)} = \frac{7089 \text{ lb/day}}{4440 \text{ mg/L} \times 8.34 \text{ lb/gal}} = 0.19 \text{ MGD}$$

$$\text{Waste (gpm)} = \frac{0.19 \text{ MGD} \times 1{,}000{,}000 \text{ gpd/MGD}}{1440 \text{ min/day}} = 132 \text{ gpm}$$

24.7.1.5 Gould Biosolids Age

Biosolids age refers to the average number of days a particle of suspend solids remains under aeration. It is a calculation used to maintain the proper amount of activated biosolids in the aeration tank. This calculation is sometimes referred to as *Gould biosolids age* so it is not confused with similar calculations such as the solids retention time (or mean cell residence time). When considering sludge age, in effect we are asking, "How many days of suspended solids are in the aeration tank?" For example, if 3000 lb of suspended solids enter the aeration tank daily, the aeration tank contains 12,000 lb of suspended solids when 4 days of solids are in the aeration tank—a sludge age of 4 days:

$$\text{Sludge age (days)} = \frac{\text{SS in tank (lb)}}{\text{SS added (lb/day)}} \qquad (24.40)$$

■ EXAMPLE 24.47

Problem: A total of 2740 lb/day suspended solids enters an aeration tank in the primary effluent flow. If the aeration tank has a total of 13,800 lb of mixed liquor suspended solids, what is the biosolids age in the aeration tank?

Solution:

$$\text{Sludge age} = \frac{\text{MLSS (lb)}}{\text{SS added (lb/day)}} = \frac{13{,}800 \text{ lb}}{2740 \text{ lb/day}} = 5 \text{ days}$$

24.7.1.6 Mean Cell Residence Time

Mean cell residence time (MCRT), sometimes called *sludge retention time*, is another process control calculation used for activated biosolids systems. MCRT represents the average length of time an activated biosolids particle remains in the activated biosolids system. It can also be defined as the length of time required at the current removal rate to remove all the solids in the system:

$$\text{MCRT (days)} = \frac{\text{MLSS (mg/L)} \times (\text{Aeration volume} + \text{Clarifier volume}) \times 8.34 \text{ lb/mg/L/MG}}{\left[\text{WAS (mg/L)} \times \text{WAS flow} \times 8.34\right] + (\text{TSS out} \times \text{flow out} \times 8.34)} \quad (24.41)$$

Key Point: MCRT can be calculated using only the aeration tank solids inventory. When comparing plant operational levels to reference materials, we must determine which calculation the reference manual uses to obtain its example values. Other methods are available to determine the clarifier solids concentrations; however, the simplest method assumes that the average suspended solids concentration is equal to the solids concentration of the aeration tank.

■ EXAMPLE 24.48

Problem: Given the following data, what is the MCRT?

Aerator volume	1,000,000 gal
Final clarifier	600,000 gal
Flow	5.0 MGD
Waste rate	0.085 MGD
MLSS	2500 mg/L
Waste	6400 mg/L
Effluent TSS	14 mg/L

Solution:

$$\text{MCRT} = \frac{2500 \text{ mg/L} \times (1.0 \text{ MG} + 0.60 \text{ MG}) \times 8.34}{\left[6400 \text{ mg/L} \times 0.085 \text{ MGD} \times 8.34\right] + (14 \text{ mg/L} \times 5.0 \text{ MGD} \times 8.34)} = 6.5 \text{ days}$$

24.7.1.6.1 Waste Quantities/Requirements

Mean cell residence time for process control requires determining the optimum range for MCRT values. This is accomplished by comparison of the effluent quality with MCRT values. When the optimum MCRT is established, the quantity of solids to be removed (wasted) is determined by

$$\text{Solids (lb/day)} = \left(\frac{\text{MLSS} \times (\text{Aerator vol.} + \text{Clarifier vol.}) \times 8.34}{\text{Desired MCRT}}\right) - (\text{TSS}_{\text{out}} \times \text{flow} \times 8.34) \quad (24.42)$$

■ EXAMPLE 24.49

$$\text{Solids} = \frac{3400 \text{ mg/L} \times (1.4 \text{ MG} + 0.50 \text{ MG}) \times 8.34}{8.6 \text{ days}} - (10 \text{ mg/L} \times 5.0 \text{ MGD} \times 8.34) = 5848 \text{ lb/day}$$

24.7.1.6.2 Waste Rate in Million Gallons per Day

When the quantity of solids to be removed from the system is known, the desired waste rate in million gallons per day can be determined. The unit used to express the rate (MGD, gpd, and gpm) is a function of the volume of waste to be removed and the design of the equipment:

$$\text{Waste (MGD)} = \frac{\text{Waste (lb/day)}}{\text{WAS concentration (mg/L)} \times 8.34} \quad (24.43)$$

$$\text{Waste (gpm)} = \frac{\text{Waste (MGD)} \times 1,000,000 \text{ gpd/MGD}}{1440 \text{ min/day}} \quad (24.44)$$

■ EXAMPLE 24.50

Problem: Given the following data, determine the required waste rate to maintain a MCRT of 8.8 days:

MLSS	2500 mg/L
Aeration volume	1.20 MG
Clarifier volume	0.20 MG
Effluent TSS	11 mg/L
Effluent flow	5.0 MGD
Waste concentration	6000 mg/L

Solution:

$$\text{Waste (lb/day)} = \frac{2500 \text{ mg/L} \times (1.20 + 0.20) \times 8.34}{8.8 \text{ days}} - (11 \text{ mg/L} \times 5.0 \text{ MGD} \times 8.34)$$

$$= 3317 \text{ lb/day} - 459 \text{ lb/day}$$

$$= 2858 \text{ lb/day}$$

$$\text{Waste (MGD)} = \frac{2858 \text{ lb/day}}{6000 \text{ mg/L} \times 8.34 \text{ gal/day}} = 0.057 \text{ MGD}$$

$$\text{Waste (gpm)} = \frac{0.057 \text{ MGD} \times 1,000,000 \text{ gpd/MGD}}{1440 \text{ min/day}} = 40 \text{ gpm}$$

24.7.1.7 Estimating Return Rates from SBV_{60}

Many methods are available for estimation of the proper return biosolids rate. A simple method described in the *Operation of Wastewater Treatment Plants: Field Study Programs*, published by California State University, Sacramento, uses the 60-min percent settled biosolids (sludge) volume. The $\%SBV_{60}$ test results can provide an approximation of the appropriate return activated biosolids rate. This calculation assumes that the SBV_{60} results are representative of the actual settling occurring in the clarifier. If this is true, the return rate in percent should be approximately equal to the SBV_{60}. To determine the approximate return rate in million gallons per day (MGD), the influent flow rate, the current return rate, and the SBV_{60} must be known. The results of this calculation can then be adjusted based upon sampling and visual observations to develop the optimum return biosolids rate.

Note: The $\%SBV_{60}$ must be converted to a decimal percent and total flow rate. (Wastewater flow and current return rate in million gallons per day must be used.)

$$\text{Est. return rate (MGD)} = \left[\text{Influent flow (MGD)} + \text{Current return flow (MGD)}\right] \times \%SBV_{60} \quad (24.45)$$

$$\text{RAS rate (GPM)} = \frac{\text{Return biosolids rate (gpd)}}{1440 \text{ min/day}} \qquad (24.46)$$

where it is assumed that

- $\%SBV_{60}$ is representative.
- Return rate in percent equals $\%SBV_{60}$.
- Actual return rate is normally set slightly higher to ensure that organisms are returned to the aeration tank as quickly as possible.

The rate of return must be adequately controlled to prevent the following:

- Aeration and settling hydraulic overloads
- Low MLSS levels in the aerator
- Organic overloading of aeration
- Septic return activated biosolids
- Solids loss due to excessive biosolids blanket depth

■ EXAMPLE 24.51

Problem: The influent flow rate is 5.0 MGD, and the current return activated sludge flow rate is 1.8 MGD. The $\%SBV_{60}$ is 37%. Based on this information, what should be the return biosolids rate in million gallons per day?

Solution:

$$\text{Return (MGD)} = (5.0 \text{ MGD} + 1.8 \text{ MGD}) \times 0.37 = 2.5 \text{ MGD}$$

24.7.1.8 Biosolids Volume Index (BVI)

The biosolids volume index (BVI) is a measure (an indicator) of the settling quality (a quality indicator) of the activated biosolids. As the BVI increases, the biosolids settle more slowly, do not compact as well, and are likely to result in an increase in effluent suspended solids. As the BVI decreases, the biosolids become denser, settling is more rapid, and the biosolids age. BVI is the volume in milliliters occupied by 1 g of activated biosolids. For the settled biosolids volume (mL/L) and the mixed liquor suspended solids (MLSS) calculation, mg/L are required. The proper BVI range for any plant must be determined by comparing BVI values with plant effluent quality:

$$\text{Biosolids Volume Index (SBI)} = \frac{\text{SBV (mL)} \times 1000}{\text{MLSS (mg/L)}} \qquad (24.47)$$

■ EXAMPLE 24.52

Problem: The SBV_{30} is 250 mL/L, and the MLSS is 2425 mg/L. What is the BVI?

Solution:

$$\text{Biosolids Volume Index (SBI)} = \frac{350 \text{ mL} \times 1000}{2425 \text{ mg/L}} = 144$$

The biosolids volume index equals 144. What does this mean? It means that the system is operating normally with good settling and low effluent turbidity. How do we know this? We know this because we compare our result (144) with the following parameters to obtain the expected condition (the result):

BVI	Expected Conditions
Less than 100	Old biosolids; possible pin floc
	Effluent turbidity increasing
100–250	Normal operation; good settling
	Low effluent turbidity
Greater than 250	Bulking biosolids; poor settling
	High effluent turbidity

24.7.1.9 Mass Balance: Settling Tank Suspended Solids

Solids are produced whenever biological processes are used to remove organic matter from wastewater. Mass balance for anaerobic biological process must take into account both the solids removed by physical settling processes and the solids produced by biological conversion of soluble organic matter to insoluble suspended matter organisms. Research has shown that the amount of solids produced per pound of BOD removed can be predicted based on the type of process being used. Although the exact amount of solids produced can vary from plant to plant, research has developed a series of K factors that can be used to estimate the solids production for plants using a particular treatment process. These average factors provide a simple method to evaluate the effectiveness of a facility's process control program. The mass balance also provides an excellent mechanism to evaluate the validity of process control and effluent monitoring data generated.

24.7.1.9.1 Mass Balance Calculation

$$BOD_{in} \text{ (lb)} = BOD \text{ (mg/L)} \times Flow \text{ (MGD)} \times 8.34 \text{ lb/gal}$$

$$BOD_{out} \text{ (lb)} = BOD \text{ (mg/L)} \times Flow \text{ (MGD)} \times 8.34 \text{ lb/gal}$$

$$\text{Solids produced (lb/day)} = [BOD_{in} \text{ (lb)} - BOD_{out} \text{ (lb)}] \times K$$

$$\text{TSS out (lb/day)} = \text{TSS out (mg/L)} \times Flow \text{ (MGD)} \times 8.34 \text{ lb/gal} \qquad (24.48)$$

$$\text{Waste (lb/day)} = \text{Waste (mg/L)} \times Flow \text{ (MGD)} \times 8.34 \text{ lb/gal}$$

$$\text{Solids removed (lb/day)} = \text{TSS out (lb/day)} + \text{Waste (lb/day)}$$

$$\% \text{ Mass balance} = \frac{(\text{Solids produced} - \text{solids removed}) \times 100}{\text{Solids produced}}$$

24.7.1.10 Biosolids Waste Based on Mass Balance

$$\text{Waste rate (MGD)} = \text{Solids produced (lb/day)}/(\text{waste concentration} \times 8.34 \text{ lb/gal}) \quad (24.49)$$

■ EXAMPLE 24.53

Problem: Given the following data, determine the mass balance of the biological process and the appropriate waste rate to maintain current operating conditions.

Extended aeration (no primary)

Influent flow	1.1 MGD	*Effluent flow*	1.5 MGD	*Waste flow*	24,000 gpd
BOD	220 mg/L	BOD	18 mg/L	TSS	8710 mg/L
TSS	240 mg/L	TSS	22 mg/L		

Solution:

$$BOD_{in} = 220 \text{ mg/L} \times 1.1 \text{ MGD} \times 8.34 \text{ lb/gal} = 2018 \text{ lb/day}$$

$$BOD_{out} = 18 \text{ mg/L} \times 1.1 \text{ MGD} \times 8.34 \text{ lb/gal} = 165 \text{ lb/day}$$

$$BOD \text{ removed} = 2018 \text{ lb/day} - 165 \text{ lb/day} = 1853 \text{ lb/day}$$

$$Solids \text{ produced} = 1853 \text{ lb/day} \times 0.65 \text{ lb/lb BOD} = 1204 \text{ lb/day}$$

$$Solids \text{ out (lb/day)} = 22 \text{ mg/L} \times 1.1 \text{ MGD} \times 8.34 \text{ lb/gal} = 202 \text{ lb/day}$$

$$Sludge \text{ out (lb/day)} = 8710 \text{ mg/L} \times 0.024 \text{ MGD} \times 8.34 \text{ lb/gal} = 1743 \text{ lb/day}$$

$$Solids \text{ removed (lb/day)} = 202 \text{ lb/day} + 1743 \text{ lb/day} = 1945 \text{ lb/day}$$

$$\% \text{ Mass balance} = \frac{1204 \text{ lb/day} - 1945 \text{ lb/day}) \times 100}{1204 \text{ lb/day}} = 62\%$$

The mass balance indicates the following:

1. The sampling points, collection methods, or laboratory testing procedures are producing nonrepresentative results.
2. The process is removing significantly more solids than is required. Additional testing should be performed to isolate the specific cause of the imbalance.

To assist in the evaluation, the waste rate based on the mass balance information can be calculated:

$$\text{Waste (gpd)} = \frac{\text{Solids produced (lb/day)}}{\text{Waste TSS (mg/L)} \times 8.34} \tag{24.50}$$

Thus,

$$\text{Waste (gpd)} = \frac{1204 \text{ lb/day} \times 1{,}000{,}000}{8710 \text{ mg/L} \times 8.34} = 16{,}575 \text{ gpd}$$

24.7.1.11 Aeration Tank Design Parameters

The two design parameters of aeration tanks are food-to-microorganism (F/M) ratio and aeration period (similar to detention time). F/M ratio (BOD loading) is expressed as pounds of BOD per day per pound of MLSS:

$$\text{F/M ratio} = \frac{133{,}690(BOD)Q_o}{(MLSS)V} \tag{24.51}$$

where
   133,690 = Factor to convert units.
   BOD = Settled BOD from primary tank (mg/L).
   Q_o = Average daily wastewater flow (MGD).
   MLSS = Mixed liquor suspended solids (mg/L).
   V = Volume of tank (ft^3).

■ EXAMPLE 24.54

Problem:

Using the data below, design a conventional aeration tank:

MGD = 1 million gal/day
BOD from primary clarifier = 110 mg/L
MLSS = 2000 mg/L
Design F/M = 0.5 per day
Design aeration period, t = 6 hr

Solution:

$$0.50 = \frac{(133,690)(110)(1)}{(2000)V}$$

$$V = 14,706 \text{ ft}^3$$

$$\text{Aeration tank volume } (V) = Q_t = (1 \times 10^6 \text{ gpd})(6 \text{ hr}) \left(\frac{1}{7.48 \text{ ft}^3/\text{gal}} \right) \left(\frac{1 \text{ day}}{24 \text{ hr}} \right) = 33,422 \text{ ft}^3$$

Assume a depth of 10 ft and a length of twice the width:

$$A = \frac{33,422}{10} = 3342 \text{ ft}^2$$

$$(2w)(w) = 3342$$

$$w = 41 \text{ ft}$$

$$l = 82 \text{ ft}$$

24.7.1.12 Lawrence and McCarty Design Model

Over the years, numerous design criteria utilizing empirical and rational parameters based on biological kinetic equations have been developed for suspended-growth systems. Based on practice, the basic Lawrence and McCarty (1970) model is widely used in the industry. The basic Lawrence and McCarty design equations used for sizing suspended-growth systems are provided below.

24.7.1.12.1 Complete Mix with Recycle

For a complete mix system, the mean hydraulic retention time (HRT) θ for the aeration basin is

$$\theta = V/Q \qquad (24.52)$$

where
θ = Hydraulic retention time (day).
V = Volume of aeration tank (m³).
Q = Influent wastewater flow (m³/day).

The mean cell residence time (θ_c) is expressed as

$$\theta_c = \frac{X}{X/t} \qquad (24.53)$$

$$\theta_c = \frac{VX}{\left(Q_{wa}X + Q_cX_c\right)} = \frac{\text{Total mass SS in reactor}}{\text{SS wasting rate}} \tag{24.54}$$

where

θ_c = Mean cell residence time based on solids in the tank (day).
X = Concentration of MLVSS maintained in the tank (mg/L).
$\Delta X/\Delta t$ = Growth of biological sludge over time period Δt (mg/L/day).
V = Volume of aeration tank (m³).
Q_{wa} = Flow of waste sludge removed from the aeration tank (m³/day).
Q_c = Flow of treated effluent (m³/day).
X_c = Microorganism concentration (VSS in effluent, mg/L).

The mean cell residence time for system-drawn sludge from the return line would be

$$Q_c = \frac{VX}{\left(Q_{wr}X_r + Q_cX_c\right)} \tag{24.55}$$

where

Q_{wr} = Flow of waste sludge from return sludge line (m³/d).
X_r = Microorganism concentration in return sludge line (mg/L).

24.7.1.12.2   Microorganism Mass Balance

The mass balance for the microorganisms in the entire activated biosolids system is expressed as (Metcalf & Eddy, 1991)

$$V\left(\frac{dX}{dt}\right) = QX_o + V\left(r_g'\right) - \left(Q_{wa}X + Q_cX_c\right) \tag{24.56}$$

where

V = Volume of aeration tank (m³).
dX/dt = Rate of change of microorganism concentration (VSS) (mg/L/m³/day).
Q = Flow (m³/day).
X_o = Microorganism concentration (VSS) in influent (mg/L).
X = Microorganism concentration in tank (mg/L).
r_g' = Net rate of microorganism growth (VSS) (mg/L/day).

The net rate of bacterial growth (r_g') is expressed as

$$r_g' = Yr_{su} - K_dX \tag{24.57}$$

where

Y = Maximum yield coefficient over finite period of log growth (mg/mg).
r_{su} = Substrate utilization rate (mg/m³).
K_d = Endogenous decay coefficient (per day).
X = Microorganisms concentration in tank (mg/L).

Assuming the cell concentration in the influent is zero and steady-state conditions, Equation 24.58 can be used:

$$\frac{Q_{wa}X + Q_eX_e}{VX} = -Y^r\frac{su}{X} - K_d \tag{24.58}$$

The net specific growth rate can be determined using

$$\frac{1}{\theta_c} = -Y\frac{r_{su}}{X} - K_d \tag{24.59}$$

The term r_{su} can be computed from the following equation:

$$r_{su} = \frac{Q}{V}(S_o - S) = \frac{S_o - S}{\theta} \tag{24.60}$$

where
  S_o = Substrate concentration in influent (mg/L).
  S = Substrate concentration in effluent (mg/L).
  $S_o - S$ = Mass concentration of substrate utilized (mg/L).
  θ = Hydraulic retention time.

24.7.1.13 Effluent Microorganism and Substrate Concentrations

The mass concentration of microorganisms X in the aeration basin can be computed from the following equation:

$$X = \frac{\theta_c Y(S_o - S)}{\theta(1 + K_d\theta_c)} = \frac{\mu_m(S_o - S)}{K(1 + K_d\theta_c)} \tag{24.61}$$

Aeration basin volume can be computed from the following equation:

$$V = \frac{\theta_c QY(S_o - S)}{X(1 + K_d\theta_c)} \tag{24.62}$$

The substrate concentration in effluent S can be determined by the following equation:

$$S = \frac{K_s(1 + K_d\theta_c)}{\theta_c(Yk - K_d) - 1} \tag{24.63}$$

where
  S = Effluent substrate (soluble BOD) concentration (mg/L).
  K_s = Half-velocity constant, substrate concentration at one-half the maximum growth rate (mg/L).
  K_d = Endogenous decay coefficient (per day).
  θ_c = Mean cell residence time based on solids in the tank (day).
  Y = Maximum yield coefficient over finite period of log growth (mg/mg).
  k = Maximum rate of substrate utilization per unit mass of microorganism per day.

Observed yield in the system can be determined by using the following equation:

$$Y_{obs} = \frac{Y}{1 + Q_{ct}} \qquad (24.64)$$

where
  Y_{obs} = Observed yield in the system with recycle (mg/mg).
  Y = Maximum yield coefficient over finite period of log growth (mg/mg).
  Q_{ct} = Mean of all residence times based on solids in aeration tank and in secondary clarifier (days).

24.7.1.13.1 Process Design and Control Relationships

The specific substrate utilization rate (closely related to F/M ratio, which is widely used in practice) can be computed by

$$U = \frac{r_{su}}{X} \qquad (24.65)$$

$$U = \frac{Q(S_o - S)}{VX} = \frac{S_o - S}{\theta X} \qquad (24.66)$$

The net specific growth rate can be computed by

$$\frac{1}{\theta_c} = YU - K_d \qquad (24.67)$$

The flow rate of waste sludge from the sludge return line will be approximately

$$Q_{wt} = \frac{VX}{\theta_c X_r} \qquad (24.68)$$

where X_r is the concentration of sludge in the sludge return line (mg/L).

24.7.1.13.2 Sludge Production

The amount of sludge generated per day can be calculated by

$$P_x = Y_{obs} Q(S_o - S)(8.34) \qquad (24.69)$$

where
  P_x = Net waste activated sludge (VSS) (kg/day, lb/day).
  Y_{obs} = Observed yield (g/g or lb/lb).
  Q = Influent wastewater flow (m³/day, MGD).
  S_o = Influent soluble BOD concentration (mg/L).
  S = Effluent soluble BOD concentration (mg/L).
  8.34 = Conversion factor [(lb/MG):(mg/L)].

24.7.1.13.3 Oxygen Requirements

The theoretical oxygen requirement to remove the carbonaceous organic matter in wastewater for an activated biosolids process is expressed by Metcalf & Eddy (1991) in SI units and the British system:

Mass of O_2 per day = Total mass of BOD_u used − 1.42(mass of organisms wasted, P_x)

$$\text{Kilograms O}_2 \text{ per day} = \frac{Q(S_o - S)}{(1000 \text{ g/kg})f} - 1.42 p_x \qquad (24.70)$$

$$\text{Kilograms O}_2 \text{ per day} = \frac{Q(S_o - S)}{(1000 \text{ g/kg})}\left(\frac{1}{f} - 1.42 Y_{obs}\right) \qquad (24.71)$$

$$\text{Pounds O}_2 \text{ per day} = Q(S_o - S) \times 8.34\left(\frac{1}{f} - 1.42 Y_{obs}\right) \qquad (24.72)$$

where
  BOD_u = Ultimate BOD.
  P_x = Net waste activated sludge (VSS) (kg/day, lb/day).
  Q = Influent flow (m³/day, MGD).
  S_o = Influent soluble BOD concentration (mg/L).
  S = Effluent soluble BOD concentration (mg/L).
  f = Conversion factor for converting BOD to BOD_u.
  Y_{obs} = Observed yield (g/g, lb/lb).
  8.34 = Conversion factor [(lb/MG):(mg/L)].

24.8 OXIDATION DITCH DETENTION TIME

Oxidation ditch systems may be used where the treatment of wastewater is amendable to aerobic biological treatment and the plant design capacities generally do not exceed 1.0 MGD. The oxidation ditch is a form of aeration basin where the wastewater is mixed with return biosolids. The oxidation ditch is essentially a modification of a completely mixed activated biosolids system used to treat wastewater from small communities. This system can be classified as an extended aeration process and is considered to be a low loading rate system. This type of treatment facility can remove 90% or more of influent BOD. Oxygen requirements will generally depend on the maximum diurnal organic loading, degree of treatment, and suspended solids concentration to be maintained in the aerated channel mixed liquor suspended solids. Detention time is the length of time required for a given flow rate to pass through a tank. Detention time is not normally calculated for aeration basins, but it is calculated for oxidation ditches.

Note: When calculating detention time, it is essential that the time and volume units used in the equation be consistent with each other.

$$\text{Detention time (hr)} = \frac{\text{Oxidation ditch volume (gal)}}{\text{Flow rate (gph)}} \qquad (24.73)$$

■ **EXAMPLE 24.55**

Problem: An oxidation ditch has a volume of 160,000 gal. If the flow to the oxidation ditch is 185,000 gpd, what is the detention time in hours?

Solution: Because detention time is desired in hours, the flow must be expressed as gph:

$$\frac{185,000 \text{ gpd}}{24 \text{ hr/day}} = 7708 \text{ gph}$$

Now calculate detention time:

$$\text{Detention time (hr)} = \frac{\text{Oxidation ditch volume (gal)}}{\text{Flow rate (gph)}} = \frac{160,000 \text{ gal}}{7708 \text{ gph}} = 20.8 \text{ hr}$$

24.9 TREATMENT PONDS

The primary goals of wastewater treatment ponds focus on simplicity and flexibility of operation, protection of the water environment, and protection of public health. Moreover, ponds are relatively easy to build and manage, they accommodate large fluctuations in flow, and they can also provide treatment that approaches conventional systems (producing a highly purified effluent) at much lower cost. It is the cost (the economics) that drives many managers to decide on the pond option of treatment. The actual degree of treatment provided in a pond depends on the type and number of ponds used. Ponds can be used as the sole type of treatment, or they can be used in conjunction with other forms of wastewater treatment—that is, other treatment processes followed by a pond or a pond followed by other treatment processes. Ponds can be classified based on their location in the system, by the type of wastes they receive, and by the main biological process occurring in the pond. First, we look at the types of ponds based on their location and the type of wastes they receive: raw sewage stabilization ponds, oxidation ponds, and polishing ponds.

24.9.1 TREATMENT POND PARAMETERS

Before we discuss process control calculations, it is important first to describe the calculations for determining the area, volume, and flow rate parameters that are crucial in making treatment pond calculations.

- Determining pond area in acres:

$$\text{Area (ac)} = \frac{\text{Area (ft}^2)}{43,560 \text{ ft}^2/\text{ac}} \tag{24.74}$$

- Determining pond volume in acre-feet:

$$\text{Volume (ac-ft)} = \frac{\text{Volume (ft}^3)}{43,560 \text{ ft}^2/\text{ac-ft}} \tag{24.75}$$

-
  Determining flow rate in acre-feet/day:

$$\text{Flow (ac-ft/day)} = \text{Flow (MGD)} \times 3069 \text{ ac-ft/MG} \tag{24.76}$$

Key Point: Acre-feet (ac-ft) is a unit that can cause confusion, especially for those not familiar with pond or lagoon operations. One ac-ft is the volume of a box with a 1-ac top and 1 ft of depth—but the top does not have to be an even number of acres in size to use ac-ft.

- Determining flow rate in acre-inches/day

$$\text{Flow (ac-in./day)} = \text{Flow (MGD)} \times 36.8 \text{ ac-in./MG} \tag{24.77}$$

24.9.2 TREATMENT POND PROCESS CONTROL CALCULATIONS

Although there are no recommended process control calculations for the treatment pond, several calculations may be helpful in evaluating process performance or identifying causes of poor performance. These include hydraulic detention time, BOD loading, organic loading rate, BOD removal efficiency, population loading, and hydraulic loading rate. In the following, we provide a few calculations that might be helpful in pond performance evaluation and identification of causes of poor process performance along with other calculations and equations that may be helpful.

24.9.2.1 Hydraulic Detention Time (Days)

$$\text{Hydraulic detention time (days)} = \frac{\text{Pond volume (ac-ft)}}{\text{Influent flow (ac-ft/day)}} \tag{24.78}$$

Key Point: Hydraulic detention time for stabilization ponds normally ranges from 30 to 120 days.

■ EXAMPLE 24.56

Problem: A stabilization pond has a volume of 54.5 ac-ft. What is the detention time in days when the flow is 0.35 MGD?

Solution:

$$\text{Flow} = 0.35\ \text{MGD} \times 3069\ \text{ac-ft/MG} = 1.07\ \text{ac-ft/day}$$

$$\text{Detention time (days)} = \frac{54.5\ \text{ac-ft}}{1.07\ \text{ac-ft/day}} = 51\ \text{days}$$

24.9.2.2 BOD Loading

When calculating BOD loading on a wastewater treatment pond, the following equation is used:

$$\text{BOD (lb/day)} = \text{BOD (mg/L)} \times \text{Flow (MGD)} \times 8.34\ \text{lb/gal} \tag{24.79}$$

■ EXAMPLE 24.57

Problem: Calculate the BOD loading (lb/day) on a pond if the influent flow is 0.3 MGD with a BOD of 200 mg/L.

Solution:

$$\text{BOD (lb/day)} = \text{BOD} \times \text{Flow} \times 8.34\ \text{lb/gal} = 200\ \text{mg/L} \times 0.3\ \text{MGD} \times 8.34\ \text{lb/gal} = 500\ \text{lb/day}$$

24.9.2.3 Organic Loading Rate

Organic loading can be expressed as pound of BOD per acre per day (most common), pounds of BOD per acre-foot per day, or people per acre per day.

$$\text{Organic loading rate (lb BOD/ac/day)} = \frac{\text{BOD (mg/L)} \times \text{influent flow (MGD)} \times 8.34}{\text{Pond area (ac)}} \tag{24.80}$$

Key Point: The organic loading rate normally ranges from 10 to 50 lb BOD per day per acre.

■ **EXAMPLE 24.58**

Problem: A wastewater treatment pond has an average width of 370 ft and an average length of 730 ft. The influent flow rate to the pond is 0.10 MGD with a BOD concentration of 165 mg/L. What is the organic loading rate to the pond in pounds per day per acre (lb/day/ac)?

Solution:

$$730 \text{ ft} \times 370 \text{ ft} \times \frac{1 \text{ ac}}{43{,}560 \text{ ft}^2} = 6.2 \text{ ac}$$

$$0.10 \text{ MGD} \times 165 \text{ mg/L} \times 8.34 \text{ lb/gal} = 138 \text{ lb/day}$$

$$\frac{138 \text{ lb/day}}{6.2 \text{ ac}} = 22.3 \text{ lb/day/ac}$$

24.9.2.4 BOD Removal Efficiency

The efficiency of any treatment process is its effectiveness in removing various constituents from the water or wastewater. BOD removal efficiency is therefore a measure of the effectiveness of the wastewater treatment pond in removing BOD from the wastewater:

$$\% \text{ BOD removed} = \frac{\text{BOD removed (mg/L)}}{\text{BOD total (mg/L)}} \times 100$$

■ **EXAMPLE 24.59**

Problem: The BOD entering a waste treatment pond is 194 mg/L. If the BOD in the pond effluent is 45 mg/L, what is the BOD removal efficiency of the pond?

Solution:

$$\% \text{ BOD removed} = \frac{\text{BOD removed (mg/L)}}{\text{BOD total (mg/L)}} \times 100 = \frac{149 \text{ mg/L}}{194 \text{ mg/L}} \times 100 = 77\%$$

24.9.2.5 Population Loading

$$\text{Population loading (people/ac/day)} = \frac{\text{BOD (mg/L)} \times \text{influent flow (MGD)} \times 8.34}{\text{Pond area (ac)}} \quad (24.81)$$

24.9.2.6 Hydraulic Loading (In./Day) (Overflow Rate)

$$\text{Hydraulic loading (in./day)} = \frac{\text{Influent flow (ac-in./day)}}{\text{Pond area (ac)}} \quad (24.82)$$

24.9.3 Aerated Ponds

Depending on the hydraulic retention time, the effluent from an aerated pond will contain from one-third to one-half the concentration of the influent BOD in the form of cell tissue (Metcalf & Eddy, 1991). These solids must be removed by settling before the effluent is discharged. The mathematical relationship for BOD removal in a complete-mix-activated pond is derived from the following equation:

$$QS_o - QS - kSV = 0 \tag{24.83}$$

Rearranged:

$$\frac{S}{S_o} = \frac{1}{1 + k(V/Q)} = \frac{\text{Effluent BOD}}{\text{Influent BOD}} \tag{24.84}$$

$$= \frac{1}{1 + k\theta} \tag{24.85}$$

where

Q = Wastewater flow (m³/day, MGD).
S_o = Influent BOD concentration (mg/L).
S = Effluent BOD concentration (mg/L).
k = Overall first-order BOD removal rate (per day).
θ = Hydrauslic retention time.

The temperature in the aerated pond resulting from the influent wastewater temperature, air temperature, surface area, and flow can be computed using the following equation (Mancini and Barnhart, 1968):

$$T_i - T_w = \frac{(T_w - T_a)fA}{Q} \tag{24.86}$$

where

T_i = Influent wastewater temperature (°C, °F).
T_w = Lagoon water temperature (°C, °F).
T_a = Ambient air temperature (°C, °F).
f = Proportionality factor = 12×10^{-6} (British system) or 0.5 (SI units).
A = Surface area of lagoon (m², ft²).
Q = Wastewater flow (m³/day, MGD).

Using Equation 24.86 rearranged, the pond water temperature is

$$T_w = \frac{AfT_a + QT_i}{Af + Q} \tag{24.87}$$

24.10 CHEMICAL DOSAGE*

Chemicals are used extensively in wastewater and water treatment operations. Plant operators add chemicals to various unit processes for slime-growth control, corrosion control, odor control, grease removal, BOD reduction, pH control, sludge-bulking control, ammonia oxidation, and bacterial reduction, among other reasons. To apply any chemical dose correctly it is important to be able to make certain dosage calculations. One of the most frequently used calculations in wastewater mathematics is the dosage or loading. The general types of mg/L to lb/day or lb calculations are for chemical dosage, BOD, chemical oxygen demand (COD), suspended solids (SS) loading/removal, pounds of solids under aeration, and waste activated sludge (WAS) pumping rate. These calculations are usually made using either of the following equations:

---

* In Chapter 23, we discussed calculations relative to the chlorination process used to treat potable water. Similar information appears here.

$$\text{Concentration (mg/L)} \times \text{Flow (MGD)} \times 8.34 \text{ lb/gal} = \text{lb/day} \qquad (24.88)$$

$$\text{Concentration (mg/L)} \times \text{Volume (MG)} \times 8.34 \text{ lb/gal} = \text{lb} \qquad (24.89)$$

Note: If mg/L concentration represents a concentration in a flow, then million gallons per day (MGD) flow is used as the second factor; however, if the concentration pertains to a tank or pipeline volume, then million gallons (MG) volume is used as the second factor.

Note: In previous years, it was normal practice to use the expression *parts per million* (ppm) as an expression of concentration, as 1 mg/L = 1 ppm; however, current practice is to use mg/L as the preferred expression of concentration.

24.10.1   CHEMICAL FEED RATE

In chemical dosing, a measured amount of chemical is added to the wastewater (or water). The amount of chemical required depends on the type of chemical used, the reason for dosing, and the flow rate being treated. The two expressions most often used to describe the amount of chemical added or required are

- Milligrams per liter (mg/L)
- Pounds per day (lb/day)

A milligram per liter is a measure of concentration. As shown below, if a concentration of 5 mg/L is desired, then a total of 15 mg chemical would be required to treat 3 L:

$$\frac{5 \text{ mg} \times 3}{\text{L} \times 3} = \frac{15 \text{ mg}}{3 \text{ L}}$$

The amount of chemical required therefore depends on two factors:

- The desired concentration (mg/L)
- The amount of wastewater to be treated (normally expressed as MGD)

To convert from mg/L to lb/day, Equation 24.88 is used.

■ EXAMPLE 24.60

Problem: Determine the chlorinator setting (lb/day) required to treat a flow of 5 MGD with a chemical dose of 3 mg/L.

Solution:

$$\text{Chemical} = \text{Chemical} \times \text{Flow} \times 8.34 \text{ lb/gal} = 3 \text{ mg/L} \times 5 \text{ MGD} \times 8.34 \text{ lb/gal} = 125 \text{ lb/day}$$

■ EXAMPLE 24.61

Problem: The desired dosage for a dry polymer is 10 mg/L. If the flow to be treated is 2,100,000 gpd, how many lb/day polymer will be required?

Solution:

$$\text{Polymer} = \text{Polymer} \times \text{Flow} \times 8.34 \text{ lb/day} = 10 \text{ mg/L} \times 2.10 \text{ MGD} \times 8.34 \text{ lb/day} = 175 \text{ lb/day}$$

Key Point: To calculate chemical dose for tanks or pipelines, a modified equation must be used. Instead of MGD flow, MG volume is used:

$$\text{Chemical (lb)} = \text{Chemical (mg/L)} \times \text{Tank volume (MG)} \times 8.34 \text{ lb/gal} \qquad (24.90)$$

■ EXAMPLE 24.62

Problem: To neutralize a sour digester, 1 lb of lime is added for every pound of volatile acids in the digester biosolids. If the digester contains 300,000 gal of biosolids with a volatile acid level of 2200 mg/L, how many pounds of lime should be added?

Solution: Because the volatile acid concentration is 2200 mg/L, the lime concentration should also be 2200 mg/L, so

$$\text{Lime required} = \text{Lime concentration} \times \text{Digester volume} \times 8.34 \text{ lb/gal}$$

$$= 2200 \text{ mg/L} \times 0.30 \text{ MG} \times 8.34 \text{ lb/gal} = 5504 \text{ lb}$$

24.10.2 CHLORINE DOSE, DEMAND, AND RESIDUAL

Chlorine is a powerful oxidizer that is commonly used in wastewater and water treatment for disinfection and in wastewater treatment for odor and bulking control, among other applications. When chlorine is added to a unit process, we want to ensure that a measured amount is added, obviously. Chlorine dose depends on two considerations: the chlorine demand and the desired chlorine residual:

$$\text{Chlorine dose} = \text{Chlorine demand} + \text{Chlorine residual} \qquad (24.91)$$

24.10.2.1 Chlorine Dose

To describe the amount of chemical added or required, we use Equation 24.92:

$$\text{Chemical (lb/day)} = \text{Chemical (mg/L)} \times \text{Flow (MGD)} \times 8.34 \text{ lb/day} \qquad (24.92)$$

■ EXAMPLE 24.63

Problem: Determine the chlorinator setting (lb/day) required to treat a flow of 8 MGD with a chlorine dose of 6 mg/L.

Solution:

$$\text{Chemical} = \text{Chemical} \times \text{Flow} \times 8.34 \text{ lb/day} = 6 \text{ mg/L} \times 8 \text{ MGD} \times 8.34 \text{ lb/gal} = 400 \text{ lb/day}$$

24.10.2.2 Chlorine Demand

The chlorine demand is the amount of chlorine used in reacting with various components of the water such as harmful organisms and other organic and inorganic substances. When the chlorine demand has been satisfied, these reactions cease.

■ EXAMPLE 24.64

Problem: The chlorine dosage for a secondary effluent is 6 mg/L. If the chlorine residual after a 30-min contact time is found to be 0.5 mg/L, what is the chlorine demand expressed in mg/L?

Solution:

$$\text{Chlorine dose} = \text{Chlorine demand} + \text{chlorine residual}$$

$$6 \text{ mg/L} = x \text{ mg/L} + 0.5 \text{ mg/L}$$

$$6 \text{ mg/L} - 0.5 \text{ mg/L} = x \text{ mg/L}$$

$$x = 5.5 \text{ mg/L chlorine demand}$$

24.10.2.3   Chlorine Residual

Chlorine residual is the amount of chlorine remaining after the demand has been satisfied.

■ **EXAMPLE 24.65**

Problem: What should the chlorinator setting (lb/day) be to treat a flow of 3.9 MGD if the chlorine demand is 8 mg/L and a chlorine residual of 2 mg/L is desired?

Solution: First calculate the chlorine dosage in mg/L:

$$\text{Chlorine dose} = \text{Chlorine demand} + \text{Chlorine residual} = 8 \text{ mg/L} + 2 \text{ mg/L} = 10 \text{ mg/L}$$

Then calculate the chlorine dosage (feed rate) in lb/day:

$$\text{Chlorine (mg/L)} \times \text{Flow (MGD)} \times 8.34 \text{ lb/gal} = \text{lb/day chlorine}$$

$$10 \text{ mg/L} \times 3.9 \text{ MGD} \times 8.34 \text{ lb/gal} = 325 \text{ lb/day chlorine}$$

24.10.3   Hypochlorite Dosage

Hypochlorite is less hazardous than chlorine; therefore, it is often used as a substitute chemical for elemental chlorine. Hypochlorite is similar to strong bleach and comes in two forms: dry calcium hypochlorite (often referred to as HTH) and liquid sodium hypochlorite. Calcium hypochlorite contains about 65% available chlorine; sodium hypochlorite contains about 12 to 15% available chlorine (in industrial strengths).

Note: Because neither type of hypochlorite is 100% pure chlorine, more lb/day must be fed into the system to obtain the same amount of chlorine for disinfection. This is an important economical consideration for those facilities thinking about substituting hypochlorite for chlorine. Some studies indicate that such a substitution can increase operating costs, overall, by up to three times the cost of using chlorine.

To calculate the lb/day hypochlorite required, a two-step calculation is required:

$$\text{Chlorine (mg/L)} \times \text{Flow (MGD)} \times 8.34 \text{ lb/gal} = \text{lb/day chlorine}$$

$$\frac{\text{Chlorine (lb/day)}}{\dfrac{\% \text{ Available}}{100}} = \text{Hypochlorite (lb/day)} \qquad (24.93)$$

■ **EXAMPLE 24.66**

Problem: A total chlorine dosage of 10 mg/L is required to treat a particular wastewater. If the flow is 1.4 MGD and the hypochlorite has 65% available chlorine, how many lb/day of hypochlorite will be required?

Solution: First calculate the lb/day chlorine required using the mg/L to lb/day equation:

$$\text{Chlorine (mg/L)} \times \text{Flow (MGD)} \times 8.34 \text{ lb/gal} = \text{lb/day chlorine}$$

$$10 \text{ mg/L} \times 1.4 \text{ MGD} \times 8.34 \text{ lb/gal} = 117 \text{ lb/day}$$

Then calculate the lb/day hypochlorite required. Because only 65% of the hypochlorite is chlorine, more than 117 lb/day will be required:

$$\frac{117 \text{ lb/day chlorine}}{\dfrac{65 \text{ available chlorine}}{100}} = 180 \text{ lb/day hypochlorite}$$

■ **EXAMPLE 24.67**

Problem: A wastewater flow of 840,000 gpd requires a chlorine dose of 20 mg/L. If sodium hypochlorite (15% available chlorine) is to be used, how many lb/day of sodium hypochlorite are required? How many gallons per day of sodium hypochlorite is this?

Solution: First calculate the lb/day chlorine required:

$$\text{Chlorine (mg/L)} \times \text{Flow (MGD)} \times 8.34 \text{ lb/gal} = \text{lb/day chlorine}$$

$$20 \text{ mg/L} \times 0.84 \text{ MGD} \times 8.34 \text{ lb/gal} = 140 \text{ lb/day chlorine}$$

Then calculate the lb/day sodium hypochlorite:

$$\frac{140 \text{ lb/day chlorine}}{\dfrac{15 \text{ available chlorine}}{100}} = 933 \text{ lb/day hypochlorite}$$

Now calculate the gpd sodium hypochlorite:

$$\frac{933 \text{ lb/day}}{8.34 \text{ lb/gal}} = 112 \text{ gpd sodium hypochlorite}$$

■ **EXAMPLE 24.68**

Problem: How many pounds of chlorine gas are necessary to treat 5,000,000 gal of wastewater at a dosage of 2 mg/L?

Solution: First calculate the pounds of chlorine required:

$$\text{Volume } (10^6 \text{ gal}) \times \text{Chlorine concentration (mg/L)} \times 8.34 \text{ lb/gal} = \text{lb chlorine}$$

Then substitute:

$$(5 \times 10^6 \text{ gal}) \times 2 \text{ mg/L} \times 8.34 \text{ lb/gal} = 83 \text{ lb chlorine}$$

24.10.4 CHEMICAL SOLUTIONS

A *water solution* is a homogeneous liquid consisting of the *solvent* (the substance that dissolves another substance) and the *solute* (the substance that dissolves in the solvent) (Figure 24.7). Water is the solvent. The solute (whatever it may be) may dissolve up to a certain point. This is called its *solubility*—that is, the solubility of the solute in the particular solvent (water) at a particular temperature and pressure. Remember, in chemical solutions, the substance being dissolved is called the *solute*, and the liquid present in the greatest amount in a solution (and that does the dissolving) is called the *solvent*. We should also be familiar with another term, *concentration*—the amount of solute dissolved in a given amount of solvent. Concentration is measured as

$$\% \text{ Strength} = \frac{\text{Weight of solute}}{\text{Weight of solution}} \times 100 = \frac{\text{Weight of solute}}{\text{Weight of solute} + \text{solvent}} \times 100 \qquad (24.94)$$

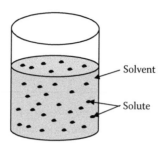

FIGURE 24.7   Solution with two components: solvents and solute.

■ **EXAMPLE 24.69**

Problem: If 30 lb of chemical is added to 400 lb of water, what is the percent strength (by weight) of the solution?

Solution:

$$\% \text{ Strength} = \frac{30 \text{ lb solute}}{\text{Weight of solution}} \times 100 = \frac{30 \text{ lb solute}}{30 \text{ lb solute} + 400 \text{ lb water}} \times 100 = 7.0\%$$

Important to making accurate computations of chemical strength is a complete understanding of the dimensional units involved; for example, it is important to understand exactly what *milligrams per liter* (mg/L) signifies:

$$\text{Milligrams per liter (mg/L)} = \frac{\text{Milligrams of solute}}{\text{Liters of solution}} \qquad (24.95)$$

Another important dimensional unit commonly used when dealing with chemical solutions is *parts per million* (ppm):

$$\text{Parts per million (ppm)} = \frac{\text{Parts of solute}}{\text{Million parts of solution}} \qquad (24.96)$$

Note: A part is usually a weight measurement.

For example:

$$8 \text{ ppm} = \frac{8 \text{ lb solids}}{1,000,000 \text{ lb solution}} = \frac{8 \text{ mg solids}}{1,000,000 \text{ mg solution}}$$

24.10.5   MIXING SOLUTIONS OF DIFFERENT STRENGTH

When different percent strength solutions are mixed, we use the following equations, depending upon the complexity of the problem:

$$\% \text{ Strength of mixture} = \frac{\text{Chemical in mixture (lb)}}{\text{Solution mixture (lb)}} \times 100 \qquad (24.97)$$

$$\% \text{ Strength of mixture} = \frac{\text{Solution 1 chemical (lb)} + \text{Solution 2 chemical (lb)}}{\text{Solution 1 (lb)} + \text{Solution 2 (lb)}} \times 100 \quad (24.98)$$

$$\text{% Strength of mixture} = \frac{\dfrac{\text{Sol. 1 (lb)} \times (\text{% Strength Sol. 1})}{100} + \dfrac{\text{Sol. 2 (lb)} \times (\text{% Strength Sol. 2})}{100}}{\text{Solution 1 (lb)} + \text{Solution 2 (lb)}} \times 100 \qquad (24.99)$$

■ **EXAMPLE 24.70**

Problem: If 25lb of a 10% strength solution are mixed with 40 lb of a 1% strength solution, what is the percent strength of the solution mixture?

Solution:

$$\text{% Strength of mixture} = \frac{\dfrac{\text{Sol. 1 (lb)} \times (\text{% Strength Sol. 1})}{100} + \dfrac{\text{Sol. 2 (lb)} \times (\text{% Strength Sol. 2})}{100}}{\text{Solution 1 (lb)} + \text{Solution 2 (lb)}} \times 100$$

$$= \frac{(25 \text{ lb})(0.1) + (40 \text{ lb})(0.01)}{25 \text{ lb} + 40 \text{ lb}} \times 100$$

$$= \frac{2.5 \text{ lb} + 0.4 \text{ lb}}{65 \text{ lb}}$$

$$= 4.5\%$$

Key Point: Percent strength should be expressed in terms of pounds chemical per pounds solution. That is, when solutions are expressed, for example, in terms of gallons, the gallons should be expressed as pounds before continuing with the percent strength calculations.

24.10.6 Solution Mixtures and Target Percent Strength

When two different percent strength solutions are mixed in order to obtain a desired quantity of solution and a target percent strength, we use Equation 24.99 and fill in the given information. Then, we find for the unknown, x.

■ **EXAMPLE 24.71**

Problem: What weights of a 3% solution and a 6% solution must be mixed to make 800 lb of a 4% solution?

Solution:

$$\text{% Strength of mixture} = \frac{\dfrac{\text{Sol. 1 (lb)} \times (\text{% Strength Sol. 1})}{100} + \dfrac{\text{Sol. 2 (lb)} \times (\text{% Strength Sol. 2})}{100}}{\text{Solution 1 (lb)} + \text{Solution 2 (lb)}} \times 100$$

$$4 = \frac{(x \text{ lb})(0.03) + (800 - x \text{ lb})(0.06)}{800 \text{ lb}} \times 100$$

$$\frac{4}{100}(800) = 0.03x + 48 - 0.06x$$

$$32 = -0.03x + 48$$

$$0.03x = 14$$

$$x = 467 \text{ lb of 3%solution}$$

$$800 - 467 = 333 \text{ lb of 6% solution}$$

24.10.7 CHEMICAL SOLUTION FEEDER SETTING (GPD)

Calculating a gallon-per-day feeder setting depends on how the solution concentration is expressed: lb/gal or percent. If the solution strength is expressed as lb/gal, use the following equation:

$$\text{Solution (gpd)} = \frac{\text{Chemical (mg/L)} \times \text{flow (MGD)} \times 8.34 \text{ lb/gal}}{\text{Chemical solution (lb)}} \qquad (24.100)$$

In water and wastewater operations, a standard, trial-and-error method known as *jar testing* is conducted to determine optimum chemical dosage. Jar testing has been the accepted bench testing procedure for many years. After jar testing results are analyzed to determine the best chemical dosage, the actual calculations are made, as demonstrated by the following example problems.

■ EXAMPLE 24.72

Problem: Jar tests indicate that the best liquid alum dose for a water is 8 mg/L. The flow to be treated is 1.85 MGD. Determine the gpd setting for the liquid alum chemical feeder if the liquid alum contains 5.30 lb of alum per gallon of solution.

Solution: First, calculate the lb/day of dry alum required, using the mg/L to lb/day equation:

$$\text{Alum} = \frac{\text{Dose (mg/L)} \times \text{flow (MGD)} \times 8.34 \text{ lb/gal}}{\text{Chemical solution (lb)}}$$

$$= 8 \text{ mg/L} \times 1.85 \text{ MGD} \times 8.34 \text{ lb/gal} = 123 \text{ lb/day}$$

Then, calculate the gpd solution required:

$$\text{Alum solution} = \frac{123 \text{ lb/day alum}}{5.30 \text{ lb alum per gal solution}} = 23 \text{ gpd}$$

The feeder setting, then, is 23 gpd alum solution. If the solution strength is expressed as a percent, we use the following equation:

$$\begin{bmatrix} \text{Chemical (mg/L)} \times \text{Flow treated (MGD)} \times 8.34 \text{ lb/gal} \end{bmatrix}$$
$$= \begin{bmatrix} \text{Solution (mg/L)} \times \text{Solution flow (MGD)} \times 8.34 \text{ lb/gal} \end{bmatrix} \qquad (24.101)$$

■ EXAMPLE 24.73

Problem: The flow to a plant is 3.40 MGD. Jar testing indicates that the optimum alum dose is 10 mg/L. What should the gpd setting be for the solution feeder if the alum solution is a 52% solution?

Solution: A solution concentration of 52% is equivalent to 520,000 mg/L:

$$\text{Desired dose (lb/day)} = \text{Actual dose (lb/day)}$$
$$\text{Chemical} \times \text{Flow treated} \times 8.34 \text{ lb/gal} = \text{Solution} \times \text{Solution flow} \times 8.34 \text{ lb/gal}$$
$$10 \text{ mg/L} \times 3.40 \text{ MGD} \times 8.34 \text{ lb/gal} = 520,00 \text{ mg/L} \times x \text{ MGD} \times 8.34 \text{ lb/gal}$$

$$x = \frac{10 \times 3.40 \times 8.34}{520,000 \times 8.34} = 0.0000653 \text{ MGD}$$

This can be expressed as gpd flow:

$$0.0000653 \text{ MGD} = 65.3 \text{ gpd flow}$$

24.10.8 CHEMICAL FEED PUMP: PERCENT STROKE SETTING

Chemical feed pumps are generally positive displacement pumps (also called *piston pumps*). This type of pump displaces, or pushes out, a volume of chemical equal to the volume of the piston. The length of the piston, called the *stroke*, can be lengthened or shortened to increase or decrease the amount of chemical delivered by the pump. As mentioned, each stroke of a piston pump displaces or pushes out chemical. When calculating the percent stroke setting, use the following equation:

$$\text{Stroke setting } (\%) = \frac{\text{Required feed (gpd)}}{\text{Maximum feed (gpd)}} \qquad (24.102)$$

■ EXAMPLE 24.74

Problem: The required chemical pumping rate has been calculated as 8 gpm. If the maximum pumping rate is 90 gpm, what should the percent stroke setting be?

Solution: The percent stroke setting is based on the ratio of the gpm required to the total possible gpm:

$$\text{Stroke setting } (\%) = \frac{\text{Required feed (gpd)}}{\text{Maximum feed (gpd)}} = \frac{8 \text{ gpm}}{90 \text{ gpm}} \times 100 = 8.9\%$$

24.10.9 CHEMICAL SOLUTION FEEDER SETTING (ML/MIN)

Some chemical solution feeders dispense chemical as milliliters per minute (mL/min). To calculate the mL/min solution required, use the following equation:

$$\text{Solution (mL/min)} = \frac{\text{gpd} \times 3785 \text{ mL/gal}}{1440 \text{ min/day}} \qquad (24.103)$$

■ EXAMPLE 24.75

Problem: The desired solution feed rate was calculated to be 7 gpd. What is this feed rate expressed as mL/min?

Solution: Because the gpd flow has already been determined, the mL/min flow rate can be calculated directly:

$$\text{Solution} = \frac{\text{gpd} \times 3785 \text{ mL/gal}}{1440 \text{ min/day}} = \frac{7 \text{ gpd} \times 3785 \text{ mL/gal}}{1440 \text{ min/day}} = 18 \text{ mL/min}$$

24.10.10 CHEMICAL FEED RATE CALIBRATION

Routinely, to ensure accuracy, we need to compare the actual chemical feed rate with the feed rate indicated by the instrumentation. To accomplish this, we use calibration calculations. To calculate the actual chemical feed rate for a dry chemical feed, place a container under the feeder, weigh the container when empty, and then weigh the container again after a specified length of time, such as 30 min. Then actual chemical feed rate can then be determined as

$$\text{Feed rate (lb/min)} = \frac{\text{Chemical applied (lb)}}{\text{Length of application (min)}} \qquad (24.104)$$

■ **EXAMPLE 24.76**

Problem: Calculate the actual chemical feed rate (lb/day) if a container is placed under a chemical feeder and a total of 2.2 lb is collected during a 30-min period.

Solution: First calculate the lb/min feed rate:

$$\text{Feed rate (lb/min)} = \frac{\text{Chemical applied (lb)}}{\text{Length of application (min)}} = \frac{2.2 \text{ lb}}{30 \text{ min}} = 0.07 \text{ lb/min}$$

Then calculate the lb/day feed rate:

$$\text{Chemical feed rate (lb/day)} = (0.07 \text{ lb/min} \times 1440 \text{ min/day}) = 101 \text{ lb/day}$$

■ **EXAMPLE 24.77**

Problem: A chemical feeder is to be calibrated. The container to be used to collect chemical is placed under the chemical feeder and weighed (0.35 lb). After 30 min, the weight of the container and chemical is found to be 2.2 lb. Based on this test, what is the actual chemical feed rate in lb/day?

Note: The chemical applied is the weight of the container and chemical minus the weight of the empty container.

Solution: First calculate the lb/min feed rate:

$$\text{Feed rate (lb/min)} = \frac{\text{Chemical applied (lb)}}{\text{Length of application (min)}}$$

$$= \frac{2.2 \text{ lb} - 0.35 \text{ lb}}{30 \text{ min}} = \frac{1.85 \text{ lb}}{30 \text{ min}} = 0.062 \text{ lb/min}$$

Then calculate the lb/day feed rate:

$$0.062 \text{ lb/min} \times 1440 \text{ min/day} = 89 \text{ lb/day}$$

When the chemical feeder is used for a solution, the calibration calculation is slightly more difficult than that for a dry chemical feeder. As with other calibration calculations, the actual chemical feed rate is determined and then compared with the feed rate indicated by the instrumentation. The calculations used for solution feeder calibration are as follows:

$$\text{Flow rate (gpd)} = \frac{(\text{mL/min}) \times 1440 \text{ min/day}}{3785 \text{ mL/gal}} \qquad (24.105)$$

Then calculate chemical dosage (lb/day):

$$\text{Chemical (lb/day)} = \text{Chemical (mg/L)} \times \text{Flow (MGD)} \times 8.34 \text{ lb/day} \qquad (24.106)$$

■ **EXAMPLE 24.78**

Problem: A calibration test is conducted for a solution chemical feeder. During 5 min, the solution feeder delivers a total of 700 mL. The polymer solution is a 1.3% solution. What is the lb/day feed rate? (Assume that the polymer solution weighs 8.34 lb/gal.)

Solution: The mL/min flow rate is calculated as

$$700 \text{ mL} \div 5 \text{ min} = 140 \text{ mL/min}$$

Then convert the mL/min flow rate to a gpd flow rate:

$$\frac{140 \text{ mL/min} \times 1440 \text{ min/day}}{3785 \text{ mL/gal}} = 53 \text{ gpd}$$

Now calculate the lb/day feed rate:

$$\text{Chemical (mg/L)} \times \text{Flow (MGD)} \times 8.34 \text{ lb/day} = \text{Chemical (lb/day)}$$

$$13{,}000 \text{ mg/L} \times 0.000053 \text{ MGD} \times 8.34 \text{ lb/day} = 5.7 \text{ lb/day polymer}$$

Actual pumping rates can be determined by calculating the volume pumped during a specified time frame; for example, if 120 gal are pumped during a 15-min test, the average pumping rate during the test is 8 gpm. The gallons pumped can be determined by measuring the drop in tank level during the timed test:

$$\text{Flow (gpm)} = \frac{\text{Volume pumped (gal)}}{\text{Duration of test (min)}} \tag{24.107}$$

The actual flow rate (gpm) is then calculated using

$$\text{Flow rate (gpm)} = \frac{0.785 \times (\text{Diameter})^2 \times \text{Drop in level (ft)} \times 7.48 \text{ gal/ft}^3}{\text{Duration of test (min)}} \tag{24.108}$$

■ **EXAMPLE 24.79**

Problem: A pumping rate calibration test is conducted for a 5-min period. The liquid level in the 4-ft-diameter solution tank is measured before and after the test. If the level drops 0.4 ft during the 5-min test, what is the pumping rate in gpm?

Solution:

$$\text{Pumping rate (gpm)} = \frac{0.785 \times (\text{Diameter})^2 \times \text{Drop in level (ft)} \times 7.48 \text{ gal/ft}^3}{\text{Duration of test (min)}}$$

$$= \frac{0.785 \times (4 \text{ ft} \times 4 \text{ ft}) \times 0.4 \text{ ft} \times 7.48 \text{ gal/ft}^3}{5 \text{ min}}$$

$$= 38 \text{ gpm}$$

24.10.11 AVERAGE USE CALCULATIONS

During a typical shift, operators log in or record several parameter readings. The data collected are important in monitoring plant operation as they provide information on how to best optimize plant or unit process operation. One of the important parameters monitored each shift or each day is actual use of chemicals. From the recorded chemical use data, expected chemical use can be forecasted. These data are also important for inventory control, because an estimate can be made of when additional chemical supplies will be required. To determine average chemical use, we first must determine the average chemical use:

$$\text{Average use (lb/day)} = \frac{\text{Total chemical used (lb)}}{\text{Number of days}} \qquad (24.109)$$

or

$$\text{Average use (gpd)} = \frac{\text{Total chemical used (gal)}}{\text{Number of days}} \qquad (24.110)$$

Then calculate the days supply in inventory:

$$\text{Days supply in inventory} = \frac{\text{Total chemical in inventory (lb)}}{\text{Average use (lb/day)}} \qquad (24.111)$$

or

$$\text{Days supply in inventory} = \frac{\text{Total chemical in inventory (gal)}}{\text{Average use (gpd)}} \qquad (24.112)$$

■ EXAMPLE 24.80

Problem: The chemical used for each day during a week is given below. Based on these data, what was the average lb/day chemical use during the week?

Monday	92 lb/day
Tuesday	94 lb/day
Wednesday	92 lb/day
Thursday	88 lb/day
Friday	96 lb/day
Saturday	92 lb/day
Sunday	88 lb/day

Solution:

$$\text{Average use (lb/day)} = \frac{\text{Total chemical used (lb)}}{\text{Number of days}} = \frac{642 \text{ lb}}{7 \text{ days}} = 91.7 \text{ lb/day}$$

■ EXAMPLE 24.81

Problem: The average chemical use at a plant is 83 lb/day. If the chemical inventory in stock is 2600 lb, how many days supply is this?

Solution:

$$\text{Days supply in inventory} = \frac{\text{Total chemical in inventory (lb)}}{\text{Average use (lb/day)}} = \frac{2600 \text{ lb}}{83 \text{ lb/day}} = 31.3 \text{ days}$$

24.11 BIOSOLIDS PRODUCTION AND PUMPING

24.11.1 PROCESS RESIDUALS

The wastewater unit treatment processes remove solids and biochemical oxygen demand from the wastestream before the liquid effluent is discharged to its receiving waters. What remains to be disposed of is a mixture of solids and wastes, called *process residuals*, more commonly referred to as *biosolids* (or *sludge*).

Note: *Sludge* is the commonly accepted term for wastewater residual solids; however, if wastewater sludge is used for beneficial reuse (i.e., as a soil amendment or fertilizer), it is commonly called *biosolids*. I choose to refer to process residuals as biosolids in this text.

The most costly and complex aspect of wastewater treatment can be the collection, processing, and disposal of biosolids. The quantity of biosolids produced may be as high as 2% of the original volume of wastewater, depending somewhat on the treatment process being used. Because the biosolids can be as much as 97% water content and because cost of disposal will be related to the volume of biosolids being processed, one of the primary purposes or goals (along with stabilizing it so it is no longer objectionable or environmentally damaging) of biosolids treatment is to separate as much of the water from the solids as possible.

24.11.2 PRIMARY AND SECONDARY SOLIDS PRODUCTION CALCULATIONS

It is important to point out that when making calculations pertaining to solids and biosolids, the term *solids* refers to dry solids and the term *biosolids* refers to the solids and water. The solids produced during primary treatment depend on the solids that settle in, or are removed by, the primary clarifier. When making primary clarifier solids production calculations, we use the mg/L to lb/day equation for suspended solids as shown below:

$$\text{SS removed (lb/day)} = \text{SS removed (mg/L)} \times \text{Flow (MGD)} \times 8.34 \text{ lb/gal} \qquad (24.113)$$

24.11.3 PRIMARY CLARIFIER SOLIDS PRODUCTION CALCULATIONS

■ EXAMPLE 24.82

Problem: A primary clarifier receives a flow of 1.80 MGD with suspended solids concentrations of 340 mg/L. If the clarifier effluent has a suspended solids concentration of 180 mg/L, how many pounds of solids are generated daily?

Solution:

$$\text{SS removed (lb/day)} = \text{SS removed (mg/L)} \times \text{Flow (MGD)} \times 8.34 \text{ lb/gal}$$

$$= 160 \text{ mg/L} \times 1.80 \text{ MGD} \times 8.34 \text{ lb/gal}$$

$$= 2402 \text{ lb/day}$$

■ EXAMPLE 24.83

Problem: The suspended solids content of the primary influent is 350 mg/L and the primary influent is 202 mg/L. How many pounds of solids are produced during a day when the flow is 4,150,000 gpd?

Solution:

$$\text{SS removed (lb/day)} = \text{SS removed (mg/L)} \times \text{Flow (MGD)} \times 8.34 \text{ lb/gal}$$

$$= 148 \text{ mg/L} \times 4.15 \text{ MGD} \times 8.34 \text{ lb/gal}$$

$$= 5122 \text{ lb/day}$$

24.11.4 SECONDARY CLARIFIER SOLIDS PRODUCTION CALCULATION

Solids produced during secondary treatment depend on many factors, including the amount of organic matter removed by the system and the growth rate of the bacteria (Figure 24.8). Because precise calculations of biosolids production is complex, we use a rough estimate method of solids production that uses an estimated growth rate (unknown) value. We use the BOD removed (lb/day) equation shown below:

$$\text{BOD removed (lb/day)} = \text{BOD removed (mg/L)} \times \text{Flow (MGD)} \times 8.34 \text{ lb/day} \quad (24.114)$$

■ EXAMPLE 24.84

Problem: The 1.5-MGD influent to the secondary system has a BOD concentration of 174 mg/L. The secondary effluent contains 22 mg/L BOD. If the bacteria growth rate for this plant is 0.40 lb SS per lb BOD removed, how many pounds of dry biosolids solids are produced each day by the secondary system?

Solution:

$$\text{BOD removed (lb/day)} = \text{BOD removed (mg/L)} \times \text{Flow (MGD)} \times 8.34 \text{ lb/day}$$

$$= 152 \text{ mg/L} \times 1.5 \text{ MGD} \times 8.34 \text{ lb/gal} = 1902 \text{ lb/day}$$

Then use the unknown *x* value to determine lb/day solids produced:

$$\frac{0.44 \text{ lb SS produced}}{1 \text{ lb BOD removed}} = \frac{x \text{ lb SS produced}}{1902 \text{ lb/day BOD removed}}$$

$$\frac{0.44 \times 1902}{1} = x$$

$$837 \text{ lb/day solids produced} = x$$

Key Point: Typically, for every pound of food consumed (BOD removed) by the bacteria, between 0.3 and 0.7 lb of new bacteria cells are produced; these are solids that have to be removed from the system.

24.11.5 PERCENT SOLIDS CALCULATIONS

Biosolids are composed of water and solids. The vast majority of biosolids is water, usually in the range of 93 to 97%. To determine the solids content of biosolids, a sample of the biosolids is dried overnight in an oven at 103 to 105°F. The solids that remain after drying represent the total solids content of the biosolids. Solids content may be expressed as a percent or as a mg/L. Either of two equations is used to calculate percent solids:

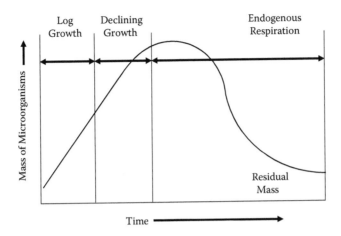

FIGURE 24.8 Bacteria growth curve.

$$\% \text{ Solids} = \frac{\text{Total solids (g)}}{\text{Biosolids sample (g)}} \times 100 \qquad (24.115)$$

$$\% \text{ Solids} = \frac{\text{Solids (lb/day)}}{\text{Biosolids (lb/day)}} \times 100 \qquad (24.116)$$

■ **EXAMPLE 24.85**

Problem: The total weight of a biosolids sample (sample only, not the dish) is 22 g. If the weight of the solids after drying is 0.77 g, what is the percent total solids of the biosolids?

Solution:

$$\% \text{ Solids} = \frac{\text{Total solids (g)}}{\text{Biosolids sample (g)}} \times 100 = \frac{0.77 \text{ g}}{22 \text{ g}} \times 100 = 3.5\%$$

24.11.6 BIOSOLIDS PUMPING CALCULATIONS

While on shift, wastewater operators are often required to make various process control calculations. An important calculation involves biosolids pumping. The biosolids pumping calculations the operator may be required to make are covered in this section.

24.11.6.1 Estimating Daily Biosolids Production

The calculation for estimation of the required biosolids pumping rate provides a method to establish an initial pumping rate or to evaluate the adequacy of the current withdrawal rate:

$$\text{Estimated pump rate} = \frac{(\text{Influent TSS conc.} - \text{Effluent TSS conc.}) \times \text{Flow} \times 8.34}{\% \text{ Solids in sludge} \times 8.34 \times 1440 \text{ min/day}} \qquad (24.117)$$

■ **EXAMPLE 24.86**

Problem: The biosolids withdrawn from the primary settling tank contain 1.4% solids. The unit influent contains 285 mg/L TSS, and the effluent contains 140 mg/L TSS. If the influent flow rate is 5.55 MGD, what is the estimated biosolids withdrawal rate in gallons per minute (assuming the pump operates continuously)?

Solution:

$$\text{Biosolids rate} = \frac{(285 \text{ mg/L} - 140 \text{ mg/L}) \times 5.55 \times 8.34}{0.014 \times 8.34 \times 1440 \text{ min/day}} = 40 \text{ gpm}$$

24.11.6.2 Biosolids Production in Pounds per Million Gallons

A common method of expressing biosolids production is in pounds of biosolids per million gallons of wastewater treated.

$$\text{Biosolids (lb/MG)} = \frac{\text{Total biosolids production (lb)}}{\text{Total wastewater flow (MG)}} \qquad (24.118)$$

■ **EXAMPLE 24.87**

Problem: Records show that a plant has produced 85,000 gal of biosolids during the past 30 days. The average daily flow for this period was 1.2 MGD. What was the plant's biosolids production in pounds per million gallons?

Solution:

$$\text{Biosolids} = \frac{\text{Total biosolids production (lb)}}{\text{Total wastewater flow (MG)}} = \frac{85,000 \text{ gal} \times 8.34 \text{ lb/gal}}{1.2 \text{ MGD} \times 30 \text{ days}} = 19,692 \text{ lb/MG}$$

24.11.6.3 Biosolids Production in Wet Tons per Year

Biosolids production can also be expressed in terms of the amount of biosolids (water and solids) produced per year. This is normally expressed in wet tons per year:

$$\frac{\text{Biosolids}}{\text{(wet tons/yr)}} = \frac{\text{Biosolids prod. (lb/MG)} \times \text{average daily flow (MGD)} \times 365 \text{ days/yr}}{2000 \text{ lb/ton}} \qquad (24.119)$$

■ **EXAMPLE 24.88**

Problem: The plant is currently producing biosolids at the rate of 16,500 lb/MG. The current average daily wastewater flow rate is 1.5 MGD. What will be the total amount of biosolids produced per year in wet tons per year?

Solution:

$$\frac{\text{Biosolids}}{\text{(wet tons/yr)}} = \frac{\text{Biosolids prod. (lb/MG)} \times \text{Average daily flow (MGD)} \times 365 \text{ days/yr}}{2000 \text{ lb/ton}}$$

$$= \frac{16,500 \text{ lb/MG} \times 1.5 \text{ MGD} \times 365 \text{ days/yr}}{2000 \text{ lb/ton}}$$

$$= 4517 \text{ wet tons/yr}$$

24.11.6.4 Biosolids Pumping Time

The biosolids pumping time is the total time the pump operates during a 24-hr period in minutes:

$$\text{Pump operating time} = \text{Time/cycle (min)} \times \text{Frequency (cycles/day)} \qquad (24.120)$$

The following information is used for Examples 24.89 to 24.94:

Frequency = 24 times/day
Pump rate = 120 gpm
Solids = 3.70%
Volatile matter = 66%

■ EXAMPLE 24.89

Problem: What is the pump operating time?

Solution:
$$\text{Pump operating time} = 15 \text{ min/hr} \times 24 \text{ cycles/day} = 360 \text{ min/day}$$

24.11.6.5 Biosolids Pumped per Day in Gallons

$$\text{Biosolids (gpd)} = \text{Operating time (min/day)} \times \text{Pump rate (gpm)} \qquad (24.121)$$

■ EXAMPLE 24.90

Problem: What is the biosolids pumped per day in gallons?

Solution:
$$\text{Biosolids (gpd)} = 360 \text{ min/day} \times 120 \text{ gpm} = 43,200 \text{ gpd}$$

24.11.6.6 Biosolids Pumped Per day in Pounds

$$\text{Biosolids (lb/day)} = \text{Gallons of biosolids pumped} \times 8.34 \text{ lb/gal} \qquad (24.122)$$

■ EXAMPLE 24.91

Problem: What is the biosolids pumped per day in pounds?

Solution:
$$\text{Biosolids (lb/day)} = 43,200 \text{ gpd} \times 8.34 \text{ lb/gal} = 360,000 \text{ lb/day}$$

24.11.6.7 Solids Pumped per Day in Pounds

$$\text{Solids pumped (lb/day)} = \text{Biosolids pumped (lb/day)} \times \% \text{ Solids} \qquad (24.123)$$

■ EXAMPLE 24.92

Problem: What are the solids pumped per day?

Solution:
$$\text{Solids pumped (lb/day)} = 360,300 \text{ lb/day} \times 0.0370 = 13,331 \text{ lb/day}$$

24.11.6.8 Volatile Matter Pumped per Day in Pounds

$$\text{Volatile matter (lb/day)} = \text{Solids pumped (lb/day)} \times \% \text{ Volatile matter} \qquad (24.124)$$

■ **EXAMPLE 24.93**

Problem: What is the volatile matter in pounds per day?

Solution:

$$\text{Volatile matter (lb/day)} = 13{,}331 \text{ lb/day} \times 0.66 = 8798 \text{ lb/day}$$

24.11.6.9 Pounds of Solids per Pounds of Volatile Solids per Day

If we wish to calculate the pounds of solids or the pounds of volatile matter removed per day, the individual equations demonstrated above can be combined into single calculations:

$$\text{Solids (lb/day)} = \text{Pump time (min/cycle)} \times \text{Frequency (cycles/day)}$$
$$\times \text{Rate (gpm)} \times 8.34 \text{ lb/gal} \times \% \text{Solids} \tag{24.125a}$$

$$\text{Volatile matter (lb/day)} = \text{Pump time (min/cycle)} \times \text{Frequency (cycles/day)}$$
$$\times \text{Rate (gpm)} \times 8.34 \text{ lb/gal} \times \% \text{Solids} \times \% \text{Volatile matter} \tag{24.125b}$$

■ **EXAMPLE 24.94**

$$\text{Solids} = 15 \text{ min/cycle} \times 24 \text{ cycles/day} \times 120 \text{ gpm} \times 8.34 \text{ lb/gal} \times 0.0370 = 13{,}331 \text{ lb/day}$$

$$\text{Volatile matter} = 15 \text{ min/cycle} \times 24 \text{ cycles/day} \times 120 \text{ gpm} \times 8.34 \text{ lb/gal} \times 0.0370 \times 0.66 = 8798 \text{ lb/day}$$

24.12 BIOSOLIDS THICKENING

Biosolids thickening (or concentration) is a unit process used to increase the solids content of the biosolids by removing a portion of the liquid fraction. In other words, biosolids thickening is all about volume reduction. Increasing the solids content allows more economical treatment of the biosolids. Biosolids thickening processes include the following:

- Gravity thickeners
- Flotation thickeners
- Solids concentrators

Biosolids thickening calculations are based on the concept that the solids in the primary or secondary biosolids are equal to the solids in the thickened biosolids. The solids are the same. It is primarily water that has been removed to thicken the biosolids and result in a higher percent solids. In this unthickened biosolids, the solids might represent 1 to 4% of the total pounds of biosolids, but when some of the water is removed those same solids might represent 5 to 7% of the total pounds of biosolids.

Note: The key to biosolids thickening calculations is that solids remain constant.

24.12.1 GRAVITY/DISSOLVED AIR FLOTATION THICKENER CALCULATIONS

Biosolids thickening calculations are based on the concept that the solids in the primary or secondary biosolids are equal to the solids in the thickened biosolids; that is, assuming a negligible amount of solids are lost in the thickener overflow, the solids are the same. Note that the water is removed to thicken the biosolids and results in higher percent solids.

24.12.1.1 Estimating Daily Biosolids Production

The calculation for estimation of the required biosolids pumping rate provides a method to establish an initial pumping rate or to evaluate the adequacy of the current pump rate:

$$\text{Estimated pump rate} = \frac{(\text{Influent TSS conc.} - \text{Effluent TSS conc.}) \times \text{Flow} \times 8.34}{\% \text{ Solids in biosolids} \times 8.34 \times 1440 \text{ min/day}} \quad (24.126)$$

■ EXAMPLE 24.95

Problem: The biosolids withdrawn from the primary settling tank contain 1.5% solids. The unit influent contains 280 mg/L TSS, and the effluent contains 141 mg/L TSS. If the influent flow rate is 5.55 MGD, what is the estimated biosolids withdrawal rate in gallons per minute (assuming that the pump operates continuously)?

Solution:

$$\text{Biosolids withdrawal rate} = \frac{(280 \text{ mg/L} - 141 \text{ mg/L}) \times 5.55 \text{ MGD} \times 8.34}{0.015 \times 8.34 \times 1440 \text{ min/day}} = 36 \text{ gpm}$$

24.12.1.2 Surface Loading Rate (gal/day/ft²)

The surface loading rate (surface settling rate) is hydraulic loading—the amount of biosolids applied per square foot of gravity thickener:

$$\text{Surface loading (gal/day/ft}^2) = \frac{\text{Biosolids applied (gpd)}}{\text{Thickener area (ft}^2)} \quad (24.127)$$

■ EXAMPLE 24.96

Problem: A 70-ft-diameter gravity thickener receives 32,000 gpd of biosolids. What is the surface loading in gallons per square foot per day?

Solution:

$$\text{Surface loading} = \frac{32,000 \text{ gpd}}{0.785 \times 70 \text{ ft} \times 70 \text{ ft}} = 8.32 \text{ gpd/ft}^2$$

24.12.1.3 Solids Loading Rate (lb/day/ft²)

The solids loading rate is the pounds of solids per day being applied to 1 ft² of tank surface area. The calculation uses the surface area of the bottom of the tank. It assumes that the floor of the tank is flat and has the same dimensions as the surface:

$$\text{Solids loading rate (lb/day/ft}^2) = \frac{\% \text{ Biosolids solids} \times \text{Biosolids flow (gpd)} \times 8.34 \text{ lb/gal}}{\text{Thickener area (ft}^2)} \quad (24.128)$$

■ EXAMPLE 24.97

Problem: The thickener influent contains 1.6% solids. The influent flow rate is 39,000 gpd. The thickener is 50 ft in diameter and 10 ft deep. What is the solid loading in pounds per day?

Solution:

$$\text{Solids loading rate} = \frac{0.016 \times 39,000 \text{ gpd} \times 8.34 \text{ lb/gal}}{0.785 \times 50 \text{ ft} \times 50 \text{ ft}} = 2.7 \text{ lb/day/ft}^2$$

24.12.2 Concentration Factor Calculations

The concentration factor (CF) represents the increase in concentration resulting from the thickener; it is a means of determining the effectiveness of the gravity thickening process:

$$CF = \frac{\text{Thickened biosolids concentration (\%)}}{\text{Influent biosolids concentration (\%)}} \qquad (24.129)$$

■ EXAMPLE 24.98

Problem: The influent biosolids contain 3.5% solids. The thickened biosolids solids concentration is 7.7%. What is the concentration factor?

Solution:

$$CF = \frac{7.7\%}{3.5\%} = 2.2$$

24.12.3 Air-to-Solids Ratio Calculations

The air-to-solids ratio is the ratio of the pounds of solids entering the thickener to the pounds of air being applied:

$$\text{Air/solids ratio} = \frac{\text{Air flow (ft}^3/\text{min)} \times 0.0785 \text{ lb/ft}^3}{\text{Biosolids flow (gpm)} \times \% \text{ Solids} \times 8.34 \text{ lb/gal}} \qquad (24.130)$$

■ EXAMPLE 24.99

Problem: The biosolids pumped to the thickener are comprised of 0.85% solids. The airflow is 13 cfm. What is the air-to-solids ratio if the current biosolids flow rate entering the unit is 50 gpm?

Solution:

$$\text{Air/solids ratio} = \frac{13 \text{ ft}^3/\text{min} \times 0.0785 \text{ lb/ft}^3}{50 \text{ gpm} \times 0.0085 \times 8.34 \text{ lb/gal}} = 0.28$$

24.12.4 Recycle Flow in Percent Calculations

The amount of recycle flow can be expressed as a percent:

$$\% \text{ Recycle} = \frac{\text{Recycle flow rate (gpm} \times 100)}{\text{Sludge flow (gpm)}} \qquad (24.131)$$

■ EXAMPLE 24.100

Problem: The sludge flow to the thickener is 80 gpm. The recycle flow rate is 140 gpm. What is the percent recycle?

Solution:

$$\% \text{ Recycle} = \frac{140 \text{ gpm} \times 100}{80 \text{ gpm}} = 175\%$$

24.12.5   Centrifuge Thickening Calculations

A centrifuge exerts a force on the biosolids thousands of times greater than gravity. Sometimes polymer is added to the influent of the centrifuge to help thicken the solids. The two most important factors that affect the centrifuge are the volume of the biosolids put into the unit (gpm) and the pounds of solids put in. The water that is removed is the *centrate*. Normally, hydraulic loading is measured as flow rate per unit of area; however, because of the variety of sizes and designs, hydraulic loading to centrifuges does not include area considerations. It is expressed only as gallons per hour. The equations to be used if the flow rate to the centrifuge is given as gallons per day or gallons per minute are

$$\text{Hydraulic loading (gph)} = \frac{\text{Flow (gpd)}}{24 \text{ hr/day}} \qquad (24.132)$$

$$\text{Hydraulic loading (gph)} = \frac{\text{Flow (gpm)} \times 60 \text{ min}}{\text{hr}} \qquad (24.133)$$

■ EXAMPLE 24.101

Problem: A centrifuge receives a waste activated biosolids flow of 40 gpm. What is the hydraulic loading on the unit in gal/hr?

Solution:

$$\text{Hydraulic loading} = \frac{40 \text{ gpm} \times 60 \text{ min}}{\text{hr}} = 2400 \text{ gph}$$

■ EXAMPLE 24.102

Problem: A centrifuge receives 48,600 gal of biosolids daily. The biosolids concentration before thickening is 0.9%. How many pounds of solids are received each day?

Solution:

$$\frac{48,600 \text{ gal}}{\text{day}} \times \frac{8.34 \text{ lb/gal}}{\text{gal}} \times \frac{0.9}{100} = 3648 \text{ lb/day}$$

24.13   STABILIZATION

24.13.1   Biosolids Digestion

A major problem in designing wastewater treatment plants is disposing of biosolids into the environment without causing damage or nuisance. Untreated biosolids are even more difficult to dispose of. Untreated raw biosolids must be stabilized to minimize disposal problems. In many cases, the term *stabilization* is considered synonymous with digestion.

Note: The stabilization of organic matter is accomplished biologically using variety of organisms. The microorganisms convert the colloidal and dissolved organic matter into various gases and into protoplasm. Because protoplasm has a specific gravity slightly higher than that of water, it can be removed from the treated liquid by gravity.

Biosolids digestion is a process in which biochemical decomposition of the organic solids occurs; in the decomposition process, the organics are converted into simpler and more stable substances. Digestion also reduces the total mass or weight of biosolids solids, destroys pathogens, and makes it easier to dry or dewater the biosolids. Well-digested biosolids have the appearance and characteristics of a rich potting soil. Biosolids may be digested under aerobic or anaerobic conditions. Most large municipal wastewater treatment plants use anaerobic digestion. Aerobic digestion finds application primarily in small activated biosolids treatment systems.

24.13.2 AEROBIC DIGESTION PROCESS CONTROL CALCULATIONS

The purpose of aerobic digestion is to stabilize organic matter, to reduce volume, and to eliminate pathogenic organisms. Aerobic digestion is similar to the activated biosolids process. Biosolids are aerated for 20 days or more. Volatile solids are reduced by biological activity.

24.13.2.1 Volatile Solids Loading (lb/ft³/day)

Volatile solids (organic matter) loading for the aerobic digester is expressed in pounds of volatile solids entering the digester per day per cubic foot of digester capacity:

$$\text{Volatile solids loading (lb/day/ft}^3) = \frac{\text{Volatile solids added (lb/day)}}{\text{Digester volume (ft}^3)} \tag{24.134}$$

■ EXAMPLE 24.103

Problem: An aerobic digester is 20 ft in diameter and has an operating depth of 20 ft. The biosolids that are added to the digester daily contain 1500 lb of volatile solids. What is the volatile solids loading in pounds per day per cubic foot?

Solution:

$$\text{Volatile solids loading} = \frac{1500 \text{ lb/day}}{0.785 \times 20 \text{ ft} \times 20 \text{ ft} \times 20 \text{ ft}} = 0.24 \text{ lb/day/ft}^3$$

24.13.2.2 Digestion Time (Days)

The theoretical time the biosolids remain in the aerobic digester is calculated by

$$\text{Digestion time (days)} = \frac{\text{Digester volume (gal)}}{\text{Biosolids added (gpd)}} \tag{24.135}$$

■ EXAMPLE 24.104

Problem: The digester volume is 240,000 gal. Biosolids are added to the digester at the rate of 15,000 gpd. What is the digestion time in days?

Solution:

$$\text{Digestion time} = \frac{240,000 \text{ gal}}{15,000 \text{ gpd}} = 16 \text{ days}$$

24.13.2.3 pH Adjustment

In many instances, the pH of the aerobic digester will fall below the levels required for good biological activity. When this occurs, the operator must perform a laboratory test to determine the amount of alkalinity required to raise the pH to the desired level. The results of the lab test must then be converted to the actual quantity required by the digester:

$$\text{Chemical required (lb)} = \frac{\text{Lab test chemical (mg)} \times \text{Digester volume} \times 3.785}{\text{Sample volume (L)} \times 454 \text{ g/lb} \times 1000 \text{ mg/g}} \quad (24.136)$$

■ **EXAMPLE 24.105**

Problem: 240 mg of lime will increase the pH of a 1-L sample of the aerobic digester contents to pH 7.1. The digester volume is 240,000 gal. How many pounds of lime will be required to increase the digester pH to 7.3?

Solution:

$$\text{Lime required} = \frac{240 \text{ mg} \times 240,000 \text{ gal} \times 3.785 \text{ L/gal}}{1 \text{ L} \times 454 \text{ g/lb} \times 1000 \text{ mg/g}} = 480 \text{ lb}$$

24.13.3 Aerobic Tank Volume Calculations

The aerobic tank volume can be computed, where no significant nitrification will occur, by the following equation (WPCF, 1985):

$$V = \frac{Q_i (X_i + YS_i)}{X \left(K_d P_v + 1/\theta_c \right)} \quad (24.137)$$

where
V = Volume of aerobic digester (ft³).
Q_i = Influent average flow rate to digester (ft³/d).
X_i = Influent suspended solids concentration (mg/L).
Y = Fraction of the influent BOD consisting of raw primary sludge (in decimals).
S_i = Influent BOD (mg/L).
X = Digester suspended solids concentration (mg/L).
K_d = Reaction-rate constant (per day).
P_v = Volatile fraction of digester suspended solids (in decimals).
θ_c = Solids retention time (days).

■ **EXAMPLE 24.106**

Problem: The pH of an aerobic digester is found to have declined to 6.1. How much sodium hydroxide must be added to raise the pH to 7.0? The volume of the digester is 370 m³. Results from jar tests show that 34 mg of caustic soda will raise the pH to 7.0 in a 2-L jar.

Solution:

$$\text{NaOH required} = (34 \text{ mg}) \div (2 \text{ L}) = 17 \text{ mg/L} = 17 \text{ g/m}^3$$

$$\text{NaOH to be added} = 17 \text{ g/m}^3 \times 370 \text{ m}^3 = 6290 \text{ g} = 6.3 \text{ kg} = 13.9 \text{ lb}$$

24.13.4 Anaerobic Digestion Process Control Calculations

The purpose of anaerobic digestion is the same as aerobic digestion: to stabilize organic matter, to reduce volume, and to eliminate pathogenic organisms. Equipment used in anaerobic digestion includes an anaerobic digester of either the floating or fixed cover type. These include biosolids pumps for biosolids addition and withdrawal, as well as heating equipment such as heat exchangers, heaters and pumps, and mixing equipment for recirculation. Typical ancillaries include gas storage, cleaning equipment, and safety equipment such as vacuum relief and pressure relief devices, flame traps, and explosion-proof electrical equipment. In the anaerobic process, biosolids enter the sealed digester where organic matter decomposes anaerobically. Anaerobic digestion is a two-stage process:

1. Sugars, starches, and carbohydrates are converted to volatile acids, carbon dioxide, and hydrogen sulfide.
2. Volatile acids are converted to methane gas.

Key anaerobic digestion process control calculations are covered in the sections that follow.

24.13.4.1 Required Seed Volume in Gallons

$$\text{Seed volume (gal)} = \text{Digester volume (gal)} \times \% \text{ Seed} \qquad (24.138)$$

■ **EXAMPLE 24.107**

Problem: The new digester requires 25% seed to achieve normal operation within the allotted time. If the digester volume is 280,000 gal, how many gallons of seed material will be required?

Solution:

$$\text{Seed volume (gal)} = 280,000 \times 0.25 = 70,000 \text{ gal}$$

24.13.4.2 Volatile Acids to Alkalinity Ratio

The ratio of volatile acids to alkalinity can be used to control an anaerobic digester:

$$\text{Ratio} = \frac{\text{Volatile acids concentration}}{\text{Alkalinity concentration}} \qquad (24.139)$$

■ **EXAMPLE 24.108**

Problem: The digester contains 240 mg/L volatile acids and 1840 mg/alkalinity. What is the volatile acids/alkalinity ratio?

Solution:

$$\text{Volatile acids/alkalinity ratio} = \frac{240 \text{ mg/L}}{1840 \text{ mg/L}} = 0.13$$

Key Point: Increases in the ratio normally indicate a potential change in the operating condition of the digester.

24.13.4.3 Biosolids Retention Time

The length of time the biosolids remain in the digester is calculated as

$$\text{Retention time} = \frac{\text{Digester volume (gal)}}{\text{Biosolids volume added (gpd)}} \qquad (24.140)$$

■ **EXAMPLE 24.109**

Problem: Biosolids are added to a 520,000-gal digester at the rate of 12,600 gal/day. What is the biosolids retention time?

Solution:

$$\text{Biosolids retention time} = \frac{520{,}000 \text{ gal}}{12{,}600 \text{ gpd}} = 41.3 \text{ days}$$

24.13.4.4 Estimated Gas Production in Cubic Feet/Day

The rate of gas production is normally expressed as the volume of gas (ft^3) produced per pound of volatile matter destroyed. The total cubic feet of gas a digester will produce per day can be calculated by

$$\text{Gas production } (ft^3/\text{day}) = \text{Volatile matter in } (lb/\text{day}) \times \% \text{ Volatile matter reduction} \tag{24.141}$$
$$\times \text{Production rate } (ft^3/lb)$$

Key Point: Multiplying the volatile matter added to the digester per day by the percent volatile matter reduction (in decimal percent) gives the amount of volatile matter being destroyed by the digestion process per day.

■ EXAMPLE 24.110

Problem: The digester reduces 11,500 lb of volatile matter per day. Currently, the volatile matter reduction achieved by the digester is 55%. The rate of gas production is 11.2 ft^3 of gas per pound of volatile matter destroyed.

Solution:

$$\text{Gas production } (ft^3/\text{day}) = 11{,}500 \text{ lb/day} \times 0.55 \times 11.2 \text{ } ft^3/lb = 70{,}840 \text{ } ft^3/\text{day}$$

24.13.4.5 Volatile Matter Reduction (Percent)

Because of the changes occurring during biosolids digestion, the calculation used to determine percent volatile matter reduction is more complicated:

$$\% \text{ Reduction} = \frac{(\%VM_{in} - \%VM_{out}) \times 100}{[\%VM_{in} - (\%VM_{in} \times \%VM_{out})]} \tag{24.142}$$

■ EXAMPLE 24.111

Problem: Using the digester data provided here, determine the percent volatile matter reduction for the digester with raw biosolids volatile matter of 71% and digested biosolids volatile matter of 54%.

Solution:

$$\% \text{ Volatile matter reduction} = \frac{0.71 - 0.54}{0.71 - (0.71 \times 0.54)} = 52\%$$

24.13.4.6 Percent Moisture Reduction in Digested Biosolids

$$\%\text{Moisture reduction} = \frac{(\%\text{Moisture}_{in} - \%\text{Moisture}_{out}) \times 100}{[\%\text{Moisture}_{in} - (\%\text{Moisture}_{in} \times \%\text{Moisture}_{out})]} \tag{24.143}$$

Key Point: % Moisture = 100% – percent solids.

■ **EXAMPLE 24.112**

Problem: Using the digester data provided below, determine the percent moisture reduction and percent volatile matter reduction for the digester.

Raw biosolids
  % Solids   9%
  % Moisture   91% (100% – 9%)
Digested biosolids
  % Solids   15%
  % Moisture   85% (100% – 15%)

Solution:

$$\%\text{Moisture reduction} = \frac{(0.91-0.85)\times100}{0.91-(0.91\times0.85)} = 44\%$$

24.13.4.7 Gas Production

When measuring the performance of a digester, gas production is one of the most important parameters. Typically, gas production ranges from 800 to 1125 L of digester gas per kilogram volatile solids destroyed. Gas produced from a properly operated digester contains approximately 68% methane and 32% carbon dioxide. If carbon dioxide exceeds 35%, the digestion system is operating incorrectly. The quantity of methane gas produced can be calculated in SI and British units, respectively, by the following equations(McCarty, 1964):

$$V = 350\left[Q(S_o - S)/(1000 \text{ g/kg}) - 1.42P_x\right] \tag{24.144}$$

$$V = 5.62\left[Q(S_o - S)8.34 - 1.42P_x\right] \tag{24.145}$$

where
  V = Volume of methane produced at standard conditions (0°C, 32°F and 1 atm), L/d or ft³/d).
  350 = Theoretical conversion factor for the amount of methane produced per kg of ultimate BOD oxidized (L/kg).
  5.62 = Theoretical conversion factor for the amount of methane produced per lb of ultimate BOD oxidized (ft³/lb).
  Q = Flow rate (m³/day, MGD).
  S_o = Influent ultimate BOD (mg/L).
  S = Effluent ultimate BOD (mg/L).
  8.34 = Conversion factor (lb/MG/mg/L).
  P_x = Net mass of cell tissue produced (kg/day, lb/day).

For a complete-mix, high-rate, two-stage anaerobic digester (without recycle), the mass of biological solids synthesized daily (P_x) can be estimated by the following equations (in SI and British system units, respectively):

$$P_x = \frac{Y[Q(S_o - S)]}{1 + K_d\theta_c} \tag{24.146}$$

$$P_x = \frac{Y[Q(S_o - S)8.34]}{1 + K_d\theta_c} \tag{24.147}$$

where
Y = Yield coefficient (kg/kg, lb/lb).
K_d = Endogenous coefficient (per day).
θ_c = Mean cell residence time (days).

Other terms are as defined previously.

■ EXAMPLE 24.113

Problem: Determine the amount of methane generated per kilogram of ultimate BOD stabilized. Use glucose $(C_6H_{12}O_6)$ as BOD and the following information:

Molecular weight of glucose = 180
Molecular weight of methane and carbon dioxide = 48
48/180 = 0.267
Oxidation of methane and carbon dioxide and water = 1.07 kg

Solution: Calculate the rate of the amount of methane generated per kg of BOD converted.

$$\frac{0.267}{1.07} = \frac{0.25}{1.0}$$

Thus, 0.25 kg of methane is produced by each kilogram of BOD stabilized. Now calculate the volume equivalent of 0.25 kg of methane at the standard conditions (0°C and 1 atm):

$$\text{Volume} = (0.25 \times 1000 \text{ g})(1 \text{ mol}/16 \text{ g})(22.4 \text{ L/mol}) = 350 \text{ L}$$

24.14 BIOSOLIDS DEWATERING AND DISPOSAL

24.14.1 BIOSOLIDS DEWATERING

The process of removing enough water from a liquid biosolids to change its consistency to that of damp solid is called *biosolids dewatering*. Although the process is also referred to as *biosolids drying*, the dry or dewatered biosolids may still contain a significant amount of water, often as much as 70%. But, at moisture contents of 70% or less, the biosolids no longer behave as a liquid and can be handled manually or mechanically. Several methods are available to dewater biosolids. The particular types of dewatering techniques or devices used best describe the actual processes used to remove water from biosolids and change their form from a liquid to damp solid. The commonly used techniques and devices include the following:

- Filter presses
- Vacuum filtration
- Sand drying beds

Note: Centrifugation is also used in the dewatering process; however, in this text we concentrate on those unit processes listed above that are traditionally used for biosolids dewatering.

An ideal dewatering operation would capture all of the biosolids at minimum cost and the resultant dry biosolids solids or cake would be capable of being handled without causing unnecessary problems. Process reliability, ease of operation, and compatibility with the plant environment would also be optimized.

24.14.2   Pressure Filtration Calculations

In pressure filtration, the liquid is forced through the filter media by a positive pressure. Several types of presses are available, but the most commonly used types are plate and frame presses and belt filter presses.

24.14.2.1   Plate and Frame Press

The plate and frame press consists of vertical plates that are held in a frame and that are pressed together between a fixed and moving end. A cloth filter medium is mounted on the face of each individual plate. The press is closed, and the biosolids are pumped into the press at pressures up to 225 psi. They then pass through feed holes in the trays along the length of the press. Filter presses usually required a precoat material, such as incinerator ash or diatomaceous earth, to aid in solids retention on the cloth and to allow easier release of the cake. Performance factors for plate and frame presses include feed biosolids characteristics, type and amount of chemical conditioning, operating pressures, and the type and amount of precoat. Filter press calculations (and other dewatering calculations) typically used in wastewater solids handling operations include solids loading rate, net filter yield, hydraulic loading rate, biosolids feed rate, solids loading rate, flocculant feed rate, flocculant dosage, total suspended solids, and percent solids recovery.

24.14.2.1.1   Solids Loading Rate

The solids loading rate is a measure of the lb/hr solids applied per square foot of plate area, as shown in Equation 24.148:

$$\text{Solids loading rate (lb/hr/ft}^2) = \frac{\text{Biosolids (gph)} \times 8.34 \text{ lb/gal} \times (\%\text{Solids}/100)}{\text{Plate area (ft}^2)} \qquad (24.148)$$

■ **EXAMPLE 24.114**

Problem: A filter press used to dewater digested primary biosolids receives a flow of 710 gal during a 2-hr period. The biosolids has a solids content of 3.3%. If the plate surface area is 120 ft^2, what is the solids loading rate in lb/hr/ft^2?

Solution: The flow rate is given as gallons per 2 hours. First express this flow rate as gallons per hour:

$$710 \text{ gal/2 hr} = 355 \text{ gal/hr}$$

Then,

$$\text{Solids loading rate} = \frac{\text{Biosolids (gph)} \times 8.34 \text{ lb/gal} \times (\%\text{Solids}/100)}{\text{Plate area (ft}^2)}$$

$$= \frac{355 \text{ gph} \times 8.34 \text{ lb/gal} \times (3.3/100)}{120 \text{ ft}^2}$$

$$= 0.81 \text{ lb/hr/ft}^2$$

Key Point: The solids loading rate measures the lb/hr of solids applied to each sq ft of plate surface area. However, this does not reflect the time when biosolids feed to the press is stopped.

24.14.2.1.2   Net Filter Yield

Operated in the batch mode, biosolids are fed to the plate and frame filter press until the space between the plates is completely filled with solids. The flow of biosolids to the press is then stopped and the plates are separated, allowing the biosolids cake to fall into a hopper or conveyor below. The net filter yield, measured in lb/hr/ft², reflects the runtime as well as the downtime of the plate and frame filter press. To calculate the net filter yield, simply multiply the solids loading rate (in lb/hr/ft²) by the ratio of filter runtime to total cycle time as follows:

$$\text{Net filter yield} = \frac{\text{Filter run time}}{\text{Total cycle time}} \qquad (24.149)$$

■ EXAMPLE 24.115

Problem: A plate and frame filter press receives a flow of 660 gal of biosolids during a 2-hr period. The solids concentration of the biosolids is 3.3%. The surface area of the plate is 110 ft². If the downtime for biosolids cake discharge is 20 min, what is the net filter yield in lb/hr/ft²?

Solution: First, calculate solids loading rate, then multiply that number by the corrected time factor:

$$\text{Solids loading rate} = \frac{\text{Biosolids (gph)} \times 8.34 \text{ lb/gal} \times (\% \text{ solids}/100)}{\text{Plate area (ft}^2)}$$

$$= \frac{330 \text{ gph} \times 8.34 \text{ lb/gal} \times (3.3/100)}{100 \text{ ft}^2}$$

$$= 0.83 \text{ lb/hr/ft}^2$$

Next, calculate net filter yield, using the corrected time factor:

$$\text{Net filter yield (lb/hr/ft}^2) = \frac{0.83 \text{ lb/hr/ft}^2 \times 2 \text{ hr}}{2.33 \text{ hr}} = 0.71 \text{ lb/hr/ft}^2$$

24.14.2.2 Belt Filter Press

The belt filter press (Figure 24.9) consists of two porous belts. The biosolids are sandwiched between the two porous belts. The belts are pulled tight together as they are passed around a series of rollers to squeeze water out of the biosolids. Polymer is added to the biosolids just before it gets to the unit. The biosolids are then distributed across one of the belts to allow for some of the water to drain by gravity. The belts are then put together with the biosolids in between.

24.14.2.2.1 Hydraulic Loading Rate

Hydraulic loading for belt filters is a measure of gpm flow per foot or belt width:

$$\text{Hydraulic loading rate (gpm/ft)} = \frac{\text{Flow (gpm)}}{\text{Belt width (ft)}} \qquad (24.150)$$

■ EXAMPLE 24.116

Problem: A 6-ft-wide belt press receives a flow of 110 gpm of primary biosolids. What is the hydraulic loading rate in gpm/ft?

Solution:

$$\text{Hydraulic loading rate (gpm/ft)} = \frac{\text{Flow (gpm)}}{\text{Belt width (ft)}} = \frac{110 \text{ gpm}}{6 \text{ ft}} = 18.3 \text{ gpm/ft}$$

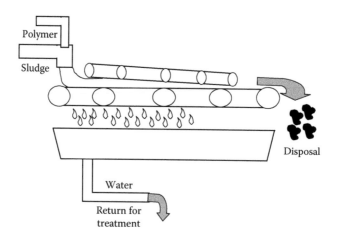

FIGURE 24.9   Belt filter press.

■ **EXAMPLE 24.117**

Problem: A belt filter press 5 ft wide receives a primary biosolids flow of 150 gpm. What is the hydraulic loading rate in gpm/ft²?

Solution:

$$\text{Hydraulic loading rate (gpm/ft)} = \frac{\text{Flow (gpm)}}{\text{Belt width (ft)}} = \frac{150 \text{ gpm}}{5 \text{ ft}} = 30 \text{ gpm/ft}$$

24.14.2.2.2   Biosolids Feed Rate

The biosolids feed rate to the belt filter press depends on several factors, including the biosolids (lb/day) that must be dewatered, the maximum solids feed rate (lb/hr) that will produce an acceptable cake dryness, and the number of hours per day the belt press is in operation. The equation used in calculating biosolids feed rate is

$$\text{Biosolids feed rate (lb/hr)} = \frac{\text{Dewatered biosolids (lb/day)}}{\text{Operating time (hr/day)}} \qquad (24.151)$$

■ **EXAMPLE 24.118**

Problem: The amount of biosolids to be dewatered by the belt filter press is 20,600 lb/day. If the belt filter press is to be operated 10 hr each day, what should the biosolids feed rate in lb/hr be to the press?

Solution:

$$\text{Biosolids feed rate (lb/hr)} = \frac{\text{Dewatered biosolids (lb/day)}}{\text{Operating time (hr/day)}} = \frac{20,600 \text{ lb/day}}{10 \text{ hr/day}} = 2060 \text{ lb/hr}$$

24.14.2.2.3   Solids Loading Rate

The solids loading rate may be expressed as lb/hr or as tons/hr. In either case, the calculation is based on biosolids flow (or feed) to the belt filter press and percent of mg/L concentration of total suspended solids (TSS) in the biosolids. The equation used in calculating solids loading rate is

$$\text{Solids loading rate (lb/hr)} = \text{Feed (gpm)} \times 60 \text{ (min/hr)} \times 8.34 \text{ lb/gal} \times (\%\text{TSS}/100) \qquad (24.152)$$

■ EXAMPLE 24.119

Problem: The biosolids feed to a belt filter press is 120 gpm. If the total suspended solids concentration of the feed is 4%, what is the solids loading rate in lb/hr?

Solution:

$$\text{Solids loading rate (lb/hr)} = \text{Feed (gpm)} \times 60 \text{ (min/hr)} \times 8.34 \text{ lb/gal} \times (\%TSS/100)$$

$$= 120 \text{ gpm} \times 60 \text{ min/hr} \times 8.34 \text{ lb/gal} \times (4/100) = 2402 \text{ lb/hr}$$

24.14.2.2.4 Flocculant Feed Rate

The flocculant feed rate may be calculated like all other mg/L to lb/day calculations and then converted to a lb/hr feed rate, as follows:

$$\text{Flocculant feed (lb/day)} = \frac{\text{Flocculant (mg/L)} \times \text{feed rate (MGD)} \times 8.34 \text{ lb/gal}}{24 \text{ hr/day}} \quad (24.153)$$

■ EXAMPLE 24.120

Problem: The flocculant concentration for a belt filter press is 1% (10,000 mg/L). If the flocculant feed rate is 3 gpm, what is the flocculant feed rate in lb/hr?

Solution: First calculate lb/day flocculant using the mg/L to lb/day calculation. Note that the gpm feed flow must be expressed as MGD feed flow:

$$\frac{3 \text{ gpm} \times 1440 \text{ min/day}}{1,000,000} = 0.00432 \text{ MGD}$$

$$\text{Flocculant feed (lb/day)} = \text{Flocculant (mg/L)} \times \text{feed rate (MGD)} \times 8.34 \text{ lb/gal}$$

$$= 10,000 \text{ mg/L} \times 0.00432 \text{ MGD} \times 8.34 \text{ lb/gal}$$

$$= 360 \text{ lb/day}$$

Then convert lb/day flocculant to lb/hr:

$$\frac{360 \text{ lb/day}}{24 \text{ hr/day}} = 15 \text{ lb/hr}$$

24.14.2.2.5 Flocculant Dosage

When the solids loading rate (tons/hr) and flocculant feed rate (lb/hr) have been calculated, the flocculant dose in lb/ton can be determined. The equation used to determine flocculant dosage is

$$\text{Flocculant dosage (lb/ton)} = \frac{\text{Flocculant (lb/hr)}}{\text{Solids treated (ton/hr)}} \quad (24.154)$$

■ EXAMPLE 24.121

Problem: A belt filter has solids loading rate of 3100 lb/hr and a flocculant feed rate of 12 lb/hr. Calculate the flocculant dose in lb per ton of solids treated.

Solution: First convert lb/hr solids loading to tons/hr solids loading:

$$\frac{3100 \text{ lb/hr}}{2000 \text{ lb/ton}} = 1.55 \text{ tons/hr}$$

Now calculate pounds flocculant per ton of solids treated:

$$\text{Flocculant dosage (lb/ton)} = \frac{\text{Flocculant (lb/hr)}}{\text{Solids treated (ton/hr)}} = \frac{12 \text{ lb/hr}}{1.55 \text{ tons/hr}} = 7.8 \text{ lb/ton}$$

24.14.2.2.6 Total Suspended Solids

The feed biosolids solids are comprised of two types of solids: suspended solids and dissolved solids. *Suspended solids* will not pass through a glass fiber filter pad. Suspended solids can be further classified as total suspended solids (TSS), volatile suspended solids, or fixed suspended solids and can also be separated into three components based on settling characteristics: settleable solids, floatable solids, and colloidal solids. Total suspended solids in wastewater is normally in the range of 100 to 350 mg/L. *Dissolved solids* will pass through a glass fiber filter pad. Dissolved solids can also be classified as total dissolved solids (TDS), volatile dissolved solids, and fixed dissolved solids. Total dissolved solids are normally in the range of 250 to 850 mg/L.

Two lab tests can be used to estimate the total suspended solids concentration of the feed biosolids to the filter press: *total residue test* (measures both suspended and dissolved solids concentrations) and *total filterable residue test* (measures only the dissolved solids concentration). By subtracting the total filterable residue from the total residue, we obtain the total nonfilterable residue (total suspended solids), as shown in Equation 24.155:

Total residue (mg/L) – Total filterable residue (mg/L) = Total nonfilterable residue (mg/L) (24.155)

■ EXAMPLE 24.122

Problem: Lab tests indicate that the total residue portion of a feed biosolids sample is 22,000 mg/L. The total filterable residue is 720 mg/L. On this basis, what is the estimated total suspended solids concentration of the biosolids sample?

Solution:

Total residue (mg/L) – Total filterable residue (mg/L) = Total nonfilterable residue (mg/L)

22,000 mg/L – 720 mg/L = 21,280 mg/L Total suspended solids

24.14.3 Rotary Vacuum Filter Dewatering Calculations

The *rotary vacuum filter* (Figure 24.10) is a device used to separate solid material from liquid. The vacuum filter consists of a large drum with large holes in it covered with a filter cloth. The drum is partially submerged and rotated through a vat of conditioned biosolids. This filter is capable of excellent solids capture and high-quality supernatant or filtrate; solids concentrations of 15 to 40% can be achieved.

24.14.3.1 Filter Loading

The filter loading for vacuum filters is a measure of lb/hr of solids applied per square foot of drum surface area. The equation to be used in this calculation is shown below:

$$\text{Filter loading (lb/hr/ft}^2) = \frac{\text{Solids to filter (lb/hr)}}{\text{Surface area (ft}^2)}$$ (24.156)

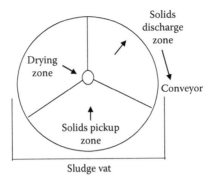

FIGURE 24.10   Vacuum Filter.

■ **EXAMPLE 24.123**

Problem: Digested biosolids are applied to a vacuum filter at a rate of 70 gpm, with a solids concentration of 3%. If the vacuum filter has a surface area of 300 ft², what is the filter loading in lb/hr/ft²?

Solution:

$$\text{Filter loading rate (lb/hr/ft}^2) = \frac{\text{Biosolids (gpm)} \times 60 \text{ min/hr} \times 8.34 \text{ lb/gal} \times (\% \text{ solids}/100)}{\text{Surface area (ft}^2)}$$

$$= \frac{70 \text{ gpm} \times 60 \text{ min/hr} \times 8.34 \text{ lb/gal} \times (3/100)}{300 \text{ ft}^2}$$

$$= 3.5 \text{ lb/hr/ft}^2$$

24.14.3.2   Filter Yield

One of the most common measures of vacuum filter performance is filter yield. It is the lb/hr of dry solids in the dewatered biosolids (cake) discharged per square foot of filter area. It can be calculated using Equation 24.157:

$$\text{Filter yield (lb/hr/ft}^2) = \frac{\text{Wet cake flow (lb/hr)} \times (\% \text{ Solids in cake}/100)}{\text{Filter area (ft}^2)} \qquad (24.157)$$

■ **EXAMPLE 24.124**

Problem: The wet cake flow from a vacuum filter is 9000 lb/hr. If the filter areas is 300 ft² and the percent solids in the cake is 25%, what is the filter yield in lb/hr/ft²?

Solution:

$$\text{Filter yield (lb/hr/ft}^2) = \frac{\text{Wet cake flow (lb/hr)} \times (\% \text{ Solids in cake}/100)}{\text{Filter area (ft}^2)}$$

$$= \frac{9000 \text{ lb/hr} \times (25/100)}{300 \text{ ft}^2} = 7.5 \text{ lb/hr/ft}^2$$

24.14.3.3   Vacuum Filter Operating Time

Use Equation 24.157 to calculate the vacuum filter operating time required to process a given lb/day solids. The vacuum filter operating time is the unknown factor, designated by *x*.

■ **EXAMPLE 24.125**

Problem: A total of 4000 lb/day primary biosolids solids is to be processed by a vacuum filter. The vacuum filter yield is 2.2 lb/hr/ft². The solids recovery is 95%. If the area of the filter is 210 ft², how many hours per day must the vacuum filter remain in operation to process these solids?

Solution:

$$\text{Filter yield (lb/hr/ft}^2\text{)} = \frac{\left(\dfrac{\text{Solids to filter (lb/day)}}{\text{Filter operation (lb/day)}}\right)}{\text{Filter area (ft}^2\text{)}} \times \frac{\% \text{ Recovery}}{100}$$

$$2.2 \text{ lb/hr/ft}^2 = \frac{4000 \text{ lb/day}}{x \text{ hr/day}} \times \frac{1}{210 \text{ ft}^2} \times \frac{95}{100}$$

$$x = \frac{4000 \times 1 \times 95}{2.2 \times 210 \times 100}$$

$$x = 8.2 \text{ hr/day}$$

24.14.3.4　Percent Solids Recovery

As mentioned, the function of the vacuum filtration process is to separate the solids from the liquids in the biosolids being processed; therefore, the percent of feed solids recovered (sometimes referred to as the *percent solids capture*) is a measure of the efficiency of the process. Equation 24.158 is used to determine percent solids recovery:

$$\% \text{ Solids recovery} = \left[\frac{\text{Wet cake flow (lb/hr)} \times (\% \text{ Solids in cake}/100)}{\text{Biosolids feed (lb/hr)} \times (\% \text{ Solids in feed}/100)}\right] \times 100 \quad (24.158)$$

■ **EXAMPLE 24.126**

Problem: The biosolids feed to a vacuum is 3400 lb/day, with a solids content of 5.1%. If the wet cake flow is 600 lb/hr with a 25% solids content, what is the percent solids recovery?

Solution:

$$\% \text{ Solids recovery} = \left[\frac{\text{Wet cake flow (lb/hr)} \times (\% \text{ solids in cake}/100)}{\text{Biosolids feed (lb/hr)} \times (\% \text{ solids in feed}/100)}\right] \times 100$$

$$= \frac{600 \text{ lb/hr} \times (25/100)}{3400 \text{ lb/hr} \times (5.1/100)} \times 100 = \frac{150 \text{ lb/hr}}{173 \text{ lb/hr}} \times 100 = 87\%$$

24.14.4　SAND DRYING BED CALCULATIONS

Drying beds are generally used for dewatering well-digested biosolids. Biosolids drying beds consist of a perforated or open joint drainage system in a support media, usually gravel or wire mesh. Drying beds are usually separated into workable sections by wood, concrete, or other materials. Drying beds may be enclosed or opened to the weather. They may rely entirely on natural drainage and evaporation processes or may use a vacuum to assist the operation. *Sand drying beds* are the oldest biosolids dewatering technique. They consist of 6 to 12 in. of coarse sand underlain by layers of graded gravel ranging from 1/8 to 1/4 in. at the top and 3/4 to 1-1/2 in. at the bottom. The total gravel thickness is typically about 1 ft. Graded natural earth (4 to 6 in.) usually makes up the bottom with a web of drain tile placed on 20- to 30-ft centers. Sidewalls and partitions between bed sections are usually made of wooden planks or concrete and extend about 14 in. above the sand surface. Typically, three calculations are used to monitor sand drying bed performance: total biosolids applied, solids loading rate, and biosolids withdrawal to drying beds.

24.14.4.1 Total Biosolids Applied

The total gallons of biosolids applied to sand drying beds may be calculated using the dimensions of the bed and depth of biosolids applied, as shown by Equation 24.159:

$$\text{Volume (gal)} = \text{Length (ft)} \times \text{Width (ft)} \times \text{Depth (ft)} \times 7.48 \text{ gal/ft}^3 \qquad (24.159)$$

■ EXAMPLE 24.127

Problem: A drying bed is 220 ft long and 20 ft wide. If biosolids are applied to a depth of 4 in., how many gallons of biosolids are applied to the drying bed?

Solution:

$$\text{Volume} = \text{Length (ft)} \times \text{width (ft)} \times \text{depth (ft)} \times 7.48 \text{ gal/ft}^3$$

$$= 220 \text{ ft} \times 20 \text{ ft} \times 0.33 \text{ ft} \times 7.48 \text{ gal/ft}^3$$

$$= 10,861 \text{ gal}$$

24.14.4.2 Solids Loading Rate

The solids loading rate may be expressed as lb/yr/ft^2. The loading rate is dependent on biosolids applied per application (lb), percent solids concentration, cycle length, and square feet of sand bed area. The equation for solids loading rate is given below:

$$\text{Solids loading rate (lb/yr/ft}^2) = \frac{\left[\left(\dfrac{\text{Biosolids applied (lb)}}{\text{Days of application}}\right) \times (365 \text{ days/yr}) \times (\% \text{ Solids/100})\right]}{\text{Length (ft)} \times \text{Width (ft)}} \qquad (24.160)$$

■ EXAMPLE 24.128

Problem: A biosolids bed is 210 ft long and 25 ft wide. A total of 172,500 lb of biosolids is applied during each application of the sand drying bed. The biosolids have a solids content of 5%. If the drying and removal cycle requires 21 days, what is the solids loading rate in lb/yr/ft^2?

Solution:

$$\text{Solids loading rate (lb/yr/ft}^2) = \frac{\left[\left(\dfrac{\text{Biosolids applied (lb)}}{\text{Days of application}}\right) \times \dfrac{365 \text{ days}}{\text{yr}} \times \dfrac{\% \text{ solids}}{100}\right]}{\text{Length (ft)} \times \text{width (ft)}}$$

$$= \frac{\left(\dfrac{172,500 \text{ lb}}{21 \text{ days}} \times \dfrac{365 \text{ days}}{\text{yr}} \times \dfrac{5}{100}\right)}{210 \text{ ft} \times 25 \text{ ft}}$$

$$= 37.5 \text{ lb/yr/ft}^2$$

24.14.4.3 Biosolids Withdrawal to Drying Beds

Pumping digested biosolids to drying beds is one method among many for dewatering biosolids, thus making the dried biosolids useful as a soil conditioner. Depending on the climate of a region, the drying bed depth may range from 8 to 18 in. The area covered by these drying beds may be substantial. For this reason, the use of drying beds is more common for smaller plants than for larger plants. When calculating biosolids withdrawal to drying beds, use Equation 24.161:

$$\text{Biosolids withdrawn (ft}^3) = 0.785 \times (\text{Diameter})^2 \times \text{Drawdown (ft)} \qquad (24.161)$$

■ EXAMPLE 24.129

Problem: Biosolids are withdrawn from a digester that has a diameter of 40 ft. If the biosolids are drawn down 2 ft, how many ft^3 will be sent to the drying beds?

Solution:

$$\text{Biosolids withdrawn (ft}^3) = 0.785 \times (\text{Diameter})^2 \times \text{Drawdown (ft)}$$

$$= 0.785 \times (40 \text{ ft} \times 40 \text{ ft}) \times 2 \text{ ft}$$

$$= 2512 \text{ ft}^3 \text{ withdrawn}$$

24.14.5 BIOSOLIDS DISPOSAL CALCULATIONS

In the disposal of biosolids, land application, in one form or another, has become not only a necessity (because of the banning of ocean dumping in the United States in 1992 and the shortage of landfill space since then) but also quite popular as a beneficial reuse practice. Beneficial reuse means that the biosolids are disposed of in an environmentally sound manner by recycling nutrients and soil conditions. The application of biosolids is occurring throughout the United States on agricultural and forest lands. For use in land applications, the biosolids must meet certain conditions. Biosolids must comply with state and federal biosolids management and disposal regulations and must also be free of materials dangerous to human health (e.g., toxins, pathogenic organisms) or dangerous to the environment (e.g., pesticides, heavy metals). Biosolids can be land applied by direct injection, by application, by incorporation (plowing in), or by composting.

24.14.5.1 Land Application

Land application of biosolids requires precise control to avoid problems. Use of process control calculations is part of the overall process control process. Calculations include determining disposal cost, plant available nitrogen (PAN), application rates (dry tons and wet tons/acre), metals loading rates, maximum allowable applications based on metals loading, and site life based on metals loading.

24.14.5.1.1 Disposal Cost

The cost of disposal of biosolids can be determined by using Equation 24.162:

$$\text{Cost} = \text{Wet tons biosolids produced per year} \times \% \text{ Solids} \times \text{Cost per dry ton} \qquad (24.162)$$

■ EXAMPLE 24.130

Problem: The treatment system produces 1925 wet tons of biosolids for disposal each year. The biosolids are comprised of 18% solids. A contractor disposes of the biosolids for $28 per dry ton. What is the annual cost for biosolids disposal?

Solution:

$$\text{Cost} = 1925 \text{ wet tons per year} \times 0.18 \times \$28/\text{dry ton} = \$9702$$

24.14.5.1.2 Plant Available Nitrogen

One factor considered when land applying biosolids is the amount of nitrogen in the biosolids available to the plants grown on the site. This includes ammonia nitrogen and organic nitrogen. The organic nitrogen must be mineralized for plant consumption. Only a portion of the organic nitrogen is mineralized per year. The mineralization factor (f^1) is assumed to be 0.20. The amount of ammonia nitrogen available is directly related to the time elapsed between applying the biosolids and incorporating (plowing) the biosolids into the soil. Volatilization rates are presented in the example below:

$$\text{PAN (lb/dry ton)} = \left[\begin{array}{l} \left(\text{Organic nitrogen (mg/kg)} \times f^1 \right) \\ + \left(\text{Ammonia nitrogen (mg/kg)} \times V_1 \right) \end{array} \right] \times 0.002 \text{ lb/dry ton} \qquad (24.163)$$

where
   PAN = Plant available nitrogen.
   f^1 = Mineral rate for organic nitrogen (assume 0.20).
   V_1 = Volatilization rate ammonia nitrogen.
      = 1.00 if biosolids are injected.
      = 0.85 if biosolids are plowed in within 24 hr.
      = 0.70 if biosolids are plowed in within 7 days.

■ EXAMPLE 24.131

Problem: Biosolids contain 21,000 mg/kg of organic nitrogen and 10,500 mg/kg of ammonia nitrogen. The biosolids is incorporated into the soil within 24 hr after application. What is the plant available nitrogen (PAN) per dry ton of solids?

Solution:

$$\text{PAN} = \left[(21,000 \text{ mg/kg} \times 0.20) + (10,500 \times 0.85) \right] \times 0.002 = 26.3 \text{ lb/dry ton}$$

24.14.5.1.3 Application Rate Based on Crop Nitrogen Requirement

In most cases, the application rate of domestic biosolids to crop lands will be controlled by the amount of nitrogen the crop requires. The biosolids application rate based on the nitrogen requirement is determined by the following:

1. Use an agriculture handbook to determine the nitrogen requirement of the crop.
2. Determine the amount of biosolids in dry tons required to provide this much nitrogen:

$$\text{Dry tons/ac} = \frac{\text{Plant nitrogen requirement (lb/ac)}}{\text{Plant available nitrogen (lb/dry ton)}} \qquad (24.164)$$

■ EXAMPLE 24.132

Problem: The crop to be planted on the land application site requires 150 lb of nitrogen per acre. What is the required biosolids application rate if the PAN of the biosolids is 30 lb/dry ton?

Solution:

$$\text{Dry tons/ac} = \frac{150 \text{ lb/ac}}{30 \text{ lb/dry ton}} = 5 \text{ dry tons/ac}$$

24.14.5.1.4   Metals Loading

When biosolids are land applied, metals concentrations are closely monitored and the loading on land application sites is calculated:

$$\text{Loading (lb/ac)} = \text{Metal conc. (mg/kg)} \times 0.002 \text{ lb/dry ton} \times \text{Application rate (dry tons/ac)} \quad (24.165)$$

■ **EXAMPLE 24.133**

Problem: Biosolids contain 14 mg/kg of lead and are currently being applied to the site at a rate of 11 dry tons/ac. What is the metals loading rate for lead in pounds per acre?

Solution:

$$\text{Loading rate} = 14 \text{ mg/kg} \times 0.002 \text{ lb/dry ton} \times 11 \text{ dry tons} = 0.31 \text{ lb/ac}$$

24.14.5.1.5   Maximum Allowable Applications Based on Metals Loading

If metals are present, they may limit the total number of applications a site can receive. Metals loadings are normally expressed in terms of the maximum total amount of metal that can be applied to a site during its use:

$$\text{Application} = \frac{\text{Maximum allowable cumulative load (lb/ac)}}{\text{Metal loading (lb/ac)}} \quad (24.166)$$

■ **EXAMPLE 24.134**

Problem: The maximum allowable cumulative lead loading is 48 lb/ac. Based on the current loading of 0.35 lb/ac, how many applications of biosolids can be made to this site?

Solution:

$$\text{Application} = \frac{48.0 \text{ lb/ac}}{0.35 \text{ lb/ac}} = 137$$

24.14.5.1.6   Site Life Based on Metals Loading

The maximum number of applications based on metals loading and the number of applications per year can be used to determine the maximum site life:

$$\text{Site life (yr)} = \frac{\text{Maximum allowable applications}}{\text{Number of applications per year}} \quad (24.167)$$

■ **EXAMPLE 24.135**

Problem: Biosolids are currently being applied to a site twice annually. Based on the lead content of the biosolids, the maximum number of applications is determined to be 135 applications. Based on the lead loading and the applications rate, how many years can this site be used?

Solution:

$$\text{Site life} = \frac{135 \text{ applications}}{2 \text{ applications per year}} = 68 \text{ yr}$$

Key Point: When more than one metal is present, the calculations must be performed for each metal. The site life would then be the lowest value generated by these calculations.

24.14.5.2 Biosolids to Compost

The purpose of composting biosolids is to stabilize the organic matter, reduce volume, eliminate pathogenic organisms, and produce a product that can be used as a soil amendment or conditioner. Composting is a biological process. In a composting operation, dewatered solids are usually mixed with a bulking agent (e.g., hardwood chips) and stored until biological stabilization occurs. The composting mixture is ventilated during storage to provide sufficient oxygen for oxidation and to prevent odors. After stabilization of the solids, they are separated from the bulking agent. The composted solids are then stored for curing and are applied to farm lands or other beneficial uses. Expected performance of the composting operation for both percent volatile matter reduction and percent moisture reduction ranges from 40 to 60%.

Performance factors related to biosolids composting include moisture content, temperature, pH, nutrient availability, and aeration. The biosolids must contain sufficient moisture to support the biological activity. If the moisture level is too low (40% less), biological activity will be reduced or stopped. At the same time, if the moisture level exceeds approximately 60%, it will prevent sufficient airflow through the mixture. The composting process operates best when the temperature is maintained within an operating range of 130 to 140°F; biological activities provide enough heat to increase the temperature well above this range. Forced air ventilation or mixing is used to remove heat and maintain the desired operating temperature range. The temperature of the composting solids, when maintained at the required levels, will be sufficient to remove pathogenic organisms. The influent pH can affect the performance of the process if extreme (less than 6.0 or greater than 11.0). The pH during composting may have some impact on the biological activity but does not appear to be a major factor. Composted biosolids generally have a pH in the range of 6.8 to 7.5. The critical nutrient in the composting process is nitrogen. The process works best when the ratio of nitrogen to carbon is in the range of 26 to 30 carbon to one nitrogen. Above this ratio, composting is slowed. Below this ratio, the nitrogen content of the final product may be less attractive as compost. Aeration is essential to provide oxygen to the process and to control the temperature. In forced air processes, some means of odor control should be included in the design of the aeration system.

Pertinent composting process control calculations include determination of percent of moisture of compost mixture and compost site capacity.

24.14.5.2.1 Blending Dewatered Biosolids with Composted Biosolids

Blending composted material with dewatered biosolids is similar to blending two different percent solids biosolids. The percent solids (or percent moisture) content of the mixture will always fall somewhere between the percent solids (or percent moisture) concentrations of the two materials being mixed. Equation 24.168 is used to determine percent moisture of mixture:

$$\begin{array}{c} \%\,\text{Moisture} \\ \text{of mixture} \end{array} = \frac{\left[\text{Biosolids (lb/day)} \times \left(\%\text{Moist.}/100\right)\right] + \left[\text{Compost (lb/day)} \times \left(\%\text{Moist.}/100\right)\right]}{\text{Biosolids (lb/day)} + \text{Compost (lb/day)}} \quad (24.168)$$

■ EXAMPLE 24.136

Problem: If 5000 lb/day dewatered biosolids is mixed with 2000 lb/day compost, what is the percent moisture of the blend? The dewatered biosolids has a solids content of 25% (75% moisture) and the compost has a 30% moisture content.

Solution:

$$\begin{aligned}
\frac{\%\text{Moisture}}{\text{of mixture}} &= \frac{\left[\text{Biosolids (lb/day)} \times (\%\text{Moist.}/100)\right] + \left[\text{Compost (lb/day)} \times (\%\text{Moist.}/100)\right]}{\text{Biosolids (lb/day)} + \text{Compost (lb/day)}} \\[6pt]
&= \frac{\left[5000 \text{ lb/day} \times (75/100)\right] + \left[2000 \text{ lb/day} \times (30/100)\right]}{5000 \text{ lb/day} + 2000 \text{ lb/day}} \\[6pt]
&= \frac{3750 \text{ lb/day} + 600 \text{ lb/day}}{7000 \text{ lb/day}} \\[6pt]
&= 62\%
\end{aligned}$$

24.14.5.2.2 Compost Site Capacity Calculation

An important consideration in compost operation is the solids processing capability (fill time), expressed in lb/day or lb/wk. Equation 24.169 is used to calculate site capacity:

$$\text{Fill time days} = \frac{\text{Total available capacity (yd}^3)}{\left(\dfrac{\text{Wet compost (lb/day)}}{\text{Compost bulk density (lb/yd}^3)}\right)} \tag{24.169}$$

■ EXAMPLE 24.137

Problem: A composting facility has an available capacity of 7600 yd³. If the composting cycle is 21 days, how many lb/day wet compost can be processed by this facility? Assume a compost bulk density of 900 lb/yd³.

Solution:

$$\frac{\text{Fill time}}{(\text{days})} = \frac{\text{Total available capacity (yd}^3)}{\left(\dfrac{\text{Wet compost (lb/day)}}{\text{Compost bulk density (lb/yd}^3)}\right)} = \frac{7600 \text{ yd}^3}{\left(\dfrac{x \text{ lb/day}}{900 \text{ lb/yd}^3}\right)}$$

$$21 \text{ days} = \frac{7600 \text{ yd}^3 \times 900 \text{ lb/yd}^3}{x \text{ lb/day}}$$

$$x \text{ lb/day} = \frac{7600 \text{ yd}^3 \times 900 \text{ lb/yd}^3}{21 \text{ days}}$$

$$x \text{ lb/day} = \frac{6{,}840{,}000 \text{ lb}}{21 \text{ days}}$$

$$= 325{,}714 \text{ lb/day}$$

24.15 WASTEWATER LABORATORY CALCULATIONS

24.15.1 Wastewater Lab

Waterworks and wastewater treatment plants are sized to meet current needs, as well as those of the future. No matter the size of the treatment plant, some space or area within the plant is designated as the lab area, which can range from being the size of a closet to being fully equipped and staffed environmental laboratories. Water and wastewater laboratories usually perform a number of different tests. Lab test results provide the operator with the information necessary to operate the treatment facility at optimal levels. Laboratory testing usually includes determining service line flushing time, solution concentration, pH, COD, total phosphorus, fecal coliform count, chlorine residual, and BOD (seeded), to name a few. The standard reference for performing wastewater testing is contained in *Standard Methods for Examination of Water & Wastewater* (Clesceri et al., 1999).

In this section, the focus is on standard water/wastewater lab tests that involve various calculations. Specifically, the focus is on calculations used to determine the proportioning factor for composite sampling, flow from a faucet estimation, service line flushing time, solution concentration, biochemical oxygen demand, molarity and moles, normality, settleability, settleable solids, biosolids total, fixed and volatile solids, suspended and volatile suspended solids, and biosolids volume index and biosolids density index.

24.15.2 COMPOSITE SAMPLING PROCEDURES AND CALCULATIONS

When preparing oven-baked food, the cook sets the correct oven temperature and then usually moves on to some other chore. The oven thermostat maintains the correct temperature, and that is that. Unlike the cook, in water and wastewater treatment plant operations the operator does not have the luxury of setting a plant parameter and then walking off and forgetting about it. To optimize plant operations, various adjustments to unit processes must be made on an ongoing basis.

The operator makes unit process adjustments based on local knowledge (experience) and on lab test results; however, before lab tests can be performed, samples must be taken. The two basic types of samples are grab samples and composite samples. The type of sample taken depends on the specific test, the reason the sample is being collected, and requirements in the plant discharge permit.

A *grab sample* is a discrete sample collected at one time and one location. It is primarily used for any parameter whose concentration can change quickly (e.g., dissolved oxygen, pH, temperature, total chlorine residual) and is representative only of the conditions at the time of collection.

A *composite sample* consists of a series of individual grab samples taken at specified time intervals and in proportion to flow. The individual grab samples are mixed together in proportion to the flow rate at the time the sample was collected to form the composite sample. The composite sample represents the character of the water/wastewater over a period of time. Because knowledge of the procedure used in processing composite samples is important (a basic requirement) to the water/wastewater operator, the actual procedure used is covered in this section:

1. Determine the total amount of sample required for all tests to be performed on the composite sample.
2. Determine the average daily flow of the treatment system.

Key Point: Average daily flow can be determined by using several months of data, which will provide a more representative value.

3. Calculate a proportioning factor:

$$\text{Proportioning factor (PF)} = \frac{\text{Total sample volume required (mm)}}{\text{No. of samples} \times \text{Average daily flow (MGD)}} \qquad (24.170)$$

Key Point: Round the proportioning factor to the nearest 50 units (e.g., 50, 100, 150) to simplify calculation of the sample volume.

4. Collect the individual samples in accordance with the schedule (e.g., once/hr, once/15 min).
5. Determine flow rate at the time the sample was collected.

6. Calculate the specific amount to add to the composite container:

$$\text{Required volume (mL)} = \text{Flow}_T \times \text{Proportioning factor} \qquad (24.171)$$

where T is the time the sample was collected.
7. Mix the individual sample thoroughly, measure the required volume, and add to composite storage container.
8. Keep the composite sample refrigerated throughout the collection period.

■ EXAMPLE 24.138

Problem: The effluent testing will require 3825 mL of sample. The average daily flow is 4.25 MGD. Using the flows given below, calculate the amount of sample to be added at each of the times shown:

8 a.m.	3.88 MGD
9 a.m.	4.10 MGD
10 a.m.	5.05 MGD
11 a.m.	5.25 MGD
12 noon	3.80 MGD
1 p.m.	3.65 MGD
2 p.m.	3.20 MGD
3 p.m.	3.45 MGD
4 p.m.	4.10 MGD

Solution:

$$\text{Proportioning factor (PF)} = \frac{3825 \text{ mL}}{9 \text{ samples} \times 4.25 \text{ MGD}} = 100$$

$\text{Volume}_{8\text{a.m.}} = 3.88 \times 100 = 388 \ (400) \text{ mL}$
$\text{Volume}_{9\text{a.m.}} = 4.10 \times 100 = 410 \ (410) \text{ mL}$
$\text{Volume}_{10\text{a.m.}} = 5.05 \times 100 = 505 \ (500) \text{ mL}$
$\text{Volume}_{11\text{a.m.}} = 5.25 \times 100 = 525 \ (530) \text{ mL}$
$\text{Volume}_{12\text{noon}} = 3.80 \times 100 = 380 \ (380) \text{ mL}$
$\text{Volume}_{1\text{p.m.}} = 3.65 \times 100 = 365 \ (370) \text{ mL}$
$\text{Volume}_{2\text{p.m.}} = 3.20 \times 100 = 320 \ (320) \text{ mL}$
$\text{Volume}_{3\text{p.m.}} = 3.45 \times 100 = 345 \ (350) \text{ mL}$
$\text{Volume}_{4\text{p.m.}} = 4.10 \times 100 = 410 \ (410) \text{ mL}$

24.15.3 BIOCHEMICAL OXYGEN DEMAND CALCULATIONS

Biochemical oxygen demand (BOD$_5$) measures the amount of organic matter that can be biologically oxidized under controlled conditions (5 days at 20°C in the dark). Several criteria determine which BOD$_5$ dilutions should be used for calculating test results. Consult a laboratory testing reference manual (such as *Standard Methods*) for this information. Two basic calculations are used for BOD$_5$. The first is used for samples that have not been seeded, and the second must be used whenever BOD$_5$ samples are seeded. Both methods are introduced and examples are provided below.

24.15.3.1 BOD$_5$ (Unseeded)

$$\text{BOD}_5 \text{ (unseeded)} = \frac{\left(\left[\text{DO}_{\text{start}} \text{ (mg/L)} - \text{DO}_{\text{final}} \text{ (mg/L)} \right] \times 300 \text{ mL} \right)}{\text{Sample volume (mL)}} \qquad (24.172)$$

■ EXAMPLE 24.139

Problem: A BOD_5 test has been completed. Bottle 1 of the test had dissolved oxygen (DO) of 7.1 mg/L at the start of the test. After 5 days, bottle 1 had a DO of 2.9 mg/L. Bottle 1 contained 120 mg/L of sample. Determine the unseeded BOD_5.

Solution:

$$BOD_5 \text{ (unseeded)} = \frac{(7.1 \text{ mg/L} - 2.9 \text{ mg/L}) \times 300 \text{ mL}}{120 \text{ mL}} = 10.5 \text{ mg/L}$$

24.15.3.2 BOD_5 (Seeded)

If the BOD_5 sample has been exposed to conditions that could reduce the number of healthy, active organisms, the sample must be seeded with organisms. Seeding requires the use of a correction factor to remove the BOD_5 contribution of the seed material:

$$\text{Seed correction} = \frac{\text{Seed material } BOD_5 \times \text{Seed in dilution (mL)}}{300 \text{ mL}} \quad (24.173)$$

$$BOD_5 \text{ (seeded)} = \frac{\left(\left[DO_{start} \text{ (mg/L)} - DO_{final} \text{ (mg/L)} - \text{Seed correction}\right] \times 300 \text{ mL}\right)}{\text{Sample volume (mL)}} \quad (24.174)$$

■ EXAMPLE 24.140

Problem: Using the data provided below, determine the BOD_5:

  BOD_5 of seed material = 90 mg/L
  Seed material = 3 mL
  Sample = 100 mL
  Start DO = 7.6 mg/L
  Final DO = 2.7 mg/L

Solution:

$$\text{Seed correction} = \frac{90 \text{ mg/L} \times 3 \text{ mL}}{300 \text{ mL}} = 0.90 \text{ mg/L}$$

$$BOD_5 \text{ (seeded)} = \frac{(7.6 \text{ mg/L} - 2.7 \text{ mg/L} - 0.90) \times 300}{\text{Sample volume (mL)}} = 12 \text{ mg/L}$$

24.15.3.3 BOD 7-Day Moving Average

Because the BOD characteristic of wastewater varies from day to day, even hour to hour, operational control of the treatment system is most often accomplished based on trends in data rather than individual data points. The BOD 7-day moving average is a calculation of the BOD trend.

Key Point: The 7-day moving average is called a moving average because a new average is calculated each day by adding the new day's value to the six previous days' values.

$$\text{7-day average BOD} = \frac{\left(\begin{array}{c} BOD_{day1} + BOD_{day2} + BOD_{day3} + BOD_{day4} \\ + BOD_{day5} + BOD_{day6} + BOD_{day7} \end{array}\right)}{7} \quad (24.175)$$

■ **EXAMPLE 24.141**

Problem: Given the following primary effluent BOD test results, calculate the 7-day average:

June 1	200 mg/L
June 2	210 mg/L
June 3	204 mg/L
June 4	205 mg/L
June 5	222 mg/L
June 6	214 mg/L
June 7	218 mg/L

Solution:

$$\text{7-day average BOD} = \frac{\left(200 + 210 + 204 + 205 + 222 + 214 + 218\right)}{7} = 210 \text{ mg/L}$$

24.15.4 MOLE AND MOLARITY CALCULATIONS

Chemists have defined a very useful unit called the *mole*. Moles and molarity, a concentration term based on the mole, have many important applications in water/wastewater operations. A mole is defined as a gram molecular weight; that is, the molecular weight expressed as grams. For example, a mole of water is 18 g of water, and a mole of glucose is 180 g of glucose. A mole of any compound always contains the same number of molecules. The number of molecules in a mole is called *Avogadro's number* and has a value of 6.022×10^{23}.

Note: How big is Avogadro's number? An Avogadro's number of soft drink cans would cover the surface of the Earth to a depth of over 200 miles.

Key Point: Molecular weight is the weight of one molecule. It is calculated by adding the weights of all the atoms that are present in one molecule. The units are atomic mass units (amu). A mole is a gram molecular weight—that is, the molecular weight expressed in grams. The molecular weight is the weight of 1 molecule in daltons. All moles contain the same number of molecules (Avogadro's number), equal to 6.022×10^{23}. The reason all moles have the same number of molecules is because the value of the mole is proportional to the molecular weight.

24.15.4.1 Moles

A mole is a quantity of a compound equal in weight to its formula weight; for example, the formula weight for water can be determined using the periodic table of elements (see Figure 24.11):

$$\begin{array}{r} \text{Hydrogen } (1.008) \times 2 = 2.016 \\ + \text{ Oxygen} = 16.000 \\ \hline \text{Formula weight of H}_2\text{O} = 18.016 \end{array}$$

Because the formula weight of water is 18.016, a mole is 18.016 units of weight. A *gram-mole* is 18.016 grams of water. A *pound-mole* is 18.016 pounds of water. For our purposes in this text, the term *mole* will be understood to mean gram-mole. The equation used to determine moles is shown below:

$$\text{Moles} = \frac{\text{Grams of chemical}}{\text{Formula weight of chemical}} \qquad (24.176)$$

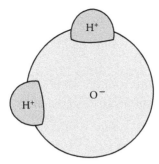

FIGURE 24.11   A molecule of water.

■ **EXAMPLE 24.142**

Problem: The atomic weight of a certain chemical is 66. If 35 g of the chemical are used in making up a 1-L solution, how many moles are used?

Solution:

$$\text{Moles} = \frac{\text{Grams of chemical}}{\text{Formula weight of chemical}} = \frac{66 \text{ g}}{35 \text{ g/mol}} = 1.9 \text{ moles}$$

The molarity of a solution is calculated by taking the moles of solute and dividing by the liters of solution:

$$\text{Molarity} = \frac{\text{Moles of solute}}{\text{Liters of solution}} \qquad (24.177)$$

■ **EXAMPLE 24.143**

Problem: What is the molarity of 2 moles of solute dissolved in 1 L of solvent?

Solution:

$$\text{Molarity} = \frac{2 \text{ moles}}{1 \text{ L}} = 2 \ M$$

Key Point: Measurement in moles is a measurement of the amount of a substance. Measurement in molarity is a measurement of the concentration of a substance—the amount (moles) per unit volume (liters).

24.15.4.2 Normality

The *molarity* of a solution refers to its concentration (the solute dissolved in the solution). The *normality* of a solution refers to the number of *equivalents* of solute per liter of solution. The definition of chemical equivalent depends on the substance or type of chemical reaction under consideration. Because the concept of equivalents is based on the reacting power of an element or compound, it follows that a specific number of equivalents of one substance will react with the same number of equivalents of another substance. When the concept of equivalents is taken into consideration, it is less likely that chemicals will be wasted as excess amounts. Keeping in mind that normality is a measure of the reacting power of a solution (i.e., 1 equivalent of a substance reacts with 1 equivalent of another substance), we use the following equation to determine normality:

$$\text{Normality } (N) = \frac{\text{No. of equivalents of solute}}{\text{Liters of solution}} \tag{24.178}$$

■ **EXAMPLE 24.144**

Problem: If 2.0 equivalents of a chemical are dissolved in 1.5 L of solution, what is the normality of the solution?

Solution:

$$\text{Normality } (N) = \frac{\text{No. of equivalents of solute}}{\text{Liters of solution}} = \frac{2.0 \text{ equivalents}}{1.5 \text{ L}} = 1.33 \, N$$

■ **EXAMPLE 24.145**

Problem: A 800-mL solution contains 1.6 equivalents of a chemical. What is the normality of the solution?

Solution: First convert 800 mL to liters:

$$800 \text{ mL} \div 1000 \text{ mL} = 0.8 \text{ L}$$

Then calculate the normality of the solution:

$$\text{Normality} = \frac{1.6 \text{ equivalents}}{0.8 \text{ L}} = 2 \, N$$

24.15.5   Settleability (Activated Biosolids Solids) Calculations

The settleability test is a test of the quality of the activated biosolids solids—or activated sludge solids (mixed liquor suspended solids). Settled biosolids volume (SBV)—or settled sludge volume (SSV)—is determined at specified times during sample testing. For control, 30- and 60-minute observations are made. Subscripts (SBV_{30} or SSV_{30} and SBV_{60} or SSV_{60}) indicate settling time. A sample of activated biosolids is taken from the aeration tank, poured into a 2000-mL graduated cylinder, and allowed to settle for 30 or 60 min. The settling characteristics of the biosolids in the graduated cylinder give a general indication of the settling of the MLSS in the final clarifier. From the settleability test, the percent settleable solids can be calculated using the following equation:

$$\% \text{ Settleable solids} = \frac{\text{Settled solids (mL)}}{\text{2000-mL sample}} \times 100 \tag{24.179}$$

■ **EXAMPLE 24.146**

Problem: The settleability test is conducted on a sample of MLSS. What is percent settleable solids if 420 mL settle in the 2000-mL graduated cylinder?

Solution:

$$\% \text{ Settleable solids} = \frac{420 \text{ mL}}{2000 \text{ mL}} \times 100 = 21\%$$

■ **EXAMPLE 24.147**

Problem: A 2000-mL sample of activated biosolids is tested for settleability. If the settled solids are measured as 410 mL, what is the percent settled solids?

Solution:

$$\% \text{ Settleable solids} = \frac{410 \text{ mL}}{2000 \text{ mL}} \times 100 = 20.5\%$$

24.15.6 SETTLEABLE SOLIDS CALCULATIONS

The settleable solids test is an easy, quantitative method to measure sediment found in wastewater. An Imhoff cone (plastic or glass 1-L cone; see Figure 24.12) is filled with 1 L of sample wastewater, stirred, and allowed to settle for 60 min. The settleable solids test, unlike the settleability test, is conducted on samples from the sedimentation tank or clarifier influent and effluent to determine percent removal of settleable solids. The percent settleable solids is determined by

$$\% \text{ Settleable solids removed} = \frac{\text{Settled solids removed (mL/L)}}{\text{Settled solids in influent (mL/L)}} \times 100 \qquad (24.180)$$

■ **EXAMPLE 24.148**

Problem: Calculate the percent removal of settleable solids if the settleable solids of the sedimentation tank influent is 15 mL/L and the settleable solids of the effluent is 0.4 mL/L.

Solution: First determine removed settleable solids:

$$15.0 \text{ mL/L} - 0.4 \text{ mL/L} = 14.6 \text{ mL/L}$$

Next, insert the parameters into Equation 24.180:

$$\% \text{ Settleable solids removed} = \frac{14.6 \text{ mL/L}}{15.0 \text{ mL/L}} \times 100 = 97\%$$

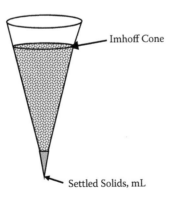

FIGURE 24.12 One-liter Imhoff cone.

■ **EXAMPLE 24.149**

Problem: Calculate the percent removal of settleable solids if the settleable solids of the sedimentation tank influent are 13 mL/L and the settleable solids of the effluent are 0.5 mL/L.

Solution: First determine removed settleable solids:

$$13 \text{ mL/L} - 0.5 \text{ mL/L} = 12.5 \text{ mL/L}$$

Next, insert the parameters into Equation 24.180:

$$\% \text{ Settleable solids removed} = \frac{12.5 \text{ mL/L}}{13.0 \text{ mL/L}} \times 100 = 96\%$$

24.15.7   BIOSOLIDS TOTAL SOLIDS, FIXED SOLIDS AND VOLATILE SOLIDS CALCULATIONS

Wastewater consists of both water and solids. The *total solids* may be further classified as either *volatile solids* (organics) or *fixed solids* (inorganics). Normally, total solids and volatile solids are expressed as percents, whereas suspended solids are generally expressed as mg/L. To calculate either percents or mg/L concentrations, certain concepts must be understood:

- *Total solids*—The residue left in the vessel after evaporation of liquid from a sample and subsequent drying in an oven at 103 to 105°C
- *Fixed solids*—The residue left in the vessel after a sample is ignited (heated to dryness at 550°C)
- *Volatile solids*—The weight loss after a sample is ignited (heated to dryness at 550°C)

Key Point: When the term *biosolids* is used, it may be understood to mean a semiliquid mass composed of solids and water. The term *solids*, however, is used to mean dry solids after the evaporation of water.

The percent total solids and percent volatile solids are calculated as follows:

$$\% \text{ Total solids} = (\text{Total solids weight/biosolids sample weight}) \times 100 \qquad (24.181)$$

$$\% \text{ Volatile solids} = (\text{Volatile solids weight/total solids weight}) \times 100 \qquad (24.182)$$

■ **EXAMPLE 24.150**

Problem: Given the information below, determine the percent solids in the sample and the percent volatile solids in the biosolids sample:

	Biosolids Sample	After Drying	After Burning (Ash)
Weight of sample and dish	73.43 g	24.88 g	22.98 g
Weight of dish (tare weight)	22.28 g	22.28 g	22.28 g

Solution: To calculate percent total solids, the grams total solids (solids after drying) and grams biosolids sample must be determined:

Total solids
  24.88 g (weight of total solids and dish)
  –22.28 g (weight of dish)
  ―――――――――――――――――――
    2.60 g (weight of total solids)

Biosolids sample
  73.43 g (weight of biosolids and dish)
  –22.28 g (weight of dish)
  ―――――――――――――――――――
  51.15 g (weight of biosolids sample)

$$\% \text{ Total solids} = \frac{\text{Total solids weight}}{\text{Biosolids sample weight}} \times 100 = \frac{2.60 \text{ g}}{51.15 \text{ g}} \times 100 = 5\%$$

To calculate the percent volatile solids, the grams total solids and grams volatile solids must be determined. Because the total solids value has already been calculated (above), only volatile solids must be calculated:

Volatile solids
  24.88 g (weight of sample and dish before burning)
  –22.98 g (weight of sample and dish after burning)
  ――――――――――――――――――――――――――――
    1.90 g (weight of solids lost in burning)

$$\% \text{ Volatile solids} = \frac{\text{Volatile solids weight}}{\text{total solids weight}} \times 100 = \frac{1.90 \text{ g}}{2.60 \text{ g}} \times 100 = 73\%$$

24.15.8 WASTEWATER SUSPENDED SOLIDS AND VOLATILE SUSPENDED SOLIDS CALCULATIONS

Total suspended solids (TSS) are the amount of filterable solids in a wastewater sample. Samples are filtered through a glass fiber filter. The filters are dried and weighed to determine the amount of total suspended solids in mg/L of sample. *Volatile suspended solids* (VSS) are those solids lost on ignition (heating to 500°C). They are useful to the treatment plant operator because they give a rough approximation of the amount of organic matter present in the solid fraction of wastewater, activated biosolids, and industrial wastes. With the exception of the required drying time, the suspended solids and volatile suspended solids tests of wastewater are similar to those of the total and volatile solids performed for biosolids. Calculations of suspended solids and volatile suspended solids are demonstrated in the example below.

Key Point: The total and volatile solids of biosolids are generally expressed as percents, by weight. The biosolids samples are 100 mL and are unfiltered.

■ EXAMPLE 24.151

Problem: Given the following information regarding a primary effluent sample, calculate (1) the mg/L suspended solids, and (2) the percent volatile suspended solids of the sample.

	After Drying (Before Burning)	After Burning (Ash)
Weight of sample and dish	24.6268 g	24.6232 g
Weight of dish (tare weight)	24.6222 g	24.6222 g

Sample volume = 50 mL

Solution: To calculate the milligrams suspended solids per liter of sample (mg/L), we must first determine grams suspended solids:

24.6268 g (weight of dish and suspended solids)
−24.6222 g (weight of dish)
_____

00.0046 g (weight of suspended solids)

Next, we calculate mg/L suspended solids (using a multiplication factor of 20, a number that will vary with sample volume) to make the denominator equal to 1 L (1000 mL):

$$\frac{0.0046 \text{ g SS}}{50 \text{ mL}} \times \frac{1000 \text{ mg}}{1 \text{ g}} \times \frac{20}{20} = \frac{92 \text{ mg}}{1000 \text{ mL}} = 92 \text{ mg/L SS}$$

To calculate percent volatile suspended solids, we must know the weight of both total suspended solids (calculated above) and volatile suspended solids:

24.6268 g (weight of dish and SS *before* burning)
−24.6232 g (weight of dish and SS *after* burning)
_____

0.0036 g (weight of solids lost in burning)

$$\% \text{ Volatile solids} = \frac{\text{Volatile solids weight}}{\text{Total solids weight}} \times 100 = \frac{0.0036 \text{ g}}{0.0046 \text{ g}} \times 100 = 78\%$$

24.15.9   BIOSOLIDS VOLUME INDEX AND BIOSOLIDS DENSITY INDEX CALCULATIONS

Two variables are used to measure the settling characteristics of activated biosolids and to determine what the return biosolids pumping rate should be. These are the *biosolids volume index* (BVI) and the *biosolids density index* (BDI):

$$\text{BVI} = \frac{\% \text{ MLSS volume after 30 min}}{\% \text{ MLSS (mg/L MLSS)}} = \text{Settled biosolids (mL)} \times 1000 \qquad (24.183)$$

$$\text{BDI} = \frac{\% \text{ MLSS}}{\% \text{ MLSS volume after 30 min}} \times 100 \qquad (24.184)$$

These indices relate the weight of biosolids to the volume the biosolids occupies. They show how well the liquid–solids separation part of the activated biosolids system is performing its function on the biological floc that has been produced and is to be settled out and returned to the aeration tanks or wasted. The better the liquid–solids separation is, the smaller will be the volume occupied by the settled biosolids and the lower the pumping rate required to keep the solids in circulation.

■ EXAMPLE 24.152

Problem: The settleability test indicates that after 30 min, 220 mL of biosolids settle in the 1-L graduated cylinder. If the mixed liquor suspended solids (MLSS) concentration in the aeration tank is 2400 mg/L, what is the biosolids volume?

Solution:

$$\text{BVI} = \frac{\text{Volume}}{\text{Density}} = \frac{220 \text{ mL/L}}{2400 \text{ mg/L}} = \frac{220 \text{ mL}}{2400 \text{ mg}} = \frac{220 \text{ mL}}{2.4 \text{ g}} = 92$$

The biosolids density index is also a method of measuring the settling quality of activated bio-solids, yet it, like the BVI parameter, may or may not provide a true picture of the quality of the biosolids in question unless compared with other relevant process parameters. It differs from the BVI in that the higher the BDI value, the better the settling quality of the aerated mixed liquor. Similarly, the lower the BDI, the poorer the settling quality of the mixed liquor. BDI is the concentration in percent solids that the activated biosolids will assume after settling for 30 min. BDI will range from 2.00 to 1.33, and biosolids with values of one or more are generally considered to have good settling characteristics. To calculate the BDI, we simply invert the numerator and denominator and multiply by 100.

■ EXAMPLE 24.153

Problem: The MLSS concentration in the aeration tank is 2500 mg/L. If the activated biosolids settleability test indicates 225 mL settled in the 1-L graduated cylinder, what is the biosolids density index?

Solution:

$$\text{BDI} = \frac{\text{Density}}{\text{Volume}} \times 100 = \frac{2500 \text{ mg}}{225 \text{ mL}} \times 100 = \frac{2.5 \text{ g}}{225 \text{ mL}} \times 100 = 1.11 \text{ mL}$$

REFERENCES AND RECOMMENDED READING

Camp, T.R. (1946). Grit chamber design. *Sewage Works Journal*, 14, 368–389.
Lawrence, A.W. and McCarty, P.L. (1970). Unified basis for biological treatment design and operation. *Journal of the Sanitary Engineering Division*, 96(3), 757–778.
Mancini, J.L. and Barnhart, E.L. (1968). Industrial waste treatment in aerated lagoons. In: Gloyna, E.R. and Eckenfelder, Jr., W.W., Eds., *Advances in Water Quality Improvement*. University of Texas Press, Austin.
Metcalf & Eddy. (1991). *Wastewater Engineering Treatment, Disposal, and Reuse*. McGraw-Hill, New York.
Russo, R. (2001). Empire Falls. Knopf, New York.
Spellman, F.R. (2010). *Spellman's Standard Handbook for Wastewater Operators*, Vol 1. CRC Press, Boca Raton, FL.
Spengel, D.B. and Dzombak, D.A. (1992). Biokinetic modeling and scale-up considerations for biological contractors. *Water Environment Research*, 64(3), 223–234.
WPCF. (1985). *Sludge Stabilization*, Manual of Practice FD-9. Water Pollution Control Federation, Alexandria, VA.

Section XII

Math Concepts:
Stormwater Engineering

"Come Watson, Come! The Game is afoot!' ..." (Doyle, 1930). Wayne County has operated an illicit Connection and Discharge Elimination Program for over 15 years. Its staff has gained valuable investigative expertise by experimenting with many different methods, committing lots of trial and error, and having a little bit of luck. Investigating for illicit discharges in the field is very similar to Holmes and Watson solving a case—it requires a mix of science, detection, deduction, and persistence.

—Dean Tuomari and Susan Thompson
Wayne County Department of Environment, Wayne, MI

As human activities alter the watershed landscape, adverse impacts to receiving waters my result from changes in the quantity and quality of stormwater runoff. Rain (and snow) falling onto the surface of unmanaged urbanizing watersheds results in a predictable increase in the quantity of runoff (and snow-melt) volume flowing to receiving waters. If left unmanaged, the hydraulic impacts (e.g., flooding, erosion, channelization) associated with the increased water volumes may be several orders of magnitude higher than that of the undisturbed watershed. In addition to causing runoff volume impacts, stormwater can also be a major nonpoint pollution source in many watersheds.

—U.S. Environmental Protection Agency

25 Stormwater Calculations

25.1 INTRODUCTION*

For the environmental practitioner involved with stormwater compliance programs, March 10, 2003, was a very significant date: It was the deadline for compliance with new National Pollutant Discharge Elimination System (NPDES) permit applications for municipal separate storm sewer systems (MS4s) that had previously been exempt, but after this date no longer are. The affected MS4s include federal and state regulated operators—serving less than 100,000 people—for areas such as military installations, prisons, hospitals, universities, and other, who are required, after March 10, 2003, to have complied with the Storm Water Phase II Rule, published December 8, 1999. Additionally, state regulators may subject certain other sources to regulations, such as municipally owned industrial sources, construction sites disturbing less than 1 acre, and sources that contribute to a significant degradation of water quality.

To comply with the new stormwater regulations, environmental engineers are called upon to design stormwater discharge control systems. In this design phase, several mathematical computations are made to ensure that the finished stormwater discharge control system meets the regulatory requirements.

This chapter provides guidelines for performing various engineering calculations associated with the design of stormwater management facilities such as extended detention and retention basins and multistage outlet structures. The prerequisite information for using these calculations is the determination of the hydrologic characteristic of the contributing watershed in the form of the peak discharge (in cubic feet per second, *cfs*), or a runoff hydrograph, depending on the hydrologic and hydraulic routing methods used. Thus, before discussing the various math computations used in engineering a stormwater discharge system, we begin by defining general stormwater terms and acronyms and present a detailed discussion of hydrologic methods.

25.2 STORMWATER TERMS AND ACRONYMS

Anti-seep collar—A device constructed around a pipe or other conduit and placed into a dam, levee, or dike for the purpose of reducing seepage losses and piping failures along the conduit it surrounds.

Anti-vortex device—A device placed at the entrance of a pipe conduit structure to help prevent swirling action and cavitation from reducing the flow capacity of the conduit system.

Aquatic bench—A 10- to 15-foot wide bench around the inside perimeter of a permanent pool that ranges in depth from zero to 12 inches. Vegetated with emergent plants, the bench augments pollutant removal, provides habitats, protects the shoreline from the effects of water fluctuations, and enhances safety.

---

* Much of the information contained in this chapter is adapted from Spellman, F.R. and Drinan, J.E., *Stormwater Discharge Management: A Practical Guide to Compliance*, Government Institutes, Lanham, MD, 2003; excerpted from *Federal and State Regulations*, Soil Conservation Service (SCS) Technical Release Nos. 20 and 55 (TR-20 and TR-55), and *Virginia Stormwater Management Handbook* (1999) and U.S. Corps of Engineers.

Aquifer—A porous, water bearing geologic formation generally restricted to materials capable of yielding an appreciable supply of water.

Atmospheric deposition—The process by which atmospheric pollutants reach the land surface either as dry deposition or as dissolved or particulate matter contained in precipitation.

Average land cover condition—The percentage of impervious cover considered to generate an equivalent amount of phosphorus as the total combined land uses within the watershed.

Bankfull flow—Condition where flow fills a stream channel to the top of bank and at a point where the water begins to overflow onto a floodplain.

Base flow—Discharge of water independent of surface runoff conditions, usually a function of groundwater levels.

Basin—A facility designed to impound stormwater runoff.

Best management practice (BMP)—Structural or nonstructural practice that is designed to minimize the impacts of changes in land use on surface and groundwater systems. Structural BMP refers to basins or facilities engineered for the purpose of reducing the pollutant load in stormwater runoff, such as Bioretention, constructed stormwater wetlands, etc. Nonstructural BMP refers to land use or development practices which are determined to be effective in minimizing the impact on receiving stream systems, such as preservation of open space and stream buffers, disconnection of impervious surfaces, etc.

Biochemical oxygen demand (BOD)—An indirect measure of the concentration of biologically degradable material present in organic wastes. It usually reflects the amount of oxygen consumed in 5 days by biological processes breaking down organic waste.

Biological processes—A pollutant removal pathway in which microbes break down organic pollutants and transform nutrients.

Bioretention basin—Water quality BMP engineered to filter the water quality volume through an engineered planting bed, consisting of a vegetated surface layer (vegetation, mulch, ground cover), planting soil, and sand bed (optional), and into the in-situ material. Also called rain gardens.

Bioretention filter—A bioretention basin with the addition of a sand layer and collector pipe system beneath the planting bed.

Catch basin—An inlet chamber usually built at the curb line of a street or low area, for collection of surface runoff and admission into a sewer or subdrain. These structures commonly have a sediment sump at its base, below the sewer or subdrain discharge elevation designed to retain solids below the point of overflow.

Channel stabilization—The introduction of natural or manmade materials placed within a channel so as to prevent or minimize the erosion of the channel bed and/or banks.

Check dam—Small dam constructed in a channel for the purpose of decreasing the flow velocity, minimizing channel scour, and promoting deposition of sediment. Check dams are a component of grassed swale BMPs.

Chemical oxygen demand (COD)—A measure of the oxygen required to oxidize all compounds, both organic and inorganic, in water.

Chute—A high-velocity, open channel for conveying water to a lower level without erosion.

COE—United States Army Corps of Engineers

Compaction—The process by which soil grains are rearranged so as to decrease void space and bring them in closer contact with one another, thereby reducing the permeability and increasing the soils unit weight, and shear and bearing strength.

Constructed stormwater wetlands—Areas intentionally designed and created to emulate the water quality improvement function of wetlands for the primary purpose of removing pollutants from stormwater.

Contour—A line representing a specific elevation on the land surface or a map.

Cradle—A structure usually made of concrete shaped to fit around the bottom and sides of a conduit to support the conduit, increase its strength and, in dams, to fill all voids between the underside of the conduit and soil.

Crest—The top of a dam, dike, spillway or weir, frequently restricted to the overflow portion.

Curve number (CN)—A numerical representation of a given area's hydrologic soil group, plant cover, impervious cover, interception and surface storage derived in accordance with Natural Resource Conservation Service methods. This number is used to convert rainfall depth into runoff volume. It is sometimes referred to as the *runoff curve number (RCN)*.

Cut—A reference to an area or material that has been excavated in the process of a grading operation.

Design storm—A selected rainfall hyetograph of specified amount, intensity, duration, and frequency that is used as a basis for design.

Detention basin—A stormwater management facility which temporarily impounds runoff and discharges it through a hydraulic outlet structure to a downstream conveyance system. While a certain amount of outflow may also occur via infiltration through the surrounding soil, such amounts are negligible when compared to the outlet structure discharge rates and, therefore, are not considered in the facility's design. Since an extended detention basin impounds runoff only temporarily, it is normally dry during nonrainfall periods.

Disturbed area—An area in which the natural vegetative soil cover or existing surface treatment has been removed or altered and, therefore, is susceptible to erosion.

Diversion—A channel or dike constructed to direct water to areas where it can be used, treated, or disposed of safely.

Drainage basin—An area of land that contributes stormwater runoff to a designated point. Also called a drainage area or, on a larger scale, a watershed.

Drop structure—A device constructed to transition water to a lower elevation.

Duration—The length of time over which precipitation occurs.

Embankment—A manmade deposit of soil, rock or other material used to form an impoundment.

Energy dissipator—A device used to reduce the velocity or turbulence of flowing water.

Erosion—The wearing away of the land surface by running water, wind, ice, or other geological agent.

- *Accelerated erosion*—Erosion in excess of what is presumed or estimated to be naturally occurring levels and which is a direct result of human activities.
- *Gully erosion*—Erosion process whereby water accumulates in narrow channels and removes the soil to depths ranging from a few inches to 1 or 2 feet to as much as 75 to 100 feet.
- *Rill erosion*—Erosion process in which numerous small channels only several inches deep are formed.
- *Sheet erosion*—Spattering of small soil particles caused by the impact of raindrops on wet soils. The loosened and spattered particles may subsequently be removed by surface runoff.

Exfiltration—The downward movement of runoff through the bottom of a stormwater facility and into the soil.

Extended detention basin—A stormwater management facility that temporarily impounds runoff and discharges it through a hydraulic outlet structure over a specified period of time to a downstream conveyance system for the purpose of water quality enhancement or stream channel erosion control. While a certain amount of outflow may also occur via infiltration through the surrounding soil, such amounts are negligible when compared to the outlet structure discharge rates and, therefore, are not considered in the facility's design. Since an extended detention basin impounds runoff only temporarily, it is normally dry during nonrainfall periods.

Extended detention basin–enhanced—An extended detention basin modified to increase pollutant removal by providing a shallow marsh in the lower stage of the basin.

Filter bed—The section of a constructed filtration device that houses the filtering media.

Filter strip—An area of vegetation, usually adjacent to a developed area, constructed to remove sediment, organic matter, and other pollutants from runoff in the form of sheet flow.

First flush—The first portion of runoff, usually defined as a depth in inches, considered to contain the highest pollutant concentration resulting from a rainfall event.

Floodplain—For a given flood event, that area of land adjoining a continuous water course which has been covered temporarily by water.

Flow splitter—An engineered hydraulic structure designed to divert a portion of storm flow to a BMP located out of the primary channel, or to direct stormwater to a parallel pipe system, or to bypass a portion of baseflow around a BMP.

Forebay—Storage space, commonly referred to as a sediment forebay, located near a stormwater BMP inlet that serves to trap incoming coarse sediments before they accumulate in the main treatment area.

Freeboard—Vertical distance between the surface elevation of the design high water and the top of a dam, levee, or diversion ridge.

Frequency (design storm frequency)—The recurrence interval of storm events having the same duration and volume. The frequency of a specified design storm can be expressed either in terms of exceedance probability or return period.

  • *Exceedance probability*—The probability that an event having a specified volume and duration will be exceeded in one time period, usually assumed to be one year. If a storm has a one- percent chance of occurring in any given year, then it has an exceedance probability of 0.01.

  • *Return period*—The average length of time between events having the same volume and duration. If a storm has a one- percent chance of occurring in any given year, then it has a return period of 100 years.

Gabion—A flexible woven wire basket composed of rectangular cells filled with large cobbles or riprap. Gabions may be assembled into many types of structures such as revetments, retaining walls, channel liners, drop structures, diversions, check dams, and groins.

Geographic Information System (GIS)—A method of overlaying spatial land and land use data of different kinds. The data are referenced to a set of geographical coordinates and encoded in a computer software system. GIS is used by many localities to map utilities and sewer lines and to delineate zoning areas.

Grassed swale—An earthen conveyance system that is broad and shallow with check dams and vegetated with erosion resistant and flood tolerant grasses, engineered to remove pollutants from stormwater runoff by filtration through grass and infiltration into the soil.

Head—The height of water above any plane or object of reference; also used to express the energy, either kinetic or potential, measured in feet, possessed by each unit weight of a liquid.

Hydraulic Engineering Circular 1 (HEC-1)—A rainfall–runoff event simulation computer model sponsored by the U.S. Corps of Engineers.

Hydric soil—A soil that is saturated, flooded, or ponded long enough during the growing season to develop anaerobic conditions in the upper part.

Hydrodynamic structure—An engineered flow-through structure that uses gravitational settling to separate sediments and oils from stormwater runoff.

Hydrograph—A plot showing the rate of discharge, depth, or velocity of flow vs. time for a given point on a stream or drainage system.

Hydrologic cycle—A continuous process by which water is cycled from the oceans to the atmosphere to the land and back to the oceans.

Hydrologic soil group (HSG)—SCS classification system of soils based on the permeability and infiltration rates of the soils. A-type soils are primarily sandy in nature with a high permeability while D-type soils are primarily clayey in nature with a low permeability.

Hyetograph—A graph of the time distribution of rainfall over a watershed.

Impervious cover—A surface composed of any material that significantly impedes or prevents natural infiltration of water into soil. Impervious surfaces include, but are not limited to, roofs, buildings, streets, parking areas, and any concrete, asphalt, or compacted gravel surface.

Impoundment—An artificial collection or storage of water, such as a reservoir, pit, dugout, sump, etc.

Industrial stormwater permit—NPDES permit issued to a commercial industry for regulating the pollutant levels associated with industrial stormwater discharges. The permit may specify on-site pollution control strategies.

Infiltration facility—A stormwater management facility which temporarily impounds runoff and discharges it via infiltration through the surrounding soil. While an infiltration facility may also be equipped with an outlet structure to discharge impounded runoff, such discharge is normally reserved for overflow and other emergency conditions. Because an infiltration facility impounds runoff only temporarily, it is normally dry during nonrainfall periods. Infiltration trench, infiltration dry well, and porous pavement are considered infiltration facilities.

Initial abstraction—The maximum amount of rainfall that can be absorbed under specific conditions without producing runoff. It is also called *initial losses.*

Intensity—The depth of rainfall divided by duration.

Invert—The lowest flow line elevation in any component of a conveyance system, including storm sewers, channels, weirs, etc.

Kjeldahl nitrogen—A measure of the ammonia and organic nitrogen present in a water sample.

Lag time—The interval between the center of mass of the storm precipitation and the peak flow of the resultant runoff.

Low impact development (LID)—Hydrologically functional site design with pollution prevention measures to reduce impacts and compensates for development impacts on hydrology and water quality.

Manning's formula—Equation used to predict the velocity of water flow in an open channel or pipeline.

Micropool—A smaller permanent pool which is incorporated into the design of larger stormwater ponds to avoid resuspension of particles, provide varying depth zones, and minimize impacts to adjacent natural features.

Modified rational method—A variation of the rational method used to calculate the critical storage volume whereby the storm duration can vary and does not necessarily equal the time of concentration.

Nonpoint source pollution—Contaminants such as sediment, nitrogen and phosphorous, hydrocarbons, heavy metals, and toxins whose sources cannot be pinpointed but rather are washed from the land surface in a diffuse manner by stormwater runoff.

Normal depth—Depth of flow in an open conduit during uniform flow for the given conditions.

Offline—Stormwater management system designed to manage a portion of the stormwater that has been diverted from a stream or storm drain. A flow splitter is typically used to divert the desired portion of the flow.

Online—Stormwater management system designed to manage stormwater in its original stream or drainage channel.

Peak discharge—The maximum rate of flow associated with a given rainfall event or channel.

Percolation rate—The velocity at which water moves through saturated, granular material.

Point source—The discernible, confined and discrete conveyance, including but not limited to, any pipe, ditch, channel, tunnel, conduit, well, container, concentrated animal feeding operation, or landfill leachate collection system from which pollutants may be discharged. This term does not include return flows from irrigated agriculture or agricultural storm water runoff.

Porosity—The ratio of pore or open space volume to total solids volume.

Principal spillway—The primary spillway or conduit for the discharge of water from an impoundment facility; generally constructed of permanent material and designed to regulate the rate of discharge.

Rational method—Means of computing peak storm drainage flow rates based on average percent imperviousness of the site, mean rainfall intensity, and drainage area.

Recharge—Replenishment of groundwater reservoirs by infiltration and transmission of water through permeable soils.

Redevelopment—Any construction, alteration, or improvement on existing development.

Retention—Permanent storage of stormwater.

Retention basin—A stormwater management facility which includes a permanent impoundment, or normal pool of water, for the purpose of enhancing water quality and, therefore, is normally wet, even during nonrainfall periods. Storm runoff inflows may be temporarily stored above this permanent impoundment for the purpose of reducing flooding or stream channel erosion.

Rip-rap—Broken rock, cobbles, or boulders placed on earth surfaces such as the face of a dam or the bank of a stream for the protection against erosive forces such as flow velocity and waves.

Riser—A vertical structure which extends from the bottom of an impoundment facility and houses the control devices (weirs/orifices) to achieve the desired rates of discharge for specific designs.

Roughness coefficient—A factor in velocity and discharge formulas representing the effect of channel roughness on energy losses in flowing water. Manning's 'n' is a commonly used roughness coefficient.

Routing—A method of measuring the inflow and outflow from an impoundment structure while considering the change in storage volume over time.

Runoff—The portion of precipitation, snow melt, or irrigation water that runs off the land into surface waters.

Runoff coefficient—The fraction of total rainfall that appears as runoff. Represented as *C* in the rational method formula.

Safety bench—A flat area above the permanent pool and surrounding a stormwater pond designed to provide a separation to adjacent slopes.

Sand filter—A contained bed of sand which acts to filter the first flush of runoff. The runoff is then collected beneath the sand bed and conveyed to an adequate discharge point or infiltrated into the in-situ soils.

SCS—Soil Conservation Service (now called Natural Resource Conservation Service, NRCS), a branch of the U.S. Department of Agriculture.

Sediment forebay—A settling basin or plunge pool constructed at the incoming discharge points of a stormwater facility.

Soil test—Chemical analysis of soil to determine the need for fertilizers or amendments for species of plants being grown.

Stage—Water surface elevation above any chosen datum.

Storm sewer—A system of pipes, separate from sanitary sewers, that only carries runoff from buildings and land surfaces.

Stormwater filtering (or filtration)—A pollutant removal method to treat stormwater runoff in which stormwater is passed through a filter media such as sand, peat, grass, compost, or other materials to strain or filter pollutants out of the stormwater.

Stormwater hot spot—An area where the land use or activities are considered to generate runoff with concentrations of pollutants in excess of those typically found in stormwater.

Stream buffers—The zones of variable width, which are located along both sides of a stream and are designed to provide a protective natural area along a stream corridor.

Surcharge—Flow condition occurring in closed conduits when the hydraulic grade line is above the crown of the sewer. This condition usually results in localized flooding or stormwater flowing out the top of inlet structures and manholes.

SWMM (Storm Water Management Model)—Rainfall-runoff event simulation model sponsored by the USEPA.

Technical Release No. 20 (TR-20), Project Formulation: Hydrology—SCS watershed hydrology computer model that is used to compute runoff volumes and route storm events through stream valleys and/or impoundments.

Technical Release No. 55 (TR-55), Urban Hydrology for Small Watersheds—SCS watershed hydrology computation model that is used to calculate runoff volumes and provide a simplified routing for storm events through stream valleys and/or ponds.

Time of concentration—The time required for water to flow from the hydrologic most distant point (in time of flow) of the drainage area to the point of analysis (outlet). This time will vary, generally depending on the slope and character of the surfaces.

Trash rack—A structural device used to prevent debris from entering a spillway or other hydraulic structure.

Travel time—The time required for water to flow from the outlet of a drainage sub-basin to the outlet of the entire drainage basin being analyzed. Travel time is normally concentrated flow through an open or closed channel.

Ultimate condition—Full watershed build-out based on existing zoning.

Ultra-urban—Densely developed urban areas in which little pervious surface exists.

Urban runoff—Stormwater from city streets and adjacent domestic or commercial properties that carries nonpoint source pollutants into the sewer systems and receiving waters.

Water quality window—The volume equal to the first 1/2 inch of runoff multiplied by the impervious surface of the land development project.

Water surface profile—Longitudinal profile assumed by the surface of a stream flowing in an open channel; hydraulic grade line.

Water table—Upper surface of the free groundwater in a zone of saturation.

Watershed—A defined land area drained by a river, stream, or drainage way, or system of connecting rivers, streams, or drainage ways such that all surface water within the area flows through a single outlet.

Wet weather flow—Combination of dry weather flows and stormwater runoff.

Wetted perimeter—The length of the wetted surface of a natural or artificial channel.

25.3 HYDROLOGIC METHODS

Hydrology is the study of the properties, distribution, and effects of water on the earth's surface, and in the soils, underlying rocks, and atmosphere. The hydrologic cycle (see Figure 25.1) is the closed loop through which water travels as it moves from one phase, or surface, to another. Water, lost from the Earth's surface to the atmosphere either by evaporation from the surface of lakes, rivers, and oceans or through the transpiration of plants, forms clouds that condense to deposit moisture on the land and sea. A drop of water may travel thousands of miles between the time it evaporates and the time it falls to Earth again as rain, sleet, or snow. The water that collects on land flows to the ocean in streams and rivers of seeps into the earth, joining groundwater. Even groundwater eventually flows toward the ocean for recycling (see Figure 25.1). When humans intervene in the natural water cycle, they generate artificial water cycles or urban water cycles (local subsystems of the water cycle, or integrated water cycles; see Figure 25.2) (Spellman and Drinan, 2000).

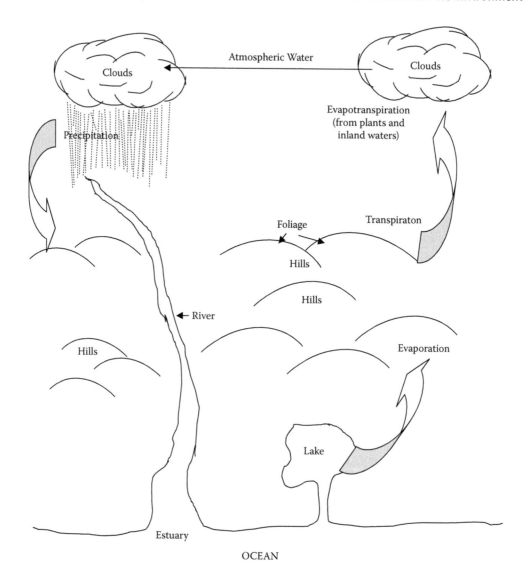

FIGURE 25.1 Natural water cycle.

The hydrologic cycle is complex, and to simulate just a small portion of it, such as the relationship between precipitation and surface runoff, can be an inexact science. Many variables and dynamic relationships must be accounted for and, in most cases, reduced to basic assumptions. However, these simplifications and assumptions make it possible to develop solutions to the flooding, erosion, and water quality impacts associated with changes in land cover and hydrologic characteristics.

Proposed engineering solutions typically involve identifying a storm frequency as a benchmark for controlling these impacts. The 2-year, 10-year, and 100-year frequency storms have traditionally been used for hydrological modeling, followed by an engineered solution designed to offset increased peak flow rates. The hydraulic calculations inherent in this process are dependent upon the engineer's ability to predict the amount of rainfall and its intensity. Recognizing that the frequency of a specific rainfall depth or duration is developed from a statistical analysis of historical rainfall data, the engineer cannot presume to accurately predict the characteristics of a future storm event. This section provides guidance for preparing acceptable calculations for various elements of the hydrologic and hydraulic analysis of a watershed.

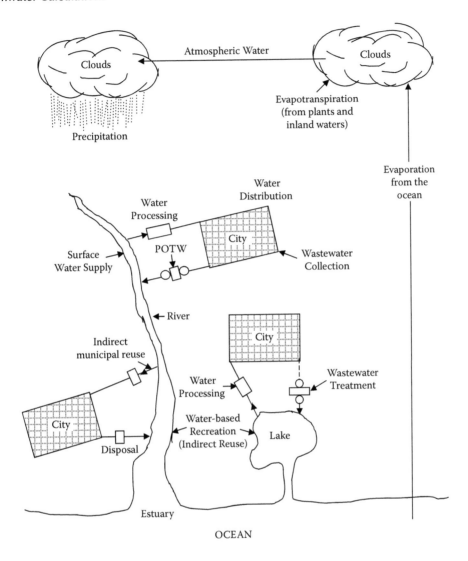

FIGURE 25.2 Urban water cycle.

25.3.1 Precipitation

Precipitation is a random event that cannot be predicted based on historical data. However, any given precipitation event has several distinct and independent characteristics which can be quantified as follows:

- *Duration*—Length of time over which precipitation occurs (hours)
- *Depth*—Amount of precipitation occurring throughout the storm duration (inches)
- *Frequency*—Recurrence interval of events having the same duration and volume
- *Intensity*—Depth divided by the duration (inches per hour)

A specified amount of rainfall may occur from many different combinations of intensities and durations, as shown, for example, in Table 25.1. Note that the peak intensity of runoff associated with each combination will vary widely. Moreover, storm events with the same intensity may have significantly different volumes and durations if the specified storm frequency (2-year, 10-year, 100-year)

TABLE 25.1

Variations of Duration and Intensity for a Given Volume

Duration (hr)	Intensity (in./hr)	Volume (in.)
0.5	3.0	1.5
1.0	1.5	1.5
1.5	1.0	1.5
6.0	0.25	1.5

Source: *Virginia Stormwater Management Handbook*, Virginia Department of Conservation and Recreation, Division of Soil and Water Conservation, Richmond, 1999.

TABLE 25.2

Variations of Volume, Duration, and Return Frequency for a Given Intensity

Duration (hr)	Volume (in.)	Intensity (in./hr)	Frequency (yr)
1.0	1.5	1.5	2
2.0	3.0	1.5	10
3.0	4.5	1.5	100

Source: *Virginia Stormwater Management Handbook*, Virginia Department of Conservation and Recreation, Division of Soil and Water Conservation, Richmond, 1999.

is different, as shown, for example, in Table 25.2. It, therefore, becomes critical for any regulatory criteria to specify the volume (or intensity) and the duration for a specified frequency design storm. Although specifying one combination of volume and duration may limit the analysis, with regard to what is considered to be the critical variable for any given watershed (erosion, flooding, water, quality, etc.), it does establish a baseline from which to work. (This analysis supports the SCS 24-hour design storm because an entire range of storm intensities is incorporated into the rainfall distribution.) Localities may choose to establish criteria based on specific watershed and receiving channel conditions, which will dictate the appropriate design storm.

25.3.1.1 Frequency

The frequency of a specified design storm can be expressed either in terms of exceedence probability or return period. *Exceedence probability* is the probability that an event having a specified volume and duration will be exceeded in one time period, which is most often assumed to be one year. *Return period* is the average length of time between events having the same volume and duration. If a storm of a specified duration and volume has a 1% chance of occurring in any given year, then it has an exceedance probability of 0.1 and a return period of 100 years. The return period concept is often misunderstand in that it implies that a 100-year flood will occur only once in a 100-year period. This will not always hold true because storm events cannot be predicted deterministically. Because storm events are random, the exceedence probability indicates that there is a finite probability (.01 for this example) that the 100-year storm may occur in any given year or consecutive years, regardless of the historic occurrence of that storm event.

25.3.1.2 Intensity–Duration–Frequency (IDF) Curves

To establish the importance of the relationship between average intensity, duration, and frequency, the U.S. Weather Bureau compiled intensity–duration–frequency (IDF) curves based on historic rainfall data for most localities across the country. The rational method uses the IDF curves directly, while SCS methods generalize the rainfall data taken from the IDF curves and create rainfall distributions for various regions of the country. There is debate concerning which combinations of storm durations and intensities are appropriate to use in a hydrologic analysis for a typical urban development. Working within the limitations of the methodology as described later in this section, small drainage areas (1 to 20 acres) in an urban setting can be accurately modeled using ether SCS or rational methods. The belief that the short, very intense storm generates the greatest need for stormwater management often leads engineers to use the rational method for stormwater management design, because this method is based on short duration storms. However, the SCS 24-hour storm is also appropriate for short duration storms because it includes short storm intensities within the 24-hour distribution.

25.3.1.3 SCS 24-Hour Storm Distribution

The SCS 24-hour storm distribution curve was derived from the National Weather Bureau's Rainfall Frequency Atlases of compiled data for areas less than 400 square miles, for durations up to 24 hours, and for frequencies from 1 to 100 years. Data analysis resulted in four regional distributions: Type I and IA for use in Hawaii, Alaska, and the Coastal side of the Sierra Nevada and Cascade Mountains in California, Washington, and Oregon; Type II distribution for most of the remainder of the United States; and Type III for the Gulf of Mexico and Atlantic coastal areas. The Type III distribution represents the potential impact of tropical storms which can produce large 24-hour rainfall amounts.

Note: For a more detailed description of the development of dimensionless rainfall distributions, refer to Section 4 of the Soil Conservation Service's *National Engineering Handbook*.

The SCS 24-hour storm distributions are based on the generalized rainfall depth–duration–frequency relationships collected for rainfall events lasting from 30 minutes up to 24 hours. Working in 30-minute increments, the rainfall depths are arranged with the maximum rainfall depth assumed to occur in the middle of the 24-hour period. The next largest 30-minute incremental depth occurs just after the maximum depth; the third largest rainfall depth occurs just prior to the maximum depth, etc. This continues with each decreasing 30-minute incremental depth until the smaller increments fall at the beginning and end of the 24-hour rainfall (see Figure 25.3). Note that this process includes all of the critical storm intensities within the 24-hour distributions. The SCS 24-hour storm distributions are, therefore, appropriate for rainfall and runoff modeling for small and large watersheds for the entire range of rainfall depths.

One of the stated disadvantages of using the *TR-55* method for hydrologic modeling is its restriction to the use of the 24-hour storm. The following discussion, taken directly from Appendix B of the *TR-55* manual (USDA, 1986), addresses this limitation:

> To avoid the use of a different set of rainfall intensities for each drainage area's size, a set of synthetic rainfall distributions having "nested" rainfall intensities was developed. The set "maximizes" the rainfall intensities by incorporating selected short-duration intensities within those needed for larger durations at the same probability level.
>
> For the size of the drainage areas for which SCS usually provides assistance, a storm period of 24 hours was chosen for the synthetic rainfall distributions. The 24-hour storm, while longer than that needed to determine peaks for these drainage areas, is appropriate for determining runoff volumes. Therefore, a single storm duration and associated synthetic rainfall distribution can be used to represent not only the peak discharges but also the runoff volumes for a range of drainage are sizes.

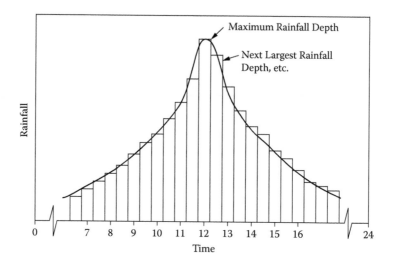

FIGURE 25.3 Typical 24-hr rainfall distribution. (From USSCS, *National Engineering Handbook*. Section 5. *Hydraulics*, U.S. Soil Conservation Service, Washington, DC, 1956.)

Figure 25.4 shows the SCS 24-hour rainfall distribution, which is a graph of the fraction of total rainfall at any given time, *t*. Note that the peak intensity for the Type II distribution occurs between time $t = 11.5$ hours and $t = 12.5$ hours.

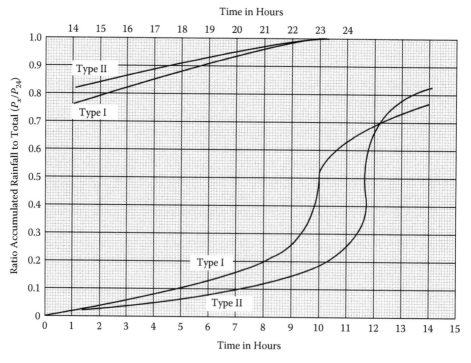

TYPE I - Coastal side of Sierra Nevada and Cascade Mountains in California, Oregon
and Washington; the Hawaiian Islands and Alaska
TYPE II - Remaining United States, Puerto Rico, and Virgin Islands

FIGURE 25.4 SCS 24-hr rainfall distribution. (From USSCS, *Urban Hydrology for Small Watersheds*, Technical Release No. 55, U.S. Soil Conservation Service, Washington, DC, 1986.)

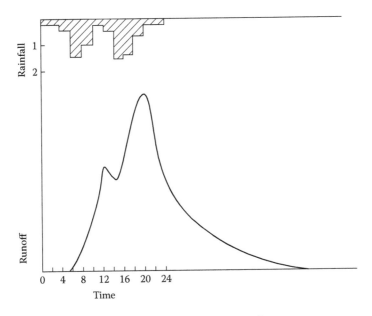

FIGURE 25.5   Rainfall hyetograph and associated runoff hydrograph.

25.3.1.4   Synthetic Storms

The alternative to a given rainfall "distribution" is to input a custom design storm into the model. This can be compiled from data gathered from a single rainfall event in a particular area, or a synthetic storm created to test the response characteristics of a watershed under specific rainfall conditions. Note, however, that a single historic design storm of known frequency is inadequate for such design work. It is better to synthesize data from the longest possible grouping of rainfall data and derive a frequency relationship as described with the IDF curves.

25.3.1.5   Single Event vs. Continuous Simulation Computer Models

The fundamental requirement of a stormwater management plan is a quantitative analysis of the watershed hydrology, hydraulics, and water quality, with consideration for associated facility costs. Computers have greatly reduced the time required to complete such an analysis. Computers have also greatly simplified the statistical analysis of compiled rainfall data. In general, there are two main categories of hydrologic computer models: single-event computer models and continuous-simulation models. Single-event computer models require a minimum of one design-storm hyetograph as input. A *hyetograph* is a graph of *rainfall intensity* on the vertical axis vs. *time* on the horizontal axis, as shown in Figure 25.5. A hyetograph shows the volume of precipitation at any given time as the area beneath the curve, and the time-variation of the intensity. The hyetograph can be a *synthetic hyetograph* or an historic *storm hyetograph*. When a frequency or recurrence interval is specified for the input hyetograph, it is assumed that the resulting output runoff has the same recurrence interval. (This is one of the general assumptions that is made for most single-event models.)

Continuous simulation models, on the other hand, incorporate the entire meteorological record of a watershed as their input, which may consist of decades of precipitation data. The data are processed by the computer model, producing a continuous runoff hydrograph. The continuous hydrograph output can be analyzed using basic statistical analysis techniques to provide discharge-frequency relationships, volume–frequency relationships, flow-duration relationships, etc. The extent to which the output hydrograph may be analyzed is dependent upon the input data available. The principal advantage of the continuous simulation model is that it eliminates the need to choose a design storm, instead providing long-term response data for a watershed that can then be statistically analyzed for the desired frequency storm.

Computer advances have greatly reduced the analysis time and related expenses associated with continuous models. It can be expected that future models, which combine some features of continuous modeling with the ease of single-event modeling, will offer quick and more accurate analysis procedures. The hydrologic methods discussed in this text are limited to single-event methodologies, based on historic data. Further information regarding the derivation of the IDF curves and the SCS 24-hour rainfall distribution can be found in NEH, Section 4, Hydrology.

25.4   RUNOFF HYDROGRAPHS

A runoff hydrograph is a graphical plot of the runoff or discharge from a watershed with respect to time. Runoff occurring in a watershed flows downstream in various patterns which are influenced by many factors, such as the amount and distribution of the rainfall, rate of snowmelt, stream channel hydraulics, infiltration capacity of the watershed, and others, that are difficult to define. No two flood hydrographs are alike. Empirical relationships, however, have been developed from which complex hydrographs can be derived. The critical element of the analysis, as with any hydrologic analysis, is the accurate description of the watershed's rainfall–runoff relationship, flow paths, and flow times. From this data, runoff hydrographs can be generated. Some of the types of hydrographs used for modeling include

- *Natural hydrographs* are obtained directly from the flow records of a gauged stream.
- *Synthetic hydrographs* are obtained by using watershed parameters and storm characteristics to simulate a natural hydrograph.
- *Unit hydrographs* are natural or synthetic hydrographs adjusted to represent one inch of direct runoff.
- *Dimensionless unit hydrographs* are made to represent many unit hydrographs by using the time to peak rates as basic units and plotting the hydrographs in ratios of these units.

25.5   RUNOFF AND PEAK DISCHARGE

Despite its simplification of the complex rainfall–runoff process, the practice of estimating runoff as a fixed percentage of rainfall has been used in the design of storm drainage systems for many years. It can be accurate when drainage areas are subdivided into homogeneous units, and when the designer has enough data and experience to use the appropriate factors. For watersheds or drainage areas comprised primarily of pervious cover such as open space, woods, lawns, or agricultural land uses, the rainfall/runoff analysis becomes much more complex. Soil conditions and types of vegetation are two of the variables that play a larger role in determining the amount of rainfall that becomes runoff. In addition, other types of flow have a larger effect on stream flow (and measured hydrograph) when the watershed is less urbanized. These factors include

1. *Surface runoff* occurs only when the rainfall rate is greater than the infiltration rate and the total volume of rainfall exceeds the interceptions, infiltration, and surface detention capacity of the watershed. The runoff flows on the land surface, collecting in the stream network.
2. *Subsurface flow* occurs when infiltrated rainfall meets an underground zone of low transmission and travels above the zone to the soil surface to appear as a seep or spring.
3. *Base flow* occurs when there is a fairly steady flow into a stream channel from natural storage. The flow comes from lakes or swamps, or from an aquifer replenished by infiltrated rainfall or surface runoff.

In watershed hydrology, it is customary to deal separately with base flow and to combine all other types of flow into direct runoff. Depending upon the requirements of the study, the designer can calculate the *peak flow rate*, in cfs (cubic feet per second), of the direct runoff from the watershed, or determine the *runoff hydrograph* for the direct runoff from the watershed. A hydrograph shows the volume of runoff as the area beneath the curve, and the time-variation of the discharge rate. If the purpose of a hydrologic study is to measure the impact of various developments on the drainage network within a watershed or to design flood control structures, then a hydrograph is needed. If the purpose of a study is to design a roadway culvert or other simple drainage improvement, then only the peak rate of flow is needed. Therefore, the purpose of a given study will dictate the methodology that should be used. Note that procedures such as the rational method and *TR-55* graphical peak discharge method do not generate a runoff hydrograph; however, the *TR-55* tabular method and the modified rational method do generate runoff hydrographs.

25.6 CALCULATION METHODS

Because they require different types of input and generate different types of results, environmental engineers responsible for stormwater design should be familiar with the different methods for calculating runoff from a watershed. The methods covered here are the rational method, modified rational method, and various *TR-55* methods (estimating runoff, graphical peak discharge, and tabular hydrograph). Note that many computer programs are available that develop these methodologies, utilizing the rainfall–runoff relationship described previously. Many of these programs also route the runoff hydrograph through a stormwater management facility, calculating the peak rate of discharge and a discharge hydrograph. Examples provided later utilize *TR-20, Project Formulation: Hydrology* (USSCS, 1982). Other readily available computer programs also utilize SCS methods. The accuracy of the computer model is based on the accuracy of the input that is typically generated through the rational or SCS methodologies covered here. Again, the engineer should be familiar with all of the methods covered here because any one may be appropriate for the specific site on the watershed being modeled.

Note: All of the methods presented below make assumptions and have limitations on the accuracy. Simply put, however, when these methods are used correctly, they will all provide a reasonable estimate of the peak rate of runoff from a drainage area or watershed.

Key Point: For small storm events (<2 inches rainfall) *TR-55* tends to underestimate the runoff, while it has been shown to be fairly accurate for larger storm events. Similarly, the rational formula has been found to be fairly accurate on smaller homogeneous watersheds, while tending to lose accuracy in the larger more complex watersheds. The following discussion provides further explanation of these methods, including assumptions, limitations, and information needed for the analysis.

25.6.1 RATIONAL METHOD

The rational method was devised for determining the peak discharges from drainage areas. Though frequently criticized for its simplistic approach, this same simplicity has made the rational method one of the most widely used techniques today. The rational formula estimates the peak rate of runoff at any location in a drainage area as a function of the runoff coefficient, mean rainfall intensity, and drainage area. The rational formula is expressed as

$$Q = C \times I \times A \tag{25.1}$$

where
    Q = Maximum rate of runoff (cfs).
    C = Dimensionless runoff coefficient, dependent upon land use.
    I = Design rainfall intensity, in inches per hour, for a duration equal to the time of concentration of the watershed.
    A = Drainage area (ac).

25.6.1.1 Assumptions

The rational method is based on the following assumptions:

- *Under steady rainfall intensity, the maximum discharge will occur at the watershed out-let at the time when the entire area above the outlet is contributing runoff. This time is commonly known as the time of concentration, tc, and is defined as the time required for runoff to travel from the most hydrologically distant point in the watershed to the outlet.*

  The assumption of steady rainfall dictates that even during longer events, when factors such as increasing soil saturation are ignored, the *maximum discharge* occurs when the entire watershed is contributing to the peak flow, at time $t = t_c$. Furthermore, this assumption limits the size of the drainage area that can be analyzed using the rational method. In large watersheds, the time of concentration may be so long that constant rainfall intensities may not occur for long periods. Also, shorter, more intense bursts of rainfall that occur over portions of the watershed may produce large peak flows. The time of concentration is equal to the minimum duration of peak rainfall. The time of concentration reflects the minimum time required for the entire watershed to contrib-ute to the peak discharge as stated above. The rational method assumes that all dis-charge does not increase as a result of soil saturation, decreased conveyance time, etc. Therefore, the time of concentration is not necessarily intended to be a measure of the actual storm duration, but simply the critical time period used to determine the average rainfall intensity from the IDF curves.

- *The frequency or return period of the computed peak discharge is the same as the frequency or return period of rainfall intensity (design storm) for the given time of concentration.*

  Frequencies of peak discharges depend not only on the frequency of rainfall intensity, but also the response characteristics of the watershed. For small and mostly impervious areas, rainfall frequency is the dominant factor since response characteristics are rela-tively constant. However, for larger watersheds, the response characteristics will have a much greater impact on the frequency of the peak discharge due to drainage struc-tures, restrictions within the watershed, and initial rainfall losses from interception and depression storage.

- *The fraction of rainfall that becomes runoff is independent of rainfall intensity or volume.*

  This assumption is reasonable for impervious areas, such as streets, rooftops, and parking lots. For pervious areas, the fraction of rainfall that becomes runoff varies with rainfall intensity and the runoff will increase. This fraction is represented by the dimensionless runoff coefficient, C. Therefore, the accuracy of the rational method is dependent on the careful selection of a coefficient that is appropriate for the storm, soil, and land use condi-tions. It is easy to see why the rational method becomes more accurate as the percentage of impervious cover in the drainage area approaches 100%.

- *The peak rate of runoff is sufficient information for the design of stormwater detention and retention facilities.*

25.6.1.2 Limitations

Because of the assumptions discussed above, the rational method should only be used when the following criteria are met:

1. The given watershed has a time of concentration, t_c, less than 20 minutes.
2. The drainage area is less than 20 acres.

For larger watersheds, attenuation of peak flows through the drainage network begins to be a factor in determining peak discharge. While there are ways to adjust runoff coefficients (C' factors) to account for the attenuation, or routing effects, it is better to use a hydrograph method or computer simulation for these more complex situations. Similarly, the presence of bridges, culverts, or storm sewers may act as restrictions that ultimately impact the peak rate of discharge from the watershed. The peak discharge upstream of the restriction can be calculated using a simple calculation procedure, such as the rational method; however, a detailed storage routing procedure which considers the storage volume above the restriction should be used to accurately determine the discharge downstream of the restriction.

25.6.1.3   Design Parameters

The following is a brief summary of the design parameters used in the rational method.

25.6.1.3.1   Time of Concentration (t_c)

The most consistent source of error in the use of the rational method is the oversimplification of the time of concentration calculation procedure. Because the origin of the rational method is rooted in the design of culverts and conveyance systems, the main components of the time of concentration are *inlet time* (or overland flow) and *pipe or channel flow time*. The *inlet overland flow time* is defined as the time required for runoff to flow overland from the furthest point in the drainage area over the surface to the inlet or culvert. The *pipe or channel flow time* is defined as the time required for the runoff to flow through the conveyance system to the design point. In addition, when an inlet time of less than 5 minutes is encountered, the time is rounded up to 5 minutes, which is then used to determine the rainfall intensity (I) for that inlet. Variations in the time of concentration can impact the calculated peak discharge. When the procedure for calculating the time of concentration is oversimplified, as mentioned above, the accuracy of the rational method is greatly compromised. To prevent this oversimplification, it is recommended that a more rigorous procedure for determining the time of concentration be used, such as the one presented in Chapter 15, Section 4, of the *National Engineering Handbook*.

Many procedures are available for estimating the time of concentration. Some were developed with a specific type or size watershed in mind, while others were based on studies of a specific watershed. The selection of any given procedure should include a comparison of the hydrologic and hydraulic characteristics used in the formation of the procedure, vs. the characteristics of the watershed under study. The engineer should be aware that if two of more methods of determining time of concentration were applied to a given watershed, there would likely be a wide range in results. The SCS method is recommended because it provides a means of estimating overland sheet flow time and shallow concentrated flow time as a function of readily available parameters such as land slope and land surface conditions. Regardless of which method is used, the result should be reasonable when compared to an average flow time over the total length of the watershed.

25.6.1.3.2   Rainfall Intensity (I)

The rainfall intensity (I) is the average rainfall rate, in inches per hour, for a storm duration equal to the time of concentration for a selected return period (e.g., 1-year, 2-year, 10-year, 25-year). Once a particular return period has been selected, and the time of concentration has been determined for the drainage area, the rainfall intensity can be read from the appropriate rainfall IDF curve for the geographic area in which the drainage area is located. These charts were developed from data furnished by the National Weather Service for regions of the U.S.

25.6.1.3.3   Runoff Coefficients (C)

The runoff coefficients for different land uses within a watershed are used to generate a single, weighted coefficient that will represent the relationship between rainfall and runoff for that watershed. Recommended values can be found in Table 25.3. In an attempt to make the rational method

TABLE 25.3

Rational Equation Runoff Coefficients

Land Use	C Value
Business, industrial and commercial	0.90
Apartments	0.75
Schools	0.60
Residential	
Lots of 10,000 ft^2	0.50
Lots of 12,000 ft^2	0.45
Lots of 17,000 ft^2	0.45
Lots of 1/2 acre or more	0.40
Parks, cemeteries and unimproved areas	0.34
Paved and roof areas	0.90
Cultivated areas	0.60
Pasture	0.45
Forest	0.30
Steep grass slopes (2:1)	0.70
Shoulder and ditch areas	0.50
Lawns	0.20

Source: USDOT, *Urban Drainage Design Manual*, Department of Transportation, Washington, DC, 2001.

more accurate, efforts have been made to adjust the runoff coefficients to represent the integrated effects of drainage basin parameters: *land use, soil type,* and *average land slope.* Table 25.3 provides recommended coefficients based on urban land use only. A good understanding of these parameters is essential in choosing an appropriate coefficient. As the slope of a drainage basin increases, runoff velocities increase for both sheet flow and shallow concentrated flow. As the velocity increases, the ability of the surface soil to absorb the runoff decreases. This decrease in infiltration results in an increase in runoff. In this case, the designer should select a higher runoff coefficient to reflect the increase due to slope.

Soil properties influence the relationship between runoff and rainfall even further since soils have differing rates of infiltration. Historically, the rational method was used primarily for the design of storm sewers and culverts in urbanizing areas; soil characteristics were not considered, especially when the watershed was largely impervious. In such cases, a conservative design simply meant a larger pipe and less headwater. For stormwater management purposes, however, the existing condition (prior to development, usually with large amounts of pervious surfaces) often dictates the allowable post-development release rate, and therefore, must be accurately modeled. Soil properties can change throughout the construction process due to compaction, cut, and fill operations. If these changes are not reflected in the runoff coefficient, the accuracy of the model will decrease. Some localities arbitrarily require an adjustment in the runoff coefficient for pervious surfaces due to the effects of construction on soil infiltration capacities.

25.6.1.3.4 Adjustment for Infrequent Storms

The rational method has undergone further adjustment to account for infrequent, higher intensity storms. This adjustment is in the form of a frequency factor (C_f) which accounts for the reduced impact of infiltration and other effects on the amount of runoff during larger storms. With the adjustment, the rational formula is expressed as follows:

TABLE 25.4

Rational Equation Frequency Factors

Frequency Factor C_f	Storm Return Frequency
1.0	10 yr or less
1.1	25 yr
1.1	50 yr
1.25	100 yr

$$Q = C \times C_f \times I \times A \qquad (25.2)$$

The C_f values are listed in Table 25.4. The product of $C_f \times C$ should not exceed 1.0.

25.6.2 MODIFIED RATIONAL METHOD

The modified rational method is a variation of the rational method, developed mainly for the sizing of detention facilities in urban areas. The modified rational method is applied similarly to the rational method except that it utilizes a fixed rainfall duration. The selected rainfall duration depends on the requirements of the user. For example, the designer might perform an iterative calculation to determine the rainfall duration, which produces the maximum storage volume requirement when sizing a detention basin.

25.6.2.1 Assumptions

The modified rational method is based on the following assumptions:

1. All of the assumptions used with the rational method apply. The most significant difference is that the time of concentration for the modified rational method is equal to the rainfall intensity averaging period rather than the actual storm duration. This assumption means that any rainfall, or any runoff generated by the rainfall, that occurs before or after the rainfall averaging period is unaccounted for. Thus, when used as a basin sizing procedure, the modified rational method may seriously underestimate the required storage volume (Walesh, 1989).
2. The runoff hydrograph for a watershed can be approximated as triangular or trapezoidal in shape. This assumption implies a linear relationship between peak discharge and time for any and all watersheds.

25.6.2.2 Limitations

All of the limitations listed for the rational method apply to the modified rational method. The key difference is the assumed shape of the resulting runoff hydrograph. The rational method produces a triangular shaped hydrograph, which then modified can generate triangular or trapezoidal hydrographs for a given watershed, as shown in Figure 25.6.

25.6.2.3 Design Parameters

The equation $Q = C \times I \times A$ (rational equation) is used to calculate the peak discharge for all three hydrographs shown in Figure 25.6. Notice that the only difference between the rational method and the modified rational method is the incorporation of the *storm duration* (*d*) into the modified rational method to generate a *volume* of runoff in addition to the peak discharge. The rational method generates the peak discharge that occurs when the entire watershed is contributing to the peak (at a time $t = t_c$) and ignores the effects of a storm which lasts longer than time t. The modified rational method, however, considers storms with a longer duration than the watershed t_c, which may have a

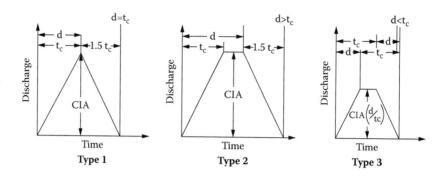

Type 1 - Storm duration, d, is equal to the time of concentration, t_c.
Type 2 - Storm duration, d, is greater than the time of concentration, t_c.
Type 3 - Storm duration, d, is less than the time of concentration, t_c.

FIGURE 25.6 Modified rational method runoff hydrographs.

smaller or larger peak rate or discharge, but will produce a greater volume of runoff (area under the hydrograph) associated with the longer duration of rainfall. Figure 25.7 shows a family of hydrographs representing storms of different durations. The storm duration that generates the greatest volume of runoff may not necessarily produce the greatest peak rate of discharge.

Note that the duration of the receding limb of the hydrograph is set to equal the time of concentration (t_c), or 1.5 times t_c. The direct solution, which will be discussed later, uses $1.5t_c$ as the receding limb. This is justified since it is more representative of actual storm and runoff dynamics. (It is also more similar to the SCS unit hydrograph where the receding limb extends longer than the risking limb.) Using 1.5 times t_c in the direct solution methodology provides for a more conservative design and will be used in this text.

The modified rational method allows the designer to analyze several different storm durations to determine the one that requires the greatest storage volume with respect to that allowable release rate. This storm duration is referred to as the critical storm duration and is used as a basin-sizing tool. The technique is discussed in more detail in later.

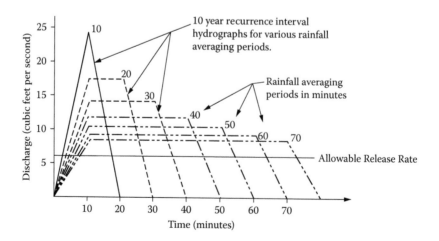

FIGURE 25.7 Modified rational method family of runoff hydrographs.

25.6.3   *TR-55* Estimating Runoff Method

The U.S. Soil Conservation Service published the second edition of *TR-55* in 1986, entitled *Urban Hydrology for Small Watersheds*. The techniques outlined in *TR-55* require the same basic data as the rational method: drainage area, time of concentration, land use, and rainfall. The SCS approach, however, is more sophisticated in that it allows the designer to manipulate the time distribution of the rainfall, the initial rainfall losses to interception and depression storage, and the moisture condition of the soils prior to the storm. The procedures developed by SCS are based on a dimensionless rainfall distribution curve for a 24-hour storm. *TR-55* presents two general methods for estimating peak discharges from urban watersheds: the graphical method and the tabular method. The *graphical method* is limited to watersheds whose runoff characteristics are fairly uniform and whose soils, land use, and ground cover can be represented by a single runoff curve number (*RCN*). The graphical method provides a peak discharge only and is not applicable for situations where a hydrograph is required.

The *tabular method* is a more complete approach and can be used to develop a hydrograph at any point in a watershed. For large areas it may be necessary to divide the area into sub-watersheds to account for major land use changes, analyze specific study points within sub-watersheds, or locate stormwater drainage facilities and assess their effects on peak flows. The tabular method can generate a hydrograph for each sub-watershed for the same storm event. The hydrographs can then be *routed* through the watershed and combined to produce a partial composite hydrograph at the selected study point. The tabular method is particularly useful in evaluating the effects of an altered land use in a specific area within a given watershed.

Prior to using either the graphical or tabular methods, the designer must determine the volume of runoff resulting from a given depth of precipitation and the time of concentration, t_c, for the watershed being analyzed. The methods for determining these values will be discussed briefly in this section. However, the reader is strongly encouraged to obtain a copy of the *TR-55* manual from the Soil Conservation Service to gain more insight into the procedures and limitations.

The runoff curve number (RCN) method is used to estimate runoff (described in detail in Section 4 of the *National Engineering Handbook*). The runoff equation (found in *TR-55* and discussed later in this section) provides a relationship between runoff and rainfall as a function of the CN. The CN is a measure of the land's ability to infiltrate or otherwise detain rainfall, with the excess becoming runoff. The CN is a function of the land cover (woods, pasture, agricultural use, percent impervious, etc.), hydrologic condition, and soils.

25.6.3.1   Limitations

- *TR-55* has simplified the relationship between rainfall and runoff by reducing all of the initial losses before runoff begins, or initial abstractions, to the term I_a, and approximating the soil and cover conditions using the variable S, potential maximum retention. Both of these terms, I_a and S, are functions of the runoff curve number. Runoff curve numbers describe average conditions that are useful for design purposes. If the purpose of the hydrologic study is to model a historical storm event, average conditions may not be appropriate.
- The designer should understand the assumption reflected in the initial abstraction term, I_a. I_a represents interception, initial infiltration, surface depression storage, evapotranspiration, and other watershed factors and is generalized as a function of the runoff curve number based on data from agricultural watersheds. This can be especially important in an urban application because the combination of impervious area with pervious area can imply a significant initial loss that may not take place. On the other hand, the combination of impervious and pervious area can underestimate initial losses if the urban area has significant surface depression storage. (To use a relationship other than the one established in *TR-55*, the designer must redevelop the runoff equation by using the original rainfall–runoff data to establish new curve number relationships for each cover and hydrologic soil group. This would represent a large data collection and analysis effort.)

- Runoff from snowmelt or frozen ground cannot be estimated using these procedures.
- The runoff curve number method is less accurate when the runoff is less than 0.5 inch. As a check, use another procedure to determine runoff.
- The SCS runoff procedures apply only to surface runoff and do not consider subsurface flow or high groundwater.
- Manning's kinematic solution (discussed later) should not be used to calculate the time of concentration for sheet flow longer than 300 feet. This limitation will affect the time of concentration calculations. Note that many jurisdictions consider 150 feet to be the maximum length of sheet flow before shallow concentrated flow develops.
- The minimum t_c used in *TR-55* is 0.1 hour.

25.6.3.2 Information Needed

Generally, a good understanding of the physical characteristics of the watershed is necessary to solve the runoff equation and determine the time of concentration. Some features, such as topography and channel geometry can be obtained from topographic maps such as the USGS 1 inch = 2000 inch quadrangle maps. Various sources of information may be accurate enough for a watershed study, however, the accuracy of the study will be directly related to the accuracy and level of detail of the base information. Ideally, a site investigation and filed survey should be conducted to verify specific features such as channel geometry and material, culvert sizes, drainage divides, ground cover, etc. Depending on the size and scope of the study, however, a site investigation may not be economically feasible. The data needed to solve the runoff equation and determine the watershed time of concentration (t_c) are listed below. These items are discussed in more detail later.

1. Soil information (to determine the hydrologic soil group
2. Ground cover type (impervious, woods, grass, etc.)
3. Treatment (cultivated or agricultural land)
4. Hydrologic conditions (for design purposes, the hydrologic condition should be considered "good" for the pre-developed condition)
5. Urban impervious area modifications (connected, unconnected, etc.)
6. Topography detailed enough to accurately identify divides, t_c and T_t flow paths and channel geometry, and surface condition (roughness coefficient)

25.6.3.3 Soil Design Parameters

In hydrograph applications, runoff is often referred to as *rainfall excess* or *effective rainfall*, and is defined as the amount of rainfall that exceeds the land's capability to infiltrate or otherwise retain the rainwater. The soil type or classification, the land use and land treatment, and the hydrologic condition of the cover are the watershed factors that will have the most significant impact on estimating the volume of rainfall excess, or runoff.

25.6.3.3.1 Hydrologic Soil Group Classification

The SCS soil classification system consists of four groups, identified as A, B, C, and D. Soils are classified into one of these categories based upon their *minimum infiltration rate*. By using information obtained from local SCS offices, soil and water conservation district offices, or from SCS Soil Surveys (published for many counties across the country), the soils in a given area can be identified. Preliminary soil identification is especially useful for watershed analysis and planning in general. When preparing a stormwater management plan for a specific site, it is recommended that soil borings be taken to verify the hydrologic soil classification. Soil characteristics associated with each hydrologic soil group are generally described as follows:

- *Group A*—Soils with low runoff-potential due to high infiltration rates even when thoroughly wetted. These soils consist primarily of deep, well to excessively drained sands and gravels with high water transmission rates (0.30 in./hr.). Group A soils include sand, loamy sand, or sandy loam.
- *Group B*—Soils with moderately low runoff potential due to moderate infiltration rates when thoroughly wetted. These soils consist primarily of moderately deep to deep and moderately well to well drained soils. Group B soils have moderate water transmission rates (0.15 to 0.30 in./hr.) and include silt loam or loam.
- *Group C*—Soils with moderately high runoff potential due to slow infiltration rates when thoroughly wetted. These soils typically have a layer near the surface that impacts the downward movement of water or soils. Group C soils have low water transmission rates (0.15 to 0.15 in./hr.) and include sandy clay loam.
- *Group D*—Soils with high runoff potential due to very slow infiltration rates. These soils consist primarily of clays with high swelling potential, soils with permanently high water tables, soils with a claypan or clay layer at or near the surface and shallow soils over nearly impervious parent material. Group D soils have very low water transmission rates (0 to 0.05 in./hr.) and include clay loam, silty clay loam, sandy clay, silty clay, or clay.

Any disturbance of a soil profile can significantly alter the soil's infiltration characteristics. With urbanization, the hydrologic soil group for a given area can change due to soil mixing, introduction of fill material from other areas, removal of material during mass grading operations, or compaction from construction equipment. A layer of topsoil may typically be saved and replaced after the earthwork is completed, but the native underlying soils have been dramatically altered. Therefore, any disturbed soil should be classified by its physical characteristics as given above for each soil group.

Some jurisdictions require all site developments to be analyzed using an HSG classification that is one category below the actual pre-developed HSG. For example, a site with a pre-developed HSG classification of B, as determined from the soil survey, would be analyzed in its developed state using an HSG classification of C.

25.6.3.3.2 *Hydrologic Condition*

Hydrologic condition represents the effects of *cover type* and *treatment* on infiltration and runoff. It is generally estimated from the density of plant and residue cover across the drainage area. *Good hydrologic condition* indicates that the cover has a low runoff potential, while *poor hydrologic condition* indicates that the cover has a high runoff potential. Hydrologic condition is used in describing non-urbanized lands such as woods, meadow, brush, agricultural land, and open spaces associated with urbanized areas, such as lawns, parks, golf courses, and cemeteries. *Treatment* is a cover type modifier to describe the management of cultivated agricultural lands. Tables 25.5A and 25.5B provide an excerpt from Table 2-2 in *TR-55* which shows the treatment and hydrologic condition for various land uses. When a watershed is being analyzed to determine the impact of proposed development, many stormwater management regulations require the designer to consider all existing or undeveloped land to be *hydrologically good conditions*. This results in lower existing condition peak runoff rates which, in turn, results in greater post-development peak control. In most cases, undeveloped land is in good hydrologic condition unless it has been altered in some way. Since the goal of most stormwater programs is to reduce the peak flows from developed or altered areas to their pre-developed or pre-altered rates, this is a reasonable approach. In addition, this approach eliminates any inconsistencies in judging the condition of undeveloped land or open space.

TABLE 25.5A

Runoff Curve Numbers for Urban Areas[a]

Cover Description		Curve Numbers for Hydrologic Soil Group			
Cover Type and Hydrologic Condition	**Average Percent Impervious Area**[b]	**A**	**B**	**C**	**D**
Fully developed urban areas (vegetation established)					
Good condition (grass cover >75%)		39	61	74	80
Impervious areas					
Paved parking lots, roofs, driveways, etc. (excluding right-of-way)		98	98	98	98
Streets and roads					
Paved; curbs and storm sewers (excluding right-of-way)		98	98	98	98
Paved; open ditches (including right-of-way)		83	89	92	93
Gravel (including right-of-way)		76	85	89	91
Dirt (including right-of-way)		72	82	87	89
Urban districts					
Commercial and business	85	89	92	94	95
Industrial	72	81	88	91	93
Residential districts by average lot size					
1/8 acre or less (town houses)	65	77	85	90	92
1/4 acre	38	61	75	83	87
1/3 acre	30	57	72	81	86
1/2 acre	25	54	70	80	85
1 acre	20	51	68	79	84
2 acres	12	46	65	77	82
Developing urban areas					
Newly graded areas (pervious areas only, no vegetation)		77	86	91	94
Idle lands (CNs are determined using cover types similar to those in *TR-55* Table 2-2c)					

[a] Refer to *TR-55* for additional cover types and general assumptions and limitations.

[b] Average runoff conditions and $I_a = 0.2S$.

Source: Adapted from *TR-55* Table 2-2a—Runoff Curve Numbers for Urban Areas.

25.6.3.3.3   Runoff Curve Number (RCN) Determination

The soil group classification, cover type, and the hydrologic condition are used to determine the runoff curve number (RCN). The RCN indicates the runoff potential of an area when the ground is not frozen. Table 25.5 provides the RCNs for various land use types and soil groups. A more complete table can be found in *TR-55*. Several factors should be considered when choosing an RCN for a given land use. First, the designer should realize that the curve numbers in Table 25.5 and *TR-55* are for the *average antecedent runoff* or *moisture condition* (ARC). The ARC is the index of runoff potential before a storm event and can have a major impact on the relationship between rainfall and runoff for a watershed. Average ARC runoff curve numbers can be converted to dry or wet values; however, the average antecedent runoff condition is recommended for design purposes. It is important to consider the list of assumptions made in developing the runoff curve numbers as provided in the Table 25.5 and in *TR-55*. Some of these assumptions are outline below.

Note: The decision to use "wet" or "dry" antecedent runoff conditions should be based on thorough field work, such as carefully monitored rain gauge data.

TABLE 25.5B
Runoff Curve Numbers for Other Agricultural Areas[a]

Cover Description		Curve Numbers for Hydrologic Soil Group			
Cover Type	Hydrologic Condition[b]	A	B	C	D
Pasture, grassland, or range—continuous forage for grazing	Good	39	61	74	80
Meadow—continuous grass, protected from grazing and generally mowed for hay	—	30	58	71	78
Brush—brush-weed-grass mixture with brush the major element	Good	230	48	65	73
Woods—grass combination (orchard or tree farm)	Good	32	58	72	79
Woods	Good	230	55	70	77
Farmsteads—buildings, lanes, driveways, and surrounding lots	—	59	74	82	86

[a] Refer to *TR-55* for additional cover types and general assumptions and limitations.
[b] Average runoff conditions and $I_a = 0.2S$.

Source: Adapted from *TR-55* Table 2-2c—Runoff Curve Numbers for Other Agricultural Lands.

The RCN determination assumptions include the following:

- The urban curve numbers, for such land uses as residential, commercial, and industrial, are computed with the percentage of imperviousness are as shown. A composite curve number should be re-computed using the actual percentage of imperviousness if it differs from the value shown.
- The impervious areas are directly connected to the drainage system.
- Impervious areas have a runoff curve number of 98.
- Pervious areas are considered equivalent to open space in good hydraulic condition.

Note: These assumptions, as well as others, are footnoted in *TR-55*, Table 2-2. *TR-55* provides a graphical solution for modification of the given RCNs if any of these assumptions do not hold true.

The environmental engineer should become familiar with the definition of *connected* vs. *unconnected* impervious areas along with the graphical solutions and the impact that their use can have on the resulting *RCN*. After some experience in using this section of *TR-55*, the designer will be able to make field evaluations of the various criteria used in the determination of the *RCN* for a given site. In addition, the designer will need to determine if the watershed contains sufficient diversity in land use to justify dividing the watershed into several sub-watersheds. If a watershed or drainage area cannot be adequately described by one weighted curve number, then the designer must divide the watershed into subarea and analyze each one individually, generate individual hydrographs, and add those hydrographs together to determine the composite peak discharge for the entire watershed. Figure 25.8 shows the decision making process for analyzing a drainage area. The flow chart can be used to select the appropriate tables or figures in *TR-55* from which to then choose the runoff curve numbers. Worksheet 2 in *TR-55* is then used to compute the weighted curve number for the area or subarea.

25.6.3.3.4   Runoff Equation
The SCS runoff equation is used to solve for runoff as a function of the initial abstraction, I_a, and the potential maximum retention, S, of a watershed, both of which are functions of the *RCN*. This equation attempts to quantify all the losses before runoff begins, including infiltration, evaporation,

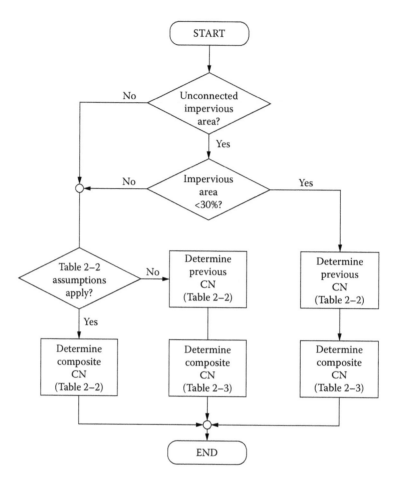

FIGURE 25.8   Runoff curve number selection flowchart.

depression storage, and water intercepted by vegetation. *TR-55* provides a graphical solution for the runoff equation. The graphical solution is found in Chapter 2 of *TR-55: Estimating Runoff.* Both the equation and graphical solution solve for depth of runoff that can be expected from a watershed or sub-watershed, of a specified *RCN*, for any given frequency storm. Additional information can be found in Section 4 of the *National Engineering Handbook.* These procedures, by providing the basic relationship between rainfall and runoff, are the basis for any hydrological study based on SCS methodology. Therefore, it is essential that the designer conduct a thorough site visit and consider the entire site features and characteristics, such as soil types and hydrologic condition, when analyzing a watershed or drainage area.

25.6.3.3.5   *Time of Concentration and Travel Time*

The time of concentration (t_c) is the length of time required for a drop of water to travel from the most hydraulically distant point in the watershed or sub-watershed to the point of analysis. The travel time (T_t) is the time it takes that same drop of water to travel from the study point at the bottom of the sub-watershed to the study point at the bottom of the whole watershed. The travel time (T_t) is descriptive of the sub-watershed by providing its location relative to the study point of the entire watershed. Similar to the rational method, the time of concentration (t_c) plays an important role in developing the peak discharge for a watershed. Urbanization usually decreases the t_c, which results in an increase in peak discharge. For this reason, to accurately model the watershed, the

TABLE 25.6A
Roughness Coefficient (*n*) for the Manning Equation—Sheet Flow

Surface Description	*n* Value
Smooth surfaces (concrete, asphalt, gravel, or bare soil)	0.011
Fallow (no residue)	0.05
Cultivated soils	
Residue cover <20%	0.06
Residue cover >20%	0.17
Grass	
Short grass prairie	0.15
Dense grasses[a]	0.24
Bermuda grass	0.41
Range (natural)	0.13
Woods[b]	
Light underbrush	0.40
Dense underbrush	0.80

[a] Includes species such as weeping lovegrass, bluegrass, buffalo grass, blue grama grass, and native grass mixtures.

[b] When selecting it, consider cover to a height of about 0.1 ft. This is the only part of the plant cover that will obstruct sheet flow.

Source: Adapted from *TR-55* Table 2-C-1—Roughness Coefficients (Manning's *n*) for Sheet Flow.

engineer must be aware of any conditions that may act to decrease the flow time, such as chan-nelization and channel improvements. On the other hand, the engineer must also be aware of the conditions within the watershed that may actually lengthen the flow time, such as surface ponding above undersized conveyance systems and culverts.

1. *Heterogeneous watersheds*—A heterogeneous watershed is one that has two or more hydrologically defined drainage areas of differing land uses, hydrologic conditions, times of concentration, or other runoff characteristics, contributing to the study point.
2. *Flow segments*—The time of concentration is the sum of the time increments for each flow segment present in the t_c flow path, such as *overland* or *sheet flow, shallow concentrated flow*, and *channel flow*. These flow types are influenced by surface roughness, channel slope, flow patterns, and slope.
   - Overland (sheet) flow is shallow flow over plane surfaces. For the purposes of deter-mining time of concentration, overland flow usually exists in the upper reaches of the hydraulic flow path. *TR-55* utilizes Manning's kinematic solution to compute t_c for overland sheet flow. The roughness coefficient is the primary culprit in the misap-plication of the kinematic t_c equation. Care should be taken to accurately identify the surface conditions for overland flow. Table 25.6A in this text and Table 3-1 in *TR-55* provide selected coefficients for various surface conditions. Refer to *TR-55* for the use of Manning's kinematic equation.
   - Shallow concentrated flow usually begins where overland flow converges to form small rills or gullies. Shallow concentrated flow can exist in small manmade drainage ditches (paved and unpaved) and in curb and gutters. *TR-55* provides a graphical solu-tion for shallow concentrated flow. The input information needed to solve for this flow segment is the land slope and the surface condition (paved or unpaved).

TABLE 25.6B
Roughness Coefficient (*n*) for the Manning Equation—
Pipe Flow

| | *n* Value Range | |
Material	From	To
Coated cast-iron	0.010	0.014
Uncoated cast-iron	0.011	0.015
Vitrified sewer pipe	0.010	0.017
Concrete pipe	0.010	0.017
Common clay drainage tile	0.011	0.017
Corrugated metal (2-2/3 × 1/2)	0.023	0.026
Corrugated metal (3 × 1 and 6 × 1)	0.026	0.029
Corrugated metal (6 × 2 structural plate)	0.030	0.033

Source: Adapted from Brater, E. and King, H., *Handbook of Hydraulics*,
            6th ed., McGraw-Hill, New York, 1976.

TABLE 25.6C
Roughness Coefficient (*n*) for the Manning Equation—
Constructed Channels

| | *n* Value Range | |
Lining Material	From	To
Concrete lined	0.012	0.016
Cement rubble	0.017	0.025
Earth, straight and uniform	0.017	0.022
Rock cuts, smooth and uniform	0.025	0.033
Rock cuts, jagged and irregular	0.035	0.045
Winding, sluggish canals	0.022	0.027
Dredged earth channels	0.025	0.030
Canals with rough stony beds, weeds on earth banks	0.025	0.035
Earth bottom, rubble sides	0.028	0.033
Small grass channels		
Long grass, 13-inch	0.042	—
Short grass, 3-inch	0.034	—

Source: Adapted from Brater, E. and King, H., *Handbook of Hydraulics*,
            6th ed., McGraw-Hill, New York, 1976.

- Channel flow occurs where flow converges in gullies, ditches or swales, and natural or artificial water conveyances (including storm drainage pipes). Channel flow is assumed to exist in perennial streams or wherever there is a well-defined channel cross-section. The Manning equation is used for open channel flow and pipe flow, and usually assumes full flow or bank-full velocity. Manning coefficients can be found in Table 25.6B–D for open-channel flow (natural and artificial channels) and closed channel flow. Coefficients can also be obtained from standard textbooks such as *Open Channel Hydraulics* (Chow, 1959) or *Handbook of Hydraulics* (King and Brater, 1976).

TABLE 25.6D
Roughness Coefficient (*n*) for the Manning Equation— Natural Stream Channels

		n Value Range	
Channel Lining		**From**	**To**
1.	Clean, straight bank, full stage, no rifts or deep pools	0.025	0.030
2.	Same as #1, but some weeds and stones	0.030	0.035
3.	Winding, some pools and shoals, clean	0.033	0.040
4.	Same as #3, lower stages, more ineffective slope and sections	0.040	0.050
5.	Same as #3, some weeds and stones	0.035	0.045
6.	Same as #4, stony sections	0.045	0.055
7.	Sluggish river reaches, rather weedy with very deep pools	0.050	0.070
8.	Very weedy reaches	0.075	0.125

Source: Adapted from Brater, E. and King, H., *Handbook of Hydraulics*, 6th ed., McGraw-Hill, New York, 1976.

25.6.4   *TR-55* GRAPHICAL PEAK DISCHARGE METHOD

The graphical peak discharge method was developed from hydrograph analyses using *TR-20, Computer Program for Project Formulation: Hydrology* (USSCS, 1983). The graphical method develops the peak discharge in cubic feet per second (cfs) for a given watershed.

25.6.4.1   Limitations

The engineer should be aware of several limitations before using the *TR-55* graphical method:

1. The watershed being studied must be hydrologically homogeneous; that is, the land use, soils, and cover are distributed uniformly throughout the watershed and can be described by one curve number.
2. The watershed may have only one main stream or flow path. If more than one is present they must have nearly equal t_c values so that one t_c represents the entire watershed.
3. The analysis of the watershed cannot be part of a larger watershed study, which would require adding hydrographs since the graphical method does not generate a hydrograph.
4. For the same reason, the graphical method should not be used if a runoff hydrograph is to be routed through a control structure.
5. When the initial abstraction/rainfall ratio (I_a/P) falls outside the range of the unit peak discharge curves (0.1 to 0.5), the limiting value of the curve must be used.

The reader is encouraged to review *TR-55* to become familiar with these and other limitations associated with the graphical method.

The graphical method can be used as a planning tool to determine the impact of development or land use changes within a watershed, or to anticipate or predict the need for stormwater management facilities or conveyance improvements. Sometimes the graphical method can be used in conjunction with the *TR-55* short-cut method for estimating the storage volume required for post-developed peak discharge control. This short-cut method is found in Chapter 6 of *TR-55*. However, it should be noted that a more sophisticated computer model such as *TR-20* or *HEC-1* or even the *TR-55* tabular hydrograph method should be used for complex, urbanizing watersheds.

25.6.4.2   Information Needed

The following represents a brief list of the parameters needed to compute the peak discharge of a watershed using the *TR-55* graphical peak discharge method:

1. The drainage area, in square miles
2. Time of concentration (t_c) in hours
3. Weighted runoff curve number (RCN)
4. Rainfall amount (P) for specified design storm, in inches
5. Total runoff (Q), in inches
6. Initial abstraction (I_a) for each subarea
7. Ratio of I_a/P for each subarea
8. Rainfall distribution (Type I, IA, II, or III)

25.6.4.3   Design Parameters

The *TR-55* peak discharge equation is

$$q_p = q_u \times A_m \times Q \times F_p \tag{25.3}$$

where

q_p = Peak discharge (cfs).
q_u = Unit peak discharge (cfs/mi²/in., csm/in.).
A_m = Drainage area (mi²).
Q = Runoff (in.).
F_p = Pond and swamp adjustment factor.

All of the required information has been determined earlier except for the unit peak discharge (q_u) and the pond and swamp adjustment factor (F_p).

The unit peak discharge (q_u) is a function of the initial abstraction (I_a), precipitation (P), and the time of concentration (t_c) and can be determined from the unit peak discharge curves in *TR-55*. The unit peak discharge is expressed in cubic feet per second per square mile per inch of runoff. Initial abstraction, as discussed previously, is a measure of all the losses that occur before runoff begins, including infiltration, evaporation, depression storage, and water intercepted by vegetation, and can be calculated from empirical equations or Table 4-1 in *TR-55*. The pond and swamp adjustment factor is an adjustment in the peak discharge to account for pond and swamp areas if they are spread throughout the watershed and are not considered in the t_c computation. Refer to *TR-55* for more information on pond and swamp adjustment factors.

The unit peak discharge (q_u) is obtained by using t_c and the I_a/P ratio with Exhibit 4-I, 4-IA, 4-II, or 4-III (depending on rainfall distribution type) in *TR-55*. As limitation number 5 above indicates, the ratio of I_a/P must fall between 0.1 and 0.5. The engineer must use the limiting value on the curves when the computed value is not within this range. The unit peak discharge is determined from these curves and entered into the above equation to calculate the peak discharge.

25.6.5   *TR-55* Tabular Hydrograph Method

The tabular hydrograph method can be used to analyze large heterogeneous watersheds. The tabular method can develop partial composite flood hydrographs at any point in a watershed by dividing the watershed into homogeneous subareas. The method is especially applicable for estimating the effects of land use change in a portion of a watershed. The tabular hydrograph method provides a tool to efficiently analyze several sub-watersheds to verify the combined impact at a downstream study point. It is especially useful to verify the timing of peak discharges. Sometimes the use of detention in a lower sub-watershed may actually increase the combined peak discharge at the study point. This procedure allows a quick check to verify the timing of the peak flows and to decide if a more detailed study is necessary.

25.6.5.1 Limitations

Some of the basic limitations of which the engineer should be aware of before using the *TR-55* tabular method include the following:

1. The travel time (T_t) must be less than 3 hours (largest T_t in *TR-55*, Exhibit 5).
2. The time of concentration (t_c) must be less than 2 hours (largest t_c in *TR-55*, Exhibit 5).
3. The acreage of the individual sub-watersheds should not differ by a factor of 5 or more.

When these limitations cannot be met, the engineer should use the *TR-20* computer program or other available computer models that will provide more accurate and detailed results. The reader is encouraged to review the *TR-55* manual to become familiar with these and other limitations associated with the tabular method.

25.6.5.2 Information Needed

The following parameters are needed to compute the peak discharge of a watershed using the *TR-55* tabular method:

1. Subdivision of the watershed into areas that are relatively homogeneous
2. The drainage area of each subarea (square miles)
3. Time of concentration (t_c) for each subarea (hours)
4. Travel time (T_t) for each routing reach (hours)
5. Weighted runoff curve number (RCN) for each subarea
6. Rainfall amount (P) for each specified design storm (inches)
7. Total runoff (Q) for each subarea (inches)
8. Initial abstraction (I_a) for each subarea
9. Ratio of I_a/P for each subarea
10. Rainfall distribution (I, IA, II, or III)

25.6.5.3 Design Parameters

The use of the tabular method requires that the engineer determine the travel time through the entire watershed. As stated previously, the entire watershed is divided into smaller sub-watersheds that must be related to one another and to the whole watershed with respect to time. The result is that the time of peak discharge is known for any one sub-watershed relative to any other sub-watershed or for the entire watershed. Travel time T_t represents the time for flow to travel from the study point at the bottom of a sub-watershed to the bottom of the entire watershed. This information must be compiled for each sub-watershed.

Note: The data for up to 10 sub-watersheds can be compiled on one *TR-55* worksheet (*TR-55* Worksheets 5a and 5b).

To obtain the peak discharge using the graphical method, the unit peak discharge is read off of a curve; however, the tabular method provides this information in the form of a table of values, found in *TR-55*, Exhibit 5. These tables are arranged by rainfall type (I, IA, II, and III), I_a/P, t_c, and T_t. In most cases, the actual values for these variables, other than the rainfall type, will be different from the values shown in the table. Therefore, a system of rounding these values has been established in the *TR-55* manual. The I_a/P term is simply rounded to the nearest table value. The t_c and T_t values are rounded together in a procedure that is outlined on pages 5-2 and 5-3 of the *TR-55* manual. The accuracy of the computed peak discharge and time of peak discharge is highly dependent on the proper use of these procedures.

The following equation, along with the information compiled on *TR-55* Worksheet 5b, is then used to determine the flow at any time:

$$q = q_t \times A_m \times Q \tag{25.4}$$

where

q = Hydrograph coordinate (cfs) at hydrograph time t.
q_t = Tabular hydrograph unit discharge at hydrograph time t from *TR-55* Exhibit 5 (csm/in.).
A_m = Drainage area of individual subarea (mi²).
Q = Runoff (in.).

The product $A_m \times Q$ is multiplied by each table value in the appropriate unit hydrograph in *TR-55* Exhibit 5 (each sub-watershed may use a different unit hydrograph) to generate the actual hydrograph for the sub-watershed. This hydrograph is tabulated on *TR-55* Worksheet 5b and then added together with the hydrographs from the other sub-watersheds, being careful to use the same time increment of each sub-watershed. The result is a composite hydrograph at the bottom of the worksheet for the entire watershed.

Note: The preceding discussion on the tabular method is taken from *TR-55* and is not complete. The engineer should obtain a copy of *TR-55* and learn the procedures and limitations as outlined in that document. Examples and worksheets are provided in *TR-55* that lead the reader through the procedures for each chapter.

25.7 GENERAL STORMWATER ENGINEERING CALCULATIONS

This section provides guidelines for performing various engineering calculations associated with the design of stormwater management facilities such as extended detention and retention basins and multistage outlet structures.

25.7.1 Detention, Extended Detention, and Retention Basin Design Calculations

Based on stormwater management regulations in general, a stormwater management basin may be designed to control water quantity (for flood control and channel erosion control) and to enhance (or treat) water quality. The type of basin selected (extended detention, retention, infiltration, etc.) and the relationship between its design components (design inflow, storage volume, and outflow) will dictate the size of the basin and serve as the basis for its hydraulic design. Some design component parameters such as design storm return frequency, allowable discharge rates, etc., may be specified by the local regulatory authority, based upon the specific needs of certain watersheds or stream channels within that locality. Occasionally, as in stream channel erosion control, it may be up to the engineer to document and analyze the specific needs of the downstream channel and establish the design parameters.

The design inflow is either the peak flow or the runoff hydrograph from the developed watershed. This inflow becomes the input data for the basin sizing calculations, often called *routings*. Various routing methods are available. Note that the format of the hydrologic input data will usually be dictated by whatever routing method is chosen. (The methods discussed in this text require the use of a peak discharge or an actual runoff hydrograph.) Generally, larger and more complex projects will require a detailed analysis, which includes a runoff hydrograph. Preliminary studies and small projects may be designed using simpler, shortcut techniques that only require a peak discharge. For all projects, the designer must document the hydrologic conditions to support the inflow portion of the hydraulic relationship.

Manipulation of the site grades and strategic placement of the permanent features such as buildings and parking lots can usually provide adequate storage volume within a basin. Sometimes the site topography and available outfall location will dictate the location of a stormwater facility.

25.7.2   ALLOWABLE RELEASE RATES

The allowable release rates for a stormwater facility are dependent on the proposed function(s) of that facility, such as flood control, channel erosion control, or water quality enhancement. For example, a basin used for water quality enhancement is designed to detain the water quality volume and slowly release it over a specified amount of time. This water quality volume is the first flush of runoff, which is considered to contain the largest concentration of pollutants (Schueler 1987). In contrast, a basin used for flood or channel erosion control is designed to detain and release runoff from a given storm event at a predetermined maximum release rate. This release rate may vary from one watershed to another based on pre-developed conditions. Localities, through stormwater management and erosion control ordinances, have traditionally set the allowable release rates for given frequency storm events to equal the watershed's pre-developed rates. This technique has become a convenient and consistent mechanism to establish the design parameters for a stormwater management facility, particularly as it relates to flood control or stream channel erosion control. Depending on location, the allowable release rate for controlling steam channel erosion or flooding may be established by ordinance using the state's minimum criteria, or by analyzing specific downstream topographic, geographic, or geologic conditions to select alternate criteria. Obviously, the engineer should be aware of what the local requirements are before beginning the design. The design examples and calculations in this text use minimum requirements for illustrative purposes.

25.7.3   STORAGE VOLUME REQUIREMENT ESTIMATES

Stormwater management facilities are designed using a trial and error process. The engineer does many iterative routings to select a minimum facility size with the proper outlet controls. Each iterative routing requires that the facility size (stage–storage relationship) and the outlet configuration (stage–discharge relationship) be evaluated for performance against the watershed requirements. A graphical evaluation of the inflow hydrograph vs. an approximation of the outflow-rating curve provides the engineer with an estimate of the required storage volume. Starting with this assumed required volume, the number of iterations is reduced. The graphical hydrograph analysis requires that the evaluation of the watershed's hydrology produce a runoff hydrograph for the appropriate design storms. Generally, local stormwater management regulations allow the use of SCS methods or the modified rational method (critical storm duration approach) for analysis. Many techniques are available to generate the resulting runoff hydrographs based on these methods. It is the engineer's responsibility to be familiar with the limitations and assumptions of the methods as they apply to generating hydrographs.

Graphical procedures can be time consuming, especially when dealing with multiple storms, and are therefore not practical when designing a detention facility for a small site development. Shortcut procedures have been developed to allow the engineer to approximate the storage volume requirements. Such methods described in *TR-55* include storage volume for detention basins (Section 5-4.2) and critical storm duration—modified rational method—direct solution, (Section 5-4.4), which can be used as planning tools. Final design should be refined using a more accurate hydrograph routing procedure. Sometimes these shortcut methods may be used for final design, but they must be used with caution because they only approximate the required storage volume.

It should be noted that the *TR-55* tabular hydrograph method does not produce a full hydrograph. The tabular method generates only the portion of the hydrograph that contains the peak discharge and some of the time steps just before and just after the peak. The missing values must be

TABLE 25.7

Hydrologic Summary, SCS Methods

Condition	Area	Runoff Curve Number (RCN)	t_c	Q_2	Q_{10}
		TR-55 **Graphical Peak Discharge**			
Pre-development	25 ac	64	0.87 hr	8.5 cfs[a]	26.8 cfs[a]
Post-development	25 ac	75	0.35 hr	29.9 cfs	70.6 cfs
		TR-20 **Computer Run**			
Pre-development	25 ac	64	0.87 hr	8.0 cfs[a]	25.5 cfs[a]
Post-development	25 ac	75	0.35 hr	25.9 cfs	61.1 cfs

[a] Allowable release rate.

extrapolated, thus potentially reducing the accuracy of the hydrograph analysis. It is recommended that if SCS methods are to be used, a full hydrograph be generated using one of the available computer programs. The accuracy of the analysis can only be accurate as the hydrograph used.

25.7.4   GRAPHICAL HYDROGRAPH ANALYSIS—SCS METHODS

The following analysis represents a graphical hydrograph analysis that results in the approximation of the required storage volume for a proposed stormwater management basin. The following procedure is presented here to illustrate this technique. See Table 25.7 for a summary of the hydrology. The *TR-20* computer-generated hydrograph is used for this example. The allowable discharge from the proposed basin has been established by ordinance (based on pre-developed watershed discharge).

25.7.4.1   Procedure

The pre- and post-developed hydrology, which includes the pre-developed peak rate of runoff (allowable release rate) and the post-developed runoff hydrograph (inflow hydrograph) is required for hydrograph analysis (see Table 25.7). Refer to Figure 25.9 for the 2-year developed inflow hydrograph and Figure 25.10 for the 10-year developed inflow hydrograph:

1. Commencing with the plot of the 2-year developed inflow hydrograph (discharge vs. time), the 2-year allowable release rate ($Q_2 = 8$ cfs) is plotted as a horizontal line starting at time $t = 0$ and continuing to the point where it intersects the falling limb of the hydrograph.
2. A diagonal line is then drawn from the beginning of the inflow hydrograph to the intersection point described above. This line represents the hypothetical rating curve of the control structure and approximates the rising limb of the outflow hydrograph for the 2-year storm.
3. The storage volume is then approximated by calculating the area under the inflow hydrograph, less the area under the rising limb of the outflow hydrograph. This is shown as the shaded area in Figure 25.9. The storage volume required for the 2-year storm, measuring the shaded area with a planimeter, could approximate S_2.

The vertical scale of a hydrograph is in cubic feet per second (cfs) and the horizontal scale is in hours (hr). Therefore, the area, as measured in square inches (in.^{2}), is multiplied by scale conversion factors of cfs per inch, hours per inch, and 3600 seconds per hour, to yield an area in cubic feet (ft^3). The conversion is

$$S_2 = (0.398 \text{ in.}^2) \times (10 \text{ cfs/in.}) \times (2.5 \text{ hr/in.}) \times (3600 \text{ sec/hr}) = 35{,}820 \text{ ft}^3 = 0.82 \text{ ac-ft}$$

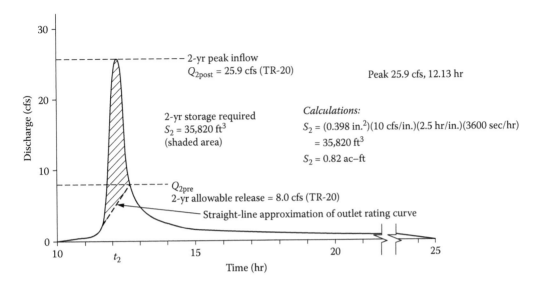

FIGURE 25.9 SCS runoff hydrograph, 2-yr post-developed.

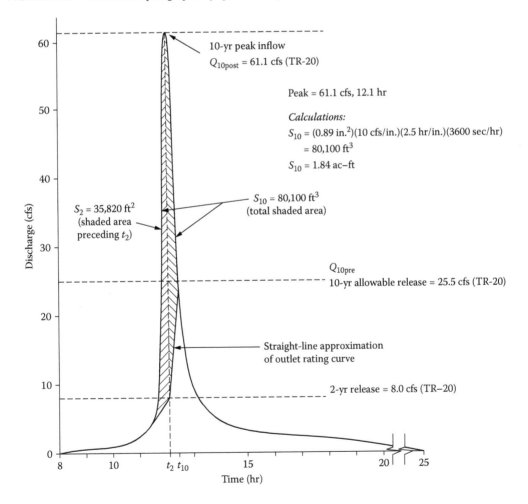

FIGURE 25.10 SCS runoff hydrograph, 10-yr post-developed.

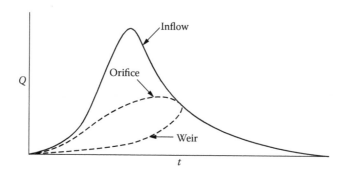

FIGURE 25.11 Typical outlet rating curves for orifice and weir outlet devices.

4. On a plot of the 10-year inflow hydrograph, the 10-year allowable release rate (Q_{10}) is plotted as a horizontal line extending from time zero to the point where it intersects the falling limb of the hydrograph.

5. By trial and error, the time (t_2) at which the S_2 volume occurs while maintaining the 2-year release is determined by planimeter. The shaded area represents this to the left of t_2 in Figure 25.10. From the intersection point of t_2 and the 2-year allowable release rate (Q_2), a line is drawn to connect to the intersection point of the 10-year allowable release rate and the falling limb of the hydrograph. This intersection point is t_{10}, and the connecting line is a straight-line approximation of the *outlet-rating curve*.

6. The area under the inflow hydrograph from time t_2 to time t_{10}, less the area under the rising limb of the hypothetical rating curve, represents the additional volume (shaded area to the right of t_2 in Figure 25.10) needed to meet the 10-year storm storage requirements.

7. The total storage volume, S_{10}, required, can be computed by adding this additional storage volume to S_2. The total shaded area under the hydrograph represents this:

$$S_{10} = (0.89 \text{ in.}^2) \times (10 \text{ cfs/in.}) \times (2.5 \text{ hr/in.}) \times (3600 \text{ sec/hr}) = 80,100 \text{ ft}^3 = 1.84 \text{ ac-ft}$$

These steps may be repeated if storage of the 100-year storm, or any other design frequency storm, is required by ordinance of downstream conditions.

In summary, the total volume of storage required is the area under the runoff hydrograph curve and above the basin outflow curve. It should be noted that the outflow-rating curve is approximated as a straight line. The actual shape of the outflow-rating curve will depend on the type of outlet device used. Figure 25.11 shows the typical shapes of outlet rating curves for orifice and weir outlet structures. The straight-line approximation is reasonable for an orifice outlet structure. However, this approximation will likely underestimate the storage volume required when a weir outlet structure is used. Depending on the complexity of the design and the need for an exact engineered solution, the use of a more rigorous sizing technique, such as a storage indication routing, may be necessary.

25.7.5 TR-55 STORAGE VOLUME FOR DETENTION BASINS (SHORTCUT METHOD)

The *TR-55* storage volume for detention basins procedure (*TR-55* shortcut procedure) provides results similar to those for graphical analysis. This method is based on average storage and routing effects for many structures. *TR-55* can be used for single-stage or multistage outflow devices. The only constraints are that (1) each stage requires a design storm and a computation of the storage required for it, and (2) the discharge of the upper stages includes the discharge of the lower stage(s). Refer to *TR-55* for more detailed discussions and limitations.

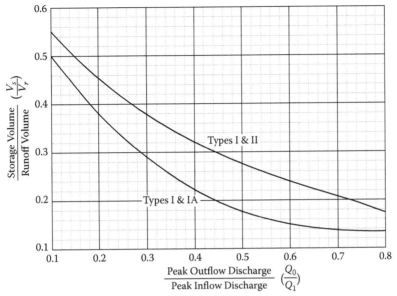

From *TR-55* Approximate Detention Basin Routing for Rainfall Types I, IA, II, and III.

FIGURE 25.12 Approximate detention basin routing for rainfall Types I, IA, II, and III.

25.7.5.1 Information Needed

To calculate the required storage volume using *TR-55*, the pre-developed and post-developed hydrology per SCS methods is needed. This includes the watershed's pre-developed peak rate of discharge, or allowable release rate (Q_o); the watershed's post-developed peak rate of discharge, or inflow (Q_i) for the appropriate design storms; and the watershed's post-developed runoff (Q), in inches. (Note that this method does not require a hydrograph.) Once these parameters are known, the *TR-55* manual can then be used to approximate the storage volume required for each design storm. The following procedure summarizes the *TR-55* shortcut method using the 25-acre watershed.

25.7.5.2 Procedure

1. Determine the peak-developed inflow (Q_i) and the allowable release rate (Q_o) from the hydrology for the appropriate design storm. The 2-year storm flow rates given below are based on *TR-55* graphical peak discharge:

$$Q_{o2} = 8.5 \text{ cfs}$$

$$Q_{i2} = 29.9 \text{ cfs}$$

Use the ratio of the allowable release rate (Q_o) to the peak developed inflow (Q_i), or Q_o/Q_i, for the appropriate design storm and refer to Figure 25.12 (or Figure 6-1 in *TR-55*) to obtain the ratio of storage volume:

$$Q_{o2}/Q_{i2} = 8.5/29.9 = 0.28$$

From Figure 25.12 or *TR-55* Figure 6-1, the ratio of storage volume (V_{s2}) to runoff volume (V_{r2}) is

$$V_{s2}/V_{r2} = 0.39$$

TABLE 25.8

Storage Volume Requirements

Method	2-yr. Storage Required	10-yr Storage Required
Graphical hydrograph analysis	0.82 ac-ft.	1.84 ac-ft.
TR-55 shortcut method	1.05 ac-ft.	1.96 ac-ft.

2. Determine the runoff volume (V_r) in acre-ft, from the *TR-55* worksheets for the appropriate design storm:

$$V_r = Q \times A_m \times 53.33$$

where
Q = Runoff (in.) from *TR-55* Worksheet 2 = 1.30 in.
A_m = Drainage area (mi²) (25 ac ÷ 640 ac/mi² = 0.039 mi²).
V_r = 1.30 × 0.39 × 53.33 = 2.70 ac-ft.

3. Multiply the V_s/V_r ratio from step 1 by the runoff volume (V_r) from step 2 to determine the volume of storage required (V_s) in ac-ft:

$$\left(\frac{V_s}{V_r} \right) V_r = V_s$$

$$0.39 \times 2.70 \text{ ac-ft} = 1.05 \text{ ac-ft}$$

4. Repeat these steps for each additional design storm as required to determine the approximate storage requirements. The 10-year storm storage requirements are presented here:
   a. Q_o = 26.8 cfs
   b. Q_i = 70.6 cfs
   c. Q_o/Q_i = 26.8/70.6 = 0.38; from Figure 25.12, V_s/V_r = 0.33
   d. $V_s = (V_s/V_r) \times V_r$ = 0.33 × 5.93ac-ft. = 1.96 ac-ft
   This volume represents the total storage required for the 2-year storm and the 10-year storm.

Note: The volume from step 4 above may need to be increased if additional storage is required for water quality purposes or channel erosion control.

The design presented above should be used with *TR-55* Worksheet 6a. The worksheet includes an area to plot the *stage–storage curve*, from which actual elevations corresponding to the required storage volumes can be derived. Table 25.8 provides a summary of the required storage volumes using the graphical SCS hydrograph analysis and the *TR-55* shortcut method.

25.7.6 GRAPHICAL HYDROGRAPH ANALYSIS, MODIFIED RATIONAL METHOD, CRITICAL STORM DURATION

The modified rational method uses the *critical storm duration* to calculate the *maximum storage volume* for a detention facility. This critical storm duration is the storm duration that generates the greatest volume of runoff and, therefore, requires the most storage. In contrast, the rational method produces a triangular runoff hydrograph that gives the peak inflow at time = t_c and falls to zero flow at time = $2.5t_c$. In theory, this hydrograph represents a storm whose duration equals the time of concentration (t_c), resulting in the greatest peak discharge for the given return frequency storm. The volume of runoff, however, is of greater consequence in sizing a detention facility. A storm whose duration is

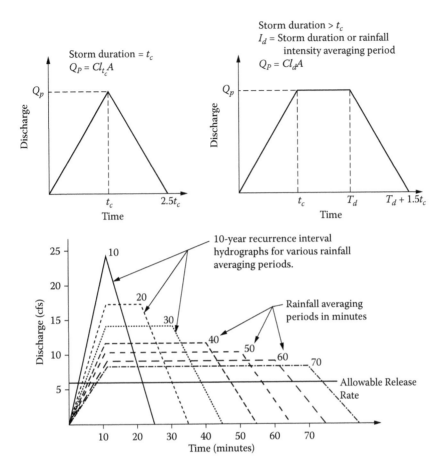

FIGURE 25.13 Modified rational method hydrographs.

longer than the t_c may not produce as large a peak rate of runoff, but it may generate a greater volume of runoff. By using the modified rational method, the designer can evaluate several different storm durations to verify which one requires the greatest volume of storage with respect to the allowable release rate. It is this maximum storage volume that the basin must be designed to detain.

The first step in determining the critical storm duration is to use the post-developed time of concentration (t_c) to generate a post-developed runoff hydrograph. Rainfall intensity averaging periods (T_d), representing time periods incrementally longer than the t_c, are then used to generate a "family" of runoff hydrographs for the same drainage area. These hydrographs will be trapezoidal with the peak discharges (Q_p) based upon the intensity (I) of the averaging period (T_d). Figure 25.13 shows the construction of a typical triangular and trapezoidal hydrograph using the modified rational method, and a family of trapezoidal hydrographs representing storms of different durations.

Note that the duration of the receding limb of the trapezoidal hydrograph (Figure 25.13) is set to equal 1.5 times the time of concentration (t_c). Also, the total hydrograph duration is $2.5t_c$ vs. $2t_c$. This longer duration is considered more representative of actual storm and runoff dynamics. It is also more analogous to the SCS unit hydrograph where the receding limb extends longer than the rising limb.

The modified rational method assumes that the rainfall intensity averaging period is equal to the actual storm duration. This means that the rainfall and runoff that occur before and after the rainfall-averaging period are not accounted for. Therefore, the modified rational method may underestimate the required storage volume for any given storm event.

TABLE 25.9

Hydrologic Summary, Rational Method

Condition	Area	Developed Runoff Coefficient (C)	t_c	Q_2	Q_{10}
Pre-developed	25 ac	0.38	0.87 hr (52 min)	17 cfs	24 cfs
Post-developed	25 ac	0.59	0.35 hr (21 min)	49 cfs	65 cfs

The rainfall intensity averaging periods are chosen arbitrarily; however, the designer should select periods for which the corresponding intensity–duration–frequency (IDF) curves are available (e.g., 10 min, 20 min, 30 min). The shortest period selected should be the time of concentration (t_c). A straight line starting at $Q = 0$ and $t = 0$ and intercepting the inflow hydrograph on the receding limb at the allowable release rate (Q_o) represents the outflow rating curve. The time averaging period hydrograph that represents the greatest storage volume required is the one with the largest area between the inflow hydrograph and outflow-rating curve. This determination is made by a graphical analysis of the hydrographs.

The next procedure presents a graphical analysis very similar to the one described earlier. Note that the rational and modified rational methods should normally be used in homogeneous drainage areas of less than 20 acres, with a t_c of less than 20 minutes. Although the watershed in our example has a drainage area of 25 acres and a t_c of greater than 20 minutes, it will be used here for illustrative purposes. Note that the pre- and post-developed peak discharges are much greater than those calculated using the SCS method applied to the same watershed. This difference may be the result of the large acreage and t_c values. A summary of the hydrology is found in Table 25.9. Note that the t_c calculations were performed using the more rigorous SCS *TR-55* method.

25.7.6.1 Information Needed

The modified rational method-critical storm duration approach is very similar to SCS methods because it requires pre- and post-developed hydrology in the form of a pre-developed peak rate of runoff (allowable release rate) and a post-developed runoff hydrograph (inflow hydrograph), as developed using the rational method.

25.7.6.2 Procedure

Refer to Figures 25.14 and 25.15.

1. Plot the 2-year developed condition inflow hydrograph (triangular) based on the developed condition (t_c).
2. Plot a family of hydrographs, with the time averaging period (T_d) of each hydrograph increasing incrementally from 21 minutes (developed condition t_c) to 60 minutes, as shown in Figure 25.14. Note that the first hydrograph is a Type 1 modified rational method triangular hydrograph, where the storm duration (d) or (T_d) is equal to the time of concentration (t_c). The remaining hydrographs are trapezoidal, or Type 2 hydrographs. The peak discharge for each hydrograph is calculated using the rational equation, $Q = CIA$, where intensity I from the IDF curve is determined using the rainfall intensity averaging period as the storm duration.
3. Superimpose the outflow-rating curve on each inflow hydrograph. The area between the two curves then represents the storage volume required, as shown in Figure 25.14. Similar cautions, as described in the SCS methods, regarding the straight-line approximation of the outlet discharge curve apply here as well. The actual shape of the outflow curve depends on the type of outlet device.

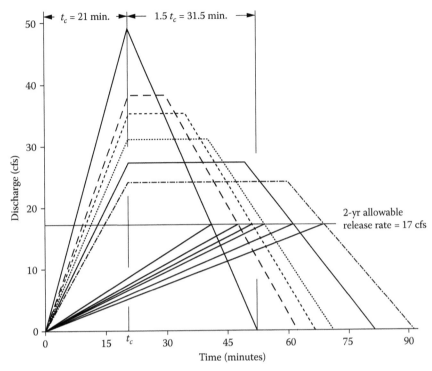

Conversion of in.^{2} to ft^3 = 10 cfs/in × 0.25 hr/in. × 3600 sec/hr = 9000 ft^3/in.^{2}

Rainfall Avg. Period (T)	Intensity I (in./hr)	Q Peak (cfs)	Area (in^2)	Storage Volume (ft^3)
21 min	3.3	49	5.62	50,580*
30 min	2.6	38	5.51	49,590
35 min	2.4	35	5.53	49,770
40 min	2.1	31	5.29	47,610
50 min	1.8	27	5.25	47,250
60 min	1.6	24	5.24	47,160

FIGURE 25.14 Modified rational method runoff hydrograph, 2-year post-developed condition.

4. Compute and tabulate the required storage volume for each of the selected durations or time-averaging periods (T_d). The storm duration that requires the maximum storage is the *critical storm* and is used for the sizing of the basin. (Storm duration equal to the t_c produces the largest storage volume required for the 2-year storm presented here.)
5. Repeat Steps 1 through 4 above for the analysis of the 10-year storage area requirements. (Figure 25.15 shows this procedure repeated for the 10-year design storm.)

Note: Conveyance systems should still be designed using the rational method, as opposed to the modified rational method, to ensure their design for the peak rate of runoff.

25.7.7 MODIFIED RATIONAL METHOD, CRITICAL STORM DURATION—DIRECT SOLUTION

A direct solution to the modified rational method, critical storm duration has been developed to eliminate the time intensive, iterative process of generating multiple hydrographs. This direct solution takes into account the storm duration and allows the engineer to solve for the time at which the storage volume curve has a slope equal to zero, which corresponds to maximum storage. The basic derivation of this method is provided below, followed by the procedure as applied to our examples.

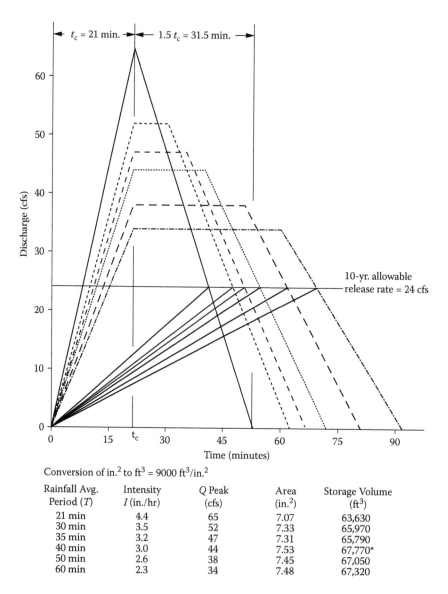

Conversion of in.^{2} to ft^3 = 9000 ft^3/in.^{2}

Rainfall Avg. Period (T)	Intensity I (in./hr)	Q Peak (cfs)	Area (in.^{2})	Storage Volume (ft^3)
21 min	4.4	65	7.07	63,630
30 min	3.5	52	7.33	65,970
35 min	3.2	47	7.31	65,790
40 min	3.0	44	7.53	67,770*
50 min	2.6	38	7.45	67,050
60 min	2.3	34	7.48	67,320

FIGURE 25.15   Modified ration method runoff hydrograph, 10-year post-developed condition.

25.7.7.1   Storage Volume

The runoff hydrograph developed with the modified rational method, critical storm duration will be either triangular or trapezoidal in shape. The outflow hydrograph of the basin is approximated by a straight line starting at 0 cfs at the time $t = 0$ and intercepting the receding leg of the runoff hydrograph at the allowable discharge (Q_o).

Note: The straight-line representation of the outflow hydrograph is a conservative approximation of the shape of the outflow hydrograph for an orifice control release structure. This method should be used with caution when designing a weir control release structure.

The area between the inflow hydrograph and the outflow hydrograph in Figure 25.16 represents the required storage volume. This area, the trapezoidal hydrograph storage volume, can be approximated using the following equation:

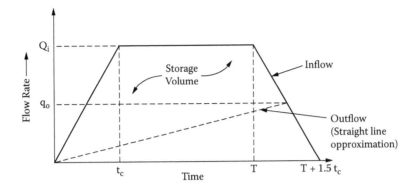

FIGURE 25.16   Trapezoidal hydrograph storage volume estimate.

$$V = \left[Q_i T_d + \frac{Q_i t_c}{4} - \frac{Q_o T_d}{2} - \frac{3 Q_o t_c}{4} \right] 60 \qquad (25.5)$$

where
V = Required storage volume (ft^3).
Q_i = Inflow peak discharge (cfs) for the critical storm duration (T_d).
T_c = Post-developed time of concentration (min).
Q_o = Allowable peak outflow (cfs).
T_d = Critical storm duration (min).

Regulatory ordinance or downstream conditions establish the allowable peak outflow. The critical storm duration (T_d) is an unknown and must be determined to solve for intensity I and to ultimately calculate the peak inflow (Q_i). Therefore, a relationship between rainfall intensity I and critical storm duration T_d must be established.

25.7.7.2   Rainfall Intensity

The rainfall intensity as taken from the IDF curves is dependent on the time of concentration (t_c) of a given watershed. Setting the storm duration (T_d) equal to t_c will provide the maximum peak discharge. As stated previously, however, it does not necessarily generate the maximum volume of discharge. Since this maximum volume of runoff is of interest, and the storm duration is unknown, rainfall intensity I must be represented as a function of time, frequency, and location. The relationship is expressed by the modified rational method intensity equation as follows:

$$I = \frac{a}{b + T_d} \qquad (25.6)$$

where
I = Rainfall intensity (in./hr).
T_d = Rainfall duration or rainfall intensity averaging period (min).
a, b = Rainfall constants developed for storms of various recurrence intervals and various geographic locations (Table 25.10).

The rainfall constants, a and b, were developed from linear regression analyses of the IDF curves and can be generated for any area where such curves are available. The limitations associated with the IDF curves, such as duration, or return frequency, will also limit development of the constants. Table 25.10 provides rainfall constants for various regions in Virginia. Substituting Equation 25.6 into the rational equation results in the rearranged rational equation as follows:

TABLE 25.10
Rainfall Constants for Virginia
(Duration, 5 min to 2 hr)

Station	Rainfall Frequency	Constants[a]	
1	2	117.7	19.1
	5	168.6	23.8
	10	197.8	25.2
2	2	118.8	17.2
	5	158.9	20.6
	10	189.8	22.6
3	2	130.3	18.5
	5	166.9	20.9
	10	189.2	22.1
4	2	126.3	17.2
	5	173.8	22.7
	10	201.0	23.9
5	2	143.2	21.0
	5	173.9	22.7
	10	203.9	24.8

[a] Constants are based on linear regression analyses of the IDF curves contained in the Virginia Department of Transportation (VDOT) *Drainage Manual*.

$$Q = C \left(\frac{a}{b + T_d} \right) A \tag{25.7}$$

where

Q = Peak rate of discharge (cfs).

a, b = Rainfall constants developed for storms of various recurrence intervals and various geographic locations (Table 25.10).

T_d = Critical storm duration (min).

C = Runoff coefficient.

A = Drainage area (acres).

Substituting this relationship for Q, Equation 25.5 then becomes:

$$V = \left[\left[C \left(\frac{a}{b + T_d} \right) A \right] \left(T_d + \frac{\left[C \left(\frac{a}{b + T_d} \right) A \right] t_c}{4} - \frac{Q_o T_d}{2} - \frac{3 Q_o t_c}{4} \right) \right] 60 \tag{25.8}$$

25.7.7.3 Maximum Storage Volume

The first derivative of this storage volume equation, Equation 25.8, with respect to time is an equation that represents the slope of the storage volume curve plotted vs. time. When this equation is set to equal zero and solved for T_d, the slope of the storage volume curve is zero, or at a maximum, as

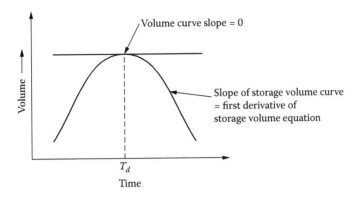

FIGURE 25.17 Storage volume vs. time curve.

shown in Figure 25.17. Equation 25.9 represents the first derivative of the storage volume equation with respect to time and can be solved for the critical storm duration (T_d):

$$T_d = \sqrt{\frac{2CAa\left(b-t_c/4\right)}{Q_o}} - b \qquad (25.9)$$

where

T_d = Critical storm duration (min).
C = Runoff coefficient.
A = Drainage area (ac).
a, b = Rainfall constants developed for storms of various recurrence intervals and various geographic locations (Table 25.10).
t_c = Time concentration (min).
Q_o = Allowable peak outflow (cfs).

Equation 25.9 is solved for T_d. Then, T_d is substituted into Equation 25.7 to solve for Q_i, and when Q_i is known, the outlet structure and the stormwater facility can be sized. This method provides a direct solution to the graphical analysis of the family of hydrographs and is quicker to use. The following procedure illustrates this method.

25.7.7.4 Information Needed

The modified rational method-direct solution is similar to the previous methods because it requires determination of the pre- and post-developed hydrology, resulting in a pre-developed peak rate of runoff (*allowable release rate*) and a post-developed *runoff hydrograph*. Table 25.9 provides a summary of the hydrology. The rainfall constants a and b for the watershed are determined from Table 25.10.

25.7.7.5 Procedure

1. Determine the 2-year critical storm duration by solving Equation 25.9:

$$T_{d2} = \sqrt{\frac{2CAa\left(b-t_c/4\right)}{Q_{o2}}} - b$$

Given:

T_{d2} = 2-year critical storm duration, (min).
C = Developed condition runoff coefficient = 0.59.
A = Drainage area = 25.0 ac.
t_c = Post-developed time of concentration = 21 min.
Q_{o2} = Allowable peak outflow = 17 cfs (pre-developed peak rate of discharge).
a_2 = 2-year rainfall constant = 130.3.
b_2 = 2-year rainfall constant = 18.5.

$$T_{d2} = \sqrt{\frac{2(0.59)(25.0)(130.3)(18.5 - 21/4)}{17}} = \sqrt{2955.0} - 18.5 = 36.2 \text{ min}$$

2. Solve for the 2-year critical storm duration intensity (I_2) using Equation 25.6 and the 2-year critical storm duration (T_{d2}):

$$I_2 = \frac{a}{b + T_{d2}}$$

where

T_{d2} = Critical storm duration = 36.2 min.
a = 2-year rainfall constant = 130.3.
b = 2-year rainfall constant = 18.5.

$$I_2 = \frac{130.3}{18.5 + 36.2} = 2.38 \text{ in./hr}$$

3. Determine the 2-year peak inflow (Q_{i2}) using the rational equation and the critical storm duration intensity (I_2):

$$Q_{i2} = C \times I_2 \times A$$

where

Q_{i2} = 2-year peak inflow (cfs).
C = Developed condition runoff coefficient = 0.59.
I_2 = Critical storm intensity = 2.38 in./hr.
A = Drainage area = 25 ac.

$$Q_{i2} = 0.59 \times 2.38 \times 25 = 35.1 \text{ cfs}$$

4. Determine the 2-year required storage volume for the 2-year critical storm duration (T_{d2}) using Equation 25.5:

$$V_2 = \left[Q_{i2}T_{d2} + \frac{Q_{i2}t_c}{4} - \frac{Q_{o2}T_{d2}}{2} - \frac{3Q_{o2}t_c}{4} \right] 60$$

where

V_2 = 2-year required storage (ft³).
Q_{i2} = 2-year peak inflow for critical storm = 35.1 cfs.
C = Developed runoff coefficient = 0.59.
A = Area = 25.0 ac.

T_{d2} = Critical storm duration = 36.2 min.
t_c = Developed condition time of concentration = 21 min.
Q_{o2} = 2-year allowable peak outflow = 17 cfs.

$$V_2 = \left[(35.1)(36.2) + \left(\frac{(35.1)(21)}{4} \right) - \left(\frac{(17)(36.2)}{2} \right) - \left(\frac{3(17)(21)}{4} \right) \right] 60 = 52,764 \text{ ft}^3 = 1.21 \text{ ac-ft}$$

Repeat Steps 2 through 4 for the 10-year storm, as follows:

5. Determine the 10-year critical storm duration T_{d10}, using Equation 25.9.

Given:

T_{d10} = 10-year critical storm duration (min).
C = Developed condition runoff coefficient = 0.59.
A = Drainage area = 25 ac.
t_c = Post-developed time of concentration = 21 min.
Q_{o10} = 24 cfs.
a_{10} = 189.2.
b_{10} = 22.1.

$$T_{d10} = \sqrt{\frac{2(0.59)(25.0)(189.2)(22.1 - 21/4)}{24}} - 22.1 = \sqrt{3918.6} - 22.1 = 40.5 \text{ min.}$$

6. Solve for the 10-year critical storm duration intensity (I_{10}) using Equation 25.6 and the 10-year critical storm duration (T_{d10}):

$$I_{10} = \frac{a}{b + T_{d10}} = \frac{189.2}{22.1 + 40.5} = 3.02 \text{ in./hr}$$

7. Determine the 10-year peak inflow (Q_{i10}) using the rational equation and the critical storm duration intensity (I_{10}).

Given:

Q_{i10} = 10-year peak inflow.
C = Developed condition runoff coefficient = 0.59.
I_{10} = Critical storm intensity = 3.02 in./hr.
A = Drainage area = 25.0 ac.

$$Q_{i10} = C \times I_{10} \times A = 44.5 \text{ cfs}$$

8. Determine the required 10-year storage volume for the 10-year critical storm duration (T_{d10}) using Equation 25.5.

Given:

V_{10} = Required storage (ft^3).
Q_{i10} = 44.5 cfs.
C = 0.59.
A = 25 ac.
T_{d10} = 40.5 min.
t_c = 21 min.
Q_{o10} = 24 cfs.

TABLE 25.11

Summary of Results: Storage Volume Requirement Estimates

Method	2-yr. Storage Required	10-yr. Storage Required
Graphical hydrograph analysis	0.82 ac-ft	1.84 ac-ft
TR-55 shortcut method	1.05 ac-ft	1.96 ac-ft
Modified rational method	1.16 ac-ft	1.56 ac-ft
Modified rational method—critical	1.21 ac ft	1.61 ac-ft
Storm duration—direct solution	$T_d = 36.2$ min.	$T_d = 40.5$ min.

$$V_{10} = \left[(44.5)(40.5) + \frac{(44.5)(21)}{4} - \frac{(24)(40.5)}{2} - \frac{(3)(24)(21)}{4} \right] 60 = 70{,}308 \text{ ft}^3 = 161 \text{ ac-ft}$$

V_2 and V_{10} represent the total storage volume required for the 2-year and 10-year storms, respectively. Table 25.11 provides a summary of the four different sizing procedures used in this section. The engineer should choose one of these methods based on the complexity and size of the watershed and the chosen hydrologic method. Using the stage–storage curve, a multistage riser structure can then be designed to control the appropriate storms and, if required, the water quality volume.

25.7.8 STAGE–STORAGE CURVE

By using one of the above methods for determining the storage volume requirements, the engineer now has sufficient information to place and grade the proposed stormwater facility. Remember, this is a preliminary sizing which needs to be refined during the actual design. Trial and error can achieve the approximate required volume achieved by designing the basin to fit the site geometry and topography. The storage volume can be computed by planimetering the contours and creating a stage–storage curve.

25.7.8.1 Storage Volume Calculations

For retention/detention basins with vertical sides, such as tanks and vaults, the storage volume is simply the bottom surface area times the height. For basins with graded (2H:1V, 3H:1V, etc.) side slopes or an irregular shape, the stored volume can be computed by the following procedure. Figure 25.18 represents the stage–storage computation worksheet completed for our example. (Note that other methods for computing basin volumes are available, such as the conic method for reservoir volumes, but they are not presented here.)

25.7.8.2 Procedure

1. Planimeter or otherwise compute the area enclosed by each contour and enter the measured value into columns 1 and 2 of Figure 25.18. The invert of the lowest control orifice represents zero storage. This will correspond to the bottom of the facility for extended detention or detention facilities, or the permanent pool elevation for retention basins.
2. Convert the planimetered area (often in square inches) to units of square feet in column 3 of Figure 25.18.
3. The average area between two contours is computed by adding the area planimetered for the first elevation, column 3, to the area planimetered for the second elevation, also column 3, and then dividing their sum by 2. This average is then written in column 4 of Figure 25.18. From this figure,

1	2	3	4	5	6	7	
ELEV.	AREA (in^2)	AREA (ft^2)	AVG. AREA (ft^2)	INTERVAL	VOL. (ft^3)	TOTAL VOLUME	
						(ft^3)	$(ac.ft.^3)$
81	0	0				0	0
			900	1	900		
82	2.0	1800				900	0.02
			2520	2	5040		
84	3.0	3240				5940	0.14
			4207	2	8414		
86	5.75	5175				14354	0.33
			7614	2	15228		
88	11.17	10053				29582	0.68
			12991	2	25982		
90	17.7	15930				55564	1.28
			20700	2	41400		
92	28.3	25470				96964	2.23
			31102	1	31102		
93	40.8	36734				128066	2.94
			38105	1	38105		
94	43.9	39476				166171	3.81

PROJECT: EXAMPLE 1 SHEET OF
COUNTY: COMPUTED BY: DATE:
DESCRIPTION:
ATTACH COPY OF TOPO: SCALE – 1" = 30 ft.

FIGURE 25.18 Stage–storage computation worksheet.

$$\text{Average area, elevation 81–82} \quad \frac{0+1800}{2} = 900 \text{ ft}^2$$

$$\text{Average area, elevation 82–84} \quad \frac{1800+3240}{2} = 2520 \text{ ft}^2$$

$$\text{Average area, elevation 84–86} \quad \frac{3240+5175}{2} = 4207 \text{ ft}^2$$

This procedure is repeated to calculate the average area found between any two consecutive contours.

4. Calculate the volume between each contour by multiplying the average area from step 3 (column 4) by the contour interval and placing the product in column 6. From Figure 25.18,

Contour interval between 81 and 82 = 1 ft × 900 ft² = 900 ft³

Contour interval between 82 and 84 = 2 ft × 2520 ft² = 5040 ft³

This procedure is repeated for each measured contour interval.

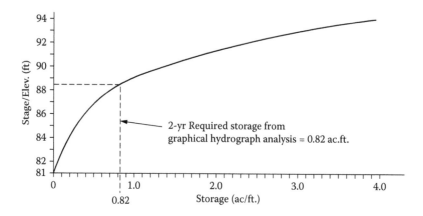

FIGURE 25.19   Stage–storage curve.

5. Sum the volume for each contour interval in column 7. Using Figure 25.18, this is simply the sum of the volumes computed in the previous step:

Contour 81 volume = 0
Contour 82 volume = 0 + 900 = 900 ft³
Contour 84 volume = 900 + 5040 = 5940 ft³
Contour 86 volume = 5940 + 8414 = 14,354 ft³

Column 8 allows for the volume to be tabulated in units of acre-feet: ft³ + 43,560 ft²/ac. This procedure is then repeated for each measured contour interval.

6. Plot the stage–storage curve with stage on the y-axis vs. storage on the x-axis. Figure 25.19 represents the stage–storage curve for our example in units of feet (stage) vs. acre-feet (storage).

The stage–storage curve allows the engineer to estimate the design high water elevation for each of the design storms if the required storage volume has been determined. This allows for a preliminary design of the riser orifice sizes and their configuration.

25.7.9   WATER QUALITY AND CHANNEL EROSION CONTROL VOLUME CALCULATIONS

Various local stormwater management regulations (the reader is advised to check with their local requirements) require that the first flush of runoff, or the water quality volume, be treated to enhance water quality. The water quality volume (V_{wq}) is the first 0.5 inches of runoff from the impervious area of development. The water quality volume must be treated using one or a combination of best management practices (BMPs) depending on the total size of the contributing watershed, amount of impervious area, and site conditions.

Calculate the water quality volume:

$$V_{wq} = \text{Impervious area (ft}^2) \times (1/2 \text{ in.})/(12 \text{ in./ft}) \tag{25.10}$$

$$V_{wq} \text{ (ac-ft)} = V_{wq} \text{ (ft}^3)/43,560 \text{ ft}^2/\text{ac} \tag{25.11}$$

The water quality volume for a wet BMP may be dependent on the specific design criteria for that BMP based on the watershed's imperviousness or the desired pollutant removal efficiency (using performance-based or technology-based criteria, respectively). This discussion is focused on the calculations associated with the control of the water quality volume in extended detention and retention basins.

Various local stormwater management regulations allow for the control of downstream channel erosion by detaining a specified volume of runoff for a period of time. Specifically, 24-hour extended detention of the runoff from the 1-year frequency storm is proposed as alternate criteria to the 2-year peak rate reduction. The channel erosion control volume (V_{ce}) is calculated by first determining the depth of runoff (in inches) based on the fraction of rainfall to runoff (runoff curve number) and then multiplying the runoff depth by the drainage area to be controlled.

25.7.9.1 Retention Basins—Water Quality Volume

The permanent pool feature of a retention basin allows for settling of particulate pollutants, such as sediment and other pollutants that adsorb to these particulates. Therefore, it is essential that the volume of the pool be both large enough and properly configured to prevent *short-circuiting*. (Short-circuiting results when runoff enters the pool and exits without sufficient time for the settling process to occur.) The permanent pool, or "dead" storage volume, of a retention facility is a function of the water quality volume. For example, a permanent pool sized to contain four times the water quality volume provides greater pollutant removal capacity than a permanent pool sized to contain two times the water quality volume. Our example analyzes a 25-acre watershed. The water quality volume and permanent pool volume calculations for a retention basin serving this watershed are calculated next.

25.7.9.1.1 Procedure
1. Calculate the water quality volume (V_{wq}) for the given watershed. Assume the commercial/industrial development disturbs 11.9 acres, with 9.28 acres (404,236 ft²) of impervious cover after development.

$$V_{wq} = 404,236 \text{ ft}^2 \times 1/2 \text{ in./12 in./ft} = 16,843 \text{ ft}^3$$

$$16,843 \text{ ft}^3/43,560 \text{ ft}^2/\text{ac} = 0.38 \text{ ac-ft}$$

2. Size the permanent pool based on the desired *pollutant removal efficiency* or the drainage area *impervious cover*. The pool volume will be sized based upon the desired pollutant removal efficiency. The permanent pool must be sized for $4 \times V_{wq}$ for a pollutant removal efficiency of 65%.

$$\text{Permanent pool volume} = V_{wq} \times 4.0 = 0.38 \text{ ac-ft} \times 4.0 = 1.52 \text{ ac-ft}$$

25.7.9.2 Extended Detention Basins—Water Quality Volume and Orifice Design

A water quality extended detention basin treats the water quality volume by detaining it and releasing it over a specified amount of time. In theory, extended detention of the water quality volume will allow the particulate pollutants to settle out of the first flush or runoff, functioning similarly to a permanent poll. Various local stormwater management regulations pertaining to water quality specify a 30-hour drawdown time for the water quality volume. This is a *brim drawdown* time, beginning at the time of peak storage of the water quality volume. Brim-drawdown time means the time required for the entire calculated volume to drain out of the basin. This assumes that the brim volume is present in the basin prior to any discharge. In reality, however, water is flowing out of the basin prior to the full or brim volume being reached. Therefore, the extended detention orifice can be sized using either of the following methods:

1. Use the maximum *hydraulic head* associated with the water quality volume (V_{wq}) and calculate the orifice size needed to achieve the required drawdown time, and route the water quality volume through the basin to verify the actual storage volume used and the drawdown time.
2. Approximate the orifice size using the *average hydraulic head* associated with the water quality volume (V_{wq}) and the required drawdown time.

The two methods for calculating the required size of the extended detention orifice allow for a quick and conservative design (Method 2, above) and a similarly quick estimation with a routing to verify the performance of the design (Method 1). Method 1, which uses the maximum hydraulic head and maximum discharge in the calculation, results in a slightly larger orifice than the same procedure using the average hydraulic head (Method 2). The routing allows the designer to verify the performance of the calculated orifice size. As a result of the routing effect however, the actual basin storage volume used to achieve the drawdown time will be less than the computed brim draw-down volume. It should be noted that the routing of the extended detention of the runoff from storms larger than the water quality storm (such as the 1-year frequency storm for channel erosion control) will result in a proportionately larger reduction in the actual storage volume needed to achieve the required extended detention. The procedure used to size an extended detention orifice includes the first steps of the design of a multistage riser for a basin controlling water quality and/or channel erosion, and peak discharge. These steps are repeated for sizing the 2-year and 10-year release openings. Other design storms may be used as required by ordinance or downstream conditions.

25.7.9.2.1 Method 1. Water Quality Orifice Design Using Maximum Hydraulic Head and Routing of Water Quality Volume

A water quality extended detention basin sized for two times the water quality volume will be used here to illustrate the sizing procedure for an extended detention orifice.

1. Calculate the water quality volume (V_{wq}) required for treatment. Assume the following:

$$V_{wq} = 404{,}236 \text{ ft}^2 \times 1/2 \text{ in./12 in./ft} = 16{,}843 \text{ ft}^3$$

$$16{,}843 \text{ ft}^3 \div 43{,}560 \text{ ft}^2/\text{ac} = 0.38 \text{ ac-ft}$$

   For extended detention basins, $2 \times V_{wq} = 2(0.38 \text{ ac-ft}) = 0.76 \text{ ac-ft} = 33{,}106 \text{ ft}^3$.
2. Determine the maximum hydraulic head (h_{max}) corresponding to the required water quality volume. Assume from our example stage vs. storage curve (Figure 25.19) that 0.76 ac-ft occurs at elevation 88 ft (approximate). Therefore, $h_{max} = 88 - 81 = 7$ ft.
3. Determine the maximum discharge (Q_{max}) resulting from the 30-hour drawdown requirement. The maximum discharge is calculated by dividing the required volume, in ft^3, by the required time, in seconds, to find the average discharge, and then multiplying by 2, to determine the maximum discharge. Assume the following:

$$Q_{avg} = \frac{33{,}106 \text{ ft}^3}{(30 \text{ hr})(3600 \text{ sec/hr})} = 0.30 \text{ cfs}$$

$$Q_{max} = 2 \times 0.30 \text{ cfs} = 0.60 \text{ cfs}$$

4. Determine the required orifice diameter by rearranging the orifice equation, Equation 25.12, to solve for the orifice area (ft^2) and then diameter (ft). Insert the values Q_{max} and h_{max} into the rearranged orifice equation, Equation 25.13, solve for the orifice area, and then solve for the orifice diameter:

$$Q = Ca\sqrt{2gh} \tag{25.12}$$

$$a = \frac{Q}{C\sqrt{2gh}} \tag{25.13}$$

where

Q = Discharge (cfs).

C = Dimensionless coefficient = 0.6.

a = Area of the orifice (ft²).

g = Gravitational acceleration (32.2 ft/sec²).

h = Head (ft).

Assume for orifice area:

$$a = \frac{0.6}{0.6\sqrt{(2)(32.2)(7.0)}}$$

For orifice diameter:

$$a = 0.047 \text{ ft}^2 = \pi d^2/4$$

$$d = \sqrt{\frac{4a}{\pi}} = \sqrt{\frac{4(0.047 \text{ ft}^2)}{\pi}} = 0.245 \text{ ft} = 2.94 \text{ in.}$$

Use a 3-inch diameter water quality orifice.

Routing the water quality volume (V_{wq}) of 0.76 ac-ft, occurring at elevation 88 ft, through a 3-in. water quality orifice will allow the engineer to verify the drawdown line, as well as the maximum elevation of 88 ft. This calculation will give the engineer the *inflow–storage–outflow relationship* in order to verify the actual storage volume needed for the extended detention of the water quality volume. The routing procedure takes into account the discharge that occurs before maximum or brim storage of the water quality volume, as opposed to the brim drawdown described in Method 2. The routing procedure is simply a more accurate analysis of the storage volume used while water is flowing into and out of the basin. Therefore, the actual volume of the basin used will be less than the volume as defined by the regulation. This procedure will come in handy if the site to be developed is tight and the area needed for the stormwater basin must be "squeezed" as much as possible. The routing effect of water entering and discharging from the basin simultaneously will also result in the actual drawdown time being less than the calculated 30 hours. Judgment should be used to determine whether the orifice size should be reduced to achieve the required 30 hours or if the actual time achieved will provide adequate pollutant removal.

Note: The designer will notice a significant reduction in the actual storage volume used when routing the extended detention of the runoff from the 1-year frequency storm (channel erosion control).

Routing the water quality volume depends on the ability to work backwards from the design runoff volume of 0.5 inches to find the rainfall amount. Using SCS methods, the rainfall needed to generate 0.5 inches of runoff from an impervious surface (RCN = 98) is 0.7 in. The SCS design storm is the Type II, 24-hour storm. Therefore, the water quality storm using SCS methods is defined as the SCS Type II, 24-hour storm, with a rainfall depth of 0.7 in. The rational method does not provide a design storm from a specified rainfall depth. Its rainfall depth depends on the storm duration (watershed t_c) and the storm return frequency. Since the water quality storm varies with runoff amount, not the design storm return frequency, an input runoff hydrograph representing the water quality volume cannot be generated using rational method parameters. Therefore, Method 1, routing of the water quality volume, must use SCS methods.

TABLE 25.12

Stage–Discharge Table:

Water Quality Orifice Design

Elevation	h (ft)	Q (cfs)
81	0	0
82	1	0.2
83	2	0.3
84	3	0.4
85	4	0.4
86	5	0.5
87	6	0.5
88	7	0.6

5. Calculate a stage–discharge relationship using the orifice equation, Equation 25.12, and the orifice size determined in step 4. Using the 3-inch-diameter orifice, the calculation is as follows:

$$Q = Ca\sqrt{2gh} = 0.6(0.047)\sqrt{(2)(32.2)(h)} = 0.22\sqrt{h}$$

where h is water surface elevation minus the orifice's centerline elevation, in ft.

Note: If the orifice size is small relative to the anticipated head (h), values of h may be defined as the water surface elevation minus the invert of the orifice elevation.

6. Complete a stage–discharge table for the range of elevations in the basin, as shown in Table 25.12.
7. Determine the time of concentration for the impervious area. From our example, the developed time of concentration (t_c) = 0.46 hr. The impervious area time of concentration (t_{cimp}) = 0.09 hr, or 5.4 min.
8. Using t_{cimp}, the stage–discharge relationship, the stage–storage relationship, and the impervious acreage (RCN = 98), route the water quality storm through the basin. The water quality storm for this calculation is the SCS Type 2, 24-hour storm, rainfall depth = 0.7 in. (Note that the rainfall depth is established as the amount of rainfall required to generate 0.5 in. of runoff from the impervious area.) The water quality volume may be routed using a variety of computer programs such as *TR-20* HEC-1 or other storage indication routing programs.
9. Evaluate the discharge hydrograph to verify that the drawdown time from maximum storage to zero discharge is a least 30 hours. (Note that the maximum storage corresponds to the maximum rate of discharge on the discharge hydrograph.) The routing of the water quality volume using *TR-20* results in a maximum storage elevation is 85.69 ft vs. the approximated 88.0 ft. The brim drawdown time is 17.5 hr (peak discharge occurs at 12.5 hr and 0.01 discharge occurs at 30 hr). For this example, the orifice size may be reduced to provide a more reasonable drawdown time and another routing performed to find the new water quality volume elevation.

25.7.9.2.2 *Method 2: Water Quality Orifice Design Using Average Hydraulic Head and Average Discharge*

For the previous example, Method 2 results in a 2.5-in. orifice (vs. a 3.0-in. orifice), and the design extended detention water surface elevation is set at 88 ft (vs. 85.69 ft). (It should be noted that trial two of the method noted above might result in a design water surface elevation closer to 88 ft.) If the basin is to control additional storms, such as 2-year or 10-year storms, the additional storage volume would be "stacked" just above the water quality volume. The invert for the 2-year control, for example, would be set at 88.1 ft.

25.7.9.3   Extended Detention Basins—Channel Erosion Control Volume and Orifice Design

Extended detention of a specified volume of stormwater runoff can also be incorporated into a basin design to protect downstream channels from erosion. Virginia's Stormwater Management Regulations, for example, recommend 24-hour extended detention of the runoff from the 1-year frequency storm as an alternative to the 2-year peak rate reduction. The discussion presented here is for the design of a channel erosion control extended detention orifice. The design of a channel erosion control extended detention orifice is similar to the design of the water quality orifice in that two methods can be employed:

1. Use the *maximum hydraulic head* associated with the specified channel erosion control (V_{ce}) storage volume and calculate the orifice size needed to achieve the required drawdown time, and route the 1-year storm through the basin to verify the storage volume and the drawdown time, or
2. Approximate the orifice size using the *average hydraulic head* associated with the channel erosion control volume (V_{ce}) and drawdown time.

The routing procedure takes into account the discharge that occurs before maximum or *brim* storage of the channel erosion control volume (V_{ce}). The routing procedure simply provides a more accurate accounting of the storage volume used while water is flowing into and out of the basin, and results in less storage volume associated with the maximum hydraulic head. The actual storage volume needed for extended detention of the runoff generated by the 1-year frequency storm will be approximately 60 percent of the calculated volume (V_{ce}) of runoff for curve numbers between 75 and 95 and time of concentration between 0.1 and 1 hour. The following procedure illustrates the design of the extended detention orifice for channel erosion control.

25.7.9.3.1   Procedure

1. Calculate the channel erosion control volume (V_{ce}). Determine the rainfall amount (inches) of the 1-year frequency storm for the local area where the project is located. With the rainfall amount and the runoff curve number (RCN), determine the corresponding runoff depth using the runoff equation (see Chapter 4 in *TR-55*) or rainfall–runoff depth charts.

   Given:

   1-year rainfall = 2.7 in., RCN = 75
   1-year frequency depth of runoff = 0.8 in.

   $$V_{ce} = 25 \text{ ac} \times 0.8 \text{ in.} \times 1 \text{ in.}/12 \text{ ft} = 1.66 \text{ ac-ft.}$$

   To account for the routing effect, reduce the channel erosion control volume:

   $$V_{ce} = 0.6 \times 1.66 \text{ ac-ft} = 1.0 \text{ ac-ft} = 43,560 \text{ ft}^3$$

2. Determine the average hydraulic head (h_{ave}) corresponding to the required channel erosion control volume:

   $$h_{avg} = (89 - 81) \div 2 = 4.0 \text{ ft}$$

3. Determine the average discharge (Q_{avg}) resulting from the 24-hour drawdown requirement. Calculate the average discharge by dividing the required volume, in cubic feet, by the required time, in seconds, to find the average discharge:

$$Q_{avg} = \frac{43{,}560 \text{ ft}^3}{(24 \text{ hr})(3600 \text{ sec/hr})} = 0.5 \text{ cfs}$$

4. Determine the required orifice diameter by rearranging the orifice equation, Equation 25.12, to solve for the orifice area (ft²) and then diameter (ft). Insert the values for Q_{avg} and h_{avg} into the rearranged orifice equation to solve for the orifice area and then solve for the orifice diameter:

$$Q = Ca\sqrt{2gh}$$

$$a = \frac{Q}{C\sqrt{2gh}}$$

where
Q = Discharge (cfs).
C = Dimensionless coefficient = 0.6.
a = Area of the orifice (ft²).
g = Gravitational acceleration = 32.2 ft/sec².
h = Head (ft).

For orifice area:

$$a = \frac{0.5}{0.6\sqrt{(2)(32.2)(4.0)}} = 0.052 \text{ ft}^2$$

For orifice diameter:

$$d = \sqrt{\frac{4a}{\pi}} = \sqrt{\frac{4(0.052 \text{ ft}^2)}{\pi}} = 0.257 \text{ ft} = 3.09 \text{ in.}$$

Use a 3.0-in.-diameter channel erosion extended detention orifice.

Method 1 results in a 3.7-in.-diameter orifice and a routed water surface elevation of 88.69 ft. Additional storms may be "stacked" just above this volume if additional controls are desired.

REFERENCES AND RECOMMENDED READING

King, H.W. and Brater, E.F. (1976). *Handbook of Hydraulics*, 6th ed. McGraw-Hill, New York.
Linsley, R.K., Kohler, M.A., and Paulhus, J.L. (1982). *Hydrology for Engineers*, 3rd ed. McGraw-Hill, New York.
Morris, H. M. and Wiggert, J.M. (1972). *Applied Hydraulics in Engineering*. John Wiley & Sons, New York.
Schueler, T. (1987). *Controlling Urban Runoff: A Practical Manual for Planning and Designing*. Metropolitan Washington Council of Governments, Washington, DC.
USDOT. (1984). *Hydrology*, Hydraulic Engineering Circular No. 19. U.S. Department of Transportation, Washington, DC.

USSCS. (1956). *National Engineering Handbook.* Section 5. *Hydraulics.* U.S. Soil Conservation Service, Washington, DC.

USSCS. (1982). *Project Formulation—Hydrology*, Technical Release No. 20. U.S. Soil Conservation Service, Washington, DC.

USSCS. (1984). *Engineering Field Manual.* U.S. Soil Conservation Service, Washington, DC.

USSCS. (1985). *National Engineering Handbook.* Section 4. *Hydrology.* U.S. Soil Conservation Service, Washington, DC.

USSCS. (1986). *Urban Hydrology for Small Watersheds*, Technical Release No. 55. U.S. Soil Conservation Service, Washington, DC.

Walesh, S.G. (1989). *Urban Surface Water Treatment.* John Wiley & Sons, New York.

Index

A

abatement period, 385
absolute pressure, 87, 408, 410, 411, 449
absolute temperature, 87, 388, 418, 491, 534
absorbed dose, 438, 444
absorbent, defined, 452
absorbers, 541–542
absorption, 385, 452–469, 487
  gaseous emissions, and, 541–542
  noise reduction by, 433
accelerated erosion, 761
acceleration, 234, 238, 246
accident, 109, 114, 197, 198, 203, 214, 391
  analysis, 386
  Boolean analysis of, 217–219
  defined, 385–386
  environmental, 213, 232
  noise, and, 422, 427
  prevention, 386
  risk, and, 394
accommodation, 386
acid rain, 399, 449
acid surge, 449
acids, 350, 449
  in water, 355, 356
acoustic trauma, 422
acoustics, 386
action level, 386, 388, 422, 428
activated biosolids, 687–701, 750
activated carbon, 47, 469, 471, 472, 477, 478, 649
activated charcoal, 469
activation energy, 364
actual cubic feet per minute (acfm), 410
acute, defined, 386
acute toxicity, 387
adjusted means, 178–179
administrative controls, 423
adsorbate, 469
adsorbents, 469
adsorber bed depth, 474–475
adsorption, 387, 469–478, 542, 605, 638
  carbon, 487
  chemical vs. physical, 470–471
  equilibrium relationships, 471–472
  factors affecting, 474–478
advective transport, 90
aerated ponds, 704–705
aeration, 564, 670, 678, 688
  basin, 699–700
  tank, 687–692, 755
    design, 696–697
aerobic decomposition, 561, 568, 569, 682
aerobic digestion, 726–727
aerobic tank volume, 727
aerodynamic diameter, 493

aerosols, 387
  biological, 388
age-adjusted mortality rate, 200–201
age-specific mortality rate, 197
agricultural effluents, 561
agricultural runoff, 581
air
  changes per hour, 410
  composition of, 403
  contamination, defined, 387
  density, 410, 498
  molecular weight, 418, 482, 550, 554
  monitoring, 407–409
  pollutants, 395–399, 491
  pollution, 387, 395–409
  pressure, 369, 370, 565, 589, 590, 607
  sampling, 387
  weight of, 369, 370, 590
air-to-filter ratio, 517, 518
air-to-solids ratio, 724
airborne contaminants, 393, 407, 414, 415, 449
albedo, 449
algae, 561, 569, 638, 678
  control, 613–614
algebra, 75–77
  Boolean, 213–219
algorithms, 93–97
  vs. computations, 93
alkalinity, 355, 357, 630–633, 728
allergens, 387
allocation, sample, 127–128
allowable loads, 250
allowable release rates, 791
allowable stress, 242
alpha radiation, 438, 440, 441, 442
alternate authorized user, 438
alternating current (AC), 289–297
alternator, 290
alum, 605, 615, 616, 618, 630–632, 712
ambient
  air concentrations, 97, 407
  air quality, 397
  concentration standards, 405
  defined, 387
  sound, 387
  temperature, 433, 705
    normal and standard, 406
American Wire Gauge (AWG), 285
ammonia gas, 453
ammonia nitrogen, 741
anaerobic digestion, 726, 727–731
analysis of covariane, 176–179
analysis of variance, 139–160, 163
  vs. t test, 141
AND gate, 214
anemometer, 410, 416

817

anions, 361
annualized installed capital cost (AICC), 524
annual payment, 226–227
antecedent runoff condition, 782
antifreeze, 351
anti-seep collar, 759
anti-vortex device, 759
apparent color, 353
aquatic bench, 759
aquifers, 581–587, 760, 772
area, 49, 60, 61, 63–66, 410
    basal, 338–340
    circle, 286–287, 597
    defined, 42
    duct, 416
    of circle, 286
    pipe, 372, 373
    treatment pond, 702
    velocity, and, 597–598
arithmetic mean, 50–52, 449
armature, 290
artesian well, 582
asbestos, 382, 397
asphyxiation, 387
association, measures of, 206–210
atmosphere, defined, 449
atmospheric deposition, 760
atmospheric pressure, 27, 29, 88, 369, 370, 371, 400, 409,
        410, 411, 413–414, 453, 580, 590, 594
atmospheric standards, 406
atom, 450
atomic number, 450
atomic weight, 450, 749
attack rate, 187, 189, 190
attenuation, 387, 423
    real-world, 425
Atterburg limits, 315
attributable proportion, 210–211
audible range, 387, 423
audiogram, 387, 423
audiologist, 423
authorized person, 387
authorized user, 438
auto-ignition temperature, 387
average, 50–52, 112
average land cover condition, 760
average use, of chemicals, 716
Avogadro's number, 748
A-weighting, 426, 427–428
axial strain, 243
axial stress, 249

B

background noise, 423
backward reactions, 364
backwash
    percent product water used for, 645–646
    rate, 642–644, 645, 647, 649, 651, 652
    troughs, 652
    water required, 644
    water tank depth, 644–645
bacteria, 561, 638, 678
bag diameter, 517, 519, 520, 521

baghouses, 387, 492, 499, 505, 516–524
bankfull flow, 760
bare rock succession, 308–309
barometric pressure, 410, 418, 452
    normal and standard, 406
barrels of oil consumed, 32–33
bar screen, head loss through, 670
Bartlett's test, 133–134
basal area, 338–340
base, triangle, 42, 60, 450
base flow, 760, 772
baseline audiogram, 423
baseline data, 387–388
bases, 350, 450
    in water, 355, 356
basin
    aeration, 699–700
    bioretention, 760
    detention, 761, 790–791
        extended, 761–762, 790–791, 809–815
        storage volume, 794–796, 806–808
    retention, 764, 790–791
        storage volume, 806–808
        water quality volume, 809
    stormwater, 760, 790–791
        storage volume, 791–806
basis for a subspace, 75
battery, 257, 258, 299
    schematic symbol for, 258
beams, 249–254
    types of, 250
bearing force, 233
behavior-based safety (BBS) management models, 388
bell-shaped distribution curve, 115
belt filter press, 733–736
benching, excavation, 318
benchmarking, 388
bending, 233, 252–253
    moment, 250–251
    strength, 244, 326
Bernoulli's equation, 598, 599–600, 670, 673–674
best management practice (BMP), 760, 762, 808
beta radiation, 438, 440, 441–442
binomial distribution, 129
bioassay, 438
biochemical oxygen demand (BOD), 51, 54, 365, 561,
        565–567, 674, 680, 682, 689, 700, 760
    calculations, 566–567, 746–748
    loading, 688, 696–697, 703
    removal calculations, 677–678, 681, 704–705, 718
    seven-day moving average, 747–748
    soluble, 684–685
    test procedure, 565–566
    types of, 571
biochemical reactions, 363
biodegradation, 363
biohazard, 388
biological aerosols, 388
biological half-life, 438, 440
biological processes, 760
biological stressors, 384
biomass expansion factors (BEFs), 332
biomass, forest-based, 325–339
    equations, 333–335

bioretention basin, 760
biosolids, 717–722
    age, 691
    composting, 743–744
    density index (BDI), 754–755
    dewatering, 731–740
    digestion, 725–731
    disposal, 740–744
    feed rate, 734
    pumping calculations, 719–722, 723
    pumping rate, 677
    retention time, 728–729
    thickening, 722–725
    total applied, 739
    volume index (BVI), 694–695, 754–755
    vs. solids, 717, 752
    withdrawal to drying beds, 740
biostatistics, 109
blinding, adsorption site, 478
board foot, 335, 337
body heat, 434–435
boiler code, 388
boiling point, 351
bolt-and-plate assembly, 248
Boltzmann's constant, 531
Boolean algebra, 213–219
    example, 217–219
bottom slope, still water body, 577
Boussinesq equation, 586
bowl efficiency, 611–612
bowl horsepower, 610, 611
Boyle's law, 388, 401, 403, 408–409
brake horsepower (BHP), 376–378
branch, duct, 410
breaking strength, 233
breakpoint chlorination, 656–657
Bremsstrahlung, 439, 442
brim drawdown, 809
brittleness, 244
Brownian diffusion, 530
bubble sort algorithm, 95
buckling, 242, 249
    stress, 254
buoyant line and point source model, 96
Byzantine Generals algorithm, 95

C

C factor, 373–374, 601
CAL3QHC/CAL3QHCR model, 97
calcium carbonate, 348, 357
calcium hypochlorite (HTH), 45, 366, 654, 658–661, 662, 708–709
calibration, 80
    dry chemical feeder, 621–623
CALINE3 model, 96
calorie, 434
CALPUFF model, 96
Camp–Shields equation, 673
canceling fraction terms, 56–59
canopy hood, 410
cantilever beams, 250
capacitance, 298
capital-recovery factor, 224–225

capture hood, 417
capture velocity, 410, 417
carbon, 330, 343, 469, 479, 480, 743
    accumulation, by forests, 35–36
    activated, 47, 469, 471, 472, 477, 478, 649
    adsorption, 474–475, 487
    atoms, 349
    capacity, 472, 476
    density, 476
    molecular weight, 32, 33, 35, 38, 39
    sequestration, 34–36, 38
    volume of in adsorber, 477
    working charge, 477
carbon dioxide, 28, 79, 345, 354, 357, 388, 401, 405, 449, 451, 478, 479, 480, 591, 678, 728, 730, 731
    emissions, 31–39
    molecular weight, 32, 33, 35, 38, 39
    percent, 406
carbon monoxide, 28, 397, 399, 451, 478
carbonaceous biochemical oxygen demand, 571
carcinogen, 388
carpal tunnel syndrome, 388
carrier gas, 452
case-control study, 209–210
case-fatality rate, 187, 202
catalysts, 363, 364, 388
catastrophe, 388
catch basin, 760
cations, 361
causal factor, 388
cause-specific mortality rate, 196–197
cave-ins, 318
CDV-700, 441, 442
Celsius scale, 24–25
census, 327
central tendency, 50, 111
centrate, 725
centrifugal force, 246–247, 504–505, 506, 670
centrifugal pumps, 375, 376
centrifuge thickening calculations, 725
centripetal force, 246–247
chaining, 63, 64
channel flow, 785, 786
    time, 775
channel stabilization, 760
characteristic polynomial of a matrix, 75
Charles' law, 388, 401–402, 403, 408–409
check dam, 760
chelation, 362
chemical bond, 450, 471
chemical change, 388, 398
chemical dosage, 365–369, 608, 613–614, 616, 622, 630–633, 655–656, 659, 660, 705–716
chemical equations, water/wastewater, 365
chemical equilibrium, 85
chemical feed rate, 706–707
    calibration, 713–714
chemical hazards, 388
chemical oxygen demand (COD), 365, 680, 689, 760
    loading, 688
chemical reactions, 361–365, 450
    types of, 361–363
chemical solution feeder setting, 618–619, 712–713
chemical spill, 388

chemical stressors, 383
chemical transport systems, 89–90
chemical usage, 623–625
    calculations, 662
chemisorption, 471
chemistry, 343–369
chemists, types of, 343
chi-square tests, 131–134
chlorination, 653–663
    breakpoint, 656–657
    chemistry, 662–663
chlorine, 462, 608, 653–663
    breakpoint, 656
    demand, 366, 655–656, 707
    dosage, 365–366, 655–656, 707–708
    feed rate, 654–655
    residual, 654, 655–657, 707, 708
chronic
    defined, 388
    diseases, 186, 191, 194, 204, 382, 383
chute, 760
circle area, 286–287, 597
circular mil, 286–287
circular-mil-foot, 287
circular tank
    circumference, 63
    surface area, 65
    volume, 68
circumference, 42, 60, 61–63
clarifiers, 625, 627, 629, 633, 637, 674, 675, 676, 678, 680,
        687, 692, 693, 717–718, 750, 751
    weir overflow rate, 674
clay, 309, 310, 311, 314, 315, 316, 317
Clean Air Act (CAA), 397
climate, 450
cloud formation, 405
cluster sampling, 130–131
c-multipliers, 171
coagulants, 605, 615
coagulation, 356, 615–625, 630
coal, railcars of burned, 38–39
coal-fired power plant emissions, 39
coefficient of determination, 164, 171
coefficient of entry, 410, 420
coefficient of friction, 245, 388, 601
coefficient of roughness, 601, 604, 764, 780, 785; *see also*
        C factor
coefficient of variation, 118
cohesion, soil, 314–315, 316
cohort study, 190, 191
coil diameter, 300
cold hazards, 437–438
collection efficiency, 493, 503
    baghouse, 518, 519, 521, 522–523
    cyclone, 504–510
    electrostatic precipitator, 510–516
    gravity settler, 499–504
    wet scrubber, 529, 530–548
colloidal solids, 352, 353
color, water, 353
column space of a matrix, 75
column stress, 252
column buckling stress, 254

columns, 249–250
combination reactions, 362, 363
combined gas law, 402–403
combustible gas indicator, 388
combustible liquid, 388
combustion, 388, 478–482
    emissions, 405–406
    limits, 480
common factor, 41
common regressions, 174–175
commutative law, 216–217
compaction, 760
    soil, 316
competent person, 388–389
complete mix with recycle, 697–698
complete mixing, 615–616, 704
complete randomization, 139–141, 144, 147, 159
COMPLEX 1 model, 97
composite number, 41
composite sample, 745–746
compost, 743–744
compressibility, soil, 316
compression force, 233
compression stress, 247
compressive strength, 243
compressive stress, 252
computations vs. algorithms, 93
concentration, 16–18, 81, 85, 90, 349–350, 706, 709
    contaminant, 396
    dilutions, and, 358–359
    factor (CF), 724
    gradient, 89, 90, 491
    in decimal percent, 45
    inlet/outlet gas, 548, 549, 552, 553
    reactant, 364
    solute, 351–352
    standards, 405
    units, 81
condensation, 483–489
conduction, heat, 434
conductive hearing loss, 424
conductivity
    hydraulic, 584, 585
    vs. resistivity, 288
conductors, 257–258, 285–288, 289, 298, 299
    unit size of, 285
cone of depression, 583
cone, volume of, 68
confidence interval, 124, 129, 131, 164–166
confidence level, 130
confidence limits, 110, 119, 124, 127, 129, 164–166, 171
confined aquifers, 581–582, 586–587
confined space, 389, 407
coniferous trees, carbon sequestration and, 34
consistency, soil, 315
constructed stormwater wetlands, 760
contact condenser, 483–485
contact power theory, 538–540, 546
containment, 389
contaminants, 59, 90, 353, 387, 397, 451, 452, 454, 469, 527
    adsorption of, 470, 474, 477
    airborne, 393, 407, 414, 415
    concentration of, 396, 399–400, 463, 464

condensation, and, 483
oxidation of, 478
contingency plan, 389
continuity equation, 372–373, 597–598
continuously supported beams, 250
continuous noise, 423
continuous simulation models, 771–772
contour, elevation, 760
controlled area, 439
controls, administrative/engineering, 423
convection, 389, 434
conversion factors, 4–39
    air pollution measurements, 25–29
    examples, 15–25
        water/wastewater, 18–24
    greenhouse gas emissions, 31–39
    soil test results, 29–31
copper, 285
copper sulfate dosing, 613–614
correlation coefficient, 120–121, 164, 171, 531
corrosion, 242, 356
corrosive material, 389
cosmic rays, 440
Coulomb's law, 257
counterelectromotive force (cemf), 299
covalent bond, 361, 362, 450
covariance, 119–120
    in randomized block design, 176–179
cradle, 761
creep, 242
crest, 761
criterion sound level, 423
critical flow, 602
critical storm duration, 778, 791, 796–806
cropland carbon stock, 37
cross-multiplying, 52–53
cross-sectional area, 63
crude mortality rate, 196
Cryptosporidium, 638
CTDMPLUS model, 97
CTSCREEN model, 97
cubic feet per minute to gallons per minute conversion, 640
cubic feet per second to gallons per minute conversion, 597
cubic feet per second to million gallons per day conversion, 22
cubic feet to cubic yards conversion, 671
cubic feet to gallon conversion, 18, 375, 639
cubic units, 42, 60
cumulative injury, 389
cumulative trauma disorder, 389
Cunningham correction factor, 495, 496, 497, 530, 531, 532, 534, 535, 542, 543, 544
current, electrical, 256–303
    average, 295
    direction of flow, 258
    effective, 294
    instantaneous, 293
    Ohm's law, and, 259–263, 270–271, 276–279
    parallel circuit, 276–279
    peak, 292, 293, 294, 295
    peak-to-peak, 293
    resistance, and, 260, 277, 280
    series circuit, 265–266

curve number (CN), 761; *see also* runoff curve number
curvilinear regression, 171–173
cut diameter, 493, 506–509, 514, 515, 537–538
cut power method, 537–538
cut set, 215–216
C-weighting, 426
cyclones, 492, 499, 504–510
    factors affecting performance of, 506–510

D

daily noise dose, 432
Dalton's law, 87, 88–89, 389
Darcy–Weisbach equation, 600–601
Darcy's law, 585, 586, 587
data, types of, 114
days supply in inventory, 623–624, 662, 716
deactivation, adsorption site, 478
dead loads, 255
death-to-case ratio, 184, 201–202
DeBruin–Keijman equation, 580
decibel (dB), 389
decimal percent, 45–46
declared pregnant woman, 439
decomposition, 563, 725
    aerobic, 561, 568, 569, 682
    forest, 35
    reaction, 362, 363
    pollutant, 479
decontamination, 389
deep well, 582
deep well turbine pump calculations, 608
defective matrix, 75
degrees of freedom, 139, 152, 153, 158, 170, 174
denominator, 52, 56–59, 182, 183, 184, 186, 189, 190, 191, 192, 202
density, 18, 246, 344–346, 451, 591–592
    air, 550, 551, 552
    correction factor, 410, 418, 419, 420
    defined, 389
    gas, 492, 506, 509, 532, 533, 534, 536, 554, 555
    liquid, 532, 533, 535, 536
    particle, 493, 494, 495, 496, 498, 500–501, 502, 503–504, 506, 507, 509, 530, 532, 534, 543, 544
    water, 82, 246, 332, 344–345, 552, 634
    wood, 339
deoxygenation, 568–569, 571
depth, defined, 42, 60
derivatives, 39
dermatitis, 389
design load, 389
design storm, 761, 771, 788, 789, 791, 794–796, 799, 808, 810, 811
    frequency, 761, 762, 768, 771, 774, 790, 811
detention basin, 761, 790–791
    extended, 761–762, 790–791, 813–815
        water quality volume, 809–813
    storage volume, 794–796, 806–808
detention time, 615–616, 617, 626, 636–637, 669, 676
    hydraulic, 674, 703
    oxidation ditch, 701–702
Deutsch–Anderson equation, 510, 511–512, 513–514, 515
dewpoint, 451, 484, 485

diagonalizable matrix, 75
diameter, 42, 60, 62–63, 65, 67, 286, 336, 372, 597
    at breast height (DBH), 161, 332, 333, 338–339
    bag, 517, 519, 520, 521
    cut, 506–509, 514, 515, 537–538
    droplet, 530, 531, 532, 533, 534, 535, 536, 543, 544
    packed column, 551–555
    packed tower absorber, 459–462
    particle, 530, 531, 532, 537, 543, 544, 649
    particulate matter, 493–494, 495, 496, 497, 498, 500,
        502, 503, 504, 506–509, 510, 515, 517
    plate tower absorber, 465–467
diffusion, 89–90, 491, 530–531
    particle, 492
digestion
    biosolids, 725–731
    time, aerobic digester, 726
dike, 389
dilution, 358–359, 561–562
    factor, 358–359
    ventilation, 410, 415
dimensional analysis, 55–59
dimensions, 15
direct current (DC)
    disadvantages, 290
    parallel, 274–284
    series, 264–274
    series–parallel, 284–285
dirt vs. soil, 309–310
discharge, 372–373, 568, 584, 604, 673–674
    capacity, 607
    coefficient, 670, 674
    elevation, 71, 595
    head, 375, 609, 610
    measurement, 562
    permit compliance, 571, 745
    peak, 763, 774–779, 783, 784, 787–791, 795–798, 801,
        810, 812
      runoff, and, 772–773
    pressure, 375, 610
    rate, 596–598, 652
    sewage, 561, 568
    stormwater, 759–814
    total, running water body, 562
    wastewater treatment plant, 564, 569
discrete variables, 129–130
disinfection, 581, 605
    chlorine, 20, 365, 367, 653, 654, 656, 707
    hypochlorite, 658, 660, 708
    hypochlorous acid, 663
    turbidity, and, 353
    well casing, 606, 608
dispersion models, 96–97
dispersive transport, 90
displacement, soil, 315
disposal cost, of biosolids, 740–741
dissolved oxygen (DO), 354, 561, 563–565, 571
    biochemical oxygen demand, and, 567, 568–569
    correction factor, 565
    saturation values, 564–565
dissolved solids, 353, 736
distribution, of data, 115–117
distributive law, 217
disturbed area, 761

diversion, 761
dividend, 42
dividing fractions, 55–59
dividing units, 55
divisor, 41, 42
dose, 389; *see also* chemical dosage, chlorine: dosage
    equivalent, 439, 444
    limits, 439
    radiation, 15, 444
    rate, radiation, 440
dosimeter, 389, 423, 439
    noise, 425, 427
    radiation, 442
    thermoluminescent, 439, 442
double hearing protection, 423
double replacement reactions, 362
drag coefficient, 536, 634–636
drainage basin, 761
drawdown, 606–608
    time, 809
drift velocity, 510, 512–513
drop structure, 761
dry chemical feeder calibration, 621–623
dry chemical feeder setting, 618
drying beds, 738–740
dry weight basis, 330
duality, 216
duct
    diameter, 416–417, 421
    electrostatic precipitator, 512, 513, 514
    noise reduction in, 432
    velocity measurements, 416–417
ductility, 244
dusts, 389
dynamic shape factor, 494
dynamics, 248

E

echelon form of a matrix, 75
economics, environmental, 223–228
effective half-life, 439, 440
effective porosity, 583
effective size (ES), 648–649
effective temperature, 437
effectiveness, 210
    vaccine, 211
efficacy, 210
    vaccine, 211
efficiency
    absorption, 462, 463, 465, 468–469
    adsorber, 470, 474, 477, 478
    air control, 450
    BOD removal, 704
    collection, 493, 503
        baghouse, 518, 519, 521, 522–523
        cyclone, 504–510
        electrostatic precipitator, 510–516
        gravity settler, 499–504
        wet scrubber, 529, 530–548
    cut diameter, and, 493
    fan, 421, 522, 523
    filtration, 647, 648, 653
    hood, 410

motor, 47, 377
Murphree, 468
overall, 468
packed tower, 453
pump, 376–378, 611–612
radiation detector, 444
removal, 453, 456, 462, 467, 468, 808, 809
effluent, 560, 562, 675–676, 678–687, 689–695, 698–702,
        704–705, 707, 717–718, 720, 723, 730, 751–753
    agricultural, 561
    flow rate, 20
    gas, 452, 469
    sewage treatment plant, 561, 667
    turbidity, 694–695
eigenspace, 75
eigenvalue, 75, 76
eigenvector, 75
elastic buckling, 249
elastic limit, 242
elastic stability, 249
electrical circuit, 257–261
electrical energy, 263
electrical grounding, 389
electrical power, 261–263
electricity, 255–303
    defined, 256
electricity use, home, 33
electromagnetic radiation, 442
electromotive force (emf), 257, 289
    self-induced, 299–301
electrons, 361, 362
electrostatic precipitator, 492, 499, 505, 510–516
    calculations, 512–516
element, defined, 451
elementary matrix, 75
embankment, 761
emergency response plan, 389
emissions, 39
    carbon dioxide, 31–39
    combustion, 405
    control
        gas, 449–489
        particulate, 491–524
        wet scrubber, 527–555
    greenhouse gas, 31
    pollution, measurement parameters, 405
    reporting units, 25
    ventilation, and, 414, 417, 418, 420
    standards, 25, 399, 405, 451
Emissions & Generation Resource Integrated Database
        (eGRID), 31
emissivity, 451
empirical rule, 116
emulsions, 352
endothermic reaction, 351, 361
energize, 390
energy, 451
    budget model, 579
    defined, 263, 390
    dissipator, 761
    electrical, 263
    kinetic, 390, 392, 419, 584, 598, 762
    potential energy, 390, 584–585, 598, 762
Engelmann spruce trees, 333

engineering, 231–303, 390
    controls, 410
        contaminants, 390
        industrial noise, 423, 429
enthalpy, 480–481
environmental air pollution, defined, 397
environmental economics, 223–228
environmental engineering, 93–97
    models, 96–97
    screening tools, 97
environmental heat, 434
environmental modeling, 79–91
    steps for developing, 79–80
environmental risk managers, 181
epidemiological theory, 390
equal allocation, 128
equal-energy rule, 423
equilibrium, 85–89
    graphs, adsorption, 471–472
Equivalencies Calculator, 31
equivalent, acid/base, 350
equivalent, chemical, 749
equivalent circuit, 281, 284
equivalent diameter, 493, 494
equivalent dose, 444
equivalent length of pipe, 374
equivalent resistance, 281–284
ergonomics, 385, 390, 392
erosion, 308, 314, 325, 332, 760, 764, 766, 791
    control, 761, 790, 791, 796, 808–814
    types of, 761
estimation, 39, 80, 109–110, 328,
    biomass, 331–332
    biosolids pumping rate, 719, 723
    return biosolids rate, 693
    transmissivity, 584
etiology, 390
Euclid, 93–94
eutrophication, 575
evaporation, surface, 578–580
evaporative cooling, 434
evapotranspiration, 579, 580
evase, 410
even number, 41
excavation, 316–318
exceedence probability, 768
excess air, 479
    percent, 405–406
excess risk, 211
exchange rate, sound, 423
exfiltration, 761
exhaust ventilation system, 412–413
exothermic reaction, 351, 361
exponents, 49–50, 57
exposure, 383–385, 386–391, 393–395, 407, 424; *see also*
        personal exposure level
    ceiling, 390
    control, 410, 418, 423
    disease, and, 206–210
    limits, 407
    thermal, 435, 438
    to asbestos, 382
    to noise, 424–428, 432
    to radiation, 384, 439, 442, 444

extended detention basin, 761–762, 790–791, 813–815
    water quality volume, 809–813
extensive property, 83
external radiation dose, 444
extrapolations, 80

F

F test, 135, 140–144, 146, 149, 157–158, 163–164, 166, 179
    with single degree of freedom, 141–142
fabric filters, 492, 499, 516–524
face velocity, 416
factor, defined, 41
factorial experiments, 149–154
Fahrenheit scale, 24–25
failure mode and effect analysis (FMEA), 214
fall arresting system, 390
fan, 421–422
    curve, 410, 421
    efficiency, 421, 522, 523
    laws, 410
    total pressure, 422
    ventilation, 410, 412
fatigue strength, 244
fault-tree analysis (FTA), 214–216
feet to pounds per square inch conversion, 72–74
field efficiency, 611–612
field head, 609, 610
field horsepower, 609, 611
filter bed, 762
    expansion, 647
    mixed media, 649
filter efficiency, 653
filter loading
    rate, 648
    rotary vacuum, 736–737
filter medium grain size, 648–649
filter operating time, rotary vacuum, 737–738
filter press calculations, 732–733
filter strip, 762
filter yield, 732–733, 737–738
filtration, 583, 605
    calculations, 638–653
    rate, 640–641
fine particles, 493
finite population correction, 119
fire, defined, 390
first flush, 762
first law of conservation of energy, 480
first law of thermodynamics, 451
fixed beams, 250
fixed-bed adsorber, 469
fixed-film rotating biological contactor, 682
fixed solids, 752–753
flame combustion, 480
flame ionization detectors (FIDs), 407
flammable liquid, 390
flammable solid, 390
flash point, 390
flexure, 252–253
float and stopwatch, 672
flocculant, 634
    dosage, 735–736
    feed rate, 732, 735

flocculation, 605, 615–625
flocculator calculations, 624–625
flooding velocity, 459–462
floodplain, 762
floor load, 254–255
flotation thickeners, 722–723
flow, 16–18
    horsepower, and, 377
    line, 585
    net, 585, 587
    open-channel, 601–603
    rate, 410, 419–420, 596–598, 610, 615, 622
        air, 517
        gas, 457, 459, 460, 462, 463, 467, 475–477, 479,
            481, 482, 484, 485, 502–504, 512–514, 518–521,
            523, 524, 534, 539, 545, 546, 549, 552–554
        liquid, 456, 458, 459, 460, 463, 467, 484, 535, 553,
            554
        mass, 456, 460, 481, 482, 484, 485
        peak, 772–773
        through filter, 638–640
        treatment pond, 702
    regime, particulate matter, 494–499
    segments, 785
    splitter, 762
fluid mechanics, 248
fluidized bed, 651–652
food/microorganism ratio, 54, 689–691, 696–697
food-specific attack rate, 189
foot-candle, 390
force, 246
    calculations, 69–75
    centrifugal, 246–247
    centripetal, 246–247
    defined, 233
    friction, and, 245
    inclined plane, and, 238–241
    magnetomotive, 288
    pressure, and, 593–594
    resolution of, 232–235
    slings, and, 235–238
    vs. weight, 593
forebay, sediment, 762
forest-based biomass computations, 325–339
forest fires, 399, 404, 491, 560
forests, 161, 309
    carbon accumulation, and, 35–36
    preserved from conversion to cropland, 36–38
    sampling, 126, 128
formula weight (FW), 359–360
forward reactions, 364
fractions, 52, 182
    converting to percentages, 45
    dividing by, 55–56
fracture, 242
free radical reaction, 362, 363
freeboard, 762
freezing point, 351
frequency, 182–187, 423
    alternating current, 291–292
    design storm, 762, 768, 771, 774, 790, 811
    distribution, 115
    octave band, 430
    years of potential life lost, and, 205

frequency factor, storm, 776–777
freshwater, sources of, 606
friction, 239, 245
    head, 70–77, 595
    loss, 373, 410, 602
    pipe, 373–374, 600–601
frostbite, 438
frostnip, 438
fuel economy, 31
fume, 390
fungi, 561, 678
fuse, 258
future dollar amount, 227
future value, 226, 228

G

gabion, 762
gallons of gasoline consumed, 32
gallons per day to million gallons per day conversion, 22
gallons per minute to million gallons per day conversion,
    22, 639, 655
gallons to cubic feet conversion, 19, 668, 671
gallons to pounds conversion, 19, 346, 592
gamma radiation, 439, 440, 442–443
gas, 391, 399–403
    chromatographs, 407
    emission control, 449–489
        absorption, 452–469
        adsorption, 469–478
        condensation, 483–489
        incineration, 478–482
    flow rate, 457, 459, 460, 462, 463, 467, 475–477, 479,
        481, 482, 484, 485, 502–504, 512–514, 518–521,
        523, 524, 534, 539, 545, 546, 549, 552–554
    laws, 399, 400–403
    particulate matter, and, 491–492
    phase, 362, 452, 455, 472, 474
        concentrations, 84, 454, 485, 542
        equilibrium, 86–87
        mass concentration, 84
        oxidation, 478
        pressure, 453
        resistance, 463
    production, 730–731
        estimated, 729
    transfer units, *see* transfer units
gauge pressure, 371, 410, 594–595
Gaussian dispersion models, 96–97
Gay-Lussac's law, 402
Geiger–Mueller counter, 441, 442, 443
general exhaust ventilation, 410
geometrical measurements, 60–68
Giardia lamblia, 638
global warming, 451
globe thermometers, 436
Gould biosolids age, 691
grab sample, 745
gram equivalent weight, 350
gram-mole, 748
grams to pounds conversion, 620
graphical hydrograph analysis, 792–794, 796–799
graphical peak discharge method, 787–788
grassed swale, 762

gravity settlers, 492, 499–504
gravity thickeners, 722–723
greatest common divisor, 41
greenhouse effect, 405, 451
greenhouse gas emissions, 79, 181, 451
    calculations, 31–39
Greenhouse Gas Equivalencies Calculator, 31
grit channel length required, 673
grit channel velocity, 672
grit removal, 667
    calculations, 670–671
ground-fault circuit interrupter (GFCI), 391
grounding, electrical, 389
groundwater, 581–587, 606
    flow, 585–587
    quality, 583
groundwater under the direct influence of surface water
    (GUDISW), 581, 583
group regression, 173–175
gully erosion, 761

H

half-life, 439, 440
hardness, steel, 244
hardness, water, 357
hazard
    defined, 391
    contributing to accident, 219, 386
    health, due to metals in water, 354
    job, analysis of, 392
    noise, 384, 386, 422, 423, 424, 425, 427, 429
    radiant heat, 384
    reactivity, 394
    soil failure, 317
hazard and operability (HAZOP) analysis, 391
Hazard Communication Standard (HAZCOM), 391
hazardous material, 391
hazardous task inventory, 424
hazardous waste, 391
Hazen–Williams equation, 373, 601, 602, 603
head, 370, 410, 413, 594–596, 609, 762
    calculations, 69–75, 375–376
    discharge, 609, 610
    field, 609, 610
    hydraulic, 584–585, 587, 809–813
    loss, 373, 374, 602
        calculations, 375, 600–601
        fixed bed flow, 649–651
        fluidized bed, 651–652
        through bar screen, 670
    pressure, and, 70, 71–73, 590, 593, 595–596, 599–600
    total, 595, 599–600
        pumping, 609
healthy worker effect, 203
hearing conservation, 391
    record, 424
    standard, 386
hearing damage risk criteria, 424
hearing handicap, 424
hearing loss, 424
    noise-induced, 425, 427
    prevention program audit, 424
hearing threshold level (HTL), 424

heat
balance calculations, 480, 484–485, 488
content, 329, 480
motor gasoline, 33
waste gas stream, 480–481
wood, 329–330
cramps, 391
exchanger, 485–489
exhaustion, 391
incineration, and, 480–481
stress, 435, 436
index, 437
stroke, 391
transfer coefficient, 485, 487, 488
heating, ventilation, and air conditioning (HVAC), 411
height, 42, 60
Henry's law, 87, 89, 454, 456, 548, 550, 551, 552, 553, 554, 663
higher heating value (HHV), 328–329
home energy use, 34
Homeland Security, 391
homogeneous variance, 131, 133–134, 136, 147, 166, 172
homoscedasticity, 166
hood, 410
capture, 417
entry loss factors, 410, 421
face velocities, 416
static pressure, 410, 416, 420–421
horizontal velocity, 615, 636–637, 674, 676
horizontal washwater troughs, 652–653
horsepower, 376–378, 609–612
hot work, 392
Howard settling chamber, 501
human factor engineering, 392
Hyatt Regency Skywalk disaster, 319–321
hydration factor, 350
hydraulic depth, 603
hydraulic detention time, 674
treatment pond, 703
Hydraulic Engineering Circular 1 (HEC-1), 762
hydraulic gradient, 584–586
hydraulic head, 584–585, 587, 809–813
hydraulic horsepower (WHP), 376
hydraulic loading
centrifuge, 725
gravity thickener, 723
rate
belt filter press, 733–734
rotating biological contactor, 683–684
trickling filter, 679–680
treatment pond, 704
trickling filter design, and, 682
hydraulic radius, 601, 602, 603, 604
hydraulic retention time
aerated pond, 704–705
complete mix with recycle, 697
microorganism mass balance, and, 699
hydraulics, 248, 369–378, 589–604
hydric soil, 762
hydrocarbons, 469, 477
hydrochloric acid, 662–663
hydrogen bond, 361
hydrogen chloride, 453, 454

hydrogen fluoride, 454
hydrogen ions, 355–357, 362
hydrogen sulfide, 453
hydrograph, 759, 762, 771, 772, 773, 775, 777, 779, 783, 787, 792, 803
discharge, 812
graphical, 792–794, 796–799
inflow, 791, 792, 794, 798, 800
outflow, 792, 800
runoff, 759, 771, 772, 773, 777, 778, 780, 787, 790, 791, 794, 800, 803, 811
tabular, 773, 787, 788–790, 791
types of, 772
hydrologic condition, 781
hydrologic cycle, 762, 765–766
hydrologic soil group (HSG), 763
hydrology, 765–772
hydrolysis, 362
hydrostatic pressure, 594
hyetograph, 763, 771
hypochlorite, 662, 663
dosage, 366–368, 708–709
feed rate, 658–660
solution, percent strength, 660–661
hypochlorous acid, 662–663
hypotenuse, 103
hypothermia, 438
hypothesis testing, 109, 110
hypothesized count, test of, 132–133

I

ideal gas, 84, 400
law, 87–88, 89, 403, 405, 418, 487, 498
identity, Boolean algebra, 217
ignition temperature, 392
illumination, 392
Imhoff cone, 751
impact strength, 244
impaction, 492, 529, 542
impervious cover, 763, 774, 783
impoundment, 763
impulse noise, 392, 424
incidence proportion, 187, 188–190
incidence rate, 187, 188, 190–195, 208
incidence vs. prevalence, 193, 194
incident, 392
incineration, 405, 450, 478–482, 487
calculations, 481–482
factors affecting, 478–481
oxygen requirement, 479
inclined plane, 238–241
incomplete combustion, 479, 480
independence, test of, 131–132
indoor air quality, 392, 395–396, 411
induced voltage, 290–291
inductance, 298–303
mutual, 301–302
total, 302
industrial hygiene, 381–383, 392, 407, 415
industrial stormwater permit, 763
industrial ventilation, 409, 411, 412, 416
inertia, 298

inertial impaction parameter, 542, 543, 544
infant mortality rate, 184, 197
infiltration facility, 763
infinite throat model, 532–537
inflow–storage–outflow relationship, 811
ingestion, 392
initial abstraction, 763
inorganic matter, 355, 752
in phase, 295
insertion sort algorithm, 95
insolation, 451
instantaneous amplitude, 293
integers, 41, 102
integrals, 39
intensity–duration–frequency curves, 769, 771, 774, 798
intensity, 387
    critical storm, 804, 805
    dose, and, 423
    forest utilization, 333
    magnetic field, 289
    of stress, 242
    radiation, 440
    rainfall, 763, 764, 766, 767, 769, 771, 773–774, 775,
        777, 797–798, 801–802
    sound, 389, 423, 426, 432–433
intensive property, 83
interception, particle, 492, 530
interface, gas phase, 452
interlock, 392
internal radiation dose, 444
International 1/4-Inch rule, 335, 336–337
International System of Units, 3–4
interpolations, 80
inventory, forest, 327
inverse element, 217
invert, 763
ion, defined, 451
ionic bond, 361, 362
ionization, 440, 441
ionizing radiation, 392
irritant, 392
ISC3 model, 97
isobar, adsorption, 472
isostere, adsorption, 472
isotherm, adsorption, 471–472

J

jar testing, 630, 712
job hazard analysis, 392
Johnstone's equation, 531–532, 542, 543–544

K

K value, 494, 495, 496, 497, 498, 515, 684–685
Kelvin scale, 399, 400, 401
kinetic energy, 390, 392, 419, 584, 598, 762
kinetic theory, 491–492
Kirchhoff's current law, 279–280
Kirchhoff's voltage law, 271–274
Kjeldahl nitrogen, 763
Knudsen number, 495
Kozeny equation, 650

L

laboratory horsepower, 610
Laboratory Safety Standard, 392
lakes, 574–575
laminar flow, 411, 492, 497, 498, 601
laminar regime, 494, 495, 496
land application, of biosolids, 740–743
land cover, 760
landfilled waste, 39
Laplace's equation, 585, 586, 587
Lapple cut diameter equation, 506, 508, 510
lapse rate, 451
Latin square design, 147–149, 160
law of conservation of energy, 599
law of conservation of mass, 362, 586
law of continuity, 372–373, 597–598
Lawrence and McCarty design model, 697–698
lead, 399
Leaning Tower of Pisa, 310
least squares, 76, 162, 164, 167
Le Chatelier principle, 85
legs, triangle, 103
length, defined, 60
Lenz's law, 299
life expectancy, 204
lift, soil, 316
lime dosage, 630–633
limits, dose, 439
linear equations, 75, 76, 100
linear functions, variance of, 121–122
linear transformation, 76
linear units, 42, 60
lines of force, 288, 289, 298
liquid
    combustible, 388
    limit, 315
    phase, 452, 454, 455, 462, 483, 542
liquid-to-gas ratio, 531, 532, 533, 535, 536, 541, 542, 543,
        544, 546
live loads, 255
loading
    baghouse, 520–521
    chemical, 365
    engineering, 235, 236, 249, 250–253
    inlet/outlet, 508, 510, 512, 514, 516, 520, 521, 522, 545,
        547
    point, 459
    rate, *see* organic loading rate, solids loading rate,
        surface loading rate
local efficiency, plate, 468
local exhaust ventilation, 411, 414–415
lockout/tagout procedure, 392
Log and Summary of Occupational Injuries and Illnesses
        (OSHA-300 Log), 392
log rules, 335–339
LONGZ model, 97
loudness, 424
lower explosive limit (LEL), 393, 480
lower flammability limit (LFL), 480
lower heating value (LHV), 328–329
lowest common multiple, 41
low impact development, 763

M

magnetic field, 288–289, 290–291
magnetomotive force (mmf), 288
major head loss, 600–601
major plot error, 158
malleability, 244
Manning's equation, 601, 603–604, 763, 786
Manning's kinematic solution, 780, 785
manometer, 411, 414, 416, 417
mass, 81, 246
    balance, 456–459, 695–696, 698–699
    concentration, 82–85
    flow rate, 456, 460, 481, 482, 484, 485
    per unit volume, 81
    transfer zone, 474–475
material balance, 456–459, 553
material hearing impairment, 424
Material Safety Data Sheet (MSDS), 393
materials, properties of, 232–233, 241–247, 247–255
maternal mortality rate, 198
math operations
    equilibrium, and, 86–89
    sequence of, 42–44
math terminology, 41–42
matrix, 75, 76, 77
matter, defined, 451
maximum storage volume, 777, 796–797, 802–803
Maxwell (Ma), 288
mean, 50–52, 112–113, 116, 449
    adjusted, 178–179
    depth, still water body, 576–577
    flow velocity, 627–628
    free path, 492
    square residual, 165, 166, 171, 174, 175
    squares, 140–141, 142, 144, 146, 147, 152, 158, 160, 171, 174, 175, 178
    standard error of, 118–119
mean cell residence time (MCRT), 692–693, 697–698, 731
mean sea level (MSL), 565
mechanical energy, 585
mechanical overload, 244
mechanics, 242, 247–255
media material content, 81–85
    liquid phase, 82–85
media, total area, 686
median, 50–52, 112, 113
medical monitoring, 393
medium, filter, 648–649
mesosphere, 451
metabolic heat, 393, 434
metabolic rate, 435
metals, in water, 354
metals loading, 742
meteorology, 451
methane, 28, 345, 401–402, 451, 591, 728
    combustion, 479
    digester production of, 730–731
    emissions, 31, 32
metric system, 3, 25, 63
micelles, 352
microorganisms, 561, 565, 654, 678, 682, 687
    effluent, 699–700
    mass balance, 698–699

micropool, 763
migration velocity, 510–511
milligrams per liter to kilograms per day conversion, 20
milligrams per liter to pounds conversion, 19–20, 365, 368, 706
milligrams per liter to pounds per day conversion, 365–366, 367, 623, 632, 706, 717
million gallons per day to cubic feet per second conversion, 21, 59
million gallons per day to gallons per day conversion, 21, 713
million gallons per day to gallons per minute conversion, 21
minimum particle size, 500–504
minimum transport velocity, 411
missing plots, 159–161
mists, 393
mixed liquor suspended solids (MLSS), 688, 689, 694, 750, 754, 755
mixed liquor volatile suspended solids (MLVSS), 689–691, 698
mixed media, filter bed, 649
mixing, 605, 615–625
    complete, 479, 562, 615, 616
    incineration, and, 479
    wastestream, 562
mixture, 451
mode, 112, 113
modeling, 575
    dispersion, 96–97
    environmental, 79–91
    groundwater, 587
    hydrologic, 766, 769, 772
    rotating biological contactor, 686
    still water body, 575
    storm events, 771–772
models, uses for, 80
modified rational method, 763, 773, 777–778, 791, 796–806
modulus of elasticity, 243
moisture content
    air, 580
    biosolids compost, 743–744
    soil, 313
    wood, 329–330, 339
moisture reduction percent, in digested biosolids, 729–730
molality, 82, 349, 358
molar concentration, 82–85
molar solutions, 359–360
molar volume of a gas, 87
molarity, 81–83, 349–350, 357, 358, 360, 748–750
    to percent solution conversion, 360
molds, 393
mole calculations, 748–750
mole fraction, 82–85
moles, 81, 82, 349, 350
molecular weight, 83, 84, 85, 359, 400, 409, 461, 748
    air, 418, 482, 550, 554
    carbon, 32, 33, 35, 38, 39
    carbon dioxide, 32, 33, 35, 38, 39
    carbon tetrachloride, 475
    glucose, 731
    mole, and, 748
    sodium carbonate, 357

sulfur dioxide, 27
toluene, 476
water, 550, 554
moment, 248–251
of inertia, 250
monitoring, 393
morbidity, 188–195
mortality, 196–205
proportionate, 186, 203
rate, 184, 196–201, 208
motor efficiency, 611–612
motor horsepower, 609
moving averages, 687
7-day, 747–748
mudballs, 646–647
multiple-bed adsorbers, 469
multiple comparisons, 141–144
multiple of a number, 41
multiple regression, 166–171
Murphree efficiency, 468
mutual inductance, 301–302
mycotoxins, 393

N

natality, 205
National Ambient Air Quality Standards (NAAQS), 397
National Pollutant Discharge Elimination System, 759, 763
natural gas, 32, 34
negative terminal, 258, 272
neonatal mortality rate, 197
net filter yield, 732–733
neutralization, 362
neutron, 440
Newton's equation for terminal settling velocity of a spherical particle, 633–635
Newton's law, 494, 495, 497, 498
Newton's second law of motion, 233, 246
Newton's third law of motion, 235
nitric oxide, 398
nitrogen, 741
nitrogen dioxide, 398
nitrogen oxides, 397, 398, 453
nitrogenous biochemical oxygen demand, 571
noise, 384, 422–433
background, 423
continuous, 423
dose, 425, 432
dosimeter, 425, 427
exposure to, 424–428, 432
hazard, 384, 386, 422, 423, 424, 425, 427, 429
area, 425
work practice, 425
impulse, 392, 424
-induced hearing loss, 425, 427
industrial, 423, 429
level measurement, 425
reduction rating, 425
nonionizing radiation, 393
nonpoint source pollution, 560, 763
nonpolar substances, 348, 349, 353, 471, 477
nonuniform flow, 602
nonvolatile solids, 353
normal distribution, 115–117

normal temperature and pressure (NTP), 406
normality, 349–350, 357, 358, 749–750
NP-complete problems, 95
Nukiyama–Tanasawa equation, 532
null hypothesis, 135
null space, 76
number of transfer units (NTUs), 463–464
numerator, 52, 56–59, 182, 183, 184, 186, 190, 194, 202
nutrients, 561

O

occupational noise exposure, 426–427
Occupational Safety and Health Act (OSH Act), 382, 393
OCD model, 97
octave band
frequency, 430
noise analyzers, 428
odd number, 41
odds ratio, 209–210
odor, Secondary Maximum Contaminant Level, 59
Ohm's law, 259–263, 266–267, 270–271, 272, 276–278, 293, 295, 296
opacity standards, 405
open-channel flow, 601–603
calculations, 603–604
operating line, 456, 459, 467
optimization, 80
optimum allocation, 128
organic loading rate, 680–681
rotating biological contactor, 685
treatment pond, 703–704
organic matter, 354, 478, 479, 752
decomposition, 561, 565, 569
eutrophication, and, 575
stabilization, 725–731
organic nitrogen, 741
OR gate, 214–215
orifice design, 809–814
osmotic pressure, 351–352
ototoxic, 425
ototraumatic, 425
out of phase, 296
outdoor air, 411
overall attack rate, 189, 190
overall collection efficiency, 501, 502, 504, 507, 508, 514, 515–516, 530
overall efficiency, 611–612
overall heat transfer coefficient, 485, 488–489
overflow rate, sedimentation, 636–637
overhanging beams, 250
overland flow, 785
overrun, 335–336
oxidation, 362, 393, 478–482
oxidation ditch detention time, 701–702
oxidizer, 393
oxygen; *see also* biochemical oxygen demand (BOD)
-deficient atmospheres, 393
deficit, 564, 568, 569, 571
requirement, incineration, 479
oxygen sag, 561, 568–569
curve, 568
ozone, 397–398, 451
depletion, stratospheric, 452

P

pacing, 63, 64
packed tower, 453
    absorber height, 462–470
    diameter, 459–462
paddle flocculator calculations, 624–625
Papadakis equation, 580
parallel circuit, 274–284
parallelogram law, 235
parent material, soil, 310
particulate emission control, 491–524
    equipment calculations, 499–524
particulate matter (PM), 393, 398, 404–405, 491–524,
        527–528
    adsorber efficiency, and, 478
    characteristics, 493–494
    collection, 492
    flow regime, 494–499
parts per billion (ppb), 82
parts per million (ppm), 25, 46, 47, 82, 706, 710
    conversion to micrograms per cubic meter, 26–28
parts per trillion (ppt), 82
passenger vehicles per year, 31–32
path set, 215, 216
peak amplitude, 292
peak discharge, 763, 774–779, 783, 784, 787–791,
        795–798, 801, 810, 812
    runoff, and, 772–773
peak-to-peak amplitude, 293
Peclet number, 530–531
pedologists, 309
penetrating radiation, 442
penetration, 521–522, 529, 532, 536, 537–538
Penman equation, 580
percent, 27, 44–47
    by mass, 81
    by volume, 81
    excess air, 405–406
    mudball volume, 646–647
    settled biosolids, 629
    solids, 718–719
    solids recovery, 738
    solutions, 360
    strength, solution, 619–621, 660–661
    stroke setting, 713
    total solids, 677
    volatile matter, 689
percolation rate, 763
perimeter, 42, 60, 61–63
period, alternating current, 291–292
permanent dipole, 471
permanent threshold shift, 426
permeability, 300, 581, 584
permissible exposure limit (PEL), 386, 393
permitted worker, 439
person-time rate, 190–193, 208
personal protective equipment (PPE), 394
pH, 451
    aerobic digester, 726–727
    water, 356, 357
phase angle, 296–297
phase difference, 296
phase equilibrium, 85–86

phase separation, 364
phase transfer, 364
phase transition, 364
phasor, 297
photochemical smog, 451
photoionization detectors (PIDs), 407
photolysis, 363
photon, 439
photosynthesis, 35, 348, 451, 564, 568–569
physical reactions, types of, 364
physical stressors, 384
pipe flow time, 775
pipe friction, 373–374, 600–601
piston pumps, 713
Pitot tube, 411, 414, 416, 417
plant available nitrogen (PAN), 741
plastic limit, 315
plasticity index, 315
plate and frame press, 732–733
plate tower, 453
    number of absorber plates, 467–469
    sizing, 465–467
plenum, 411
plug flow, 615–616, 636
plume dispersion models, 96–97
pneumatics, 248
point source pollution, 560, 764
Poisson distribution, 111
Poisson's ratio, 243, 314
polar substances, 348, 349, 471
polarity
    AC voltage, 290
    voltage drop, 272–273
    water molecule, 347
pollutant, 353, 451, 542
    air, 395–399, 491
    concentration of, 399–400
    mass rate standards, 405
    primary, 451
    secondary, 452
    stream, 560–561
polymerization, 363
population
    correlation coefficient, 120–121
    defined, 327
    loading, 704
    mean estimate, 123, 127
    variance estimate, 123
    vs. sample, 114
porosity, 312, 581, 583, 764
positive displacement pumps, 375, 376, 713
positive terminal, 258, 272
postneonatal mortality rate, 197–198
potential energy, 390, 584–585, 598, 762
pound force, 233–234
pound-mole, 748
pounds per day to milligrams per liter conversion, 20, 368
pounds per square foot to pounds per square inch
        conversion, 371
pounds per square inch to feet conversion, 72–74, 375,
        593–594, 609
pounds to flow conversion, 21
pounds to gallons conversion, 19
pounds to milligrams/liter conversion, 20

power
   electrical, 261–263
      parallel circuit, 284
      series circuit, 268–271
   sound, 426, 429–430, 431
powers, mathematical, 49–50
precipitation, 767–773
   characteristics, 767
   solids, 630
prediction, 80
preliminary assessment, 394
preliminary treatment, 667–674
presbycusis, 425
present value, 224, 227, 228
present worth, 225
pressure, 85, 394, 400, 411, 413–414, 451
   air, 369, 370, 565, 589, 590, 607
   adsorption, and, 474
   atmospheric, 27, 29, 88, 369, 370, 371, 400, 409, 410,
        411, 413–414, 453, 580, 590, 594
   calculations, 69–75, 418–422
   drop, 411, 510
      baghouse, 516, 519, 521, 522, 523
      cyclone, 506
      electrostatic precipitator, 510
      wet scrubber, 528, 540–541, 545, 546, 552, 555
   fan, 421–422
   filtration, 732–736
   gas volume, and, 401, 408
   head, and, 70, 71–73, 585, 590, 593, 595–596, 599–600
   hydraulic head, and, 584–585
   loss, 420–421
      wet scrubber, 538–539, 546
   mixture of gases, 88–89
   osmotic, 351–352
   reaction rates, and, 364
   solubility, and, 351, 663
   sound, 423, 424, 425, 426, 427, 430–432
      additions, 431
      level, 426, 430
      temperature, and, 402
   velocity, and, 598–600
   water, 589–596
      force, and, 593–594
prevalence rate, 187, 193–196
prevalence vs. incidence, 193, 194
Priestly–Taylor equation, 579
primary air quality standards, 397
primary pollutants, 451
primary porosity, 583
primary treatment, 561, 674–678, 717–718
prime number, 41
priming, plate tower, 465
probability, 110–111, 213, 214–215
process rate standards, 405
process residuals, 717
product, math, 42
product over the sum method, 283–284
propane cylinders, 38
proportion, 53–54, 182, 184–186, 197, 202
   attributable, 210–211
proportional allocation, 128
proportionate mortality, 186, 203
   ratio, 203

protozoans, 561, 638, 678
public health impact, 210–211
puff dispersion model, 96
pump column, 608
pump efficiency, 376–378
pumping rate
   backwash, 645
   biosolids, 677, 719, 723, 754
   calculations, 374–375
   chemical feeder, 623, 713, 715
   cone of depression, and, 583
   waste activated sludge, 705
pumping water level, 606–607
purification, stream, 569–572

Q

quadratic equations, 99–102
qualitative data, 114
quantitative data, 114
quartz, 314
quotient, 42

R

race-specific mortality rate, 198
radiant heat, 384, 394, 434
radiation, 384, 392, 393, 394, 438–444
   alpha, 441
   beta, 441–442
   energy budget, and, 579
   gamma, 442–443
   ionizing, 440
radioactive decay, 439
   equations, 443–444
radioisotope, 439
radiological half-life, 440
radius, 42, 60, 64–65, 67
   of curvature, 252
radon, 451
railcars of coal burned, 38–39
rain shadow effect, 452
rainfall intensity, 763, 764, 766, 767, 769, 771, 773–774,
        775, 777, 797–798, 801–802
Raleigh scattering, 452
randomization, complete, 139–141
randomized block design, 144–147, 147, 155–156, 159
   analysis of covariance in, 176–179
random sampling, 123, 129
range, 112, 113
Raoult's law, 87, 89
rapid mixing, 615
rate, statistical, 182, 187
   case-fatality, 202
   mortality, 184, 196–201, 208
   ratio, 208
rated sling loads, 238
ratio, 52–54, 81, 182–184, 197
   air-to-solids, 724
   death-to-case, 201–202
   drainage area to still water body capacity, 577–578
   food-to-microorganism, 689–691
   odds, 209–210
   Poisson's, 243, 314

proportion, and, 186
proportionate mortality, 203
rate, 208
risk, 206–208
trigonometric, 103–105
volatile acids to alkalinity, 728
rational method, 764, 773, 773–778
modified, 763, 773, 777–778, 791, 796–806
reactance, 298
reaction force, 253–254
reaction rates, chemical, 363–364
reactive substance, 394
reactivity hazard, 394
reaeration, 568–570
real ear attenuation at threshold (REAT), 425
real-world attenuation, 425
receiving hood, 410
recharge, 764
reciprocal method, 282–283
recirculation ratio, 681–682
rectangle, area of, 64–66
recycle flow, 724
recycled wastes, 39
redevelopment, 764
reduction, 362
reduction standards, 405
regression, 161–179
coefficient, 110, 162, 164, 166
curvilinear, 171–173
equations, biomass, 332, 333
group, 173–175
line, 163–164
multiple, 166–171
relative humidity, 452
relative velocity, 624, 634
reoxygenation, 568–569
replacement air, 411
replacement reactions, 362
replicates required, 138
replications, unequal, 143–144
reportable quantity (RQ), 394
residence time; see also mean cell residence time
combustion, 481–482
contaminant stream, 474
incineration, 478–479
particulate matter in atmosphere, 404
resistance, 257–303
alternating current, 295–296
current, and, 260, 277, 280
gas phase, 463
heat, 485
Ohm's law, and, 259–263, 270–271
parallel circuit, 280–282
series circuit, 264–265
resistivity, 285, 288
vs. conductivity, 288
resistors, 257, 258, 259, 260, 264, 265, 266, 267, 268, 269, 270, 272–273, 275, 276, 277, 278, 280, 281, 282, 283, 284, 290, 298
power rating of, 262
resolution of forces, 232–235
Resource Conservation and Recovery Act (RCRA), 394
restrained beams, 250

restricted area, 439
resultant force, 235
retention basin, 764, 790–791
storage volume, 806–808
water quality volume, 809
retention time
biosolids, 728–729
hydraulic
aerated pond, 704–705
complete mix with recycle, 697
microorganism mass balance, and, 699
solids, 691, 692, 727
water, 577, 578
return air, 411
return biosolids rate, 693–694
return period, storm events, 768
Reynolds number, 492, 533–534, 536, 601
drag coefficient, and, 634–636
ribbon test, 318
rich mixture, 479
rill erosion, 761
rip-rap, 764
rise rate, backwash, 643–644
riser, 764, 810
risk, 181–211, 394, 395
assessment, 394
characterization, 394
difference, 211
management, 394
ratio, 206–208
root-mean-square (RMS), 294
rotary vacuum filter, 736–738
rotating biological contactor (RBC)
calculations, 682–686
modeling performance of, 686
roughness coefficient, 601, 604, 764, 780, 785; see also C factor
roughness, pipe wall, 373–374, 601, 604
rounding numbers, 47–49
routings, design inflow, 790
running waters, 559–572
dilution capacity of, 562
dissolved oxygen levels in, 563–565
measurement of discharges in, 562
time of travel, 563
runoff, 560, 764
annual, 578
agricultural, 581
calculations, 773–790
coefficient, 764, 773, 774, 775, 775–776, 802, 803, 804, 805
curve number, 761, 779, 780, 782–783, 788, 789, 809, 813
equation, 783–784
hydrograph, 759, 771, 772, 773, 777, 778, 780, 787, 790, 791, 794, 800, 803, 811
peak discharge, and, 772–773
peak intensity, 767–768
stormwater, 564, 565, 760, 761, 762
stream, 569
surface, 578, 760, 761, 772
urban, 560, 565, 765
rupture, 242

S

Safe Drinking Water Act, 583
safety bench, 764
safety factor, 233, 394
safety standard, 394
salts, in water, 355
sample
    allocation, 127–128
    correlation coefficient, 120–121
    defined, 327
    size, 124–126, 128–129, 130, 137
    vs. population, 114
sampling, 326, 327, 332, 333
    air, 387, 407–409
    methods, 114
    thermal exposure, 435–437
    units, 327
    variables, 123–129
    wastewater, 745–746
sand, 310, 314, 317
sand drying beds, 738–740
sand filter, 764
saturated vapor pressure, 580
saturation, 349
    soil, 313
scaling, timber, 335–339
Scheffe's test, 143–144
schematic diagrams, 258
scintillators, 441, 442
scour, velocity of, 673–674
SCREEN3 model, 97
screening, 605, 667
    pit capacity calculations, 669
    removal calculations, 667–668
screening tools, 97
Scribner Decimal C rule, 335, 336, 338
scrubbers, wet, 492, 499, 527–555
scrubbing systems, 452–469
secondary air quality standards, 397
secondary attack rate, 189, 190
secondary containment, 394
Secondary Maximum Contaminant Level (SMCL), 59
secondary pollutants, 452
secondary porosity, 583
secondary treatment, 561, 686, 687, 718
second law of thermodynamics, 452
section modulus, 250
security assessment, 394
sediment forebay, 762, 764
sediment porosity, 583
sedimentation, 605, 633–636, 638
    calculations, 625–637, 676–677
    diameter, 493, 494
    types of, 634
    test, 317
seeded biochemical oxygen demand, 746–747
    procedure, 567
seed viability, 130
seed volume required, 728
self-inductance, 299–301
self-purification, stream, 559, 560, 561, 569–572
sensitivity, soil, 315

sensitizers, 394
sensorineural hearing loss, 424, 425
sequence of math operations, 42–44
serial dilutions, 358–359
series aiding/opposing, 273
series circuit, 264–274
series–parallel circuit, 284–285
sets, 213, 215
settleability, 750–751
    test, 754, 755
settleable solids, 353, 674, 750–752
settled biosolids volume, 693–694, 750
settled sludge volume, 750
settlement, soil, 316
settling, 605, 633–636
    velocity, 493, 494–496, 637, 672–673
    time, 672–673, 750
seven-day moving average, 747–748
sewage treatment plant effluent, 561, 667
sex-specific mortality rate, 198
shallow concentrated flow, 785
shallow well water, 581
shear, 252
shear force, 233
shear strength, 244
shear stress, 247, 252, 570
sheet erosion, 761
sheet flow, 785
sheeting, shoring, 318
shielding
    excavation, 318
    radiation, 384, 440, 442, 443
shoreline development index, 576
shoring, excavation, 318
short circuit, 258
short-term exposure limit (STEL), 394
SHORTZ model, 97
shrinkage limit, 315
shrub characterization, 327
SI units, 3–4
sick-building syndrome, 411
sidewall depth (SWD), 42, 60
significant digits, 47–49
silica, 395
silt, 310, 314, 315, 317, 353
siltation, 575
simple random sampling, 123
single-event computer models, 771–772
singular matrix, 77
sinking fund, 226–227
site life, metals loading and, 742
six-and-three rule, 417
SKI calculus, 94
slime, 678, 680, 682
slings, 235–238
slope, open-channel, 603
sloping, excavation, 318
slot velocity, 411
sloughings, 682
sludge, 54, 345, 365, 696, 698, 717, 724, 750; *see also* biosolids
    age, 691
    generated per day, 700–701

retention time, 692
return activated, 694
volume, 693
vs. biosolids, 717
waste activated, 365, 687, 690, 705
slugs, 234
Smalian Cubic Volume rule, 336, 337
smog, 399, 451
sociacusis, 425
sodium hypochlorite, 366, 654, 659–660, 661, 708–709
soil
  carbon stock, 37
  characteristics, 311–314, 776
  classification, 317–318, 780–781
  compaction, 316
  compressibility, 316
  failure, 316–318
  hydric, 762
  mechanics, 307–321
  particles, 314–315
  phase mass concentration, 84
  physics, 319
  porosity, 312
  solids, specific gravity, 313–314
  stress and strain, 315
  test, 764
  void ratio, 312
  volume, 311–313
  vs. dirt, 309–310
  water content, 315
  weight, 311–313
Soil Conservation Service (SCS), 764, 775, 779, 792–794
  storm durations, 769–770
solids
  concentrators, 722
  in water, 353
  inventory, 688
  loading rate
    filter press, 732–733, 734–735
    sand drying bed, 739
    thickener, 723
  percent, 718–719, 721, 722, 723
  percent recovery, 738
  production calculations, 717–718
  pumped per day, 721–722
  soil, 313–314
  suspended, *see* suspended solids
  types of, 736
  vs. biosolids, 717, 752
solubility, 452, 453–456, 709
  predicting, 350–351
soluble biochemical oxygen demand, 684–685
solute, 81, 82, 90, 347–350, 351, 352, 354, 358, 452, 619, 660, 709
solutions, 359–360, 452, 709–713
  acidic, 449
  hydrophilic, 352
  percent strength of, 619–621, 660–661, 710–711
  properties of, 351–352
  vs. suspensions, 352
  water, 347–350
solvent, 81, 82, 347–350, 351, 352, 358, 359, 660, 709
sorting, algorithms for, 95

sound
  speed of, 430
  intensity, 426
    level, 432–433
  level meter (SLM), 426, 427–428
  power, 426, 429–430, 431
  pressure, 423, 424, 425, 426, 427, 430–432
    additions, 431
    level, 426, 430
    temperature, and, 402
specific gravity, 23–24, 246, 344–346, 395, 452, 496, 498, 502, 507, 508, 591–592, 649
  protoplasm, 725
  soil solids, 313–314
  wood, 332, 339
specific heat, 480, 481, 484, 485, 488
specific resistance, 285, 288
specific weight, 18, 375–376
  water, 344, 589
specific yield, 584, 586, 607–608
sphere, 42, 60, 61
  surface area, 421
  volume of, 68
spillway, 764
split-plot design, 154–159, 160
spring water, 582
square mil, 285–286
square units, 42, 60, 63
stabilization, 725–731, 743
  channel, 760
  ponds, 702, 703
stack emissions, 405
stack height, 417
stack, ventilation, 411
stage–storage curve, 806–808
standard conditions, 400
standard cubic feet per minute, 411
standard deviation, 116, 117–118, 121, 131
standard error, 118–119, 121, 122, 123, 127, 129, 131
standard temperature and pressure (STP), 27, 84, 87, 400, 406, 418, 589, 590
standard threshold shift (STS), 426
static electricity, 441
static head, 70–77, 595
static pressure, 411, 413, 414, 418–420, 422, 590
static water level, 606–607, 609
statics, 247–248
statistics, 109–179
  forest biomass, 328
steady state, 85–86
Stevin's law, 371, 591
still waters, 573–580
  calculations, 575–578
  surface evaporation, and, 578–580
Stokes' diameter, 493, 534
Stokes' law, 494, 495, 497, 498–499, 501, 502, 503, 634, 635
storage volume, basin, 791–808
  maximum, 777, 796, 797, 802–803
storm duration, 777, 778, 791, 796–806, 811
storm frequency factor, 776–777
storm sewer, 764
  design of, 776

Storm Water Management Model (SWMM), 765
stormwater
    calculation methods, 773–790
    calculations, 759–814
        engineering, 790–814
    filtration, 764
    hot spot, 765
    management systems, 763
    runoff, 564, 565, 760, 761, 762
strain, 242, 243, 247
    soil, 315
straining, 638
stratified random sampling, 126–129
stratosphere, 452
stream buffers, 765
stream flow, annual, 605
stream purification, 569–572
Streeter–Phelps equation, 569–572
stress, 233, 235, 242, 247, 249, 252
    allowable, 242
    intensity of, 242
    soil, 315
    ultimate, 242
stressors, workplace, 383–385
structural failure, 316, 319–321
subplot error, 156, 158
subsoil, 310
subspace, 75, 76, 77
substrate concentration, 699–700
substrate utilization rate, 700
subsurface flow, 772
suction pressure, 411
sulfur dioxide, 398–399, 453–457, 462, 466–467
sulfur oxide gases, 398
sum of squares, 112, 133, 136, 140, 142, 145–146, 148, 151, 152, 153, 155, 156–157, 162, 163, 164, 168, 170, 171, 176–177, 179
supply ventilation system, 413
surcharge, 765
surface area, 60, 63, 67, 476–477
    activated charcoal, 469
    bag, 520, 523, 524
    basin, 637, 806
    circle, 64
    condenser, 487, 488
    duct, 514
    filter, 640, 641, 648, 679, 737
    media, 683–686
    lagoon, 705
    plate, 512, 513, 732, 733
    reaction rates, and, 364
    smoothness, 245
    solid, 364
    sphere, 421
    still water body, 576–577, 614
    stream, 570
    tank, 65, 675, 723
surface condenser, 483, 485–489
surface evaporation, 578–580
surface loading rate, gravity thickener, 723
surface overflow rate, 627, 675, 676
surface runoff, 578, 760, 761, 772
surface water, 581, 583, 606

Surface Water Treatment Rule, 583, 638
suspended solids, 353, 365, 674, 717, 736, 752, 753–754
    fixed, 736
    loading, 688
    mass balance, 695–696
    mixed liquor, 688, 689, 694, 750, 754, 755
    mixed liquor volatile, 689–691, 698
    removal calculations, 677–678
    removed, 681
    total, 353, 734–735, 736, 753–754
suspension, 352
sustainability, 223
swale, grassed, 762
synthetic storms, 771
system effect loss, 417
system operating point, 421

T

t test, 135–138, 146, 158, 159
    for paired plots, 137–138
    for unpaired plots, 135–137
    vs. analysis of variance, 141
tabular hydrograph, 773, 787, 791
    method, 788–790
tallow, 487
tanker truck, gasoline, 33
tank volume, 625–626, 644–645
Technical Release No. 20 (TR-20), 765
Technical Release No. 55 (TR-55), 765, 769, 773, 779–790
temperature, 400–401
    adsorption, and, 474
    changes in, 82, 85, 89
    conversions, 4, 24–25
    density, and, 344, 345, 591
    dissolved oxygen saturation concentration, and, 564
    energy budget, and, 579
    filter bags, and, 518
    incineration, and, 478
    molality, and, 82
    molarity, and 82
    particle energy, and, 400
    Peclet number, and, 531
    pressure, and, 402
    reaction rates, and, 364
    solubility, and, 349, 350, 351
    specific heat, and, 485
    standard, 27; see also standard temperature and
        pressure
    stream, 569
    surface heat exchanger, and, 486–487
    volume, and, 401–402, 408–409
    workplace, 433–438
temporary threshold shift, 426
tensile force, 233
tensile strength, 243
tensile stress, 247, 252
tension, 247
terminal settling velocity, 494, 495, 496, 498–499, 503, 633–636
tertiary treatment, 561
theoretical air, 406
theoretical detention time, 636–637

thermal pollution, 351
thermal stress, 433–438
thermoluminescent dosimeter (TLD), 439
thermosphere, 452
therms of natural gas, 32
thread test, 317–318
threshold limit value (TLV), 395
    heat stress, 437
threshold odor number (TON), 59–60
threshold shift, 426
tight buildings, 395, 411
timber scaling, 335–339
time of concentration, 765, 774, 775, 777–780, 784–789,
    796–798, 801, 804, 805, 812, 813
time of travel, running water, 563
time-weighted average (TWA), 395
topsoil, 310
torque, 248
torsional force, 233
torsional strength, 244
total brake horsepower, 609
total differential head, 377
total dynamic head, 71
total effective dose equivalent (TEDE), 439
total energy, 598–600
total filterable residue test, 736
total head, 595, 599–600
total media area, 686
total pressure, 411, 413–414
total pumping head, 609
total quality management (TQM), 395
total residue test, 736
total solids percent, 752–753
total suspended solids (TSS), 353, 734–735, 736, 753–754
toughness, 244
toxicity, 395
toxicology, 395
toxins, airborne, 449
transfer units, 539, 540, 546, 547, 548, 549, 550, 551, 552,
    554
transformations, 131
transformer, 290
transition regime, 494, 495, 497
transmissivity, 584
transverse strain, 243
trash rack, 765
Traveling Salesman problem, 95
travel time, water, 765, 784–786, 789
treatment pond, 702–705
    area, 702
    flow rate, 702
    hydraulic detention time, 703
    volume, 702
trench boxes, 318
trenchfoot, 438
trend, 327
triangle law, 235
trickling filter
    calculations, 678–682
    design, 682
trigonometry, 103–105
triple bottom line, 223, 326
troposphere, 452
true color, 353

turbidity, 353, 583, 655, 694–695
turbine pump calculations, 609–613
turbulence, 601
    incineration, and, 479
    stream, 561, 568
        reaeration constant for, 570
turbulent flow, 411, 601, 634
turbulent regime, 494, 495
two-way table, 151, 157
Type A, B, C soil, 317
Type I/II errors, 111

U

ultimate stress, 242
unconfined aquifer, 581, 586, 587
underrun, 335–336
undissolved solids, 353
Unified System of Classification, 315
uniform flow, 602
uniformity coefficient (UC), 648–649
unit filter run volume (UFRV), 641–642
unit peak discharge, 788
units, 15
universal solvent, 347
unsafe condition, 395
unseeded biochemical oxygen demand calculations,
    746–747
unseeded biochemical oxygen demand procedure, 566
upper explosive limit (UEL), 395, 480
uprights, shoring, 318
urban runoff, 560, 565, 765

V

vaccine efficacy/effectiveness, 211
vacuum
    filtration, 731, 736–738
    perfect, 370, 371, 590, 594
valence, 361
VALLEY model, 97
van der Waals' adsorption, 471, 474
vapor pressure, 89, 351
variance, 121–122, 123, 125–126, 128, 166
    analysis of, 139–160
    Bartlett's test of homogeneity of, 133–134
    within-group, 136
variation, source of, 139
varied flow, 602
vector diagrams, 297
vectors, 76, 77
vehicle miles traveled (VMT), 31
velocity, 411, 595
    area, and, 597–598
    capture, 410, 417
    duct, 416–417
    filtration, 517
    flooding, 459–462
    fluidizing, 651
    gas, 529, 530, 532, 533, 534, 535, 536, 541, 543, 544,
        548, 551, 554, 555
    gas, adsorption and, 474
    grit channel, 672
    head, 70–77, 595, 599, 600

hood-face, 416
horizontal, 615, 636–637, 674, 676
liquid, 554
liquid droplet, 529
mean flow, 627–628
migration, 510–511
molecular, 491
open-channel, 602
pressure, 410, 411, 413, 414, 416–420, 422
pressure, and, 598–600
relative, 624, 634
scour, 673–674
settling, 493, 494–496, 637, 672–673
sound, 426
stream, 562, 568, 570
terminal settling, 494, 495, 496, 498–499, 503, 633
throughput, 512
water, 596–598, 624
velocity pressure, 411, 413, 418–420, 422
vena contracta, 421
ventilation, 409–422
good practices, 417
measurements, 416–417
system drawings, 417
Venturi scrubber, 453, 492, 499, 527, 528, 530
calculations, 542–555
packed column height and diameter, 551–555
packed tower, 549–551
scrubber design, 542–548
spray tower, 548–549
efficiency, 531–541
viscosity, 601, 461, 462, 468, 555, 601
air, 496, 498, 503, 508
gas, 391, 491, 492, 494, 495, 500, 506, 507, 510, 511, 530, 531, 532, 533, 534, 535, 536, 542, 543, 544
solvent, 551
water, 461, 552, 634, 650, 651
VISCREEN model, 97
void ratio, 312
volatile acids to alkalinity ratio, 728
volatile matter
percent, 689
pumped per day, 721–722
reduction by anaerobic digestion, 729
volatile organic compounds (VOCs), 397, 398, 528
volatile solids, 353
loading, aerobic digester, 726
percent, 752–753
volatile suspended solids (VSS), 54
percent, 753–754
voltage, 257–303
average, 295
drop, 266–268, 271–273, 285
polarity, 272–273
effective, 294
effective value, 294
instantaneous, 293, 294
migration velocity, and, 511
Ohm's law, and, 259–263, 267–268, 270–271, 276
parallel circuit, 275–276
peak, 292, 293, 294, 295
peak-to-peak, 293
series circuit, 266–268
sources, series aiding and opposing, 273

volume, 16–18, 23–24, 42, 49, 60, 61, 66–68, 81, 87, 400
basin, 616–617
development, still water body, 577
flow rate, 411
over volume test, 629
percent, 81
mudball, 646–647
pond, 613
still water body, 575
tank, 625–626
temperature, and, 401–402, 408–409
treatment pond, 702
volumetric flow rate, 502, 503, 504, 512, 513, 514, 517, 518, 519, 520, 521, 523, 524
vulnerability assessment, 395

W

washwater troughs, 652–653
waste activated sludge (WAS), 365, 687, 690, 705
waste, recycled vs. landfilled, 39
waste gas stream, 479, 480, 481
waste rate
F/M ratio and, 690–691
mass balance, and, 695–696
mean cell residence time, and, 692–693
Waste Reduction Model (WARM), 39
wastewater
calculations, 667–755
laboratory calculations, 744–755
treatment plants, 561, 564
water
at rest, 371
budget model, 578
chemistry, 346–369
constituents, 353–358
cycle, 560, 765–766
density, 82, 246, 332, 344–345, 552, 634
discharge, 372–373
gauge, 413
horsepower, 610–611
hydraulics, 589–604
molecular weight, 550, 554
molecules, 346–347
pH, 356, 357
phase mass concentration, 84
pressure, 589–596
quality volume, 808–815
quality window, 765
retention time, 577, 578
solutions, 347–348, 709
source calculations, 606–608
specific gravity, 345, 592
specific weight, 589
storage, 612–614
surface profile, 765
table, 765
aquifer, 581
treatment, 605–663
chemicals, 347
viscosity, 461, 552, 634, 650, 651
weight of, 72, 344, 345, 370, 590, 591, 592, 593
watershed, 765
heterogeneous, 785

wavelength, 426, 430
    alternating current, 292
weight, 16–18
weighted measurements, 426
weights, ingredient, 16
weir overflow rate, 628–629, 636–637, 674, 675–677
welding, 248
well drawdown, 606–608
well pump calculations, 609–613
well yield, 607–608
wet bulb globe temperature (WBGT), 435–436, 437
wetlands, 760
wet scrubbers, 492, 499, 527–555
wet shaking test, 317
wet weather flow, 765
wet weight basis, 329–330
wetted perimeter, 765
width, 42, 60
wood
    density, 339
wood columns, 254

work, 234, 239, 261, 376
work-load assessment, 435
workers' compensation, 395
working stress, 242
workplace stressors, 383–385

X

x-rays, 440, 442–443

Y

years of potential life lost (YPLL), 204–205
yield point, 242

Z

zero energy state, 395
zone of influence, 583
zone of saturation, 581
zoogleal slime, 682